Some Useful Values

Earth

Mean radius	6.37×10^6 m
Mass	5.97×10^{24} kg
Mean radius of orbit about sun	1.49×10^{11} m
Period of orbit about sun	3.16×10^7 s

Moon

Radius	1.74×10^6 m
Mass	7.35×10^{22} kg
Mean distance from earth (center-to-center)	3.80×10^8 m
Period of orbit about earth	2.36×10^6 s

Sun

Radius	6.96×10^8 m
Mass	1.99×10^{30} kg

Physics for Students of Science and Engineering

Physics for Students
of Science and Engineering

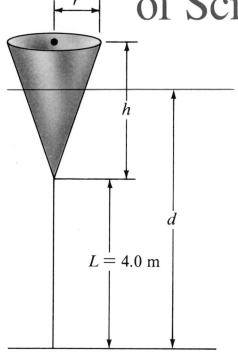

A. L. Stanford
J. M. Tanner

Georgia Institute of Technology

Academic Press, Inc.
(Harcourt Brace Jovanovich, Publishers)
Orlando   San Diego   San Francisco   New York   London
Toronto   Montreal   Sydney   Tokyo   São Paulo

Cover photo by Carroll Morgan.

Photo Credits: Figure 21.1, Fundamental Photographs, New York;
Figure 21.4, Yoav/Phototake; Figure 21.11(c), *Physics*, by
Edward R. McCliment, Harcourt Brace Jovanovich ©1984;
Figure 21.13, Vincent Mallette; Figure 21.21, Yoav/Phototake;
Figure 23.9, Yoav/Phototake; Figure 24.11, photo courtesy of
AT&T Bell Laboratories.

Academic Press, Inc.
Orlando, Florida 32887

United Kingdom Edition Published by Academic Press, Inc.
(London) Ltd., 24/28 Oval Road, London NW1 7DX

ISBN: 0-12-663380-0

Library of Congress Catalog Card Number: 84-70476

Printed in the United States of America

Contents

Contents

Preface

This book is a calculus-based textbook of introductory physics designed for a one-year or three-semester program. It is intended for students who are concurrently studying calculus. The use of elementary calculus, while limited in early chapters, increases as the text proceeds.

Why, one may reasonably ask, is there a need for yet another textbook in introductory physics? Our experience at Georgia Tech has compelled the conclusion that a textbook appropriate to the modern student of science or engineering has not been available.

Students of science or engineering, when taking their first course in general physics, have considerable demands upon their time. Consequently, they are acutely aware of the need for a textbook that provides what they need to learn—what they must learn—and provides it efficiently and effectively. The topics in the main body of this text are limited to those that we have judged to be practically appropriate to a first course in physics. Therefore, the text is not encyclopedic: not every "pet topic" of every instructor is within the primary text. Many of the traditional peripheral topics of general physics are, however, available for consideration in the "mini-lectures" of the Group B problem sets. Thus, the text itself focuses on essentials. The text takes, we hope, a pithy, efficient, no-nonsense approach to the basic issues and was designed with the student in mind at every juncture.

Students quickly perceive that the critical issue for them is problem solving. This book treats the issue of problem solving carefully and extensively. First, the text provides a problem-solving strategy applicable to all word problems and amplifies that strategy specifically for certain categories that are particularly difficult for students of beginning physics. Further, at the end of each chapter is a Problem-Solving Summary. Chapter summaries of many textbooks merely restate principles and list some of the formulas of each chapter. In contrast, summaries in this book point out how to identify the category into which each problem falls, warn of the common pitfalls, remind the reader of critical sign conventions, and suggest how the problem-solving strategy may be most effectively applied.

The problem sets at the end of each chapter are designed, as is the entire book, for students. The problems are not separated into categories but are arranged roughly in order of increasing difficulty. As students progress successfully through this arrangement, their confidence in their problem-solving ability increases. Equally importantly, this arrangement recognizes that, for a student, determination of what the problem is about is part of the problem! (To assist instructors in preparing assignments and examinations, a categorical arrangement of problems is provided in the Instructor's Manual that accompanies the textbook.)

The problem sets for each chapter are divided into two parts. Group A problems are those that a student should find instructive and appropriate in the thorough preparation for examinations. Group B problems are, in general, more challenging: they provide more extensive use of the calculus, introduce many topics of physics that are not the basic material of beginning courses, occasionally offer the opportunity to utilize programmable calculators or microcomputers, and provide "mini-lectures" on advanced topics that certain instructors will choose to include in introductory courses.

The text of this book is brief. We have tried to provide an exposition of physical principles that is efficient without sacrificing precision or clarity. Elaboration of particular issues is provided where our experience indicates that students have difficulty.

The text is interspersed with Exercises, which are, in general, one-step problems designed to entice the student to use pencil and paper as a part of the study process. These Exercises also offer students an occasion to determine whether or not they understand the immediately preceding material in the text. Numerous worked-out problems are included at appropriate points in the text. Some of these are embedded within the text where special techniques need to be described in detail. Most, however, are set apart and labeled Examples. These Examples within each chapter are intended as models for students, models that illustrate the application of the problem-solving strategy to specific topics.

The last chapter of the text is an introduction to wave mechanics and may be omitted without affecting the completeness of the text for most curricula. This chapter is included because modern curricula in science and engineering are beginning to require that students become familiar with tunneling, band theory, and similar wave mechanical phenomena at an early stage in their education.

We are pleased to thank the numerous people who have helped us prepare and test this textbook. Ms Audrey Ralston typed the text and graciously suffered our continual repairs. Parts of the manuscript were read by various members of the Georgia Tech faculty, and we thank them for their advice and suggestions. A host of reviewers from a large number of institutions provided continual critical attention throughout the preparation of the manuscript. Their contributions to the technical accuracy, pedagogy, and clarity of the text and problem sets are greatly appreciated. We especially thank the many students at Georgia Tech who used and criticized the manuscript and who have, over the years, made us aware that the time has come for a new textbook in general physics . . . for students.

A. L. Stanford
J. M. Tanner
Atlanta, Georgia

1 Introduction

1.1 Physics and the Scientific Method

Physics is a natural science. It is one of humankind's responses to its curiosity about how nature works, about how the universe is ordered.

Like other modern natural sciences, physics has evolved to become a logical process based on the *scientific method*. This method is rooted in a philosophy that recognizes no truths and embraces no dogma but seeks to be completely objective and practical. The scientific method may be considered an investigative process composed of three parts:

1. Physical processes are observed and measured both quantitatively and qualitatively. This step necessarily includes the conception and definition of appropriate quantities by which measurements may be made.

2. A hypothesis is offered, usually in the form of a general principle or a mathematical statement of relationships between physical quantities (time and distance, for example). These principles or relationships can be used to predict the results of other similar physical processes.

3. The hypothesis is subjected to experimental tests of its validity. Its predictions are compared to actual measured values.

Hypotheses proposed according to the scientific method are retained only if they enjoy continued and unfailing success. A single instance in which a hypothesis fails to predict successfully the outcome of a pertinent, repeatable experiment requires either rejection of the hypothesis or its modification to rectify that failure. Throughout the history of science many hypotheses have been discarded, and many have been changed. Those that have enjoyed some measure of success but are without extensive experimental verification over a long period of time are referred to as theories (those that have not had some success are not referred to at all). Hypotheses that have withstood successfully the repeated and diverse trials of experiment are accorded the title *law*, but even the most venerated laws of physics are not considered "true" by scientists. Laws are, along with all the tenets of science, acceptable only as long as they continue to coincide with measurements of physical processes. Scientists do not "believe" the laws of physics; they merely use them in very practical ways, maintaining a healthy skepticism that permits continual checking of current laws and theories and encourages speculation about new hypotheses. In this way the scientific method provides a rational approach to an intellectual and logical comprehension of natural phenomena.

1.2 Units

Physics is a science of relationships and measurements of the physical world, and understanding physics requires an understanding of the measurement process. The measurements of physical quantities are determined quantitatively, in terms of some units like feet, meters, miles per hour, or kilograms.

Standards and Nomenclature

Measurements in terms of units require standards. For example, in the metric system of units the standard unit of length is the meter; a table that is 2.7 meters long, for example, is 2.7 times the length of the standard meter.

The standard unit of time is the *second* (s). Originally defined in terms of a fraction of a mean solar day on earth, the second is now defined in terms of certain electromagnetic emissions from the element cesium. Another basic standard in the metric system is the *kilogram* (kg), a unit of mass defined to be equal to the mass of a particular body of metal kept in France. The concept of mass will be considered later, but for now it is sufficient to note that the mass of a given object is an expression of the quantitity of matter contained in that body. The third basic unit of the metric system is the *meter* (m). Over the years the meter has been defined successively in terms of a quadrant of the surface of the earth, in terms of the distance between the marks on a metal bar, and in terms of a specified number of wavelengths of certain electromagnetic emissions from a particular species of atom. In 1983 the *standard meter* was redefined by international agreement to be the distance that light travels through a vacuum in $1/299,792,458$ of a second. Thus the basic unit of length is defined in terms of our best measured value of the speed of light c. In 1983 the accepted value of the speed of light in vacuum was taken to be

$$c = 299,792,458 \text{ m/s}$$

The wide range of magnitudes of measurements encountered in physics makes it convenient to use multiples and submultiples of the standard units. The metric system is a decimal system, that is, it is based on powers of 10. This system is particularly amenable to using prefixes to specify multiplying factors that can be associated with units. Table 1.1 lists some of the common prefixes and those that will be used throughout this book. The prefixes or their abbreviations may be used with any metric unit or its abbreviation. For example, the kilowatt, or kW, is 10^3 watts, and the nanosecond, or ns, is 10^{-9} second.

The metric system of units, known as the SI (for *Système Internationale*), will be used in this book along with the British Engineering system, often called the English system. The metric system uses the kilogram as the standard unit of mass, the meter for length, and the second for time. In the English system the

TABLE 1.1   **Common Prefixes and Their Multiplying Factors Associated with Physical Units**

Factor	Prefix	Prefix Abbreviation
10^9	giga	G
10^6	mega	M
10^3	kilo	k
10^{-2}	centi	c
10^{-3}	milli	m
10^{-6}	micro	μ
10^{-9}	nano	n
10^{-12}	pico	p

standard force (the choice of force or mass as fundamental is arbitrary) is the *pound* (lb), defined to be that force with which the earth pulls on a mass of 0.45359237 kilogram at a certain location on the surface of the earth. The standard length in the English system is the *yard* (yd), which is specified in terms of the meter such that

$$1 \text{ yd} = 0.91440183 \text{ m}$$

It follows that

$$1 \text{ inch (in.)} = 2.54 \text{ cm}$$

The unit of time, the second, is the same in the English and SI systems.

Conversion of Units

A basic skill necessary to the successful solution of many physics problems is the conversion of units between the metric and English systems. It may be necessary to determine, for example, the number of inches in a half mile (mi) or the number of meters in six feet. In any case, confusion and error can be avoided by using a simple procedure based on the principle that a given measure (including units) is not changed when multiplied by unity, that is, by the number 1. Of course, unity can be represented by any fraction in which the numerator and the denominator are equal or equivalent. The fractions 7/7, 3 ft/1 yd, and 2.54 cm/1 in., for example, are all equal to unity. A half mile can be converted to inches without changing its measure (that is, without changing the magnitude of its given length) by starting with the given value and multiplying it by appropriate fractions, each of which is equal to unity, as many times as needed:

$$0.50 \text{ mi} \times \left(\frac{5280 \text{ ft}}{1 \text{ mi}}\right) \times \left(\frac{12 \text{ in.}}{1 \text{ ft}}\right) = 31{,}680 \text{ in.} = 3.2 \times 10^4 \text{ in.}$$

The key to this procedure is finding the appropriate fractions equal to unity that should be used. Most people probably do not know offhand the number of inches in a mile, but many people know that 5280 feet are equal to a mile. Thus, the fraction that converts miles to feet is used, anticipating the next step, which uses

12 in. in a foot. Notice that in constructing the first fraction, 1 mi is placed in the denominator so that its unit cancels algebraically with that of the given 0.50 mi. At this stage, 0.50 mi has been converted to feet. Similarly, the second fraction is formed so that the unit feet cancels the previously obtained feet, leaving the final product expressed in inches. This process accomplishes the desired conversion. For a given conversion, an arbitrary number of fractions, each equal to unity, may be used as necessary to convert units step by step to provide the given measure in the desired units.

Here we have been treating units as if they were algebraic quantities, and indeed they may be manipulated algebraically. As a further example, the area of a rectangle with sides of lengths 3 m and 2 m is obtained by multiplying its height h by its width w.

$$\text{Area} = hw = (3 \text{ m})(2 \text{ m}) = 6 \text{ m}^2$$

in which multiplication of the unit of height by the unit of width gives the unit of area, m^2. A similar algebraic treatment of units applies to division. For example, an automobile traveling a distance of 120 mi on 3.0 gallons (gal) of fuel achieves a mileage rating of

$$\frac{120 \text{ mi}}{3.0 \text{ gal}} = 40 \text{ mi/gal}$$

Addition and subtraction of physical quantities require particular care with respect to units. Any terms to be added algebraically must be expressed in identical units before the usual rules of algebraic addition may be applied. The procedure for adding 3 ft and 2 in., for example, is

$$3 \text{ ft} + 2 \text{ in.} = 3 \text{ ft}\left(\frac{12 \text{ in.}}{1 \text{ ft}}\right) + 2 \text{ in.}$$
$$= 36 \text{ in.} + 2 \text{ in.} = 38 \text{ in.}$$

Note to the Student

The following exercises and those throughout the book are integral parts of the text. At times, new and essential material will be introduced *only* in the exercises. These exercises are, in general, one-step problems designed to ensure that the student understands the preceding textual material. Answers to the exercises usually accompany each problem or question. Any difficulty with an exercise should prompt the student to review the appropriate section of the text before proceeding with new material.

E 1.1   A person's height is measured at 5 ft, 8 in. Express this height in centimeters.   Answer: 1.7×10^2 cm

E 1.2   A world-class sprinter can run the 100-m dash in about 10 s, which corresponds to an average speed of 10 m/s. Express this sprinter's speed in miles per hour.   Answer: 22 mi/h

E 1.3   An acre is the area of a square plot of land measuring approximately 209 ft along each side. How many square meters correspond to an acre?   Answer: 4.1×10^3 m^2

E 1.4   The *density* of an object is equal to its mass divided by its volume.

A cube of lead 4.2 cm along an edge has a mass of 865 grams (g). Calculate the density of lead in kg/m^3.   Answer: 1.1×10^4 kg/m^3

E 1.5   The mass of the earth is 6.0×10^{24} kg, and its radius is 4.0×10^3 mi. Determine the average density of the earth in g/cm^3.   Answer: 5.5 g/cm^3

1.3 Vectors

Mathematics is the working language of physics. The material in this text requires facility with algebra and elementary plane geometry, a basic knowledge of analytic geometry, and a growing familiarity with the fundamental concepts of the calculus. Certain aspects of trigonometry are so frequently required that considerable facility with the manipulation of trigonometric relationships is needed. A review of trigonometry is provided in the Appendix. Similarly, necessary to the mastery of physics is a comfortable working knowledge of vector analysis. Some familiarity with vector notations and operations is necessary for handling the physical principles and their applications presented in this book. For these reasons, the more commonly used definitions and operations associated with vectors are presented here for review or study, as required. This section has been made compact intentionally. A student studying vectors for the first time may wish to consult a mathematics textbook.

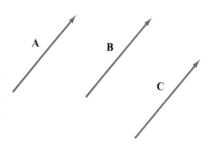

Figure 1.1
Three equal vectors. The vectors **A**, **B**, and **C** all have the same magnitude and direction.

Vector and Scalar Quantities

Certain physical quantities have a direction associated with them. One cannot, for example, exert a push, that is, a force, on an object without pushing in a specific direction. The push must be in some definite direction, such as vertically upward, eastward, or in a direction 30° north of westward. Such quantities also have a magnitude, of course. The push in our example must have associated with it some number of pounds of force applied in the specified direction. Such physical quantities, those that have both magnitude and direction, are called *vector quantities*. In contrast, physical quantities that have no direction associated with them, like volume, are called *scalar quantities*, and such quantities have associated magnitudes (such as a volume of 2.0 liters) but no associated directions.

Here we shall consider some of the basic operations involving *vectors*, the graphical or analytical representations of vector quantities. All the operations considered here will be used extensively throughout this book.

A vector quantity is represented graphically by an arrow, the length of which represents the magnitude of the quantity and the direction of which represents the direction of the quantity. Vectors having identical magnitudes and identical directions are mathematically equal. Thus, in Figure 1.1, the vectors **A**, **B**, and **C** are equal, or **A** = **B** = **C**. (Note that vector quantities are represented by boldface letters.) The magnitude of a vector **A** is represented by either A or |**A**|. In Figure 1.2 the vectors **A**, 2**A**, and −**A** are shown graphically. The vector 2**A** is twice the length of **A** and is in the same direction, and the vector −**A** is the same length as **A** but is in the opposite, or antiparallel, direction.

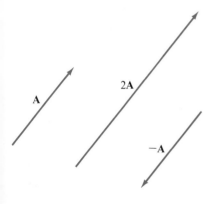

Figure 1.2
Graphical representation of the effect of multiplying a vector by a scalar value. The vector 2**A** is twice the length of **A** and is in the same direction as **A**. The vector −**A** is the same length as **A** but is in an opposite direction.

Vector Addition and Subtraction in Polar Form

A vector may be represented in *polar form* by giving its magnitude and direction, such as $\mathbf{A} = 12$ m $\underline{/212°}$. Two vectors $\mathbf{A}$ and $\mathbf{B}$ may be added graphically, as shown in Figure 1.3. By placing the tail of $\mathbf{B}$ at the head of $\mathbf{A}$, while maintaining the length and direction of both arrows, the sum $\mathbf{C}$ is formed by the vector from the tail of $\mathbf{A}$ to the head of $\mathbf{B}$. Then,

$$\mathbf{A} + \mathbf{B} = \mathbf{C} \tag{1-1}$$

symbolically represents the addition operation. The magnitude and direction of $\mathbf{C}$ is determined by applying trigonometry to the triangle thus formed, that is, by using the laws of sines and cosines.

Figure 1.3
The graphical addition of two vectors. The vector $\mathbf{B}$ is added to $\mathbf{A}$ by placing the tail of $\mathbf{B}$ to the head of $\mathbf{A}$ without changing the magnitude (length) or direction of either vector. The sum of $\mathbf{A}$ and $\mathbf{B}$ is the vector $\mathbf{C}$, formed by drawing a vector from the tail of $\mathbf{A}$ to the head of $\mathbf{B}$.

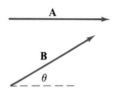
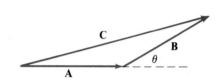

Example 1.1

PROBLEM In Figure 1.3 let $A = 4$, $B = 6$, and $\theta = 60°$. Calculate the magnitude of $\mathbf{C}$ and the angle between $\mathbf{A}$ and $\mathbf{C}$.

SOLUTION Figure 1.4 shows the triangle relating the known quantities A, B, and θ to the unknown quantities C, α, β, and γ. (Notice that the figure shows the known values for A, B, and θ and that the unknown quantities specifically asked for in the problem are labeled with question marks. Other unknown quantities in the triangle, like the angles α and γ, are labeled in case they are needed.)

The law of cosines relates C to A, B, and γ, giving

$$C^2 = A^2 + B^2 - 2AB \cos\gamma$$

$$C^2 = 4^2 + 6^2 - 2(4)(6) \cos\gamma$$

$$C^2 = 52 - 48 \cos\gamma$$

Because γ and θ are supplementary angles, or $\gamma + \theta = 180°$, it follows that

$$\gamma + 60° = 180°$$

$$\gamma = 120°$$

Thus, C is now determined by

$$C^2 = 52 - 48 \cos 120°$$

$$C^2 = 76$$

$$C = \sqrt{76}$$

The angle β is determined using the law of sines:

$$\frac{\sin\beta}{B} = \frac{\sin\gamma}{C}$$

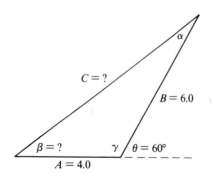

Figure 1.4   Example 1.1

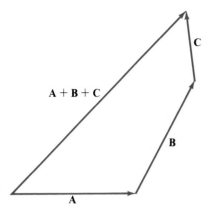

Figure 1.5
The graphical addition of three vectors. The vector sum of **A**, **B**, and **C** is obtained by placing the tail of **B** to the head of **A**, the tail of **C** to the head of **B**, and the sum **A** + **B** + **C** is drawn from the tail of **A** to the head of **C**.

$$\sin\beta = \frac{B}{C}\sin\gamma$$

$$\sin\beta = \frac{6}{\sqrt{76}}\sin 120°$$

$$\sin\beta = 0.596$$

$$\beta = 37°$$

A simple check for this solution can be accomplished by calculating the angle α using the law of sines and then adding α, β, and γ to make sure they add to 180°. Thus, the law of sines gives

$$\frac{\sin\alpha}{A} = \frac{\sin\gamma}{C}$$

$$\sin\alpha = \frac{A}{C}\sin\gamma$$

$$\sin\alpha = \frac{4}{\sqrt{76}}\sin 120°$$

$$\sin\alpha = 0.397$$

$$\alpha = 23°$$

and the sum of the interior angles of the triangle is

$$\alpha + \beta + \gamma = 23° + 37° + 120° = 180° \quad \blacksquare$$

E 1.6   If **F** = **C** + **D**, determine the magnitude of **F** and the angle between **F** and **C**.   Answer: 9.0, 39°

Exercise 1.6

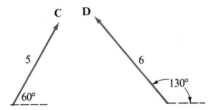

E 1.7   Convince yourself that **A** + **B** = **B** + **A**, that is, that vector addition is commutative. (*Hint:* The sum of two vectors can be represented graphically as the diagonal of a parallelogram.)

Three or more vectors may be added graphically by extending the procedure of placing the tail of a vector to be added to the head of the last one added. Thus, in Figure 1.5, the sum **A** + **B** + **C** is drawn from the tail of the first to the head of the last. Because vector addition is commutative, the order in which the vectors **A**, **B**, and **C** are added does not affect the result. It follows then, for example, that

$$\mathbf{A} + \mathbf{B} + \mathbf{C} = \mathbf{B} + \mathbf{A} + \mathbf{C} \tag{1-2}$$

Vector subtraction is defined by

$$\mathbf{A} - \mathbf{B} = \mathbf{A} + (-\mathbf{B}) \tag{1-3}$$

The vector **B** is subtracted from the vector **A** by adding the negative of **B** to **A**. Figure 1.6 illustrates vector subtraction.

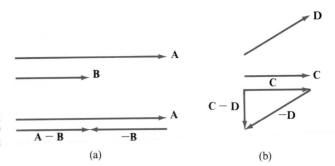

Figure 1.6
Graphical representation of vector subtraction. **(a)** **A** − **B** is formed by adding the vector −**B** to **A**. **(b)** To subtract **D** from **C**, the negative of **D** is formed and added to **C**.

(a)   (b)

E 1.8   If **C** = **A** − **B**, determine the magnitude of **C** and the angle between **C** and **A**.   Answer: 6.1, 25°

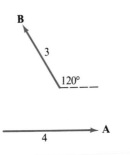

Exercise 1.8

E 1.9   If **F** = **E** − **D**, determine the magnitude of **F** and the angle between **F** and **E**.   Answer: 11, 75°

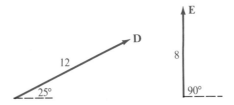

Exercise 1.9

Vector Multiplication

We have already seen that a vector may be multiplied by a scalar; in Figure 1.2 the vector 2**A** has twice the magnitude of **A** and the same direction as **A**. There are two distinct ways to multiply a vector by a vector. One way yields a product that is a scalar; the other way yields a product that is a vector.

The *scalar product* of two vectors **A** and **B** (written **A** · **B** and sometimes called the "dot product") is defined by

$$\mathbf{A} \cdot \mathbf{B} = AB \cos\phi \qquad (1\text{-}4)$$

where A and B are the magnitudes of **A** and **B**, and ϕ is the smaller angle between **A** and **B** when they are drawn with their tails at a common point. Notice that the result of this multiplication operation is not a vector but a real number with an algebraic sign. The following example and exercises will consider this aspect of the scalar product.

Example 1.2

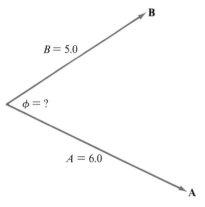

$B = 5.0$

$\phi = ?$

$A = 6.0$

Figure 1.7   Example 1.2

PROBLEM   Two vectors **A** and **B** have magnitudes that are given by $A = 6$ and $B = 5$. Calculate the angle between the two vectors if (a) $\mathbf{A} \cdot \mathbf{B} = 18$, and (b) $\mathbf{A} \cdot \mathbf{B} = -21$.

SOLUTION   The unknown angle ϕ between **A** and **B** is shown in Figure 1.7. Using Equation (1-4) gives

$$\mathbf{A} \cdot \mathbf{B} = AB \cos\phi$$

$$\cos\phi = \frac{\mathbf{A} \cdot \mathbf{B}}{AB}$$

(a) If $\mathbf{A} \cdot \mathbf{B} = 18$, then it follows that

$$\cos\phi = \frac{18}{(6)(5)} = 0.60$$

$$\phi = 53°$$

(b) But if $\mathbf{A} \cdot \mathbf{B} = -21$, then the calculation for ϕ is

$$\cos\phi = \frac{-21}{(6)(5)} = -0.70$$

$$\phi = 134° \quad \blacksquare$$

E 1.10   Calculate the scalar product for each of the illustrated pairs of vectors.   Answers: (a) 41; (b) 0; (c) −20; (d) −12

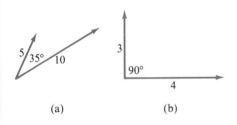

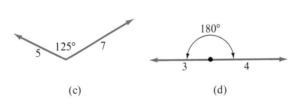

(a)   (b)   (c)   (d)

Exercise 1.10

E 1.11   In the previous exercise you saw examples of pairs of vectors with scalar products that are positive, zero, or negative. See if you can generalize these results by completing the following sentences:

(a) The scalar product of two vectors is positive if the angle between the two is _____.
                                              Answer: ≥0 and <90°

(b) The scalar product of two nonzero vectors is zero if the angle between the two is _____.   Answer: 90°, i.e., the vectors are perpendicular

(c) The scalar product of two vectors is negative if the angle between the two is _____.   Answer: >90° and ≤180°

(d) If $\mathbf{a} \cdot \mathbf{b} = ab$, then the two vectors are _____.
                          Answer: parallel, i.e., in the same direction

(e) If $\mathbf{a} \cdot \mathbf{b} = -ab$, then the two vectors are _____.
                                          Answer: in opposite directions

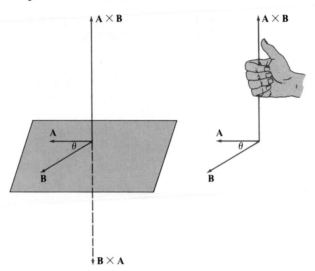

Figure 1.8
The right-hand rule for specifying the direction of **A** × **B**. When two vectors **A** and **B** are positioned with their tails at a common point, the vector **A** × **B** is perpendicular to the plane containing **A** and **B** in the direction of the thumb of the right hand when the fingers of that hand curve in the direction from **A** toward **B** through θ, the smaller of the two angles between **A** and **B**.

E 1.12 You know that real number multiplication is commutative, that is, if r_1 and r_2 are real numbers, then $r_1 r_2 = r_2 r_1$. Is this also true for the scalar product of two vectors? In other words, does **A** · **B** = **B** · **A**?

The *vector product* of two vectors **A** and **B** is written **A** × **B** and is sometimes called the "cross product." This vector product is defined such that the magnitude of the product is given by

$$|\mathbf{A} \times \mathbf{B}| = AB \sin\phi \qquad (1\text{-}5)$$

where, again, A and B are magnitudes of **A** and **B**, and ϕ is the smaller angle between the vectors **A** and **B** when they are drawn with their tails at a common point. The direction of the product **A** × **B** is perpendicular to both **A** and **B**, that is, it is perpendicular to the plane containing **A** and **B**. The direction of **A** × **B** is further specified by the so-called right-hand rule, in which the fingers of the right hand are curved in a direction from **A** toward **B** through the smaller angle between **A** and **B**. Then the extended thumb of the right hand is perpendicular to the plane of **A** and **B** and is in the direction of **A** × **B**. The right-hand rule determination of the direction of **A** × **B** is shown in Figure 1.8. **B** × **A** is a vector of the same magnitude as **A** × **B** but is in the opposite direction. Thus we conclude that **B** × **A** = −**A** × **B**; so the vector product is not commutative.

E 1.13 Calculate **a** × **b** for each of the vector pairs shown.
Answers: (a) 8.5, out of the page; (b) 8.5, into the page; (c) 8.5, out of the page; (d) 12, into the page; (e) zero

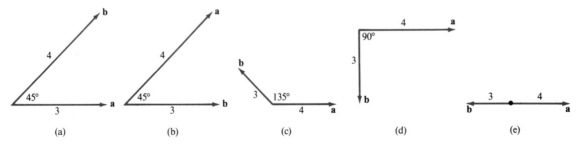

Exercise 1.13

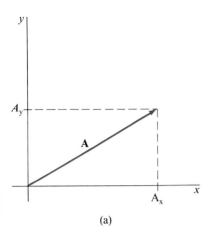

(a)

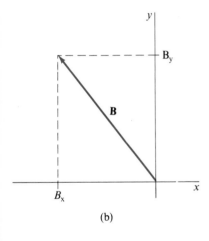

(b)

Figure 1.9
Rectangular components of vectors. **(a)** The rectangular components A_x and A_y of a vector **A** in the first quadrant are both positive numbers. **(b)** For a vector **B** lying in the second quadrant, its rectangular components are B_x, a negative number, and B_y, a positive number.

E 1.14   Complete the following sentences:

(a) The vector product of two nonzero vectors is zero if _____.
> Answer: (a) the two are parallel or antiparallel

(b) The magnitude of the vector product of two vectors of specified magnitudes is a maximum if _____.
> Answer: the two are perpendicular

(c) If $|\mathbf{a} \times \mathbf{b}| = ab$, then _____.
> Answer: the two are perpendicular

(d) If $|\mathbf{a} \times \mathbf{b}| = 0$ and $a \neq 0$, $b \neq 0$, then _____.
> Answer: the two are parallel or antiparallel

E 1.15   Explain why $\mathbf{a} \cdot (\mathbf{a} \times \mathbf{b}) = 0$ for any two vectors **a** and **b**.

Rectangular Components of Vectors

A vector can always be specified by its magnitude and direction, but it is often necessary or convenient to specify one by its *rectangular components* A_x and A_y (and A_z if the vector is three dimensional). A_x, the *x component* of the vector **A**, is the *x* coordinate of the head of **A** when the vector **A** is drawn with its tail at the origin of a coordinate system. The *y component* of **A**, A_y, is similarly defined. A_x and A_y are illustrated in Figure 1.9(a). Note that the rectangular components of a vector are real numbers that may be positive, zero, or negative. Figure 1.9(b) illustrates a vector **B** having a negative *x* component and a positive *y* component.

A vector may be represented by giving its magnitude and direction (polar representation) or by giving its *x* and *y* components (rectangular representation). The equivalence of these two representations and the process of changing from one representation to the other may be seen by considering the vector **C** of Figure 1.10(a). The magnitude of **C** is 10, and its direction is 30° (measured counterclockwise from the positive *x* axis, according to convention). In the polar representation, **C** may be written as $\mathbf{C} = 10\underline{/30°}$. The rectangular components of **C** are obtained by applying right-triangle trigonometry to Figure 1.10(a), giving

$$C_x = C \cos\theta \quad \text{and} \quad C_y = C \sin\theta \tag{1-6}$$

where $C = 10$ and $\theta = 30°$. Then the components are given by

$$C_x = 10 \cos30° = 8.7 \quad \text{and} \quad C_y = 10 \sin30° = 5.0 \tag{1-7}$$

Figure 1.10
(a) Determination of rectangular components of a vector **C** that is specified in the polar representation. **(b)** Determination of the polar representation of a vector **D** with rectangular components that are known.

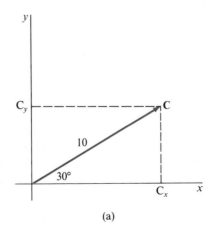

(a)

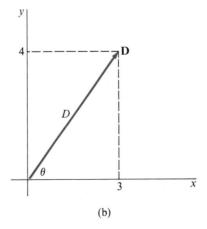

(b)

Alternatively, the magnitude and direction of a vector may be determined if its rectangular components are known. In Figure 1.10(b) the vector **D** has components $D_x = 3$ and $D_y = 4$. Using the Pythagorean theorem and trigonometry, we see that

$$D = \sqrt{D_x^2 + D_y^2} \text{ and } \tan\theta = \frac{D_y}{D_x} \tag{1-8}$$

so that

$$D = \sqrt{3^2 + 4^2} = 5$$

$$\tan\theta = \frac{4}{3} = 1.33 \text{ or } \theta = \tan^{-1}(1.33) = 53°$$

E 1.16 Calculate the rectangular components for each of the vectors shown.
Answers: $a_x = -1.7$, $a_y = 4.7$; $b_x = -3.0$, $b_y = -5.2$; $c_x = -2.7$, $c_y = 7.5$

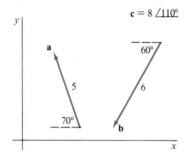

Exercise 1.16

E 1.17 Calculate the magnitude and direction of each of the vectors for which rectangular components are given. Answer in the form $\mathbf{a} = a\underline{/\theta}$, where θ is measured counterclockwise from the positive x direction.
(a) $a_x = 3.7$ $a_y = 2.9$ Answer: $4.7\underline{/38°}$
(b) $b_x = -2.2$ $b_y = -1.3$ (Be careful with the direction!)
                                                              Answer: $2.5\underline{/210°}$
(c) $c_x = 1.9$ $c_y = -2.2$ Answer: $2.9\underline{/310°}$

Vector Operations in Component Notation

So far vectors have been referred to as representations of quantities having both magnitude and direction. A more frequently useful representation in physical problems and in three-dimensional situations is *component notation*. In this form a vector is written in terms of its components

$$\mathbf{A} = A_x\hat{\mathbf{i}} + A_y\hat{\mathbf{j}} + A_z\hat{\mathbf{k}} \tag{1-9}$$

in which A_x, A_y, and A_z are the x, y, and z components of **A**. The symbols $\hat{\mathbf{i}}$, $\hat{\mathbf{j}}$, and $\hat{\mathbf{k}}$ are *unit vectors* in the positive x, y, and z directions, respectively, as shown in Figure 1.11. A unit vector has a length of one unit and is used simply to indicate direction. A vector **A** with components $A_x = 3$, $A_y = 4$, and $A_z = 2$ could, for example, be represented by

$$\mathbf{A} = 3\hat{\mathbf{i}} + 4\hat{\mathbf{j}} + 2\hat{\mathbf{k}} \tag{1-10}$$

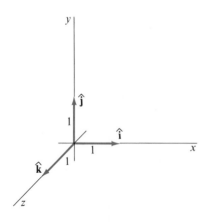

Figure 1.11
The unit vectors, $\hat{\mathbf{i}}$, $\hat{\mathbf{j}}$, and $\hat{\mathbf{k}}$ of a rectangular coordinate system.

The vector **A** is represented in Equation (1-10) in component notation and is shown in Figure 1.12. The *component vectors* of **A**, namely $3\hat{\mathbf{i}}$, $4\hat{\mathbf{j}}$, and $2\hat{\mathbf{k}}$, are seen in Figure 1.12(a) to be the projections of **A** onto the x, y, and z axes. Figure 1.12(b) illustrates how the vector **A** is the sum of its component vectors.

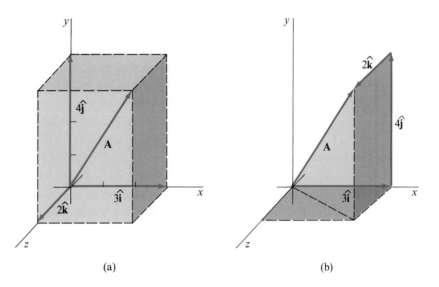

Figure 1.12
The rectangular component vectors of a vector $\mathbf{A} = 3\mathbf{i} + 4\mathbf{j} + 2\mathbf{k}$. **(a)** The component vectors are the projections of **A** onto the rectangular axes. **(b)** The vector **A** is the sum of its component vectors.

(a)   (b)

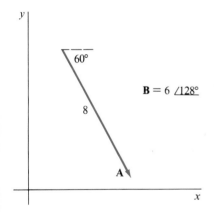

Exercise 1.18

E 1.18   Use the unit vectors $\hat{\mathbf{i}}$ and $\hat{\mathbf{j}}$ to write in rectangular form each of the vectors shown here.   Answers: $\mathbf{A} = 4.0\hat{\mathbf{i}} - 6.9\hat{\mathbf{j}}$; $\mathbf{B} = -3.7\hat{\mathbf{i}} + 4.7\hat{\mathbf{j}}$

E 1.19   Write in polar form each of the vectors that are specified here: $\mathbf{a} = 2.7\hat{\mathbf{i}} - 3.2\hat{\mathbf{j}}$; $\mathbf{b} = -1.3\hat{\mathbf{i}} - 4.8\hat{\mathbf{j}}$.   Answers: $\mathbf{a} = 4.2\underline{/311°}$; $\mathbf{b} = 5.0\underline{/260°}$

Component notation makes it particularly easy to add or subtract vectors. The sum of $\mathbf{A} = A_x\hat{\mathbf{i}} + A_y\hat{\mathbf{j}} + A_z\hat{\mathbf{k}}$ and $\mathbf{B} = B_x\hat{\mathbf{i}} + B_y\hat{\mathbf{j}} + B_z\hat{\mathbf{k}}$ is obtained by algebraically adding the individual components:

$$\mathbf{A} + \mathbf{B} = (A_x + B_x)\hat{\mathbf{i}} + (A_y + B_y)\hat{\mathbf{j}} + (A_z + B_z)\hat{\mathbf{k}} \qquad (1\text{-}11)$$

Vector subtraction is similarly accomplished, subtracting algebraically the individual components:

$$\mathbf{A} - \mathbf{B} = (A_x - B_x)\hat{\mathbf{i}} + (A_y - B_y)\hat{\mathbf{j}} + (A_z - B_z)\hat{\mathbf{k}} \qquad (1\text{-}12)$$

E 1.20   For the vectors **A** and **B** of Exercise 1.18, calculate $\mathbf{A} - \mathbf{B}$ and express the result in polar form.   Answer: $14\underline{/300°}$

E 1.21   Extending vector addition and subtraction to include three or more vectors is accomplished simply using component notation. For example,

$$\mathbf{A} + \mathbf{B} - \mathbf{C} = (A_x + B_x - C_x)\hat{\mathbf{i}} + (A_y + B_y - C_y)\hat{\mathbf{j}}$$

For the vectors $\mathbf{a} = 2.7\hat{\mathbf{i}} - 3.2\hat{\mathbf{j}}$, $\mathbf{b} = -1.3\hat{\mathbf{i}} - 4.8\hat{\mathbf{j}}$, and $\mathbf{c} = 2.2\hat{\mathbf{i}} + 1.8\hat{\mathbf{j}}$, calculate $\mathbf{b} - \mathbf{a} - \mathbf{c}$ and express the result in rectangular form.
   Answer: $3.6\hat{\mathbf{i}} - 6.2\hat{\mathbf{j}}$

Scalar multiplication and vector multiplication in component notation require that we take particular note of the unit vectors $\hat{\mathbf{i}}$, $\hat{\mathbf{j}}$, and $\hat{\mathbf{k}}$. Because these three unit vectors are mutually perpendicular, the definition of a scalar product (Equation [1-4], $\mathbf{A} \cdot \mathbf{B} = AB \cos\phi$) demands that

$$\hat{\mathbf{i}} \cdot \hat{\mathbf{i}} = 1 \qquad \hat{\mathbf{j}} \cdot \hat{\mathbf{i}} = 0 \qquad \hat{\mathbf{k}} \cdot \hat{\mathbf{i}} = 0$$
$$\hat{\mathbf{i}} \cdot \hat{\mathbf{j}} = 0 \qquad \hat{\mathbf{j}} \cdot \hat{\mathbf{j}} = 1 \qquad \hat{\mathbf{k}} \cdot \hat{\mathbf{j}} = 0 \qquad (1\text{-}13)$$
$$\hat{\mathbf{i}} \cdot \hat{\mathbf{k}} = 0 \qquad \hat{\mathbf{j}} \cdot \hat{\mathbf{k}} = 0 \qquad \hat{\mathbf{k}} \cdot \hat{\mathbf{k}} = 1$$

A right-handed coordinate system, which will be used exclusively in this book, is an arrangement of the x, y, and z axes such that $\hat{\mathbf{i}} \times \hat{\mathbf{j}} = \hat{\mathbf{k}}$ according to the right-hand rule of the vector cross product. Therefore, the definition of the vector product requires that

$$\hat{\mathbf{i}} \times \hat{\mathbf{i}} = 0 \qquad \hat{\mathbf{j}} \times \hat{\mathbf{i}} = -\hat{\mathbf{k}} \qquad \hat{\mathbf{k}} \times \hat{\mathbf{i}} = \hat{\mathbf{j}}$$
$$\hat{\mathbf{i}} \times \hat{\mathbf{j}} = \hat{\mathbf{k}} \qquad \hat{\mathbf{j}} \times \hat{\mathbf{j}} = 0 \qquad \hat{\mathbf{k}} \times \hat{\mathbf{j}} = -\hat{\mathbf{i}} \qquad (1\text{-}14)$$
$$\hat{\mathbf{i}} \times \hat{\mathbf{k}} = -\hat{\mathbf{j}} \qquad \hat{\mathbf{j}} \times \hat{\mathbf{k}} = \hat{\mathbf{i}} \qquad \hat{\mathbf{k}} \times \hat{\mathbf{k}} = 0$$

Using the scalar products of unit vectors of Equation (1-13), the scalar product of $\mathbf{A} = A_x\hat{\mathbf{i}} + A_y\hat{\mathbf{j}} + A_z\hat{\mathbf{k}}$ and $\mathbf{B} = B_x\hat{\mathbf{i}} + B_y\hat{\mathbf{j}} + B_z\hat{\mathbf{k}}$ becomes

$$\mathbf{A} \cdot \mathbf{B} = (A_x\hat{\mathbf{i}} + A_y\hat{\mathbf{j}} + A_z\hat{\mathbf{k}}) \cdot (B_x\hat{\mathbf{i}} + B_y\hat{\mathbf{j}} + B_z\hat{\mathbf{k}})$$
$$\mathbf{A} \cdot \mathbf{B} = A_xB_x + A_yB_y + A_zB_z \qquad (1\text{-}15)$$

All other terms of the scalar product, like $A_xB_y\hat{\mathbf{i}} \cdot \hat{\mathbf{j}}$, are zero because, according to Equation (1-13), only the $\hat{\mathbf{i}} \cdot \hat{\mathbf{i}}$, $\hat{\mathbf{j}} \cdot \hat{\mathbf{j}}$, and $\hat{\mathbf{k}} \cdot \hat{\mathbf{k}}$ terms are not zero.

Example 1.3

PROBLEM If $\mathbf{A} = 4\hat{\mathbf{i}} - 6\hat{\mathbf{j}} + 5\hat{\mathbf{k}}$ and $\mathbf{B} = 3\hat{\mathbf{i}} + 2\hat{\mathbf{j}} - 7\hat{\mathbf{k}}$, calculate (a) $\mathbf{A} \cdot \mathbf{B}$ and (b) the angle between $\mathbf{A}$ and $\mathbf{B}$.

SOLUTION (a) Because $\mathbf{A}$ and $\mathbf{B}$ are given in component form, Equation (1-15) gives

$$\mathbf{A} \cdot \mathbf{B} = A_xB_x + A_yB_y + A_zB_z$$
$$\mathbf{A} \cdot \mathbf{B} = (4)(3) + (-6)(2) + (5)(-7) = -35$$

(b) Because the definition of $\mathbf{A} \cdot \mathbf{B}$ is

$$\mathbf{A} \cdot \mathbf{B} = AB \cos\phi$$

the angle ϕ between $\mathbf{A}$ and $\mathbf{B}$ may be calculated according to

$$\cos\phi = \frac{\mathbf{A} \cdot \mathbf{B}}{AB}$$

$$\cos\phi = \frac{-35}{\sqrt{4^2 + 6^2 + 5^2} \, \sqrt{3^2 + 2^2 + 7^2}} = -0.507$$

$$\phi = 120° \quad \blacksquare$$

E 1.22 If $\mathbf{a} = 2\hat{\mathbf{i}} - 4\hat{\mathbf{j}} + 3\hat{\mathbf{k}}$, $\mathbf{b} = -6\hat{\mathbf{i}} + 4\hat{\mathbf{k}}$, and $\mathbf{c} = 5\hat{\mathbf{j}} - 7\hat{\mathbf{k}}$, calculate: (a) $\mathbf{a} \cdot \mathbf{b}$, (b) $\mathbf{a} \cdot \mathbf{c}$, (c) $\mathbf{b} \cdot \mathbf{c}$, (d) $\mathbf{a} \cdot (\mathbf{b} + \mathbf{c})$.

Answers: (a) 0; (b) -41; (c) -28; (d) -41

E 1.23 For $\mathbf{a}$, $\mathbf{b}$, and $\mathbf{c}$ of Exercise 1.22, which of the vectors $\mathbf{b}$, $\mathbf{c}$, or $(\mathbf{b} + \mathbf{c})$ is perpendicular to $\mathbf{a}$? Why?

Answer: $\mathbf{b}$; because $\mathbf{a} \cdot \mathbf{b} = 0 = ab \cos\phi$, $\cos\phi = 0$ and $\phi = 90°$

E 1.24   If $\mathbf{a} = 2\hat{\mathbf{i}} + 2\hat{\mathbf{j}} - \hat{\mathbf{k}}$, $\mathbf{b} = 9\hat{\mathbf{i}} + 20\hat{\mathbf{j}} + 12\hat{\mathbf{k}}$, and $\mathbf{c} = 2\hat{\mathbf{i}} - 2\hat{\mathbf{j}} + \hat{\mathbf{k}}$, calculate the angle between (a) $\mathbf{a}$ and $\mathbf{b}$, (b) $\mathbf{a}$ and $\mathbf{c}$, and (c) $\mathbf{b}$ and $\mathbf{c}$.

Answers: (a) 52°; (b) 96°; (c) 98°

The vector product of **A** and **B** in component form is somewhat more complicated.

$$\mathbf{A} \times \mathbf{B} = (A_y B_z - A_z B_y)\hat{\mathbf{i}} + (A_z B_x - A_x B_z)\hat{\mathbf{j}} + (A_x B_y - A_y B_x)\hat{\mathbf{k}} \qquad \textbf{(1-16)}$$

may be remembered more easily in the mnemonic form of a determinant:

$$\mathbf{A} \times \mathbf{B} = \begin{vmatrix} \hat{\mathbf{i}} & \hat{\mathbf{j}} & \hat{\mathbf{k}} \\ A_x & A_y & A_z \\ B_x & B_y & B_z \end{vmatrix} \qquad \textbf{(1-17)}$$

Example 1.4   **PROBLEM**   If the vectors **A** and **B** are those given in Example 1.3, namely, $\mathbf{A} = 4\hat{\mathbf{i}} - 6\hat{\mathbf{j}} + 5\hat{\mathbf{k}}$ and $\mathbf{B} = 3\hat{\mathbf{i}} + 2\hat{\mathbf{j}} - 7\hat{\mathbf{k}}$,
(a) calculate $\mathbf{A} \times \mathbf{B}$.
(b) verify that $|\mathbf{A} \times \mathbf{B}| = AB \sin\phi$, using ϕ as determined in Example 1.3.
(c) show that $\mathbf{A} \times \mathbf{B}$ is perpendicular to both **A** and **B**.
   SOLUTION   (a) Equation (1-16) applied to **A** and **B** gives

$$\mathbf{A} \times \mathbf{B} = (A_y B_z - A_z B_y)\hat{\mathbf{i}} + (A_z B_x - A_x B_z)\hat{\mathbf{j}} + (A_x B_y - A_y B_x)\hat{\mathbf{k}}$$

$$\mathbf{A} \times \mathbf{B} = [(-6)(-7) - (5)(2)]\hat{\mathbf{i}} + [(5)(3) - (4)(-7)]\hat{\mathbf{j}} + [(4)(2) - (-6)(3)]\hat{\mathbf{k}}$$

$$\mathbf{A} \times \mathbf{B} = 32\hat{\mathbf{i}} + 43\hat{\mathbf{j}} + 26\hat{\mathbf{k}}$$

Alternatively, Equation (1-17) may be used to obtain

$$\mathbf{A} \times \mathbf{B} = \begin{vmatrix} \hat{\mathbf{i}} & \hat{\mathbf{j}} & \hat{\mathbf{k}} \\ 4 & -6 & 5 \\ 3 & 2 & -7 \end{vmatrix}$$

$$\mathbf{A} \times \mathbf{B} = \hat{\mathbf{i}} \begin{vmatrix} -6 & 5 \\ 2 & -7 \end{vmatrix} - \hat{\mathbf{j}} \begin{vmatrix} 4 & 5 \\ 3 & -7 \end{vmatrix} + \hat{\mathbf{k}} \begin{vmatrix} 4 & -6 \\ 3 & 2 \end{vmatrix}$$

$$\mathbf{A} \times \mathbf{B} = 32\hat{\mathbf{i}} + 43\hat{\mathbf{j}} + 26\hat{\mathbf{k}}$$

(b) The magnitude of $\mathbf{A} \times \mathbf{B}$ is given by

$$|\mathbf{A} \times \mathbf{B}| = |32\hat{\mathbf{i}} + 43\hat{\mathbf{j}} + 26\hat{\mathbf{k}}|$$

$$|\mathbf{A} \times \mathbf{B}| = \sqrt{32^2 + 43^2 + 26^2} = 60$$

In Example 1.3 the angle between **A** and **B** was found to be 120°. It follows, then, that

$$AB \sin\phi = \sqrt{4^2 + 6^2 + 5^2}\, \sqrt{3^2 + 2^2 + 7^2}\, \sin 120° = 60$$

(c) To show that $\mathbf{A} \times \mathbf{B}$ is perpendicular to **A** and to **B**, we need only demonstrate that $\mathbf{A} \cdot (\mathbf{A} \times \mathbf{B})$ and $\mathbf{B} \cdot (\mathbf{A} \times \mathbf{B})$ are each equal to zero. This is accomplished by using the result of part (a) to get

$$\mathbf{A} \cdot (\mathbf{A} \times \mathbf{B}) = (4)(32) + (-6)(43) + (5)(26) = 0$$

$$\mathbf{B} \cdot (\mathbf{A} \times \mathbf{B}) = (3)(32) + (2)(43) + (-7)(26) = 0 \quad \blacksquare$$

E 1.25   If $\mathbf{A} = 4\hat{\mathbf{i}} - 7\hat{\mathbf{j}}$, $\mathbf{B} = 5\hat{\mathbf{i}} + 3\hat{\mathbf{j}}$, and $\mathbf{C} = -2\hat{\mathbf{i}} - 4\hat{\mathbf{j}}$, calculate:
(a) $\mathbf{A} \times \mathbf{B}$; (b) $\mathbf{B} \times \mathbf{C}$; (c) $\mathbf{A} \times \mathbf{C}$.   Answers: (a) $47\hat{\mathbf{k}}$; (b) $-14\hat{\mathbf{k}}$; (c) $-30\hat{\mathbf{k}}$

E 1.26   Using the vectors **a**, **b**, and **c**, as given in Exercise 1.24, calculate **a** × **b**, **b** × **c**, and **c** × **a**.

Answers: $44\hat{i} - 33\hat{j} + 22\hat{k}$; $44\hat{i} + 15\hat{j} - 58\hat{k}$; $4\hat{j} + 8\hat{k}$

1.4 Problem-Solving: A Strategy

For beginning students, the study of physics often amounts to solving problems. Indeed, learning basic physics means learning to solve "word problems," that is, problems stated in words. Of course, a knowledge of the principles, definitions, and relationships of physics is a necessary prerequisite to the successful study of physics. A considerable portion of this book is devoted to those fundamentals. This section, however, and part of each subsequent chapter, is concerned with the presentation and development of a strategy designed to assist the student in the successful solution of problems. Here, then, we present an overall strategy for solving problems, a process intended to be applicable to any word problem.

Word problems generally lend themselves to analysis in four steps, which may be characterized as follows:

Step 1.   Understanding and visualizing the problem. Identifying known and unknown quantities.

Step 2.   Relating unknown quantities to known quantities.

Step 3.   Executing the mathematics to obtain the solution.

Step 4.   Checking the solution.

Let us examine the steps individually. Step 1 may seem obvious, because it is necessary to understand what a problem is about in order to attempt its solution with any chance of success. But understanding a problem means understanding all the words used to state it, not only their literal meanings but also their implications for the problem at hand. For example, the word *frictionless* appears frequently in physics problems. Most people would say this word suggests smoothness, but that association is not enough. In a physics problem, a frictionless surface can push on objects only in specific directions, and this implication may be crucial in the analysis of a given problem. Therefore, the technical definitions and implications of the words in a problem should be clearly understood before tackling the problem.

A second aspect of understanding a word problem hinges on recognizing common assumptions and agreements between the writer of the problem and the problem solver. For example, few problems specifically state that the physical situation of the problem takes place on or near the surface of the earth. It is generally understood that if the problem involves an alien planet, that condition would be specified in the statement of the problem. A perfectly unambiguous problem that details every possible contingency would require unreasonable space and effort for its statement. Common sense is the appropriate guide in this respect.

Step 1 also suggests visualizing the problem. Picturing the problem in one's mind is not usually sufficient; a graphic representation, such as a drawing or sketch, however crude, is an immense aid in understanding a word problem. The process of translating the literal meanings of the words of a problem into an accurate mental image of what is taking place is a crucial aspect of problem-solving. Making a drawing, if possible, is the surest means of making a successful transition from words to a useful and effective mental image. In particular, it is helpful to identify the known and unknown quantities in the problem (the last task in Step 1) by putting appropriate labels with quantities directly on the drawing. Thus, a known distance may be labeled $x = 12$ m and an unknown force indicated by $F = ?$ This technique makes the issues of the problem clearer and more meaningful.

Step 2 offers the opportunity to use the principles, definitions, and relationships that have been learned. Every problem asks the solver to find some unknown quantity. To find it, the solver must relate that unknown quantity to those that are known or given in the problem. Here is where the student calls on the resources acquired during the study of principles and relationships. This step, relating unknown quantities to known quantities, usually becomes easier with experience and practice. In physics problems, the process of relating "unknowns" to "knowns" is almost always accomplished in the form of equations in which symbols are used to represent physical quantities.

Step 3, solving the equations obtained in Step 2, is not usually a concern of students prepared to study physics. It is assumed that students using this text can carry out the appropriate mathematical operations necessary to obtain the solution.

Step 4 serves as a check on both the mathematical execution of Step 3 and the selection of appropriate relationships in Step 2. Usually it is not possible to check the solution of a physics problem by direct substitution, in the sense that the solution of most algebra problems may be checked, but some procedures are often available that provide confidence in the reasonableness of a solution. The most obvious check is the examination of the units associated with the solution. If an answer that is supposed to be a distance turns up with units of time, it is immediately clear that an error has been made. Less obvious procedures, like estimations, are often appropriate. This type of check is illustrated in the four-step-strategy example that follows. Finally, solutions to physics problems may frequently be checked by what are sometimes called "limiting cases." The ability to make these checks effectively usually depends on the problem solver's experience in recognizing what occurs in simple physical situations and relating or comparing a particular solution to the simple situation. As an example, suppose we have found a solution that describes the motion of an object moving along a horizontal surface. We could examine the solution to see if it predicts what we would expect if the surface were frictionless. Although such procedures do not guarantee that a solution is correct, they afford a measure of confidence in the answer and provide the further benefit of exercising and strengthening our physical insight into the problem.

Let us now look at the use of the four-step strategy with a simple problem:

How long after midnight does it take for the hands of a clock to coincide again?

Step 1. The only word in the problem that may need clarifying is *coincide*. The hands of a clock coincide when they point in the same direction. What are the assumptions made in the statement of the problem? Without belaboring this issue, the problem obviously assumes that we know the functional principles of how a clock indicates the passage of time. We are supposed to know that both hands of a clock coincide at midnight, that the minute hand moves through an

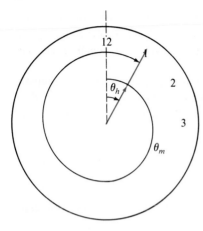

Figure 1.13
The hands of a clock coinciding for the first time after midnight. The figure shows the angles θ_h and θ_m through which the hour hand and minute hand have turned since midnight.

angle of 360° every hour, and that the hour hand moves only 1/12 as fast, that is, through an angle of 30° every hour. To visualize the pertinent aspects of the problem, construct a simple diagram (*see* Fig. 1.13) that shows the positions of both hands of a clock when they first coincide after midnight. The figure also shows the angle through which each hand has turned since midnight. Let θ_m be the angle through which the minute hand has turned and θ_h be the angle through which the hour hand has turned.

Step 2. This problem requires that we understand the notion of rate, that the minute hand moves at an angular rate of 360°/h, and the hour hand turns at the rate of 30°/h. The mathematical relationship that we can establish is suggested by the units in the problem. An angular rate R has the unit (deg/h). If we multiply R by the time t with the unit (h), we obtain an angle θ, measured in (deg). Thus, the appropriate relationship here is

$$\theta(\text{deg}) = R\left(\frac{\text{deg}}{\text{h}}\right) t(\text{h})$$

or

$$\theta(\text{deg}) = Rt(\text{deg}) \tag{1-18}$$

We now apply Equation (1-18) to each hand of the clock, recognizing certain facts from the figure. We do not know the angle θ_h through which the hour hand has moved in the time t that has elapsed since midnight, so we write

$$\theta_h = 30t \tag{1-19}$$

Neither θ_h nor t is a known value. Obviously we cannot solve one equation with two unknown quantities. We proceed, therefore, by considering further information in the figure, the information associated with the minute hand. The minute hand has moved through an angle θ_m, which is clearly $360° + \theta_h$, in the same elapsed time t. Then for the minute hand, we may write

$$\theta_m = 360t$$

$$360 + \theta_h = 360t \tag{1-20}$$

Because t is the time elapsed since midnight, both Equations (1-19) and (1-20) include the same unknown quantities, θ_h and t. We now have a sufficient number of relationships to determine t, the desired quantity representing the time elapsed between midnight and the instant the hands next coincide.

Step 3. The execution of the mathematical solution is, in this case, a simple algebraic substitution and manipulation. Substituting Equation (1-19) for θ_h into Equation (1-20) gives

$$360 + 30t = 360t$$

$$360 = 330t$$

$$t = \frac{360}{330} = \frac{12}{11} = 1\frac{1}{11} \text{ h} \tag{1-21}$$

Because t is the elapsed time since midnight, the result of Equation (1-21) is the value required in the statement of the problem.

Step 4. How may we check the result? We may estimate what we might have expected without working the problem, namely in somewhat more than 1.0 h. When the minute hand has completed one revolution from its position at midnight, the hour hand has moved to 1 on the clock. Five minutes later, when

$t = 1\frac{1}{12}$ h, the minute hand has reached 1 on the clock, but the hour hand has moved forward slightly. It should then not take much longer for the hands to coincide. Because $1\frac{1}{11}$ h is only slightly greater than $1\frac{1}{12}$ h, we conclude that our answer is probably correct.

A second example, in which the steps are not discussed as extensively, demonstrates further the utility of this problem-solving process.

Example 1.5

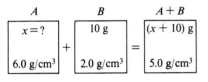

Figure 1.14   Example 1.5

PROBLEM   How many grams of a powder A with a density of 6.0 g/cm^3 must be added to 10 g of a powder B, the density of which is 2.0 g/cm^3, to give a mixture of density 5.0 g/cm^3?

SOLUTION   Let x be the number of grams in powder A. Figure 1.14 depicts the situation. If the volume of A is added to the volume of B, we obtain the volume of $A + B$, or

$$V_A + V_B = V_{A+B} \tag{1-22}$$

The units suggest that the volumes of A, B, and A + B may be obtained by dividing the number of grams of each material by the density of that material. Then Equation (1-22) becomes

$$\frac{x \text{ g}}{6 \text{ g/cm}^3} + \frac{10 \text{ g}}{2 \text{ g/cm}^3} = \frac{(x + 10) \text{ g}}{5 \text{ g/cm}^3}$$

$$\frac{x}{6} \text{ cm}^3 + 5 \text{ cm}^3 = \frac{x + 10}{5} \text{ cm}^3 \tag{1-23}$$

Equation (1-23) may be solved algebraically:

$$5x + 150 = 6x + 60$$

$$x = 90 \text{ g}$$

Thus, 90 g of A (6.0 g/cm^3) must be added to 10 g of B (2.0 g/cm^3) to produce 100 g of a mixture with density 5.0 g/cm^3. This result may be checked by direct substitution into Equation (1-23):

$$\frac{90}{6} + 5 = \frac{90 + 10}{5}$$

$$20 = 20 \quad \blacksquare$$

This problem-solving approach will be utilized throughout this book in three ways, with *Problems*, *Examples*, and *Problem-Solving Summaries:*

1. Each chapter contains illustrative *Problems* within the text. In these problems, extended discussions relate the principles, definitions, and relationships being studied to the problem-solving process. These problems emphasize specialized techniques and hints that are applicable to similar problems. Problem-solving hazards are pointed out in these discussions, and appropriate safeguards are recommended.

2. Each chapter includes *Examples*, typical problems associated with the topics of that chapter. These examples, which use the problem-solving strategy without extended discussion, are intended as models for the student.

3. Each chapter contains a *Problem-Solving Summary*, which reiterates the principles and relationships of the chapter in terms of problem-solving. The summary offers suggestions on how problem categories may be identified: When is a

problem about energy, or momentum, or rotation, for example. Each chapter summary suggests how the basic problem-solving strategy may be modified or specialized to apply most effectively to the specific topical material of that chapter. Finally, the problem-solving summary of each chapter attempts to identify common pitfalls of problems and suggests how these hazards may be recognized and avoided.

Problems

Note to the Student
   The problems at the end of each chapter are separated into two groups, A and B. Group A problems are based on material contained in the preceding text. Group B problems are, in general, more challenging and frequently present new material.

GROUP A

1.1 The density of mercury is 13.6 g/cm³.

(a) Calculate the density of mercury in kg/m³.

(b) One kilogram of mass weighs approximately 2.2 pounds. Could you lift a suitcase filled with mercury if the dimensions of the suitcase are 15 cm × 50 cm × 60 cm?

1.2 A person who weighs 150 lb has a mass of approximately 70 kg. Using the fact that an average person floats in water with very little exposed volume and, therefore, has approximately the same density as water, namely 1.0 g/cm³, estimate the volume in m³ and ft³ of a person weighing 150 lb.

1.3 If $\mathbf{A} = 8\hat{\mathbf{i}} - 9\hat{\mathbf{j}}$ and $\mathbf{B} = 12\underline{/300°}$,

(a) express **A** in polar form.

(b) express **B** in rectangular form.

(c) calculate $\mathbf{A} + \mathbf{B}$.

(d) calculate $\mathbf{A} \cdot \mathbf{B}$.

1.4 If **k** and **m** are the vectors shown,

(a) express **k** in polar form.

(b) express **m** in rectangular form.

(c) calculate $|2\mathbf{k} + \mathbf{m}|$.

(d) calculate $\mathbf{k} \cdot \mathbf{m}$.

1.5 Point B is 15 m east of point A, and point C is 10 m south of point B.

(a) How far is point C from point A?

(b) What is the direction from point A to point C?

(c) Let $\hat{\mathbf{i}}$ point to the east and $\hat{\mathbf{j}}$ to the north. Express the vector **d** drawn from point A to point C in rectangular and polar forms.

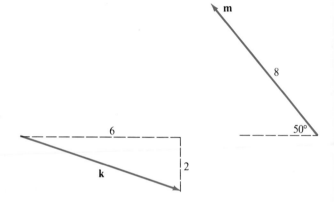

Figure 1.15   Problem 1.4

1.6 If $\mathbf{r} = 6\underline{/110°}$ and $\mathbf{s} = 8\underline{/340°}$, calculate

(a) $\mathbf{r} - \mathbf{s}$.

(b) $2\mathbf{r} + 3\mathbf{s}$.

(c) $\mathbf{r} \cdot \mathbf{s}$.

(d) $\mathbf{r} \times \mathbf{s}$.

1.7 If $\mathbf{K} = 7\hat{\mathbf{i}} - 3\hat{\mathbf{j}} + 2\hat{\mathbf{k}}$ and $\mathbf{L} = 4\hat{\mathbf{i}} + 5\hat{\mathbf{j}} - 3\hat{\mathbf{k}}$, calculate

(a) the magnitudes of **K** and **L**.

(b) $|\mathbf{K} + \mathbf{L}|$.

(c) $\mathbf{K} \cdot \mathbf{L}$.

(d) the cosine of the angle between **K** and **L**.

(e) the angle between **K** and **L**.

1.8 If $\mathbf{D} = -5\hat{\mathbf{i}} + 6\hat{\mathbf{j}} - 3\hat{\mathbf{k}}$ and $\mathbf{E} = 7\hat{\mathbf{i}} + 8\hat{\mathbf{j}} + 4\hat{\mathbf{k}}$, determine

(a) the magnitudes of **D** and **E**.

(b) $\mathbf{D} \cdot \mathbf{E}$.

(c) $\mathbf{D} \times \mathbf{E}$.

(d) the angle between **D** and **E**.

1.9 Using vectors **R**, **S**, and **T** (*see* Fig. 1.16), evaluate

(a) **R** · (**S** × **T**).

(b) **R** × (**S** × **T**).

(c) (**S** · **T**) **R**.

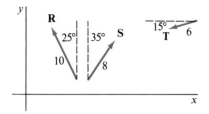

Figure 1.16   Problem 1.9

1.10 Given that **r** = 12/70°, **s** = −8**î** + 6**ĵ**, **t** = 4**î** + 5**ĵ**, and **u** = 2**r** − 3**s** + **t**, express **u** in rectangular form.

1.11 Points B and C are located relative to the origin, as shown in Figure 1.17.

(a) Calculate the length of the line segment connecting points B and C.

(b) Calculate the angle between the line segment connecting B and C and the line segment connecting O and B.

(c) If **d** is a vector starting at B and ending at C, write **d** in polar form and in component form.

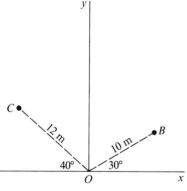

Figure 1.17   Problem 1.11

1.12 For the vectors **A**, **B**, and **C** in Figure 1.18, calculate

(a) **A** · (**B** − **C**).

(b) (**A** · **B**) **C**.

(c) |**A** − **B**|.

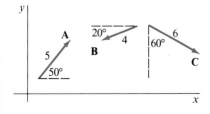

Figure 1.18   Problem 1.12

1.13 Given that **A** = 5/40°, **B** = 8/230°, **C** = 4/340°, and **D** = 2**A** + **B** − 3**C**, express **D** in polar form.

1.14 An airplane starting from airport A flies 300 km east, then 350 km 30° west of north, and then 150 km north to arrive finally at airport B. There is no wind on this day.

(a) In what direction should the pilot fly to travel directly from A to B?

(b) How far will the pilot travel in this direct flight?

1.15 A racetrack consists of two parallel sides with semicircular ends, as shown in Figure 1.19. The vector **d** connects track positions A and B. Express **d** in polar form.

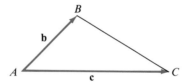

Figure 1.19   Problem 1.15

1.16 Consider the triangle ABC with vectors **b** and **c** constructed, as shown in Figure 1.20. Show that $\frac{1}{2}$|**b** × **c**| equals the area of the triangle.

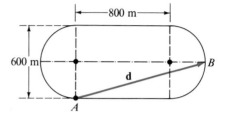

Figure 1.20   Problem 1.16

1.17 Consider the 2 ft × 3 ft × 5 ft rectangular volume shown in Figure 1.21.

(a) Calculate the length of a vector drawn diagonally from corner A to corner B.

(b) Find the angle between the diagonal vector of part (a) and a vector drawn from point A to point C.

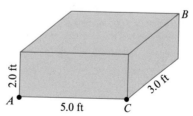

Figure 1.21   Problem 1.17

1.18 A person walks from A to B to C along the path shown in Figure 1.22. The distance R is 35 ft.

(a) Find the total distance walked.

(b) If **D** is a vector from A to C, express **D** in terms of **î** and **ĵ**.

(c) Express **D** in polar form.

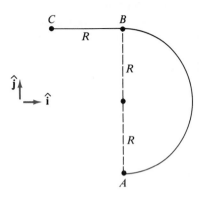

Figure 1.22   Problem 1.18

GROUP **B**

1.19 Unit vectors are used to indicate direction. Given any vector **v**, a unit vector parallel to **v**, written **v̂**, is given by **v**/v. If **a** = $2\hat{\imath} - \hat{\jmath} + 2\hat{k}$ and **b** = $9\hat{\imath} + 20\hat{\jmath} + 12\hat{k}$, find

(a) a unit vector **â** in the same direction as **a**.

(b) a unit vector **b̂** in the same direction as **b**.

(c) a unit vector **ĉ** perpendicular to both **a** and **b**.

(d) Is **ĉ** unique? Why?

1.20 Any vector **b** can be written as the sum of two mutually perpendicular vectors $\mathbf{b}_{\parallel}$ and $\mathbf{b}_{\perp}$ that are parallel (or antiparallel) to and perpendicular to any other vector **a**, as depicted in Figure 1.23.

(a) Show that $\mathbf{b}_{\parallel} = (\hat{\mathbf{a}} \cdot \mathbf{b})\hat{\mathbf{a}}$ where $\hat{\mathbf{a}} = \mathbf{a}/a$ is a unit vector parallel to **a**.

(b) Show that $\mathbf{b}_{\perp} = \mathbf{b} - (\hat{\mathbf{a}} \cdot \mathbf{b})\hat{\mathbf{a}}$.

(c) Verify directly that $\mathbf{b}_{\parallel}$ and $\mathbf{b}_{\perp}$ are perpendicular.

(d) If **a** = $12\hat{\imath} - 16\hat{\jmath} + 15\hat{k}$ and **b** = $10\hat{\imath} + 10\hat{\jmath} - 5\hat{k}$, calculate $\mathbf{b}_{\parallel}$ and $\mathbf{b}_{\perp}$.

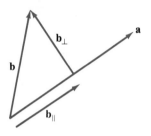

Figure 1.23   Problem 1.20

1.21 If **A** = $6\hat{\imath} - 8\hat{\jmath}$, **B** = $-8\hat{\imath} + 3\hat{\jmath}$, and **C** = $26\hat{\jmath} + 19\hat{\jmath}$, determine a and b so that

$$a\mathbf{A} + b\mathbf{B} + \mathbf{C} = 0$$

2 Particle Kinematics

Kinematics is the branch of mechanics that describes the motion of physical bodies. It is distinguished from *dynamics*, which considers relationships between the motions of physical bodies and the forces (pushes or pulls) that are associated with those motions.

For the present we shall be interested in the kinematics of *particles*, models of physical bodies that may be considered mathematical points for the purpose of mathematical and physical analyses. A physical body may be regarded as a particle when only its *translational* motion is being considered, that is, when any vibrations or rotations of the body are of no consequence to the problem at hand. Large or small bodies may be treated as particles. The earth in its orbit or an automobile moving along a highway may each be thought of as a particle provided we are interested in its translational motion only. Thus, in describing translational motion of physical bodies, we may use the model of a point particle. In this chapter, we shall consider the kinematics of particles.

To be physically meaningful a description of motion must be measurable relative to some frame of reference, such as a coordinate system that could be established within a laboratory. With this in mind, a description of the motion of a particle at a given instant should answer the following questions:

1. Where is the particle relative to a chosen reference frame?

2. How fast is the particle moving, and in what direction is it moving?

3. To what extent and in which direction is its motion changing? In this chapter we shall consider how these questions may be answered quantitatively and how the answers are interpreted and applied.

2.1 Motion Along a Straight Line (Rectilinear Motion)

Position, Velocity, and Acceleration

Perhaps the simplest example of kinematics is the motion of a particle constrained to move along a straight line. Suppose such a particle is moving along

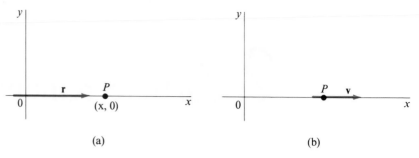

Figure 2.1
(a) The position vector **r** locating a particle P lying on the axis. **(b)** The instantaneous velocity vector **v** for a particle P moving along the x axis.

the x axis, as shown in Figure 2.1. In Figure 2.1(a) the *position* of the particle is specified by a position vector **r**, a vector that originates at the coordinate origin and terminates at the particle P. As the particle moves, the vector **r** changes, so **r** is a function of time and may be written as **r**(t) to emphasize its dependence on time. Then for this one-dimensional case, the position vector **r** is specified by

$$\mathbf{r}(t) = x(t)\hat{\mathbf{i}} \tag{2-1}$$

The average velocity $\bar{\mathbf{v}}$ of a particle as it moves from a position $\mathbf{r}_1$ at time t_1 to a position $\mathbf{r}_2$ at time t_2 is defined to be

$$\bar{\mathbf{v}} = \frac{\mathbf{r}_2 - \mathbf{r}_1}{t_2 - t_1} = \frac{x_2 - x_1}{t_2 - t_1}\hat{\mathbf{i}} = \frac{\Delta x}{\Delta t}\hat{\mathbf{i}} \tag{2-2}$$

where $\Delta x\hat{\mathbf{i}} = (x_2 - x_1)\hat{\mathbf{i}}$ is the displacement of the particle during the time interval $\Delta t = t_2 - t_1$. Both displacement and average velocity are vector quantities.

The instantaneous velocity **v** of the particle at point P in Figure 2.1 is defined to be the limit of the average velocity as the time interval Δt approaches zero while the displacement always includes the point P. In the notation of the calculus, instantaneous velocity is

$$\mathbf{v}(t) = \lim_{\Delta t \to 0} \frac{\Delta x}{\Delta t}\hat{\mathbf{i}} = \frac{dx}{dt}\hat{\mathbf{i}} = v_x(t)\hat{\mathbf{i}} \tag{2-3}$$

Thus, the instantaneous velocity $\mathbf{v}(t)$, a vector quantity illustrated in Figure 2.1(b), is the rate at which the position of a particle is changing with respect to time. The magnitude of **v** at a given instant is $|dx/dt| = |v_x|$, the *speed* of the particle; the motion of the particle is in the positive x direction if dx/dt is positive and in the negative x direction if dx/dt is negative. The speed of the particle at any instant is how fast the particle is moving at that instant. Its speed at a given instant is equal to the distance the particle would traverse each second if it proceeded at the rate it is moving at the given instant.

The rate at which the velocity of a particle is changing at any instant is the *instantaneous acceleration* **a** of that particle. Then the acceleration

$$\mathbf{a}(t) = \lim_{\Delta t \to 0} \frac{\Delta \mathbf{v}}{\Delta t} = \frac{d\mathbf{v}}{dt} = \frac{dv_x}{dt}\hat{\mathbf{i}} = a_x(t)\hat{\mathbf{i}} \tag{2-4}$$

is a vector quantity. Here the magnitude of the instantaneous acceleration is $|dv_x/dt|$, which specifies how much the speed is changing per unit time at a given instant. If the rate dv_x/dt is positive, the acceleration is in the positive x direction; if dv_x/dt is negative, the acceleration of the particle is in the negative x direction.

The unit of position is the meter or the foot, depending on the system of units being used. Similarly, the unit of velocity or speed is m/s (or ft/s), and the unit of

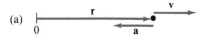

(a)

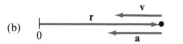

(b)

(c)

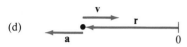

(d)

Exercise 2.1

acceleration is m/s^2 (or ft/s^2). The unit of acceleration may become more mean-ingful if it is rewritten (m/s)/s, which suggests that a constant acceleration of 3 m/s^2 means that the velocity is changing 3 m/s every second.

Problem-solving and analysis of rectilinear motion can be simplified by using the position, velocity, and acceleration vectors rather than the unit vector no-tation, because x, v_x, and a_x contain equivalent information to $\mathbf{r}$, $\mathbf{v}$, and $\mathbf{a}$. For example, suppose that at a given instant the position, velocity, and acceleration of a particle are $\mathbf{r} = 3\,\hat{\mathbf{i}}$ m, $\mathbf{v} = 4\,\hat{\mathbf{i}}$ m/s, and $\mathbf{a} = -2\,\hat{\mathbf{i}}$ m/s^2. It follows that $x = +3$ m (the particle is 3 m from the origin in the positive x direction), $v_x = +4$ m/s (the particle is moving in the positive x direction with a speed of 4 m/s), and $a_x = -2$ m/s^2 (the acceleration of the particle is 2 m/s^2 in the negative x direction). Further, it can be deduced that at that given instant the distance from the origin to the particle (3 m) is increasing because x and v_x have the same sign. Similarly, the speed of the particle is decreasing at the given instant because v_x and a_x have different signs.

E 2.1   Each of the accompanying figures shows position, velocity, and acceleration vectors for an object in rectilinear motion. For each figure indicate: (i) whether the distance from the origin to the object is increasing or decreasing, and (ii) whether the speed of the object is increasing or decreasing.

Answers: (a) increasing, decreasing; (b) decreasing, increasing; (c) increasing, increasing; (d) decreasing, decreasing

E 2.2   An electric car moves airport commuters back and forth along a straight path from point A to B, a point 2000 ft away from A. Point O, halfway between A and B, is chosen as a coordinate origin with the direction from O to B taken as positive.
(a) If the car is 600 ft from B and is moving 15 ft/s toward B with its speed increasing at 2 ft/s^2, what are x, v_x, and a_x?

Answers: 400 ft, 15 ft/s, 2 ft/s^2
(b) At a given instant when the position, velocity, and acceleration of the car are $\mathbf{r} = -200\hat{\mathbf{i}}$ ft, $\mathbf{v} = -12\hat{\mathbf{i}}$ ft/s, and $\mathbf{a} = 3.0\hat{\mathbf{i}}$ ft/s^2, how far is the car from B, how fast is it moving, and at what rate is its speed changing and is it increas-ing or decreasing?   Answers: 1200 ft, 12 ft/s, decreasing at 3.0 ft/s^2

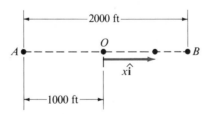

Exercise 2.2

E 2.3   A particle is moving along the x axis. At one instant $x = -4.0$ m, $v_x = 2.0$ m/s, and $a_x = 1.0$ m/s^2.
(a) Write expressions for $\mathbf{r}$, $\mathbf{v}$, and $\mathbf{a}$ at that instant.

Answers: $-4.0\hat{\mathbf{i}}$ m, $2.0\hat{\mathbf{i}}$ m/s, $1.0\hat{\mathbf{i}}$ m/s^2
(b) State whether the distance from the origin to the particle is increasing or decreasing at that instant. At what rate?   Answers: decreasing, 2.0 m/s
(c) State whether the speed of the particle is increasing or decreasing at that instant. At what rate?   Answers: increasing, 1.0 m/s^2

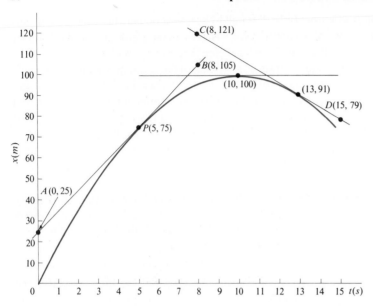

Figure 2.2   Example 2.1

Example 2.1

PROBLEM   The position of a particle moving along the x axis is shown in Figure 2.2 as a function of time for $0 \leq t \leq 15$ s. Using graphical techniques, determine the velocity v_x of the particle at $t = 5$ s, 10 s, and 13 s.

SOLUTION   The velocity v_x at any time t_1 is dx/dt, which is equal to the slope of a line drawn tangent to the curve $x(t)$ at the point (t_1, x_1) where $x_1 = x(t_1)$. The figure shows these tangent lines for the times $t = 5$ s, 10 s, and 13 s. The velocities of the particle are equal to the slopes of these lines:

$$v_x(5) = \frac{x_B - x_A}{t_B - t_A} = \frac{105 \text{ m} - 25 \text{ m}}{8 \text{ s} - 0} = 10 \text{ m/s}$$

$$v_x(10) = 0$$

$$v_x(13) = \frac{x_D - x_C}{t_D - t_C} = \frac{79 \text{ m} - 121 \text{ m}}{15 \text{ s} - 8 \text{ s}} = -6 \text{ m/s}$$

Thus, at $t = 5$ s the particle is moving in the positive x direction with a speed of 10 m/s. At $t = 10$ s, it is momentarily at rest. At $t = 13$ s its motion is in the negative direction at a speed of 6 m/s. ∎

E 2.4   Figure 2.2 shows the position of a particle as a function of time. The curve is a graph of

$$x(t) = (20t - t^2) \text{ m}$$

for $0 \leq t \leq 15$ s. Use the calculus to verify the results obtained in Example 2.1 for $v_x(5)$, $v_x(10)$, and $v_x(13)$.

E 2.5   The position of a motorcycle as a function of time is expressed by $x = 24t^2 - t^3$, where x is measured in feet and t in seconds for $0 \leq t \leq 10$ s. (a) Determine the velocity of the motorcycle as a function of time.

Answer: $(48t - 3t^2)$ ft/s

(b) Determine its acceleration as a function of time.   Answer: $(48 - 6t)$ ft/s^2
(c) When does the speed reach its maximum value? What is this maximum speed?   Answer: 8 s, 190 ft/s = 130 mi/h

Constant Acceleration in Rectilinear Motion

Several special cases arise in the study of kinematics. These are limited situations, so commonly encountered and of such practical value that they receive particular attention in introductory studies. Students are usually expected to be familiar with such special cases and their applications. Here and in the following section we shall consider two important situations that are special cases of rectilinear motion: particles moving with constant acceleration and particles in free-fall.

When the acceleration of a particle does not change during our span of interest, we say that the acceleration of the particle is constant. Then Equation (2-4) may be written (suppressing the vector notation) as

$$a_x = \frac{dv_x}{dt} \quad ; \quad a_x \text{ constant} \tag{2-5}$$

The derivative form of Equation (2-5) suggests that there is an antiderivative form (indefinite integral) that may be written

$$v_x = \int a_x \, dt + C \quad ; \quad a_x \text{ constant} \tag{2-6}$$

where C is a constant of integration. Then,

$$v_x = a_x \int dt + C$$

$$v_x = a_x t + C \tag{2-7}$$

The constant of integration C is determined by specifying that when $t = 0$, the particle has a velocity v_{ox}, usually called the initial velocity. Then Equation (2-7) becomes, when $t = 0$,

$$v_{ox} = a_x \cdot (0) + C$$

$$C = v_{ox}$$

which, when substituted into Equation (2-7) gives

$$v_x = v_{ox} + a_x t \tag{2-8}$$

This equation predicts the velocity v_x of a particle moving along a straight line with constant acceleration a_x at any time t if we know its velocity v_{ox} at time $t = 0$.

In a similar way, $v_x = dx/dt$ has a corresponding antiderivative relationship,

$$x = \int v_x \, dt + C' \tag{2-9}$$

Here we can substitute the expression obtained in Equation (2-8) for v_x:

$$x = \int (v_{ox} + a_x t) \, dt + C'$$

Because a_x is a constant (as is v_{ox}), x becomes

$$x = v_{ox}t + \tfrac{1}{2}a_x t^2 + C' \qquad (2\text{-}10)$$

Here again, the constant of integration is determined by specifying the position of the particle at $t = 0$ to be x_o. Then Equation (2-10) determines C' to be

$$x_o = v_{ox} \cdot (0) + \tfrac{1}{2}a_x \cdot (0)^2 + C'$$

or

$$C' = x_o \qquad (2\text{-}11)$$

and Equation (2-10) becomes

$$x = x_o + v_{ox}t + \tfrac{1}{2}a_x t^2 \qquad (2\text{-}12)$$

In this relationship x_o, v_{ox}, and a_x are constants; and Equation (2-12) predicts the position of the particle at any time t if the constants are known. Equations (2-8) and (2-12) are the key relationships in rectilinear kinematics when the acceleration of a particle is constant. Note that these equations relate velocity to time and position to time. To relate velocity to position, solve Equation (2-8) for t and substitute that solution into Equation (2-12). The result is

$$v_x^2 = v_{ox}^2 + 2a_x(x - x_o) \qquad (2\text{-}13)$$

The special case of constant acceleration is important in elementary mechanics. These relationships, Equations (2-8), (2-12), and (2-13), will be used frequently, and each should be clearly understood. They are listed here:

$$v_x = v_{ox} + a_x t$$

$$x = x_o + v_{ox}t + \tfrac{1}{2}a_x t^2 \qquad (2\text{-}14)$$

$$v_x^2 = v_{ox}^2 + 2a_x(x - x_o)$$

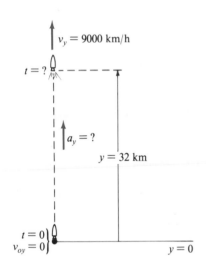

$v_y = 9000$ km/h

$t = ?$

$a_y = ?$

$y = 32$ km

$t = 0$
$v_{oy} = 0$

$y = 0$

Figure 2.3   Example 2.2

Example 2.2

PROBLEM   A rocket launched from rest at ground level rises vertically and reaches a speed of 9000 km/h (2.5 km/s) at a height of 32 km above the launch point. Assuming the acceleration of the rocket to be constant, determine this acceleration and the time required for this portion of the flight.

SOLUTION   Figure 2.3 shows the known quantities v_{oy}, y, and v_y and the unknown quantities a_y and t. The constant acceleration equation, which relates the known values of v_{oy}, y, and v_y to the desired acceleration, is

$$v_y^2 = v_{oy}^2 + 2a_y(y - y_o)$$

Because $v_{oy} = 0$, $y_o = 0$, $v_y = 9000$ km/h $= 2.5 \times 10^3$ m/s, and $y = 32$ km $= 3.2 \times 10^4$ m, we have

$$v_y^2 = 2a_y y$$

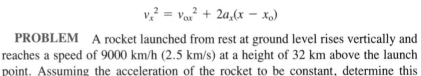

$$a_y = \frac{v_y^2}{2y} = \frac{(2.5 \times 10^3 \text{ m/s})^2}{2(3.2 \times 10^4 \text{ m})} = 98 \text{ m/s}^2$$

The time required for this period of acceleration is obtained using the relationship

$$v_y = v_{oy} + a_y t$$

Because $v_{oy} = 0$ and $a_y = v_y^2/2y$ (from above), we obtain

$$v_y = \frac{v_y^2}{2y} t$$

$$t = \frac{2y}{v_y} = \frac{2 \times 3.2 \times 10^4 \text{ m}}{2.5 \times 10^3 \text{ m/s}} = 26 \text{ s} \quad \blacksquare$$

E 2.6   A certain not-too-swift automobile accelerates from rest to 20 m/s (45 mi/h) in 15 s. Assuming constant acceleration,
(a) what is this acceleration?   Answer: 1.3 m/s²
(b) how far does the vehicle travel during this time?   Answer: 150 m

E 2.7   A minimum landing distance of 800 ft is required for a small airplane, which touches down at a speed of 50 mi/h (73 ft/s). Assuming constant acceleration,
(a) what is this acceleration?   Answer: -3.4 ft/s² (Why is it negative?)
(b) how much time is required for the plane to stop?   Answer: 22 s

E 2.8   The muzzle velocity of a bullet is observed to be 640 m/s after traveling the length of a 1.2-m rifle barrel. Assuming constant acceleration, find (a) the acceleration of the bullet, and (b) the time the bullet spends in the barrel after firing.   Answers: (a) 1.7×10^5 m/s²; (b) 3.8×10^{-3} s = 3.8 ms

Although emphasis so far has been on the kinematic special case in which acceleration of a particle is constant, it should be noted that Equation (2-4) has an antiderivative form in which a_x is not constant but is a function of time:

$$v_x(t) = \int a_x(t) \, dt + C$$

This case of nonconstant acceleration will not be treated here.

Free-Fall

A particle that is released in the vicinity of the surface of the earth is said to be in *free-fall*, whether it is ascending, descending, or momentarily at rest. Experiments show that such a particle experiences a downward (toward the center of the earth) acceleration of magnitude 9.8 m/s² = 32 ft/s², for which we use the symbol g, the magnitude of the gravitational acceleration. We shall adopt the convention that *g is a positive value*. Usually the positive y direction is designated as the upward direction, and the negative y direction as the downward. With these conventions, then, a particle in free-fall is described by Equations (2-14) with, of course, y substituted for x:

$$y = y_o + v_{oy}t - \tfrac{1}{2}gt^2$$

$$v_y = v_{oy} - gt$$

$$v_y^2 = v_{oy}^2 - 2g(y - y_o) \qquad (2\text{-}15)$$

In Equations (2-15) the acceleration is $a_y = -g$ because of the sign conventions.

In this study of free-fall and throughout this text in the treatment of particles moving through the air, the resistance of the air on the particle is not considered unless that effect is specifically identified.

Example 2.3

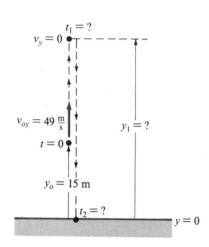

Figure 2.4   Example 2.3

PROBLEM   A rock is thrown with a velocity of 49 m/s upward from a point 15 m above the ground.
(a) When does the rock reach its maximum height?
(b) Calculate the maximum height of the rock above the ground.
(c) When does the rock hit the ground?

SOLUTION   The position and velocity of the rock are given as functions of time by

$$y = y_o + v_{oy}t - \frac{1}{2}gt^2 = (15 + 49t - 4.9t^2) \text{ m}$$
$$v_y = v_{oy} - gt = (49 - 9.8t) \text{ m/s}$$

(a) As shown in Figure 2.4, at the position of maximum height the rock is momentarily at rest, that is,

$$v_y = 0$$
$$49 - 9.8t = 0$$
$$t_1 = 5.0 \text{ s}$$

(b) Then the maximum height y_1 is equal to $y(5)$:

$$y_1 = y(5) = 15 + 49(5) - 4.9(5)^2 = 140 \text{ m}$$

(c) At the instant the rock returns to the ground, y is equal to zero, so that

$$15 + 49t - 4.9t^2 = 0$$

The two roots of this quadratic equation are -0.30 s and 10 s. The positive root is, of course, the one desired:

$$t_2 = 10 \text{ s}$$

Can you offer an interpretation of the negative root? ∎

E 2.9   A rock is launched vertically from the ground at $t = 0$ with an initial speed of 85 ft/s.
(a) When does the rock reach its maximum height?   Answer: 2.6 s
(b) What is this maximum height?   Answer: 110 ft
(c) What are its velocity and acceleration at this maximum height?
                                          Answers: zero, -32 ft/s^2
(d) What are its speed and direction of motion 3.0 s after launch?
                                          Answer: 12 ft/s, downward
(e) What is its velocity when it is 80 ft above the ground?
                                          Answer: ± 46 ft/s. (Be sure you understand the physical significance of the two signs.)

E 2.10   A ball is dropped at $t = 0$ from the top of a building. The top is 45 m above the ground.
(a) When does the ball strike the ground?   Answer: 3.0 s
(b) What is the speed of the ball just before hitting the ground?
                                          Answer: 30 m/s

2.2 Motion in a Plane

The description of the motion of a particle that moves in more than one dimension is necessarily more complicated than that of a particle in rectilinear motion. Nevertheless, the concepts and treatments of position, velocity, and acceleration are similar. In this section we shall consider the kinematics of particles in two dimensions.

Position, Velocity, and Acceleration

As a particle moves through space, it traces out a path called its *trajectory*. The position of the particle on its trajectory at any instant is specified by a position vector **r** that begins at the origin of the coordinate system and terminates at the particle *P* (*see* Fig. 2.5[a]). The instantaneous velocity of the particle has a direction tangential to the trajectory in the direction the particle is moving, as shown in Figure 2.5(b).

Figure 2.5
(a) The position vector **r** locating the particle *P* on its trajectory. **(b)** The instantaneous velocity **v** of the particle *P*. When the vector **v** is drawn originating from the particle, it is tangential to the trajectory of the particle.

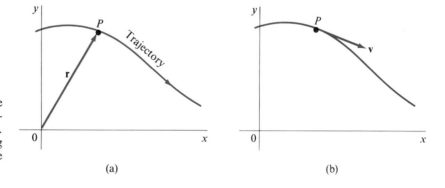

(a)   (b)

For a particle moving in a plane, the analytical expressions for position, velocity, and acceleration have, in general, two rectangular components:

$$\mathbf{r} = x\hat{\mathbf{i}} + y\hat{\mathbf{j}} \tag{2-16}$$

$$\mathbf{v} = \frac{d\mathbf{r}}{dt} = \frac{dx}{dt}\hat{\mathbf{i}} + \frac{dy}{dt}\hat{\mathbf{j}} = v_x\hat{\mathbf{i}} + v_y\hat{\mathbf{j}} \tag{2-17}$$

$$\mathbf{a} = \frac{d\mathbf{v}}{dt} = \frac{dv_x}{dt}\hat{\mathbf{i}} + \frac{dv_y}{dt}\hat{\mathbf{j}} = a_x\hat{\mathbf{i}} + a_y\hat{\mathbf{j}} \tag{2-18}$$

Position, velocity, and acceleration vectors may be written in two forms: the rectangular form used in Equations (2-16), (2-17), and (2-18) or the polar form. In their polar forms, $\mathbf{r} = r\,\underline{/\alpha}$, $\mathbf{v} = v\,\underline{/\beta}$, and $\mathbf{a} = a\,\underline{/\gamma}$, the magnitudes and the directions of these vectors are expressly emphasized. These component forms, $\mathbf{r} = x\hat{\mathbf{i}} + y\hat{\mathbf{j}}$, $\mathbf{v} = v_x\hat{\mathbf{i}} + v_y\hat{\mathbf{j}}$, and $\mathbf{a} = a_x\hat{\mathbf{i}} + a_y\hat{\mathbf{j}}$, are particularly useful in applications in which it is convenient to think of the simultaneous motions of two points, one moving along the *x* axis described by the kinematic variables x, v_x, and a_x, and the other moving along the *y* axis described by y, v_y, and a_y. This application of component forms of vectors will be useful in studies of projectile motion.

E 2.11   At a certain instant the position, velocity, and acceleration of a particle are $\mathbf{r} = (12\hat{\mathbf{i}} - 27\hat{\mathbf{j}})$ m, $\mathbf{v} = (21\hat{\mathbf{i}} + 19\hat{\mathbf{j}})$ m/s, and $\mathbf{a} = (-15\hat{\mathbf{i}} + 11\hat{\mathbf{j}})$ m/s². At that instant,

(a) write $\mathbf{r}$, $\mathbf{v}$, and $\mathbf{a}$ in polar form.

Answers: 30 m $\underline{/294°}$, 28 m/s $\underline{/42°}$, 19 m/s² $\underline{/144°}$

(b) how far is the particle from the origin?   Answer: 30 m
(c) how fast is the particle traveling?   Answer: 28 m/s
(d) what is its direction of motion?   Answer: 42°

Equations (2-16), (2-17), and (2-18) provide the analytical mechanism for specifying the velocity of a particle in terms of its position and for specifying its acceleration in terms of that velocity. But these questions do not, in this form, lend themselves directly to a clear interpretation of the physical relationships between these quantities. Let us see, then, how they are related physically.

Suppose a position vector $\mathbf{r}$ locates a particle moving with velocity $\mathbf{v}$ at a particular instant. Figure 2.6(a) shows $\mathbf{r}$ and $\mathbf{v}$ at that instant with the velocity vector drawn originating at the position of the particle. Instead of examining the x and y components of the velocity vector, let us look at the components of $\mathbf{v}$ that are parallel and perpendicular to the position vector $\mathbf{r}$. The component parallel to $\mathbf{r}$, $\mathbf{v}_\parallel$, tells us, in effect, how rapidly $\mathbf{r}$ is changing in magnitude independently of any change in the direction of $\mathbf{r}$. In other words, $\mathbf{v}_\parallel$ is a measure of how the position of the particle is changing radially; if $\mathbf{v}_\parallel$ is away from the origin, the distance of the particle from the origin is increasing, and if $\mathbf{v}_\parallel$ is toward the origin, the vector $\mathbf{r}$ is decreasing in magnitude. In Figure 2.6(a), for example, the magnitude of $\mathbf{r}$ is increasing. On the other hand, $\mathbf{v}_\perp$ indicates the extent to which the direction of $\mathbf{r}$ is changing independently of any change in the magnitude (or length) of $\mathbf{r}$. The direction of $\mathbf{v}_\perp$ also indicates whether θ is increasing or decreasing; that is, the direction of $\mathbf{v}_\perp$ determines which way $\mathbf{r}$ is turning about the origin (clockwise or counterclockwise). In Figure 2.6(a), for example, $\mathbf{r}$ is turning clockwise.

Now look at the similar situation in Figure 2.6(b) that relates the velocity and acceleration of the same particle. Note that the velocity vector has been drawn at the coordinate origin, and the acceleration vector has been drawn originating at the head of the velocity vector. The component of the particle's acceleration that is parallel to the velocity $\mathbf{v}$, $\mathbf{a}_\parallel$, indicates the rate at which the magnitude of $\mathbf{v}$ is changing independently of any change in the direction of $\mathbf{v}$. Thus, if $\mathbf{a}_\parallel$ lies in the

Figure 2.6
(a) Resolution of a velocity vector $\mathbf{v}$ of a particle into component vectors parallel and perpendicular to the position vector $\mathbf{r}$ of the particle. (b) Resolution of the acceleration vector $\mathbf{a}$ of a particle into component vectors parallel and perpendicular to the velocity vector $\mathbf{v}$ of the particle.

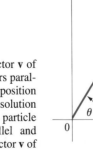

(a)

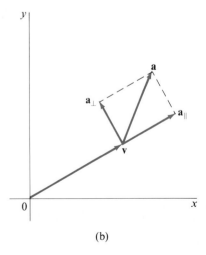

(b)

same direction as **v**, the speed (the magnitude of **v**) of the particle is increasing; and if $\mathbf{a}_\parallel$ is antiparallel to **v**, the speed of the particle is decreasing. Is the speed increasing or decreasing in Figure 2.6(b)? The component of **a** that is perpendicular to **v**, $\mathbf{a}_\perp$, indicates the extent to which the velocity is changing direction independently of any change in speed. Here again, the direction of $\mathbf{a}_\perp$ indicates the sense (clockwise or counterclockwise) in which **v** is turning. Which way is **v** turning in Figure 2.6(b)?

These physical considerations will be especially useful in this chapter when we consider circular motion.

E 2.12   At a certain time the position, velocity, and acceleration of a particle are given by $\mathbf{r} = (3\hat{\mathbf{i}} + 7\hat{\mathbf{j}})$ m, $\mathbf{v} = 4\hat{\mathbf{j}}$ m/s, and $\mathbf{a} = (2\hat{\mathbf{i}} - 3\hat{\mathbf{j}})$ m/s^2. At this instant,

(a) how is the magnitude of **r** changing?   Answer: increasing
(b) is **r** turning in a clockwise (CW) or a counterclockwise (CCW) direction?
                                                                    Answer: CCW
(c) is the speed of the particle increasing or decreasing?   Answer: decreasing
(d) how is **v** turning?   Answer: CW

E 2.13   If $\mathbf{r} = 15$ m $\underline{/300°}$, $\mathbf{v} = 10$ m/s $\underline{/180°}$, and $\mathbf{a} = 8.0$ m/s^2 $\underline{/90°}$,

(a) how is **r** changing at this instant?
                            Answer: decreasing in magnitude and turning CW
(b) how is **v** changing at this instant?
                            Answer: not changing in magnitude and turning CW

Projectile Motion

A particle released or projected near the surface of the earth with a horizontal component of velocity executes what is called *projectile motion*. If such a motion is described within a coordinate system oriented so that the x axis is horizontal and the y axis vertical, the particle experiences constant acceleration of magnitude g in the negative y direction during its flight. Again we will disregard the effects of air resistance on the motion of the particle. In this idealization, the particle experiences no acceleration in a horizontal direction. Thus the particle experiences an acceleration such that $a_x = 0$ and $a_y = -g$ during its entire flight.

Suppose, then, we consider a particle launched with an initial velocity $\mathbf{v}_o$ that has a direction in the plane of the trajectory at an angle θ above the positive x axis, as seen in Figure 2.7. The trajectory may be analyzed by placing the origin

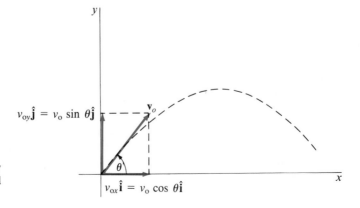

Figure 2.7
The resolution of the initial velocity vector of a particle into its vertical and horizontal component vectors.

of a horizontal-vertical rectangular coordinate system so that it coincides with the initial ($t = 0$) position of the particle. Such an orientation permits an analysis in terms of the separate vertical and horizontal components of the motion. Thus, the initial velocity vector $\mathbf{v}_o$ has rectangular components

$$v_{ox} = v_o \cos\theta \qquad (2\text{-}19)$$

$$v_{oy} = v_o \sin\theta \qquad (2\text{-}20)$$

Consider the horizontal direction first. Because $a_x = 0$, Equation (2-8) becomes

$$v_x = v_{ox} \qquad (2\text{-}21)$$

for all $t \geq 0$ during the flight. Because of our choice for the placement of the coordinate origin, x_o is also zero, and Equation (2-12) becomes

$$x = v_{ox}t \qquad (2\text{-}22)$$

Similarly, the vertical motion, in which $y_o = 0$ and $a_y = -g$, is described by the free-fall relationships, Equation (2-15),

$$y = v_{oy}t - \tfrac{1}{2}gt^2 \qquad (2\text{-}23)$$

$$v_y = v_{oy} - gt \qquad (2\text{-}24)$$

for all $t \geq 0$.

By eliminating t from Equations (2-22) and (2-23), we may see that the trajectory of a projectile is a parabola. That parabolic path is an idealization that results from our having neglected air resistance. In actuality, a projectile moving through the air has an acceleration with both horizontal and vertical components. As a consequence, an actual projectile moving through air will neither rise as high nor travel as far as our analysis indicates.

Equations (2-19) through (2-24) describe projectile motion, and the examples and exercises that follow will suggest how specific information about projectile motion may be obtained. Here let us indicate how the velocity of the particle is determined at any instant during the flight of a projectile. Figure 2.8 shows the particle at an arbitrary point P in its trajectory. The velocity of the particle at this particular time is the vector sum of the horizontal and vertical components of velocity,

$$\mathbf{v} = v_x\hat{\mathbf{i}} + v_y\hat{\mathbf{j}} \qquad (2\text{-}25)$$

where v_x and v_y are given by Equations (2-21) and (2-24). The magnitude of the velocity is

$$v = \sqrt{v_x^2 + v_y^2} \qquad (2\text{-}26)$$

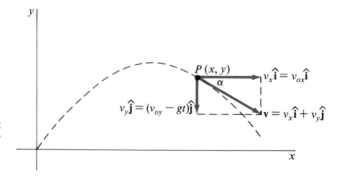

Figure 2.8
The horizontal and vertical component vectors of the velocity $\mathbf{v}$ at an arbitrary point P on the trajectory of a projectile.

Its direction is tangent to the trajectory but is usually specified relative to one of the axes, such as the angle α measured from the horizontal direction in Figure 2.8 in which

$$\alpha = \tan^{-1}\left(\frac{v_y}{v_x}\right) \tag{2-27}$$

Example 2.4

PROBLEM   A ball is thrown at $t = 0$ with an initial velocity $\mathbf{v}_o = 100$ ft/s $\diagup 60°$.

(a) When does the ball reach its maximum height?
(b) What is the maximum height attained by the ball?
(c) What are the position (relative to its $t = 0$ position), velocity, and acceleration of the ball at the position of maximum height?

SOLUTION   The components of the initial velocity are

$$v_{ox} = v_o \cos 60° = (100 \text{ ft/s})(0.5) = 50 \text{ ft/s}$$

$$v_{oy} = v_o \sin 60° = (100 \text{ ft/s})(0.866) = 87 \text{ ft/s}$$

Thus, the position, velocity, and acceleration components of the ball are

$$x = v_{ox}t = (50t) \text{ ft} \qquad y = v_{oy}t - \frac{1}{2}gt^2 = (87t - 16t^2) \text{ ft}$$

$$v_x = v_{ox} = 50 \text{ ft/s} \qquad v_y = v_{oy} - gt = (87 - 32t) \text{ ft/s}$$

$$a_x = 0 \qquad a_y = -g = -32 \text{ ft/s}^2$$

(a) At the position of maximum height, the y component of the velocity of the ball is momentarily zero, that is,

$$0 = v_y = 87 - 32t$$

$$t = 2.7 \text{ s}$$

(b) The maximum height h, then, is equal to the value of y at $t = 2.7$ s:

$$h = 87(2.7) - 16(2.7)^2 = 120 \text{ ft}$$

(c) At the instant of maximum height, the position, velocity, and acceleration of the ball, shown in Figure 2.9, are

$$\mathbf{r} = (50)(2.7)\hat{\mathbf{i}} + 120\hat{\mathbf{j}} = (130\hat{\mathbf{i}} + 120\hat{\mathbf{j}}) \text{ ft}$$

$$\mathbf{v} = 50\hat{\mathbf{i}} + 0\hat{\mathbf{j}} = 50\hat{\mathbf{i}} \text{ ft/s}$$

$$\mathbf{a} = 0\hat{\mathbf{i}} - 32\hat{\mathbf{j}} = -32\hat{\mathbf{j}} \text{ ft/s}^2$$

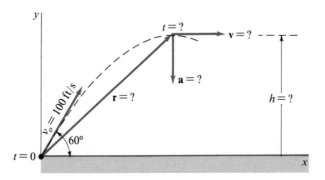

Figure 2.9   Example 2.4

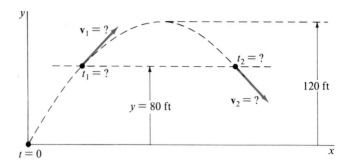

Figure 2.10   Example 2.5

Example 2.5

PROBLEM   Consider the trajectory of the ball described in Example 2.4.
(a) When is the ball 80 ft above its initial position?
(b) What is the velocity of the ball when it is 80 ft above its initial position?

  SOLUTION   Figure 2.10 shows the trajectory of the ball with the maximum height of 120 ft, as determined in Example 2.4. We expect to find two times when the ball is 80 ft above the initial position.
(a) Setting the height y of the ball equal to 80 ft gives

$$80 = 87t - 16t^2$$

$$16t^2 - 87t + 80 = 0$$

The roots of this quadratic equation are

$$t_1 = 1.2 \text{ s} \quad ; \quad t_2 = 4.2 \text{ s}$$

which are the times when the ball is 80 ft above the initial position.
(b) The velocity of the ball at each of these times is

$$\mathbf{v}_1 = \mathbf{v}(t_1) = 50\hat{\mathbf{i}} + [87 - (32)(1.2)]\hat{\mathbf{j}} = (50\hat{\mathbf{i}} + 48\hat{\mathbf{j}}) \text{ ft/s}$$

$$\mathbf{v}_2 = \mathbf{v}(t_2) = 50\hat{\mathbf{i}} + [87 - (32)(4.2)]\hat{\mathbf{j}} = (50\hat{\mathbf{i}} - 48\hat{\mathbf{j}}) \text{ ft/s}$$

At the earlier time t_1, the velocity $\mathbf{v}_1$ of the ball has a positive y component, that is, the ball is still ascending. The negative y component of $\mathbf{v}_2$ indicates that the ball is descending at the later time t_2. ■

E 2.14   A projectile is launched at $t = 0$ from ground level over flat terrain with an initial velocity given by 57 m/s $\underline{/60°}$.
(a) Calculate v_{ox} and v_{oy}.   Answers: 28 m/s, 49 m/s
(b) Write expressions for $v_x(t)$ and $v_y(t)$.
                                    Answers: $v_x(t) = 28$ m/s, $v_y(t) = (49 - 9.8t)$ m/s
(c) Assuming $x = y = 0$ at $t = 0$, write expressions for $x(t)$ and $y(t)$.
                                    Answers: $x(t) = (28t)$ m, $y(t) = (49t - 4.9t^2)$ m

E 2.15   Consider the projectile of Exercise 2.14.
(a) What is its velocity at the instant it reaches its maximum height?
                                                      Answer: 28 $\hat{\mathbf{i}}$ m/s
(b) Because $v_y = 0$ at the maximum height, at what time is this maximum height achieved?   Answer: 5.0 s
(c) What is this maximum height?   Answer: 120 m
(d) What is its acceleration at this maximum height?   Answer: $-9.8\hat{\mathbf{j}}$ m/s^2

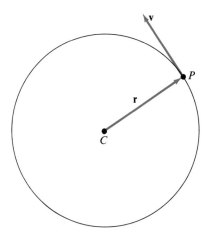

Figure 2.11
The velocity vector **v** of a particle *P* located by the position vector **r** relative to the center *C* of the circular path of the particle.

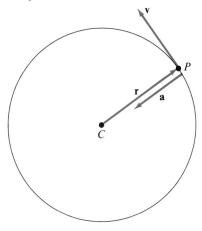

Figure 2.12
The acceleration vector **a**, velocity vector **v**, and position vector **r** for a particle *P* executing uniform circular motion about the center *C* of the circular path of the particle.

E 2.16   Consider the projectile of Exercise 2.14.
(a) Determine **v** at *t* = 3.0 s and 7.0 s.

Answers: $(28\hat{\mathbf{i}} + 20\hat{\mathbf{j}})$ m/s, $(28\hat{\mathbf{i}} - 20\hat{\mathbf{j}})$ m/s

(b) Recall that $\mathbf{a} = (-9.8\hat{\mathbf{j}})$ m/s^2 at all times during the flight of the projectile. Is the speed of the projectile increasing, decreasing, or not changing at times *t* = 3.0 s and 7.0 s?   Answers: decreasing, increasing

Uniform Circular Motion

An important special case in kinematics is that of a particle moving with uniform speed in a circular path. This situation has some unique features that we shall now consider.

A particle *P* in uniform circular motion, moving in a circular path with constant speed, is shown in Figure 2.11. The particle is located relative to the center *C* of the circle by a position vector **r**, which has a magnitude that does not change as the particle moves but with a direction that is changing continuously. Also, because the speed of the particle remains constant, the velocity vector does not change in magnitude, but the direction of **v** is continuously changing.

We have seen earlier in this chapter that the component of the acceleration perpendicular to the velocity is a measure of the rate at which the velocity vector is changing direction while the speed remains constant. Figure 2.12 illustrates the relationships among **r**, **v**, and **a** for a particle executing uniform circular motion. The position vector **r** does not change in magnitude, and **v** is perpendicular to **r**, indicating that the direction of **r** is changing (counterclockwise in Fig. 2.12). Similarly, because the magnitude of **v** is not changing (the speed of the particle is constant), **a** is perpendicular to **v**. Also, because **v** itself is turning counterclockwise, the direction of **a** is toward the center *C*. Because **a** is toward the center of the circle, it is called the *centripetal acceleration*. For a particle in uniform circular motion, the acceleration of the particle is *always* toward the center. The magnitude of the centripetal acceleration of a particle can be determined as indicated by the diagrams in Figure 2.13. As a particle in uniform circular motion moves from point *A* to point *B*, it traces out an arc length Δs in a time Δt. If Δt is a sufficiently short time, Δs is approximately a straight line. The change in velocity $\Delta \mathbf{v}$ takes place in the same time Δt, so that $\mathbf{v}_2 = \mathbf{v}_1 + \Delta \mathbf{v}$. The

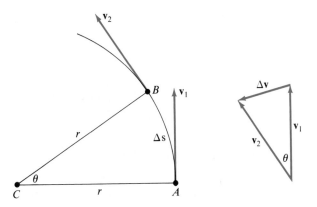

Figure 2.13
A particle in uniform circular motion.

velocity triangle is similar to the triangle *ABC* because **v** is perpendicular to **r** at all times. Then, we may write

$$\frac{|\Delta v|}{v} \cong \frac{\Delta s}{r}$$

$$|\Delta v| \cong \frac{v}{r}\Delta s \qquad (2\text{-}28)$$

where v is the magnitude $|v_1| = |v_2|$, because the speed is constant. Dividing both sides of Equation (2-28) by Δt gives

$$\frac{|\Delta v|}{\Delta t} \cong \frac{v}{r}\frac{\Delta s}{\Delta t} \qquad (2\text{-}29)$$

Taking the limit of both sides of Equation (2-29) as $\Delta t \to 0$, we obtain

$$\lim_{\Delta t \to 0}\frac{|\Delta v|}{\Delta t} = \frac{v}{r}\lim_{\Delta t \to 0}\frac{\Delta s}{\Delta t}$$

$$\left|\frac{dv}{dt}\right| = \frac{v}{r}\frac{ds}{dt}$$

$$a_c = \frac{v^2}{r} \qquad (2\text{-}30)$$

where $a_c = |dv/dt|$ and $v = ds/dt$. In the limit, as $\Delta t \to 0$, Δv is perpendicular to v; and the acceleration is, as we have already noted, toward the center of the circular path of the particle. The magnitude of the centripetal acceleration, Equation (2-30), is obtained in another way in Problem 2.39 (*see* Problems, Group B).

E 2.17   An automobile travels at a constant speed of 80 km/h (22 m/s) around a flat, circular (radius = 300 m) roadway. Calculate the magnitude of its acceleration.   Answer: 1.7 m/s²

E 2.18   A skilled pilot flies a stunt plane in a vertical circle of 1200-ft radius at a constant speed of 140 mi/h (210 ft/s). Calculate the pilot's acceleration at: (a) the highest point, and (b) the lowest point on the circle.
Answers: (a) 35 ft/s², downward; (b) 35 ft/s², upward

E 2.19   The earth rotates one revolution about its own axis every 24 h. If the mean radius of the earth is 6.4×10^6 m, calculate: (a) the speed of a point on the equator of the earth, and (b) the magnitude of the centripetal acceleration of this point.   Answers: (a) 460 m/s = 1000 mi/h; (b) 0.034 m/s²

E 2.20   The earth revolves around the sun once each year in an approximately circular path of mean radius 9.3×10^7 mi. Calculate: (a) the orbital speed of the earth, and (b) the magnitude of the earth's centripetal acceleration.
Answers: (a) 9.8×10^4 ft/s = 6.7×10^4 mi/h; (b) 0.020 ft/s² = 0.0060 m/s²

2.3 Relative Motion

In some physical situations it becomes important to realize that a particle is moving relative to more than one reference frame (coordinate system) at the same time. For example, an airplane may be moving relative to the air in which it is flying while, at the same time, that air is moving relative to the surface of the earth. Clearly, the motion of the airplane would be described differently by a balloonist floating along with the air and an observer on the earth. These descriptions are, of course, related, and such relationships are often useful. We can discover these relationships of relative motion from the following analysis.

Look at the points A, B, and C in Figure 2.14(a). The position vector $\mathbf{r}_{BC}$ locates the point B, the origin of the coordinate axes S', relative to the point C, which is the origin of another set of coordinate axes S. Similarly, the vector $\mathbf{r}_{AB}$ locates the point A relative to B, and $\mathbf{r}_{AC}$ locates A relative to C. The orientation of the arrowheads on the three position vectors tells us that the location vectors are related by

$$\mathbf{r}_{AC} = \mathbf{r}_{BC} + \mathbf{r}_{AB}$$

or

$$\mathbf{r}_{AC} = \mathbf{r}_{AB} + \mathbf{r}_{BC} \qquad (2\text{-}31)$$

By taking the time derivative of both sides of Equation (2-31), we obtain

$$\frac{d\mathbf{r}_{AC}}{dt} = \frac{d\mathbf{r}_{AB}}{dt} + \frac{d\mathbf{r}_{BC}}{dt}$$

or

$$\mathbf{v}_{AC} = \mathbf{v}_{AB} + \mathbf{v}_{BC} \qquad (2\text{-}32)$$

which relates the relative velocities. Here, again, $\mathbf{v}_{AB}$ means the velocity of the point A relative to B, and so on. Equation (2-32) means that a vector triangle such as that of Figure 2.14(b) exists for the three relative velocity vectors. Solving the vector equation is equivalent to solving for the unknown values of the vector triangle. The equation and the triangle each contain six values, three magnitudes (the lengths of the sides of the triangle) and three directions.

A memory device may be helpful in recalling the velocity relationship for relative velocity problems. Notice in Equation (2-32) that the subscripts on the resultant vector of the sum are the same as the outside subscripts of the two added vectors. The third subscript appears as the inner subscript on the added vectors. This rule may be applied to construct the appropriate relationship using any of

Figure 2.14
(a) Locating the point A relative to two sets of coordinate axes S and S'. (b) A vector triangle relating the velocity vectors $\mathbf{v}_{AB}$, $\mathbf{v}_{BC}$, and $\mathbf{v}_{AC}$ such that $\mathbf{v}_{AC} = \mathbf{v}_{AB} + \mathbf{v}_{BC}$.

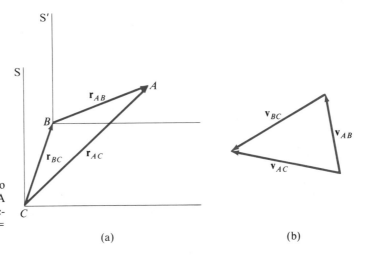

(a)   (b)

the three vectors as the sum of the other two. For example, if we are to determine $\mathbf{v}_{AB}$, we could write

$$\mathbf{v}_{AB} = \mathbf{v}_{AC} + \mathbf{v}_{CB} \qquad (2\text{-}33)$$

It should be observed, however, that in a given problem, the value of $\mathbf{v}_{BC}$ may be given when $\mathbf{v}_{CB}$ is required. The velocity of B relative to C is oppositely directed to the velocity of C relative to B:

$$\mathbf{v}_{BC} = -\mathbf{v}_{CB} \qquad (2\text{-}34)$$

Example 2.6

PROBLEM   An ocean liner moving west with a speed of 20 km/h passes an oil tanker moving with a speed of 12 km/h in a direction 30° east of north. How fast and in what direction is the tanker moving relative to the liner?

SOLUTION   Let $\mathbf{v}_{TL}$ be the velocity of the tanker relative to the liner, $\mathbf{v}_{TW}$ be the velocity of the tanker relative to the water, and $\mathbf{v}_{LW}$ be the velocity of the liner relative to the water. These velocities are related by

$$\mathbf{v}_{TW} = \mathbf{v}_{TL} + \mathbf{v}_{LW}$$

Because we seek the velocity of the tanker relative to the liner $\mathbf{v}_{TL}$, we rewrite this as

$$\mathbf{v}_{TL} = \mathbf{v}_{TW} - \mathbf{v}_{LW}$$

Figure 2.15(a) shows this vector triangle, and Figure 2.15(b) shows the corresponding "scalar" triangle with the known and unknown quantities indicated. The speed v_{TL} of the tanker relative to the liner is obtained by using the law of cosines to get

$$v_{TL}^2 = v_{TW}^2 + v_{LW}^2 - 2v_{TL}v_{LW}\cos120°$$
$$v_{TL}^2 = (12)^2 + (20)^2 - 2(12)(20)(-0.5) = 780$$
$$v_{TL} = 28 \text{ km/h}$$

Using the law of sines, we get

$$\frac{\sin\alpha}{20} = \frac{\sin120°}{28}$$

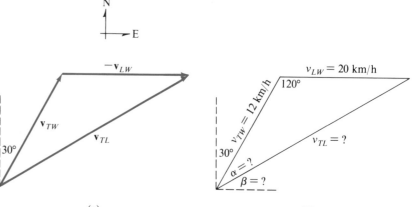

Figure 2.15   Example 2.6   (a)   (b)

$$\sin\alpha = \frac{20}{28}\frac{\sqrt{3}}{2}$$

$$\alpha = 38°$$

The angle β is obtained by noting in Figure 2.15(b) that

$$\beta + \alpha + 30° = 90°$$

$$\beta = 60° - \alpha = 60° - 38° = 22°$$

Thus, the tanker is moving relative to the liner with a velocity of 28 km/h in a direction 22° north of east. ■

E 2.21   A child runs 1.5 m/s toward the rear of a passenger car of a train moving forward with a speed of 4.0 m/s. What is the velocity of the child relative to the earth?   Answer: 2.5 m/s in the direction of the motion of the train

E 2.22   A water bug swims with a speed of 0.30 m/s toward the north relative to the surface of a stream, which flows 0.40 m/s toward the east relative to the shore. What is the velocity of the water bug relative to the shore?
Answer: 0.50 m/s, 37° north of east

2.4 Problem-Solving Summary

Kinematics describes motion. The quantities used to describe motion are position, velocity (or speed, which is the magnitude of velocity), acceleration, and time. A problem that asks us to find one of these variables in terms of two or more of the others is a kinematics problem. So far we have distinguished four types of kinematics problems: (a) motion in a straight line, (b) projectile motion, (c) uniform circular motion, and (d) relative motion.

If all the motion of a problem takes place in a straight line, say the x direction, a pertinent question to ask is, "Is the acceleration constant?" If the acceleration is *not* constant, we must use the defining relationships,

$$v_x = \frac{dx}{dt} \quad ; \quad a_x = \frac{dv_x}{dt}$$

or their antiderivative forms,

$$x = \int v_x(t)dt + C \quad ; \quad v_x = \int a_x(t)dt + C$$

Far more commonly, rectilinear problems involve constant acceleration; in that case, Equations (2-14) apply:

$$x = x_o + v_{ox}t + \frac{1}{2}a_xt^2$$

$$v_x = v_{ox} + a_xt$$

$$v_x^2 = v_{ox}^2 + 2a_x(x - x_o)$$

Listing the unknown variable and the known variables will specify which of the above equations is appropriate.

Free-fall, the situation in which an object is released or thrown *vertically*, is a case of straight-line motion with constant acceleration. Signs should receive special attention in free-fall problems. One way to avoid difficulty with signs is to choose upward as the positive direction from any chosen origin. Then, because g is a positive value, the constant acceleration has magnitude g and its direction is always downward (negative). Equations (2-15) then apply:

$$y = y_o + v_{oy}t - \frac{1}{2}gt^2$$

$$v_y = v_{oy} - gt$$

$$v_y^2 = v_{oy}^2 - 2g(y - y_o)$$

Care must be taken with the sign of every value entered into these equations. Positions above the chosen origin are positive; positions below the origin are negative. Velocities in the upward direction are positive; velocities in the downward direction are negative.

Projectile motion occurs when a body is released or thrown with a velocity having a horizontal component. In every phase of a projectile problem, which we work in two parts (the vertical part and the horizontal part), we must be aware that there is no acceleration associated with the horizontal motion. The vertical part of the problem is treated as free-fall. The mathematics of a projectile problem is simplified by choosing the origin of the x and y coordinates to be at the point from which the projectile is fired or released at $t = 0$. Two equations relate the components of the position of the projectile and the time of flight to the initial velocity components, v_{ox} and v_{oy}:

$$x = v_{ox}t$$

$$y = v_{oy}t - \frac{1}{2}gt^2$$

Similarly, the velocity of the projectile at any time during its flight is the vector sum of two components:

$$v_x = v_{ox} \text{ (constant)}$$

$$v_y = v_{oy} - gt$$

Notice that both the position and the velocity of the projectile are expressed in terms of v_{ox} and v_{oy}. Therefore, finding the appropriate relationships for solving projectile problems is usually facilitated by immediately resolving the initial velocity v_o of the projectile into its components, v_{ox} and v_{oy}, even if the magnitude or the direction of v_o is unknown.

Uniform circular motion is identified by recognizing that a particle is moving at *constant speed* v in a circular path of radius r at the instant of interest. Two facts are then pertinent: the (centripetal) acceleration is toward the center of the circular path, and its magnitude is v^2/r.

Relative velocity problems may be spotted by phrases like "velocity with respect to" and "velocity relative to." Usually three speeds and three directions are involved in relative velocity problems. Diagrams are essential in determining the known and unknown quantities, which are related through a velocity triangle:

$$\mathbf{v}_{AB} = \mathbf{v}_{AC} + \mathbf{v}_{CB}$$

Once the triangle is formed, each of the desired unknown quantities is obtained from the trigonometric relationships of the triangle. In writing the velocity triangle equation, remember that $\mathbf{v}_{AC}$ is obtained when $\mathbf{v}_{CA}$ is known by using $\mathbf{v}_{AC} = -\mathbf{v}_{CA}$.

Problems

GROUP A

2.1 An apple drops at $t = 0$ from a limb that is 3 m above the ground.

  (a) When does the apple hit the ground?

  (b) How fast is the apple moving just before it hits the ground?

2.2 A certain baseball is thrown vertically upward with an initial ($t = 0$) speed of 50 ft/s.

  (a) When will the ball be moving downward with a speed of 25 ft/s?

  (b) At the instant the ball is moving at 25 ft/s upward, what is the height of the ball above its initial ($t = 0$) position?

2.3 A projectile is given an initial ($t = 0$) velocity having a horizontal component of 20 m/s and a vertical component of 49 m/s upward.

  (a) When will the projectile reach its maximum height?

  (b) What is the speed of the projectile at $t = 3.0$ s?

  (c) How far is the projectile from its initial position at the time $t = 3.0$ s?

2.4 A soccer ball is kicked from the ground with an initial velocity of 20 m/s directed 40° above the horizontal. At what distance from the initial point will the ball hit the ground?

2.5 A child rides a Ferris wheel with a diameter of 25 m. The wheel completes one turn every 21 s. Calculate the child's acceleration at

  (a) the top of the circle.

  (b) the bottom of the circle.

2.6 Calculate the magnitude of the centripetal acceleration of the tip of a propeller if it is turning at 2000 revolutions per minute (rpm) and the tip is 3.0 ft from the propeller's center.

2.7 An electric vehicle starts from rest and accelerates for 20 s at a rate of 1.5 m/s² in a straight line. Then the vehicle slows at a constant rate of 0.30 m/s² until it stops. All the motion takes place along a straight line.

  (a) Determine the time interval from start to stop.

  (b) What is the maximum speed of the vehicle?

  (c) How far does the vehicle travel from start to stop?

2.8 As an object moves along the x axis, its position is given by

$$x(t) = (t^4 - 18t^2) \text{ m}$$

for $-4 \text{ s} \le t \le 4 \text{ s}$.

  (a) At what times and positions, if any, is the object momentarily at rest?

  (b) At what times, if any, is the speed of the object not changing?

2.9 A pilot flies an airplane pointed toward the north with an airspeed of 85 mi/h. If the wind velocity relative to the ground is 35 mi/h toward the east, determine the speed and direction of motion of the airplane relative to the ground.

2.10 If a ship moving 22 km/h through the water is steered 12° south of east, its velocity relative to the shore is observed to be 23 km/h toward the east. Find the velocity of the water relative to the shore.

2.11 A centrifuge for testing pressure suits used by pilots flying high-performance aircraft is essentially a small cockpit rotating in a horizontal circle of radius 25 ft. If the motion of the centrifuge is to cause the pilot to accelerate at 10g (320 ft/s²),

  (a) how fast is the cockpit moving?

  (b) how much time is required for each revolution?

2.12 An automobile tire has an outer diameter of 0.65 m.

  (a) How many revolutions per second does this tire turn when it is on an automobile that is moving at 89 km/h (55 mi/h)?

  (b) What is the magnitude of the centripetal acceleration of a point on the periphery of the tire in part (a)?

2.13 A car moving 60 mi/h eastward has a head-on collision with a brick wall, and, while stopping, the rear bumper travels 4.0 ft. Assuming constant acceleration, determine

  (a) the time required for the car to come to rest.

  (b) the acceleration of the car during the collision.

2.14 An object is released from a point 110 m above the ground. Two seconds later a second object is thrown downward from the same height (110 m). The two objects hit the ground simultaneously.

(a)   How much time elapses while the first object falls?

(b)   Find the initial speed of the second object.

(c)   Determine the speed of each object just before it strikes the ground.

2.15   A fast duck is flying $(20\hat{\mathbf{i}} + 40\hat{\mathbf{j}})$ mi/h at the same altitude as a slow airplane flying with a velocity of $(-80\hat{\mathbf{i}} + 40\hat{\mathbf{j}})$ mi/h.

(a)   How fast is the duck moving relative to the airplane?

(b)   In what direction is the duck moving relative to the airplane?

2.16   Particles A, B, and C, executing rectilinear motion, have positions given by

$$x_A = (3t - 4) \text{ m}$$

$$x_B = (t^2 - t) \text{ m}$$

$$x_C = (4t^3 - t + 4) \text{ m}$$

where t is measured in seconds.

(a)   When are A and B at the same x coordinate?

(b)   Which of the particles have constant acceleration?

(c)   Which particle has the greatest speed at $t = 2$ s?

(d)   When do B and C have equal accelerations?

(e)   When do B and C have equal speeds?

2.17   Two trains run on parallel tracks. Train A starts from rest at the instant train B, moving with a constant speed of 30 mi/h, passes A. Train A maintains a constant acceleration of 0.20 ft/s² until it overtakes B.

(a)   When and where will A have the same speed as B?

(b)   When and where will A overtake B? What speed will A have at this instant?

2.18   A child throws a ball vertically. Two seconds after the ball is thrown, it is caught by a friend on a balcony 16 ft above the point of release.

(a)   Determine the initial speed of the ball.

(b)   When did the ball first pass the friend on the balcony?

(c)   Find the speed of the ball when it first passed the balcony.

2.19   An airplane flying in a wind that is blowing at 55 km/h toward the northeast (45° north of east) is observed to move relative to the ground at 250 km/h, 25° north of west. Determine the velocity of the airplane through the air.

2.20   Raindrops are falling straight downward. When observed from a car traveling 55 mi/h, the drops streak the side window at an angle of 40° with the vertical. Find the speed with which the drops are falling.

2.21   Figure 2.16 shows a graph of position as a function of time for a particle moving along the x axis.

(a)   Determine the velocity of the particle at $t = 1$ s.

(b)   Determine the velocity of the particle at $t = 3$ s.

(c)   Sketch a graph of v_x as a function of t for $0 < t < 6$ s, omitting $t = 2$ s where v_x undergoes an "unphysical" instantaneous change.

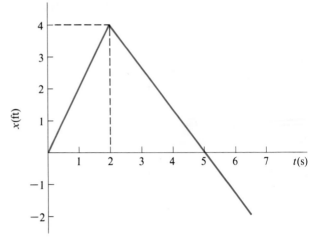

Figure 2.16   Problem 2.21

2.22   The graph in Figure 2.17 shows v_x versus t for a particle. (The abrupt changes in the slope of v_x at $t = 2$ s and $t = 4$ s are not "physically realistic.")

(a)   Calculate the acceleration of the particle at $t = 1$, 3, and 6 s.

(b)   Sketch a graph of a_x versus t for $0 < t < 8$ s, excluding $t = 2$ s and $t = 4$ s.

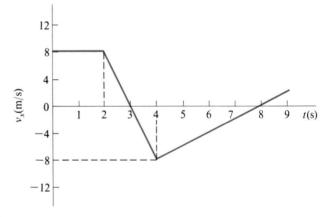

Figure 2.17   Problem 2.22

2.23   If the initial speed and direction of a projectile are as shown (*see* Fig. 2.18), determine that

(a)   the horizontal range R is given by

$$R = \frac{v_0^2}{g} \sin 2\theta.$$

(b) the maximum height H above the initial position is given by

$$H = \frac{v_0^2}{2g} \sin^2\theta.$$

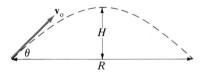

Figure 2.18   Problem 2.23

2.24 A baseball player throws a ball vertically and catches it 3.8 s later. Find

(a) the initial speed of the ball.

(b) the maximum height the ball achieves.

2.25 Object A is thrown vertically from the ground and rises to a maximum height of 220 ft above the ground. At the instant A reaches its maximum height, object B is thrown from the ground with a speed equal to the initial speed of A.

(a) At what height above the ground will A and B pass each other?

(b) Calculate the elapsed time between B's release and the time the two particles pass each other.

(c) Determine the velocity of each object at the instant they pass each other.

2.26 A ball is thrown from ground level with an initial velocity of 37 m/s, 50° above the horizontal, toward a vertical wall at a horizontal distance of 120 m.

(a) Which does it hit first, the ground or wall? Where?

(b) What initial speed is required if the ball is to hit at the base of the wall?

2.27 The velocity of a rock thrown by a child is 45 ft/s, 38° above the horizontal. When the rock leaves the child's hand, it is 5.0 ft above the level ground.

(a) When does the rock reach its maximum height?

(b) What is the maximum height of the rock above the ground?

(c) When does the rock hit the ground?

(d) How far away (horizontally) from the child is the point of impact?

(e) What is the speed of the rock just before it strikes the ground?

2.28 An airplane releases a package at an instant when the airplane and package have a velocity of 270 km/h, 30° below the horizontal. At this instant of release, the airplane is 1200 m above level ground.

(a) How much time elapses between the release and the impact with the ground?

(b) How far does the package travel horizontally after release?

(c) What is the speed of the package just before impact?

2.29 A motorcyclist's acceleration is given by

$$a_x = (20 - 2t) \text{ ft/s}^2$$

for $0 \le t \le 20$ s. At $t = 0$, the velocity and position of the cyclist are 10 ft/s and 25 ft.

(a) Determine $v_x(t)$ and $x(t)$.

(b) When does the cyclist achieve maximum speed?

(c) What is the maximum speed?

(d) What total distance does the cyclist travel during the time interval 6 s $< t <$ 15 s?

2.30 A certain artillery shell is fired with an initial velocity of 300 m/s, 55° above the horizontal. It explodes on a mountainside 42 s after firing.

(a) Calculate the position of the explosion relative to the firing point.

(b) Determine the velocity of the shell just before the explosion.

2.31 Ten seconds after being fired, a projectile strikes a hillside at a point displaced 2000 ft horizontally and 300 ft vertically above the point of launch. Calculate for the projectile its

(a) initial velocity.

(b) maximum height above its launch point.

(c) velocity just before impact.

2.32 A flowerpot is dropped from the top of a building. The top is 120 ft above the ground.

(a) How much time elapses before the pot hits the ground?

(b) What is the speed of the pot just before it hits the ground?

(c) What is its speed when it passes a window 80 ft above the ground?

(d) How much time elapses between the release of the flowerpot and its passing a window 80 ft above the ground?

2.33 A ball is thrown with an initial speed of 35 m/s. It returns to the same elevation 6.0 s after release. Find the maximum height the ball rises above the point of release.

2.34 A simple model for the kinematics of a vehicle assumes that the maximum acceleration in any gear is constant over a given speed range, as shown in the following table:

Gear	1	2	3	4
Speed range (mi/h)	0–15	15–30	30–45	45–75
Maximum acceleration (ft/s^2)	14	10	6.0	3.0

(a) Determine the minimum time for this vehicle to reach a speed of 60 mi/h starting from rest.

(b) Calculate the distance traveled by the vehicle in going from rest to 60 mi/h at maximum acceleration.

(c) If the vehicle has a speed of 35 mi/h, calculate the minimum time elapsed and the distance traveled as it accelerates to 70 mi/h.

2.35 A hunter wishes to cross a river that is 1.5 km wide and flows with a velocity of 5.0 km/h parallel to its banks, as shown in Figure 2.19. The hunter uses a small powerboat that moves at a maximum speed of 12 km/h with respect to the water.

(a) The hunter crosses the river in a minimum time by aiming the boat directly toward a point straight across the river, such that $\alpha = 90°$. Why?

(b) What is the minimum time for crossing?

(c) Determine the velocity of the boat with respect to the earth for this minimum time of crossing.

(d) What is the hunter's position relative to the starting position on reaching the opposite bank in minimum time?

(e) If the hunter wishes to land at a point opposite the launch point, determine α and the time for crossing. Assume the boat is run at maximum speed.

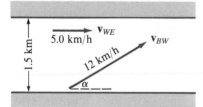

Figure 2.19   Problem 2.35

2.36 The average speed $\bar{v}$ for a particle is defined by

$$\frac{\bar{v}}{7} = \frac{\text{distance traveled}}{\text{time elapsed}}, \text{ a scalar quantity.}$$

(a) A tourist drives 100 km at 50 km/h and another 100 km at 40 km/h. Find the average speed during this 200-km trip.

(b) A motorcyclist travels on an interstate highway at 50 mi/h for 2 h and 40 mi/h for another 2 h. Determine the cyclist's average speed during this 4-h trip.

2.37 A particle moves around a circle centered on the origin. The radius of the circle is 15 ft. At $t = 0$ the position of the particle is $15\ \hat{\mathbf{i}}$ ft; it reaches a position of $-15\ \hat{\mathbf{i}}$ ft at $t = 5.0$ s. The particle returns for the first time to its original position ($t = 0$ position) at $t = 12$ s. Assume its motion to be counterclockwise at all times during this time interval of 12 s.

(a) Calculate the average velocity of the particle during the $0 < t < 5.0$ s time interval.

(b) Calculate its average speed (*see* Problem 2.36) for the $0 < t < 5.0$ s time interval.

(c) Repeat (a) and (b) for $0 < t < 12$ s.

GROUP B

2.38 An object is dropped from a height h above the ground. Point A is 70 m above the ground, and Point B is 60 m above the ground. A time of 0.40 s elapses between the instant the object passes A and the instant it passes B. Calculate the initial height h.

2.39 A particle moves according to

$$x = R\cos(\omega t) \qquad y = R\sin(\omega t)$$

where R and ω are positive constants. By combining vector components and by differentiation show that

(a) $\mathbf{r} = R(\hat{\mathbf{i}}\cos\omega t + \hat{\mathbf{j}}\sin\omega t)$

(b) $\mathbf{v} = \omega R(-\hat{\mathbf{i}}\sin\omega t + \hat{\mathbf{j}}\cos\omega t)$

(c) $\mathbf{a} = -\omega^2 R(\hat{\mathbf{i}}\cos\omega t + \hat{\mathbf{j}}\sin\omega t)$

(d) the particle is executing uniform circular motion with a speed given by $v = R\omega$

(e) the (centripetal) acceleration is radially inward with a magnitude given by $a_c = v^2/R$

2.40 The initial speed of a projectile is v_0, and its initial direction is θ above the horizontal. Show that

(a) for a fixed initial speed, the maximum horizontal range is equal to v_0^2/g when $\theta = \pi/4$ radians $= 45°$.

(b) for a fixed initial speed, the maximum height equals $v_0^2/2g$ when $\theta = \pi/2$ radians $= 90°$.

(c) if $v_0^2 > gR$, there are always two initial angles for a given horizontal range R and that these two angles are complementary. Sketch these two trajectories for both $R = 200$ m and $v_0 = 60$ m/s. Is the time of flight the same for both trajectories?

(d) there is a θ such that the horizontal range is equal to twice the maximum height. Find that value of θ.

2.41 Because integration is the inverse operation of differentiation and $v_x = dx/dt$, it follows that

$$x(t_2) = x(t_1) + \int_{t_1}^{t_2} v_x(t)dt$$

where $x(t_1)$ and $x(t_2)$ are positions at times t_1 and t_2, respectively. Thus, we have

$$\int_{t_1}^{t_2} v_x(t)dt = x(t_2) - x(t_1)$$

or the (definite) integral of the velocity of a particle over any time interval is equal to its change in position during that time interval. Just as the derivative operation has graphical interpretation, the slope of a tangent line, so does the integral operation have a graphical meaning. More specifically, the integral of v_x for $0 < t < 2$s is equal to the area of the shaded rectangle shown in Figure 2.20, that is,

$$\int_0^2 v_x dt = (8 \text{ m/s})(2\text{s}) = 16 \text{ m}$$

Thus, we have $x(2) - x(0) = 16$ m. If the position of the particle at $t = 0$ is $x = -10$ m, then we get

$$x(2) = x(0) + \int_0^2 v_x dt = -10 \text{ m} + 16 \text{ m} = 6 \text{ m}$$

From the graph in Figure 2.20 and the data given, determine

(a) $\int_2^4 v_x dt$
(d) $x(4)$
(b) $\int_4^5 v_x dt$
(e) $x(5)$
(c) $\int_5^8 v_x dt$
(f) $x(8)$

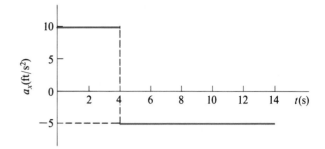

Figure 2.21   Problem 2.43

2.45 The polar unit vectors $\hat{r}$ and $\hat{\theta}$ are shown in Figure 2.22.

(a) Show that

$$\hat{r} = \hat{i} \cos\theta + \hat{j} \sin\theta$$
$$\hat{\theta} = -\hat{i} \sin\theta + \hat{j} \cos\theta$$

(b) Verify that $\hat{r}$ and $\hat{\theta}$ are perpendicular. (Remember that $\mathbf{a} \cdot \mathbf{b} = ab \cos\phi$.)

(c) Show that $\hat{\theta} = d\hat{r}/d\theta$ and that $\hat{r} = -d\hat{\theta}/d\theta$.

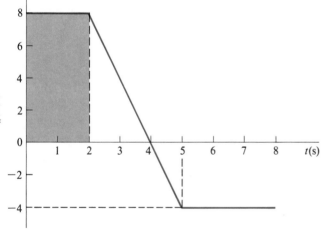

Figure 2.20   Problem 2.41

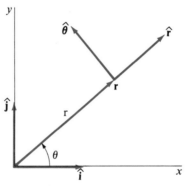

Figure 2.22   Problem 2.45

2.42 Using the results of Problem 2.41, construct a graph of x for $0 < t < 8.0$ s. Be sure to pay attention to intervals on the graph where the slope is constant, zero, and so on.

2.43 The acceleration of a particle is shown as a function of time (see Fig. 2.21). If the position and velocity of the particle at $t = 0$ are -80 ft and 20 ft/s, calculate:

(a) $\int_0^4 a_x dt$
(c) $v_x(4)$
(b) $\int_4^{12} a_x dt$
(d) $v_x(12)$

2.44 Using the data from Problem 2.43, graph v_x for the time interval $0 < t < 12$ s and determine the position of the particle at $t = 12$ s. Graph x for $0 < t < 12$ s.

2.46 Consider a particle moving with constant speed v in a circle of radius R centered on the origin. Use the polar unit vectors $\hat{r}$ and $\hat{\theta}$, as defined in Problem 2.45.

(a) Show that the position of the particle is $\mathbf{r} = R\hat{r}$.

(b) Show that the velocity of the particle is $\mathbf{v} = R\dot{\theta}\hat{\theta}$, where the "dot" notation is used to denote differentiation with respect to time, or $\dot{\theta} \equiv d\theta/dt$.

(c) Show that $\mathbf{v}$ is counterclockwise (clockwise) if θ is increasing (decreasing) with respect to time.

(d) Show that the speed of the particle is given by $v = R|\dot{\theta}|$.

(e) Show that, for uniform circular motion, θ changes at a constant rate, that is, $\dot{\theta} = $ constant and $\ddot{\theta} \equiv d^2\theta/dt^2 = 0$.

(f) Show that $\mathbf{a} = -R\dot{\theta}^2\hat{r}$.

(g) Verify that the magnitude of the centripetal acceleration is given by $a_c = v^2/R$.

2.47 If the speed of a particle is changing while the particle is executing circular motion (radius = R), then the particle experiences a tangential acceleration. In this case show that the acceleration is given by

$$\mathbf{a} = -a_c\hat{\mathbf{r}} + a_t\hat{\boldsymbol{\theta}}$$

where

$$a_c = R\dot{\theta}^2 = v^2/R$$
$$a_t = R\ddot{\theta}$$

2.48 Consider a particle moving in the plane. Use the polar unit vectors $\hat{\mathbf{r}}$ and $\hat{\boldsymbol{\theta}}$, as defined in Problem 2.45.

(a) Show that $\mathbf{r} = r\hat{\mathbf{r}}$.

(b) Show that $\mathbf{v} = \dot{r}\hat{\mathbf{r}} + r\dot{\theta}\hat{\boldsymbol{\theta}}$, where the "dot" notation is used to denote differentiation with respect to time, that is, $\dot{r} \equiv dr/dt$ and $\dot{\theta} \equiv d\theta/dt$.

(c) Show that the particle's speed is given by

$$v = \sqrt{(\dot{r})^2 + (r\dot{\theta})^2}$$

(d) Show that $\mathbf{v}_\parallel = \dot{r}\hat{\mathbf{r}}$ and $\mathbf{v}_\perp = r\dot{\theta}\hat{\boldsymbol{\theta}}$. Note that in this form it is clear that $\mathbf{v}_\parallel$ results from $\mathbf{r}$'s changing magnitude ($\dot{r}$), while $\mathbf{v}_\perp$ results from the changing direction ($\dot{\theta}$) of $\mathbf{v}$.

(e) Show that $\mathbf{a} = (\ddot{r} - r\dot{\theta}^2)\hat{\mathbf{r}} + (r\ddot{\theta} + 2\dot{r}\dot{\theta})\hat{\boldsymbol{\theta}}$.

(f) Show that Problems 2.39 and 2.46 are special cases of the more general results in this problem.

2.49 The hunter of Problem 2.35 wishes to start at point A to reach point B (*see* Fig. 2.23). To reach point B the hunter may land at any point on the opposite bank and drive a snowmobile with a maximum speed of 20 km/h to B. In what direction should the boat be pointed, that is, what is the value of α, to minimize the transit time from A to B? What is the minimum time?

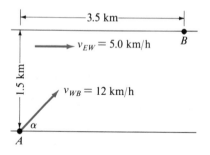

Figure 2.23   Problem 2.49

2.50 At $t = 0$ observers A and B are at rest with respect to the earth. For $t > 0$, B is rocketed vertically from the surface of the earth with an acceleration given by

$$\mathbf{a} = \frac{gt}{1 + t^2}\hat{\mathbf{j}}$$

where t is measured in seconds. A remains at rest on the earth. Both A and B observe a projectile C whose position relative to A is given by

$$\mathbf{r} = [200t\hat{\mathbf{i}} + (300t - 16t^2)\hat{\mathbf{j}}]\text{ ft}$$

(a) Determine the velocity of B with respect to the earth as a function of time.

(b) Determine the velocity of C relative to B as a function of time.

2.51 It is desired to aim a projectile with an initial speed v_o so that it hits a target displaced a distance d horizontally and a distance h vertically upward.

(a) Show that this can be accomplished only if

$$v_o^2 \geq g(d + 2h)$$

(b) For $v_o^2 = g(d + 2h)$, show that the angle between the initial velocity and the horizontal is

$$\theta = \tan^{-1}(1 + 2h/d)$$

(c) Show that for $v_o^2 > g(d + 2h)$ there are two initial angles given by

$$\theta_1 = \tan^{-1}\left\{\frac{v_o^2}{gd} + \sqrt{\left(\frac{v_o^2}{gd}\right)^2 - \frac{v_o^2 h}{gd^2} - 1}\right\}$$

$$\theta_2 = \tan^{-1}\left\{\frac{v_o^2}{gd} - \sqrt{\left(\frac{v_o^2}{gd}\right)^2 - \frac{v_o^2 h}{gd^2} - 1}\right\}$$

(d) If $d = 2000$ ft and $h = 400$ ft, determine the minimum initial speed and the corresponding initial angle.

(e) If $d = 2000$ ft, $h = 400$ ft, and $v_o = 360$ ft/s, find the initial angles.

2.52 A projectile is launched with an initial speed v_o and an initial direction θ above the horizontal. Show that the distance from the initial position to the projectile always increases with time if $\theta < \sin^{-1}(2\sqrt{2}/3)$.

3 Force and Motion: Particle Dynamics

We have been considering the motion and changes in motion of particles without regard to what might have influenced those changes. In Chapter 2 we learned to recognize a change in motion as acceleration. Any influence that can cause a body to accelerate will now be called a *force*. Force, which we recognize intuitively as a push or a pull, is a fundamental interaction between physical bodies that can produce changes in the velocities of the bodies. The study of the motion of particles in terms of the forces associated with that motion is called *particle dynamics*, the subject of this chapter.

The principles that relate force and motion were formulated by Isaac Newton (1642–1727). These principles, now referred to as Newton's laws of motion, are profound insights into the natural processes involving force and motion. Here we shall state and discuss each of Newton's laws and illustrate their usefulness in a number of physical situations.

3.1 Newton's First Law

A body has a constant velocity unless it is acted on by a net force.

To say that a body has a constant velocity is to say that the body experiences no acceleration. The words *net force* in the first law remind us that many individual forces may act on a body, but the resultant or net force must be equal to zero if the body is not to accelerate. We should also recognize that because force is inherently a vector quantity (one may not push or pull on a body without doing so in a particular direction), the net force on a body is the vector sum of all the forces acting on that body.

Consider one implication of the first law. The assertion that a body on which there is no net force experiences no acceleration implies that a moving body will maintain a constant velocity forever unless acted on by a net force. We intuitively expect a body at rest to remain at rest when no forces act on it, but the first law asserts that constant velocity is an equally natural state of motion. Of course, none of us has ever experienced the situation in which no forces are present on a body. We must, therefore, imagine, just as Newton did, the idealized situation in which there is no gravity, no friction, no forces of any kind. In such a situation, a body at rest would remain at rest forever, and a body moving with a given velocity would move with that velocity forever.

Newton's first law, however, does not apply to all *reference frames*, the coordinate systems in which we specify motion. Suppose, for example, one chooses as a reference frame an automobile that is accelerating rapidly in a forward direction. A passenger seated in the car would feel the force of the seat pushing forward on her back but would not experience any acceleration relative to the automobile, the chosen reference frame. This situation, a net force on the passenger without her accelerating, seems to be a violation of the first law. What then are the restrictions on Newton's first law? We may assert generally that the first law is valid only in an unaccelerated frame, which we shall call an *inertial frame*. (The concept of an inertial frame is complex and has some implications that will not be included in our discussion here. An extended and interesting treatment of inertial and accelerated reference frames is in *Mechanics*, vol. 1 of *Berkeley Physics Course*, 2nd ed. [New York: McGraw-Hill, 1973], pp. 102–120.)

Where do we find an inertial frame? The earth itself is accelerating, because it moves in a curved trajectory about the sun, and even a laboratory on the surface of the earth is accelerated by the rotation of the earth. So how can we justify experimental measurements made in our laboratory? In Chapter 2 (Exercises 2.19 and 2.20) we calculated the magnitude of the accelerations that result from the revolution and rotation of the earth. Those accelerations are small (less than 0.040 m/s^2) compared with values we commonly measure. Therefore, we are justified in disregarding these effects in most experiments. For most purposes, then, and for purposes of our studies here, we shall assume that the earth is an adequate inertial frame.

Another implication of the first law is that every physical body possesses an intrinsic characteristic called *inertia*, the tendency of a body at rest to remain at rest or the tendency of a body in motion to continue that motion. The first law, in fact, implies that a force must be applied to a body in order to overcome such tendencies. As an example, a passenger in a bus has a tendency to proceed in a straight line when the bus turns suddenly. She remains in the bus because the bus exerts a force on her in the direction of the turn to provide the change in her velocity.

Newton's first law, then, establishes the notions of inertia and inertial frames. It also asserts that constant velocity is a natural state of a body, a state that changes only if a net force acts on the body. These concepts are thought-provoking, useful, and, of course, necessary to the understanding of mechanics. Usefulness is epitomized by Newton's second law, which we now consider.

3.2 Newton's Second Law

The acceleration of a body is proportional to the net force acting on that body, and the acceleration of that body is in the direction of that net force.

This statement of Newton's second law can be expressed mathematically. The proportionality between the net force $\Sigma \mathbf{F}$ on a body and the acceleration $\mathbf{a}$ of that body may be expressed by $\Sigma \mathbf{F} \propto \mathbf{a}$. By writing $\Sigma \mathbf{F}$, we emphasize that the net force on a body is the vector sum of all external forces acting on that body. The symbol ∝ may be read "is proportional to." The vector relation implies that the acceleration of the body is in the direction of the net, or resultant, force on the body. In order to express the proportionality as an equation, we introduce a constant of proportionality m, a scalar quantity:

$$\sum \mathbf{F} = m\mathbf{a} \qquad (3\text{-}1)$$

This is a mathematical statement of Newton's second law, where m is the *mass* of the body. Mass is seen from Equation (3-1) to be the ratio of the magnitudes (because the net force and acceleration are in the same direction) of the net force on a body and the acceleration of that body:

$$m = \frac{\left|\sum \mathbf{F}\right|}{a} \qquad (3\text{-}2)$$

For a specific body, the ratio $|\sum \mathbf{F}|/a$ is always the same. If the magnitude of the net force is made to have different values, the magnitude of the acceleration of the body will assume appropriate values that maintain the constant ratio $|\sum \mathbf{F}|/a$. In other words, the mass of the body remains unchanged. Thus the mass of a given body is an intrinsic property characteristic of that body.

Although mass is properly associated with the quantity of matter in a body, Equation (3-2) offers a clearer interpretation of the physical significance of the concept of mass. The relationship $m = |\sum \mathbf{F}|/a$ implies that, for a given acceleration, a large mass requires a greater net force acting on it than does a lesser mass. So mass is a measure of the sluggishness, or the reluctance, of a body to accelerate under the influence of a net force. Whereas the first law implies the existence of inertia, the second law establishes mass as the measure of inertia.

The unit associated with mass in the English system is specified by the second law, $\sum \mathbf{F} = m\mathbf{a}$. The pound (lb) is the defined unit of force in the English system (see Section 1.2), and the unit of acceleration is ft/s^2. The unit of mass is named the *slug*, and Equation (3-2) relates the slug to more fundamental units:

$$m = \frac{\left|\sum \mathbf{F}\right|}{a}$$

$$\text{slug} = \frac{\text{lb}}{\text{ft/s}^2} = \frac{\text{lb} \cdot \text{s}^2}{\text{ft}} \qquad (3\text{-}3)$$

The slug is a descriptive name for the unit of mass. We have already noted that the mass of a body is a measure of its sluggishness to accelerate under the influence of a net force.

In a similar way, $\sum \mathbf{F} = m\mathbf{a}$ dictates the unit of force, named the *newton* (N), that must be used in the metric system. Recall that the unit of mass is defined to be the kilogram (kg) and the unit of acceleration is m/s^2. Then

$$\left|\sum \mathbf{F}\right| = ma$$

$$\text{N} = \frac{\text{kg} \cdot \text{m}}{\text{s}^2} \qquad (3\text{-}4)$$

specifies that 1 N is the net force that causes an acceleration of 1 m/s^2 when that force acts on a mass of 1 kg. (The defined units in the SI system are *chosen* to be those of mass and acceleration, and the defined units in the English system are *chosen* to be those of force and acceleration.)

E 3.1   Any object that is in free-fall near the surface of the earth accelerates 9.8 m/s^2 downward. What is the resultant force on a 70-kg object falling
freely?   Answer: 690 N downward

E 3.2   During launch, an 80-kg astronaut is accelerated upward at a rate
of 20 m/s^2. What is the resultant force on this astronaut?
                                                            Answer: 1600 N upward

E 3.3   Forces F_1 and F_2 are the only forces acting on an object of mass m.
If $F_1 = 4.0$ N, $F_2 = 3.0$ N, $m = 2.0$ kg, and F_1 and F_2 are directed as shown for
three cases, determine the resulting acceleration in each case.
                    Answers: $0.50\hat{i}$ m/s^2; $3.5\hat{i}$ m/s^2; $(2.0\hat{i} + 1.5\hat{j})$ m/s^2

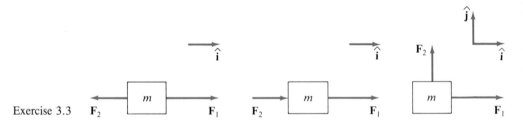

Exercise 3.3

Newton's second law in its equation form, $\Sigma F = ma$, is the fundamental tool
of dynamics. We shall have numerous occasions to use this relationship and
recall its interpretation. The final tool, Newton's third law, remains to be
considered.

3.3 Newton's Third Law

**Whenever two bodies interact, the force exerted by the first body on the second body
is equal in magnitude and opposite (antiparallel) in direction to the force exerted by
the second body on the first.**

Forces come in pairs. *A* pushes on *B* only if *B* pushes back on *A*. These are
some of the facts implicit in Newton's third law. In order to state the law in an
analytically useful way, suppose that F_{12} represents the force that body 1 exerts
on body 2. Then there exists a force F_{21}, the force that body 2 exerts on body 1,
and the forces are related by

$$F_{12} = -F_{21} \qquad\qquad (3\text{-}5)$$

The negative sign in Equation (3-5) points out that the pair of forces are oppositely directed. The subscripts emphasize that the forces under consideration
are on different bodies.

Two important physical ideas are embodied in the third law. First, you cannot
exert a force on an object unless the object exerts a force of equal magnitude on
you. This fact provides the physical reasoning for a boxer to "roll with a punch."
A boxing glove cannot exert more force on your chin than you permit your chin
to exert on the glove. Second, it is important to remember that two third-law

forces, which are equal in magnitude and oppositely directed, *never* act on the same body.

E 3.4   For each of the following forces, Newton's third law requires the existence of a second force. Describe that second force: (a) an upward force of 160 lb on a person's feet by the floor, (b) an upward buoyant force of 600 N by the water on a person floating on the water's surface, (c) the force of attraction on the moon by the earth, and (d) a downward gravitational force of 170 lb on a person by the earth.

Answers: (a) 160 lb force downward on the floor by that person's feet; (b) 600 N force downward on the water by the person; (c) an oppositely directed attractive force on the earth by the moon; (d) 170 lb upward force on the earth by the person

3.4   Weight and Mass

We saw in Chapter 2 that a body that is released in space near the surface of the earth accelerates downward with magnitude g. Newton's second law, $\Sigma\mathbf{F} = m\mathbf{a}$, then requires that there must be a net downward force on the body. The second law further prescribes that the downward force has magnitude mg. This force is the gravitational pull of the earth on the body. The force of the gravitational attraction of the earth on a body is called the *weight* $\mathbf{W}$ of the body. Under these conditions $|\Sigma\mathbf{F}| = m\mathbf{a}$ becomes, for this body,

$$W = mg$$

or

$$m = \frac{W}{g} \qquad\qquad (3\text{-}6)$$

which relates the mass and weight of the body at or near the surface of the earth. Notice that weight is a force. Therefore, in the English system the unit of weight is the pound (lb); in the metric system the unit of weight is the newton (N).

E 3.5   Recall that an object that weighs 1 lb near the surface of the earth has a mass of approximately 454 grams.
(a) What is its weight in newtons?   Answer: 4.5 N
(b) What is its mass in slugs?   Answer: $\frac{1}{32}$ slug = 0.031 slug
(c) How many pounds are equivalent to 1 newton?   Answer: 0.23 lb
(d) How many kilograms are equivalent to 1 slug?   Answer: 15 kg

E 3.6   What is the mass in kilograms and slugs of (a) a person weighing 175 lb? (b) a car weighing 2400 lb? and (c) a book weighing 10 N?

Answers: (a) 80 kg = 5.4 slugs; (b) 1.1×10^3 kg = 75 slugs; (c) 1.0 kg = 0.070 slug

E 3.7   What is the weight in newtons and pounds of (a) a person whose mass is 74 kg? (b) 105 grams of salt? (c) a portable TV with a mass of 1.0 slug?

Answers: (a) 730 N = 160 lb; (b) 1.0 N = 0.23 lb; (c) 14 N = 32 lb

3.5 Applications of Newton's Laws

Although Newton's laws can be stated briefly, their implications are profound, and their applications extensive. The application of Newton's laws usually means solving dynamics problems, which are commonly stated as word problems. This section is intended to assist the student in making a logical, systematic approach to the solution of dynamics problems. Here we shall deal with process and technique as well as with principles and definitions. But first, we should stress that no techniques or procedures are adequate substitutes for a thorough knowledge of the basic principles and a clear understanding of the definitions of the quantities used in solving problems.

The basic issue in dynamics problems is solving the fundamental dynamic equation, $\Sigma \mathbf{F} = m\mathbf{a}$. Perhaps the most common form of problem in dynamics requires that the acceleration of a given mass be determined when the mass is subjected to known forces. Sometimes more than one mass is involved; occasionally the mass is the unknown quantity to be determined. In any case, the problem ultimately becomes finding the solution of $\Sigma \mathbf{F} = m\mathbf{a}$.

Let us first consider an uncomplicated one-body problem and discuss some of its problem-solving aspects that are applicable generally to dynamics problems.

A horizontal string with a constant tension of 20 N pulls a 10-kg block that is on a horizontal frictionless surface, as shown in Figure 3.1. What is the acceleration of the block?

It may be tempting to treat such a simple dynamics problem casually: $F = ma$, so $20 = 10a$, and $a = 2$. Right? Definitely not. Let us use this simple example to emphasize some important problem-solving techniques that are particularly useful in the study of dynamics.

Acceleration is a vector quantity, and acceleration of a particle is not specified without a magnitude and a direction. Furthermore, a numerical answer is meaningless without units except for some special quantities that are real numbers, ratios, or similar values that clearly do not require units. Therefore, we may contend that, in general, the answer to every physics problem should have three parts unless there is a specific reason for its having fewer. Each answer should include a magnitude, usually a number or an algebraic expression; a direction specified according to coordinate axes or directions used in the problem; and a unit consistent with the system of units in the statement of the problem. Thus, the answer suggested above is unsatisfactory, and the approach to the problem is even worse.

Plugging values for a given force and mass into $F = ma$ and solving algebraically for a is a naive and overly simplistic approach that should be avoided. True, the problem under consideration is simple, but the procedure should be methodological. Because the object is to learn something of the problem-solving process, let us utilize the problem-solving strategy developed in Section 1.4 to demonstrate in detail how it may be applied to a dynamics problem—however simple.

We must first read and understand any problem under consideration. Do you, for example, know what *frictionless* means? If you do, have you considered what this word implies that is meaningful and relevant to the problem at hand? Do you

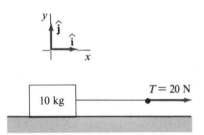

Figure 3.1
A horizontal string with a tension of 20 N attached to a 10-kg block resting on a horizontal frictionless surface.

know what *tension* means in a technical sense? Let us clear up these points before proceeding.

The block is said to be on a frictionless surface. This means that the surface is incapable of exerting forces parallel to itself. A *frictionless* surface can exert only forces that are *normal*, or perpendicular, to itself. The problem also speaks of tension in a string. When a flexible line, like a string, rope, or chain, is pulled taut, the *tension* in the line is the magnitude of the force with which the line pulls on any object attached to either of its ends. Of course, a flexible line can only pull; one cannot push with a flexible line. Tension has the units of force. Although tension is a scalar quantity, it results in a vector force on any object to which a taut line is attached. The direction of that force is along the taut line, away from the attached object.

An important aid in addressing word problems is a pictorial representation of the physical situation. A drawing of some kind is useful if the problem can be represented in a graphical way. Figure 3.2(a) is a drawing that represents the physical situation of the problem we are considering.

Now we may consider what is known and what is to be determined in this problem. The mass of the block and a force acting on it are given. We are looking for the acceleration of the block. At this point we *assume* the direction in which we expect the block to accelerate. This is a guess, but it is one that does not require much intuition. Because the block is being pulled toward the right (in the positive *x* direction), we expect it to accelerate to the right. In Figure 3.2(a) we indicate the assumed direction of acceleration by the vector **a**. Now consider what relationship connects or relates the known force and mass to the unknown acceleration. Because the problem involves motion caused by force (that is, it is a dynamics problem), we should expect to use Newton's second law. But the second law, $\Sigma \mathbf{F} = m\mathbf{a}$, requires that we use the *net* force on the block. Let us, therefore, set about finding the net, or resultant, force on the block in a systematic way.

Obviously the correct resultant force can be found only if we have identified and included in our considerations *all* the external forces acting on the block. How can we find all the forces on the block and be assured that none has been omitted? We may do this by first considering the body under consideration as if it were in isolation, the block in this case. This isolation is suggested in Figure 3.2(a) by the dashed circle around the block. Figure 3.2(b) shows the isolated body with all the forces that act on it.

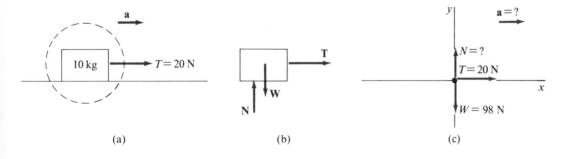

(a)   (b)   (c)

Figure 3.2
Constructing a force diagram. **(a)** The dashed circle assists in identifying all contact forces on the block. **(b)** The isolated body is drawn with all the forces shown acting on the block. **(c)** The block is represented in the force diagram as a particle at the origin of a rectangular coordinate system. All the forces acting on the block are drawn from the origin.

The first and the mandatory force that we include acting on every physical body near the surface of the earth is the weight **W** of the body. (Here we assume the problem is describing a situation on the surface of the earth. Otherwise, the location would have been specifically mentioned.) Because the mass of the block is given, the magnitude of its weight must be calculated from Equation (3-6):

$$W = mg$$

$$W = (10 \text{ kg})(9.8 \text{ m/s}^2) = 98 \text{ N} \tag{3-7}$$

A force vector labeled **W** is drawn pointing downward on the body of Figure 3.2(b). Now we proceed to the remaining forces.

All mechanical forces other than the weight of the block must result from something in contact with the body. By observing the dashed circle of Figure 3.2(a), we can identify everything that comes in physical contact with the block. In this problem the only things touching the block are the horizontal surface and the string. We must account, therefore, for all contact forces the surface and string may exert on the block. Because the surface is frictionless, it can exert only a force normal to the surface (upward, in this case). The normal force **N** exerted by the surface on the block is drawn on the isolated body. Notice that the drawing includes only the forces acting *on* the block, never the forces exerted *by* the block. Finally, the string exerts a force on the block to the right. The magnitude of this force was given to be 20 N (the tension in the string). When the vector representing the force on the block by the string has been drawn on the figure, we are confident that all external forces on the block have been included. We have included the weight of the block and every contact force acting on the block.

Look now at the *force diagram* (sometimes called a *free-body diagram*) of Figure 3.2(c). Here all the forces on the body, which is represented as a particle at the origin, are drawn originating at the origin. We have oriented a rectangular coordinate system so that one axis is parallel to the assumed direction of acceleration. We expect no acceleration in the *y* direction, because the block does not leap off the surface or fall through it. Then the sum of the *y* components of the forces must be equal to zero, or

$$\sum F_y = 0$$

$$N - W = 0$$

$$N = W = 98 \text{ N} \tag{3-8}$$

that is, the magnitude of the normal force is, therefore, equal to that of the weight of the block.

The resultant of all three external forces acting on the block is now seen to be

$$\sum \mathbf{F} = 20\hat{\mathbf{i}} \text{ N} \tag{3-9}$$

According to Newton's second law, the acceleration of the block is in the direction of the resultant external force on the block. From Equation (3-9), the net force is in the positive *x* direction, the direction we assumed for the acceleration **a** of the block. Consequently, Newton's second law gives

$$\sum \mathbf{F} = m\mathbf{a}$$

$$20\hat{\mathbf{i}} \text{ N} = (10 \text{ kg})a\hat{\mathbf{i}}$$

$$a = \frac{20 \text{ N}}{10 \text{ kg}} = \frac{20 \text{ kg} \cdot \text{m/s}^2}{10 \text{ kg}} = 2.0 \text{ m/s}^2 \tag{3-10}$$

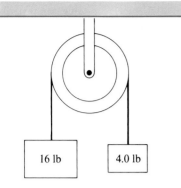

Figure 3.3
Two blocks, weighing 16 lb and 4.0 lb, suspended from a weightless, frictionless pulley.

The answer, correctly stated, is that the acceleration of the block is 2.0 m/s² in the positive x direction, or simply $\mathbf{a} = 2.0\hat{\mathbf{i}}$ m/s².

The foregoing problem dealt with the dynamics of a single body in one dimension. The processes and techniques that have been described are, however, generally applicable. Other useful problem-solving techniques are demonstrated by the following example that involves more than one body.

Two blocks, one weighing 16 lb and the other weighing 4.0 lb, are connected by a weightless string and suspended over a fixed, weightless, frictionless pulley, as shown in Figure 3.3. When the two-block system is released, what is the acceleration of each block and what is the tension in the string?

In this problem, the weightless string and the weightless, frictionless pulley are idealizations that are often used in physics studies. Such a string offers no inertia. Such a pulley offers no inertia and exerts no force tangential to its periphery. The pulley merely serves to change the direction of the string in this problem. The system may be represented by a drawing like that of Figure 3.4(a), in which we have indicated the assumed direction of the acceleration of each body in the problem. The assumed direction of the acceleration of each body is again a guess, but, as we shall see, no harm is done if the guess is wrong. Here we have assumed that the 16-lb block will accelerate downward, and the 4.0-lb block will accelerate upward. The acceleration of each block has the same magnitude a because we assume that the string is inextensible (cannot be stretched). Next we mentally isolate the 16-lb block and construct its force diagram, as in Figure 3.4(b). Its weight $\mathbf{W}_1$ is a 16-lb downward force, and the only contact force on this block is that provided by the string, which exerts an upward force $\mathbf{T}$ on the block. No other forces act on the 16-lb block. First we obtain the mass of the 16-lb block from Equation (3-6):

$$m_1 = \frac{W_1}{g} \qquad (3\text{-}11)$$

When writing Newton's second law in component form, forces parallel to the assumed acceleration are entered as positive quantities, and those antiparallel to

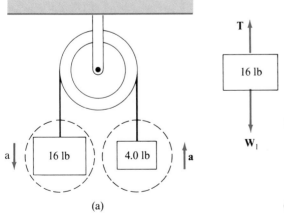

(a)

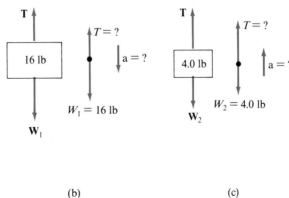

(b)   (c)

Figure 3.4
Constructing force diagrams for a two-body system. **(a)** The dashed circles help to identify the contact forces on the blocks. The directions of the assumed accelerations of the blocks are indicated. **(b)** The 16-lb block is isolated and its force diagram is constructed. **(c)** The 4.0-lb block is isolated and its force diagram is constructed.

the assumed acceleration are entered as negative quantities. Because W_1 is parallel to $\mathbf{a}$, and $\mathbf{T}$ is antiparallel to $\mathbf{a}$ in Figure 3.4(b), we write

$$W_1 - T = m_1 a$$

$$W_1 - T = \frac{W_1}{g}a \qquad (3\text{-}12)$$

Similarly, when considering the force diagram for the 4-lb block of weight W_2, as shown in Figure 3.4(c), we write

$$T - W_2 = m_2 a$$

$$T - W_2 = \frac{W_2}{g}a \qquad (3\text{-}13)$$

E 3.8   (a) Solve Equations (3-12) and (3-13) for the unknowns a and T to get

$$a = \left(\frac{W_1 - W_2}{W_1 + W_2}\right)g$$

$$T = \frac{2W_1 W_2}{W_1 + W_2}$$

(b) Show that the results of part (a) have the correct units.
(c) Do the results of part (a) agree with what you would expect for the special cases: (i) $W_2 = 0$; (ii) $W_1 = W_2$; and (iii) $W_1 = 0$?
(d) For the values of W_1 and W_2 given, determine a and T.

Answers: 19 ft/s^2; 6.4 lb

That a turns out to be a positive value means that we guessed the directions of the accelerations of the blocks correctly. A negative value of a would indicate that the blocks accelerate oppositely to the assumed directions. Notice that the tension T turned out to be a value between 4 lb and 16 lb, which assures that the net force is upward on the 4-lb block and downward on the 16-lb block.

E 3.9   What is the resultant (net) force on (a) the 16-lb block? and (b) the 4-lb block?   Answers: (a) 9.6 lb downward; (b) 2.4 lb upward

In this problem, Newton's second law was applied to each of the two bodies. It is generally true that dynamics problems involving n connected bodies having a common magnitude of acceleration will require that n equations obtained from $\Sigma \mathbf{F} = m\mathbf{a}$ in order to determine the acceleration.

One more example will be discussed here. In this problem we consider a single body moving on an inclined plane.

Determine the acceleration of a block of mass m sliding down a frictionless plane inclined at an angle θ from the horizontal, as shown in Figure 3.5.

This problem is represented by Figure 3.6(a), in which the acceleration is assumed to be down the plane. Isolation of the block with all the external forces acting on it is indicated in Figure 3.6(b). Here again, a frictionless plane can exert only a normal force $\mathbf{N}$ on the block. The plane is the only thing touching the block, so the weight of the block, $\mathbf{W} = m g$ downward, and the normal force $\mathbf{N}$ are all the external forces on the block. These forces are transferred to the force

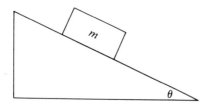

Figure 3.5
A block of mass m sliding down a frictionless plane inclined at an angle θ from the horizontal.

Figure 3.6
Constructing a force diagram for a block moving down an inclined plane. **(a)** The dashed circle helps in identifying contact forces on the block. **(b)** The isolated body is drawn with all the forces acting on the block. **(c)** The force diagram shows all the forces on the block resolved into component vectors parallel and perpendicular to the assumed direction of the acceleration of the block.

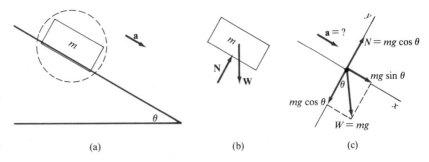

(a)   (b)   (c)

diagram of Figure 3.6(c). In this case, we set one of the rectangular axes along the plane because the direction of the assumed acceleration is down the plane. We resolve all forces along these axes to facilitate finding the resultant force, which is in the direction of the (assumed) acceleration. The weight has a component down the plane of magnitude $mg \sin\theta$. (Convince yourself that the angle θ in Figure 3.6[c] is the same angle θ at which the plane is inclined.) The component of the weight perpendicular to the plane has magnitude $mg \cos\theta$. In the y direction, we expect no acceleration, so the y components of the forces must sum to zero:

$$\sum F_y = 0$$

$$N - mg \cos\theta = 0$$

$$N = mg \cos\theta \tag{3-14}$$

In the x direction, Newton's second law is $\sum F_x = ma_x$. Because $mg \sin\theta$ is the only force in the x direction, it follows that

$$\sum F_x = ma_x$$

$$mg \sin\theta = ma$$

$$a = g \sin\theta \tag{3-15}$$

E 3.10   A 100-lb block of ice slides down a frictionless plane inclined 30° from the horizontal.
(a) What is the acceleration of the block?   Answer: 16 ft/s² down the plane
(b) What is the resultant force on the block?   Answer: 50 lb down the plane

Let us use Equation (3-15) to illustrate how students may check the solutions of some problems. First, the units should be checked. We expect the units of our result to be those of acceleration. Since g has the units of acceleration (m/s² or ft/s²) and $\sin\theta$ is unitless, we have no difficulty concerning units. Now let us look at some "limiting cases." Look first at the solution of Equation (3-15) and the drawing of Figure 3.6(a). What would our solution become if θ were zero (the plane becomes horizontal)? In this case, $\sin\theta$ is zero, and Equation (3-15) reduces to $a = 0$. Our solution makes sense in this limiting case, for when $\theta = 0$, the plane is horizontal; and we would not expect the block to accelerate. Similarly, if θ were equal to 90° (the plane would be vertical), $\sin\theta$ would be unity and a would be equal to g. This result corresponds to what we should expect in this limiting case of a vertical plane. If the frictionless plane were vertical, the block, in effect, would be in free-fall and would have a downward acceleration of magnitude g.

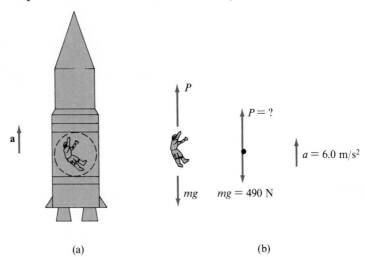

Figure 3.7   Example 3.1   (a)   (b)

Example 3.1

PROBLEM   An astronaut, whose mass is 50 kg, pilots a rocket ship as it rises from the launch pad with an upward acceleration of 6.0 m/s². What force does the rocket ship exert on the astronaut during the time of ascent?

SOLUTION   We isolate the astronaut, as shown by the dashed circle in Figure 3.7(a). The force diagram for the astronaut in Figure 3.7(b) includes the astronaut's weight, a downward force of magnitude mg, and an upward force of magnitude P, which is exerted on the astronaut by the rocket ship. The quantity P is obtained by requiring that the sum of the vertical forces must equal ma, that is,

$$P - mg = ma$$

Because **P** is in the same direction as the acceleration, its magnitude P is entered as a positive quantity; mg is entered negatively because the astronaut's weight is opposite to the direction of the acceleration. Solving for P gives

$$P = m(g + a)$$
$$P = 50(9.8 + 6) = 790 \text{ N}$$

Thus the rocket ship exerts a force of 790 N upward on the astronaut.   ■

Example 3.2

PROBLEM   A truck pulls a 2000-lb trailer. If the truck accelerates at a constant rate from rest to a speed of 30 mi/h (44 ft/s) in 25 s on level ground, calculate the horizontal force the truck exerts on the trailer during this time.

SOLUTION   Figure 3.8(a) shows the trailer as the object to be isolated, and Figure 3.8(b) shows the force diagram for the trailer. The force **N** represents the total vertical force on the trailer by the road and the trailer hitch, and P is the magnitude of the horizontal force on the trailer by the truck. Because **P** is the only force in the direction of the acceleration, we have, using Newton's second law,

$$P = ma$$

The trailer's mass m is given by

$$m = \frac{W}{g} = \frac{2000}{32} = 62.5 \text{ slugs}$$

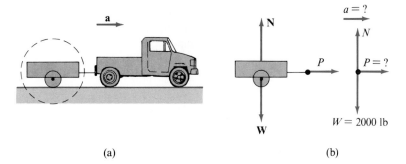

Figure 3.8   Example 3.2   (a)   (b)

The acceleration is determined from the kinematic relationship

$$v = v_o + at$$

with $v_o = 0$, $v = 44$ ft/s, and $t = 25$ s. Thus we have

$$a = \frac{v}{t} = \frac{44}{25} = 1.76 \text{ ft/s}^2$$

Consequently, the quantity P is given by

$$P = (62.5)(1.76) = 110 \text{ lb} \quad \blacksquare$$

Example 3.3

PROBLEM   A jetliner accelerates at a constant rate from rest to a speed of 60 m/s (220 km/h) while moving a distance of 600 m along a runway. Determine the force exerted on a 50-kg passenger by the airplane.

SOLUTION   Figure 3.9 shows the passenger isolated and the corresponding force diagram for this passenger. The quantities V and H are the vertical and horizontal components of the force exerted on the passenger by the airplane, and the downward force having a magnitude equal to mg is the passenger's weight. Applying Newton's second law in both the horizontal and vertical directions gives

$$H = ma$$

$$V - mg = 0$$

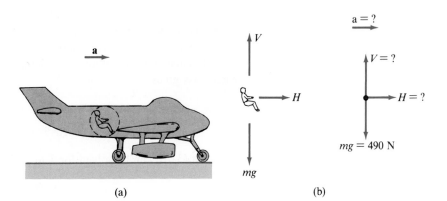

Figure 3.9   Example 3.3   (a)   (b)

These two equations contain the three unknown quantities V, H, and a. Obviously, more information is required. We use the constant acceleration relationship

$$v_x{}^2 = v_{ox}{}^2 + 2a_x(x - x_o)$$

Here, $v_{ox} = 0$, $x_o = 0$, $a_x = a$, $v_x = 60$ m/s, and $x = 600$ m, so that we have

$$v_x{}^2 = 2ax$$

$$a = \frac{v_x{}^2}{2x} = \frac{(60)^2}{2(600)} = 3 \text{ m/s}^2$$

Thus, the components H and V are given by

$$H = ma = (50)(3) = 150 \text{ N}$$

$$V = mg = (50)(9.8) = 490 \text{ N}$$

The force $\mathbf{F}$ exerted on the passenger by the airplane is then

$$\mathbf{F} = H\hat{\mathbf{i}} + V\hat{\mathbf{j}} = (150\hat{\mathbf{i}} + 490\hat{\mathbf{j}}) \text{ N} = 510 \text{ N}\underline{/73°} \quad \blacksquare$$

Example 3.4

PROBLEM   A force $\mathbf{P}$ acts on a mass m_1 (6.0 kg), which slides on a frictionless horizontal surface (*see* Fig. 3.10). A second mass m_2 (4.0 kg) is connected to m_1 by a string that passes over a pulley with negligible mass and friction. If m_2 has an upward acceleration of 4 m/s^2 and $\theta = 30°$, calculate P and the tension in the connecting string.

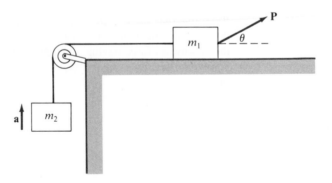

Figure 3.10   Example 3.4

SOLUTION   Figure 3.11 shows m_1 and m_2 as the objects to be isolated and the force diagram for each mass. The force $\mathbf{P}$ is resolved into components perpendicular and parallel to the acceleration. Applying Newton's second law to each mass gives

$$T - m_2g = m_2a \tag{3-16}$$

$$P\cos\theta - T = m_1a \tag{3-17}$$

Equation (3-16) can be solved for T to get

$$T = m_2(g + a) \tag{3-18}$$

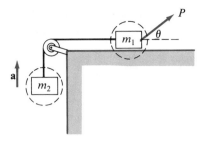

Figure 3.11   Example 3.4

A solution for P is obtained by adding Equations (3-16) and (3-17) to get

$$P \cos\theta - m_2 g = (m_1 + m_2)a$$

$$P = \frac{m_2 g + (m_1 + m_2)a}{\cos\theta} \qquad (3\text{-}19)$$

We can check these results by considering the special case of $a = 0$ (the blocks are not accelerating) and $\theta = 0$ (**P** acts horizontally). Because m_2 is not accelerating, the upward tension T must be equal to the downward weight $m_2 g$, that is, $T = m_2 g$. Similarly, because m_1 is also not accelerating, P must be equal to T, or $P = T = m_2 g$. Setting a and θ equal to zero in Equations (3-18) and (3-19) then gives

$$T = m_2 g$$

$$P = m_2 g$$

the results we expect. Finally, the numerical values for P and T are

$$P = \frac{(4)(9.8) + (10)(4)}{\cos 30°} = 91 \text{ N}$$

$$T = (4)(9.8 + 4) = 55 \text{ N}$$

Example 3.5   PROBLEM   The mass m_1 (3.0 kg) slides on a frictionless surface inclined at an angle α with the horizontal, as shown in Figure 3.12. If $P = 50$ N, the mass m_2 (2.0 kg) has an upward acceleration of 2.5 m/s². Calculate the angle α if the pulley has negligible mass and friction.

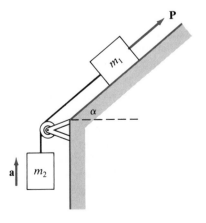

Figure 3.12   Example 3.5

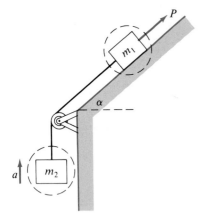

Figure 3.13   Example 3.5

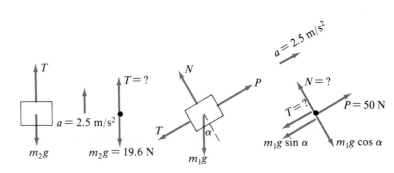

SOLUTION   Figure 3.13 shows the isolated objects m_1 and m_2 and the force diagram for each mass. The weight vector for m_1 has been resolved into components perpendicular and parallel to the acceleration. Applying Newton's second law to each mass gives

$$P - T - m_1 g \sin\alpha = m_1 a$$

$$T - m_2 g = m_2 a$$

The unknown quantities are T and α. Adding these two equations eliminates T and gives

$$P - m_1 g \sin\alpha - m_2 g = (m_1 + m_2)a$$

$$m_1 g \sin\alpha = P - m_2 g - (m_1 + m_2)a$$

$$\sin\alpha = \frac{P - m_2 g - (m_1 + m_2)a}{m_1 g}$$

$$\sin\alpha = \frac{50 - (2)(9.8) - (2 + 3)(2.5)}{(3)(9.8)} = 0.609$$

$$\alpha = 38° \quad \blacksquare$$

3.6 Problem-Solving Summary

Dynamics of particles relates the translational motions of particles to the forces that cause changes in those motions. The physical quantities that we associate with translational dynamics are force, mass, and acceleration. A dynamics problem asks us to find the value of any one of these quantities when the other two are known, given, or capable of being determined from kinematical data in the problem. The appropriate physical relationship to use in solving a problem in translational dynamics is Newton's second law: $\Sigma\mathbf{F} = m\mathbf{a}$.

A dynamics problem may be stated so that a body is characterized by either its mass (measured in kg or slug) or its weight (measured in N or lb). When the mass m is given or used in a problem, one of the forces acting on the body is its weight $\mathbf{W}$, which has magnitude $W = mg$ and is directed downward. When the weight W is given or used, the mass m of the body is given by $m = W/g$. Remember that a weight is always a force.

Solving a dynamics problem successfully often hinges on finding the net force $\Sigma\mathbf{F}$ on one or more bodies. Newton's second law requires that the net force on a body is in the same direction as the acceleration of that body. Therefore, once the problem has been pictured in a drawing (Step 1 of the problem-solving strategy of Section 1.4), it is appropriate to assume a direction for the acceleration of each body in the problem. This assumption is equivalent to assuming the direction of the net force on each body.

Finding the net force $\Sigma\mathbf{F}$ on each body is the crucial issue for most students trying to solve a dynamics problem. A careful, methodical procedure will help to avoid the omission of any forces that do act on a body or the addition of any forces that do not act on that body. The following routine for determining the net force on a body, a composite of Steps 1 and 2 of our overall problem-solving strategy, involves drawings and relationships:

1. Isolate each body in the problem by mentally tracing a path completely around the body, noting anything that touches the body. Everything in contact with the body may exert a force on the body.

2. Draw the force vectors acting on the isolated body, starting with the weight of the body. The weight of a body is *always* a force on the body. Then draw *all* contact forces on the body, indicating the direction of each force. Label the forces.

3. Construct a force diagram, representing the body as a particle at the origin of a rectangular coordinate system oriented with one axis along the direction of the assumed acceleration of the body. Draw a vector, originating at the origin, to represent each force on the body.

4. Resolve all the force vectors of the force diagram into component vectors along the coordinate axes.

5. Write equations for Newton's second law in component form. Remember that the force components perpendicular to the acceleration must add to zero, and those parallel to the acceleration must sum to *ma*.

Here it may be asked, "Why does the four-step strategy, intended to apply to any word problem, need to be so specifically enlarged for dynamics problems?" Dynamics problems are notorious for being rife with hazards for the inexperienced problem solver. This detailed procedure is an attempt to cope with these difficulties in a methodical way.

The foregoing procedure should be applied to each body in a dynamics problem. When more than one body is present, each application of this procedure will usually produce a relationship with more than one unknown quantity. Thus, the procedure is applied to different bodies in the problem until the number of equations is equal to the number of unknown quantities. The equations so obtained may then be solved.

Finally, dynamics problems occasionally require consideration of Newton's third law, namely, that forces exerted on each other by two bodies are equal in magnitude and oppositely directed. An important point in this respect is that the forces are *never* on the same body. So when a single body is isolated and the forces acting on it are considered, one must not include a force that the isolated body is exerting on another body as if it were a force on the isolated body itself. These points may be easier to remember if we recognize that the second law is about forces on a single body and the third law is about forces on two distinct bodies.

Problems

GROUP A

3.1 What force **P** must be exerted on an 8.0-kg mass in order that the mass accelerate upward at 4.0 m/s²?

3.2 At what angle must a frictionless plane be inclined from the horizontal so that a block placed on the plane experiences an acceleration down the plane of 4.0 m/s²?

3.3 An 8.0-kg mass on a frictionless horizontal surface has a horizontal force of 100$\hat{\mathbf{i}}$ N applied to it continuously. What additional force **P** must be applied if the mass is to accelerate at (a) 6.0$\hat{\mathbf{i}}$ m/s²? (b) −6.0$\hat{\mathbf{i}}$ m/s²?

3.4 A 2.0-kg block on a frictionless plane inclined at 37° from the horizontal has two forces of magnitudes 10 N and P acting on it parallel to the plane, as shown in Figure 3.14. What is the value of P if the block has

- (a) a constant velocity?
- (b) an acceleration up the plane of 4.0 m/s²?
- (c) an acceleration down the plane of 4.0 m/s²?

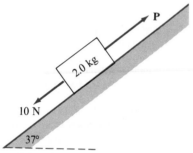

Figure 3.14   Problem 3.4

3.5 What is the magnitude of **P** and the tension T if the system shown in Figure 3.15 is accelerating

- (a) upward at 6.0 m/s²?
- (b) downward at 6.0 m/s²?

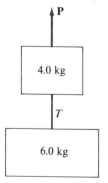

Figure 3.15   Problem 3.5

3.6 What minimum distance must an 8.0-kg mass be lowered vertically from rest in 5.0 seconds by a string that will support a maximum tension of 50 N?

3.7 Two blocks (m_1 = 5.0 kg, m_2 = 7.0 kg) connected by a string are pulled across a horizontal frictionless surface by a horizontal force **F** (*see* Fig. 3.16). If the blocks are observed to have an acceleration of 2.4$\hat{\mathbf{i}}$ m/s²,

- (a) determine F.
- (b) determine the tension in the connecting string.
- (c) what force does m_2 exert on the connecting string?
- (d) what force does m_1 exert on the connecting string?
- (e) what force does m_1 exert on the frictionless surface?

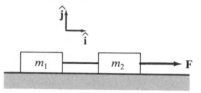

Figure 3.16   Problem 3.7

3.8 A 74-lb box is pulled across a horizontal frictionless surface by a 10-lb force acting 30° above the horizontal.

- (a) What is the magnitude of the acceleration of the box?
- (b) What force does the surface exert on the box?

3.9 A 5.0-lb weight is suspended by a string from the ceiling of an elevator. Find the tension in the string if the elevator is moving

- (a) upward with its speed decreasing at a rate of 7.0 ft/s².
- (b) downward with a constant speed of 15 ft/s.
- (c) downward with its speed increasing at a rate of 7.0 ft/s².

3.10 The inclined plane shown in Figure 3.17 is frictionless. Determine P if the 20-kg block

- (a) remains at rest.
- (b) moves with a constant speed up the plane.
- (c) moves with an acceleration of 5.0 m/s² up the plane.
- (d) moves with an acceleration of 2.0 m/s² down the plane.

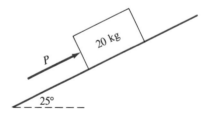

Figure 3.17   Problem 3.10

3.11 Three blocks stacked vertically (*see* Fig. 3.18) are accelerated upward at 1.8 m/s² by the vertical force **P**. Find

- (a) the magnitude of **P**.
- (b) the force exerted on the 15-kg block by the 11-kg block.

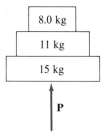

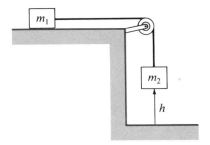

Figure 3.18   Problem 3.11

3.12 A person who normally weighs 130 lb stands on a bathroom scale in an elevator and observes the scale to read 140 lb.

(a) What is the acceleration of the elevator?

(b) Is the elevator ascending or descending?

3.13 The two objects that are shown in Figure 3.19 are released at $t = 0$ from rest with m_2 initially at a height h above the floor. Disregard friction and the mass of the pulley. Given that $m_1 = 20$ kg, $m_2 = 6.0$ kg, and $h = 2.0$ m,

(a) when will m_2 hit the floor?

(b) how fast will m_2 be moving the instant before it hits the floor?

Figure 3.19   Problem 3.13

3.14 A rocket weighing 2.4×10^6 lb rises the first 300 ft above its rest position in 6.0 s. Assuming the rocket's acceleration to be constant, determine the upward force of the rocket's thrust.

3.15 The 4.0-kg mass (*see* Fig. 3.20) is released from rest when it is 2.5 m above the floor. It hits the floor 1.8 s later. Disregarding the mass of the pulley and any frictional forces, find

(a) the tension in the connecting string.

(b) the mass M.

Figure 3.20   Problem 3.15

3.16 A horizontal force of magnitude P pushes two objects across a horizontal frictionless surface, as shown in Figure 3.21. If $P = 18$ N, $m_1 = 4.0$ kg, and $m_2 = 3.0$ kg, find

(a) the acceleration of m_2.

(b) the force exerted on m_1 by m_2.

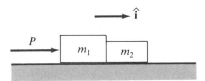

Figure 3.21   Problem 3.16

3.17 Only three horizontal forces act on a 2.4-kg mass that is moving in a horizontal plane with an acceleration $\mathbf{a} = 1.5\hat{\mathbf{j}}$ m/s^2. Two of these forces are expressed as $\mathbf{F}_1 = (3.1\hat{\mathbf{i}} - 2.7\hat{\mathbf{j}})$ N and $\mathbf{F}_2 = (2.0\hat{\mathbf{i}} + 1.5\hat{\mathbf{j}})$ N. Determine the third force $\mathbf{F}_3$.

3.18 A 160-lb person rides an elevator that is accelerating upward at a rate of 4.0 ft/s^2.

(a) What is the resultant force acting on the person?

(b) What force does the floor of the elevator exert on the person?

(c) What force does the person exert on the floor?

(d) If the elevator with its occupants weighs 1800 lb and is supported by a vertical cable, what is the tension in the cable when the elevator accelerates upward at a rate of 4.0 ft/s^2?

3.19 Two blocks are pushed across a horizontal frictionless surface by a 5.0-lb force directed 25° below the horizontal, as shown in Figure 3.22. The surface between the two blocks is frictionless.

(a) Calculate the acceleration of the blocks.

(b) Find the force exerted on the 15-lb block by the 28-lb block.

(c) What force does the horizontal surface exert on the 15-lb block?

(d) What force does the horizontal surface exert on the 28-lb block?

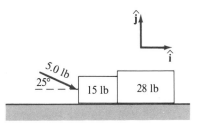

Figure 3.22   Problem 3.19

3.20   The mass of the test-driver of a rocket sled is 73 kg. At an instant when the sled is accelerating 18 m/s² horizontally, determine

(a)   the resultant force on the test-driver.

(b)   the force exerted on the driver by the seat.

3.21   Two objects connected by a string are arranged as shown in Figure 3.23. The block m_1 slides on a horizontal frictionless surface. The pulley is frictionless and has negligible mass. Then if $m_1 = 1.8$ kg, and $m_2 = 2.2$ kg, what is

(a)   the acceleration of m_2?

(b)   the tension in the string?

(c)   the resultant force on m_1?

(d)   the resultant force on m_2?

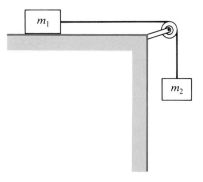

Figure 3.23   Problem 3.21

3.22   A 2400-lb car moving 55 mi/h toward the east on a level roadway brakes to a stop in 280 ft with constant acceleration.

(a)   Find the acceleration of the car.

(b)   What force does the roadway exert on the car while it is braking?

(c)   What force does the car exert on a 125-lb occupant during the braking?

3.23   The horizontal force **P** causes the 15-lb object shown in Figure 3.24 to accelerate upward at a rate of 11 ft/s². The horizontal surface on which the 10-lb block slides is smooth, and the pulley is massless and frictionless.

(a)   Calculate the magnitude of **P**.

(b)   Calculate the tension in the connecting string.

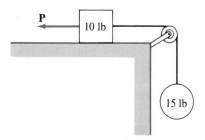

Figure 3.24   Problem 3.23

3.24   The cable holding a submerged buoy 25 ft below the surface of a lake snaps, and the buoy rises to the surface in 2.2 s. The buoy weighs 230 lb when it is out of the water. Assume the acceleration of the buoy to be constant as it rises through the water.

(a)   What is the acceleration of the buoy as it rises through the water?

(b)   What is the speed of the buoy as it emerges from the water's surface?

(c)   Determine the buoyant force on the submerged buoy.

3.25   Three masses are connected by strings and arranged as shown in Figure 3.25. The horizontal surface is smooth, and the massless pulley is frictionless. What is

(a)   the magnitude of the accelerations of the blocks?

(b)   the tension in string 1?

(c)   the tension in string 2?

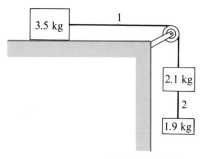

Figure 3.25   Problem 3.25

3.26   Disregard any pulley mass or friction in Figure 3.26. If $m_1 = 4.8$ kg, $m_2 = 2.1$ kg, and $m_3 = 1.2$ kg, find

(a)   the magnitude of the acceleration of the blocks.

(b)   the tension in string 1.

(c)   the tension in string 2.

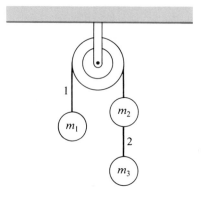

Figure 3.26   Problem 3.26

3.27 The force **P** causes the two blocks to accelerate at 8.0 ft/s² up the frictionless plane (*see* Fig. 3.27). Find

(a) the magnitude of **P**.

(b) the tension in the connecting string.

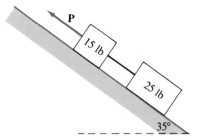

Figure 3.27 Problem 3.27

3.28 A horizontal force **P** acts on an 8.0-kg block that slides on a frictionless plane inclined 40° with the horizontal, as shown in Figure 3.28. If $P = 82$ N, find

(a) the acceleration of the block.

(b) the magnitude of the force on the block by the plane.

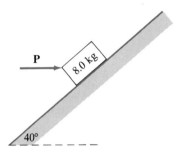

Figure 3.28 Problem 3.28

3.29 In Figure 3.29 if the tension in string 1 is 95 N and the horizontal surface and the massless pulleys shown in the figure are frictionless,

(a) determine m.

(b) find the acceleration of the 5.0-kg mass.

(c) find the tension in string 2.

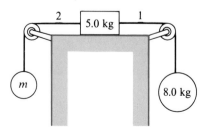

Figure 3.29 Problem 3.29

3.30 A 0.25-lb ball traveling horizontally at 80 ft/s to the left strikes a vertical wall and rebounds to the right with the same speed. High-speed movies of this collision show that the ball is in contact with the wall for 5.5 ms. Assuming constant acceleration, find

(a) the acceleration of the ball while it is in contact with the wall.

(b) the force on the ball by the wall.

3.31 If the sliding surfaces shown in Figure 3.30 are smooth, and the pulley is massless and frictionless, find

(a) the acceleration of the 4.0-kg mass.

(b) the tension in the connecting string.

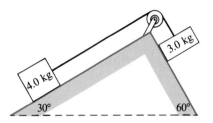

Figure 3.30 Problem 3.31

3.32 A 15-gm bullet leaves the barrel of a rifle with a speed of 840 m/s.

(a) Assuming the acceleration of the bullet to be constant, find the magnitude of the force on the bullet as it moves the length (65 cm) of the barrel.

(b) Calculate the ratio of the magnitude of this force to the weight of the bullet.

3.33 If the sliding surfaces shown in Figure 3.31 are smooth and the pulley is massless and frictionless, find

(a) the magnitude of the accelerations of the blocks.

(b) the tension in the connecting string.

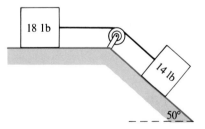

Figure 3.31 Problem 3.33

3.34 If the sliding surfaces shown in Figure 3.32 are smooth and the pulley is massless and frictionless, find

(a) the acceleration of the 18-lb block.

(b) the tension in the connecting string.

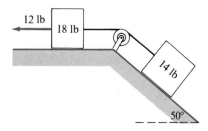

Figure 3.32   Problem 3.34

3.35 A construction worker wishes to throw a 15-lb block 20 ft vertically above the point of release from his hand. The vertical travel of the block while it is in contact with his hand is 2.8 ft. Assuming constant acceleration, what force must the worker exert on the block?

3.36 The 1.5-kg block in Figure 3.33 is observed to accelerate 2.7 m/s^2 to the left. If the sliding surfaces are smooth and the pulleys are massless and frictionless, find

(a)   the tension in the connecting string.

(b)   the mass M.

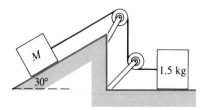

Figure 3.33   Problem 3.36

3.37 The 5.0-kg block slides down the frictionless inclined plane shown in Figure 3.34 with an acceleration of 4.3 m/s^2 down the plane. The pulley is massless and frictionless.

(a)   Determine M.

(b)   What is the tension in the connecting string?

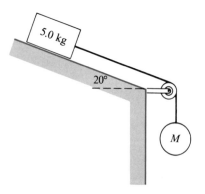

Figure 3.34   Problem 3.37

3.38 A 2600-lb car accelerates 5.0 ft/s^2 up a rough road inclined 10° from the horizontal.

(a)   What force parallel to the surface of the road is exerted on the car by the road?

(b)   What is the magnitude of the total force exerted on the car by the road?

3.39 A horizontal force of 12 N is exerted on a 5.0-kg block that slides on a frictionless horizontal surface (*see* Fig. 3.35). A 2.0-kg block on top of the 5.0-kg block stays fixed with respect to the 5.0-kg block.

(a)   What is the acceleration of the blocks?

(b)   What is the force exerted on the 2.0-kg block by the 5.0-kg block?

(c)   What force does the 2.0-kg block exert on the 5.0-kg block?

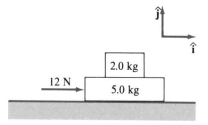

Figure 3.35   Problem 3.39

3.40 A horizontal force ($8.0\hat{\mathbf{i}}$ lb) acts on a 20-lb block sliding on a frictionless horizontal surface, as shown in Figure 3.36. A 4.0-lb block on top of the 20-lb block is observed to move with an acceleration of $4.5\hat{\mathbf{i}}$ ft/s^2.

(a)   What is the acceleration of the 20-lb block?

(b)   What force does the 20-lb block exert on the 4.0-lb block?

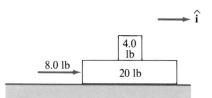

Figure 3.36   Problem 3.40

3.41 Disregard any friction and pulley mass in the configuration shown in Figure 3.37. What is

(a)   the magnitude of the acceleration of the blocks?

(b)   the tension in the connecting string?

3.42 The 10-kg block shown in Figure 3.38 moves with an acceleration of 5.2 m/s^2 to the right on the horizontal frictionless surface. The pulley is frictionless and of negligible mass. What is the magnitude of the force **F**?

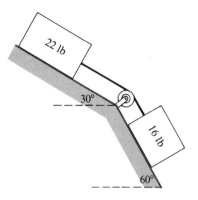

Figure 3.37   Problem 3.41

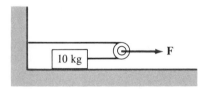

Figure 3.38   Problem 3.42

GROUP **B**

3.43   Disregard friction and pulley mass in Figure 3.39. Then if $F = 18$ lb, find

(a)   the acceleration of the 10-lb block.

(b)   the acceleration of the 20-lb block.

(c)   the tension in string 1.

(d)   the tension in string 2.

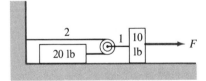

Figure 3.39   Problem 3.43

3.44   Blocks of mass m_1 and m_2 are arranged as shown in Figure 3.40. Answer the following questions in terms of m_1, m_2, and g, without concern for friction and pulley mass.

(a)   What is the acceleration of m_1?

(b)   What is the acceleration of m_2?

(c)   What is the tension in string 1?

(d)   What is the tension in string 2?

(This problem is an excellent opportunity to check your results against limiting cases. Try, for example, the case in which $m_2 = 0$.)

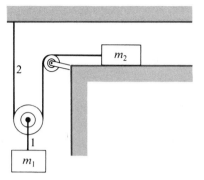

Figure 3.40   Problem 3.44

3.45   Neglect any friction and pulley mass in Figure 3.41. Determine the magnitude of the force **P** and the tension in string 2 if W_1 is to accelerate upward at a rate of $g/3$. Express your answer in terms of the weights W_1 and W_2.

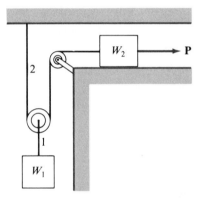

Figure 3.41   Problem 3.45

3.46   Blocks of weights W_1, W_2, and W_3 are arranged as shown in Figure 3.42. Disregard friction and pulley mass. Express your answers in terms of W_1, W_2, W_3, and g.

(a)   What is the magnitude of the accelerations of the blocks?

(b)   What is the tension in string 1?

(c)   What is the tension in string 2?

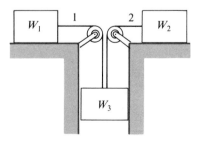

Figure 3.42   Problem 3.46

3.47 Suppose $m_1 = 5.0$ kg, $m_2 = 5.0$ kg, and $m_3 = 10$ kg in Figure 3.43. Disregard any friction and pulley mass.

(a) Convince yourself that $2a_1 = a_2 + a_3$.

(b) Determine the acceleration of each mass.

(c) Determine the tension in each string.

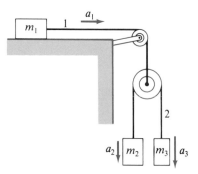

Figure 3.43   Problem 3.47

3.48 The pulleys are massless and frictionless in Figure 3.44.

(a) Convince yourself that $2a_1 = a_2 + a_3$.

(b) Calculate the acceleration of each weight.

(c) Calculate the tension in each string.

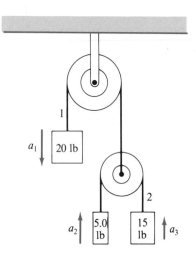

Figure 3.44   Problem 3.48

3.49 A sphere falling vertically in air experiences a resistive (upward) force proportional to its speed.

(a) Show that for such a sphere, Newton's second law may be written

$$\frac{dv}{dt} = g - \frac{g}{v_t}v$$

where v is the downward speed of the sphere and v_t is that speed for which the net force is zero. This speed v_t is called the *terminal speed*.

(b) Show that if $v = 0$ at $t = 0$, then the speed of the sphere is given by

$$v = v_t(1 - e^{-(gt)/v_t})$$

(c) Sketch a graph of the speed of the sphere as a function of time.

(d) If v_t is observed to be 240 mi/h for a certain sphere and if $v = 0$ at $t = 0$, when will the sphere achieve a speed of 120 mi/h?

(e) How far will the sphere fall during the time required for its speed to increase from zero to 120 mi/h?

(f) If $v = 3v_t$ at $t = 0$, solve for $v(t)$.

3.50 A particle moving horizontally experiences a resistive force proportional to its speed. If its speed is v_o at $t = 0$, and it travels a total distance d before stopping, show that its speed and position are given by

$$v(t) = v_o e^{-(v_o t)/d}$$

$$x(t) = d(1 - e^{-(v_o t)/d})$$

where x is measured from the $t = 0$ position.

4 Further Application of Newton's Laws

The preceding chapter introduced Newton's laws of motion, the principles that form the basis of dynamics. We described some simple dynamics problems that require the use of these principles. In this chapter we will extend the use of the same basic principles and problem-solving techniques to apply to a broader variety of problems and, in some cases, to physical situations that are not as idealized as those of Chapter 3. For example, we will no longer restrict ourselves to frictionless surfaces but will introduce a more realistic and practical representation of surfaces. Then we will consider how Newton's second law may be applied to particles in circular motion. Next, another of Newton's discoveries, the law of universal gravitation, will be discussed and applied to appropriate situations. Finally, in this chapter we will look at the special case of Newton's second law in which physical bodies are in static equilibrium. In this case of bodies at rest, we will consider physical bodies that have extension in space, that is, bodies that are not treated merely as mathematical points. These considerations provide the occasion to introduce the concept of torque and the conditions that must be satisfied in order that a body be in static equilibrium. This chapter, then, introduces several new concepts while extending the usefulness of Newton's laws of motion.

4.1 Friction

When two bodies are in contact, there are interactions at the molecular level within the region of contact. Although an adequate theoretical treatment that explains the contact forces from fundamental principles is not now feasible, we can make a number of experimental observations that permit us to analyze the macroscopic effects of contact between physical bodies. Let us look at some experimental situations using a book on a desk.

Suppose first that a book is resting on a horizontal surface, as illustrated in Figure 4.1(a). Only two forces act on the book: Its weight $\mathbf{W}$ is a downward force on the book, and the table exerts a force that, because the book is at rest, must be oppositely directed from the weight and equal in magnitude. The force that the table exerts upward on the book is perpendicular (or normal) to the horizontal contact surface and is, therefore, known as the *normal force* $\mathbf{N}$. Now

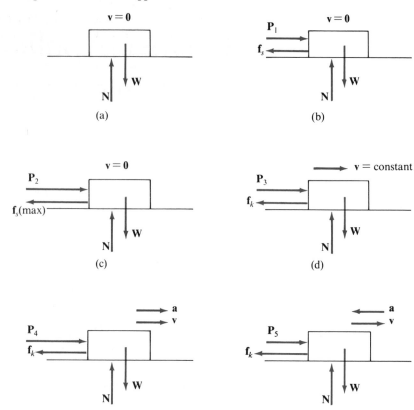

Figure 4.1
A book on a horizontal surface **(a)** resting, **(b)** resting with a small horizontal
force P_1 applied, **(c)** resting with the
greatest horizontal force P_2 that does
not cause the book to move, **(d)** moving
at a constant velocity with a horizontal
force P_3 applied, **(e)** accelerating in the
direction of **v** when P_4 is greater than
f_k, **(f)** accelerating in a direction opposite to **v** when P_5 is less than f_k.

suppose that we push on the book with a horizontal force P_1 directed toward the
right, as in Figure 4.1(b) and that P_1 is a small force. We observe that the book
does not move, even though a horizontal force has been applied. The acceleration of the book is still zero, which means the resultant force on the book must
still be zero. The normal force **N** still balances the weight **W** in the vertical
direction, so we must conclude that there is a horizontal force to the left that
balances the push P_1 to the right. Because only the table is present to produce
additional contact forces, we must further conclude that the table exerts a force
on the book parallel to the surface of contact. This force that is exerted by the
table on the book in a direction parallel to the contact surface is known as the
force of static friction f_s, because the book is static (not moving in the chosen
inertial frame).

To continue our experimental observations, assume we push harder and harder
on the book until we feel it *just* start to move. The instant before it starts to move
is the instant of *impending motion*, as shown in Figure 4.1(c). At this stage, we
must conclude that the force of static friction has reached its utmost magnitude.
In other words, the table at this moment has exerted the greatest force parallel to
the contact surface that the table can provide. Let us call this force the *maximum
static frictional force* $f_s(\text{max})$. Just before the book moves, that is, at the instant
of impending motion of the book, the horizontal forces must sum vectorially to
zero; or the book would accelerate. Next we see that $f_s(\text{max}) = -P_2$, where P_2 is
the greatest horizontal force that can be applied without setting the book in
motion. As the book starts to move, suppose we adjust the magnitude of our push

to a value $\mathbf{P}_3$ so that the book moves along the table at a constant velocity. This situation is illustrated in Figure 4.1(d). If our sense of touch is sufficiently sensitive, we notice that $\mathbf{P}_3$ is a force of lesser magnitude than that of $\mathbf{P}_2$. (Have you noticed when trying to move a heavy box across a room that it takes more force to start the box moving than to keep it in motion once started?) Moving now at constant velocity, the book again has zero acceleration, and the resultant force must again be zero. Therefore, the frictional force must now be equal in magnitude and opposite in direction to the push $\mathbf{P}_3$. We call the frictional force exerted on a moving body the *force of kinetic friction* $\mathbf{f}_k$, which is a force parallel to the contact surface and in a direction opposite to the velocity of the body.

Suppose now that the book is moving horizontally, as indicated in Figure 4.1(e). If we apply a horizontal force $\mathbf{P}_4$ (with magnitude greater than that of $\mathbf{f}_k$) in the direction of the velocity $\mathbf{v}$ of the book, a net force, $\mathbf{P}_4 + \mathbf{f}_k$, acts on the book in the direction of $\mathbf{v}$. Then the book experiences an acceleration $\mathbf{a}$ in the direction of $\mathbf{v}$, and the speed of the book increases. Finally, consider the application of a horizontal force $\mathbf{P}_5$ (with magnitude less than that of $\mathbf{f}_k$) in the direction of $\mathbf{v}$, as shown in Figure 4.1(f). In this case, the net force, $\mathbf{P}_5 + \mathbf{f}_k$, is directed oppositely to the velocity $\mathbf{v}$. The book then has an acceleration $\mathbf{a}$ that is directed oppositely to $\mathbf{v}$, and the speed of the book decreases. In general, then, we may express the acceleration $\mathbf{a}$ of the book, using Newton's second law, by

$$\mathbf{a} = \frac{\sum \mathbf{F}}{m}$$

$$\mathbf{a} = \frac{\mathbf{P} + \mathbf{f}_k}{m} \tag{4-1}$$

where $\mathbf{P}$ is a horizontal force applied to the book in a direction parallel or antiparallel to the velocity of the book.

Let us see what we may deduce from these experiments. First, frictional forces are parallel to contact surfaces. So each surface that is in contact with a body may exert a force that has two components: a normal force perpendicular to the surface and a frictional force parallel to the surface. The magnitude of the force of static friction may be zero or may assume any value between zero and some maximum value $f_s(\text{max})$. In other words, f_s is only as big as it has to be to keep the body from moving. Further, the force of kinetic friction on a body is opposite to the direction of the motion of the body; and the force of static friction on a body at rest is opposite to the direction of the tendency of the body to move, that is, opposite to the direction the body would move if there were no friction. Finally, we may observe that the force of kinetic friction is usually less in magnitude than the maximum value of the force of static friction for a given body and a given surface.

If we performed additional experiments of a more quantitative nature, we could learn even more useful properties of frictional forces. For example, it can be demonstrated that, to a good approximation, frictional forces between flat, unlubricated surfaces are independent of the area of contact between two bodies, as long as the normal force exerted on each body is constant. Therefore, in our considerations and in our problems, we shall assume that the area of contact between bodies is of no consequence in the calculation of frictional forces. Another result of quantitative measurements made on bodies and surfaces of a wide variety of materials indicates that the magnitude of the maximum static frictional force is directly proportional to the magnitude of the normal force, or

$$f_s(\text{max}) \propto N$$

This relationship may be written as an equation by introducing a coefficient of proportionality μ_s:

$$f_s(\text{max}) = \mu_s N \qquad (4\text{-}2)$$

This equation relates the maximum static force of friction to the normal force and defines the *coefficient of static friction* μ_s, a constant that is characteristic of a given pair of surfaces. A similar equation relates the magnitude of the force of kinetic friction to the magnitude of the normal force and defines a second constant μ_k, the *coefficient of kinetic friction:*

$$f_k = \mu_k N \qquad (4\text{-}3)$$

Again, the value of the coefficient is characteristic of a specific pair of surfaces.

The coefficients of static and kinetic friction are very nearly constant values for a given pair of surfaces. Actually, the coefficient of kinetic friction may vary slightly with relative velocity between the surfaces in contact, but here we will assume that μ_s and μ_k are constant for a given pair of surfaces. The numerical value of the coefficients μ_s and μ_k for actual materials can be determined by a very simple experimental procedure that illustrates how frictional forces may be treated analytically.

Suppose we wish to determine the coefficient of static friction for a brick on a wooden surface. Let us place the brick on a plane made of wood and slowly change the angle of the plane relative to the horizontal, as shown in Figure 4.2(a), until the brick begins to move. At that instant we stop increasing the inclination and note the angle θ at which the brick begins to move. This situation,

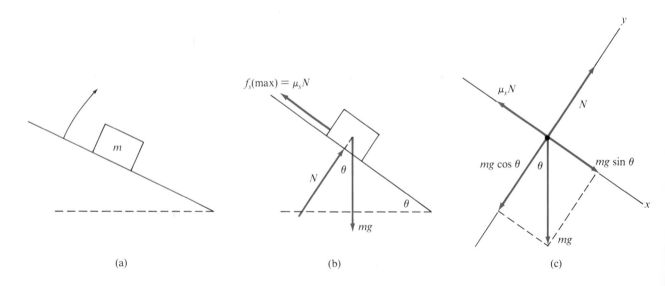

Figure 4.2
(a) Increasing the angle of an inclined plane until a brick starts to slide. (b) Forces on the brick at the moment of impending motion. (c) Force diagram for the brick at the moment of impending motion.

or more precisely the situation of impending motion the instant before the brick moves, is shown for the isolated body in Figure 4.2(b). Figure 4.2(c) shows the appropriate force diagram with rectangular axes oriented parallel and perpendicular to the plane. All forces are resolved along the chosen axes. At the instant depicted, the force of static friction has reached its maximum value and, therefore, has a magnitude equal to $\mu_s N$. The frictional force is parallel to the inclined plane and is directed up the plane, opposing the tendency of the brick to move down the plane. The magnitude N of the normal force is equal to the magnitude $mg \cos\theta$ of the y component of the weight because no y-directed acceleration occurs. At the moment of impending motion, the acceleration in the x direction, along the plane, is still zero; and the sum of x-directed forces is zero, or

$$\sum F_x = 0$$

$$mg \sin\theta - \mu_s N = 0 \tag{4-4}$$

Because N is equal to $mg \cos\theta$, we have

$$mg \sin\theta = \mu_s mg \cos\theta \tag{4-5}$$

Division of both sides of Equation (4-5) by $mg \cos\theta$ gives

$$\tan\theta = \mu_s \tag{4-6}$$

Equation (4-6) indicates that a measurement of the angle at which the motion of the brick begins specifies the coefficient of static friction μ_s between the brick and the plane.

A similar measurement that determines the angle of the incline with the horizontal when the brick moves at constant velocity yields the value of the coefficient of kinetic friction μ_k in terms of that angle. This situation is considered in the problems at the end of this chapter.

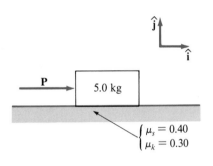

Exercise 4.1

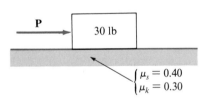

Exercise 4.3

E 4.1   A 5.0-kg body initially at rest on a horizontal surface ($\mu_s = 0.40$, $\mu_k = 0.30$) is acted upon by a horizontal force **P**, as shown in the accompanying diagram. Determine the force on the body by the surface for each of the values of P given.
(a) $P = 10$ N, (b) $P = 25$ N
                    Answers: (a) $(-10\hat{\imath} + 49\hat{\jmath})$ N; (b) $(-15\hat{\imath} + 49\hat{\jmath})$ N

E 4.2   Consider Exercise 4.1. What is the resultant force on the 5-kg body in each case?   Answers: (a) 0; (b) $10\hat{\imath}$ N

E 4.3   A horizontal force **P** acts on the 30-lb body (*see* diagram). For what range of values of P will the body remain at rest?
                    Answer: $0 \leq P \leq 12$ lb

E 4.4   If a 10-kg body is sliding across a surface, as shown in the accompanying diagram,
(a) what is the force on the body by the surface?   Answer: $(-29\hat{\imath} + 98\hat{\jmath})$ N
(b) what is the force on the body by the surface after the body stops?
                    Answer: $98\hat{\jmath}$ N

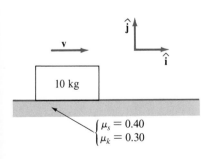

Exercise 4.4

Example 4.1

PROBLEM   A horizontal force **P** acts on a 5.0-kg block initially at rest on a surface inclined 25° from the horizontal (*see* Fig. 4.3). The coefficients of static and kinetic friction for the block and the surface are $\mu_s = 0.20$ and $\mu_k = 0.10$.
(a) For what range of values of P will the block remain at rest?
(b) If $P = 40$ N, what is the acceleration of the block?

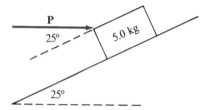

Figure 4.3   Example 4.1

SOLUTION   (a): The situation in which P assumes its minimum value P_1 is shown in Figure 4.4(a). Here the impending motion is down the plane, and the maximum force of static friction acts up the plane. Because the block remains at rest, the forces acting on it must sum to zero. In component form, this condition is

$$\sum F_x = 0$$

$$N - P_1 \sin\theta - mg\cos\theta = 0$$

$$P_1 \cos\theta + f_s(\text{max}) - mg\sin\theta = 0$$

These two equations include three unknown quantities, namely, P_1, N, and $f_s(\text{max})$. Of course, $f_s(\text{max})$ and N are related by

$$f_s(\text{max}) = \mu_s N$$

Solving these equations for P_1 gives

$$P_1 = \left(\frac{\sin\theta - \mu_s \cos\theta}{\cos\theta + \mu_s \sin\theta}\right) mg$$

$$P_1 = \left(\frac{\sin25° - (0.2)\cos25°}{\cos25° + (0.2)\sin25°}\right)(5.0)(9.8) = 12 \text{ N}$$

Figure 4.4(b) depicts the situation for which P assumes its maximum value P_2. The impending motion is up the plane, and, consequently, the force of static friction acts down the plane with a maximum magnitude. Once again the forces on the block must sum to zero, so we have

$$\sum F_x = 0$$

$$N - P_2 \sin\theta - mg\cos\theta = 0$$

$$P_2 \cos\theta - f_s(\text{max}) - mg\sin\theta = 0$$

and, again, $f_s(\text{max})$ and N are related by

$$f_s(\text{max}) = \mu_s N$$

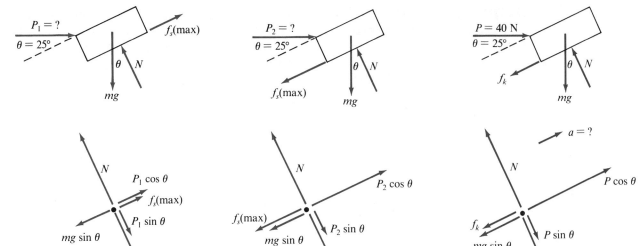

Figure 4.4   Example 4.1

Solving these equations for P_2 yields

$$P_2 = \left(\frac{\sin\theta + \mu_s \cos\theta}{\cos\theta - \mu_s \sin\theta}\right) mg$$

$$P_2 = \left(\frac{\sin 25° + (0.2) \cos 25°}{\cos 25° - (0.2) \sin 25°}\right) (5.0)(9.8) = 36 \text{ N}$$

Thus, the block remains at rest for values of P that are in the range

$$12 \text{ N} \leq P \leq 36 \text{ N}$$

(b): If $P = 40$ N, the block accelerates up the plane, and the force of kinetic friction acts down the plane, as shown in Figure 4.4(c). Applying Newton's second law in component form gives

$$\sum F_x = ma$$

$$P \cos\theta - mg \sin\theta - f_k = ma$$

$$N - P \sin\theta - mg \cos\theta = 0$$

where the unknown quantities are a, f_k, and N. Relating f_k and N gives

$$f_k = \mu_k N$$

Eliminating f_k and N and solving for a gives

$$a = \frac{P(\cos\theta - \mu_k \sin\theta) - mg(\sin\theta + \mu k \cos\theta)}{m}$$

$$a = \frac{40(\cos 25° - (0.1) \sin 25°) - (5.0)(9.8)(\sin 25° + (0.1) \cos 25°)}{5.0}$$

$$a = 1.9 \text{ m/s}^2 \quad \blacksquare$$

4.2 Dynamics of Circular Motion

In Chapter 2 we saw that a particle in uniform circular motion (moving in a circular path at constant speed) experiences an acceleration with magnitude v^2/r, where v is the speed of the particle and r is the radius of its circular path. Even though the speed v of a particle in uniform circular motion is constant, a change in direction of a vector quantity like $\mathbf{v}$ is as real and is as effective in producing acceleration as a change in its magnitude. The direction of the particle's acceleration is at every instant toward the center of the circular path. Then the presence of a centripetal acceleration

$$a_c = \frac{v^2}{r} \tag{4-7}$$

directed toward the center requires, in accordance with Newton's second law, that there must be a resultant force toward the center such that

$$\sum \mathbf{F} = m\mathbf{a}$$

$$F_c = ma_c$$

or

$$F_c = \frac{mv^2}{r} \tag{4-8}$$

Here F_c, which is the magnitude of the resultant of all external forces on the particle moving in a circle, is called the *centripetal force* on the particle. Again, at every instant the direction of this net force is toward the center of the circle. It is important to recognize that if a particle is moving in a circular path and if its speed is not changing at a particular instant, the sum of *all* external forces on the particle must be toward the center of the particle's circular path. Further, the magnitude of that resultant force is related to the speed, mass, and trajectory of the particle by Equation (4-8).

An instructive example of the dynamics of a particle in circular motion is the conical pendulum:

A particle is attached to a string that is fixed at its upper end to a ceiling, as shown in Figure 4.5(a). The particle moves in a horizontal circular path of radius r at a constant speed v. Express the angle θ the string makes with the vertical in terms of v and r.

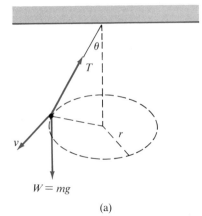

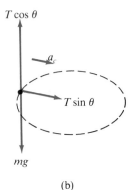

Figure 4.5
The conical pendulum.

(a)   (b)

We may assume that the particle has mass m and that the string is under tension T. The only forces on the particle are its weight W, which equals mg, directed downward, and the tension T directed along the string. In Figure 4.5(b) these forces are resolved into vertical and horizontal components. Because there is no vertical acceleration on a particle moving in a horizontal plane, we may write

$$\sum F_y = 0$$

$$T \cos\theta - mg = 0$$

$$T \cos\theta = mg \qquad \text{(4-9)}$$

The magnitude of the resultant force on the particle, therefore, is equal to the horizontal component, $T \sin\theta$, of the tension. This component of the tension is directed toward the center of the circular path of the particle. Then because $F_c = T \sin\theta$, Equation (4-8) becomes

$$T \sin\theta = \frac{mv^2}{r} \qquad \text{(4-10)}$$

If Equation (4-10) is divided by Equation (4-9), we obtain

$$\tan\theta = \frac{v^2}{gr}$$

$$\theta = \tan^{-1}\left(\frac{v^2}{gr}\right) \qquad \text{(4-11)}$$

which expresses the required angle θ in terms of the given quantities v and r.

E 4.5   In Equation (4-11) what is the unit of (v^2/gr)? Is this unit appropriate?

E 4.6   A 2.0-kg object sliding on a frictionless horizontal surface moves along a circular path with an 0.80-m radius. The centripetal force is provided by a horizontal string from the object to the center of the circle.
(a) What is the tension in the string if the object is moving 1.6 m/s?

Answer: 6.4 N

(b) If the tension in the string is 15 N, what is the speed of the object?

Answer: 2.5 m/s

E 4.7   A 150-kg motorcycle travels without sliding at a speed of 20 m/s around a level circular track. The radius of the circular track is 120 m. What frictional force (magnitude and direction) does the surface of the track exert on the motorcycle?   Answer: 500 N radially inward

Example 4.2   **PROBLEM**   A truck travels with a speed v around a circular racetrack with a radius r. The surface of the road is banked and makes an angle α with the horizontal, as shown in Figure 4.6. For what value of α (expressed in terms of v and r) will there be no tendency of the truck to slide up or down the roadway? Calculate α for $v = 55$ mi/h (81 ft/s) and $r = 800$ ft.

SOLUTION   For any banking angle $\alpha > 0$, if the vehicle is at rest or moving very slowly, it would tend to slide down the banked road surface; a force of static friction acts upon the surface on the tires to prevent this sliding, as shown in

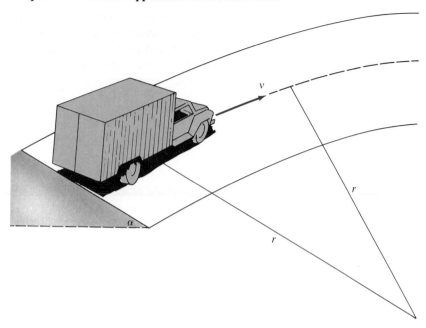

Figure 4.6   Example 4.2

Figure 4.7(a). At very high speeds, because of the inertia of the truck, it tends to slide up the banked surface; a force of static friction on the tires acts down the surface to prevent this sliding, as shown in Figure 4.7(b). For one particular speed, the truck exhibits no tendency to slide. At this speed no frictional force acts, and the only two forces on the truck are its weight and the normal force on it by the surface, as shown in Figure 4.7(c). As you can see from the force diagram of Figure 4.7(c), the vertical component $N \cos\alpha$ of **N** supports the weight of the truck, and the horizontal component $N \sin\alpha$ provides the centripetal force:

$$N \cos\alpha = mg$$
$$N \sin\alpha = \frac{mv^2}{r}$$

If the second equation is divided by the first, we get

$$\tan\alpha = \frac{v^2}{gr}$$

$$\alpha = \tan^{-1}\frac{v^2}{gr} = \tan^{-1}\left[\frac{(81)^2}{(32)(800)}\right] = 14° \quad \blacksquare$$

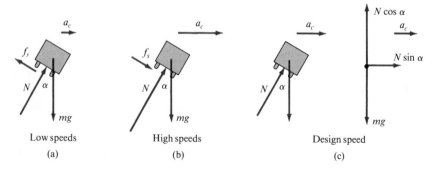

Figure 4.7   Example 4.2

Example 4.3   PROBLEM   Consider a 1.0-lb chunk of rubber on the outer edge of a tire that has a 15-in. radius. If the outer edge of the tire is moving 60 mi/h (88 ft/s) relative to the center, what centripetal force is required to keep this piece of rubber moving in a circular path?

SOLUTION   The centripetal acceleration of a point on the outer edge of the tire is radially inward and has a magnitude of

$$a_c = \frac{v^2}{r} = \frac{(88)^2}{1.25} = 6.2 \times 10^3 \text{ ft/s}^2$$

The magnitude of the centripetal force necessary to provide this acceleration to a 1.0-lb object is given by

$$F_c = \frac{w}{g}a_c = \frac{1}{32}(6.2 \times 10^3) = 190 \text{ lb}$$

This rather large inward force is provided, of course, by that part of the tire in contact with the chunk. ∎

4.3   Law of Universal Gravitation

We have been using the fact that the earth pulls on every physical body near its surface with a force directed toward the center of the earth. We have referred to this force on a body as the weight of the body. Newton's law of universal gravitation extends the notion of one body pulling on another to include all bodies in the universe. This law may be stated as follows:

Every particle of mass in the universe attracts every other particle of mass. For a given pair of particles, the force on each particle is directly proportional to the product of their masses and inversely proportional to the square of the distance between them.

This principle may be stated mathematically by

$$F_g \propto \frac{m_1 m_2}{r^2}$$

$$F_g = G\frac{m_1 m_2}{r^2} \tag{4-12}$$

In Equation (4-12) G is the constant of proportionality relating the magnitude F_g of the gravitational force, the masses m_1 and m_2 of any two particles, and the distance r separating the two particles. The value of the gravitational constant has been measured to be

$$G = 6.67 \times 10^{-11} \text{ N} \cdot \text{m}^2 \cdot \text{kg}^{-2} \tag{4-13}$$

The quantity F_g given by Equation (4-12) is the magnitude of each of a pair of forces. Newton's third law prescribes that forces come in pairs and that the two forces of such a pair are equal in magnitude, are oppositely directed, and are acting on different bodies. Because the gravitational forces on both masses m_1 and m_2 of a pair are attractive, a force of magnitude F_g on each particle is directed toward the other, as indicated in Figure 4.8.

Figure 4.8
Attractive gravitational forces on two masses.

An important and useful fact associated with gravitation is the result of a theorem. Its proof will not be given here, but the theorem may be stated as follows:

A spherically symmetrical mass distribution produces the same force on a body outside that sphere as would be produced if all the mass within the sphere were located at its center.

Notice that "spherically symmetrical" does not require that a spherical mass be uniformly dense in order to qualify as a mass to which this theorem applies. A spherical mass in which density varies, but only in the radical direction, is a spherically symmetrical mass distribution. Thus, the earth, which has a dense core and a relatively less dense crust, is a good approximation to a spherically symmetrical mass. Therefore, in gravitational problems involving the earth and a body located at or outside of the earth's surface, we may treat the earth as if it were a point mass located at the center of the earth. And now that this theorem is available to us, the value of the mass M_e of the earth may be easily calculated.

Consider a mass m located on the surface of the earth. The weight of this mass has a magnitude mg, which is the magnitude of the force with which the earth pulls on the mass. Then, according to Equation (4-12), we may write

$$F_g = mg = G\frac{mM_e}{R^2}$$

$$g = \frac{GM_e}{R^2}$$

$$M_e = \frac{gR^2}{G} \tag{4-14}$$

in which R is the radius of the earth. Because we are treating the mass of the earth as if it were concentrated at its center, the radius $R = 6.4 \times 10^6$ m of the earth separates the two masses, the earth and the mass of weight mg. Substituting numerical values for g, R, and G into Equation (4-14) yields the mass M_e of the earth:

$$M_e = \frac{gR^2}{G} = \frac{(9.8)(6.4 \times 10^6)^2}{6.67 \times 10^{-11}}$$

$$M_e = 6.0 \times 10^{24} \text{ kg} \tag{4-15}$$

The magnitude of the mass of the earth should suggest why, in mechanical problems, we do not have to consider any gravitational forces on a body by any nearby masses other than the earth. Perhaps the following exercises will convey something of the magnitudes associated with masses and forces of gravitation.

E 4.8   Estimate the gravitational force on one 80-kg person by another person 1 m away. Is the magnitude of this force comparable to the weight of either person? Is it negligible compared to the weight of either person?

Answer: 4.3×10^{-7} N

E 4.9   The previous exercise demonstrates that the gravitational force on you by another person or by any comparably sized object is negligible. What about the force of attraction on you by a nearby mountain? To answer this question, determine the gravitational force on an 80-kg person by a sphere of 1.0-km diameter uniformly filled with a material having a density of 5.0 g/cm³. Assume that the person stands at the surface of the sphere.

Answer: 5.6 × 10⁻² N

E 4.10   The mass and radius of the moon are respectively equal to 7.4×10^{22} kg and 1.7×10^{6} m. Calculate the free-fall acceleration of an object on or near the surface of the moon. See Equation (4-14).   Answer: 1.6 m/s²

In many situations that may be analyzed in terms of gravitational forces, the numerical value of G may not be required. For example, consider the following problem:

The mass of the earth is about 81 times the mass of the moon. What fraction of the distance to the moon in a straight line would a spaceship on earth have to travel in order to reach the point of so-called weightlessness? Weightlessness here is defined as the point at which the pull of the earth on the spaceship is equal to the pull of the moon in the opposite direction.

The physical situation of this problem is pictured in Figure 4.9(a). The distance from the earth to the moon is d, and x is the distance from the earth to the point P of weightlessness. It follows, then, that the distance from the weightless point to the moon is $d - x$. Let the mass of the spaceship be m and let M_e and M_m represent the masses of the earth and moon. The forces of gravitational attraction on the spaceship by the earth and the moon, shown in Figure 4.9(b), are $\mathbf{F}_e$ and $\mathbf{F}_m$. According to the law of gravitation, Equation (4-12), the magnitudes of these forces are

$$F_e = G\frac{mM_e}{x^2} \tag{4-16}$$

and

$$F_m = G\frac{mM_m}{(d-x)^2} \tag{4-17}$$

The weightless condition of the spaceship requires that $F_e = F_m$, or

$$G\frac{mM_e}{x^2} = G\frac{mM_m}{(d-x)^2}$$

$$\frac{M_e}{x^2} = \frac{M_m}{(d-x)^2} \tag{4-18}$$

Figure 4.9
(a) The point P of weightlessness between the earth and the moon.
(b) Gravitational forces by the earth and by the moon on the weightless spaceship.

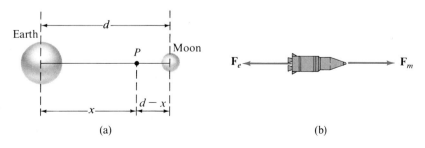

(a)   (b)

Inserting the information that $M_e = 81\ M_m$, Equation (4-18) becomes

$$\frac{81\ M_m}{x^2} = \frac{M_m}{(d-x)^2}$$

$$81\ (d-x)^2 = x^2$$

$$9\ (d-x) = x$$

$$\frac{x}{d} = \frac{9}{10} \tag{4-19}$$

Thus the spaceship would have to travel 9/10 of the distance (less the radius of the earth) to the moon in order to experience weightlessness.

In the foregoing problem, the ratio that we sought turned out to be independent of the value of G, and the answer did not depend on the mass of the spaceship (it would be weightless at the designated point regardless of its mass). In fact, although we knew the ratio of the masses of the earth and moon, neither of the values of these masses was required. These facts suggest a useful hint in approaching many kinds of physics problems: When unknown physical quantities appear in the analysis of a problem, assign symbols to those quantities and proceed.

The law of universal gravitation is applicable to other types of space-age situations. The mechanics of satellites, shuttles, and orbiting bodies depends directly on the gravitational law. The following example illustrates this use of the law of universal gravitation.

The first space shuttle orbited the earth at an altitude of about 140 mi (about 230 km) in an approximately circular orbit. Determine its speed in orbit and the time it took to negotiate each circuit of the earth.

Figure 4.10(a) illustrates the orbit of the space shuttle at a radius r, extending 6.6×10^6 m from the center of the earth. At every instant during orbiting, the shuttle (to which we assign a mass m) is traveling at a constant speed v and is,

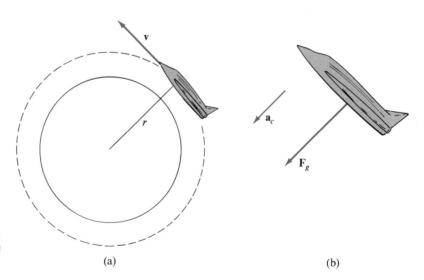

Figure 4.10
(a) Space shuttle orbiting the earth.
(b) Gravitational force $\mathbf{F}_g$ on the shuttle.

(a)

(b)

therefore, in uniform circular motion. The gravitational force F_g with which the earth attracts the shuttle is the only force on the shuttle, and this fact is indicated in the diagram of Figure 4.10(b). Then F_g is the magnitude of the net force on the spacecraft; and that force, as well as the acceleration $\mathbf{a}_c$ of the spacecraft, must be directed toward the center of the circular path of the shuttle. The gravitational force is, therefore, the centripetal force on the shuttle, so we may write

$$\frac{mv^2}{r} = G\frac{mM_e}{r^2} = F_g$$

$$v^2 = \frac{GM_e}{r} = \frac{(6.67 \times 10^{-11})(5.98 \times 10^{24})}{6.60 \times 10^6}$$

$$v = 7800 \text{ m/s} \ (= 17,000 \text{ mi/h}) \tag{4-20}$$

Knowing the speed of the shuttle to be 7.8×10^3 m/s permits us to calculate the time required for the shuttle to complete each revolution about the earth. Dividing the circumference of the circular path of the shuttle by its speed gives the time t required to complete an orbit:

$$t = \frac{2\pi r}{v} = \frac{2\pi(6.60 \times 10^6)}{7.8 \times 10^3}s = 5.3 \times 10^3 \text{ s} = 1.5 \text{ h} \tag{4-21}$$

E 4.11   The mean earth-moon distance is 3.8×10^5 km. Calculate (a) the speed of the moon in its orbit around the earth, and (b) the time for one orbit of the earth by the moon.   Answers: (a) 1.0 km/s; (b) 27 days

E 4.12   The mean earth-sun distance is 1.5×10^8 km, and the time for one earth orbit of the sun is 365 days.
(a) Calculate the speed of the earth around the sun.   Answer: 30 km/s
(b) Calculate the mass of the sun.   Answer: 2.0×10^{30} kg

Example 4.4   **PROBLEM**   The planet Venus requires 225 days to orbit the sun, which has a mass $M_s = 2.0 \times 10^{30}$ kg, in an almost circular trajectory. Calculate the radius of this orbit and the orbital speed of Venus as it circles the sun.

   SOLUTION   The centripetal force on Venus, which keeps it in its circular orbit about the sun is, of course, the gravitational force on Venus by the sun. Thus, combining Newton's second law and the law of gravitation gives

$$\frac{mv^2}{r} = \frac{GmM_s}{r^2}$$

where m is the mass of Venus, M_s is the mass of the sun, v is the orbital speed of Venus, and r is the radius of the circular trajectory Venus follows. Multiplying both sides of this equation by r/m gives

$$v^2 = \frac{GM_s}{r}$$

Another relationship is needed to solve this equation because it includes the two unknown quantities v and r. Equation (4-21) provides this relationship,

$$t = \frac{2\pi r}{v}$$

where t is the time required to complete one orbit. Solving this equation for v and substituting into the previous equation gives, after a little algebra,

$$r = \left(\frac{GM_s t^2}{4\pi^2}\right)^{1/3}$$

$$r = \left(\frac{6.67 \times 10^{-11} \times 2.0 \times 10^{30} \times (225 \times 24 \times 3600)^2}{4\pi^2}\right)^{1/3} = 1.1 \times 10^{11} \text{ m}$$

The orbital speed of Venus is then

$$v = \frac{2\pi r}{t} = \frac{2\pi(1.1 \times 10^{11})}{225 \times 24 \times 3600} = 3.5 \times 10^4 \text{ m/s} = 35 \text{ km/s} \quad \blacksquare$$

4.4 Static Equilibrium

When a physical body is in *static equilibrium*, it is not moving. Until now we have focused our attention on describing the motion of bodies and investigating how forces are related to that motion. One state of motion of a body is, of course, to be at rest; recognizing this state may permit us to use helpful information about the forces acting on such a body. We know, for example, that a body at rest has no acceleration and, therefore, can have no net force acting on it, in accordance with Newton's second law. This condition of static equilibrium, that the net force on a resting body must be zero, is usually written

$$\sum \mathbf{F} = \mathbf{0} \tag{4-22}$$

which emphasizes that, although many forces may be exerted simultaneously on a body, the vector sum of all forces must be equal to zero if the body is to remain at rest. Equation (4-22), then, is a condition of static equilibrium, a special case of $\sum \mathbf{F} = m\mathbf{a}$ in which $\mathbf{a} = \mathbf{0}$.

Let us now see how the condition of static equilibrium, Equation (4-22), may be applied to some basic static situations. Among the simplest cases of static equilibrium is a 10-lb block suspended from a ceiling by a string, as shown in Figure 4.11(a). Using the problem-solving procedures developed for dynamics, we mentally isolate the body, draw the appropriate force diagram, and observe that the tension in the string must be equal to the weight of the block. We may note, then, that when a body at rest is attached to only a vertical string, the tension in the string will be equal to the weight of the body.

Now if we complicate the suspension of the 10-lb block a bit by using strings attached to a ceiling and a wall, as in Figure 4.11(b), we focus attention on the point where the three strings meet and that point is in equilibrium. Suppose we designate the tensions in the strings by T_1, T_2, and T_3. The force diagram for the point where the three strings meet is seen in Figure 4.11(c), where the forces on that point have been resolved along rectangular axes. If there is no net force on this point, there is no net force along either rectangular axis. In other words, the force components along the x axis must sum vectorially to zero, and those along the y axis must similarly sum to zero. These facts may be succinctly expressed by

$$\sum F_x = 0 \tag{4-23}$$

$$\sum F_y = 0 \tag{4-24}$$

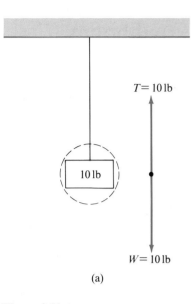

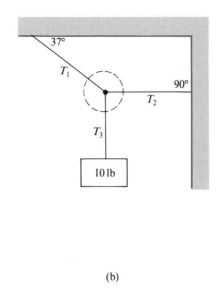

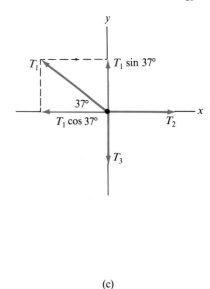

(a)   (b)   (c)

Figure 4.11
(a) A simple vertical suspension of a 10-lb block. **(b)** A static situation. The knot is in static equilibrium. **(c)** Force diagram for the knot.

Continuing with our example, we substitute the values of the force components into Equations (4-23) and (4-24):

$$T_2 - T_1 \cos 37° = 0 \tag{4-25}$$

$$T_2 \sin 37° - T_3 = 0 \tag{4-26}$$

We know that $T_3 = 10$ lb, so we can substitute the numerical values into Equation (4-26) to give

$$0.6\, T_1 - 10 = 0$$

$$T_1 = 16.7 \text{ lb} \tag{4-27}$$

which may be substituted into Equation (4-25):

$$T_2 - (16.7)(0.8) = 0$$

$$T_2 = 13 \text{ lb} \tag{4-28}$$

The tension in each string is now determined.

An interesting aspect of these results is that a 10-lb block can produce in each of the two strings supporting it a tension greater than the weight of the block.

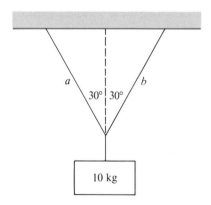

Exercise 4.13

E 4.13   A 10-kg object is suspended, as shown in the accompanying diagram. Determine the tension in string a and in string b.

Answer: $T_a = T_b = 57$ N

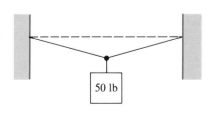

Exercise 4.14

E 4.14   A 50-lb object is suspended at the midpoint of a cable (see diagram). If the angle between the cable and the horizontal is 1.5°, what is the tension in the cable? Compare this tension to the weight of the suspended object.

Answer: 960 lb

Torque and Rotational Equilibrium

So far we have considered the motion (or lack of it) of physical bodies in terms of the particle model, in which only the translation of the body is significant. Now we must take into account the sizes of real bodies and recognize that forces acting on a given real body may cause that body to *rotate*, or turn, about some axis as well as to translate through space.

The effects of forces in causing bodies to rotate are familiar to us from everyday experience. Look, for example, at the opening or closing of a door. In such a case we are interested in causing the door to rotate about a fixed axis through its hinges. Suppose in Figure 4.12 we are looking downward on a thin door along the line of a fixed axis (through a door's hinges). Further, suppose we consider the effects of exerting forces, each of which has a magnitude of 5.0 lb. In Figure 4.12(a) we may observe that if we apply a 5.0-lb force, like F_1, F_2, or F_3 along a line passing through the axis in any direction, no rotation of the door occurs. Indeed no rotation takes place if we push on the door in a direction like that of F_4, which is also along a line that passes through the axis. Now suppose we exert our 5.0-lb force at a particular point on the door in any one of several directions, as indicated by F_5, F_6, and F_7 in Figure 4.12(b). What rotational effects would be observed? Either of the forces F_5 or F_7 would produce rotation, but neither would be as effective as F_6. We surmise, therefore, that a force applied at a particular point on the door is most effective in producing rotation when it is perpendicular to the plane of the door. We have found already from considering Figure 4.12(a) that the other extreme, forces applied parallel to the plane of the door, cause no rotation because their lines of action pass through the axis. Finally, let us examine in Figure 4.12(c) the effect of applying our 5.0-lb force on a perpendicular to the plane of the door at different distances from the axis. Try it on a door! The force F_{10} applied farthest from the axis is clearly more effective in producing rotation than either F_8 or F_9.

How may we summarize these observations? First, when a force is applied to the door along a line that does not pass through the axis, only that component of the force that is perpendicular to the plane of the door tends to rotate the door. Also, the farther away from the axis the perpendicular component of a force acts, the more effective the force is in producing rotation. To put this summary into a more quantitative form, let us introduce the concept of *torque* caused by a force applied to a body. Torque is defined so that it is a measure of the effectiveness of a force in producing rotation of a body about a chosen axis. Any axis may be chosen, even one that does not lie within the body being considered.

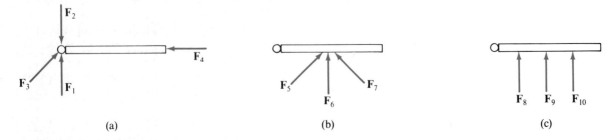

Figure 4.12
Some of the ways a 5-lb force may be applied to a door.

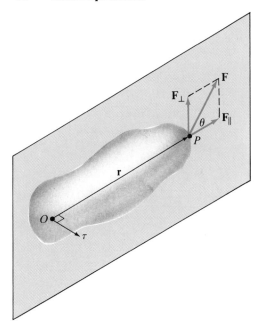

Figure 4.13
A force **F** applied to a body. The force acts at a distance r from the axis of rotation.

To define torque analytically, let us refer to Figure 4.13, in which a force **F** is applied to an object at a point P located by the vector **r** from an axis of rotation passing through the point O. The force **F**, which makes an angle θ with **r**, lies in the shaded plane and has components $\mathbf{F}_\parallel$ and $\mathbf{F}_\perp$ that are parallel and perpendicular to **r**. Since $\mathbf{F}_\parallel$ lies on a line passing through O, it produces no rotational effect about an axis through O. Then the torque produced by **F** is caused solely by the effect of $\mathbf{F}_\perp$, and the magnitude τ of the torque about an axis through O caused by the force F is

$$\tau = rF_\perp = rF \sin\theta \qquad (4\text{-}29)$$

Torque is a vector quantity with a magnitude given by Equation (4-29). The definition of the torque τ produced on an object about an axis by a force **F** applied to the object at a point located by **r** from that axis is

$$\boldsymbol{\tau} = \mathbf{r} \times \mathbf{F} \qquad (4\text{-}30)$$

The direction of $\boldsymbol{\tau}$ is perpendicular to the plane containing **r** and **F** and is specified by the right-hand rule: When the fingers of the right hand are curled from the direction of **r** toward the direction of **F**, the extended thumb of that hand points in the direction of $\boldsymbol{\tau}$.

We may note from Equation (4-29) that the magnitude of the torque produced by a force depends on: (a) the magnitude F of the force; (b) in what direction (θ) that force acts; and (c) where (r) on the body, relative to the chosen axis, that force is applied.

From either of Equations (4-29) and (4-30) we may see that the unit of torque is a product of a unit of force and a unit of length. In the SI system, the unit of torque is the newton-meter (N · M); in the English system the unit of torque is the pound-foot (lb · ft).

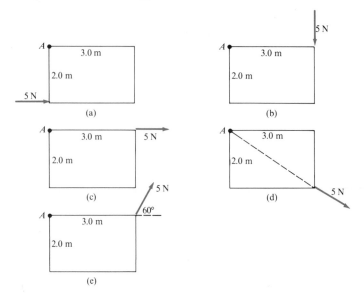

Exercise 4.15

E 4.15 A rectangular sheet of plywood is to be rotated about an axis perpendicular to the sheet and through point A, as shown in each of the accompanying figures. Each figure shows a 5.0-N force acting on the plywood. For each situation, determine the magnitude and sense (clockwise, CW; or counterclockwise, CCW) of the torque about the axis A.

Answers: (a) 10 N·m, CCW; (b) 15 N·m, CW; (c) zero; (d) zero; (e) 13 N·m, CCW

Before proceeding with further considerations of static equilibrium, it may be helpful to investigate briefly the rotational effects of several forces acting simultaneously on an extended body. Look, then, at the diagrams of Figure 4.14, each of which represents a merry-go-round. In each of the four cases two children are exerting forces of equal magnitudes on the merry-go-round. In which of the four cases do the children produce rotation of the merry-go-round about an axis through its center? Why?

In Figure 4.14(a) both forces provide torques that tend to cause the merry-go-round to rotate in a counterclockwise sense (their torques are directed out of the page), and the magnitudes of these torques are additive even though the forces are in opposite directions. The magnitudes of the torques caused by the forces in Figure 4.14(b) are also additive. On the other hand, the torques produced by the forces in Figure 4.14(c) are in opposing senses, and there is no net torque on the merry-go-round in this case, even though the forces are in the same direction. It should be clear that the forces in Figure 4.14(d) produce no net torque. We may conclude, therefore, that the magnitudes of torques about a given axis caused by two or more forces are algebraically additive if we take into account the sense (clockwise or counterclockwise) of each torque. Conventionally, we regard counterclockwise torques as positive and clockwise torques as negative.

E 4.16 A rectangular sheet of plywood is to be rotated about an axis perpendicular to the sheet and through point A, as shown on the figure. Calculate the net torque about this axis when three forces act as shown.

Answer: 12 ft·lb CW

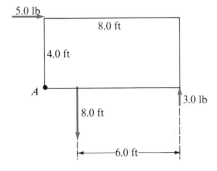

Exercise 4.16

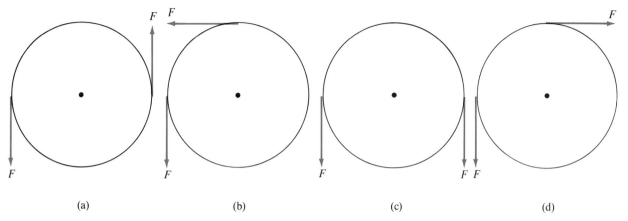

(a)   (b)   (c)   (d)

Figure 4.14
A merry-go-round viewed from above in four situations. All the applied forces have the same magnitudes.

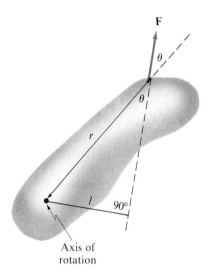

Figure 4.15
The force **F** applied to the same body as in Figure 4.13. The lever arm *l* is equal to *d* sin θ.

An additional point of technique may be made here: The calculation of torques is sometimes made easier by using an equivalent alternative form to that of Equation (4-29). But first we need to define lever arm. The *lever arm l* of a force **F** about a chosen axis is the perpendicular distance from the line along that force to the axis, as shown in Figure 4.15. The magnitude τ of the torque caused by that force of magnitude *F* is

$$\tau = Fl \tag{4-31}$$

In Figure 4.15, the same force **F** is applied to the same body as in Figure 4.13. From the diagram, the lever arm *l* is of length *r* sinθ, so Equation (4-31) becomes

$$\tau = Fl = Fr \sin\theta \tag{4-32}$$

which is equivalent to Equation (4-29).

E 4.17   The force **F** acts on an extended body, as shown in the accompanying diagram.
(a) Calculate $F_\perp$, the component of **F** that produces torque about *A*, the indicated axis of rotation.   Answer: 12 lb
(b) Use the result of part (a) to determine the torque of **F** about *A*.
                                              Answer: 24 ft · lb CCW
(c) Calculate *l*, the lever arm of **F** about *A*.   Answer: 1.6 ft
(d) Use the result of part (c) to determine the torque of **F** about *A*. Verify that this result agrees with that of part (b).   Answer: 24 ft · lb CCW

Center of Gravity

When we calculate torques on an extended body, where on the body should we assume the pull of the earth to act? Of course, the earth pulls on every particle of mass within the body. For purposes of calculating torques, however, we may assume a single point, called the *center of gravity* of the body, through which all the weight of the body acts. For a body that is symmetrical and of uniform density, its center of gravity is located at the geometrical center of the body. In Chapter 6 we shall see how to calculate the center of mass of bodies having more

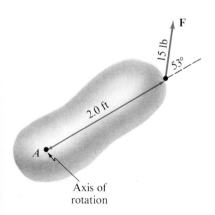

Exercise 4.17

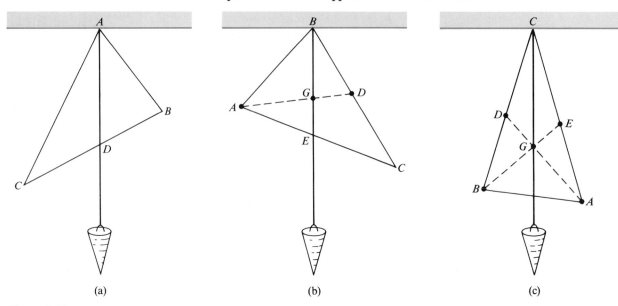

Figure 4.16
Experimental determination of the center of gravity of a two-dimensional body.

general shapes. At that time, we shall see that the center of gravity and the center of mass are identical for bodies in which the acceleration of gravity g is the same throughout the body. (For all practical purposes, this will be the case for every object on which we have occasion to calculate torques.) For our purposes here, we may assume that the location of the center of gravity of any regularly shaped extended body is at its geometrical center unless otherwise specified.

The location of the center of gravity of a two-dimensional object of arbitrary shape may be determined experimentally by a simple process illustrated in Figure 4.16. A triangular-shaped object, for example, freely suspended from one point (*see* Fig. 4.16[a]) to which a plumb bob is attached will assume a position in which the center of gravity lies along the vertical line AD of the bob. Otherwise, the weight of the object would produce a torque about an axis through the point of suspension. Now, if the object is freely suspended at another point on the triangle, as in Figure 4.16(b), we conclude that the center of gravity must lie along BE. The intersection of the lines AD and BE uniquely specifies the center of gravity G. Suspending the object freely at any other point on the triangle, as in Figure 4.16(c), will cause the center of gravity G to lie again along the vertical.

The Conditions of Static Equilibrium

Let us return to the issue of static equilibrium. Newton's second law, $\Sigma \mathbf{F} = m\mathbf{a}$, incorporates the fact that a static body, since it is not moving and therefore has no acceleration, has no net force acting on it. This is a condition of static equilibrium. We now know, however, that a body may have a net force equal to zero yet may not be static; it may, in fact, be undergoing rotation (look at Figure 4.14[a], for example). Consequently, for a body to be in static equilibrium, we must impose an additional condition: The net torque on the body must be zero. Here we may ask, "About what axis should the torques on the body be calculated? About what axis is the net torque equal to zero?" These questions

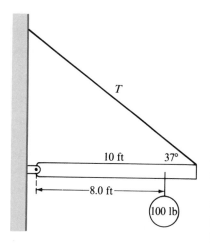

Figure 4.17
A 100-lb ball suspended from a horizontal uniform beam that weighs 50 lb.

may be answered by a question, "About what axis is the body not rotating?" Clearly, if the body is static, it is not rotating about *any* axis. Therefore, the net torque on a static body caused by all the forces on that body must be equal to zero when calculated using *any* axis. This condition, which pertains to any axis, may be expressed by

$$\sum \boldsymbol{\tau} = \mathbf{0} \tag{4-33}$$

This means that, for any given arbitrary axis, all the forces acting on a body in static equilibrium produce torques about that axis, the magnitudes of which sum algebraically to zero.

An astute reader could ask about a body in static equilibrium, "How do I know that if the torques sum to zero about a given axis and the forces sum to zero, the torques will sum to zero about any other axis?" We have not addressed this point, but Problem 4.47 provides the opportunity to convince yourself of its validity.

A body in static equilibrium, then, is subject to two conditions of equilibrium, both of which must be satisfied. These conditions are, once again, that the net force and the net torque are equal to zero on a body in static equilibrium:

$$\sum \mathbf{F} = \mathbf{0}$$

$$\sum \boldsymbol{\tau} = \mathbf{0} \tag{4-34}$$

These conditions are applied in the following problem:

> A uniform 50-lb beam 10 ft long is supported in a horizontal position by a pin and a weightless rope attached to a vertical wall, as shown in Figure 4.17. A 100-lb ball is suspended from the beam 8.0 ft from the wall. Calculate the tension T in the rope. What force does the beam exert on the pin?

In this problem, the forces of interest act on the beam. Therefore, let us isolate the beam and draw all the forces acting *on* it, as shown in Figure 4.18. In this figure, the force **T** on the beam by the string has been resolved into its x and y components. Similarly, since we do not know the magnitude or direction of the force that the wall exerts on the beam, that force is represented by its horizontal and vertical components F_H and F_V. Here the key word *uniform* tells us that we may treat the weight of the beam as if it were acting at the geometrical center of the beam, that is, at the center of gravity of the beam. Now, because the beam is stationary, we may apply the condition of static equilibrium, $\sum \mathbf{F} = \mathbf{0}$, in its component form, namely, $\sum F_x = 0$ and $\sum F_y = 0$:

$$F_H - 0.8T = 0 \tag{4-35}$$

$$F_V + 0.6T - 100 - 50 = 0 \tag{4-36}$$

Equations (4-35) and (4-36) contain three unknown quantities, T, F_H, and F_V; so another relationship is required. Let us use the remaining condition of static equilibrium, $\sum \boldsymbol{\tau} = \mathbf{0}$. We may select any axis about which the torques are to be calculated. If we choose an axis passing through the pin supporting the beam, we obtain, using $\sum \boldsymbol{\tau} = \mathbf{0}$,

$$0.6T(10) - 50(5) - 100(8) = 0$$

$$6T = 1050$$

$$T = 180 \text{ lb} \tag{4-37}$$

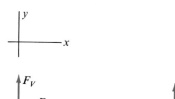

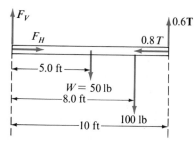

Figure 4.18
The forces acting on the beam of Figure 4.17.

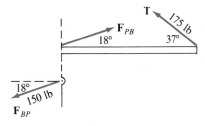

Figure 4.19
The forces **T** and **F**$_{PB}$ acting on the beam of Figure 4.17. **F**$_{PB}$ is the force the beam exerts on the pin.

where the force of magnitude $0.6T$ has a lever arm of 10 ft about the chosen axis, the 100-lb force has a lever arm of 8 ft, and the 50-ft weight has a lever arm of 5 ft. By substituting $T = 175$ lb into Equations (4-35) and (4-36), we find that

$$F_H = 140 \text{ lb}$$

$$F_V = 45 \text{ lb} \qquad \text{(4-38)}$$

That F_H and F_V are positive means that we correctly guessed the directions of the horizontal and vertical components of the force of the pin on the beam. The force **F**$_{PB}$ of the pin on the beam is, of course, the vector sum of **F**$_H$ and **F**$_V$. The magnitude of **F**$_{PB}$ is

$$|\mathbf{F}_{PB}| = \sqrt{140^2 + 45^2} = 150 \text{ lb} \qquad \text{(4-39)}$$

and **F**$_{PB}$ acts on the beam at an angle θ measured from the horizontal given by

$$\theta = \tan^{-1}\left(\frac{45}{140}\right) = 18° \qquad \text{(4-40)}$$

Figure 4.19 shows the forces **T** and **F**$_{PB}$ acting on the beam. Recall that in the statement of the problem, we were asked to determine the force **F**$_{BP}$ of the beam on the pin. Newton's third law requires that $\mathbf{F}_{BP} = -\mathbf{F}_{PB}$. Figure 4.19 shows the force **F**$_{BP}$ with a magnitude of 150 lb, directed to the left and below the horizontal at an angle of 18°.

In the foregoing problem, we selected the axis through the pin for calculating torques. Any axis could have been chosen, so why was this choice advantageous? Of the six force components acting on the beam in Figure 4.18, three forces are on lines that pass through the pin. Such forces have no lever arms and, therefore, produce no torque about an axis through the pin. A judicious choice of the axis for computing torques reduces the number of terms in the $\Sigma\tau = 0$ equation by the number of forces with lines of action that pass through the axis. Nevertheless, any axis may be chosen about which the sum of torques is equal to zero for a body in static equilibrium.

E 4.18 Using the values just obtained for T, F_V, and F_H, verify that all torques sum to zero about an axis through the right end of the beam.

4.5 Problem-Solving Summary

This chapter has introduced several new categories of problems to which Newton's laws apply. It has, therefore, extended the range of physical systems to which we may apply the techniques introduced in Chapter 3 for solving dynamics problems. Now we have the tools to consider dynamics problems involving frictional forces, the centripetal forces of circular motion, and gravitational forces. We also have been introduced to a special case of Newton's second law, namely statics, by which we may analyze the forces and the torques on bodies that are at rest.

Several points are worth remembering in solving problems involving surfaces at which there is friction. A rough surface not only may exert a normal force on an object—even a frictionless surface may do that—but may also exert a force on the object parallel to the surface. Frictional forces on an object are generally in a direction opposite to the direction of the motion of the object or opposite to the direction the object is tending to move. The magnitude f_s of the force of static friction exerted on an object at rest assumes any value between zero and $\mu_s N$ necessary to keep the object from moving. On the other hand, the magnitude f_k of the force of kinetic friction exerted on a moving object is always equal to $\mu_k N$. Notice that the magnitude N of the normal force exerted on the object by the surface is not necessarily equal to the weight of the object. Careful examination of the force diagram will help to determine the correct value for N.

A particle in uniform circular motion is identified by two characteristics: It is moving at a constant speed v and is moving in a circular path with a radius r. Then the acceleration of the particle is directed radially toward the center of the circular path and has magnitude v^2/r. Because the net force on the particle must be in the direction of the acceleration of the particle, the net force must be toward the center. That is, the net force is centripetal with magnitude mv^2/r. Mistakes associated with the forces on a particle in circular motion may be minimized by following our usual dynamics problem procedure, which may be made specific for circular motion:

1. On the force diagram for the particle, indicate the direction of the centripetal acceleration $\mathbf{a}_c$ (toward the center of the circular motion).

2. In writing the relationship expressing Newton's second law, enter the magnitude of the forces that are in the direction of $\mathbf{a}_c$ with positive signs, enter the magnitudes of the forces that are opposite to $\mathbf{a}_c$ with negative signs, and equate their sum to mv^2/r.

The gravitational force on each of two particles with masses m_1 and m_2, separated by a distance r, has the magnitude Gm_1m_2/r^2. Problems that utilize the gravitational force (always an attractive force) in this formulation may be identified by two observations:

1. A mechanical force is present without there being any physical contact, that is, a force acts on an object at a distance.

2. The object under consideration is not near the surface of the earth; near the surface of the earth we represent the gravitational force on an object by its weight $W = mg$.

Thus, problems involving the forces on planets or their satellites, like a moon, a rocket ship, or a space station, normally involve gravitational forces. Problems involving gravitational forces are usually most easily set up and solved in symbolic form, that is, using symbols to represent all the physical quantities. The solutions of these problems are often independent of one of the masses involved; sometimes they are independent of the gravitational constant G. Using symbolic relationships makes it easier to recognize that certain quantities appearing in the law of universal gravitation need not be known quantities in a particular problem situation. Therefore, when a quantity is not known or given, it is a good problem-solving procedure to assign a symbol to that quantity and proceed.

A statics problem may be identified by the object of interests being at rest and remaining at rest. Usually we are asked to find one or more of the forces on a resting body or some quantity that can be related to such forces. We know two facts about a body that remains at rest: The net force on the body is equal to zero, and the net torque on the body about any axis is equal to zero. These facts permit us to solve most simple statics problems with a straightforward procedure. Just as in dynamics problems, we isolate the body on which the forces of interest act. By resolving all the forces acting on that body into convenient x and y components, we require that the component sums, ΣF_x and ΣF_y, are each equal to zero. These conditions usually provide two equations, but these equations may contain more than two unknown quantities. A third relationship may be obtained using the fact that the net torque $\Sigma\tau$ on the body is equal to zero about *any* axis. The mathematics of a statics problem is simplified by judiciously choosing the axis about which the torques are calculated. Usually that means selecting an axis having the most forces whose lines of action pass through that axis (or, even better, selecting the axis having the lines of action of the greatest number of unknown forces passing through it). This simplification is effected, of course, because a force acting along a line that passes through an axis produces no torque about that axis.

Problems

GROUP A

4.1 A uniform beam weighing 100 lb is supported in a horizontal position by two vertical ropes, as shown in Figure 4.20.

(a) Draw a diagram showing the isolated beam and all forces acting on it.

(b) Write an equation requiring that all torques sum to zero about an axis of rotation through the left end of the beam. Solve this equation for the tension in rope 2.

(c) Using $\Sigma F_y = 0$, calculate the tension in rope 1.

(d) Verify that all torques sum to zero about an axis of rotation passing through the intersection of rope 2 and the beam.

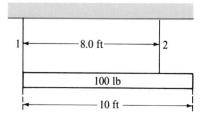

Figure 4.20   Problem 4.1

4.2 An irregularly shaped object is suspended by two vertical ropes (*see* Fig. 4.21). The tension in rope 1 is 450 N, and the tension in rope 2 is 650 N.

(a) What is the mass of the object?

(b) Determine the horizontal distance x, shown in the figure, that locates the center of gravity (c.g.).

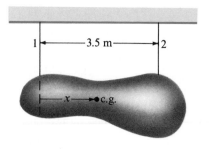

Figure 4.21   Problem 4.2

4.3 At one instant, three asteroids are positioned as shown in Figure 4.22. If $m_A = 4.0 \times 10^7$ kg, $m_B = 5.0 \times 10^8$ kg, $m_C = 6.0 \times 10^7$ kg, $a = 1.2$ km, and $b = 1.8$ km, what is the total gravitational force on asteroid B?

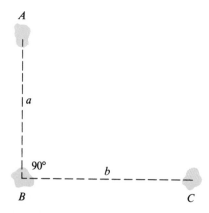

Figure 4.22   Problem 4.3

4.4 A uniform sphere (*see* Fig. 4.23) having a mass of 25 kg and a radius of 20 cm is held by a string against a frictionless vertical wall.

(a) What is the tension in the string?

(b) What is the force exerted on the sphere by the wall?

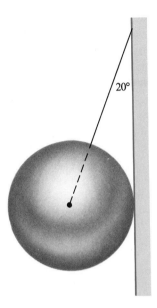

Figure 4.23   Problem 4.4

4.5 A 100-lb crate sits on the floor of a railroad car, which is accelerating 0.40 ft/s² toward the east.

(a) What frictional force does the floor exert on the crate, assuming that the crate does not slide?

(b) What is the resultant force on the crate?

(c) What force does the crate exert on the floor?

4.6 An irregularly shaped object (*see* Fig. 4.24) is supported in two positions. The tensions in strings 1, 2, and 3 are 48 N, 85 N, and 70 N, respectively. Calculate the distances x and y that locate the center of gravity.

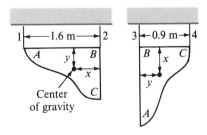

Figure 4.24   Problem 4.6

4.7 A 180-lb person stands on a uniform beam weighing 75 lb. The beam is supported at each end, as depicted in Figure 4.25.

When the person stands 8.0 ft from the left end of the beam, what vertical force does

(a) the right support exert on the beam?

(b) the left support exert on the beam?

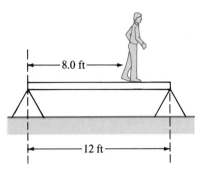

Figure 4.25   Problem 4.7

4.8 A horizontal force F acts on a 5.0-kg body initially at rest on a horizontal surface (*see* Fig. 4.26). The coefficients of friction for the block and surface are $\mu_s = 0.40$ and $\mu_k = 0.20$.

(a) For what range of values of F will the body remain at rest?

(b) If $F = 12$ N, what force does the surface exert on the body?

(c) If $F = 25$ N, what force does the surface exert on the body?

(d) What is the resultant force on the body in part (b)? in part (c)?

(e) What is the acceleration of the body in part (b)? in part (c)?

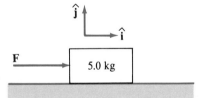

Figure 4.26   Problem 4.8

4.9 A skidding 1800-lb car comes to a stop from an initial speed of 55 mi/h in a distance of 190 ft on a level road. What is the coefficient of kinetic friction between the tires of the car and the surface of the road?

4.10 What minimum coefficient of friction between the tire and road is required if a motorcycle is to accelerate from rest to a speed of 80 km/h in 4.8 s? Assume that 60 percent of the weight is supported by the rear wheel.

4.11 A 1600-lb car travels without sliding at a speed of 60 mi/h around a level circular road. The radius of the circular path is 600 ft.

  (a) What frictional force does the road exert on the car?

  (b) What is the minimum value of μ_s between the tire and the road?

  (c) If $\mu_s = 0.25$, what is the maximum speed that the car may be driven around this curve without sliding?

4.12 A 120-lb girl riding a Ferris wheel moves at a constant speed of 15 ft/s around a vertical circular path 80 ft in diameter.

  (a) What force does the Ferris wheel exert on her at the top of the circle? at the bottom of the circle?

  (b) What is the resultant force on the girl at the top of the circle?

4.13 A uniform 20-kg beam is supported in a horizontal position by a pin and a cable, as shown in Figure 4.27. Masses m_1 and m_2 are suspended from the beam. If the masses $m_1 = 40$ kg and $m_2 = 15$ kg, calculate

  (a) the tension in the cable.

  (b) the force exerted on the pin by the beam.

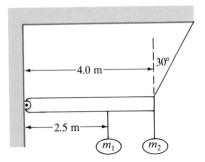

Figure 4.27   Problem 4.13

4.14 A uniform 60-kg beam with a length of 6.0 m is placed on two supports (*see* Fig. 4.28). Let x be the distance from the left end of the beam to a 70-kg person on the beam.

  (a) If $x = 4.5$ m, what vertical forces do supports 1 and 2 exert on the beam?

  (b) What is the maximum distance the person can walk from the left end without tilting the beam?

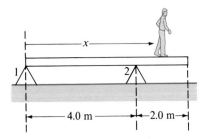

Figure 4.28   Problem 4.14

4.15 While two astronauts were on the surface of the moon, a third astronaut orbited the moon. Assume the orbit to be circular and 100 km above the surface of the moon. If the mass and radius of the moon are 7.4×10^{22} kg and 1.7×10^6 m, determine

  (a) the orbiting astronaut's acceleration.

  (b) the astronaut's orbital speed.

  (c) the period (time for one revolution) of the orbit.

4.16 Xeronians (occupants of the imaginary planet Xeron) measure the free-fall acceleration of an object near the surface of Xeron to be 6.0 m/s². They also measure their spherical planet to have a radius of 6000 km.

  (a) What is the mass of Xeron?

  (b) What is the orbital speed of a satellite circling 1500 km above the surface of Xeron?

4.17 A uniform sphere weighing 45 lb is placed in a wedge formed by two flat frictionless surfaces each tilted 15° from the vertical, as shown in Figure 4.29. What is the magnitude of the force exerted on the sphere by each surface?

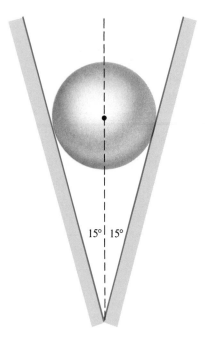

Figure 4.29   Problem 4.17

4.18 Two identical uniform spheres, each of which has a mass of 3.0 kg, are suspended, as shown in Figure 4.30. Assume the surfaces of the spheres are frictionless.

  (a) What is the tension in each string?

  (b) What force (magnitude only) does one sphere exert on the other?

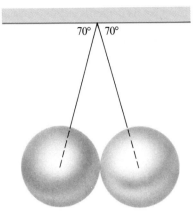

Figure 4.30   Problem 4.18

4.19 A force **P** parallel to a surface inclined 35° above the horizontal acts on an 18-lb block initially at rest on the surface (*see* Fig. 4.31). If the coefficients of friction for the block and surface are $\mu_s = 0.32$ and $\mu_k = 0.24$,

(a) for what range of values of P will the block remain at rest?

(b) what is the acceleration of the block if $P = 16$ lb?

(c) what is the acceleration of the block if $P = 4.0$ lb?

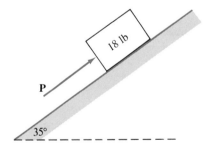

Figure 4.31   Problem 4.19

4.20 A horizontal force **P** acts on a 5.0-kg block, the first of three blocks positioned as in Figure 4.32. The blocks are initially at rest. The coefficients of friction between the blocks and the horizontal surface are $\mu_s = 0.30$ and $\mu_k = 0.20$.

(a) For what range of values of P will the blocks remain at rest?

(b) If $P = 20$ N, what force does the 5.0-kg block exert on the 3.0-kg block?

(c) If $P = 36$ N, what force does the 5.0-kg block exert on the 3.0-kg block?

(d) If $P = 36$ N, what force does the 5.0-kg block exert on the horizontal surface?

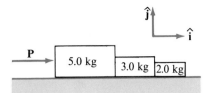

Figure 4.32   Problem 4.20

4.21 A 70-kg object sits without sliding on a horizontal platform that rotates at a constant rate of one revolution every 6 s about a fixed vertical axis (*see* Fig. 4.33). The object moves in a circle having a radius r of 2.0 m.

(a) What is the speed of the object?

(b) What is the acceleration of the object?

(c) What is the frictional force on the object by the platform?

(d) What is the minimum value of μ_s, the coefficient of static friction, between the object and the platform?

(e) If $\mu_s = 0.40$, what is the radius of the largest circle the object can traverse without sliding?

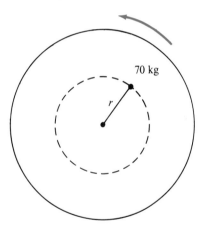

Figure 4.33   Problem 4.21

4.22 A 3.0-kg ball attached by a 2.0-m string to a vertical axis is rotated about this axis and moves in a horizontal circle. The angle between the string and the axis is 28°, as shown in Figure 4.34.

(a) What is the tension in the string?

(b) What is the resultant force on the ball?

(c) What is the ball's speed as it rotates?

4.23 A 100-lb object is suspended as shown in Figure 4.35.

(a) What is the tension in string a?

(b) What is the tension in string b?

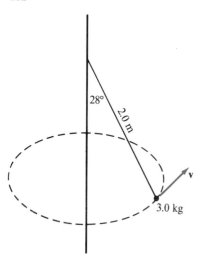

Figure 4.34   Problem 4.22

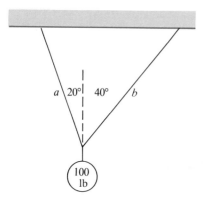

Figure 4.35   Problem 4.23

4.24   A 10-kg body is suspended as shown in Figure 4.36.

(a)   What is the tension in string a?

(b)   What is the tension in string b?

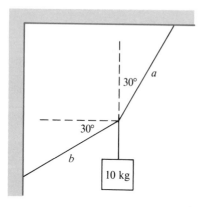

Figure 4.36   Problem 4.24

4.25   A uniform 100-lb beam is held in a vertical position by a pin at its lower end and a cable at its upper end. A horizontal 200-lb force acts as shown in Figure 4.37.

(a)   What is the tension in the cable?

(b)   What force does the pin exert on the beam?

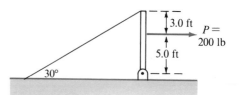

Figure 4.37   Problem 4.25

4.26   A horizontal force **P** is applied as shown in Figure 4.38. The two blocks are initially at rest on a horizontal surface. The frictional forces on both blocks by the surface are determined by $\mu_s = 0.35$ and $\mu_k = 0.25$.

(a)   For what range of values of P will the blocks remain at rest?

(b)   If $P = 15$ lb, what is the tension in the connecting string?

(c)   If $P = 25$ lb, what is the tension in the connecting string, and what is the magnitude of the accelerations of the blocks?

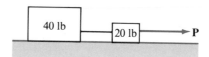

Figure 4.38   Problem 4.26

4.27   The force **P** acts on a 15-kg crate initially at rest on a horizontal surface (*see* Fig. 4.39). The coefficients of friction for the crate and the surface are $\mu_s = 0.35$ and $\mu_k = 0.20$.

(a)   For what range of values of P will the crate remain at rest?

(b)   If $P = 30$ N, what force does the surface exert on the crate?

(c)   If $P = 55$ N, what is the acceleration of the crate?

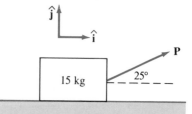

Figure 4.39   Problem 4.27

4.28 At an instant when a 2500-lb airplane is flying 130 mi/h (190 ft/s) around a horizontal circular path having a radius of 1500 ft, what is the magnitude of the force exerted on the airplane by the air?

4.29 An 8.0-lb object attached by two strings to a vertical axis, as shown in Figure 4.40, is rotated about this axis and moves in a horizontal circle. Suppose the tension in the horizontal 2.0-ft string is 24 lb.

(a) What is the tension in the 4.0-ft string?

(b) What is the resultant force on the object?

(c) What is the speed of the object as it rotates?

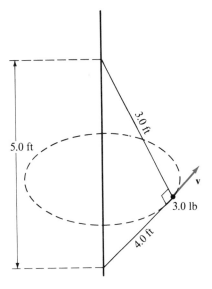

Figure 4.41   Problem 4.30

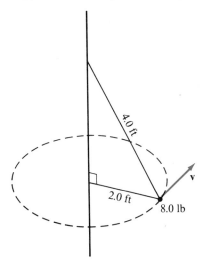

Figure 4.40   Problem 4.29

4.30 A 3.0-lb ball attached by two strings to a vertical axis (*see* Fig. 4.41) is rotated about this axis and moves in a horizontal circle.

(a) What is the minimum speed for the ball if the 4.0-ft string is to be fully extended?

(b) For the speed obtained in part (a), what is the tension in the 3.0-ft string? the 4.0-ft string?

(c) If the tension in the 4.0-ft string is 6.0 lb, what is the tension in the 3.0-ft string, and what is the speed of the ball?

4.31 A horizontal force **P** acts on a 12-kg block initially at rest (*see* Fig. 4.42). The 12-kg block slides on a horizontal surface with $\mu_s = 0.50$ and $\mu_k = 0.30$. Disregard any pulley mass or friction.

(a) For what range of values of P will the block remain at rest?

(b) If P = 150 N, what is the magnitude of the accelerations of the blocks, and what is the tension in the connecting string?

(c) Answer part (b) for P = 10 N.

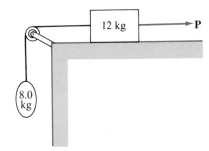

Figure 4.42   Problem 4.31

4.32 The 10-kg block, shown in Figure 4.43, slides on the horizontal surface with $\mu_s = 0.35$ and $\mu_k = 0.20$. Disregarding any pulley mass or friction, determine the force on the 10-kg block by the surface, the magnitude of the accelerations of the blocks, and the tension in the connecting string if

(a) M = 3.0 kg.

(b) M = 5.0 kg.

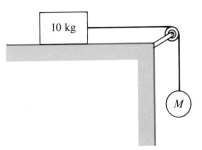

Figure 4.43   Problem 4.32

4.33 A uniform beam having a mass of 40 kg and a length of 2.8 m is held in place at its lower end by a pin. Its upper end leans against a frictionless vertical wall (*see* Fig. 4.44).

(a) What force does the wall exert on the beam?

(b) What force does the pin exert on the beam?

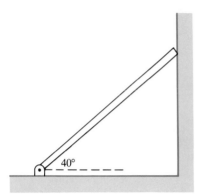

Figure 4.44   Problem 4.33

4.34 A 25-ft crane supported at its lower end by a pin is elevated by a horizontal cable, as depicted in Figure 4.45. A 500-lb load is suspended from the outer end of the crane. The center of gravity of the crane is 10 ft from the pin, and the crane weighs 200 lb.

(a) What is the tension in the horizontal cable?

(b) What force does the pin exert on the crane?

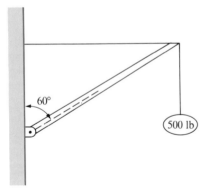

Figure 4.45   Problem 4.34

4.35 A 7.0-kg block is initially at rest on a surface inclined at an angle θ with the horizontal, as shown in Figure 4.46. The coefficients of friction for the block and surface are $\mu_s = 0.40$ and $\mu_k = 0.30$. Determine the force on the block by the surface and the resultant force on the block if

(a) $\theta = 20°$.

(b) $\theta = 30°$.

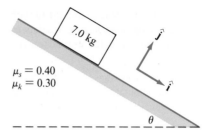

Figure 4.46   Problem 4.35

4.36 The coefficients of friction for a 25-lb body and an inclined plane (*see* Fig. 4.47) are $\mu_s = 0.45$ and $\mu_k = 0.20$. Disregard any pulley mass or friction.

(a) For what range of values of W will the 25-lb body remain at rest?

(b) If $W = 25$ lb, what is the acceleration of the weight W, and what is the tension in the connecting string?

(c) Repeat part (b) for $W = 4.0$ lb.

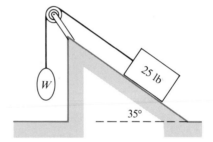

Figure 4.47   Problem 4.36

4.37 A force **F** acts on a 10-kg block sliding on a flat surface inclined 30° from the horizontal, as shown in Figure 4.48. Suppose that $F = 63$ N. Then the block slides at a constant speed up the inclined surface.

(a) What frictional force acts on the block?

(b) What is μ_k for the block and surface?

(c) If the block slides down the surface with $F = 0$, what frictional force acts on the block? What is its acceleration?

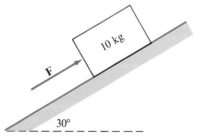

Figure 4.48   Problem 4.37

4.38 Two identical 1.4-kg masses are fixed to opposite ends of a light rigid rod that is 0.80 m long. This apparatus rotates in a horizontal plane about a vertical axis through the center of the rod. If the apparatus makes one revolution every 0.20 s, what is the tension in the rod?

4.39 The planet Neptune, with a diameter of 5.0×10^7 m, completes an orbit of the sun in 165 years. The free-fall acceleration near Neptune's surface is 12 m/s². Assume Neptune to be in a circular path about the sun ($M_s = 2.0 \times 10^{30}$ kg).

  (a) Calculate the mass of Neptune.

  (b) What is the period of a satellite orbiting 6.0×10^7 m above the surface of Neptune?

  (c) Calculate the radius of Neptune's orbit about the sun.

  (d) What is Neptune's orbital speed about the sun?

4.40 A uniform 2.5-kg sphere rests on two flat frictionless surfaces (*see* Fig. 4.49).

  (a) What force does surface A exert on the sphere?

  (b) What force does surface B exert on the sphere?

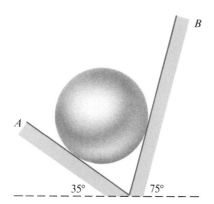

Figure 4.49   Problem 4.40

4.41 A uniform strut weighs 150 lb. It is held in place by a pin and a cable (*see* Fig. 4.50). The strut supports a 200-lb weight. Determine

  (a) the tension in the cable.

  (b) the force exerted on the beam by the pin.

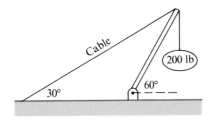

Figure 4.50   Problem 4.41

4.42 A 35-lb object slides at a constant speed down a flat surface inclined 25° with the horizontal.

  (a) What frictional force does the surface exert on the object?

  (b) What is the resultant force on the object?

  (c) If the same object slides up the inclined surface, what frictional force does the surface exert on it?

4.43 A block slides on a surface inclined 25° with the horizontal. The coefficients of friction between the block and surface are $\mu_s = 0.35$ and $\mu_k = 0.20$. Initially the block has a speed of 12 ft/s up the plane.

  (a) How far beyond its initial position will the block slide before it stops?

  (b) After the block stops, does it slide back down the plane?

  (c) What will be the speed of the block when it again passes its initial position?

4.44 An 8.0-kg block is initially at rest on a flat surface inclined at an angle θ above the horizontal. As θ is increased, the block remains at rest until θ becomes equal to 33°. If θ is held constant at this value, the block is moving at a speed of 3.5 m /s after it slides a distance of 3.0 m down the surface. Calculate μ_s and μ_k for this block on this surface.

4.45 Consider a planet with a spherical core and a concentric shell of different densities. Suppose the core has a uniform density of 9.0 g/cm³ and a diameter of 6.0×10^6 m. Assume the concentric shell to have a uniform density of 4.0 g/cm³ and a thickness of 1.0×10^6 m.

  (a) What is the mass of the planet?

  (b) What is the free-fall acceleration near the surface of the planet?

4.46

  (a) Show that the time T for one revolution by an earth satellite in a circular orbit of radius r is given by

$$T = \frac{2\pi r^{3/2}}{\sqrt{GM_e}}$$

    where M_e is the earth's mass.

  (b) Use this result to calculate the height above the surface of the earth of a satellite that appears stationary to an observer on the equator.

GROUP B

4.47

  (a) Show that

$$\tau_A = \tau_B + \mathbf{R} \times \mathbf{F}$$

where τ_A is the torque of the force $\mathbf{F}$ about point A, τ_B is the torque of **F** about B, and **R** is the position vector from A to B.

(b) Show that for any set of forces $F_1, F_2, \ldots, F_N$,

$$\sum_{n=1}^{N} \tau_{A,n} = \sum_{n=1}^{N} \tau_{B,n} + \mathbf{R} \times \sum_{n=1}^{N} \mathbf{F}_n$$

where $\tau_{A,n}$ is the torque of the force F_n about point A and $\tau_{B,n}$ is the torque of F_n about point B.

(c) Show that if the net force on a body is zero and if the net torque of these forces about any one point is zero, then the net torque about any other point is zero.

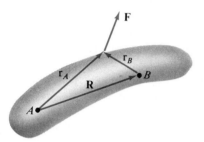

Figure 4.51   Problem 4.47

4.48 A 220-lb bureau is to be pushed across a level floor by a force **P** acting horizontally at a height h above the floor. The center of gravity of the bureau is depicted in Figure 4.52. The coefficients of friction for the bureau and the floor are $\mu_s = 0.40$ and $\mu_k = 0.30$.

(a) If $P = 50$ lb and $h = 3.0$ ft, determine the normal forces on the bureau by the floor at points A and B.

(b) Find a minimum height h_o such that if $h > h_o$ the bureau will tilt rather than slide as **P** is increased in magnitude.

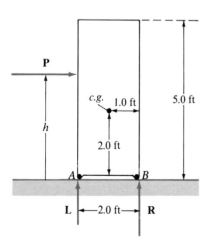

Figure 4.52   Problem 4.48

4.49 A horizontal force **P** acts on the 50-lb body, which slides on a horizontal frictionless surface, as shown in Figure 4.53. The coefficients of friction between the 15-lb body and the 50-lb bodies are $\mu_s = 0.40$ and $\mu_k = 0.20$.

(a) For what range of values of P will the two bodies move with the same acceleration?

(b) If $P = 20$ lb, what are the accelerations of the blocks, and what is the force on the 15-lb body by the 50-lb body?

(c) If $P = 35$ lb, what are the accelerations of the blocks, and what is the force on the 15-lb body by the 50-lb body?

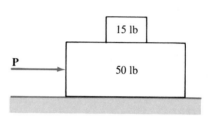

Figure 4.53   Problem 4.49

4.50 The coefficients of friction for all surfaces in Figure 4.54 are $\mu_s = 0.30$ and $\mu_k = 0.20$. Initially the 20-lb block is at rest.

(a) For what range of values of P will the 20-lb block remain at rest?

(b) If $P = 18$ lb, what is the acceleration of the 20-lb block, and what is the tension in the string tied to the wall?

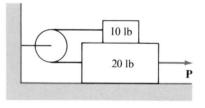

Figure 4.54   Problem 4.50

4.51 The coefficients of friction for all surfaces in Figure 4.55 are $\mu_s = 0.30$ and $\mu_k = 0.20$. Disregard any pulley mass or friction. Initially both blocks are at rest.

(a) For what range of values of P will the blocks remain at rest?

(b) If $P = 18$ lb, what is the magnitude of the accelerations of the blocks, and what is the tension in the string connecting the blocks?

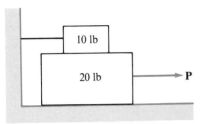

Figure 4.55   Problem 4.51

4.52 A body of mass M is initially at rest on a horizontal frictionless surface. The surface is caused to tilt so that its angle with the horizontal changes at a rate of $\pi/20$ radians/s for the time interval $0 \le t \le 10$ s.

(a) How far has the body moved along the surface at the instant its angle with the horizontal is 45°?

(b) Repeat part (a) if the surface is not frictionless and if $\mu_s = 0.30$ and $\mu_k = 0.20$.

4.53 A 1.2-lb weight swings in a vertical circle at the end of a string of negligible weight (*see* Fig. 4.56). The string is 2.0 ft long. At the lowest point of its path, the weight is moving at a speed of 10 ft/s. At the highest point, the string makes an angle of 77° with the vertical. Determine the acceleration of the weight and the tension in the string at

(a) the lowest point.

(b) the highest point.

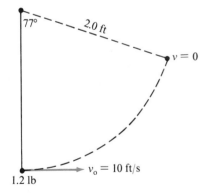

Figure 4.56   Problem 4.53

4.54 A pendulum is made using an 0.80-kg object fixed to one end of a string of negligible mass. The other end of the string is fixed so that the pendulum swings in a vertical circle with a radius of 1.2 m (*see* Fig. 4.57). Calculate the acceleration of the object and the tension in the string at position

(a) A if $v_A = 4.0$ m/s.

(b) B if $v_B = 7.9$ m/s.

(c) C if $v_C = 6.3$ m/s.

(d) D if $v_D = 5.3$ m/s.

(e) E if $v_E = 7.2$ m/s.

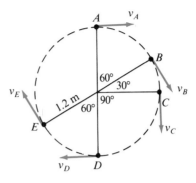

Figure 4.57   Problem 4.54

4.55 A force **P** acts as shown in Figure 4.58 on a block initially at rest on a horizontal surface with a coefficient of static friction μ_s.

(a) Show that the block will remain at rest if P satisfies

$$0 \le P < \frac{\mu_s W}{\cos\theta - \mu_s \sin\theta}$$

for $(\cos\theta - \mu_s \sin\theta) > 0$.

(b) Show that for $\theta > \tan^{-1}(1/\mu_s)$ the block remains at rest regardless of how large **P** becomes.

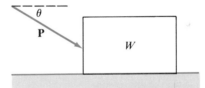

Figure 4.58   Problem 4.55

4.56 A force **P** acts on a mass m sliding across a horizontal surface with a coefficient of kinetic friction μ_k (*see* Fig. 4.59). If the mass m is to move at a constant speed,

(a) express P as a function of m, μ_k, and θ.

(b) show that P is a minimum for $\theta = \tan^{-1} \mu_k$.

(c) calculate the minimum value of P if $\mu_k = 0.30$.

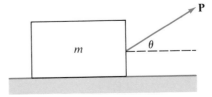

Figure 4.59   Problem 4.56

4.57 A strut of negligible weight supports a weight W, as shown in Figure 4.60. The upper end of the strut is held in place by a cable, and the lower end is held in place by the force of static friction. For what range of values of μ_s will the strut remain fixed regardless of the value of W?

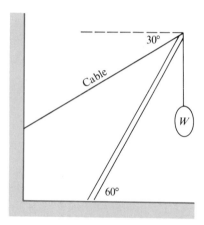

Figure 4.60   Problem 4.57

4.58 A uniform 20-ft ladder weighing 75 lb leans against a smooth vertical wall (*see* Fig. 4.61).

(a) What force does the wall exert on the ladder?

(b) What force does the horizontal surface exert on the ladder?

(c) What is the minimum value of μ_s between the ladder and the floor?

(d) If $\mu_s = 0.30$, how far can a 120-lb person climb up the ladder before the ladder begins to slip?

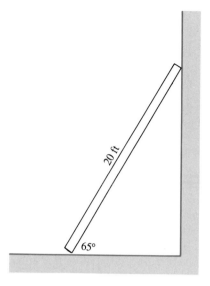

Figure 4.61   Problem 4.58

4.59 Consider a spherically symmetrical object of radius R with a mass density d that varies according to

$$d = \frac{d_o}{1 + (r/R)^3} \qquad 0 \leq r \leq R$$

where r is the radial distance from the center of the sphere and d_o is a constant.

(a) Determine the mass of the object in terms of d_o and R.

(b) Express the average mass density of the object in terms of d_o and R.

(c) If two such objects are placed with their centers separated by a distance $3R$, what force (magnitude) does each exert on the other?

5 Work, Power, and Energy

This chapter introduces one of the most important and perhaps the most far-reaching of all principles in the sciences—the conservation of energy. A conservation principle is a rule or a natural law that specifies that the value of a physical quantity does not change during the course of a physical process but remains constant. The quantity that does not change is said to be conserved. The simplicity of conservation principles makes them concise expressions of natural law and powerful tools of scientific analysis. Conservation of energy is only one of a number of conservation principles that students of science or engineering will encounter. The significance and the usefulness of these principles should become apparent to the student as they are used to analyze and interpret physical phenomena.

Of course, we have not yet defined energy. This chapter will introduce several physical quantities and concepts that are necessary for the understanding of energy and its associated conservation principle. First we will define work and describe how it may be accomplished and calculated. Then we will briefly consider power, a quantity that is useful in many practical applications. Next we will encounter energy in its mechanical forms, namely, kinetic and potential energies, and see how these quantities are related to work. Finally, we will see how kinetic and potential energies are used to express the conservation of energy principle, the essential physical relationship of this chapter.

5.1 Work

In the nonscientific world, work is often thought of in terms of some physical or mental effort. In physics, however, the term *work* is defined precisely. Doing work requires the use of force, and work on a body does not take place without displacement of that body. We will begin with a simple situation in which work occurs: work done by a constant force.

Work by a Constant Force

Suppose a constant force $\mathbf{F}$ is applied to a particle that moves in a straight line, say along the x axis from x_1 to x_2 through a displacement of $\Delta \mathbf{r} = (x_2 - x_1)\hat{\mathbf{i}} = \Delta x \hat{\mathbf{i}}$. The *work* done by the force $\mathbf{F}$ on the particle is defined to be the product of the magnitude of the displacement and the magnitude of that component of the

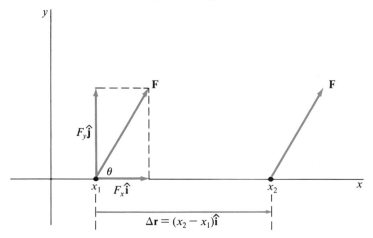

Figure 5.1
A constant force **F** applied to a particle
constrained to move along the *x* axis.

force that is in the direction of the displacement. In Figure 5.1 the force **F** is
applied to a particle at an angle θ measured from the *x* axis. As the particle is
displaced by $\Delta\mathbf{r} = \Delta x\hat{\mathbf{i}}$, the vector component of force in the direction of the
displacement is $F_x\hat{\mathbf{i}}$. Then the work *W* done by the force **F** on the particle
in displacing it by $\Delta\mathbf{r} = \Delta x\hat{\mathbf{i}}$ is

$$W = F_x\,\Delta x = F\,\Delta x\,\cos\theta \qquad (5\text{-}1)$$

Equation (5-1) may be expressed as the scalar product of the applied force **F** and
the displacement $\Delta\mathbf{r}$, or

$$W = \mathbf{F}\cdot\Delta\mathbf{r} \qquad (5\text{-}2)$$

which emphasizes that work is a scalar quantity and, therefore, has no direction
associated with it. The value of work, however, may be positive ($0 \le \theta < 90°$),
zero ($\theta = 90°$), or negative ($90° < \theta \le 180°$). Notice that the component $F_y\hat{\mathbf{j}}$ of
the applied force that is perpendicular to the displacement of the particle does no
work on the particle. In fact, no amount of force on a body results in work unless
that force has a component in the direction of the displacement of the body.

The units of work are those of force multiplied by length. In the metric system
the unit of work is the newton-meter ($\text{N}\cdot\text{m}$), which is defined to be the *joule* (J).
In the English system the unit of work is the foot-pound ($\text{ft}\cdot\text{lb}$).

E 5.1   A 10-lb object is dropped from a height of 50 ft above the ground
and falls to the ground. Determine the work done by the gravitational force on
the object as it undergoes this displacement.   Answer: 500 ft·lb

E 5.2   If the same 10-lb object is taken from the ground back to its
original position 50 ft above the ground, how much work is done by the gravitational force on the object during this displacement?   Answer: −500 ft·lb

E 5.3   Calculate the number of joules that is equivalent to 1 ft·lb of work.
Recall that 4.45 N = 1.0 lb and that 1.0 in. = 2.54 cm.   Answer: 1.4 J

Now consider further the work done on a particle by a constant force **F** for the
same special case in which the motion of the particle is along a straight line, say

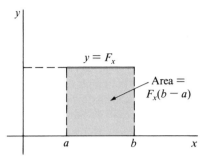

Figure 5.2
A graph of the x component of a constant force acting on a particle as it moves from $x = a$ to $x = b$. The work done by this force is equal to $F_x(b - a)$, the shaded area.

the x axis. As the particle moves from $x = a$ to $x = b$, the force **F** does work on the particle, which, according to Equation (5-1), is equal to $F_x(b - a)$, where F_x is the component of **F** in the direction of the displacement of the particle. In Figure 5.2 the x component of the force on the particle is shown as a function of position. The work $F_x(b - a)$ done on the particle by the force is equal to the area indicated in Figure 5.2 enclosed by the line $y = F_x$, the x axis, and the lines $x = a$ and $x = b$.

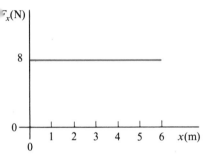

Exercise 5.4

E 5.4   As a 2.0-kg object moves along the x axis, it is acted upon by the 8.0-N force shown in the accompanying figure. Determine the work done by this force if (a) the object moves from $x = 0$ m to $x = 3.0$ m, (b) the object moves from $x = 2.0$ m to $x = 5.0$ m, and (c) the object moves from $x = 4.0$ to $x = 1.0$ m.
                                                    Answers: (a) 24 J; (b) 24 J; (c) −24 J

Work by a Variable Force

If the force exerted on a particle that is constrained to move along a straight line, say the x axis, is not a constant force but one that changes as the position of the particle changes, how is the work done by this force determined? Because any displacement of this rectilinear motion is along the x axis, only the x component F_x of the force does work. (Remember that the force components perpendicular to the displacement of the particle do no work.) The dependency of F_x upon the position of the particle is usually symbolized functionally by $F_x(x)$. Let $F_x(x)$ be represented graphically by the curve of Figure 5.3. As before, the work done by F_x is equal to the area of the region bounded by the curve $y = F_x(x)$, the x axis, and the lines $x = a$ and $x = b$. Since this area may be expressed as the definite integral of $F_x(x)$ from $x = a$ to $x = b$, the work done by the variable force F_x is given by

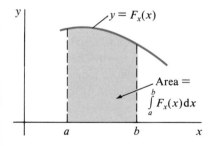

Figure 5.3
A graph of the x component of a variable force acting on a particle as it moves from $x = a$ to $x = b$. The work done by this force is equal to $\int_a^b F_x(x)dx$, the shaded area.

$$W_{a \to b} = \int_a^b F_x(x)\,dx \tag{5-3}$$

The mathematical symbols in Equation (5-3) may be interpreted physically. The quantity $F_x(x)\,dx$ represents the infinitesimal work done by the variable force F_x as the particle moves through an infinitesimal displacement from x to $x + dx$. The integral sign represents a summation, although, actually, it is the limit of a sequence of sums. The total work $W_{a \to b}$ done by F_x on the particle as it moves from $x = a$ to $x = b$ is then the sum of all the infinitesimal amounts of work done by F_x.

Example 5.1

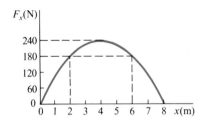

Figure 5.4   Example 5.1

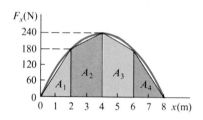

Figure 5.5   Example 5.1

PROBLEM   The force $F_x(x)$ given graphically in Figure 5.4 acts on an object as it moves along the x axis. Use graphical techniques to estimate the work done by this force as the object on which it acts moves from $x = 0$ to $x = 8.0$ m.

SOLUTION   As we have seen, the work $W_{0\to8}$ to be determined is the area enclosed between the graph $F_x(x)$ and the x axis between $x = 0$ and $x = 8.0$ m. The shaded area shown in Figure 5.5 approximates the actual area under the curve. Since the areas A_1 and A_4 are each the area of a triangle having a base of 2.0 m and a height of 180 N, it follows that

$$A_1 = A_4 = \tfrac{1}{2}(2 \text{ m})(180 \text{ N}) = 180 \text{ J}$$

Similarly, the areas A_2 and A_3 are each the area of a trapezoid having a base of 2.0 m and an average height of $(180 + 240)/2 = 210$ N, so that we have

$$A_2 = A_3 = (2 \text{ m})(210 \text{ N}) = 420 \text{ J}$$

The sum of these areas,

$$A_1 + A_2 + A_3 + A_4 = 180 + 420 + 420 + 180 = 1200 \text{ J}$$

is our estimate for $W_{0\to8}$, that is,

$$W_{0\to8} = \int_0^8 F_x(x)\,\mathrm{d}x \approx 1200 \text{ J} \quad \blacksquare$$

Can you explain why this value of 1200 J underestimates $W_{0\to8}$? When you work the following exercise, you will determine the exact answer for $W_{0\to8}$.

    E 5.5   Figure 5.4 is a graph of $F_x = (120x - 15x^2)$ N for $0 \le x \le 8$ m. Use Equation (5-3) to determine $W_{0\to8}$, the work that was estimated in Example 5.1.   Answer: 1.3 kJ

    E 5.6   An object moving along the x axis is acted upon by the force shown. Use graphical techniques to calculate the work done by this force as the object moves from $x = 6.0$ ft to $x = 12$ ft.   Answer: 110 ft·lb

    E 5.7   The force shown graphically in Exercise 5.6 may be expressed functionally as $F_x(x) = 2x$ lb, where x is measured in feet. Use Equation (5-3) to verify the result obtained for Exercise 5.6.

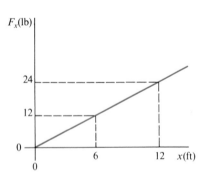

Exercise 5.6

    The following example uses Equation (5-3) to calculate the work done by a variable force. Consider a weightless spring, that is, a spring having negligible mass. One end of the spring is attached to a wall. A particle on the other end of the spring experiences a force exerted by the spring (*see* Fig. 5.6) that obeys a force rule known as *Hooke's law*,

$$F_x = -kx \tag{5-4}$$

Here x is the displacement of the spring from its equilibrium (unextended and uncompressed) position. Figure 5.6 illustrates that the force F_x exerted on the particle has a direction opposite to the displacement of the spring from its equilibrium position ($x = 0$). This is indicated by the negative sign in Equation (5-4). The constant k in Hooke's law is known as the *spring constant*, a value measured in units of force per unit length (of extension or compression). In the

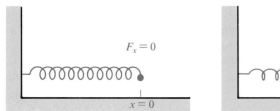 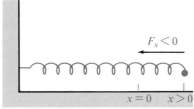 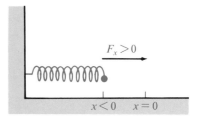

Figure 5.6
The force $\mathbf{F}_x$ exerted by a spring on a particle at the end of the spring. The spring is in equilibrium at $x = 0$.

metric system k has the unit N/m, and the English unit of k is lb/ft. A stiff spring, then, is one that has a large value of k. Actual springs, although not weightless, very nearly obey Hooke's law as long as they are neither stretched beyond the limits of extension that permanently deform them or compressed to the point where the adjacent coils are touching.

E 5.8   Coil springs are often used in automobile suspension systems. Suppose a 2600-lb auto is supported equally by four coil springs and the weight of the auto depresses each spring 4.0 in. Calculate the spring constant for each spring.   Answer: 2000 lb/ft

E 5.9   Scales for measuring the weight of produce may be constructed with a spring as the essential device. If 2.0 kg of artichokes extend the spring 3.0 cm, what is the spring constant?   Answer: 650 N/m

Now suppose we calculate the work done by a force $\mathbf{P}(x)$ as it extends a spring a distance d from its equilibrium position, as shown in Figure 5.7. The force is applied in the same direction as the extension of the spring (opposite to the direction of the force of Equation [5-4] exerted by the spring). The applied force has only an x component given by

$$P_x(x) = kx \tag{5-5}$$

and the work done by the force in extending the spring a distance d from its equilibrium position is

$$W_{0 \to d} = \int_0^d P_x(x)\,dx = \int_0^d kx\,dx = \tfrac{1}{2}kd^2 \tag{5-6}$$

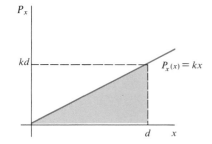

Exercise 5.10

E 5.10   The figure is a graph of Equation (5-5). Show that the shaded area is equal to the work given by Equation (5-6).

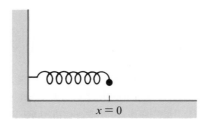 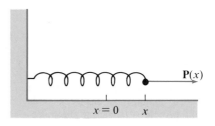

Figure 5.7
A force $\mathbf{P}(x)$ extending a spring from its equilibrium position $x = 0$.

E 5.11

(a) How much work is done in compressing each auto spring of Exercise 5.8 a distance of 4.0 inches?

(b) How much work does the spring do during this compression?

Answers: (a) $110 \text{ ft} \cdot \text{lb}$; (b) $-110 \text{ ft} \cdot \text{lb}$

E 5.12   Calculate the work done in extending the spring of Exercise 5.9 a distance of 3.0 cm.   Answer: 0.29 J

Example 5.2   PROBLEM   An object of mass m sliding on a horizontal frictionless surface is attached, as shown in Figure 5.8(a), to two ideal springs having spring constants k_1 and k_2. At the equilibrium position of the object, neither of the two springs is extended or compressed. What is the spring constant k for a single ideal spring, attached to m, as shown in Figure 5.8(b), which would cause the motion of m to be identical to its motion in Figure 5.8(a)?

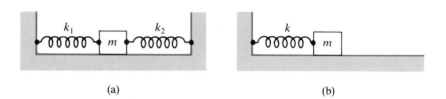

| (a) | (b) |

Figure 5.8   Example 5.2

SOLUTION   If x measures the horizontal position of m relative to its equilibrium ($x = 0$) position, the spring to the left of m in Figure 5.8(a) exerts a force on m given by

$$F_1(x) = -k_1 x$$

Similarly, the spring to the right of m exerts a force on m given by

$$F_2(x) = -k_2 x$$

Consequently, the total force ΣF_x on m is

$$\Sigma F_x = F_1(x) + F_2(x) = -k_1 x - k_2 x = -(k_1 + k_2)x$$

For the single-spring situation shown in Figure 5.8(b), the force on m, if its horizontal position is measured by x, is

$$F(x) = -kx$$

Thus if the single spring of Figure 5.8(b) is to exert the same total force on m at a given position as the two springs of Figure 5.8(a), that is, $F(x) = \Sigma F_x$, then we see

$$k = k_1 + k_2 \quad \blacksquare$$

Work by an Arbitrary Force

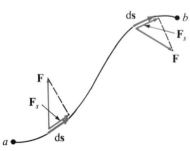

Figure 5.9
A variable force **F** acting on a particle moving along an arbitrary path from a to b. The infinitesimal displacement ds of the particle is tangent to the trajectory at every point.

Figure 5.9 represents the most general situation for calculating the work done by a force. Here a variable force **F** does work on a particle moving along an arbitrary trajectory from position a to position b. Because **F** is variable, its magnitude and direction may change continuously as the particle moves. The

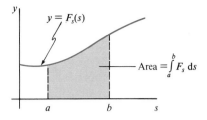

Figure 5.10
A graph of the tangential component $F_s(s)$ of a force plotted as a function of arc length s. The work done by the force on a particle that moves from a to b is equal to $\int_a^b F_s ds$, the shaded area.

work done by **F** is then the sum of the infinitesimal amounts of work dW done by **F** as the particle undergoes an infinitesimal displacement d**s**. For each infinitesimal displacement, the work done by **F** is

$$dW = \mathbf{F} \cdot d\mathbf{s} = F_s ds \tag{5-7}$$

where F_s is the component of **F** in the direction of the displacement d**s**. The summation of the infinitesimal amounts of work done as the particle moves from a to b is

$$W_{a \to b} = \int_a^b \mathbf{F} \cdot d\mathbf{s} = \int_a^b F_s ds \tag{5-8}$$

Equation (5-8) is called a *line integral*. It can be integrated only if we know the tangential component of **F** as a function of arc length s. Figure 5.10 is a graph that shows the tangential component F_s plotted as a function of arc length s along the trajectory of the particle. The work done by **F** on the particle as it moves from $s = a$ to $s = b$ is equal to the area indicated in Figure 5.10. Equation (5-8) is the most general definition of work.

E 5.13   Show that Equation (5-8) reduces to Equation (5-2) if **F** is constant. (*Hint:* Recall that an integral is essentially a sum and that a constant factor that appears in each term in a sum may be factored out of the sum.)

E 5.14   Show that Equation (5-8) reduces to Equation (5-3) if the trajectory of the particle is along the x axis.

5.2 Power

The rate at which work is done is often of interest in practical applications. If a large motor can do a given amount of work faster than a smaller motor, we say that the larger motor is the more powerful one. *Power* is the measure of the rate at which work is being done. In the case of a motor the *average power* $\overline{P}$ being developed over a period of time is defined as the quantity of work ΔW done by the motor divided by the time interval Δt required to do that work, and, in general, average power $\overline{P}$ is defined to be

$$\overline{P} = \frac{\Delta W}{\Delta t} \tag{5-9}$$

If work is expressed as a function of time, the *instantaneous power* P being developed at any instant is defined to be

$$P = \frac{dW}{dt} \tag{5-10}$$

It is sometimes convenient to express power in terms of a constant force **F** acting on an object moving at a constant velocity **v**. This is the case when, for example, an outboard motor causes the water to exert a constant force **F** on the propeller (and, therefore, on the boat) when a motorboat is moving at a constant velocity **v**. The power delivered by the motor at any instant is given by

$$P = \frac{dW}{dt} = \lim_{\Delta t \to 0} \frac{\mathbf{F} \cdot \Delta \mathbf{r}}{\Delta t} = \mathbf{F} \cdot \frac{d\mathbf{r}}{dt} \tag{5-11}$$

where **r** is the position of the boat. Because the time rate of change of position is the velocity of the boat, Equation (5-11) becomes

$$P = \mathbf{F} \cdot \mathbf{v} \qquad (5\text{-}12)$$

Equation (5-12) is correct even if the force **F** and the velocity **v** are not constant. The work dW by any force **F** acting on an object that undergoes an infinitesimal displacement d**s** is done as the object moves with a velocity **v** for an infinitesimal time dt. Then, using d**s** = **v** dt, we have

$$dW = \mathbf{F} \cdot d\mathbf{s} = \mathbf{F} \cdot \mathbf{v}\, dt$$

or

$$\frac{dW}{dt} = P = \mathbf{F} \cdot \mathbf{v} \qquad (5\text{-}13)$$

whether or not **F** and **v** are constant.

The units of power are those of work divided by time. In the metric system, the unit of power is, therefore, J/s, which is defined to be the *watt* W. In the English system, the unit of power is ft·lb/s. A common engineering unit of power is the *horsepower* (hp), which is related to standard units by

$$1 \text{ hp} = 550 \text{ ft} \cdot \text{lb/s} \cong 746 \text{ W} \qquad (5\text{-}14)$$

Sometimes work is expressed in terms of power multiplied by time. Thus the kilowatt-hour (kW·h) and the watt-second (W·s) are units of work.

E 5.15   If you push with a horizontal force of 80 lb on a car that is moving 3.0 mi/h (4.4 ft/s) in the same direction, what power are you developing?   Answer: 350 ft·lb/s = 0.64 hp = 480 W

E 5.16   A unit commonly used for measuring electrical power is the kilowatt (kW). Express the kilowatt in horsepower and ft·lb/s.   Answer: 1 kW = 1.3 hp = 740 ft·lb/s

E 5.17   Electrical work is usually measured in kW·h.
(a) Through what distance would you have to exert a 50-lb (220 N) force to do a kW·h of work?   Answer: 16 km = 10 mi
(b) How much would you earn for doing the work in part (a) if you were paid 10 cents per kW·h, a typical rate for electrical work? Is electrical work expensive?

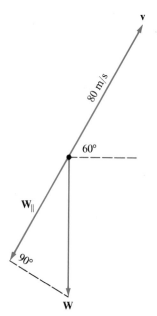

Figure 5.11   Example 5.3

Example 5.3   **PROBLEM**   At an instant when a 2.0-kg projectile is moving with a velocity of 80 m/s directed 60° above the horizontal, at what rate is the gravitational force on the projectile doing work?

SOLUTION   Figure 5.11 shows the projectile at the instant of interest. The rate at which the gravitational force **W** is then doing work, that is, the power P being delivered by this force at this instant is

$$P = \mathbf{W} \cdot \mathbf{v} = mgv \cos(90° + 60°)$$

$$= 2(80)(9.8) \cos 150° = -1360 \ W = -1.4 \text{ kW}$$

The negative sign in this result indicates that the component of **W** parallel to **v**, $W_\parallel$, is directed oppositely to **v**; at this instant, **W** is doing negative work on the projectile. ∎

5.3 Energy

We have defined work and have seen that if a force is to do work on an object, the object must undergo a displacement and the force on the object must have a component in the direction of that displacement. A particle or a system of particles that has the capacity to do work is said to possess *energy*, a physical quantity associated with the particle or system that may take many forms. Like work, energy is a word with nonscientific connotations. It is often used to refer to activity in general or merely an urge to be active. In the technical sense, however, a system that has the ability to do work is said to have energy. What is more, energy is a scalar quantity that is usually expressed in terms of the same units as those of work, specifically, J or ft·lb. In this chapter, we are primarily concerned with kinetic and potential energies. Other forms of energy, like electrical energy, will be described in later chapters.

Kinetic Energy

One of the forms of energy that may be associated with a particle is called *kinetic energy*, the measure of the ability of a particle to do work because of its motion. A moving bullet, for example, has kinetic energy because its motion ensures that if the bullet strikes an object, the bullet can exert a force on that object and move matter through a distance, thereby doing work. In order to express kinetic energy in terms of the physical quantities we know, let us consider the situation shown in Figure 5.12. A block of mass m, to which a massless spring is attached, is moving at a speed v_o toward a wall in Figure 5.12(a). The spring makes its initial contact with the wall in Figure 5.12(b). In Figure 5.12(c) the spring is being compressed as the block slows, and in Figure 5.12(d) the block has come to rest.

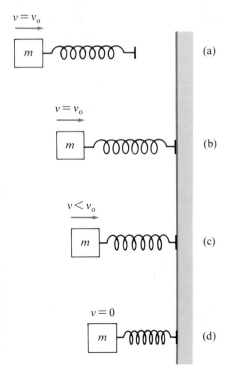

Figure 5.12
(a) A mass m with a spring attached moving with speed $v = v_o$ toward a wall. **(b)** The spring in initial contact with the wall. The speed of the mass is still $v = v_o$. **(c)** The spring being compressed. The speed of the mass is now less than v_o. **(d)** The mass momentarily at rest.

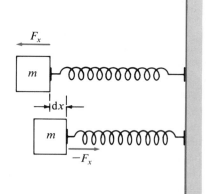

Figure 5.13
The mass of Figure 5.12 shown in two positions separated by an infinitesimal displacement dx during the compression of the spring. The force $\mathbf{F}_x$ is exerted on the block by the spring. The force $-\mathbf{F}_x$ is exerted on the spring by the block.

We have called the measure of the ability of the block to do work because of its motion its kinetic energy. Before it strikes the wall, the block is moving ($v = v_0$) and, therefore, possesses kinetic energy. After it comes to rest, it is not moving ($v = 0$) and, therefore, has no kinetic energy. What, then, was the kinetic energy of the block initially when its speed was v_0? It is the work the block does (on the spring) in coming to rest, that is, in giving up its initial kinetic energy. Let us now calculate that work.

Figure 5.13 shows the block of mass m at two positions separated by an infinitesimal displacement dx. (Since the situation here involves rectilinear motion, vector notation is not used.) The force F_x exerted on the block by the spring is to the left, in the negative x direction. According to Newton's third law, the force exerted on the spring by the block is $-F_x$. Then the work W done by the block on the spring is

$$W = \int_i^f (-F_x)dx \qquad (5\text{-}15)$$

where i and f indicate the initial and final state of the variable x over which the integration takes place. Because F_x is the resultant force on the block, we may write $F_x = ma_x$, or

$$-F_x = -ma_x = -m\frac{dv_x}{dt} \qquad (5\text{-}16)$$

Substituting Equation (5-16) into Equation (5-15) gives

$$W = -\int_i^f m\frac{dv_x}{dt}dx = -m\int_i^f \frac{dv_x}{dt}\,dx \qquad (5\text{-}17)$$

The definition of velocity provides that $v_x = dx/dt$, or $dx = v_x dt$, which, when substituted into Equation (5-17) results in

$$W = -m\int_i^f \frac{dv_x}{dt}\,v_x dt = -m\int_{v_x=v_0}^{v_x=0} v_x dv_x = \tfrac{1}{2}mv_0{}^2 \qquad (5\text{-}18)$$

This work, $\tfrac{1}{2}mv_0{}^2$, done by the block in coming to rest from its initial state of motion is equal to the kinetic energy the block had when it was moving at a speed v_0. Then we define the kinetic energy K of a body of mass m moving at a speed v to be

$$K = \tfrac{1}{2}mv^2 \qquad (5\text{-}19)$$

Equation (5-19) indicates that it is the speed of a body and not its direction of motion that is a factor in the kinetic energy of that body. Since the quantity v is squared, the sign, or direction, of the velocity of the body does not affect the ability of the body to do work.

The units of kinetic energy are the same as those of work, namely, J and ft·lb.

E 5.18   By substituting the appropriate units for m and v, show that kinetic energy as defined by Equation (5-19) has the metric unit of joule and the English unit of ft·lb.

E 5.19   Calculate the kinetic energy of
(a) a 2.0-kg aardvark running with a speed of 10 m/s, and
(b) a 5.0-oz (5/16 lb) ball moving at 100 mi/h (150 ft/s).

Answers: (a) 100 J; (b) 100 ft·lb

The Work-Energy Principle

We have just seen that the kinetic energy of an object is a measure of the work that the object can perform because of its motion. Here we shall start with Newton's second law and obtain a relationship between the work done on an object and the corresponding change in the kinetic energy of that object.

Newton's second law is usually written

$$\sum \mathbf{F} = m\mathbf{a} \tag{5-20}$$

in which $\sum \mathbf{F}$ is the vector sum of all forces acting on the object and $\mathbf{a}$ is the resulting acceleration of that object. If we form the scalar product of each side of Equation (5-20) with the velocity $\mathbf{v}$ of the particle, the result is

$$\left(\sum \mathbf{F}\right) \cdot \mathbf{v} = m\mathbf{a} \cdot \mathbf{v} \tag{5-21}$$

Now $(\sum \mathbf{F}) \cdot \mathbf{v}$ is the instantaneous rate at which all the forces acting on the mass are performing work on that mass, that is,

$$\left(\sum \mathbf{F}\right) \cdot \mathbf{v} = \frac{dW}{dt} \tag{5-22}$$

where W is the work done on the mass m by the net force acting on it. The right side of Equation (5-21), $m\mathbf{a} \cdot \mathbf{v}$, may be written

$$m\mathbf{a} \cdot \mathbf{v} = m\frac{d\mathbf{v}}{dt} \cdot \mathbf{v} = m\frac{d}{dt}\left(\frac{\mathbf{v} \cdot \mathbf{v}}{2}\right) = m\frac{d}{dt}\left(\frac{v^2}{2}\right) = \frac{d}{dt}\left(\frac{1}{2}mv^2\right) = \frac{dK}{dt} \tag{5-23}$$

Substituting Equations (5-22) and (5-23) into Equation (5-21) gives

$$\frac{dW}{dt} = \frac{dK}{dt} \tag{5-24}$$

Thus, the rate at which work is done on an object by all the forces acting on it is equal to the rate at which the kinetic energy of that object changes. If Equation (5-24) is integrated with respect to time, we find

$$W_{1 \to 2} = K_2 - K_1 \tag{5-25}$$

where $W_{1 \to 2}$ is the work done on an object by the net force on the object during a time interval beginning at the time t_1 and ending at the time t_2. The kinetic energy of the object at time t_2 is K_2; the kinetic energy of the object is K_1 at time t_1. Equation (5-25) is the *work-energy principle*, which states that during any time interval the work done by the *resultant force* on an object is equal to the change in the kinetic energy of that object during the same interval. Notice that the work done by the resultant force on an object is equal to the sum of the work done by *all* the forces acting on that object.

E 5.20 Verify each of the steps of Equation (5-23).

E 5.21 Use the work-energy principle to show that the kinetic energy of an object at a given instant is equal to the work required to bring the object from rest to the state of its motion at that given instant.

E 5.22   How much work must be done on a 5.0-kg mass to change its velocity from (a) $2.0\hat{\mathbf{i}}$ m/s to $8.0\hat{\mathbf{i}}$ m/s? (b) $2.0\hat{\mathbf{i}}$ m/s to $-8.0\hat{\mathbf{j}}$ m/s? (c) $8.0\hat{\mathbf{i}}$ m/s to $2.0\hat{\mathbf{j}}$ m/s? and (d) $2.0\hat{\mathbf{i}}$ m/s to $2.0\hat{\mathbf{j}}$ m/s?

Answers: (a) 150 J; (b) 150 J; (c) -150 J; (d) zero

E 5.23   How much work must be done on a 4.0-lb body to bring it from rest to a velocity of $(6.0\hat{\mathbf{i}} + 8.0\hat{\mathbf{j}})$ ft/s?

Answer: 6.2 ft · lb

Example 5.4

PROBLEM   The force $F_x(x)$ of Example 5.1 and Exercise 5.5 is the only force acting on a 10-kg object as it moves from $x = 0$ to $x = 8.0$ m. If the velocity of the object at $x = 0$ is $v_x = 12$ m/s, how fast is it moving at $x = 8.0$ m?

SOLUTION   According to the work-energy principle, the (total) work $W_{0\rightarrow 8}$ on the object is related to the initial and final kinetic energies by

$$W_{0\rightarrow 8} = K_8 - K_0$$

Because $K = \frac{1}{2}mv^2$, we have

$$W_{0\rightarrow 8} = \tfrac{1}{2}mv_8{}^2 - \tfrac{1}{2}mv_0{}^2$$

$$v_8{}^2 = v_0{}^2 + 2W_{0\rightarrow 8}/m$$

Recall from Exercise 5.5 that the work $W_{0\rightarrow 8}$ is 1300 J. Then we obtain

$$v_8{}^2 = 12^2 + 2(1300)/10 = 400$$

$$v_8 = 20 \text{ m/s}$$

Potential Energy

We have seen that the kinetic energy of a particle depends on the mass and speed of that particle. Although we usually measure the speed of a particle relative to the earth, the kinetic energy of that particle is associated with the particle itself. Now we consider a second form of mechanical energy called *potential energy*. Like kinetic energy, potential energy represents the ability to do work. But potential energy arises because of interactions between two or more bodies. Thus the potential energy of a system of interacting objects represents the ability of the system to do work because of its configuration.

Here we shall consider two forms of mechanical potential energy: gravitational and elastic. The gravitational potential energy that we shall associate with a particle arises from the configuration of the particle and the earth, a system that is interacting by means of the gravitational force. Elastic energy, on the other hand, arises from the configuration of a spring or some similar device that may have the ability to do work in stressed configurations. Let us examine gravitational potential energy first.

Gravitational Potential Energy

A particle anywhere outside the earth interacts with the earth through the attractive gravitational forces on the particle and the earth. We shall, for convenience, separate the special case of a particle near the surface of the earth from the more general case because many of the situations we encounter occur near the surface of the earth.

Let us consider first the near-earth situation, in which the magnitude g of the free-fall acceleration of a body is constant. We designate some reference position along the vertical to be $y = 0$. This arbitrary choice gives us a reference level from which vertical distances may be measured. Let us define the *gravitational potential energy* of a particle to be zero at the reference level $y = 0$. At any other vertical position, such as $y = h$, the gravitational potential energy is defined to be the work that the gravitational force would do on the particle if the particle were moved from a position at $y = h$ to the reference level $y = 0$.

To put this definition into a useful analytical form, consider a body of mass m at a vertical height $h > 0$, *above* the reference level $y = 0$. The work that would be done on the body by the gravitational force $\mathbf{F}_g$ as the body moves downward from $y = h$ to $y = 0$ is

$$U_g = \int_h^{\text{ref}} \mathbf{F}_g \cdot d\mathbf{s} \tag{5-26}$$

where the limits on the integral indicate that the position of the body changes from h to the reference level. Because $\mathbf{F}_g$ is $-mg\hat{\mathbf{j}}$ and $d\mathbf{s}$ is $dy\hat{\mathbf{j}}$, Equation (5-26) becomes

$$U_g = \int_{y=h}^{y=0} (-mg)\,dy = -mgy \Big|_{y=h}^{y=0}$$

$$U_g = mgh \tag{5-27}$$

Thus for a body at a height h *above* the reference level, the gravitational potential energy U_g of the body is a positive value.

By a similar analysis, a body at a vertical position *below* the reference level $y = 0$ has a negative potential energy. In this case, the force of gravity and the displacement of the body in moving to $y = 0$ are in opposite directions. The work that would be done by the gravitational force if the body were moved from $y = h < 0$ to $y = 0$ then has a negative value. Thus U_g is negative if h is negative. Equation (5-27), therefore, expresses gravitational potential energy of a body of mass m displaced by a vertical distance h (positive or negative) from an arbitrary reference level at $y = 0$.

The choice of the reference level $y = 0$ is arbitrary. If the body of mass m in Figure 5.14(a) moves, for example, a distance d from a vertical position $y = a$ to a higher position $y = b$, the change in U_g is a positive value,

$$\Delta U_g = mgb - mga = mg(b - a) = mgd \tag{5-28}$$

where $b - a$ is the distance d through which the body is raised. Suppose now that a new reference level is chosen, as shown in Figure 5.14(b). With this new reference level, the change in the height d of the body is described relative to the new reference level $y' = 0$ by

$$\Delta U_g = mg(b + c) - mg(a + c) = mg(b - a) = mgd \tag{5-29}$$

Clearly the choice of reference level does not affect the change in U_g. Therefore, we may select any reference level from which to measure vertical positions when calculating changes in gravitational potential energy.

E 5.24   A 35-lb suitcase is lifted 15 floors (150 ft) in an elevator. What is the change in the gravitational potential energy of this suitcase?

Answer: 5.3×10^3 ft $\cdot$ lb

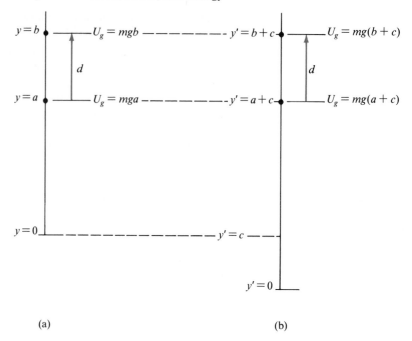

Figure 5.14
The change in gravitational potential
energy is independent of the choice of
reference level.

(a)   (b)

E 5.25   If a 1.5-kg object in free-fall descends 30 m,
(a) what is the change in its potential energy?   Answer: −440 J
(b) how much work is done on the object by the earth's gravitational force?
                                                             Answer: 440 J
(c) Why do the answers of parts (a) and (b) have opposite signs?

The arbitrariness of the choice of a reference level for U_g becomes especially useful when we consider gravitational potential energy for bodies not necessarily near the surface of the earth. Consider a body of mass m at a distance $r > R_e$, the radius of the earth, from the center of the earth. By our definition, the gravitational potential energy U_g of the body at the position r is the work that the gravitational force would do on this body if the body were moved vertically (radially) from its position r to a chosen reference level:

$$U_g = \int_r^{\text{ref}} \mathbf{F} \cdot d\mathbf{s} \tag{5-30}$$

But what reference level should we choose? Two choices that would be physically meaningful are those at which r is the radius of the earth R_e and at which r is an infinite distance from the earth. The usual choice is that for which $r \to \infty$, where we define the gravitational potential energy to be zero. At any distance $r > R_e$ from the center of the earth, a body of mass m experiences a force GmM_e/r^2 directed toward the center of the earth. Then the work that would be done by the gravitational force in moving the body to infinity is

$$U_g = \int_r^{\infty} \left(-\frac{GmM_e}{r^2}\right) dr \tag{5-31}$$

The negative sign appears in Equation (5-31) because the force of magnitude GmM_e/r^2 is radially inward, while $d\mathbf{s}$ is radially outward. Because G, m, and M_e are constants, we may write

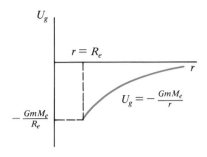

Figure 5.15
The gravitational potential energy U_g of a body of mass m as a function of radial distance r from the center of the earth. The radius of the earth is R_e.

$$U_g = -GmM_e \int_r^\infty r^{-2}dr = -GmM_e \left(-\frac{1}{r}\right)\Bigg|_r^\infty$$

$$U_g = -\frac{GmM_e}{r} \qquad (5\text{-}32)$$

The negative value of U_g for a body at any radial distance r measured from the center of the earth results from having chosen $U_g = 0$ at infinity. The gravitational potential energy U_g of a body of mass m is graphically shown as a function of r in Figure 5.15. Here we can see that the potential energy of a body increases when r increases, even though U_g is at all times a negative value. The dashed line $r = R_e$ appears because Equation (5-32) is valid only for $r \geq R_e$, where $U_g \geq -GmM_e/R_e$. Why does Equation (5-32) not apply for $r < R_e$?

E 5.26
(a) Calculate the gain in gravitational potential energy that results if a 5.0-kg object is raised from the surface of the earth to a height of 2.0×10^7 m above the surface.   Answer: 2.4×10^8 J
(b) What result do you get for part (a) if you incorrectly use the near-earth gravitational potential energy expression (mgh)?   Answer: 9.8×10^8 J

E 5.27   If a 15-kg object falls from a height of 1000 km to the surface of the earth,
(a) determine the corresponding change in gravitational potential energy.
Answer: -1.3×10^8 J
(b) calculate the work done by the gravitational force of the earth on the object.
Answer: 1.3×10^8 J

Elastic Potential Energy

The last form of mechanical energy that we shall consider is that of energy stored in a spring. When a spring is extended (or compressed) from its equilibrium position $x = 0$, it has the ability to do work as it returns to the equilibrium position. Thus a spring may have the potential for doing work because of its configuration. In a manner similar to the way we defined gravitational potential energy, we define the *elastic potential energy* U_e of a spring to be zero at its reference position $x = 0$, where the spring is in equilibrium. At any other position, say at $x = d$, either extended or compressed from equilibrium, the elastic potential energy U_e is defined to be the work that would be done by the spring in returning to its reference position:

$$U_e = \int_d^{\text{ref}} F_x dx \qquad (5\text{-}33)$$

where the integration takes place from $x = d$ to the reference level. The force exerted by an ideal spring, extended a displacement x from equilibrium, is $-kx$, where k is the spring constant. Therefore, U_e becomes

$$U_e = \int_d^{x=0} (-kx)\,dx = -\frac{kx^2}{2}\Bigg|_d^{x=0} = \frac{1}{2}kd^2 \qquad (5\text{-}34)$$

Notice that $\frac{1}{2}kd^2$ is the value we calculated earlier in this chapter to be the work done on a spring to extend (or compress) it a distance d from its equilibrium position.

E 5.28   A spring (k = 200 N/m) is compressed a distance d = 20 cm. Calculate the elastic potential energy stored in the spring.   Answer: 4.0 J

5.4 Conservation of Energy

Consider now a particle of mass m at rest at a height h above the surface (y = 0) of the earth, as shown in Figure 5.16. At that position, the particle has potential energy U_g = mgh, and its kinetic energy K is zero. Suppose we release the particle from rest. After release, the system (the particle and the earth pulling on each other) is said to be *isolated*. We say a system is isolated if no forces external to the system are doing any work on the system. As the particle falls from its initial height h to the ground, consider how the energies, both kinetic and potential, associated with the particle change.

When the particle is released at y = h, its gravitational potential energy is U_g = mgh. Because the particle is released at rest, its initial kinetic energy K is zero. Now we calculate the energies U_g and K when the particle has fallen half the distance h. At this position, y = $h/2$, U_g is equal to $mg(h/2)$. The speed of the particle at a height $h/2$ above the ground can be calculated from the kinematic relationship $v_y{}^2 = v_{oy}{}^2 - 2g(y - y_o)$, in which y_o = h, y = $h/2$, and v_{oy} = 0. Then we have

$$v_y{}^2 = 0 - 2g\left(\frac{h}{2} - h\right) = gh \qquad (5\text{-}35)$$

and the kinetic energy K at y = $h/2$ is

$$K = \frac{1}{2}mv_y{}^2 = \frac{1}{2}mgh \qquad (5\text{-}36)$$

As we see in Figure 5.16, when the particle has fallen half the distance to the ground, the total mechanical energy of the particle is equally divided between its potential and kinetic energies.

When the particle has reached a height $h/4$, its potential energy U_g is $mg(h/4)$. At this point, we find that

$$v_y{}^2 = 0 - 2g\left(\frac{h}{4} - h\right) = \frac{3gh}{2}$$

$$K = \frac{1}{2}mv_y{}^2 = \frac{3}{4}mgh \qquad (5\text{-}37)$$

Finally, immediately before the particle strikes the earth (just as y = 0) the potential energy of the particle is zero. At this point, the kinetic energy of the particle is found to be

$$v_y{}^2 = 0 - 2g(0 - h) = 2gh$$

$$K = \frac{1}{2}m(2gh) = mgh \qquad (5\text{-}38)$$

a value of kinetic energy equal to the original potential energy of the particle. It becomes obvious now that as the particle falls, its potential energy is converted to kinetic energy.

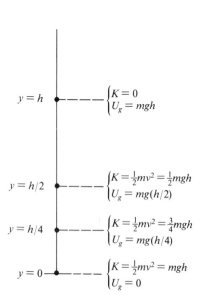

Figure 5.16
The gravitational potential energy U_g and the kinetic energy K of a particle of mass m at various heights above a reference level y = 0. The particle falls from rest at y = h.

$y = h$ $\begin{cases} K = 0 \\ U_g = mgh \end{cases}$

$y = h/2$ $\begin{cases} K = \frac{1}{2}mv^2 = \frac{1}{2}mgh \\ U_g = mg(h/2) \end{cases}$

$y = h/4$ $\begin{cases} K = \frac{1}{2}mv^2 = \frac{3}{4}mgh \\ U_g = mg(h/4) \end{cases}$

$y = 0$ $\begin{cases} K = \frac{1}{2}mv^2 = mgh \\ U_g = 0 \end{cases}$

The most important aspect of this example is that the sum of the kinetic and potential energies of the particle remains constant. The sum of the two forms of mechanical energy is conserved during the fall. The sum of the kinetic and potential energies of the particle is defined as the *total mechanical energy E* of the particle. Then, as long as an earth-particle system is isolated, we may write

$$E = K + U_g = \text{constant} \qquad (5\text{-}39)$$

Equation (5-39) is a statement of the conservation of mechanical energy for a particle interacting with the earth only.

It is reasonable to ask what happens to the mechanical energy of a body after it has fallen to earth and come to rest. The original energy of position of the body was converted during the fall to energy of motion. While the body is falling, all the particles that comprise the body are taking part in an organized motion: All the particles are moving downward together. But when the body strikes the ground, the particles in the body and those in the ground are moved about in a disorganized way. Particles rub against particles, and the energy that began as potential energy and became energy of organized motion is finally converted to *thermal energy*, the energy of disorganized motion.

Friction is an example of particles rubbing against particles. Consequently, we must recognize that when friction is present in a physical process, mechanical energy may be dissipated, or lost from the system. The total mechanical energy of the system that would otherwise remain constant is diminished by any dissipative losses of energy converted to thermal energy. We must, therefore, adjust our statement of the conservation of energy to take such losses into account. Let us see how this may be done by considering the following example.

Suppose a 10-lb cube falls from rest down a vertical shaft, as shown in Figure 5.17. What will be the speed of the cube after it has fallen through a vertical distance of 100 ft if: (a) the walls of the shaft are frictionless, and (b) the walls exert a constant frictional force of 2.0 lb on the cube?

The change in vertical position of the cube during the process of this problem suggests that the problem may be treated by consideration of the conservation of energy. For convenience, let us choose the final position of the cube to be the reference level $y = 0$, where $U_g = 0$. Then, at the initial position of the cube, $y = 100$ ft, its initial potential energy U_{gi} is $mgy = 1000$ ft·lb. Its initial kinetic energy K_i is equal to zero because the cube is at rest. We record these data on the working diagram of Figure 5.18(a). A bookkeeping procedure like this is often helpful in solving problems using the conservation of energy principle. The total mechanical energy $E = K_i + U_{gi}$ of the system is initially 1000 ft·lb. In part (a) of this problem, the walls are frictionless. They can exert forces on the cube as it falls but only in a horizontal direction. Such forces are called *constraining forces*; these horizontal forces do no work on the cube because displacements of the cube are vertical. The cube-earth system is, therefore, isolated, and $E = K + U_g$ is conserved. After the cube has fallen 100 ft, E is still equal to 1000 ft·lb. At $y = 0$, the final potential energy U_{gf} is zero, and the final kinetic energy K_f of the cube is

$$K_f = \frac{1}{2}mv^2 = \frac{1}{2}\left(\frac{10}{32}\right)v^2 = \frac{5}{32}v^2 \qquad (5\text{-}40)$$

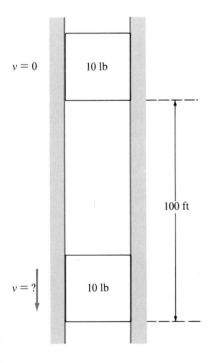

Figure 5.17
A 10-lb block falling down a vertical shaft.

$v = 0$ 10 lb

100 ft

$v = ?$ 10 lb

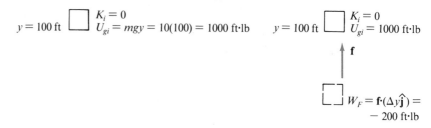

Figure 5.18
A working diagram for the kinetic and potential energies of the cube of Figure 5.17.

(a)   (b)

These values of K_f and U_{gf} are entered on the diagram of Figure 5.18(a). Because mechanical energy is conserved as the cube falls, we may equate the initial and final mechanical energies:

$$K_i + U_{gi} = K_f + U_{gf} \qquad (5\text{-}41)$$

Substituting the values we have determined for each term, we obtain

$$0 + 1000 = \frac{5}{32}v^2 + 0$$

$$v^2 = 6400$$

$$v = 80 \text{ ft/s} \qquad (5\text{-}42)$$

In part (b) of this problem, the initial total energy E of the system is again 1000 ft·lb, because $K_i = 0$ and $U_{gi} = mgy = 1000$ ft·lb. When the cube reaches its final position, $y = 0$, its potential energy U_{gf} is zero, and its kinetic energy K_f is $mv^2/2 = 5v^2/32$. But now the total mechanical energy $K_f + U_{gf}$ is no longer equal to $K_i + U_{gi}$. Mechanical energy has been lost from the cube-earth system. The energy lost, or dissipated as thermal energy, is equal to the work done by the force $\mathbf{f}$ of friction on the cube. In Figure 5.18(b) we see that the force of friction is upward, opposite to the direction of the displacement of the cube. Then the work W_f done by friction is

$$W_f = \mathbf{f} \cdot (\Delta y \hat{\mathbf{j}}) = 2(100)(-1) = -200 \text{ ft} \cdot \text{lb} \qquad (5\text{-}43)$$

Because the work W_f done by friction is negative, energy is lost from the system, and the total energy of the system finally will be less than it was initially. Thus, we must deduct the energy lost from the initial value $K_i + U_{gi}$. Adding W_f to $K_i + U_{gi}$ accomplishes this deduction:

$$K_i + U_{gi} + W_f = K_f + U_{gf} \qquad (5\text{-}44)$$

Substituting the values of our problem into Equation (5-44) gives

$$0 + 1000 - 200 = \frac{5}{32}v^2 + 0$$

$$v^2 = 5120$$

$$v = 72 \text{ ft/s} \tag{5-45}$$

a speed less than the cube attained in the frictionless situation of part (a).

E 5.29   Suppose a 2.0-kg body sliding across a horizontal surface experiences a 4.0-N frictional force on it by the surface.
(a) How much work is done on the body by the frictional force as the body slides 5.0 m?   Answer: −20 J
(b) What is the kinetic energy of the body after it slides 5.0 m beyond a point where its speed is 8.0 m/s?   Answer: 44 J

Example 5.5   PROBLEM   A body slides down a curved frictionless surface, as shown in Figure 5.19. The body is moving 15 ft/s when it is 10 ft vertically above the horizontal part of the surface. With what speed v_1 does the body move as it slides on the horizontal part of the frictionless surface?

SOLUTION   The initial energy of the body is

$$K_i + U_{gi} = \frac{1}{2}mv_0^2 + mgy_0$$

where the horizontal portion of the surface has been assigned a height of $y = 0$. The energy of the body as it slides along the horizontal surface is

$$K_f + U_{gf} = \frac{1}{2}mv_1^2 + mg(0) = \frac{1}{2}mv_1^2$$

Because no dissipative forces are acting, the final and initial energies are equal, so that

$$K_f + U_{gf} = K_i + U_{gi}$$

$$\frac{1}{2}mv_1^2 = \frac{1}{2}mv_0^2 + mgy_0$$

$$v_1^2 = v_0^2 + 2gy_0$$

$$v_1^2 = (15)^2 + 2(32)(10) = 865$$

$$v_1 = 29 \text{ ft/s} \quad \blacksquare$$

Figure 5.19   Example 5.5

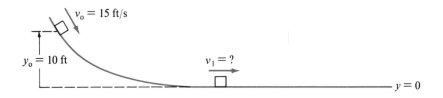

$v_0 = 15$ ft/s

$y_0 = 10$ ft

$v_1 = ?$

$y = 0$

5.5 Conservative and Nonconservative Forces

If as a particle moves from one position to another position, the work done by a force **F** acting on the particle is the same along all paths connecting the two positions, and if this statement is true for all initial and final positions, then the force **F** is said to be a *conservative force*. The significance of this statement may be seen by considering a particle of mass m that moves along an arbitrary frictionless path from point A to point B, as indicated in Figure 5.20(a). Here the y direction is vertical, and the x direction is horizontal. Instead of moving along the actual curved path, suppose the particle were to move from A to B along a series of vertical and horizontal paths, as shown in Figure 5.20(b). Then the gravitational force of the earth does work on the particle as it moves along the vertical segments 1 and 3, but it does no work as the particle moves along the horizontal segments 2 and 4. Hence, if B is at $y = 0$, the work done by the gravitational force is mgy_A as the particle traverses the segments 1, 2, 3, and 4 from A to B. Now, if the path of the particle is divided into smaller vertical and horizontal segments, as shown in Figure 5.20(c), the same argument applies: The net work done on the particle by the force of gravity is that done only along the vertical segments, namely, mgy_A. As the vertical and horizontal segments are made arbitrarily small, the stepped vertical-horizontal segments become the smooth curve from A to B. Therefore, work done by the gravitational force on the particle as it moves along the curve from A to B is mgy_A. Indeed, since the curved path we have considered connecting A and B is arbitrary, the work done on the particle by the gravitational force is independent of any path the particle may traverse from A to B.

Now let the particle of Figure 5.20 be returned from B to A. The work done on the particle by the gravitational force is $-mgy_A$ in this case, because the gravitational force is opposite in direction to each vertical displacement of the particle. Therefore, if the particle makes a round-trip from A to B *along arbitrary paths*,

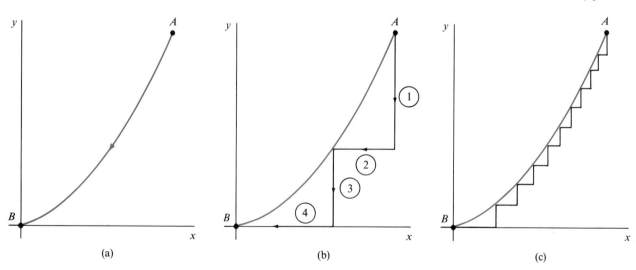

(a)   (b)   (c)

Figure 5.20
(a) An arbitrary frictionless path along which a mass m moves from A to B. (b) Vertical and horizontal paths connecting points A and B. (c) An increased number of vertical and horizontal paths between A and B.

the work done by the gravitational force is equal to zero. In general, the work done on a particle by a conservative force, such as the gravitational force, is equal to zero around any closed path.

A potential energy may be associated with each conservative force, and we may define each such potential energy in terms of work done by that conservative force. The potential energy U_A of an object at any point A is equal to the work done by the conservative force $\mathbf{F}_C$ as the object is displaced from point A *along any path* to an arbitrary reference position at which the potential energy is chosen to be equal to zero, or

$$U_A = \int_A^{\text{ref}} \mathbf{F}_C \cdot d\mathbf{s} \qquad (5\text{-}46)$$

In particular, because the gravitational force $\mathbf{F}_g$ is conservative, the gravitational potential energy U_g of an object at a vertical height h above an arbitrary reference height (where $U_g = 0$) is expressed by Equation (5-26), or

$$U_g = \int_h^{\text{ref}} \mathbf{F}_g \cdot d\mathbf{s}$$

The reference height at which $U_g = 0$ is arbitrary because choosing a different reference height is equivalent to adding equal quantities to both sides of Equation (5-41) or (5-44), expressions of the conservation of energy. In other words, only differences in potential energy are significant, so the choice of a zero position for potential energy is arbitrary.

Frictional forces are not conservative. Moving a particle around a closed path along which there is friction requires that negative work be done on the particle by the frictional force as the particle traverses each length of the path. Therefore, the work done by a frictional force around a closed path is negative. Even more important, the work done by a frictional force around a closed path is not zero. Then, because a conservative force does no work around a closed path, frictional forces cannot be conservative.

We may now state the principle of conservation of energy in a more general form than before. If, as the system evolves from its initial state to its final state, the work done by nonconservative forces is represented by W_{NC}, we may write

$$K_i + U_{gi} + U_{ei} + W_{NC} = K_f + U_{gf} + U_{ef} \qquad (5\text{-}47)$$

Here the elastic potential energy U_e has been included in the total mechanical energy. Equation (5-47) is a statement of the conservation of mechanical energy modified to account for the effects of nonconservative forces. The following points are pertinent:

1. The speeds used to determine the kinetic energies of particles are measured from an inertial frame. For most situations, this means that we measure speeds relative to the earth, which approximates an inertial frame.

2. Forces associated with potential energies are conservative. In particular, the gravitational force and forces associated with idealized springs are conservative. Because only changes in potential energies are significant, the reference position at which a potential energy is chosen to be equal to zero is arbitrary.

3. Frictional forces are nonconservative. The work by frictional forces changes the total mechanical energy of the system.

4. Constraint forces that do no work on the system do not change the total mechanical energy of the system.

A 2.0-kg block starts from rest at point A and slides to point B along a frictionless arc having a radius of 1.0 m, as shown in Figure 5.21. The block continues to slide along a horizontal surface, where the coefficient of kinetic friction between the block and surface is 0.20. An ideal spring with a spring constant of 10 N/m has one end fixed to a wall. In its equilibrium position, the other end of the spring is at point C, 2.0 m from B. If the block comes to rest at D, what is the distance d from C to D?

Consider the mechanical energy of the block at its initial and final positions (points A and D, respectively). At point A, the block is at rest, so $K_i = 0$. Choosing the level of the horizontal surface to be the zero of gravitational potential energy, we find that $U_{gi} = mgy_i = 19.6$ J. In this initial state, the spring is not compressed and, therefore, contributes nothing to the potential energy; thus, U_{ei} is equal to zero. In the final state, when the block has come to rest at point D, its kinetic energy is $K_f = 0$. Along the floor, and at D in particular, U_g is equal to zero. Let d be the positive distance that the spring is compressed from its equilibrium position, that is, the distance from C to D. The final potential energy of the spring U_{ef} is $\frac{1}{2}kd^2 = 5d^2$. One further term in Equation (5-47) remains to be considered: the work done by the nonconservative frictional force. While the block is on the horizontal surface, the frictional force $\mathbf{f}$ has a magnitude of 3.92 N and is toward the left. Then the work done by the frictional force is

$$W_{NC} = -(3.92)(2 + d) \qquad (5\text{-}48)$$

Notice that the displacement through which the force of friction acts includes the distance d from C to D. Substituting the values we have determined into Equation (5-47) gives

$$K_i + U_{gi} + U_{ei} + W_{NC} = K_f + U_{gf} + U_{ef}$$

$$0 + 19.6 + 0 + (-3.92)(2 + d) = 0 + 0 + 5d^2$$

$$5d^2 + 3.92d - 11.8 = 0 \qquad (5\text{-}49)$$

The roots of Equation (5-49) are $d = 1.2$ and $d = -1.2$. We specified d to be a positive distance, so the distance from C to D is 1.2 m.

E 5.30   Use the statement of conservation of mechanical energy and Figure 5.21 to determine the kinetic energy of the block at (a) position B, and (b) position C.   Answers: (a) 20 J, (b) 12 J

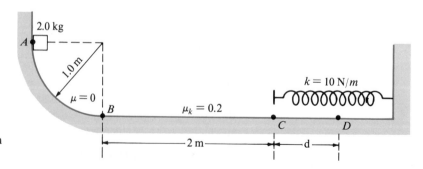

Figure 5.21
A 2.0-kg mass, which is released from rest from the position shown.

Example 5.6   PROBLEM   A vertical spring (k = 200 lb/ft) with its lower end resting on a fixed horizontal surface is compressed a distance d. An object of weight W is placed on top of the compressed spring, and the system is then released from rest. If d = 6.0 in. and W = 10 lb, calculate the speed of the object as it moves through the equilibrium position of the spring.

SOLUTION   The equilibrium, initial, and final states are shown in Figure 5.22. If the initial state is taken as the reference position (U_g = 0) for the gravitational potential energy, then the initial mechanical energy of the system is

$$K_i + U_{gi} + U_{ei} = \frac{1}{2}\frac{W}{g}(0)^2 + W(0) + \frac{1}{2}kd^2$$

Similarly, the final energy of the system is

$$K_f + U_{gf} + U_{ef} = \frac{1}{2}\frac{W}{g}v^2 + Wd + \frac{1}{2}k(0)^2$$

Since there are no dissipative forces, the initial and final energies are equal:

$$\frac{1}{2}kd^2 = \frac{1}{2}\frac{W}{g}v^2 + Wd$$

Solving for v^2 and then v gives

$$v^2 = \frac{gkd^2}{W} - 2gd$$

$$v^2 = \frac{(32)(200)(0.5)^2}{10} - 2(32)(0.5) = 128$$

$$v = 11 \text{ ft/s} \quad \blacksquare$$

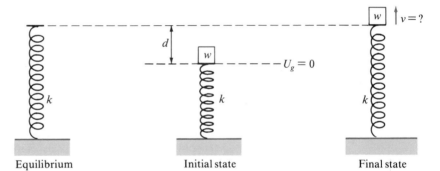

Figure 5.22   Example 5.6   Equilibrium   Initial state   Final state

Example 5.7   PROBLEM   A 4.0-kg object is released from rest at a height of 10,000 km above the surface of the earth. Calculate the kinetic energy of the body when it has fallen a vertical distance of 8000 km.

SOLUTION   The total energy of the initial state shown in Figure 5.23 is

$$K_i + U_{gi} = \frac{1}{2}(m)(0)^2 - \frac{GmM_e}{R_e + h} = -\frac{GmM_e}{R_e + h}$$

where we are given m = 4 kg, M_e = 6.0 × 10²⁴ kg, R_e = 6.4 × 10⁶ m, and h = 1.0 × 10⁷ m. For the final state, the total energy is

$$K_f + U_{gf} = K_f - \frac{GmM_e}{R_e + h - d}$$

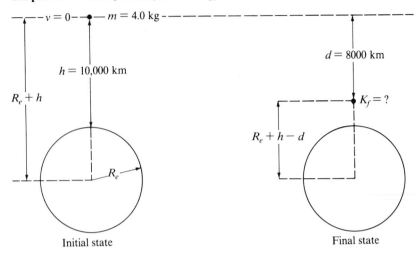

Figure 5.23   Example 5.7

where $d = 8.0 \times 10^6$ m. Because no dissipative forces act on the system, the initial and final energies are equal. We have then

$$-\frac{GmM_e}{R_e + h} = K_f - \frac{GmM_e}{R_e + h - d}$$

Solving for the final kinetic energy gives

$$K_f = \frac{GmM_e}{R_e + h - d} - \frac{GmM_e}{R_e + h} = \frac{GmM_e d}{(R_e + h - d)(R_e + h)}$$

$$K_f = \frac{(6.7 \times 10^{-11})(4.0)(6.0 \times 10^{24})(8.0 \times 10^6)}{(8.4 \times 10^6)(16.4 \times 10^6)} = 9.3 \times 10^7 \text{ J} \quad \blacksquare$$

In physical systems that involve more than a single particle, these principles of energy are still valid. But how may these multiple-particle systems be treated analytically? We may assert generally that for a system of interacting particles, the total kinetic energy K of the system is the sum of the kinetic energies of the individual particles of the system. Further, the total potential energy of the system is the sum of the potential energies associated with the configurations of the particles within the system. The energy conservation principle that we have seen applied to a single particle is also applicable to multiple-particle systems. In this book, only a few simple systems that involve the energetics of more than one particle are considered, and the computation of the potential energies in these systems is straightforward. The application of the principle of conservation of energy to simple systems with more than one particle is illustrated by the following example.

Example 5.8

PROBLEM   Figure 5.24(a) shows two objects of mass m and $3m$ connected by a string that passes over a pulley of negligible mass and friction. If this system is released from rest, how fast are the objects moving after they have each moved a distance of 1.2 m?

SOLUTION   The initial state of Figure 5.24(a) is chosen as the reference (zero) state for the gravitational potential energy of each object. Consequently, the initial total energy of the system is

$$K_i + U_{gi} = 0$$

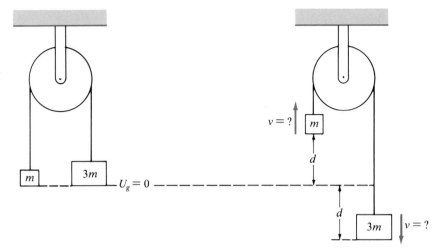

Figure 5.24   Example 5.8

The final state, when each object has moved a distance d, is shown in Figure 5.24(b). This state has a total energy given by

$$K + U_g = \frac{1}{2}mv^2 + \frac{1}{2}(3m)v^2 + mgd + 3mg(-d)$$

$$K + U_g = 2mv^2 - 2mgd$$

Notice that this energy includes the kinetic energy of each object, the positive potential energy (mgd) of the smaller mass that has risen a distance d above the reference position, and the negative potential energy ($-3mgd$) of the greater mass, which is below the reference position. Since no dissipative forces act on the system, the total energy remains constant. Thus we have

$$2mv^2 - 2mgd = 0$$

$$v^2 = gd$$

$$v = \sqrt{gd}$$

For $d = 1.2$ m, we get

$$v = \sqrt{(9.8 \text{ m/s}^2)(1.2 \text{ m})} = 3.4 \text{ m/s} \quad \blacksquare$$

Although using the conservation of energy principle is the most efficient way to solve this problem, the same solution can be obtained by using dynamics and kinematics. See if you can obtain the result by using dynamics and kinematics.

5.6 Problem-Solving Summary

Like many other physical concepts, work, power, and energy are most effectively used in solving problems by applying the definitions of these concepts strictly and carefully to the problem at hand.

The definition of work, $W_{a \to b} = \int_a^b \mathbf{F} \cdot d\mathbf{s}$, suggests a number of practical facts that are useful in problem-solving. First, work is done by a force $\mathbf{F}$ applied to an object when that object moves from point a to point b. It follows that no work can be done on an object unless the object moves. Furthermore, no work is done by $\mathbf{F}$ unless there is a component of $\mathbf{F}$ along the direction of the displacement of the object. Only the component of a force parallel to the displacement of an object can do work on that object. The integral in the definition of work suggests that the work done by the force $\mathbf{F}$ as the object moves from a to b is equal to the area enclosed by the curve of $F_s(s)$, the component of $\mathbf{F}$ parallel to ds; the s axis; and the lines $s = a$ and $s = b$. Simpler situations, like the frequently encountered case of a constant force acting on an object that moves in a straight line, say the x axis, are special cases of the general definition of work. For this case, the work by a constant force, $\mathbf{F} = F_x\hat{\mathbf{i}} + F_y\hat{\mathbf{j}}$, on an object moving along the x axis is

$$W_{a \to b} = \int_a^b F_x \, dx$$

$$W_{a \to b} = F_x(b - a)$$

A common error in computing work results from carelessness with the signs associated with the force applied to an object and the displacement of that object. The work $\mathbf{F} \cdot \mathbf{ds}$ is negative when the component of $\mathbf{F}$ parallel to the displacement of the object is oppositely directed from the displacement.

Because power P is the time rate of doing work, the power developed by a force doing work at any instant is dW/dt, whereas average power $\bar{P}$ is the work W done in a time interval Δt divided by that time interval, or $\bar{P} = W/\Delta t$. A measure of power that is frequently more useful, however, is in terms of the force $\mathbf{F}$ applied to an object moving at a velocity $\mathbf{v}$, that is, $P = \mathbf{F} \cdot \mathbf{v}$.

We have considered three forms of mechanical energy: kinetic energy, gravitational potential energy, and elastic potential energy. A clear understanding of the definition of each form of energy is a considerable aid in solving energy problems.

The kinetic energy of a body represents its ability to do work because of its motion. A body of mass m moving with speed v (in any direction) has kinetic energy K given by $K = \frac{1}{2}mv^2$. The gravitational potential energy U_g of a body measures the ability of that body to do work because of its position with respect to another body, usually the earth. Near the surface of the earth, an object of mass m has gravitational potential energy $U_g = mgh$, where h is a vertical distance measured from an arbitrary reference level $h = 0$ (at which U_g is chosen to be zero). The value of h is positive if the object is above $h = 0$ and negative if the object is below $h = 0$. An object of mass m that is not necessarily near the surface of the earth has a gravitational potential energy U_g associated with its distance r from the center of the earth given by $U_g = -(GmM_e)/r$, where M_e is the mass of the earth. But when is an object not near the surface of the earth? A rule of thumb is: If an object is 60 km (or 40 mi) above the surface of the earth, use $U_g = -(GmM_e)/r$ to be correct within approximately 1 percent (accurate to three significant figures). Finally, the elastic potential energy U_e stored in an ideal spring (one that obeys Hooke's law, $F_x = -kx$) is given by $U_e = \frac{1}{2}kd^2$, where d is the distance the spring is compressed or extended from its equilibrium position and k is the spring constant.

The most general definition of potential energy is written in terms of any conservative force $\mathbf{F}_C$. The potential energy U_A of an object at the point A is defined to be the work done by a conservative force $\mathbf{F}_C$ as the object moves from point A along any path to a reference position at which the potential energy is chosen to be zero:

$$U_A = \int_A^{ref} \mathbf{F}_C \cdot d\mathbf{s}$$

The work-energy principle relates the work done on an object by the resultant force on that object to the change in the kinetic energy of the object that takes place while that work is being done. This principle is often useful in problem-solving situations, especially when the speed of an object is to be determined after known forces have done work on the object. Care should be taken that the resultant (net) of all forces on the object is determined before calculating the work done on the object by the net force. This work done by the net force may then be equated to the change in the kinetic energy, $\frac{1}{2}mv_f^2 - \frac{1}{2}mv_i^2$, of the object.

The principle of conservation of energy is an effective tool in solving problems that may be identified by the following characteristics:

1. An object changes its vertical position and in doing so exchanges its gravitational potential energy for kinetic energy, or vice versa.

2. An object moves so as to compress or extend an ideal spring, thereby converting either its gravitational potential energy (proportional to its height) or its kinetic energy (proportional to the square of its speed), or both, into a change in the elastic potential energy of the spring.

3. A problem is presented in such a way that we can easily relate changes in speed (through kinetic energy) to changes in position (through potential energy). Thus, if an object changes position vertically and we are asked to find its change in speed, we should consider using the conservation of energy principle. On the other hand, problems that ask us to find accelerations or specific forces are probably more easily treated as dynamics problems.

Perhaps the first question that should be asked about a problem involving the principle of conservation of energy is, "Are nonconservative forces present?" If no nonconservative forces affect the problem, mechanical energy is conserved, and we may use

$$K_i + U_{gi} + U_{ei} = K_f + U_{gf} + U_{ef}$$

In using the conservation of energy relationship, remember that K and U_e are always positive values (or zero), whereas U_g may be a positive or negative value. When nonconservative forces act on the objects of interest in a problem, the conservation statement must be altered:

$$K_i + U_{gi} + U_{ei} + W_{NC} = K_f + U_{gf} + U_{ef}$$

In this case, W_{NC} is the work done by nonconservative forces as the system evolves from the initial to the final state. In this relationship, W_{NC} is usually a negative quantity (when, for example, frictional forces are the nonconservative forces in the problem). A negative value of W_{NC} reduces the initial energy of the system by the amount of energy lost to the work done by nonconservative forces.

Problems

GROUP A

5.1   A 0.30-kg object is thrown vertically from ground level and rises to a maximum height of 25 m. Calculate the initial kinetic energy of the object.

5.2   At what speed must a 0.40-kg object move if its kinetic energy is to equal its gain in gravitational potential energy when it is raised a vertical distance of 20 m?

5.3   A 5.0-kg body is accelerated from a speed of 8.0 m/s to a final speed of 14 m/s. What total work is done to accomplish this change in the motion of the body?

5.4   As a body moves along the x axis, the resultant force acting on it is given by

$$\sum \mathbf{F} = -9x^2 \hat{\mathbf{i}} \text{ N}$$

How much does the kinetic energy of the body change as it slides from $x = 2.0$ m to $x = 4.0$ m?

5.5   If a 4.0-lb object is thrown vertically upward with an initial speed of 35 ft/s, what is its kinetic energy after it has risen 12 ft?

5.6   A 2.0-kg object moves from position A to position B.

(a)   If positions A and B have the same height, how much work is done by the gravitational force of the earth on the object?

(b)   Does the answer to part (a) depend upon the path the object follows as it moves from position A to position B?

(c)   If point B is displaced 10 m horizontally and 8 m vertically (upward) from A, how much work is done by the gravitational force of the earth on the object as it moves from A to B?

5.7   A force acting on a body as it moves along the x axis is given by $(4x\hat{\mathbf{i}})$ N where x is measured in meters. Calculate the work done by this force as the body moves from

(a) $x = 1.0$ m to $x = 3.0$ m.

(b) $x = -1.0$ m to $x = 3.0$ m.

(c) $x = 2.0$ m to $x = 0$.

5.8   Sketch a graph of F_x versus x (-1 m $< x <$ 3 m) for the force given in Problem 5.7. Use this graph to verify the results of Problem 5.7.

5.9   If you push a 40-kg crate at a constant speed of 1.4 m/s across a horizontal floor ($\mu_k = 0.25$), at what rate is work being done on the crate by

(a)   you?

(b)   the frictional force?

5.10   The only horizontal forces acting on a 3.0-kg mass are $\mathbf{F}_1 = (8\hat{\mathbf{i}} + 6\hat{\mathbf{j}})$ N and $\mathbf{F}_2 = (-2\hat{\mathbf{i}} + 3\hat{\mathbf{j}})$ N. Here the unit vectors $\hat{\mathbf{i}}$ and $\hat{\mathbf{j}}$ both point in horizontal directions. If the mass is initially at rest, how fast will it be moving after it has traveled a horizontal distance of 2.0 m?

5.11   As a particle moves from the origin to $(3\hat{\mathbf{i}} - 4\hat{\mathbf{j}})$ m, it is acted upon by a constant force given by $(4\hat{\mathbf{i}} + 5\hat{\mathbf{j}})$ N. Calculate the work done by this force as the particle moves through the given displacement.

5.12   An object moving along the x axis is acted upon by the force shown in Figure 5.25.

(a)   Calculate the work done by this force as the object moves from $x = 2.0$ ft to $x = 12$ ft.

(b)   What constant force acting on the object would produce the same work as that obtained in part (a)?

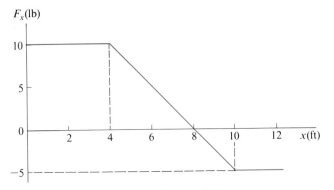

Figure 5.25   Problem 5.12

5.13   A 0.40-kg mass is suspended from the end of a string having a length of 1.5 m. If the mass is released from an initial position in which the string is horizontal, what is the kinetic energy of the mass at the lowest point of the swing?

5.14   The total work done on a 7.0-lb body during a 10-s time interval is 38 ft·lb. If the body is moving with a speed of 12 ft/s at the beginning of the time interval, what is its speed at the end of the 10-s interval?

5.15   If you pull a 20-kg crate across a horizontal surface at a constant speed of 1.5 m/s by exerting a constant force of 45 N directed 30° above the horizontal,

(a)   calculate the work you perform while moving the crate 12 m.

(b)   at what rate are you doing work on the crate?

5.16   An 0.80-kg body on a horizontal frictionless surface is attached to one end of a horizontal spring ($k = 1200$ N/m). The other end of the spring is held in a fixed position. If the body moves with a speed of 11 m/s through the equilibrium position, determine the maximum distance the spring is extended.

5.17 A 0.20-kg ball is thrown vertically from the ground with an initial speed of 20 m/s. Let y measure the height of the ball above the ground. Choose the gravitational potential energy of the ball to be zero at the ground.

(a) Write an expression for the gravitational potential energy as a function of y.

(b) Write an expression for the total mechanical energy as a function of y.

(c) Write an expression for the kinetic energy of the ball as a function of y. For what range of values of y is this expression valid?

(d) What is the maximum height above the ground attained by the ball?

5.18 A 6.0-lb body sliding on a horizontal frictionless surface is attached to one end of a horizontal spring ($k = 75$ lb/ft). The other end of the spring is held fixed. The 6.0-lb body is moved so as to stretch the spring 9.0 in., and the system is released from rest. What is the speed of the body as it passes through the equilibrium position?

5.19 A 150-kg rocket has an initial vertical speed of 5.0 km/s when its engine stops at a height of 200 km above the surface of the earth. What is the kinetic energy of the rocket when it is 1000 km above the surface of the earth?

5.20 A 25-kg projectile fired vertically from the surface of the earth rises to a maximum height of 4000 km. Disregarding any atmospheric air resistance, calculate the initial kinetic energy of the projectile.

5.21 When a pitcher releases a 5-ounce (5/16-lb) ball, it is traveling 100 ft/s. If 0.20 s elapses from the beginning of the throw until release of the ball, calculate

(a) the work the pitcher does on the ball.

(b) the average power the pitcher develops while throwing the ball.

5.22 A 1000-kg car accelerates from rest to 80 km/h in 15 s.

(a) Determine the work done on this car by the resultant force acting on it during this 15-s time interval.

(b) Calculate the average power (in kW and hp) developed by the resultant force on the car during this time interval.

5.23 A vertical spring ($k = 1200$ N/m) is compressed a distance d from its equilibrium length by a 15-kg mass, which is lowered until it is supported at rest by the spring.

(a) Calculate d.

(b) Calculate the elastic potential energy when the spring is supporting the mass.

(c) Calculate the change in gravitational potential energy as the mass is lowered through the distance d.

5.24 A 120-lb woman jogs up four flights of stairs in 20 s. As a result, she rises 50 ft above her initial position. At what average rate does she do work during this ascent?

5.25 A satellite follows a circular path as it orbits a spherically symmetric planet. Show that its kinetic energy K and the gravitational potential energy U_g of the satellite-planet system are related by

$$K = \frac{1}{2}|U_g|$$

5.26 A 4.0-kg object slides down a frictionless plane inclined 40° with the horizontal. Its initial velocity at the top of the plane is 6.0 m/s down the plane.

(a) Calculate the work by the gravitational force on the block as it slides 10 m down the inclined surface.

(b) What is the kinetic energy of the object after sliding 10 m down the surface?

5.27 A 6.0-lb object projected vertically rises 130 ft above its initial position.

(a) Use energy considerations to determine its initial speed.

(b) At what rate is the gravitational force on the projectile doing work at the instant it is 75 ft above the initial position and rising?

(c) Repeat part (b) for the instant when the projectile achieves its maximum height.

(d) Repeat part (b) for the instant when the projectile is 75 ft above its initial position and falling.

5.28 A 4.0-kg object is released from rest. After falling 10 m, its speed is 14 m/s. Calculate the work by the force of air resistance on the object during this 10-m fall.

5.29 A 15-lb block slides at a constant speed of 10 ft/s down a plane inclined 25° with the horizontal. At what rate is work being done by

(a) the gravitational force of the earth on the block?

(b) the frictional force on the block?

5.30 If an object is released from rest at a height of 4000 km above the surface of the earth, determine its speed, disregarding air resistance,

(a) at a height of 2000 km above the surface of the earth.

(b) just before hitting the surface of the earth.

5.31 A block weighing 3.0 lb is attached to one end of a horizontal spring ($k = 25$ lb/ft). The other end of the spring is fixed. The block slides on a horizontal frictionless surface, and its speed as it passes through the equilibrium position is 10 ft/s.

(a) What is the maximum extension of the spring?

(b) What is the speed of the block at an instant when the spring is compressed 0.50 ft?

(c) If x measures the displacement of the spring from equilibrium, write an equation expressing the kinetic energy of the block as a function of x.

5.32 Two objects are connected by a string that passes over a frictionless pulley of negligible mass (see Fig. 5.26). If the system is released from rest, what is the kinetic energy of the 15-lb object after it has fallen 4.0 ft?

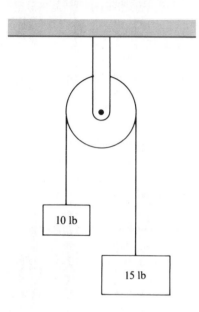

Figure 5.26   Problem 5.32

5.33   A particle from outer space is attracted to the earth from a very large, essentially infinite, distance. With what minimum speed will it strike the upper reaches of the atmosphere (100 km above the surface of the earth)?

5.34   A 0.60-kg mass at the end of a light rod swings, as shown in Figure 5.27, in a vertical circle that has a radius of 1.5 m. The mass is released from rest at position *A* with the rod horizontal.

(a)   What is the kinetic energy of the mass at position *B*?

(b)   At what rate is work being done by the gravitational force acting on the mass at position *B*?

(c)   What is the speed of the mass at position *C*?

(d)   What force does the rod exert on the mass at position *C*?

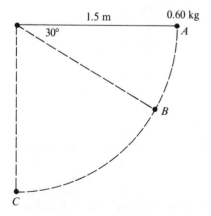

Figure 5.27   Problem 5.34

5.35   A 0.30-kg mass sliding on a horizontal frictionless surface is attached to one end of a horizontal spring ($k = 500$ N/m). The other end of the spring is held stationary. With the spring extended 0.20 m, the mass is released from rest.

(a)   What is the speed of the mass as it passes through the equilibrium position?

(b)   What is its speed when the spring is compressed 0.10 m?

(c)   At what rate is the spring doing work on the mass when the spring is compressed 0.10 m and the mass is moving away from the equilibrium position?

5.36   If **F** is the sum of two forces $\mathbf{F}_1$ and $\mathbf{F}_2$, show that the work done by **F** for any displacement equals the sum of the work done by $\mathbf{F}_1$ and $\mathbf{F}_2$ for the same displacement.

5.37   A 2.0-kg object moving along the *x* axis is acted upon by the force shown in Figure 5.28. The velocity of the object at $x = 0$ is $v_x = 5.0$ m/s. Calculate its

(a)   kinetic energy at $x = 3.0$ m.

(b)   speed at $x = 6.0$ m.

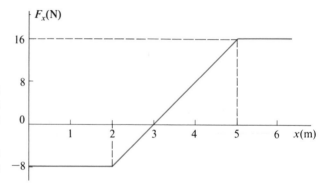

Figure 5.28   Problem 5.37

5.38   The pulley in Figure 5.29 is frictionless and has negligible mass, and the surfaces are frictionless. If the system is released from rest,

(a)   what is the combined kinetic energy of the two blocks after the 20-lb block slides 6.0 ft?

(b)   what is the speed of either block after the 20-lb block slides 6.0 ft?

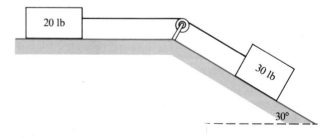

Figure 5.29   Problem 5.38

5.39 The initial speed of a 15-kg projectile launched vertically from the surface of the earth is 8.0 km/s.

(a) What is the kinetic energy of the projectile when it is a distance of 4000 km above the surface of the earth?

(b) Calculate the maximum height above the surface of the earth achieved by the projectile.

5.40 Estimate the magnitude of the force exerted on a car by air resistance if the car is moving at a constant speed of 55 mi/h (81 ft/s) and the engine is developing 100 hp. Assume that 50 percent of the energy that the engine expends is dissipated internally in the car.

5.41 The coefficient of kinetic friction between an 18-kg crate and the horizontal surface on which it slides is 0.20. The initial speed of the crate is 8.0 m/s.

(a) What is the kinetic energy of the crate after sliding 5.0 m?

(b) How far will the crate slide before stopping?

5.42

(a) Use $W_{a \to b} = \int_a^b \mathbf{F} \cdot d\mathbf{s}$ to show that any constant force is conservative.

(b) As an object slides along a uniform level surface, the force of kinetic friction has a constant magnitude. Why does it not follow from part (a) that this frictional force is conservative?

5.43 Two weights, $W_1 = 10$ lb and $W_2 = 20$ lb, are connected to a spring ($k = 20$ lb/ft) that has its lower end fixed (see Fig. 5.30). The surface on which W_1 slides is frictionless and inclined 30° with the horizontal. Disregard any pulley mass or friction. The system is released from rest with the spring at its equilibrium length.

(a) How far will W_2 fall before stopping?

(b) What is the kinetic energy of W_2 after it has fallen 1.0 ft?

(c) At an instant when W_2 has fallen 1.0 ft and is descending, at what rate is the spring doing work on W_1?

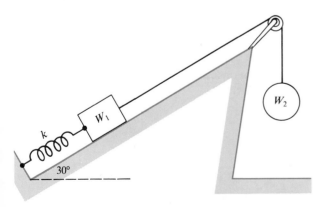

Figure 5.30   Problem 5.43

5.44 The two masses in Figure 5.31 are released from rest. After the 3.0-kg mass has fallen 1.5 m, it is moving with a speed of 4.0 m/s.

(a) What is the change in the potential energy of the two-mass system during this time interval?

(b) What is the change in the kinetic energy of the system during this time interval?

(c) Calculate the work done during this time interval by the frictional force on the 2.0-kg block.

(d) What is the magnitude of the frictional force on the 2.0-kg block?

(e) At what rate is work being done on the system by gravitational forces at an instant when the blocks are moving at a rate of 2.0 m/s?

Figure 5.31   Problem 5.44

5.45 Three objects are connected, as shown in Figure 5.32. The mass and friction of each pulley is negligible. The coefficients of friction between the 3.0-kg block and the horizontal surface are $\mu_s = 0.40$ and $\mu_k = 0.20$. If the system is released from rest and M is moving with a speed of 2.9 m/s after it has fallen 1.4 m, determine the mass M.

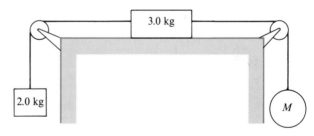

Figure 5.32   Problem 5.45

5.46 A 15-kg mass is dropped from a height of 5.0 m above the upper end of a vertical spring having its lower end sitting on a fixed horizontal surface. If the elastic constant for the spring is 1800 N/m,

(a) calculate the maximum compression of the spring.

(b) find the kinetic energy of the mass at the instant when the spring is compressed 0.50 m.

5.47 The system shown in Figure 5.33 is released from rest. Any friction in the axle of the pulley is negligible, but the mass of the pulley is not negligible. At the instant the 3.0-kg object has fallen 2.0 m, it has a speed of 2.6 m/s. What is the kinetic energy (of rotation) of the pulley at this instant?

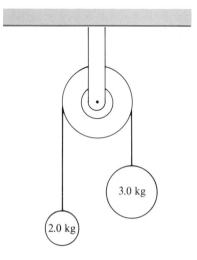

Figure 5.33   Problem 5.47

5.48 The horizontal surface on which the weight W slides is frictionless. The two springs in Figure 5.34 are neither stretched nor compressed at the equilibrium position. If $k = 40$ lb/ft and $W = 6.0$ lb, how far will the weight slide before stopping if it moves through the equilibrium position with a velocity of 20 ft/s toward the right?

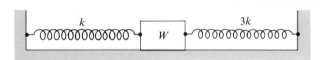

Figure 5.34   Problem 5.48

5.49 The 4.0-kg mass shown in Figure 5.35 slides on a horizontal frictionless surface. At the equilibrium position, neither of the springs is stretched or compressed. If the mass is moved to the right 0.20 m and released from rest,

(a) what is its initial acceleration?

(b) how fast is it moving when it passes through the equilibrium position?

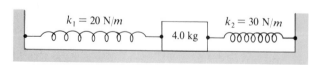

Figure 5.35   Problem 5.49

5.50 A 2.0-kg projectile is launched from ground level with an initial velocity of 150 m/s directed 30° above the horizontal.

(a) Use energy considerations to determine the maximum height of the projectile above the ground and its kinetic energy at that position.

(b) Suppose that as a result of air resistance the projectile actually reaches a maximum height of 240 m with a speed of 110 m/s. How much energy was dissipated by air resistance?

(c) If the same quantity of energy is dissipated in the descent of the projectile as in its ascent in part (b), with what speed will the projectile strike the ground?

5.51 A spring ($k = 600$ N/m) is arranged on a plane inclined 35° with the horizontal (*see* Fig. 5.36). A 1.5-kg mass is pushed against the spring until the spring is compressed 20 cm, and the system is then released from rest.

(a) If the surface is frictionless, how far will the mass slide before stopping?

(b) If $\mu_k = 0.20$ for the block and surface, repeat part (a).

(c) In part (b) how much energy is dissipated by the frictional force as the block slides from its initial position to its position of maximum height?

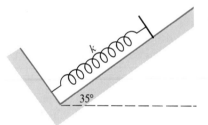

Figure 5.36   Problem 5.51

5.52 A mass m slides on a frictionless spherical surface of radius R, as shown in Figure 5.37. Its speed at its lowest position is v_0. If $m = 0.80$ kg, $R = 40$ cm, and $v_0 = 2.0$ m/s,

(a) what is its kinetic energy when α is 45°?

(b) what is the maximum value of α?

(c) write an expression for the speed of the mass as a function of α.

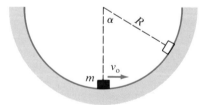

Figure 5.37   Problem 5.52

5.53 A mass m is attached to the top of a vertical spring that has an elastic constant k. The lower end of the spring is fixed. The mass is moved downward so as to compress the spring a distance d from its equilibrium length. This system is then released from rest. If m = 2.0 kg, k = 200 N/m, and d = 25 cm,

(a) what is the maximum height the mass attains above its lowest position?

(b) what is the maximum speed the mass achieves as it rises?

(c) write an expression for the kinetic energy of the mass as a function of y, its height above the lowest position.

5.54 A 20-lb block that is attached to one end of an ideal spring (k = 25 lb/ft) slides on a horizontal surface, as shown in Figure 5.38. The coefficients of friction between the block and surface have the values μ_s = 0.30 and μ_k = 0.20. Initially, the spring is at its equilibrium length, and the block is moving with a speed of 3.0 ft/s in a direction so as to extend the spring.

(a) How far will the block slide before coming to rest the first time?

(b) Will it start moving again?

(c) How fast will it be moving when it next passes through the equilibrium position?

(d) Where will the block stop and remain at rest?

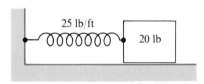

25 lb/ft

20 lb

Figure 5.38 Problem 5.54

GROUP **B**

5.55 A projectile is launched vertically with a speed v_o from the surface of a spherical planet of radius R. The free-fall acceleration near the surface of the planet is g. Show that

(a) the maximum height the projectile attains is given by

$$h = \frac{v_o^2/2g}{1 - (v_o^2/2gR)}$$

provided $v_o^2 < 2gR$.

(b) the result of part (a) reduces to the expected value for sufficiently small v_o, that is, for $v_o \ll \sqrt{2gR}$.

5.56 The *escape speed* v_e for a planet is defined as the minimum launch speed a projectile must be given at the surface of the planet such that the projectile will not return to the planet, that is, it "escapes" gravitational capture by the planet.

(a) Show that $v_e = \sqrt{2gR}$ where g is the surface free-fall acceleration for the planet and R is the radius of the planet.

(b) What is the total mechanical energy of a projectile if it is launched with escape speed?

(c) If E is the total mechanical energy of a projectile launched vertically, show that the projectile returns to the planet if $E < 0$ and escapes the planet if $E \geq 0$.

(d) If the initial vertical launch speed for the projectile is given by $v_o = \alpha v_e$ with $\alpha \geq 1$, show that, as the distance between the projectile and the planet becomes infinitely large, the projectile approaches a limiting (minimum) speed v_∞ given by $v_\infty = \sqrt{\alpha^2 - 1}\, v_e$.

5.57 Suppose a 40-kg rocket is launched vertically with a speed of 12 km/s from the surface of a uniform spherical planet. The mass and radius of the planet are equal to 3.0×10^{24} kg and 4.0×10^3 km.

(a) To what maximum height above the surface of the planet does the rocket rise?

(b) What is the minimum kinetic energy of the rocket during this flight?

(c) What is the minimum launch speed (the escape speed) for which the rocket will not return?

5.58 Consider an isolated system consisting of two particles that interact with each other but with no other objects. Assume that the interaction is conservative.

(a) Show that the rate at which work is being done on the two particles at any instant is

$$\frac{dW}{dt} = \mathbf{F}_{21} \cdot \mathbf{v}_1 + \mathbf{F}_{12} \cdot \mathbf{v}_2$$

where $\mathbf{F}_{21}$ is the force by particle 2 on particle 1, $\mathbf{v}_1$ is the velocity of particle 1, and so on.

(b) Show that at any instant

$$\frac{dU}{dt} = -\mathbf{F}_{21} \cdot \mathbf{v}_1 - \mathbf{F}_{12} \cdot \mathbf{v}_2$$

where U is the potential energy of the system.

(c) Show that

$$\frac{dK_1}{dt} = \mathbf{F}_{21} \cdot \mathbf{v}_1$$

$$\frac{dK_2}{dt} = \mathbf{F}_{12} \cdot \mathbf{v}_2$$

where K_j is the kinetic energy of particle $j (j = 1,2)$.

(d) Finally, show that the total mechanical energy E, defined by $E = K_1 + K_2 + U$, does not change with time, that is, the total energy of the system is conserved.

(e) Can you extend this treatment to a three-particle system?

5.59 Two planetoids, each having a mass of 5.0×10^{20} kg, are initially at rest and are separated by a distance of 2.0×10^6 m. If the two planetoids are in outer space so that the only force on either planetoid is the gravitational attraction by the other, calculate the speed of either when they are 1.0×10^6 m apart. Estimate the time that elapses during this acceleration.

5.60 Two 2.0-kg blocks connected by a spring ($k = 800$ N/m) are released from rest on a horizontal frictionless surface with the spring extended 0.25 m from its equilibrium length.

(a) Calculate the magnitude of the initial acceleration of either mass.

(b) Determine the speed of either mass at an instant when the spring is compressed 0.10 m from its equilibrium length.

5.61 Three identical ideal springs ($k = 800$ N/m) of negligible mass and two blocks of equal mass ($m = 2.0$ kg) are arranged as shown in Figure 5.39. The blocks slide on a horizontal frictionless surface. In the equilibrium position, each spring is neither stretched nor compressed. Suppose that the two blocks are each moved 5.0 cm so as to increase their separation by a distance of 10 cm, and the system is then released from rest in this configuration.

(a) What is the magnitude of the initial acceleration of either block?

(b) Calculate the speed of either block as it passes through its equilibrium position.

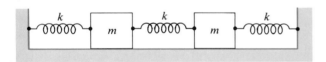

Figure 5.39   Problem 5.61

5.62 The potential energy U associated with a conservative force $\mathbf{F}_C$ can be calculated from Equation (5-46)

$$U_A = \int_A^{\text{ref}} \mathbf{F}_C \cdot d\mathbf{s}$$

This problem seeks to find the inverse relationship, that is, given the potential energy function U, how can the force be determined? Suppose the force $F_x\hat{\mathbf{i}}$ acts on a particle executing one-dimensional motion along the x axis.

(a) Show that if F_x can be expressed as a (integrable) function of x, that is, $F_x = F_x(x)$, then $F_x\hat{\mathbf{i}}$ is conservative.

(b) If $U(x)$ is the potential energy corresponding to the conservative force $F_x(x)\hat{\mathbf{i}}$, show that

$$U(x) = \int_x^{x_0} F_x(x')dx'$$

where $U(x_0) = 0$.

(c) Use the mean-value theorem for integrals, which is $\int_a^b f(x')dx' = f(\tau)(b - a)$ with $a \le \tau \le b$, to show that

$$U(x + \Delta x) - U(x) = -F_x(\tau)\Delta x; \qquad x \le \tau \le x + \Delta x$$

(d) Show that

$$F_x(x) = -\frac{dU}{dx}$$

which is, of course, just the inverse relationship to

$$U(x) = \int_x^{x_0} F_x(x')dx'$$

(e) Verify directly by differentiation that $F_x = -dU/dx$ for the gravitational potential energy $U_g(y) = mgy$ and for the elastic potential energy $U_e(k) = \frac{1}{2}kx^2$.

(f) Part (d) can be generalized to three dimensions to get

$$F_x(x,y,z) = -\frac{\partial U}{\partial x}, \; F_y(x,y,z) = -\frac{\partial U}{\partial y}, F_z(x,y,z) = -\frac{\partial U}{\partial z}$$

See if you can demonstrate this result starting with Equation (5-39).

5.63 Suppose that the only interaction for a particle executing one-dimensional motion along the x axis is given by the potential energy $U(x)$ shown in Figure 5.40(a). As the particle moves, its total energy E remains constant, as shown in Figure 5.40(b). The kinetic energy K of the particle is given as a function of x by

$$K(x) = E - U(x)$$

Figure 5.40(c) shows $K(x)$ for the total energy E in Figure 5.40(b). The positions $x = x_1$ and $x = x_2$, shown in Figure 5.40, at which the potential energy is equal to the total energy are called *turning points* of the motion.

(a) Describe the subsequent motion of a particle that has total energy E if the particle is initially at a position $x(x_1 < x < x_2)$ moving in the positive direction. Pay careful attention to what happens at $x = x_2$. Why are the positions x_1 and x_2 called "turning points"?

(b) Consider the motion of a 2.0-kg mass sliding on a horizontal frictionless surface attached to one end of a horizontal spring ($k = 50$ N/m). The other end of the spring is held fixed. If the velocity of the mass at $x = 0$ (the equilibrium position) is given by $v_x = 10$ m/s, construct a graph for this motion corresponding to Figure 5.40(c).

(c) For the mass-spring system of part (b), where are the turning points? Show that the velocity of the mass is given by

$$v_x(x) = (\pm 5\sqrt{4 - x^2}) \text{ m/s}$$

What is the physical significance of the $\pm$ sign of this result?

5.64 For a particle executing one-dimensional motion, a position $x = x_1$ is called a *position of equilibrium* if the force on the particle at that point is zero. A position $x = x_2$ is called a *position*

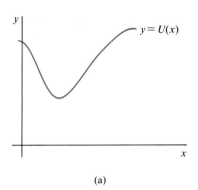

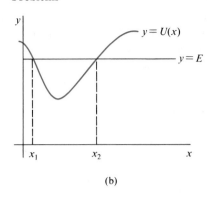

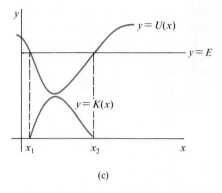

(a)   (b)   (c)

Figure 5.40   Problem 5.63

of stable equilibrium if x_2 is a position of equilibrium and if for all arbitrary small displacements from this position, the force on the particle is toward the equilibrium position, that is, the force is a *restoring force*. A position $x = x_3$ is called a *position of unstable equilibrium* if x_3 is a position of equilibrium and if for all arbitrary small displacements from this position, the force on the particle is away from the equilibrium position, that is, the force is a *destabilizing force*.

(a)   Show that if x_1 is a position of equilibrium, then it follows that $F_x(x_1) = 0$ and

$$\left(\frac{dU}{dx}\right)_{x=x_1} = 0$$

(b)   Show that if x_2 is a position of stable equilibrium, then for x near x_2,

$$F_x(x) < 0 \qquad \text{for } x > x_2$$
$$F_x(x) > 0 \qquad \text{for } x < x_2$$
$$U(x) > U(x_2) \qquad \text{for } x \neq x_2$$

(c)   Show that if x_3 is a position of unstable equilibrium, then for x near x_3,

$$F_x(x) > 0 \qquad \text{for } x > x_3$$
$$F_x(x) < 0 \qquad \text{for } x < x_3$$
$$U(x) < U(x_3) \qquad \text{for } x \neq x_3$$

(d)   Positions of stable and unstable equilibrium are depicted graphically in Figure 5.41. Study these carefully and be sure you understand how the graphs illustrate parts (b) and (c).

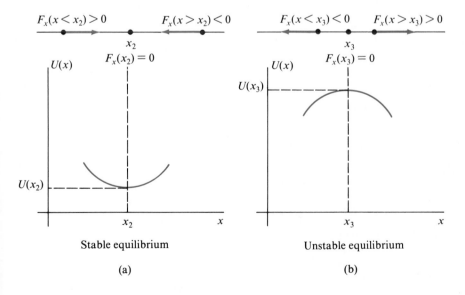

Stable equilibrium   Unstable equilibrium

(a)   (b)

Figure 5.41   Problem 5.64

5.65 A mass m attached to a light rod of length L rotates in a vertical circle centered on point P, as shown in Figure 5.42. Let s measure the arc length along the circle away from the lowest point, as shown, with $s > 0$ for CCW displacement and $s > 0$ for CW displacement.

(a) Show that the gravitational potential energy of this system is given by

$$U(s) = mgL\left(1 - \cos\frac{s}{L}\right)$$

(b) Sketch a graph of $U(s)$ for $-3\pi L < s < 2\pi L$.

(c) What values of s correspond to positions of stable equilibrium?

(d) What values of s correspond to positions of unstable equilibrium?

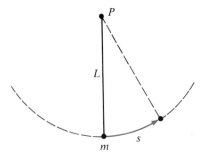

Figure 5.42   Problem 5.65

5.66 A particle is said to be in a *bound state* if its motion is localized to some finite region of space. Alternatively, if the energy of the particle is such that it can move an unlimited distance away from its initial position, the particle is said to be in a *free state*. Consider the potential energy function $U(x)$ shown in Figure 5.43. Let E be the total energy of a particle with a motion that is determined by $U(x)$.

(a) Show that values of E such that $U_o \leq E \leq U_1$ correspond to bound states.

(b) Show that values of E such that $U_1 < E < U_2$ correspond to free states.

(c) Show that values of E such that $U_2 < E$ correspond to free states.

(d) What is the essential physical difference between the free states of part (b) and those of part (c)?

(e) The *binding energy* E_B for a position of stable equilibrium is equal to the minimum kinetic energy a particle must possess at that position if it is to be in a free state.

For $U(x)$ of Figure 5.43, express E_B for the position of stable equilibrium $x = x_o$ in terms of U_0, U_1, and U_2.

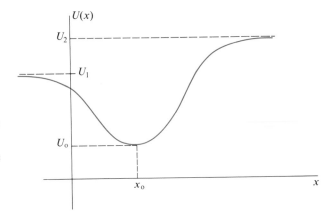

Figure 5.43   Problem 5.66

5.67 Consider the one-dimensional potential energy function

$$U(x) = U_0\left(\frac{x^2}{a^2}\right) \qquad -\infty < x < \infty$$

where U_0 and a are positive constants.

(a) Determine the force $F_x(x)$ corresponding to this potential energy function.

(b) Sketch a graph of $U(x)$ as a function of x.

(c) Determine all values of x that correspond to positions of stable or unstable equilibrium.

(d) If E is the total energy of a particle with a motion that is governed by $U(x)$, determine the ranges of values of E that correspond to bound states or to free states.

(e) For each position of stable equilibrium and values of E that correspond to bound states for this position of stable equilibrium, determine the turning points in terms of E, U_0, and a.

(f) For each position of stable equilibrium, determine the binding energy E_B for that position in terms of U_0.

5.68 Repeat Problem 5.67 for $U(x) = U_0(x^2/a^2 - 2x^4/a^4)$ and $-\infty < x < \infty$.

5.69 Repeat Problem 5.67 for $U(x) = U_0(a/x + x/a - 2)$ and $0 < x < \infty$.

5.70 Repeat Problem 5.67 for $U(x) = -U_0 e^{-x^2/a^2}$ where $-\infty < x < \infty$.

6 Momentum and Collisions

This chapter continues our focus on conservation principles. Here the conserved quantity is momentum. Like energy, which was discussed in Chapter 5, momentum is a composite of fundamental quantities, in this case, mass and velocity. In this chapter we will define momentum and develop the conservation principle associated with it. We will specify the conditions under which momentum is conserved and illustrate the usefulness of the conservation of momentum. In particular, we will use momentum conservation to investigate collisions and explosions, interactions within systems that involve more than one particle. We will also introduce the concept of impulse, which is associated with changes in momentum.

Before we consider momentum and its associated conservation principle, it is convenient to introduce the notion of center of mass. The concept of center of mass is immediately applicable to situations that are appropriately treated by considerations of momentum.

6.1 Center of Mass

If you throw a baton, dumbbell, or bowling pin, you can see that its motion is complicated (*see* Fig. 6.1). But associated with that body there is one point, known as the *center of mass* of that body, that moves along a simple parabolic path. The center of mass of an object translates in response to forces acting on it as if the object were a single particle located at the center of mass of the object.

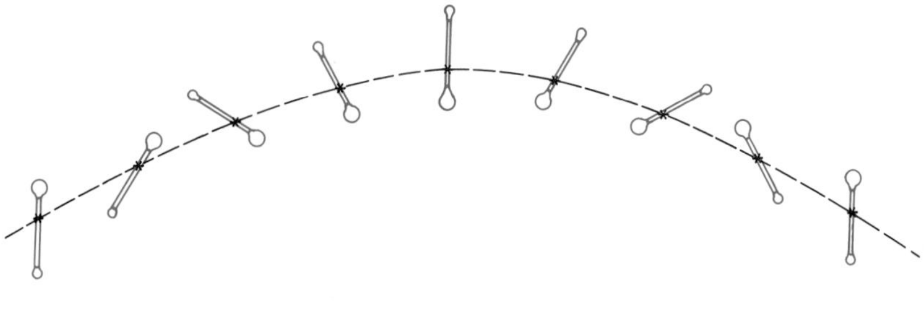

Figure 6.1
A baton with unequal masses on each end. One point associated with the baton follows the trajectory of a single particle.

The translational motion of any system of masses can, therefore, be treated as if that system were a single particle with a mass equal to the total mass of the system and with a position at the center of mass of the system. Before we prove this statement, let us see how the center of mass is defined mathematically and how we may calculate its location in a given system of masses.

An extended body is composed of many individual particles. Let us, therefore, consider first a system of particles. We define the center of mass of a collection of n particles, each with mass m_i and each located in a reference frame at (x_i, y_i, z_i), to be the point (x_{cm}, y_{cm}, z_{cm}) specified by

$$x_{cm} = \frac{\sum_{i=1}^{n} m_i x_i}{\sum_{i=1}^{n} m_i} = \frac{\sum_{i=1}^{n} m_i x_i}{M} \tag{6-1}$$

$$y_{cm} = \frac{\sum_{i=1}^{n} m_i y_i}{\sum_{i=1}^{n} m_i} = \frac{\sum_{i=1}^{n} m_i y_i}{M} \tag{6-2}$$

$$z_{cm} = \frac{\sum_{i=1}^{n} m_i z_i}{\sum_{i=1}^{n} m_i} = \frac{\sum_{i=1}^{n} m_i z_i}{M} \tag{6-3}$$

where M is the total mass of the system of particles. The following example illustrates the use of these equations.

Where on the line between the earth and the moon is the center of mass of the earth-moon system?

Because only two bodies are involved, the problem is one dimensional, and a single coordinate will suffice to specify the location of the center of mass of the two-particle system. In calculations of center of mass, the choice of a coordinate system is arbitrary. Let us then choose the center of the earth as the origin. The moon is located at the mean earth-moon distance, 3.8×10^8 m, from the center of the earth. Then with the earth at $x = 0$ and the moon at $x_m = 3.8 \times 10^8$ m, as shown in Figure 6.2, and using the mass of the earth $M_e = 6.0 \times 10^{24}$ kg and the mass of the moon $M_m = 7.4 \times 10^{22}$ kg, we may use Equation (6-1) to locate the earth-moon center of mass. According to Equation (6-1), we must sum over two particles:

$$x_{cm} = \frac{\sum_{i=1}^{2} m_i x_i}{\sum_{i=1}^{2} m_i} = \frac{M_e(0) + M_m x_m}{M_e + M_m} = \frac{0 + (7.4 \times 10^{22})(3.8 \times 10^8)}{6.0 \times 10^{24} + 7.4 \times 10^{22}}$$

$$x_{cm} = 4.6 \times 10^6 \text{ m}$$

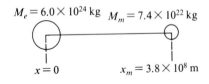

$M_e = 6.0 \times 10^{24}$ kg $M_m = 7.4 \times 10^{22}$ kg

$x = 0$ $x_m = 3.8 \times 10^8$ m

Figure 6.2
Diagram of the earth-moon system. The coordinate origin, $x = 0$, is at the center of the earth.

Because the radius of the earth is 6.4×10^6 m, we see that the center of mass of the earth-moon system lies within the earth. It is about this point, nearly 2 million meters below the surface of the earth, that the earth and the moon spin as this point revolves in a smooth trajectory about the sun.

Now look at a two-dimensional system, a set of particles lying in a plane.

Locate the center of mass of three particles, $m_1 = 2.0$ kg, $m_2 = 4.0$ kg, and $m_3 = 6.0$ kg, located at (3, 4), (−2, −1), and (4, −3), respectively, where the coordinates are in meters.

Equations (6-1) and (6-2) specify the coordinates of the center of mass of the three particles:

$$x_{cm} = \frac{\displaystyle\sum_{i=1}^{3} m_i x_i}{\displaystyle\sum_{i=1}^{3} m_i} = \frac{(2)(3) + (4)(-2) + (6)(4)}{2 + 4 + 6} = \frac{22}{12} = 1.8 \text{ m} \qquad (6\text{-}4)$$

$$y_{cm} = \frac{\displaystyle\sum_{i=1}^{3} m_i y_i}{\displaystyle\sum_{i=1}^{3} m_i} = \frac{(2)(4) + (4)(-1) + (6)(-3)}{2 + 4 + 6} = -\frac{14}{12} = -1.2 \text{ m} \qquad (6\text{-}5)$$

This point (x_{cm}, y_{cm}) is shown in Figure 6.3 as the point labeled c.m.

It should be clear from the foregoing examples how the distribution of mass affects the location of the center of mass of a collection of particles. It should be obvious, for example, that the center of mass of two particles of equal mass is halfway between them on a line passing through them. If one mass is the greater of the two, their center of mass lies closer to the more massive one.

E 6.1   Two bodies, $m_1 = 0.30$ kg and $m_2 = 0.50$ kg, are placed on the x axis so that $x_1 = 2.0$ m and $x_2 = -3.0$ m. Where should a body, $m_3 = 0.3$ kg, be placed on the x axis so that the center of mass of this system is located at the origin?   Answer: $x_3 = 3.0$ m

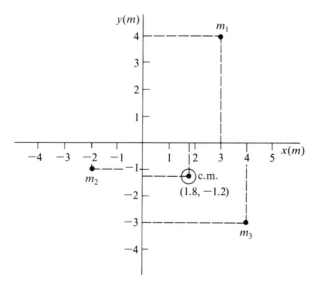

Figure 6.3
Three particles with c.m. as the center of mass.

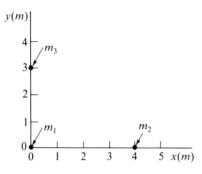

Exercise 6.2

E 6.2   If $m_1 = 0.20$ kg, $m_2 = 0.30$ kg, and $m_3 = 0.50$ kg, calculate the position of the center of mass of the system (*see* diagram).

Answer: $x_{cm} = 1.2$ m, $y_{cm} = 1.5$ m)

The center of mass of a continuous body may be determined by treating the body as if it were composed of infinitesimal masses dm, each at a point (x, y, z). The center of mass of the body is then calculated by a summation (integration) process:

$$x_{cm} = \frac{\int x \, dm}{\int dm} = \frac{\int x \, dm}{M} \tag{6-6}$$

$$y_{cm} = \frac{\int y \, dm}{\int dm} = \frac{\int y \, dm}{M} \tag{6-7}$$

$$z_{cm} = \frac{\int z \, dm}{\int dm} = \frac{\int z \, dm}{M} \tag{6-8}$$

Each integration in Equations (6-6) through (6-8) is carried out so as to include all of the body. Thus $\int dm$ is equal to the total mass M of the body. Problems in Group B at the end of this chapter indicate how such calculations are made. One important fact about the centers of mass of continuous bodies is particularly useful: The center of mass of a symmetrical body of uniform density lies at its geometrical center.

We have asserted that the center of mass of a system of particles responds to external forces acting on the system as if all of its mass were located at the center of mass and as if the net external force on the system were applied at the center of mass. That statement is equivalent to saying that we may treat a system of particles in dynamical situations as if the system were a single particle located at the center of mass of the system with a mass equal to the total mass of the system. For simplicity, let us demonstrate the validity of this assertion for a two-particle system. The following analysis may be extended to an arbitrary number of particles.

Let two particles with masses m_1 and m_2 be located at positions $\mathbf{r}_1$ and $\mathbf{r}_2$ measured from an arbitrary origin. The center of mass $\mathbf{r}_{cm}$ of the two-particle system is located at

$$\mathbf{r}_{cm} = \frac{m_1\mathbf{r}_1 + m_2\,\mathbf{r}_2}{M} \tag{6-9}$$

where $M = m_1 + m_2$ is the total mass of the system. The velocity $\mathbf{v}_{cm}$ of the center of mass may be obtained by taking the time derivative of Equation (6-9):

$$\mathbf{v}_{cm} = \frac{d\mathbf{r}_{cm}}{dt} = \frac{m_1 \dfrac{d\mathbf{r}_1}{dt} + m_2 \dfrac{d\mathbf{r}_2}{dt}}{M} = \frac{m_1\mathbf{v}_1 + m_2\mathbf{v}_2}{M} \tag{6-10}$$

Similarly, the acceleration $\mathbf{a}_{cm}$ of the center of mass is

$$\mathbf{a}_{cm} = \frac{d\mathbf{v}_{cm}}{dt} = \frac{m_1 \dfrac{d\mathbf{v}_1}{dt} + m_2 \dfrac{d\mathbf{v}_2}{dt}}{M} = \frac{m_1\mathbf{a}_1 + m_2\mathbf{a}_2}{M} \qquad (6\text{-}11)$$

from which we many obtain

$$M\mathbf{a}_{cm} = m_1\mathbf{a}_1 + m_2\mathbf{a}_2 = \Sigma\mathbf{F}_1 + \Sigma\mathbf{F}_2 = \Sigma\mathbf{F} \qquad (6\text{-}12)$$

Here, $\Sigma\mathbf{F}_1$ is the resultant of all forces acting on m_1, and $\Sigma\mathbf{F}_2$ is the resultant of all forces acting on m_2; $\Sigma\mathbf{F}$, therefore, represents the net force acting on the system. The net force $\Sigma\mathbf{F}_1$ on m_1 includes both the internal forces $\Sigma\mathbf{F}_{21}$ (any forces exerted on m_1 by m_2) and any external forces $\Sigma\mathbf{F}_{1,ext}$ exerted on m_1 by agents outside the system. In a like manner, the net force on m_2 includes the internal forces $\Sigma\mathbf{F}_{12}$ and the external forces $\Sigma\mathbf{F}_{2,ext}$. Then Equation (6-12) may be written

$$M\mathbf{a}_{cm} = \sum\mathbf{F}_{21} + \sum\mathbf{F}_{1,ext} + \sum\mathbf{F}_{12} + \sum\mathbf{F}_{2,ext} \qquad (6\text{-}13)$$

But Newton's third law assures us that $\Sigma\mathbf{F}_{12} = -\Sigma\mathbf{F}_{21}$. Then Equation (6-13) becomes

$$M\mathbf{a}_{cm} = \sum\mathbf{F}_{1,ext} + \sum\mathbf{F}_{2ext} = \sum\mathbf{F}_{ext} \qquad (6\text{-}14)$$

where $\Sigma\mathbf{F}_{ext}$ represents the net external force acting on the system. Equation (6-14) is in the form of Newton's second law for a particle of mass M. Thus, *the center of mass of a system of particles having a total mass M responds to the resultant force acting on the system exactly as would a particle of mass M located at the center of mass of the system.* This statement justifies our treating any physical body as if it were a particle when we consider its translational kinematics or dynamics.

E 6.3   At a specified time, two objects ($m_1 = 2.0$ kg, $m_2 = 3.0$ kg) have velocities $\mathbf{v}_1 = (4.0\hat{\mathbf{i}} + 8.0\hat{\mathbf{j}})$ m/s and $\mathbf{v}_2 = (-6.0\hat{\mathbf{i}} + 3.0\hat{\mathbf{j}})$ m/s. What is the velocity of the center of mass of this system at this instant?

Answer: $(-2.0\hat{\mathbf{i}} + 5.0\hat{\mathbf{j}})$ m/s

E 6.4   What is the acceleration of the center of mass of a two-particle system ($m_1 = 0.20$ kg, $m_2 = 0.30$ kg) at an instant when the resultant force on m_1 is $(5.0\hat{\mathbf{i}} - 3.0\hat{\mathbf{j}})$ N and the resultant force on m_2 is $(3.0\hat{\mathbf{i}} + 7.0\hat{\mathbf{j}})$ N?

Answer: $(16\hat{\mathbf{i}} + 8.0\hat{\mathbf{j}})$ m/s^2

6.2 Conservation of Linear Momentum

The *linear momentum* $\mathbf{p}$ of a particle of mass m moving at a velocity $\mathbf{v}$ is defined to be the product of m and $\mathbf{v}$:

$$\mathbf{p} = m\mathbf{v} \qquad (6\text{-}15)$$

Because the product of m and $\mathbf{v}$ is the product of a scalar and a vector quantity, $\mathbf{p}$

is a vector quantity. Further, the direction of the momentum $\mathbf{p}$ is the same as that of the velocity $\mathbf{v}$ of the particle.

Equation (6-15) defines linear momentum, and the physical significance of momentum is apparent from the definition. Newton, in his writings on natural science, referred to momentum as "quantity of motion"—not a bad insight into the concept of momentum. We can see, for example, that a fly buzzing along at a brisk 20 mi/h does not have much momentum because its mass is small. It would not be too difficult to stop the fly. On the other hand, a freight train moving at a mere 2.0 mi/h or a 5.0-g bullet moving at 300 m/s each has considerable momentum, and either would be difficult to stop; but a battleship at rest has no momentum at all. An understanding of the physical significance of momentum is most simply obtained by using the definition, $\mathbf{p} = m\mathbf{v}$.

Like energy, linear momentum is an important physical quantity because we can associate a conservation principle with it. Unlike energy, momentum has units that are not given a special name in either system of units. Therefore, the units of momentum are kg·m/s and slug·ft/s.

In order to consider the conservation principle associated with momentum, let us reconsider Newton's second law, $\Sigma\mathbf{F} = m\mathbf{a}$, in which m may be the mass of a single particle. In light of our discussion of center of mass, we may let M represent the mass of a system of particles in which the center of mass has acceleration $\mathbf{a}_{cm}$ and write $\Sigma\mathbf{F} = M\mathbf{a}_{cm}$. In either case, $\Sigma\mathbf{F}$ is the net external force acting on the particle or on the system. Let us now define the *momentum* $\mathbf{P}$ *of a system of particles* to be the vector sum of the momenta $\mathbf{p}$ of the individual particles within the system. From Equation (6-10), the momentum $\mathbf{P}$ of the system is given by

$$\mathbf{P} = M\mathbf{v}_{cm} \qquad (6\text{-}16)$$

Then we may state the second law in terms of linear momentum for a particle of mass m or a system of mass M:

$$\sum\mathbf{F} = m\mathbf{a} = m\frac{d\mathbf{v}}{dt} = \frac{d(m\mathbf{v})}{dt} = \frac{d\mathbf{p}}{dt} \qquad (6\text{-}17)$$

$$\sum\mathbf{F} = M\mathbf{a}_{cm} = M\frac{d\mathbf{v}_{cm}}{dt} = \frac{d(M\mathbf{v}_{cm})}{dt} = \frac{d\mathbf{P}}{dt} \qquad (6\text{-}18)$$

In both Equations (6-17) and (6-18) the next to last step is appropriate only if the particle or the system of particles has a constant mass. The case of constant mass is the only one we shall consider here. Interesting and instructive examples of variable mass situations are considered in problems of Group B at the end of the chapter.

It should now be clear that a single particle is a one-particle system. In other words, Equation (6-17) for a single particle is identical to Equation (6-18) because the single-particle mass m is equal to the system mass M and $\mathbf{p}$ is equal to $\mathbf{P}$.

E 6.5   At an instant when the velocity and acceleration of a 0.30-kg body are $\mathbf{v} = (4.0\hat{\mathbf{i}} + 6.0\hat{\mathbf{j}})$ m/s and $\mathbf{a} = (-3.0\hat{\mathbf{i}} + 4.0\hat{\mathbf{j}})$ m/s^2, calculate its momentum and the rate at which its momentum is changing.
    Answer: $\mathbf{p} = (1.2\hat{\mathbf{i}} + 1.8\hat{\mathbf{j}})$ kg·m/s, $d\mathbf{p}/dt = (-0.9\hat{\mathbf{i}} + 1.2\hat{\mathbf{j}})$ kg·m/s^2

E 6.6   Two interacting particles are isolated from any other interactions.

At an instant when the force by particle 2 on particle 1 is $(4.0\hat{\mathbf{i}} - 5.0\hat{\mathbf{j}})$ N, at what rate is

(a) the momentum of particle 1 changing?   Answer: $(4.0\hat{\mathbf{i}} - 5.0\hat{\mathbf{j}})$ kg·m/s^2

(b) the momentum of particle 2 changing?   Answer: $(-4.0\hat{\mathbf{i}} + 5.0\hat{\mathbf{j}})$ kg·m/s^2

(c) the momentum of the system changing?   Answer: zero

The conservation of linear momentum follows from Equation (6-18) for the special case in which the net force on a system is equal to zero. In that case, when $\Sigma\mathbf{F} = \mathbf{0}$, Equation (6-18) becomes

$$\frac{d\mathbf{P}}{dt} = \mathbf{0} \tag{6-19}$$

Equation (6-19) may be interpreted mathematically: If the derivative of a quantity is zero, that quantity is a constant. Physically, Equation (6-19) says the time rate at which the momentum $\mathbf{P}$ is changing is equal to zero; therefore, $\mathbf{P}$ is not changing. This, then, is the *principle of conservation of linear momentum:*

The linear momentum of a system of particles is conserved, or remains the same, as long as no net external force acts on that system.

It follows from the momentum conservation principle that the momentum of a system of particles will not change unless a net external force is applied to that system.

Within a system of particles, there may be internal forces acting among the particles. But Newton's third law ensures that for every internal force acting on a particle of the system, there is another force, equal in magnitude and opposite in direction, that acts on another particle within the system. Therefore, the sum of all the internal forces within a system is equal to zero. Consequently, the motion of the center of mass of the system is not affected by internal forces. Only an external force can alter the motion of the center of mass of a system. For example, when a system of particles, like a bomb, explodes somewhere along its trajectory, the explosion consists entirely of internal forces. If we neglect air resistance on the fragments of the bomb, no external forces that were not on the bomb before the explosion act on the fragments while they are airborne. Thus, the center of mass of the bomb continues on the same trajectory it followed before the explosion. Similarly, a person walking eastward along a floating boat causes the boat to move in a westward direction relative to the earth. Nevertheless, the motion of the center of mass of the person-boat system is unaffected if no net force acts on the system. (The person would have to walk slowly so that the resistance of the water on the boat is a negligible external force.)

The following problem illustrates the use of both conservation of linear momentum and the concept of center of mass.

In outer space, where no appreciable gravitational forces act, an astronaut in her life-support suit has a mass of 100 kg (*see* Fig. 6.4). She and a 200-kg fuel tank are at rest immediately above the antenna of her spaceship. She is attached to the fuel tank by a 50-m rope. She pushes away from the fuel tank. How far from the antenna can she move? What will be her velocity compared to that of the tank immediately before the rope is completely extended?

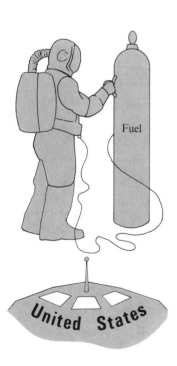

Fuel

Figure 6.4
How far from the antenna can the astronaut move, and what will be her velocity?

We assume that the system under consideration consists of two particles, the astronaut and the fuel tank, and disregard the mass of the rope. The center of mass of the system is initially at the antenna, which we may take to be at $x = 0$. Forces are acting on the particles of the system while the astronaut pushes on the tank (and the tank pushes back on her). These forces are indicated in Figure 6.5(a). The force $\mathbf{F}_{AT}$ on the tank by the astronaut is equal in magnitude and opposite in direction to the force $\mathbf{F}_{TA}$ on the astronaut by the tank. The resultant of these internal forces is zero. Similarly, in Figure 6.5(b), where the system is shown while the rope is bringing the astronaut and tank to rest, the forces $\mathbf{F}_A$ on the astronaut and $\mathbf{F}_T$ on the tank sum to zero. These are internal forces. Then at all times the net force on the astronaut-tank system is zero. Therefore, the momentum of the system cannot change, and the center of mass cannot change its position, since it was initially at rest. Measuring positions of the particles of the system relative to the antenna, we locate the center of mass initially at $x = 0$. Because there are no external forces on the system, the center of mass must remain at $x = 0$. Let d be the (positive) distance from the astronaut to the antenna in Figure 6.5(b). Then, according to Equation (6-1), the center of mass is located by

$$x_{cm} = \frac{-m_A d + m_T(50 - d)}{m_A + m_T} = 0 \qquad (6\text{-}20)$$

where m_A and m_T are the masses of the astronaut and tank. Solving Equation (6-20) for d gives

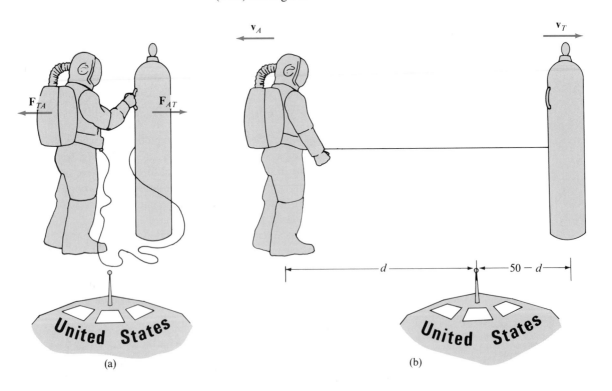

(a)   (b)

Figure 6.5
(a) An astronaut and fuel tank pushing on each other. $\mathbf{F}_{TA}$ is the force on the astronaut by the tank; $\mathbf{F}_{AT}$ is the force on the tank by the astronaut. **(b)** The astronaut and tank when the 50-m rope connecting them is almost fully extended.

$$d = \frac{50 \, m_T}{m_A + m_T}$$

$$d = \frac{50(200)}{100 + 200} = 33 \text{ m}$$

Thus the astronaut ends up 33 m from the antenna, while the tank is at a distance $50 - 33 = 17$ m from the antenna.

What is the astronaut's speed before the rope is completely extended? Initially, both the astronaut and tank were at rest, and the momentum of the system was equal to zero. No external forces acted on the system; its momentum was conserved. Therefore, the momentum of the system must be equal to zero immediately before the rope is extended. When they reach the end of their rope (as it were), the astronaut and tank must have momenta that are equal in magnitude but oppositely directed. Because the momentum of the fuel-tank astronaut system is equal to zero, the velocities of the astronaut, $\mathbf{v}_A$, and the tank, $\mathbf{v}_T$, are

$$m_A\mathbf{v}_A + m_T\mathbf{v}_T = 0$$

$$(100 \text{ kg})\mathbf{v}_A + (200 \text{ kg})\mathbf{v}_T = 0$$

$$\mathbf{v}_A = -2\mathbf{v}_T \qquad (6\text{-}21)$$

Until the rope is completely extended, the astronaut is moving with twice the speed of the fuel tank but in a direction opposite to the velocity of the tank.

6.3 Collisions

When two particles interact intensely for a brief time, we say these particles undergo a *collision*. In many mechanical systems, like a ball bouncing off a surface or a ball struck by a bat, large forces act on the ball while it is in contact with the surface or the bat. Before and after such collisions, these large contact forces are not present. Collisions between particles are characterized by having distinct periods before and after the intense forces of collision come into play. Physicists sometimes treat interactions that take place on a relatively long time scale as collisions; for example, an asteroid that enters the solar system and is bent around the sun before leaving the solar system is considered a collision. But here again, the time of intense interaction between the sun and asteroid (when the asteroid is near the sun) is small compared to the total time the system is observed. In this chapter, we will concentrate on collisions that are characterized by physical contact between bodies.

Impulse

We have seen in Equation (6-17) that Newton's second law may be written so that the net force $\Sigma\mathbf{F}$ on a particle is the time rate of change of the momentum $\mathbf{p}$ of the particle:

$$\Sigma\mathbf{F} = \frac{d\mathbf{p}}{dt} \qquad (6\text{-}22)$$

If we now consider **F** to be an intense force of interaction on a particle during a collision, we may assume that any other forces acting on the particle, its weight, for example, are small compared to **F**. Then we may treat **F** as the resultant force on the particle during the collision and write Equation (6-22) as

$$d\mathbf{p} = \mathbf{F}\,dt \qquad (6\text{-}23)$$

Using i to represent the initial instant of contact in a collision and f to represent the final instant of contact, we may integrate both sides of Equation (6-23) to give

$$\int_{\mathbf{p}_i}^{\mathbf{p}_f} d\mathbf{p} = \int_{t_i}^{t_f} \mathbf{F}\,dt$$

$$\mathbf{p}_f - \mathbf{p}_i = \int_{t_i}^{t_f} \mathbf{F}\,dt \equiv \mathbf{J} \qquad (6\text{-}24)$$

Hence $\mathbf{p}_i$ is the momentum of the particle initially, immediately before the collision, and $\mathbf{p}_f$ is its momentum finally, immediately after the collision. The integral on the right side of Equation (6-24) is called the *impulse* **J** transmitted to the particle during the collision. The left side of Equation (6-24) represents the change in momentum $\Delta\mathbf{p} = \mathbf{p}_f - \mathbf{p}_i$ that the particle undergoes as a result of the collision. Then we may write

$$\mathbf{J} = \Delta\mathbf{p} = \mathbf{p}_f - \mathbf{p}_i \qquad (6\text{-}25)$$

which says that the impulse transmitted to a particle during a collision is equal to the change in the momentum of the particle. What is more, since Equation (6-25) is a vector equation, the direction of **J** is the same as the direction of the *change* in momentum, a *vector difference* between two momenta. The units of impulse are the same as the units of momentum.

Suppose a golf ball is driven horizontally (in the x direction) from a tee by a collision with the head of a golf club. The momentum of the ball in the x direction is changed drastically by the collision, so we know that it receives considerable impulse but do not know the details of the impulse. Even if we measure how long the club head is in contact with the ball, we do not know how the force on the ball varies with time during the collision. Suppose for the moment that the force is a known function of time and $F_x(t)$ looks like the curve of Figure 6.6. Then, according to Equation (6-24), the x component of the impulse delivered to the ball by the club head is

$$J_x = \int_{t_i}^{t_f} F_x(t)dt \qquad (6\text{-}26)$$

which is the area under the curve of $F_x(t)$ versus t in Figure 6.6. This figure also shows a rectangular area that is equal to the area under the curve of $F_x(t)$. There is some value $\bar{F}_x$, the average of F_x, which, when multiplied by the time of contact $\Delta t = t_f - t_i$, gives the same value (area) as $\int_{t_i}^{t_f} F_x(t)dt$. Then

$$J_x = \bar{F}_x\Delta t = \Delta p_x \qquad (6\text{-}27)$$

relates the average force $\bar{F}_x$ exerted on the ball to the change ΔP_x in momentum of the ball.

The arguments resulting in Equation (6-27) are equally valid for y and z components. Therefore, we may generalize Equation (6-27):

$$\mathbf{J} = \bar{\mathbf{F}}\,\Delta t = \Delta\mathbf{p} \qquad (6\text{-}28)$$

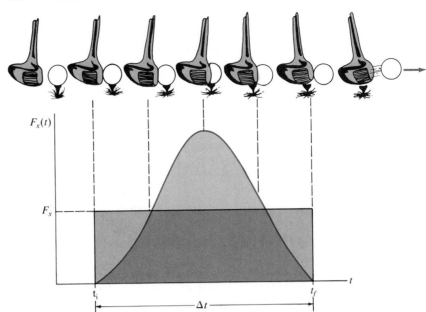

$F_x(t)$

F_x

t_i Δt t_f

t

Figure 6.6
A graph of the force $F_x(t)$ on a golf ball during a collision with a golf club head. The area under the rectangle of height $\overline{F}_x$ is equal to the area under the curve of $F_x(t)$.

Equation (6-28) incorporates the fact that, for a given time interval, the average force on a particle is in the same direction as that of the change in momentum of the particle.

Equation (6-28) provides the physical reasoning behind the golf or tennis pro's plea to "follow through" when trying to teach someone to change the momentum of a golf or tennis ball as much as possible. The longer the club head or racquet is in contact with a ball, the greater will be the impulse transmitted to the ball. By similar reasoning, we see that a long gun barrel permits the force of exploding gunpowder to act for a long time on a bullet or artillery shell.

E 6.7   Because impulse is essentially force multiplied by time, impulse is often measured in N·s or lb·s. Show that these units are the same as those of momentum.

The following problem illustrates again how the change in momentum of a body may be related to the average force exerted on the body in a collision.

A 0.50-kg ball moving at 10 m/s strikes a wall at 45°, as shown in Figure 6.7, and bounces off at 45° with a speed of 10 m/s. Calculate the average force the wall exerts on the ball during the collision if the ball and wall are in contact for 0.010 s. Compare this force to the weight of the ball.

The impulse that is imparted to the ball is equal to the change in momentum $\Delta \mathbf{p} = \mathbf{p}_f - \mathbf{p}_i$ of the ball during the collision. Let us calculate $\mathbf{p}_f$ and $\mathbf{p}_i$:

$$\mathbf{p}_f = m\mathbf{v}_f = 0.5\,(-10\,\cos45°\hat{\mathbf{i}} - 10\,\sin45°\hat{\mathbf{j}})$$

$$= (-3.5\hat{\mathbf{i}} - 3.5\hat{\mathbf{j}})\ \text{kg}\cdot\text{m/s} \tag{6-29}$$

$$\mathbf{p}_i = m\mathbf{v}_i = 0.5\,(10\,\cos45°\hat{\mathbf{i}} - 10\,\sin45°\hat{\mathbf{j}})$$

$$= (3.5\hat{\mathbf{i}} - 3.5\hat{\mathbf{j}})\ \text{kg}\cdot\text{m/s} \tag{6-30}$$

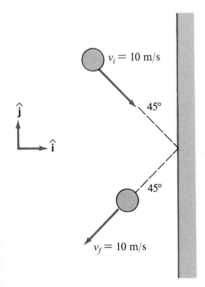

$v_i = 10$ m/s

$\hat{\mathbf{j}}$

$\hat{\mathbf{i}}$

45°

45°

$v_f = 10$ m/s

Figure 6.7
A ball striking a wall at 45° and bouncing off at 45°.

Then the change in the momentum $\Delta\mathbf{p} = \mathbf{p}_f - \mathbf{p}_i$ of the ball is $-7.0\hat{\mathbf{i}}$ kg·m/s. Because the collision time is small, we assume that the total impulse imparted to the ball results from the force exerted by the wall. Then, according to Equation (6-28), we may write

$$\bar{\mathbf{F}} = \frac{\Delta\mathbf{p}}{\Delta t} = \frac{-7.0\hat{\mathbf{i}} \text{ kg·m/s}}{0.010} = -700\hat{\mathbf{i}} \text{ N} \qquad (6\text{-}31)$$

and note that the magnitude of the average force $\bar{\mathbf{F}}$ is 700 N compared to the weight $W = 0.5(9.8) = 4.9$ N of the ball. In other words, the ratio of $|\bar{\mathbf{F}}|$ to W is

$$\frac{|\bar{\mathbf{F}}|}{W} = \frac{700}{4.9} = 140 \qquad (6\text{-}32)$$

This result, Equation (6-32), justifies our assumption that the impulse transmitted to the ball by the force of gravity is negligible compared to the impulse transmitted by the wall.

E 6.8   A 5.0-oz (5/16 lb) ball initially moving 80 ft/s toward the left is struck by a bat. Immediately after the collision, the velocity of the ball is 100 ft/s toward the right.
(a) Calculate the change in momentum of the ball during its collision with the
    bat.   Answer: 1.8 slug·ft/s, right
(b) Calculate the impulse imparted to the ball by the bat.
                                                            Answer: 1.8 lb·s, right

E 6.9   The velocity of a 2.0-kg body before it undergoes a collision with another body is $(7.0\hat{\mathbf{i}} - 8.0\hat{\mathbf{j}})$ m/s. Immediately after the collision, the velocity of the 2.0-kg body is $(12\hat{\mathbf{i}} + 2.0\hat{\mathbf{j}})$ m/s.
(a) Calculate the change in the momentum of the 2.0-kg body.
                                                Answer: $(10\hat{\mathbf{i}} + 20\hat{\mathbf{j}})$ kg·m/s
(b) What impulse is imparted to the body by this collision?
                                                Answer: $(10\hat{\mathbf{i}} + 20\hat{\mathbf{j}})$ N·s
(c) What impulse is imparted to the other body by this collision?
                                                Answer: $(-10\hat{\mathbf{i}} - 20\hat{\mathbf{j}})$ N·s

E 6.10   If the collision of E 6.9 lasts for 0.050 s, calculate the average force on the 2.0-kg body during the collision.   Answer: $(200\hat{\mathbf{i}} + 400\hat{\mathbf{j}})$ N

E 6.11   A constant resultant force of $3.0\hat{\mathbf{i}}$ N acts for 2.0 s on a 0.50-kg body. The velocity of the body at the beginning of this time interval is $8.0\hat{\mathbf{j}}$ m/s.
(a) What impulse is imparted to the body during this 2.0-s time interval?
                                                            Answer: $6.0\hat{\mathbf{i}}$ N·s
(b) What is the change in the momentum of the body during this time interval?
                                                            Answer: $6.0\hat{\mathbf{i}}$ kg·m/s
(c) What is the momentum of the body at the end of this time interval?
                                                Answer: $(6.0\hat{\mathbf{i}} + 4.0\hat{\mathbf{j}})$ kg·m/s

Classifying Collisions Energetically

Analysis of collisions involves relating the motions of particles after they collide to their motions before the collision. We now have the necessary tools for collision analysis in the form of conservation principles. Before we analyze collisions, however, let us distinguish between two kinds of collisions: elastic and inelastic.

A collision between particles is said to be *elastic* if the sum of the kinetic energies of the particles is the same after the collision as the sum of their kinetic energies before the collision. We need not consider gravitational potential energy because the energies of the particles are considered immediately before and immediately after the brief time of collision. During that time, the vertical positions of the particles do not change, and their potential energies do not change.

If a collision is not elastic, it is said to be *inelastic*. In an inelastic collision, the kinetic energy of the system of colliding particles is not conserved. The kinetic energy of a system may be decreased or increased during an inelastic collision. In a crash of two trains, for example, the initial kinetic energy of the system is decreased; this energy is dissipated in deforming the trains. In the explosion of a bomb, the kinetic energy of the system is increased because the explosion converts energy stored chemically to kinetic energy of the particles. The extreme case of an inelastic collision in which the colliding particles stick together and proceed as a single unit is called a *perfectly inelastic collision*.

Collisions in One Dimension

When two particles undergo a *head-on collision*, their motions before and after the collision are along the same straight line. Let us now consider a head-on collision in which all motion takes place along the x axis. Since only rectilinear motion is involved, we may dispense with vector notation. Remember, however, that positive values of v represent velocities in the positive x direction, and negative values of v represent velocities in the negative x direction. Suppose that two masses, m_1 and m_2, are involved in a head-on collision. Let v_{1i} and v_{2i} represent their initial velocities, that is, their velocities immediately before the collision. The initial situation is represented by the diagram of Figure 6.8.

We may assume that momentum is conserved in every collision. This statement is strictly true only when no external forces act on the colliding particles, but it is a good approximation when the impulse of external forces is small compared to the impulse of the intense forces of the collision. Then, assuming that momentum is conserved in collisions, we may write

$$\mathbf{P}_i = \mathbf{P}_f \tag{6-33}$$

or, the initial momentum of the system is equal to its final momentum. Equation (6-33) may be written

$$m_1 v_{1i} + m_2 v_{2i} = m_1 v_{1f} + m_2 v_{2f} \tag{6-34}$$

in which v_{1f} and v_{2f} are the final velocities (after the collision) of the masses m_1 and m_2. For inelastic collisions in one dimension, Equation (6-34) is generally applicable. For completely inelastic collisions, where the masses stick together, Equation (6-34) becomes

$$m_1 v_{1i} + m_2 v_{2i} = (m_1 + m_2) v_f \tag{6-35}$$

where v_f is the velocity of the combined particles after the collision.

If a head-on collision is elastic, the kinetic energy of the system is conserved, as is the momentum of the system. Then, in addition to Equation (6-34), we also use

Figure 6.8
Two masses m_1 and m_2 with velocities v_{1i} and v_{2i} immediately before a head-on collision.

4.0 lb

1.0 lb

2.0 ft/s

18 ft/s

Figure 6.9
At what velocity do the stone and putty
proceed after colliding?

Chapter 6   Momentum and Collisions

$$\frac{1}{2}m_1v_{1i}^2 + \frac{1}{2}m_2v_{2i}^2 = \frac{1}{2}m_1v_{1f}^2 + \frac{1}{2}m_2v_{2f}^2 \tag{6-36}$$

Thus, for elastic collisions in one dimension, Equations (6-34) and (6-36) simultaneously apply to the system.

The following problems illustrate the analysis of head-on collisions.

The 4.0-lb stone, shown in Figure 6.9 moving to the right at 2.0 ft/s, strikes a 1.0-lb blob of putty moving left at 18 ft/s, and the two stick together. At what velocity do they proceed after colliding?

The collision in this situation is perfectly inelastic, and the principle of conservation of momentum expressed in Equation (6-35) applies:

$$m_1v_{1i} + m_2v_{2i} = (m_1 + m_2)\,v_f$$

$$\frac{4}{32}(2) + \frac{1}{32}(-18) = \frac{5}{32}v_f$$

$$v_f = -\frac{10}{5} = -2.0 \text{ ft/s} \tag{6-37}$$

Here velocities to the right are assigned a positive value, and those to the left, a negative value. The result, Equation (6-37), indicates that the combined mass is moving to the left at a speed of 2.0 ft/s.

A comparison of the kinetic energies of this system before and after the collision is informative:

$$K_i = \frac{1}{2}m_1v_{1i}^2 + \frac{1}{2}m_2v_{2i}^2 = \frac{1}{2}\left(\frac{4}{32}\right)(4) + \frac{1}{2}\left(\frac{1}{32}\right)(324) = 5.3 \text{ ft·lb}$$

$$K_f = \frac{1}{2}(m_1 + m_2)\,v_f^2 = \frac{1}{2}\left(\frac{4+1}{32}\right)(4) = 0.31 \text{ ft·lb}$$

These values show clearly that kinetic energy is not conserved in this inelastic collision. Indeed most of the original energy of motion is spent in deforming the putty during the collision.

E 6.12   A 10-kg object moving with a velocity of $200\hat{i}$ m/s "explodes" into two parts. Immediately after the explosion, one part with a mass of 8.0-kg has a velocity of $350\hat{i}$ m/s, and the other part (mass = 2.0 kg) has a velocity of $-400\hat{i}$ m/s.
(a) Was momentum conserved during the explosion?
Answer: Yes, $\mathbf{P}_f = \mathbf{P}_i = 2000\hat{i}$ kg·m/s
(b) Determine the final kinetic energy of the system.   Answer: 6.5×10^5 J
(c) Determine the change in the kinetic energy of the system.
Answer: 4.5×10^5 J

E 6.13   Two snowmobiles, Frosty and Flakey, of equal masses suffer a rear-end collision. Just before the collision Frosty's velocity is 30 m/s toward the east and Flakey's velocity is 10 m/s in the same direction. Just after the collision, Frosty is moving 15 m/s toward the east.
(a) What is Flakey's velocity immediately after the collision?
Answer: 25 m/s, east
(b) Was this an elastic collision?   Answer: No, $K_f/K_i = 0.85$

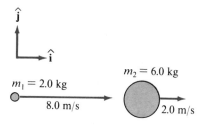

$m_2 = 6.0$ kg

$m_1 = 2.0$ kg

8.0 m/s   2.0 m/s

Figure 6.10
Determine the velocities of each of the
two masses after they collide.

Figure 6.10 shows a mass $m_1 = 2.0$ kg, moving at $8.0\hat{i}$ m/s, which overtakes and collides head-on with a mass $m_2 = 6.0$ kg moving in the same direction at $2.0\hat{i}$ m/s. If the collision is elastic, determine the velocities of each of the two masses after the collision.

Because this collision is elastic, both momentum and kinetic energy are conserved. Therefore, Equations (6-34) and (6-36) apply simultaneously:

$$m_1 v_{1i} + m_2 v_{2i} = m_1 v_{1f} + m_2 v_{2f}$$

$$(2)(8) + (6)(2) = 2v_{1f} + 6v_{2f}$$

$$14 = v_{1f} + 3v_{2f} \tag{6-38}$$

$$\frac{1}{2}m_1 v_{1i}^2 + \frac{1}{2}m_2 v_{2i}^2 = \frac{1}{2}m_1 v_{1f}^2 + \frac{1}{2}m_2 v_{2f}^2$$

$$\frac{1}{2}(2)(64) + \frac{1}{2}(6)(4) = \frac{1}{2}(2)\,v_{1f}^2 + \frac{1}{2}(6)\,v_{2f}^2$$

$$76 = v_{1f}^2 + 3v_{2f}^2 \tag{6-39}$$

Solving Equations (6-38) and (6-39) simultaneously involves a quadratic equation and, therefore, provides two solution sets:

$$\left\{ \begin{aligned} v_{1f} &= 8.0 \text{ m/s} \\ v_{2f} &= 2.0 \text{ m/s} \end{aligned} \right\} \tag{6-40}$$

and

$$\left\{ \begin{aligned} v_{1f} &= -1.0 \text{ m/s} \\ v_{2f} &= 5.0 \text{ m/s} \end{aligned} \right\} \tag{6-41}$$

Examination of the physical situation of the problem provides the meanings of the solution sets. The solutions of Equation (6-40) cannot apply to the situation after the collision, because the mass m_1, initially on the left, would have to pass through the mass m_2 in order to have a greater speed in the same direction than that of m_2. The solution of Equation (6-40) is not, therefore, physically meaningful *after* the collision. This solution does, however, reiterate the physical situation *before* the collision. In elastic collisions, the appearance of the initial velocities as one of the solutions serves as a check on the accuracy of the solutions. The physically meaningful solution for the situation after the collision is, therefore, Equation (6-41), from which we see that the 6.0-kg mass is moving (right) at $5.0\hat{i}$ m/s while the 2.0-kg mass is moving (left) at $-1.0\hat{i}$ m/s.

Collisions in Two Dimensions

When two particles collide and the collision is not head-on, the particles may approach each other from arbitrary directions, collide, and move away from each other in two- or three-dimensional space. The principles and techniques appropriate to the analysis of such collisions will be illustrated here by consideration of a collision between two particles in two dimensions. Collisions in two dimensions take place in a plane, like billiard balls colliding on a horizontal table. Just as in one-dimensional collisions, the conservation principles, the conservation of energy and the conservation of momentum, apply to elastic collisions in two dimensions. In two dimensions, if a collision is inelastic, only the principle of conservation of momentum is useful; kinetic energy is not conserved.

In a two-dimensional inelastic collision, we may apply the conservation of momentum as expressed in Equation (6-33), $\mathbf{P}_i = \mathbf{P}_f$. Because Equation (6-33) is a vector equation, we can write it as two equations for a two-dimensional situation:

$$P_{xi} = P_{xf} \tag{6-42}$$

$$P_{yi} = P_{yf} \tag{6-43}$$

In other words, the x component of momentum is conserved, and the y component of momentum is conserved. Consider how Equations (6-42) and (6-43) may be used to analyze a perfectly inelastic collision:

An 1800-lb car heading east at 60 mi/h (88 ft/s) collides with a northbound 2400-lb pick-up truck moving at 30 mi/h. The vehicles become a single mangled mass. Immediately after the collision, what is the velocity of the wrecked mass?

Because the collision here is inelastic, we can use only the conservation of momentum; we cannot assume that kinetic energy is conserved. The physical situation before the collision is suggested by Figure 6.11(a). The initial momentum of the system (the car and the truck) has components

$$P_{xi} = m_c v_{xi} = \left(\frac{1800}{32}\right)(88) = 5000 \text{ slug} \cdot \text{ft/s} \tag{6-44}$$

$$P_{yi} = m_t v_{yi} = \left(\frac{2400}{32}\right)(44) = 3300 \text{ slug} \cdot \text{ft/s} \tag{6-45}$$

Here m_c and m_t represent the masses of the car and the truck. The weights and velocities of the vehicles have been converted to appropriate units in the English system. Figure 6.11(b) represents the situation immediately after the collision, where the velocity $\mathbf{v}_f$ of the wreckage has components v_{xf} and v_{yf}. Then the final components of the momentum are

$$P_{xf} = (m_c + m_t)\, v_{xf} = \left(\frac{1800}{32} + \frac{2400}{32}\right) v_{xf} = 130\, v_{xf} \tag{6-46}$$

$$P_{yf} = (m_c + m_t)\, v_{yf} = \left(\frac{1800}{32} + \frac{2400}{32}\right) v_{yf} = 130\, v_{yf} \tag{6-47}$$

Figure 6.11
(a) A car and truck approaching an intersection where they collide. (b) The mangled mass of the car-truck after the collision. The mass is moving at a velocity $\mathbf{v}_f$ in a direction θ measured N of E immediately after the collision.

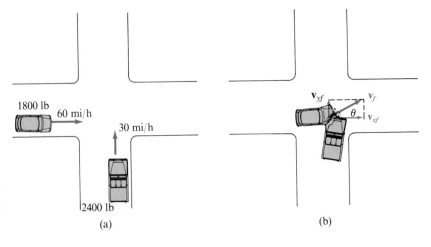

Because momentum is conserved in the collision, it follows that $P_{xf} = P_{xi}$, or

$$130 \, v_{xf} = 5000$$

$$v_{xf} = 38 \text{ ft/s} \tag{6-48}$$

and $P_{yf} = P_{yi}$, or

$$130 \, v_{yf} = 3300$$

$$v_{yf} = 25 \text{ ft/s} \tag{6-49}$$

These components of $\mathbf{v}_f$ may be combined to find v_f:

$$v_f = \sqrt{v_{xf}^2 + v_{yf}^2} = \sqrt{38^2 + 25^2} = 45 \text{ ft/s} \tag{6-50}$$

The angle θ is specified by

$$\theta = \tan^{-1}\left(\frac{25}{38}\right) = 33° \qquad \text{N of E} \tag{6-51}$$

Thus the two unknown quantities, v_f and θ, are determined by the two scalar equations that express the conservation of momentum in two dimensions.

E 6.14   Calculate the change in the kinetic energy of the car-truck system of Figure 6.11 as a result of the collision.   Answer: -1.5×10^5 ft·lb

Now consider another collision in two dimensions:

Figure 6.12(a) shows a white ball of mass 2.0 kg moving along the surface of a horizontal table with a speed of 10 m/s before striking a resting black ball of equal mass. Immediately after they collide, the white ball is moving at 5.0 m/s at an angle of 60° from its original direction of motion, as shown in Figure 6.12(b). What is the speed and direction of the black ball immediately after the collision? Was the collision elastic?

We seek v_{2f} and θ. Let the original direction of motion of the white ball be the x direction. We may then use the fact that the x component of momentum is conserved in the collision:

$$m_1 v_{1xi} + m_2 v_{2xi} = m_1 v_{1xf} + m_2 v_{2xf}$$

$$(2)(10) + (2)(0) = (2)(5 \cos 60°) + 2v_{2f} \cos\theta$$

$$20 = 5 + 2v_{2f} \cos\theta$$

$$2v_{2f} \cos\theta = 15 \tag{6-52}$$

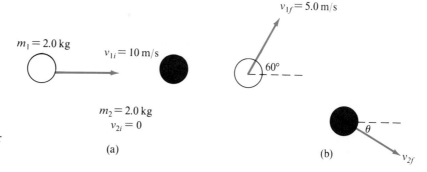

Figure 6.12
White and black balls before and after they collide.

Similarly, the y component of momentum is conserved:

$$m_1 v_{1yi} + m_2 v_{2yi} = m_1 v_{1yf} + m_2 v_{2yf}$$

$$0 + 0 = (2)(5 \sin 60°) + 2v_{2f} \sin\theta$$

$$2v_{2f} \sin\theta = -8.66 \tag{6-53}$$

Dividing Equation (6-53) by Equation (6-52) gives

$$\tan\theta = \frac{-8.7}{15} = -0.580$$

$$\theta = -30° \tag{6-54}$$

which specifies the final direction of the black ball. Its final speed v_{2f} may be determined by substituting $\theta = -30°$ into Equation (6-53):

$$2v_{2f} \sin(-30°) = -8.66$$

$$v_{2f} = 8.66 \text{ m/s} \tag{6-55}$$

Whether or not the collision was elastic is decided by checking to see if the kinetic energy is the same before and after the collision:

$$\frac{1}{2}m_1 v_{1i}{}^2 + \frac{1}{2}m_2 v_{2i}{}^2 = \frac{1}{2}(2)(10^2) = 100 \text{ ft} \cdot \text{lb} \tag{6-56}$$

$$\frac{1}{2}m_1 v_{1f}{}^2 + \frac{1}{2}m_2 v_{2f}{}^2 = \frac{1}{2}(2)(5^2) + \frac{1}{2}(2)(8.66^2) = 100 \text{ ft} \cdot \text{lb} \tag{6-57}$$

Equations (6-56) and (6-57) assure that kinetic energy was conserved in the collision. The collision, therefore, was elastic.

Example 6.1

PROBLEM   A bullet of mass m (10 g) moving vertically upward with a speed v (500 m/s) as shown in Figure 6.13, embeds itself in a wooden block of mass M (2.0 kg). To what maximum height does the block rise above its initial position?

SOLUTION   Immediately after the collision, the combined block and bullet move upward with a speed V, as shown in Figure 6.14(a). Imposing conservation of momentum in the vertical direction gives

$$P_{yf} = P_{yi}$$

or

$$(M + m)V = mv$$

$$V = \frac{m}{M + m}v = \left(\frac{0.01}{2 + 0.01}\right)500 = 2.5 \text{ m/s}$$

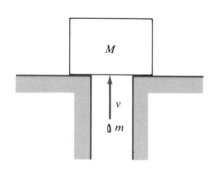

Figure 6.13   Example 6.1

The composite mass moves to a maximum height h, as shown in Figure 6.14(b), where the gravitational potential energy, $(M + m)gh$, relative to its initial position is equal to its initial kinetic energy, $\frac{1}{2}(M + m)v^2$, that is,

$$(M + m)gh = \frac{1}{2}(M + m)v^2$$

$$h = \frac{v^2}{2g} = \frac{(2.5)^2}{2(9.8)} = 0.32 \text{ m} \quad \blacksquare$$

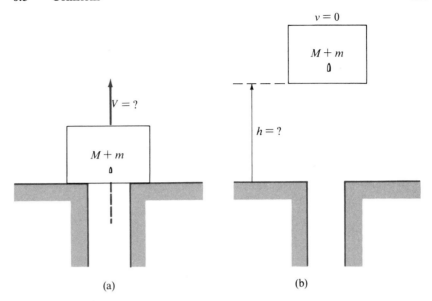

Figure 6.14   Example 6.1   (a)   (b)

E 6.15   What percent of the initial kinetic energy of the bullet is dissipated in the collision of Example 6.1? What happens to this "lost" kinetic energy?   Answer: 99.5%

Example 6.2   **PROBLEM**   Blocks m_1 and m_2 slide on a horizontal frictionless surface, as shown in Figure 6.15. A spring ($k = 800$ N/m) of negligible mass is fixed to m_2 as shown. Before the collision, m_1 is moving 8.0 m/s toward the right, and m_2 is moving 6.0 m/s toward the left. At the instant when m_1 is moving 4.0 m/s toward the right, determine the velocity of m_2 and the distance the spring is compressed.

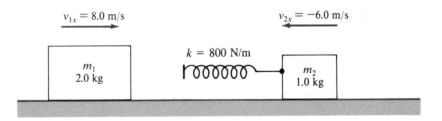

Figure 6.15   Example 6.2

SOLUTION   Figure 6.16 depicts the collision at the instant of interest. Since no net external force acts on the two-mass system, the momentum at this instant must be equal to the initial momentum, that is,

$$m_1 V_{1x} + m_2 V_{2x} = m_1 v_{1x} + m_2 v_{2x}$$

$$(2)(4) + (1)\, V_{2x} = (2)(8) + (1)(-6)$$

$$8 + V_{2x} = 16 - 6$$

$$V_{2x} = 2.0 \text{ m/s}$$

(Notice that the momentum of m_2 before the collison is negative, because we have taken the positive direction toward the right.) Thus m_2 is moving 2.0 m/s

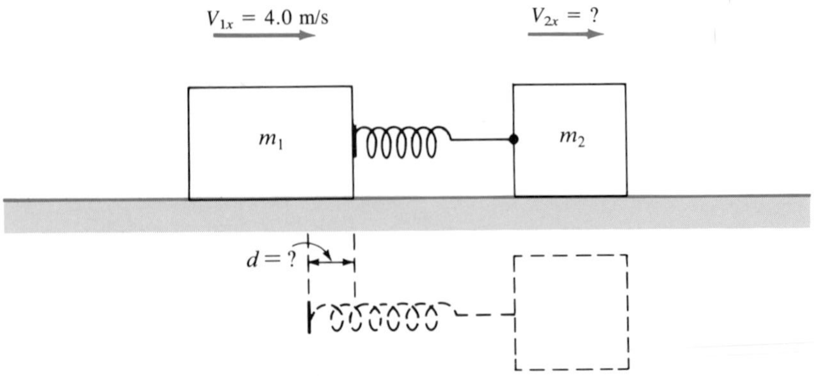

$V_{1x} = 4.0$ m/s $V_{2x} = ?$

m_1 m_2

$d = ?$

Figure 6.16   Example 6.2

toward the right at this instant. The compression distance d shown in Figure 6.16 may be determined using energy conservation (since no dissipative forces act on the system):

$$\frac{1}{2}m_1V_1^2 + \frac{1}{2}m_2V_2^2 + \frac{1}{2}kd^2 = \frac{1}{2}m_1v_1^2 + \frac{1}{2}m_2v_2^2$$

$$\frac{1}{2}(2)(4)^2 + \frac{1}{2}(1)(2)^2 + \frac{1}{2}(800)\ d^2 = \frac{1}{2}(2)(8)^2 + \frac{1}{2}(1)(6)^2$$

$$16 + 2 + 400\ d^2 = 64 + 18$$

$$d^2 = 0.16$$

$$d = 0.40 \text{ m} \quad \blacksquare$$

Example 6.3

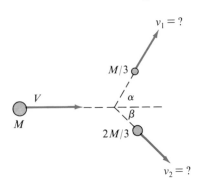

Figure 6.17   Example 6.3

PROBLEM An atomic particle of mass M traveling with a speed V disintegrates into two particles, as shown in Figure 6.17. One of these particles has a mass of $M/3$ and moves with a speed v_1 in the direction shown. The second particle has a mass of $2M/3$ and moves with a speed v_2 in the direction shown. Find expressions for v_1, v_2, and K_f/K_i, the ratio of the final kinetic energy to the initial kinetic energy. These expressions should be written in terms of α, β, and V. Finally, calculate the ratios v_1/V, v_2/V, and K_f/K_i for $\alpha = 40°$ and $\beta = 60°$.

SOLUTION Requiring momentum components (both parallel and perpendicular to the initial direction of motion, as shown in Figure 6.17) to be conserved gives

$$MV = \frac{1}{3}Mv_1\ \cos\alpha + \frac{2}{3}Mv_2\ \cos\beta$$

$$0 = \frac{1}{3}Mv_1\ \sin\alpha - \frac{2}{3}Mv_2\ \sin\beta$$

If each of these equations is multiplied by $3/M$, the result is

$$3V = v_1\ \cos\alpha + 2v_2\ \cos\beta$$

$$0 = v_1\ \sin\alpha - 2v_2\ \sin\beta$$

After applying a bit of algebra and trigonometry, we get

$$v_1 = \frac{3\ \sin\beta}{\sin(\alpha + \beta)}\ V$$

$$v_2 = \frac{3\ \sin\alpha}{2\ \sin(\alpha + \beta)}\ V$$

The final kinetic energy is

$$K_f = \frac{1}{2}\left(\frac{M}{3}\right)v_1{}^2 + \frac{1}{2}\left(\frac{2M}{3}\right)v_2{}^2$$

$$K_f = \frac{M}{6}\left[\frac{3\ \sin\beta}{\sin(\alpha + \beta)}\right]^2 V^2 + \frac{M}{3}\left[\frac{3\ \sin\alpha}{2\ \sin(\alpha + \beta)}\right]^2 V^2$$

$$K_f = \left\{\frac{3\ (\sin^2\alpha + 2\ \sin^2\beta)}{2\ \sin^2(\alpha + \beta)}\right\}\frac{1}{2}MV^2$$

or

$$\frac{K_f}{K_i} = \frac{3\ (\sin^2\alpha + 2\ \sin^2\beta)}{2\ \sin^2(\alpha + \beta)}$$

For $\alpha = 40°$ and $\beta = 60°$, we get

$$\frac{v_1}{V} = \frac{3\ \sin60°}{\sin(40° + 60°)} = 2.6$$

$$\frac{v_2}{V} = \frac{3\ \sin40°}{2\ \sin(40° + 60°)} = 0.98$$

$$\frac{K_f}{K_i} = \frac{3\ (\sin^240° + 2\ \sin^260°)}{2\ \sin^2(40° + 60°)} = 3.0 \quad ■$$

E 6.16   Do the "bit" of algebra and trigonometry necessary to obtain the expressions for v_1 and v_2 in Example 6.3. *One hint:* You will need to use the identity $\sin(\alpha + \beta) = \sin\alpha\ \cos\beta + \cos\alpha\ \sin\beta$.

Example 6.4

PROBLEM   A 5.0-kg body moving with a speed of 20 m/s explodes into a 2.0-kg mass moving at 35 m/s and a 3.0-kg mass moving at 30 m/s. Determine the direction of motion, relative to the initial direction of motion of the 5.0-kg body, of each of the two pieces after the explosion.

SOLUTION   Figure 6.18(a) shows the explosion with the unknown scattering angles α and β. The system momentum $\mathbf{P}_0$ before the explosion has a magnitude

$$P_0 = (5.0\ \text{kg})(20\ \text{m/s}) = 100\ \text{kg} \cdot \text{m/s}$$

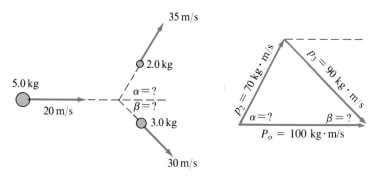

Figure 6.18   Example 6.4   (a)   (b)

After the explosion, the momentum $\mathbf{p}_2$ of the 2.0-kg mass has a magnitude

$$p_2 = (2.0 \text{ kg})(35 \text{ m/s}) = 70 \text{ kg} \cdot \text{m/s}$$

and the magnitude of the momentum $\mathbf{p}_3$ of the 3.0-kg mass is

$$p_3 = (3.0 \text{ kg})(30 \text{ m/s}) = 90 \text{ kg} \cdot \text{m/s}$$

Because momentum is conserved, $\mathbf{P}_0 = \mathbf{p}_2 + \mathbf{p}_3$. This vector triangle is shown in Figure 6.18(b). The angles α and β can be determined with the law of cosines, that is,

$$70^2 + 100^2 - 2(70)(100)\cos\alpha = 90^2$$

$$\cos\alpha = \frac{70^2 + 100^2 - 90^2}{2(70)(100)} = 0.486$$

$$\alpha = 61°$$

and

$$90^2 + 100^2 - 2(90)(100)\cos\beta = 70^2$$

$$\cos\beta = \frac{90^2 + 100^2 - 70^2}{2(90)(100)} = 0.733$$

$$\beta = 43° \quad \blacksquare$$

6.4 Problem-Solving Summary

The principle of conservation of momentum is a powerful problem-solving tool. Problems to which this principle applies may be identified by recognizing a system of interacting particles on which no net external force acts. If no net external force acts on a system, we are assured of two facts:

1. The linear momentum $\mathbf{P} = \Sigma_i\, m_i\mathbf{v}_i$ of the system remains constant, that is, neither its magnitude nor its direction can change.

2. The center of mass of the system maintains a constant velocity.

When no net external force acts on a system, linear momentum is conserved; the initial momentum $\mathbf{P}_i$ of the system is equal to its final momentum $\mathbf{P}_f$. If a net external force $\Sigma\mathbf{F}$ acts on a system, the impulse provided by $\Sigma\mathbf{F}$, $\mathbf{J} = \int_{t_i}^{t_f} \Sigma\mathbf{F}dt$, alters the initial momentum $\mathbf{P}_i$ of the system to give it a final momentum $\mathbf{P}_f$, according to $\mathbf{P}_f = \mathbf{P}_i + \mathbf{J}$. Thus, impulse may be interpreted as a change in momentum: $\mathbf{J} = \mathbf{P}_f - \mathbf{P}_i = \Delta\mathbf{P}$. Note that $\Delta\mathbf{P}$ is equal to the final value of $\mathbf{P}$ less its initial value, a vector difference that is frequently miscalculated.

The principle of conservation of linear momentum is most useful in collision problems. Then we may use $\mathbf{P}_i = \mathbf{P}_f$ in all collision problems if we remember that $\mathbf{P}_i$ represents the vector sum of the momenta of the colliding particles before the collision and $\mathbf{P}_f$ is the vector sum of their momenta after the collision.

In two-dimensional collisions, the relationships, $P_{xi} = P_{xf}$ and $P_{yi} = P_{yf}$, express the conservation of momentum. When calculating one of these components of the total momentum, be sure to include the appropriate sign for the velocity component of each particle. Care with signs is especially important in collision problems.

Although we may use the momentum conservation principle in any collision problem, the kinetic energy of the system may or may not be changed as a result of the collision. If a given problem states that the collision is elastic, the kinetic energy of the system is the same before and after the collision, and we may use the relationship $K_i = K_f$. In some problems in which the elastic condition is not specifically stated, enough additional information may be available to permit the construction of an appropriate energy relationship.

The concept of center of mass of a system of particles (including continuous bodies) is important because it permits us to treat the whole system as if it were a single particle, that is, we may use $\mathbf{P} = M\mathbf{v}_{cm}$ and $\Sigma\mathbf{F} = M\mathbf{a}_{cm}$. The center of mass is located relative to any set of coordinate axes at the point (x_{cm}, y_{cm}, z_{cm}), where x_{cm} is given by $\Sigma_{i=1}^{n} m_i x_i / \Sigma_{i=1}^{n} m_i$; similar expressions define y_{cm} and z_{cm}. Other than problems that simply locate the center of mass of a system, most problems involving center of mass use the following fact: When no net external force acts on a system, the velocity of the center of mass does not change. In the special case in which a system has no net force on it and has no net momentum (the center of mass is at rest), the position of the center of mass cannot change no matter what internal movements occur within the system. This case usually involves problems that may be solved by application of the definitions that locate the center of mass.

Problems

GROUP A

6.1 Four particles with masses $m_1 = 4.0$ kg, $m_2 = 2.0$ kg, $m_3 = 6.0$ kg, and $m_4 = 10$ kg lie along the x axis at the points $x_1 = 8.0$ m, $x_2 = 4.0$ m, $x_3 = 0$, and $x_4 = -2.0$ m. Locate the center of mass of this system.

6.2 A system of particles is composed of $m_1 = 4.0$ kg at the point $(4.0$ m, -6.0 m$)$, $m_2 = 6.0$ kg at $(-4.0$ m, 2.0 m$)$, and $m_3 = 1.0$ kg at the point $(7.0$ m, $0)$. Locate the center of mass of the system.

6.3 A 2.0-kg mass falling vertically at 20 m/s explodes into two fragments. The fragment of mass 0.50 kg is rising vertically at 4.0 m/s after the explosion. What is the velocity of the other fragment?

6.4 A 2.0-kg block and a 6.0-kg block have a compressed spring between them. The blocks are placed on a frictionless surface and released from rest. The 6.0-kg mass moves away with a velocity of $12\hat{\mathbf{i}}$ m/s.

(a) With what velocity does the 2.0-km mass move?

(b) How much work is done by the spring on the blocks?

6.5 While heading eastward at 30 mi/h, a 2400-lb car collides with a truck heading westward at 30 mi/h. Immediately after the collision, the combined wreckage is moving westward at 5 mi/h. What is the weight of the truck?

6.6 Ten seconds after starting from rest, a 200-kg motorcycle is moving 30 m/s toward the east.

(a) What is the impulse imparted to the motorcycle during this 10-s time interval?

(b) What average resultant force acts on the motorcycle during this time interval?

6.7 The only force acting on a 0.50-kg object is shown in Figure 6.19. The velocity of the object at $t = 0$ is $v_x = 40$ m/s.

(a) Calculate the impulse delivered by F_x during the time interval $0 \le t \le 20$ s.

(b) Determine the maximim speed of the object during the time interval $0 \le t \le 20$ s.

(c) Calculate the momentum of the object at $t = 20$ s.

6.8 During a 5.0-s time interval, a 12-lb body undergoes a change in velocity from $8.0\hat{\mathbf{i}}$ ft/s to $10\hat{\mathbf{j}}$ ft/s.

(a) What change in momentum of the body occurs during this time interval?

(b) What impulse is imparted to the body during this time interval?

(c) Calculate the average resultant force acting on the body during this time interval.

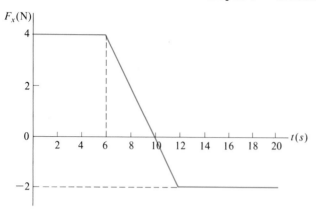

Figure 6.19   Problem 6.7

6.9   While moving with a velocity of $5.0\hat{i}$ m/s, a 3.0-kg object collides with and sticks to a 2.0-kg body moving with a velocity of $(-4.0\hat{i} + 8.0\hat{j})$ m/s.

(a)   What is the final velocity of the composite body?

(b)   What impulse is imparted to the 3.0-kg body?

(c)   What impulse is imparted to the 2.0-kg body?

6.10   Before exploding into three pieces (1.0 kg, 2.0 kg, and 3.0 kg), a 6.0-kg mass is moving with a velocity of $2.0\hat{i}$ m/s. Immediately after the explosion, the 1.0-kg mass has a velocity of $4.0\hat{i}$ m/s, and the 2.0-kg mass has a velocity of $-3.0\hat{j}$ m/s. What is the velocity of the 3.0-kg mass immediately after the explosion?

6.11   A 12-ft uniform plank weighing 80 lb rests on a horizontal surface of negligible friction. Both the plank and a 120-lb person standing on one end of the plank are initially at rest. The person walks to the opposite end of the plank and stops there. How far does the plank move relative to the surface?

6.12   A completely inelastic collision occurs between a 5.0-kg mass moving upward at 20 m/s and a 2.0-kg mass moving downward at 10 m/s. How high will the combined mass rise above the point of the collision?

6.13   An elastic collision occurs when a 5.0-kg body moving upward at 20 m/s collides head-on with a 2.0-kg body moving downward at 10 m/s. How high will each of the two bodies rise above the point of the collision?

6.14   As a result of a glancing collision with a surface, the velocity of a 2.0-lb weight changes from 40 ft/s directed horizontally to 30 ft/s directed 60° above the horizontal. If the collision lasts 0.020 s, calculate the average force on the weight by the surface.

6.15   An automatic rifle fires 10-g bullets horizontally with a speed of 400 m/s at a rate of 240 per minute. What is the magnitude of the average horizontal force the rifle exerts on the person holding the rifle while it is being fired?

6.16   While moving horizontally with a speed v_0, a bullet of mass m embeds itself in a wooden block of mass M initially at rest. The combined masses follow the trajectory shown in Figure 6.20. If $m = 0.010$ kg, $M = 2.0$ kg, $h = 1.0$ m, and $d = 1.5$ m, calculate v_0.

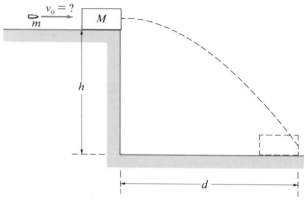

Figure 6.20   Problem 6.16

6.17   A 2.0-lb body moving 10 ft/s collides with a 6.0-lb body moving at a right angle to the direction of the 2.0-lb body. The two bodies stick together and just after the collision have a velocity that makes an angle of 30° with the original direction of motion of the 2.0-lb body. How much kinetic energy is dissipated in this collision?

6.18   Falling vertically at 8.0 m/s, a ball of mass 0.50 kg strikes a floor and bounces straight upward at 6.0 m/s.

(a)   What impulse does the ball receive from the floor?

(b)   If the ball is in contact with the floor for 0.050 s, what average force does the floor exert on the ball?

6.19   The velocity of a baseball weighing 5.0 oz is changed from 60 mi/h (88 ft/s) directed horizontally to 120 mi/h in the opposite direction when the ball is hit by a bat.

(a)   What is the magnitude of the impulse imparted to the ball by the bat?

(b)   If the bat exerts an average force of 100 lb on the ball, how long is the bat in contact with the ball?

6.20   Three particles, each having a mass of 0.25 kg, lie at the vertices of an equilateral triangle at the points $(-2.0$ m, $0)$, $(2.0$ m, $0)$, and $(0, 2\sqrt{3}$ m$)$.

(a)   Locate the center of mass of the particles.

(b)   A force of $0.25\hat{i}$ N is applied to the particle located at $(0, 2\sqrt{3}$ m$)$ for 4.0 s. Determine the location of the particle after the 4.0-s interval and locate the new center of mass of the three-particle system.

(c)   What is the acceleration of the center of mass of the system while the force acts?

(d)   Calculate the new location of the center of mass after the force has acted for 4.0 s using the result of part (c). Compare this result to that of part (b).

6.21 After falling vertically from a height of 20 m, a 0.20-kg ball collides with the floor and rises to a height of 16 m before coming to rest momentarily.

(a) What impulse does the ball receive from the floor?

(b) If the average force on the ball is 400 N while it is in contact with the floor, how long was it touching the floor?

(c) How much energy does the ball lose during the collision?

6.22 Two blocks of mass m and $3m$ start from rest with a compressed spring between them. What is the ratio of their final kinetic energies?

6.23 Two swimmers simultaneously dive off a 100-lb rowboat. Hank, who weighs 180 lb, dives toward the east at an angle of 30° above the horizontal at a speed of 6.0 ft/s. Susan, a 120-pounder, dives toward the west at 45° from the horizontal at 4.0 ft/s. Disregarding the momentum transferred to the water, what is the velocity of the boat immediately after their dives?

6.24 An elastic collision occurs between a 3.0-kg body moving with a velocity of $10\hat{i}$ m/s and a body of mass M initially at rest. After the collision the 3.0-kg body has a velocity of $-5.0\hat{i}$ m/s. Determine M and the velocity of the mass M after the collision.

6.25 A 0.40-kg body moving with a velocity of $10\hat{i}$ m/s collides elastically with a mass M initially at rest. After the collision the 0.40-kg mass has a velocity of $5.0\hat{j}$ m/s. Calculate M and the velocity of the mass M after the collision.

6.26 A bullet of mass m (15 g) moving horizontally with a speed v_1 (1000 m/s) hits and passes through a wooden block of mass M (0.50 kg), which is initially at rest on a horizontal frictionless surface. The bullet emerges from the block moving horizontally with a speed v_2 (400 m/s).

(a) What is the speed of the block after the collision?

(b) Account for all the initial kinetic energy of the bullet.

6.27 Just before embedding itself in a wooden block of mass M (1995 g) at rest on a horizontal surface, a bullet of mass m (5 g) is moving horizontally with a speed v (1000 m/s). The coefficient of kinetic friction for the block and horizontal surface is 0.20.

(a) How far will the block slide after the collision?

(b) Account for all the initial kinetic energy of the bullet.

6.28 Show that when two particles of equal mass undergo an elastic, head-on collision, they exchange speeds.

6.29 The block of mass $M = 1.0$ kg contains a coiled spring and a ball of mass $m = 0.25$ kg. This system is released from rest on a horizontal surface 2.0 m above a floor, as shown in Figure 6.21. The block is on a smooth surface, and when the ball is ejected horizontally from the block, the block moves a vertical height of 1.0 m before coming to rest. How far does the ball travel horizontally before striking the floor? How much energy is stored in the spring before the ball is ejected?

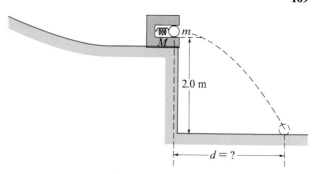

Figure 6.21   Problem 6.29

6.30 A block of mass $M = 1.0$ kg contains a coiled spring and a ball of mass $m = 0.25$ kg. The spring is released when the block-ball system is at rest at the edge of a frictionless plane inclined 20° from the horizontal (*see* Fig. 6.22). The ball, initially 2.0 m above a horizontal floor, strikes the floor a horizontal distance of 3.0 m from the release point. How far along the plane does the block move before coming to rest? How much energy was initially stored in the spring?

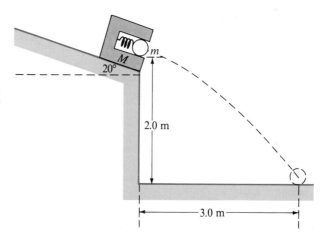

Figure 6.22   Problem 6.30

6.31 A mass m_1 suspended on a string of length L is released from rest with the string horizontal, as shown in Figure 6.23 on the following page. At the lowest point in its swing, m_1 collides elastically with a stationary mass $m_2 = 2m_1$ suspended on a string of length L. Calculate the maximum angle that the string suspending m_2 makes with the vertical following the initial collision.

6.32 The two masses ($m_1 = 3.0$ kg, $m_2 = 5.0$ kg) moving ($v_1 = 12$ m/s, $v_2 = 4.0$ m/s) (*see* Fig. 6.24) slide on a horizontal frictionless surface. A spring ($k = 1000$ N/m) is fixed to m_2.

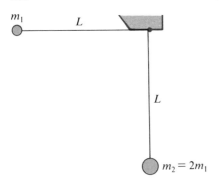

Figure 6.23   Problem 6.31

(a) Find the maximum distance of compression of the spring during the collision.

(b) At the instant m_1 is moving 8.0 m/s toward the right, find the distance the spring is compressed.

(c) Determine the velocities of the two masses after the collision.

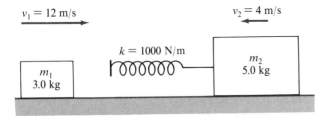

Figure 6.24   Problem 6.32

6.33 A particle of mass $3m$ moving with a velocity $v_x = v$ undergoes an elastic one-dimensional collision with a particle of mass m moving with a velocity $v_x = -v$. Answer the following in terms of m and v.

(a) What is the velocity of each particle after the collision?

(b) What is the change in momentum of each particle as a result of the collision?

(c) What are the relative velocities of the particles before and after the collision?

6.34 An elastic one-dimensional collision occurs when a particle of mass $2m$ moving with a velocity $v_x = 2v$ collides with a particle of mass m moving with a velocity $v_x = v$. Answer the following in terms of m and v:

(a) What is the velocity of each particle after the collision?

(b) What is the change in momentum of each particle as a result of the collision?

(c) What is the change in kinetic energy of each particle as a result of the collision?

6.35 Before exploding into two parts, a 6.0-kg object is moving at 20 m/s. After the explosion, a 2.0-kg piece is moving in a direction 50° away from the original direction of motion of the 6.0-kg object. The other part is observed to be moving with a speed of 25 m/s. Calculate the speed of the 2.0-kg piece, the direction of motion of the 4.0-kg piece, and the kinetic energy added to the system by the explosion.

6.36 A 2.0-kg mass moving with a velocity $v_x = 6.0$ m/s has a one-dimensional collision with a 4.0-kg mass moving with a velocity $v_x = 3.0$ m/s. As a result of the collision, an additional 18 J of kinetic energy is added to the system. Calculate the velocity of each of the masses after the collision. (You will get two solutions. Which of these is correct? Can you give a physical interpretation to the other?)

6.37 A sled with an initial mass of 250 kg is to be propelled from rest across a horizontal frozen surface of negligible friction by ejecting 50 kg of mass from the rear with a horizontal speed of 30 m/s *relative to the sled*. What is the final speed of the sled if the 50-kg mass is

(a) released in one firing?

(b) released in two successive firings of 25 kg?

6.38 A mass m_1 (0.20 kg) moving with a speed of 5.0 m/s collides elastically with a mass m_2 (0.30 kg) initially at rest. After the collision, m_1 moves in a direction 30° away from its initial direction of motion.

(a) Determine the speed of each mass after the collision.

(b) What is the direction of motion of m_2 after the collision?

6.39 The three blocks in Figure 6.25 are shown in their initial states on a horizontal frictionless surface. All collisions are head-on and elastic. Calculate the *final* speed of each of the blocks.

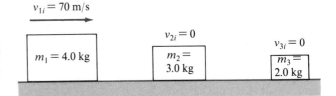

Figure 6.25   Problem 6.39

6.40 The blocks are shown (*see* Fig. 6.26) in their initial states on a horizontal frictionless surface. All collisions are elastic and head-on. What is the final velocity of each of the three blocks?

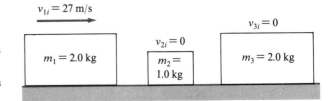

Figure 6.26   Problem 6.40

GROUP **B**

6.41 At an instant when a 20-g mass is moving 15 m/s upward, it is hit from below by a 10-g mass moving 30 m/s upward. This head-on collision is elastic and occurs 15 m above a horizontal surface. When either mass hits this horizontal surface, the mass rebounds instantly with no loss of kinetic energy. How much time elapses between the first and second collisions of these two masses?

6.42 The blocks shown in Figure 6.27 slide on a horizontal frictionless surface. The 4.0-kg block is initially at rest. All collisions are elastic and head-on.

(a) What is the *final* velocity of each block?

(b) Compare the initial and final values of the momentum and kinetic energy of the two-block system. What happened?

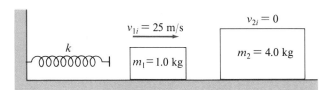

Figure 6.27   Problem 6.42

6.43 The position of the center of mass of a one-dimensional body oriented along the *x* axis is determined from

$$x_{cm} = \frac{\int x \, dm}{\int dm}$$

where *x* is the position of the infinitesimal mass d*m*. Figure 6.28 depicts just such a body extending from $x = 0$ to $x = L$. The infinitesimal quantities d*x* and d*m* are related by

$$dm = \lambda(x)dx$$

where $\lambda(x)$ is the (linear) mass density of the body at point *x*. Thus the total mass *M* and the *x* coordinate of the center of mass are given by

$$M = \int_0^L \lambda(x) \, dx$$

$$x_{cm} = \frac{1}{M} \int_0^L x \, \lambda(x) \, dx$$

(a) Determine *M* and x_{cm} for a uniform rod of linear density λ_o and length *L*.

(b) Determine *M* and x_{cm} for a rod of length *L* with a density that varies according to $\lambda_o x/L$, where λ_o is a constant.

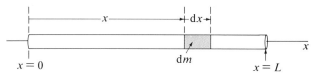

Figure 6.28   Problem 6.43

6.44 The position of the center of mass of a two-dimensional body lying in the *xy* plane is determined from

$$x_{cm} = \frac{\int x \, dm}{\int dm} \qquad y_{cm} = \frac{\int y \, dm}{\int dm}$$

where (x,y) is the position of the infinitesimal mass d*m*. Figure 6.29(a) illustrates such a body. The infinitesimal quantities d*x*, d*y*, and d*m* are related by

$$dm = \sigma(x,y)dydx$$

where $\sigma(x,y)$ is the (areal) mass density of the body at the point (x,y). The total mass *M* and the center of mass coordinates are given by

$$M = \int\int \sigma(x,y)dydx$$

$$x_{cm} = \frac{1}{M} \int\int x \, \sigma(x,y)dydx$$

$$y_{cm} = \frac{1}{M} \int\int y \, \sigma(x,y)dydx$$

The limits on the integrals are determined by the geometry of the body.

(a) Determine *M*, x_{cm}, and y_{cm} for a uniform (areal density σ_o) rectangular body shown in Figure 6.29(b).

(b) Determine *M*, x_{cm}, and y_{cm} for a uniform (areal density σ_o) triangular body shown in Figure 6.29(c).

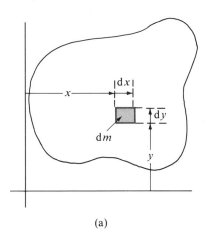

(a)

Figure 6.29(a)   Problem 6.44

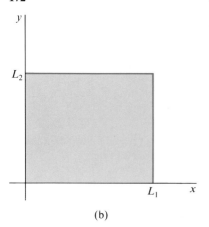

(b)

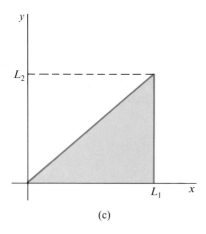

(c)

Figures 6.29b and c

6.45   The semicircular object shown in Figure 6.30 is cut from a uniform sheet of metal. Determine the coordinates of the center of mass of the object.

6.46   A uniform triangle of mass M is joined with a uniform circular quadrant of mass $2M$ as shown in Figure 6.31. Determine the coordinates of the center of mass of the composite body.

6.47   A particle collides elastically with an identical particle initially at rest. Show that the angle between the velocities of the two particles after the collision is $90°$ if the collision is not head-on.

6.48   Consider an elastic one-dimensional collision between two particles. Show that the velocity of one particle relative to the other changes its sign (direction), but not its magnitude, as a result of the collision.

6.49   A rocket is propelled by ejecting fuel with a speed v_o relative to the rocket. Consider a rocket in outer space, that is, a rocket on which no external forces act.

(a)   Use momentum conservation to show that for one-dimensional motion along the x axis

$$m \frac{dv_x}{dt} = -v_o \frac{dm}{dt}$$

where m is the mass of the rocket at the time t, v_x is the velocity of the rocket at the time t, and dm/dt is the (negative) rate at which the mass of the rocket is changing at the time t.

(b)   Show that if $v_x = 0$ and $m = m_o$ at $t = 0$, then

$$m = m_o e^{-(v_x/v_o)}$$

or

$$v_x = v_o \ln(m_o/m)$$

(c)   Suppose that (i) $dm/dt = -m_o/2T$ for $0 \le t \le T$, (ii) $v_x = 0$ at $t = 0$ and (iii) $m = m_o$ at $t = 0$. Use part (b) to determine $v_x(t)$, and determine the maximum speed attained by the rocket.

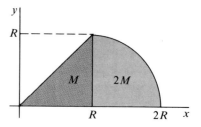

Figure 6.30   Problem 6.45

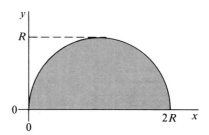

Figure 6.31   Problem 6.46

7 Rotational Motion

We saw in Chapter 4 how torques applied to an extended body may cause the body to turn, or rotate, about an axis that is fixed in space. In this chapter, we will further investigate rotation. We begin with consideration of the simplest physical situation in which rotation takes place, the pure rotation of a body about a fixed axis.

We will see how rotational motion may be described analytically for bodies undergoing pure rotation; *rotational kinematics* describes rotational motion. Next we will introduce *rotational energetics*, an analysis of the ability of bodies to do work because they are rotating. Energetics offers an occasion to introduce the important concept of rotational inertia, which will be especially useful for topics on rotational motion that will be covered later. In discussing bodies undergoing pure rotation, we will define and develop the notion of angular momentum, a fundamental quantity of considerable physical significance.

Then we will consider a more general form of motion of an extended body—the combination of rotation and translation, which involves rotation of a body about an axis that is itself moving through space. We will concentrate on two specific cases of simultaneous rotation and translation: bodies with circular symmetry that are rolling without slipping along a plane surface and bodies with circular symmetry with motions that are determined by a string wound about the body.

Finally, we will develop and illustrate a rotational conservation principle, the conservation of angular momentum. This rotational principle is of comparable stature in physics to that of the conservation of linear momentum, a principle introduced in Chapter 6.

7.1 Rotation About a Fixed Axis

An extended body is composed of many particles. If the distance between each particle of the body remains constant, we call that body a *rigid body*. Rigid bodies are an idealization—they don't exist in any state—but actual bodies can flex. We use the notion of a rigid body to facilitate the analysis of the motion of actual physical bodies. In the analysis of many physical bodies, the rigid-body assumption is an excellent approximation. In this chapter we will assume that all extended bodies are rigid bodies.

Let us now consider the description of the purely rotational motion of a rigid body turning about an axis fixed in space.

Rotational Kinematics

Consider the circular disk of Figure 7.1(a). It is a rigid body attached to a rod lying along an axis passing through the center of the disk normal to the face of the disk. An *axis of rotation* of a body is a line in space about which the particles within the body maintain a constant distance and, therefore, move in a circular path about the axis. Because the disk is a rigid body, the rotational motion of the disk may be described by the motion of an arbitrary particle within the disk. Look, then, at the particle P on the face of the disk shown in Figure 7.1(b). We will now define the variables that specify the rotational motion of the disk.

The angle θ, measured counterclockwise from a fixed reference line to a radial line through the point P, as shown in Figure 7.1(b), is the *angular position* of the particle at point P. The length s of the arc that lies at a distance r from the axis is related to θ by

$$s = r\theta \tag{7-1}$$

when θ is *measured in radians*. The rate at which s changes with respect to time is obtained by differentiating Equation (7-1) with respect to time. Recognizing that r is a constant, we obtain

$$\frac{ds}{dt} = r\frac{d\theta}{dt} \tag{7-2}$$

In Equation (7-2), ds/dt is v_t, the tangential component of the instantaneous velocity of the particle, or

$$v_t = \frac{ds}{dt} \tag{7-3}$$

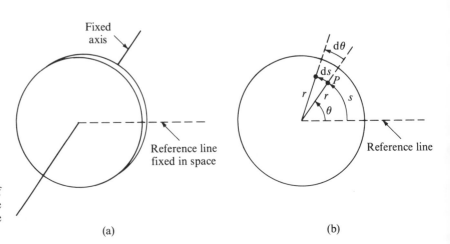

Figure 7.1
A rigid circular disk with a fixed axis of rotation passing through its center. The axis is perpendicular to the face of the disk.

(a)   (b)

We define $d\theta/dt$, the instantaneous time rate of change of angular position, to be the *instantaneous angular velocity* ω of P (and, therefore, of the disk), or

$$\omega \equiv \frac{d\theta}{dt} \qquad (7\text{-}4)$$

The unit of angular velocity is radians per second (rad/s). Using the definitions of Equations (7-3) and (7-4), we may write Equation (7-2) as

$$v_t = r\omega \qquad (7\text{-}5)$$

which relates linear velocity to angular velocity.

   The rotational motion that will be considered in this text is confined to a plane perpendicular to an axis in a fixed direction. This limitation simplifies our treatment significantly, because the rotational motion may be treated as if it were, in effect, one dimensional. It is one dimensional in the sense that the rotational variables are determined by signed ($+$ or $-$) scalar quantities. Because we will consider only the rotation of particles that are turning about an axis with a fixed direction, we may restrict our considerations to angular displacements that are counterclockwise ($\theta > 0$) or clockwise ($\theta < 0$). For the same reason, we will treat ω, which is actually a vector quantity, as a signed quantity that is either CCW ($\omega > 0$) or CW ($\omega < 0$).

   The time derivative of Equation (7-5) yields, since r is constant,

$$\frac{dv_t}{dt} = r\frac{d\omega}{dt} \qquad (7\text{-}6)$$

The left side of Equation (7-6) is a_t, the instantaneous tangential acceleration of the particle at P, or

$$a_t = \frac{dv_t}{dt} \qquad (7\text{-}7)$$

We define $d\omega/dt$ to be the *instantaneous angular acceleration* α of the particle (and of the disk):

$$\alpha \equiv \frac{d\omega}{dt} \qquad (7\text{-}8)$$

The unit of angular acceleration is radians per second per second (rad/s^2). Equation (7-6) may now be written

$$a_t = r\alpha \qquad (7\text{-}9)$$

E 7.1   Show that the centripetal acceleration a_c for point P in Figure 7.1(b) is given by $a_c = r\omega^2$.

E 7.2   Consider a point P on the face of the circular disk of Figure 7.1(b). The distance from the axis of rotation (through the center of the disk) to the point P is 0.15 m.
(a) If the disk rotates through 2.5 revolutions, what total distance does P travel?
                                                                Answer: 2.4 m
(b) At an instant when the disk is rotating at 1.5 rad/s CCW, what is the velocity of point P?   Answer: 0.23 m/s CCW
(c) At an instant when the disk is rotating at 1.5 rad/s CCW and accelerating at 0.50 rad/s^2 CW, what is the acceleration of point P?
                          Answer: $a_t = 7.5\text{cm/s}^2$ CW, $a_c = 34$ m/s^2 radially inward

E 7.3   The angular displacement of a gear is $\theta = (2t^2 - 9t + 4)$ rad where t is measured in seconds.
(a) Calculate its angular velocity and angular acceleration at $t = 2$ s.
                          Answer: $\omega = -1$ rad/s $= 1$ rad/s CW, $\alpha = 4$ rad/s^2
(b) Is the angular acceleration constant?   Answer: Yes

We see now that purely rotational motion may be characterized by the variables θ, ω, and α. Let us now examine the physical interpretation of the signs associated with the rotational quantities. A positive value of θ specifies a counterclockwise (CCW) angular displacement, and, therefore, a negative value of θ is associated with a clockwise (CW) angular displacement. Then because $\omega = d\theta/dt$, a positive value of ω represents CCW rotation; a negative value of ω indicates CW rotation. When ω is positive and its magnitude is increasing, α is positive, and the rotational speed $|\omega|$ is increasing. On the other hand, when ω is positive and its magnitude is decreasing, α is negative, and the rotational speed is decreasing. In general, if ω and α have the same sign, the rotational speed is increasing. When ω and α have different signs, the rotational speed is decreasing.

E 7.4   A wheel is turning CCW and its angular speed is decreasing at a given instant. At that instant,
(a) is the angular displacement increasing or decreasing?   Answer: Increasing
(b) in what direction (sense) is the wheel accelerating?   Answer: CW

There is a correspondence between the rotational variables θ, ω, and α for a particle or rigid body and the variables x, v_x, and a_x that describe translational motion for a particle or rigid body. In Chapter 2 we derived a set of equations (Equation [2-14]) relating x, v_x, a_x, and t for the special case of constant acceleration a_x. A similar set of equations can be derived, using the same procedures as those in Chapter 2, for the rotational variables θ, ω, α, and t for constant α. We will not repeat that procedure here but simply note that the corresponding equations of rotational kinematics for constant angular acceleration may be formed from the translational equations by making the following replacements:

Translational		Rotational
x	$\rightarrow$	θ
v_x	$\rightarrow$	ω
a_x	$\rightarrow$	α

With these substitutions, Equations (2-14) become the equations of rotational kinematics for constant angular acceleration:

Transational		Rotational	
$x = x_o + v_{ox}t + \dfrac{1}{2}a_xt^2$	$\rightarrow$	$\theta = \theta_o + \omega_o t + \dfrac{1}{2}\alpha t^2$	(7-10)
$v_x = v_{ox} + a_xt$	$\rightarrow$	$\omega = \omega_o + \alpha t$	(7-11)
$v_x^2 = v_{ox}^2 + 2a_x(x - x_o)$	$\rightarrow$	$\omega^2 = \omega_o^2 + 2\alpha(\theta - \theta_o)$	(7-12)

In Equations (7-10), (7-11), and (7-12), θ_o and ω_o represent the values of θ and ω when $t = 0$.

E 7.5 A disk, starting from rest at $t = 0$, rotates with a constant angular acceleration of 0.40 rad/s² CCW.

(a) What is the angular velocity of the disk at $t = 3.0$ s?

Answer: 1.2 rad/s CCW

(b) Through what angle does the disk rotate during the first 5.0 s of its motion?

Answer: 5.0 rad CCW

(c) What is the angular velocity of the disk after it has turned 4.0 revolutions?

Answer: 4.5 rad/s CCW

Example 7.1   **PROBLEM**   The motor driving a grinding wheel is turned off at $t = 0$ when the wheel is rotating at 60 revolutions per second (rev/s). The wheel stops rotating 80 s later. Assuming constant angular acceleration,

(a) calculate this angular acceleration.

(b) through what angle does the wheel turn while it is stopping?

(c) what is the angular velocity of the wheel after it has turned 1000 rev?

SOLUTION   Because the angular acceleration α is constant, the angular velocity at any time is given by $\omega = \omega_\text{o} + \alpha t$ where $\omega_\text{o} = 60$ rev/s $= 120\pi$ rad/s. Since $\omega = 0$ at $t = 80$ s, it follows that

$$0 = 120\pi \text{ rad/s} + \alpha(80\text{s})$$

or

$$\alpha = -\frac{120\pi \text{ rad/s}}{80 \text{ s}} = -1.5\pi \text{ rad/s}^2$$

The angle through which the wheel turns while stopping is then

$$\theta = \theta_\text{o} + \omega_\text{o}t + \frac{1}{2}\alpha t^2$$

$$\theta = 0 + (120\pi)(80) + \frac{1}{2}(-1.5\pi)(80)^2$$

$$\theta = 4800\pi \text{ rad} = 2400 \text{ rev}$$

Finally, the angular velocity when $\theta = 1000$ rev $= 2000\pi$ rad is determined by

$$\omega^2 = \omega_\text{o}^2 + 2\alpha\theta = (120\pi)^2 + 2(-1.5\pi)(2000\pi) = 8400\pi^2$$

or

$$\omega = 290 \text{ rad/s} = 46 \text{ rev/s}$$

E 7.6   How much time is required (after $t = 0$) for the grinding wheel of Example 7.1 to turn through 1000 rev?   Answer: 19 s

Rotational Energy and Moment of Inertia

A single particle of mass m rotating at a radius r about a fixed axis with angular speed $|\omega|$ has a linear speed $|v_t|$, which, from Equation (7-5), is $|v_t| = |r\omega|$. The energy that the particle possesses because of its motion is $\frac{1}{2}mv_t^2$. The fact that the particle is rotating does not change its energy of motion, because the kinetic energy of the particle depends only on its mass and speed. It is often convenient, however, to express the energy of a rotating particle in terms of its rotational variables. Then, using Equation (7-5), the kinetic energy of the particle is

$$K = \frac{1}{2} m v_t^2 = \frac{1}{2} m r^2 \omega^2 \tag{7-13}$$

We now define mr^2, the product of the mass of the particle and the square of the radius of its circular path, to be the *moment of inertia I* of the particle about its axis of rotation, that is,

$$I = mr^2 \tag{7-14}$$

The units of moment of inertia are $\text{kg} \cdot \text{m}^2$ and $\text{slug} \cdot \text{ft}^2$. Equation (7-13) may be written in terms of I so that the kinetic energy of the particle is given by

$$K = \frac{1}{2} I \omega^2 \tag{7-15}$$

E 7.7   A 1.2-kg mass moves around a circular path with a 0.50-m radius. The mass moves with a constant speed and completes one cycle every 0.40 s.
(a) What is the angular speed of the mass?   Answer: 16 rad/s
(b) Calculate the moment of inertia of the mass about the axis of rotation.
                                                             Answer: $0.30 \text{ kg} \cdot \text{m}^2$
(c) Use Equation (7-15) to determine the kinetic energy of the mass.
                                                             Answer: 37 J
(d) What is the linear speed of the mass as it moves around the circle?
                                                             Answer: 7.9 m/s
(e) Use $\frac{1}{2} m v^2$ to determine the kinetic energy of the mass. Does your answer agree with the answer for part (c)?   Answer: 37 J

Now we consider a rigid body composed of many individual particles of masses m_1, m_2, . . . at fixed distances r_1, r_2, . . . from the axis of rotation. Because each particle is turning at the angular speed ω of the body about the fixed axis, the kinetic energy K of the rigid body is

$$K = \frac{1}{2} (m_1 r_1^2 \omega^2 + m_2 r_2^2 \omega^2 + \dots) = \frac{1}{2} \left(\sum_i m_i r_i^2 \right) \omega^2 \tag{7-16}$$

Then, much like the case of a single particle, we define the moment of inertia of the rigid body to be

$$I = \sum_i m_i r_i^2 \tag{7-17}$$

Now Equation (7-15), $K = \frac{1}{2} I \omega^2$, represents the rotational kinetic energy of the rigid body having a moment of inertia I and rotating with an angular speed $|\omega|$, just as it does for a single particle.

E 7.8   A rigid body composed of two identical particles each of mass m connected by a light rod with negligible mass and of length L is rotated about an axis perpendicular to the rod and passing through the midpoint of the rod. Let $m = 0.20$ kg and $L = 1.2$ m.
(a) Calculate the moment of inertia of the body about the axis of rotation.
                                                             Answer: $0.14 \text{ kg} \cdot \text{m}^2$

(b) If the body rotates at a constant angular speed of 1.0 rev/s (2π rad/s), what is its kinetic energy?   Answer: 2.9 J

Finally, let us consider rigid bodies composed of a continuous distribution of mass. In such cases, we may assume that each differential mass dm is at a distance r from a particular rotational axis. Then the moment of inertia of the body about the chosen axis is

$$I = \int_{body} r^2 dm \tag{7-18}$$

in which the integration extends throughout the entire volume of the body. The moment of inertia for continuous rigid bodies is readily calculated for some simple geometric shapes about an axis that coincides with an axis of symmetry of the body. The calculation of I using Equation (7-18) may be more difficult for bodies with less regular shapes.

Equation (7-18) shows that the moment of inertia of a body depends on how its mass is distributed relative to a given axis of rotation. The farther away each bit of mass is from the axis, the greater is its contribution to the moment of inertia of the body.

It will be useful to calculate the moment of inertia of a continuous, uniform shape. Consider, then, a uniform thin rod of mass M and length L. *Thin* means its cross-sectional dimensions are small compared to its length. Let us calculate the moment of inertia of this rod about an axis perpendicular to the length of the rod and passing through the center of the rod, as shown in Figure 7.2. Let the rod lie along the x axis. An arbitrary element of mass dm is located a distance x from the axis. The moment of inertia of the rod about this axis is

$$I = \int_{body} r^2 dm = \int_{body} x^2 dm \tag{7-19}$$

We may express the differential mass as $dm = \rho dV$, where ρ is the constant density of the rod. The differential volume dV is equal to Adx, where A is the constant cross-sectional area of the rod. Then Equation (7-19) becomes

$$I = \rho A \int_{-L/2}^{L/2} x^2 dx = \rho A \left[\frac{x^3}{3}\right]_{-L/2}^{L/2} = \frac{\rho A L^3}{12} \tag{7-20}$$

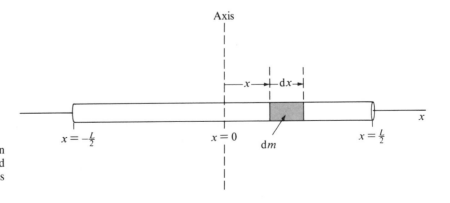

Figure 7.2
A thin rod of length L. The axis shown is perpendicular to the length of the rod and passes through the center of mass of the rod.

TABLE 7.1   Moments of Inertia of Some Common Shapes

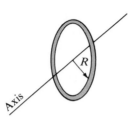

Hoop or cylindrical shell
$I = MR^2$

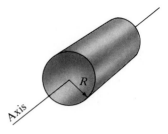

Uniform solid cylinder or disk
$I = \frac{1}{2}MR^2$

Uniform solid sphere
$I = \frac{2}{5}MR^2$

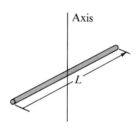

Uniform long, thin rod
$I = \frac{1}{12}MR^2$

The total mass M of the rod is equal to ρAL, so the moment of inertia of the rod about an axis through its center and perpendicular to its length may be written

$$I = \frac{1}{12}ML^2 \tag{7-21}$$

A few geometrical shapes appear frequently in rotational situations. Their moments of inertia can be calculated using Equation (7-18). These moments of inertia are so commonly encountered that they are presented individually in Table 7.1. This table gives the moments of inertia of four geometrical shapes about particular axes, each of which passes through the center of mass of the body. In each case, the total mass of the body is M.

E 7.9   Suppose that the hoop, cylinder, and sphere of Table 7.1 have equal masses and radii. Let I_h, I_c, and I_s be the corresponding moments of inertia calculated using the table.
(a) Show that $I_s < I_c < I_h$.
(b) Explain why you should expect the inequalities of part (a).

The usefulness of the concept of moment of inertia is extended by the *parallel-axis theorem:*

The moment of inertia $I_{\parallel}$ of a body of mass M about an axis parallel to another axis passing through that body's center of mass is given by

$$I_{\parallel} = I_{cm} + Mh^2 \tag{7-22}$$

where I_{cm} is the moment of inertia of the body about the axis through its center of mass and h is the perpendicular distance between the parallel axes.

This theorem will not be proved here. Instead, let us demonstrate its validity by direct calculation of the moment of inertia of a thin uniform rod of mass M and length L about an axis perpendicular to the rod through one end of the rod. In Figure 7.3 we see that the new axis is parallel to one passing through the center (of mass) of the rod. Calculating the moment of inertia $I_{\parallel}$, we again use Equation (7-18):

$$I_{\parallel} = \int_{body} r^2 dm = \rho A \int_0^L x^2 dx$$

$$I_{\parallel} = \rho A \left[\frac{x^3}{3} \right]_0^L = \frac{\rho A L^3}{3}$$

And since $M = \rho A L$, we obtain

$$I_{\parallel} = \frac{1}{3} M L^2 \tag{7-23}$$

This result may be obtained directly from the parallel-axis theorem using $h = L/2$ and Equation (7-21):

$$I_{\parallel} = I_{cm} + M h^2$$

$$I_{\parallel} = \frac{1}{12} M L^2 + M \left(\frac{L}{2} \right)^2 = \frac{1}{3} M L^2 \tag{7-24}$$

E 7.10   A uniform solid sphere has a mass of 6.0 kg and a diameter of 0.20 m. Calculate the moment of inertia of the sphere about an axis
(a) passing through the center of the sphere.
(b) tangential to the surface of the sphere.

Answers: (a) 0.024 kg·m²; (b) 0.084 kg·m²

The parallel-axis theorem confirms our physical interpretation of moment of inertia. By moving the axis of rotation a distance h from an axis passing through the center of mass of a body, we effectively shift the mass of the body farther from the axis of rotation. The moment of inertia thereby becomes correspondingly greater. In the following section we will be able to attach further physical significance to the concept of the moment of inertia of a body.

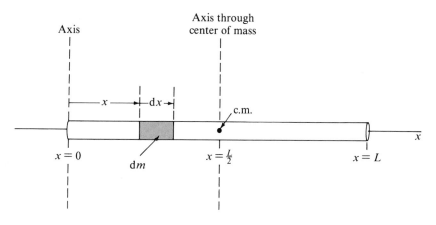

Figure 7.3
A thin rod of length L. The axis passing through the left end of the rod is perpendicular to the rod and parallel to the axis passing through the center of mass of the rod.

E 7.11   What is the radius of a uniform 10-kg sphere that has a minimum moment of inertia of 0.15 kg·m²?   Answer: 0.19 m

E 7.12   Show that the moment of inertia of a rigid body about any one of a set of parallel axes is a minimum for that axis that passes through the center of mass of the body.

Example 7.2   **PROBLEM**   A composite body built using a uniform thin rod and a uniform solid sphere is rotated about an axis perpendicular to the rod, as shown in Figure 7.4. If $M = 3.0$ kg, $L = 0.40$ m, $m = 0.50$ kg, and $r = 0.10$ m, calculate the moment of inertia of the body about the given axis.

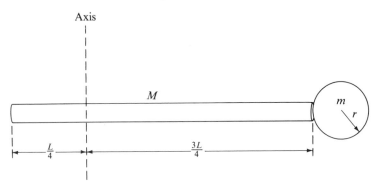

Figure 7.4   Example 7.2

SOLUTION   Because $I = \Sigma_i m_i r_i^2$, it follows that the moment of inertia about any axis of a collection of bodies is equal to the sum of the moments of inertia of the individual bodies (about that same axis). Thus, for this problem, we have $I = I_{\text{rod}} + I_{\text{sphere}}$. If we use the parallel-axis theorem and the appropriate expressions from Figure 7.3, we get

$$I_{\text{rod}} = \frac{1}{12} ML^2 + M\left(\frac{L}{4}\right)^2 = \frac{1}{12}(3)(0.4)^2 + 3(0.1)^2 = 0.070 \text{ kg·m}^2$$

$$I_{\text{sphere}} = \frac{2}{5} mr^2 + m\left(\frac{3}{4}L + r\right)^2 = \frac{2}{5}(0.5)(0.1)^2 + 0.5(0.4)^2 = 0.082 \text{ kg·m}^2$$

Combining these two values gives

$$I = I_{\text{rod}} + I_{\text{sphere}} = 0.070 + 0.082 = 0.15 \text{ kg·m}^2$$

Angular Momentum

The *angular momentum* **L** of a particle at any instant about a point O, the origin of a coordinate frame, is defined to be

$$\mathbf{L} = \mathbf{r} \times m\mathbf{v} = \mathbf{r} \times \mathbf{p} \tag{7-25}$$

where **r** is the position vector of the particle, m is its mass, **v** is its velocity, and **p** is its linear momentum. Angular momentum is a vector quantity. Its magnitude is $L = rp \sin\theta$, where θ is the angle between **r** and **p**; its direction is specified by the right-hand rule for vector products (see Chapter 1). The units of angular momentum are kg·m²/s or slug·ft²/s.

We will concentrate on simple rotational systems in which particles rotate only about axes that have fixed directions. In such cases, we may treat the angular momentum of a particle or a body as a signed ($+$ or $-$) scalar quantity. Then the angular momentum L is positive when ω, the angular velocity, is positive and negative when ω is negative. In those cases for which the rotational motion of a body is not confined to a plane, such three-dimensional motion is appropriately (and necessarily) described using the vectors $\boldsymbol{\tau}$, $\boldsymbol{\omega}$, and $\mathbf{L}$. The vector treatment of three-dimensional motion, like gyroscopic motion, for example, is considered in Problems 7.49–7.52 in the Group B problems at the end of this chapter.

Consider Figure 7.5, which shows a particle of mass m in pure rotation. At a particular instant the particle is turning at a fixed distance r about the axis with a tangential velocity v_t. At this instant, the angular momentum of the particle may be expressed as

$$L = rmv_t = rp_t \qquad (7\text{-}26)$$

where $p_t = mv_t$ is the tangential component of the linear momentum of the particle.

Angular momentum is the rotational counterpart of linear momentum in translational motion. We have seen in Chapter 6 how Newton's second law relates the net force $\Sigma\mathbf{F}$ on a body to the time rate of change, $d\mathbf{p}/dt$, of its linear momentum, that is, we know $\Sigma\mathbf{F} = d\mathbf{p}/dt$. We may expect a similar relationship in rotational motion between the net torque $\Sigma\tau$ on a body and the time rate of change of its angular momentum dL/dt. Let us determine that relationship.

Suppose a particle of mass m is constrained so that it rotates in a circular path of radius r. If a force $\mathbf{F}$ acts on this particle, only that component F_t that is tangential to the circle affects the rotation. The tangential force F_t shown in Figure 7.6 produces a torque (about the center) $\tau = F_t r$.

Now suppose several tangential forces F_{t1}, F_{t2}, . . . are acting on the particle. (Consistent with our use of signed scalar quantities, a tangential force F_t is positive if it produces a CCW torque.) Then the net torque on the particle is

$$\sum\tau = F_{t1}r + F_{t2}r + \ldots = r\sum F_t \qquad (7\text{-}27)$$

Of course, a radial force must be acting on the particle in addition to the tangential forces. This is because a particle of mass m moving in a circular path of radius r at speed v_t has a centripetal force $\Sigma\mathbf{F}_r$ toward the center of magnitude mv_t^2/r. This radial force does not change the rotation of the body because it produces no torque about the axis. The net tangential force that affects rotation is given by

$$\sum F_t = ma_t = m\frac{dv_t}{dt} = \frac{d(mv_t)}{dt} = \frac{dp_t}{dt}$$

where $a_t = dv_t/dt$ is the tangential acceleration of the particle and p_t is the tangential component of the linear momentum of the particle. Then Equation (7-27) becomes

$$\sum\tau = r\frac{dp_t}{dt} \qquad (7\text{-}28)$$

We can show that the right side of Equation (7-28) is the time rate of change of angular momentum by using Equation (7-25):

$$L = rp_t$$

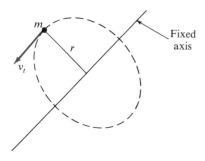

Figure 7.5
A particle of mass m moving at a fixed distance r from a fixed axis. The particle is moving with a tangential velocity v_t.

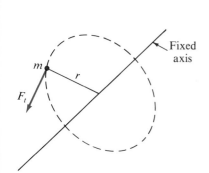

Figure 7.6
A particle of mass m turning about a fixed axis at a constant distance r. The particle is acted on by a force having a tangential component F_t.

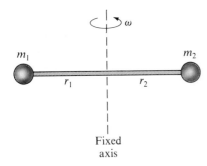

Figure 7.7
Two point masses m_1 and m_2 at fixed distances r_1 and r_2 from an axis fixed in space. The masses rotate at an angular velocity ω.

Because r is constant, we have

$$\frac{dL}{dt} = r\frac{dp_t}{dt} \tag{7-29}$$

Then, comparing Equations (7-28) and (7-29), we conclude

$$\sum \tau = \frac{dL}{dt} \tag{7-30}$$

as we had anticipated. Equation (7-30) is the rotational version of Newton's second law, and it plays a major role in rotational dynamics.

If we now apply the definition of angular momentum (Equation [7-25]) to a particle of mass m moving with velocity v_t in a circular path of radius r, we may write

$$L = rp_t = rmv_t = mr^2\omega \tag{7-31}$$

Here we have used $v_t = r\omega$, where ω is the angular velocity of the particle about the axis. Since mr^2 is the moment of inertia I of the particle about the axis, Equation (7-31) becomes

$$L = I\omega \tag{7-32}$$

Although Equations (7-30) and (7-32) were derived for a single particle, both apply to a rigid body rotating about a fixed axis. This relationship may be seen using Figure 7.7 in a simple case of two point masses m_1 and m_2 connected by a weightless rigid rod, which rotates about an axis perpendicular to the rod. The system is rotating at an angular velocity ω. The masses are at distances r_1 and r_2 from the axis. The total angular momentum L of the system is the sum of L_1 and L_2, the angular momenta of the individual masses:

$$L = L_1 + L_2 = m_1 v_1 r_1 + m_2 v_2 r_2 \tag{7-33}$$

Because $v_1 = r_1\omega$ and $v_2 = r_2\omega$, the angular momentum of the system is

$$L = m_1 r_1^2 \omega + m_2 r_2^2 \omega = (m_1 r_1^2 + m_2 r_2^2)\omega$$

$$L = I\omega \tag{7-34}$$

in which $I = m_1 r_1^2 + m_2 r_2^2$ is the moment of inertia of the rigid body made up of m_1 and m_2. Thus, because Equations (7-32) and (7-34) are identical, the equation $L = I\omega$ applies to a single particle or a rigid body composed of many particles.

E 7.13   What is the angular momentum of a cylindrical shell that has a mass of 2.0 slugs and a radius of 6.0 in. when it is turning about its cylindrical axis at 180 rpm (rev/min). Answer: $9.4 \text{ slug} \cdot \text{ft}^2/\text{s}$

Because the angular momentum L of a rigid body is the sum of angular momenta of the individual particles that comprise the system and because any torque applied to a particle in the system is a torque applied to the system, Equation (7-30), $\sum \tau = dL/dt$, applies to rigid bodies.

Rotational Dynamics

Dynamics relates the motion of objects to the agents that cause that motion. In Chapter 3 we saw that forces are the agents that produce acceleration in translational motion: Newton's second law, $\Sigma\mathbf{F} = m\mathbf{a}$, expresses the relationship between force and acceleration. We may obtain the corresponding relationship between torque and angular acceleration for rotational motion from Equation (7-30). For a system in which the moment of inertia is constant, as is the case for a rigid body constrained to rotate about its axis of symmetry, we may write

$$\Sigma\tau = \frac{dL}{dt} = \frac{d(I\omega)}{dt} = I\frac{d\omega}{dt}$$

$$\Sigma\tau = I\alpha \qquad\qquad (7\text{-}35)$$

Equation (7-35) may be thought of as the rotational form of Newton's second law. Notice that if the net torque $\Sigma\tau$ on a rigid body is zero, the angular acceleration $\alpha = d\omega/dt$ is zero, and the angular velocity ω of the rigid body is constant.

Equation (7-35) suggests how we may further extend our understanding of the physical significance of moment of inertia. Physically, moment of inertia plays a similar role in rotational dynamics to the role of mass in linear dynamics. Recall that $\Sigma\mathbf{F} = m\mathbf{a}$ permits our interpreting mass as a measure of the reluctance of a body to accelerate under the influence of a given force. Similarly, Equation (7-35) suggests that the moment of inertia of an object about a chosen axis is a measure of the reluctance of that object to undergo angular acceleration under the influence of a given torque. Further, Equation (7-17), $I = \Sigma_i m_i r_i^2$, emphasizes that the farther the mass of an object is located from the axis of rotation, the greater will be the moment of inertia about that axis. Distributing more of the mass of an object far from a chosen axis of rotation causes that object to acquire greater resistance to angular acceleration when it is subjected to a given torque about that axis.

E 7.14   A uniform solid cylinder (mass = 20 kg, radius = 0.12 m) is constrained to rotate about its axis of symmetry. The only torque about this axis of rotation results from a person's pulling on and unwinding a string (constant tension = 3.0 N) that is wound around the periphery of the cylinder.
(a) Calculate the net torque about the axis of rotation.   Answer: 0.36 N·m
(b) What is the moment of inertia of the cylinder about the axis of rotation?
                                                             Answer: 0.14 kg·m²
(c) Determine the angular acceleration of the cylinder.   Answer: 2.5 rad/s²

E 7.15   Two particles, each having a mass of 0.40 kg, are attached to opposite ends of a 1.5-m rod of negligible mass. This rigid body is constrained to rotate about a horizontal axis that is perpendicular to the rod and through a point on the rod 0.50 m from one mass, as shown in the accompanying diagram. The rod is initially horizontal.

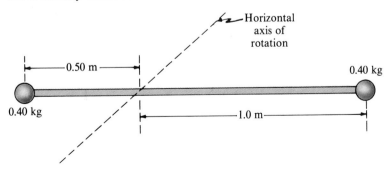

Exercise 7.15

(a) Calculate the initial net torque about the axis of rotation.

Answer: 0.39 N·m CW

(b) What is the moment of inertia about the axis?   Answer: 0.50 kg·m^2

(c) What is the initial angular acceleration?   Answer: 0.78 rad/s^2 CW

When using Equation (7-35) in problems that involve rotational dynamics, the moment of inertia is either given or must be calculated from a specified mass distribution. We may also expect that dynamics problems using Equation (7-35) will often be combined with kinematics, and we may expect to see physical situations that involve both rotation and translation. The following example illustrates typical uses of Equation (7-35) for rotational systems with fixed axes.

Two identical toy rockets, each with a mass m of 0.25 kg, are fixed at opposite ends of a meter stick ($L = 1.0$ m) so that each rocket exerts a force F perpendicular to the stick in the plane of the motion of the stick, as shown in Figure 7.8. The meter stick, which has a mass M of 0.10 kg, is mounted on a frictionless bearing about an axis perpendicular to the stick through the center of the stick. When they are firing, the rockets cause the stick to start from rest and acquire an angular speed of 60 rad/s in 6.0 s. Assume that the masses of the rockets do not change appreciably during this 6.0-s interval. What is the thrust (assumed to be constant) of each rocket?

Which words in the statement of the problem require explanation? *Thrust* refers to the force exerted on the rockets by the escaping gases. A frictionless bearing is a connection between a rotating object and its axis of rotation that exerts no torque on the object. We are seeking the forces causing rotation of the system in this problem and, therefore, should expect to relate the net torque on the stick-rocket system to the angular acceleration of the stick by using $\Sigma \tau = I \alpha$. The moment of inertia, however, is not given, which suggests that we first calculate I for the system. The moment of inertia of the system is the sum of the

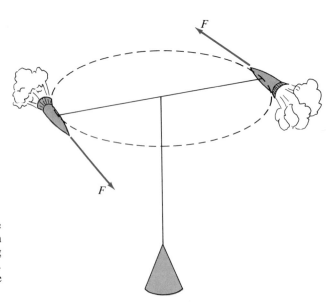

Figure 7.8
Two toy rockets attached to opposite ends of a meter stick that is free to turn about a perpendicular axis passing through the center of mass of the stick. Each rocket exerts a force of magnitude F on the system.

moments of inertia of the stick and of the two rockets, which we shall treat as point masses moving in a circle of radius $L/2$:

$$I = \frac{1}{12}ML^2 + 2m\left(\frac{L}{2}\right)^2$$

$$I = \frac{1}{12}\left(\frac{1}{10}\right)(1)^2 + 2\left(\frac{1}{4}\right)\left(\frac{1}{2}\right)^2 = \frac{2}{15} \text{ kg}\cdot\text{m}^2 \tag{7-36}$$

The angular acceleration α may be determined from the kinematic data given:

$$\omega = \omega_o + \alpha t$$

$$60 = 0 + \alpha(6)$$

$$\alpha = 10 \text{ rad/s}^2 \tag{7-37}$$

The torque produced by each rocket on the system is F times the radius, 0.50 m, of the circular path of the rocket. Each torque is in the CCW sense if we view the system from above, so the net torque $\Sigma\tau$ is the sum of these two CCW torques. We may then substitute the calculated values, $I = 2/15$ kg $\cdot$ m^2 and $\alpha = 10$ rad/s^2, into $\Sigma\tau = I\alpha$ to obtain

$$\sum\tau = I\alpha$$

$$F\left(\frac{1}{2}\right) + F\left(\frac{1}{2}\right) = \left(\frac{2}{15}\right)(10)$$

$$F = \frac{4}{3}\text{ N} \tag{7-38}$$

which is the desired thrust of each rocket.

Another example illustrates the combination of rotational and linear dynamics.

A 32-lb disk is 1.0 ft in diameter and is free to rotate about a fixed axis along the symmetry axis of the disk. A 16-lb block is attached to a string that is wrapped around the disk, as shown in Figure 7.9(a). What is the magnitude of the acceleration of the block?

The linear displacement x of the block is related to the angular displacement θ of the disk through $x = r\theta$. It follows that the linear acceleration a of the block is related to the angular acceleration of the disk by $a = r\alpha$. Because this problem involves two bodies, we may expect that by isolating each body we can obtain two dynamical expressions that are related. Let m represent the mass of the block and M represent the mass of the disk. Isolation of the block, as shown in Figure 7.9(b), gives

$$\sum \mathbf{F} = m\mathbf{a}$$

$$mg - T = ma \tag{7-39}$$

where T is the tension in the string. Equation (7-39) has two unknown quantities, T and a, which suggests that we seek a second relationship by consideration of the other body, the disk. Figure 7.9(b) shows the forces on the disk. Because the disk undergoes no translation, the net force on the disk is zero. The weight $\mathbf{W}$ of the disk and the tension T act downward on the disk, so the bearing that supports the disk must exert an upward force $\mathbf{B}$ on the disk such that $B = T + W$. Neither

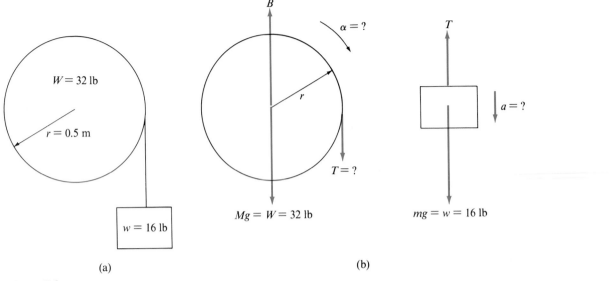

Figure 7.9
(a) A block attached to a string that is wrapped around the periphery of a disk. (b) The forces acting on the block and the disk.

B nor **W** exerts a torque on the disk about the fixed axis. Only T causes a torque, so the net torque $\Sigma\tau$ on the disk is rT, and we have

$$\sum\tau = I\alpha$$

$$rT = \left(\frac{1}{2}Mr^2\right)\left(\frac{a}{r}\right)$$

$$T = \frac{Ma}{2} \qquad\qquad (7\text{-}40)$$

Substitution of $T = Ma/2$ into Equation (7-39) gives

$$mg - \frac{Ma}{2} = ma$$

$$a = \frac{mg}{\left(m + \dfrac{M}{2}\right)} \qquad\qquad (7\text{-}41)$$

and substitution of numerical values, $m = 16/32$ slug and $M = 32/32$, slug gives

$$a = 16 \text{ ft/s}^2 \qquad\qquad (7\text{-}42)$$

E 7.16
(a) Show that the tension T of the string of Figure 7.9 is given by

$$T = \frac{mMg}{2m + M}.$$

(b) Show that the expressions for a, Equation (7-40), and T reduce to the expected values for $M = 0$.

Example 7.3   **PROBLEM**   Two weights ($W_1 = 18$ lb, $W_2 = 20$ lb) are connected by a string that passes over a pulley ($I = 0.040$ slug·ft², $r = 0.20$ ft) arranged as shown in Figure 7.10(a). The string does not slip on the pulley. Calculate the angular acceleration of the pulley.

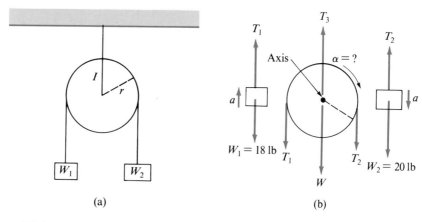

Figure 7.10   Example 7.3   (a)   (b)

SOLUTION   Figure 7.10(b) shows the force diagrams for the two weights and the pulley. Notice that even though the same string is attached to W_1 and W_2, the tension in that string is assigned different values (T_1 and T_2) on the left and right sides of the pulley. These values must be different (in fact, T_1 must be less than T_2) if the pulley is to accelerate (CW) as shown in the figure. (If $T_1 = T_2$, then the net torque about the pulley axis of rotation is zero, and no rotation occurs.) Applying $\Sigma F_y = ma_y$ to each of the weights and $\Sigma\tau = I\alpha$ to the pulley and using $a = r\alpha$ gives

$$T_1 - W_1 = \frac{W_1}{g}a = \frac{W_1}{g}r\alpha$$

$$W_2 - T_2 = \frac{W_2}{g}a = \frac{W_2}{g}r\alpha$$

$$T_2 r - T_1 r = I\alpha$$

If the third equation is divided by r and the three equations are then added, the result is

$$W_2 - W_1 = \left(\frac{W_1 + W_2}{g} + \frac{I}{r}\right)\alpha$$

or

$$\alpha = \left(\frac{W_2 - W_1}{W_1 + W_2 + Ig/r^2}\right)\frac{g}{r}$$

Substituting the values specified in the problem gives

$$\alpha = \left(\frac{20 - 18}{18 + 20 + (0.04)(32)/(0.2)^2}\right)\frac{32}{0.2} = 4.6 \text{ rad/s}^2$$

E 7.17
(a) Explain why in Example 7.3 you should expect T_1 and T_2 to satisfy

$$W_1 < T_1 < T_2 < W_2$$

(b) Calculate T_1 and T_2 for the values specified in Example 7.3 and show that the inequality of part (a) is satisfied.

E 7.18
Show that the equation obtained in Example 7.3, which expresses α in terms of W_1, W_2, I, r, and g, reduces to the expected value for $W_1 = W_2$.

7.2 Simultaneous Translation and Rotation

A physical body may undergo translation and rotation at the same time. Although in general such combined motion may be complex, we will consider only simple systems in which the centers of mass of bodies having circular symmetry move parallel to a straight line fixed in space. A wheel, a disk, and a sphere rolling along a flat surface are examples of such motion. A yo-yo unwinding on a string is another example. Further, we will consider only cases in which bodies roll or unwind without slipping. Before we deal with specific problems involving rolling bodies, however, let us investigate the special role of the center of mass in relating the translational motion of a rolling body to its rotational motion, a role that significantly simplifies the treatment of rolling bodies.

In Figure 7.11, we see a disk of radius r rolling without slipping along a flat surface so that its center of mass C is moving with a speed v_{cm} relative to the surface, which we may assume to be an inertial frame. As the disk rolls without slipping and turns through an angle θ, the arc length s along the periphery of the disk is equal to the distance x_{cm} through which the center of mass translates. If θ is measured in radians, we may write

$$x_{cm} = s = r\theta \tag{7-43}$$

from which

$$v_{cm} = \frac{ds}{dt} = r\frac{d\theta}{dt} = r\omega \tag{7-44}$$

and

$$a_{cm} = \frac{dv_{cm}}{dt} = r\frac{d\omega}{dt} = r\alpha \tag{7-45}$$

These relations, Equations (7-43), (7-44), and (7-45), relate the kinematic variables of translation of the body's center of mass to the associated rotational variables of the body.

We are no longer considering bodies that are turning about fixed axes. Now we are concerned with bodies with centers of mass that are moving, perhaps even accelerating. Can we still use the basic relationship of rotational dynamics, $\Sigma\tau = dL/dt$, when the center of mass of a body is moving? Fortunately, we can because of the following theorem, which we will state without proof:

The net torque $\Sigma\tau_{cm}$ acting on a system and calculated about an axis through the center of mass of the system is equal to the time rate of change dL_{cm}/dt of the angular momentum of the system measured relative to the same axis through the center of mass, *regardless of any motion of the center of mass.*

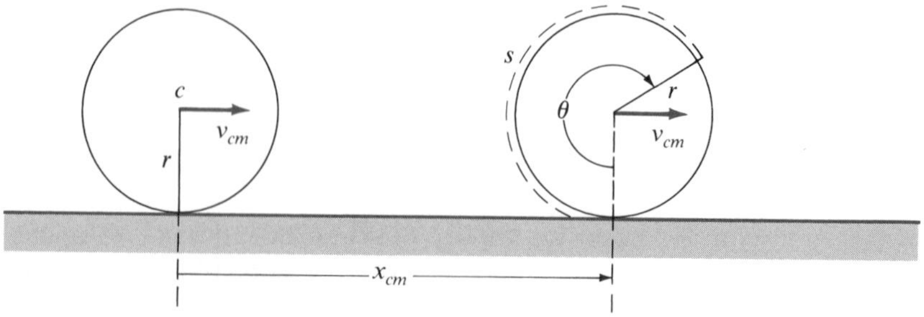

Figure 7.11
A disk of radius r shown before and after it has rolled through an angle θ along a flat surface without slipping. The center of the disk is moving parallel to the surface at a speed v_{cm}. The center moves through a linear distance x_{cm}.

Mathematically, this theorem may be stated:

$$\sum \tau_{cm} = \frac{dL_{cm}}{dt} = I_{cm}\alpha \tag{7-46}$$

where I_{cm} is the (assumed) constant moment of inertia of the system measured about an axis through the center of mass and α is the angular acceleration of the system about the same axis.

One further theorem is useful in analyzing the motion of bodies that are simultaneously translating and rotating:

The total kinetic energy K of a body is the sum of its rotational kinetic energy K_{rot}, measured relative to an axis through the center of mass, and its translational kinetic energy K_{trans}, measured relative to an inertial frame.

We may illustrate the use of this theorem by considering a cylinder of radius R, shown in Figure 7.12, rolling without slipping on a flat surface. In Figure 7.12(a), we see that the center of mass is moving relative to the surface (assumed to be an inertial frame) with speed v_{cm}, and the angular velocity of the cylinder is ω. In Figure 7.12(b) the same rolling cylinder is depicted, but now we consider the cylinder to be rotating about an instantaneous axis along the line of contact between the cylinder and the flat surface, that is, an axis passing through point A. The cylinder is, at this instant, in pure rotation about the axis through A, so its total kinetic energy K is given by

$$K = \frac{1}{2}I_A\omega^2 \tag{7-47}$$

where I_A is the moment of inertia of the cylinder about the instantaneous axis. This moment of inertia is given by the parallel-axis theorem:

$$I_A = I_{cm} + MR^2 \tag{7-48}$$

Substituting Equation (7-48) into Equation (7-47) gives

$$K = \frac{1}{2}(I_{cm} + MR^2)\omega^2$$

$$K = \frac{1}{2}I_{cm}\omega^2 + \frac{1}{2}MR^2\omega^2 \tag{7-49}$$

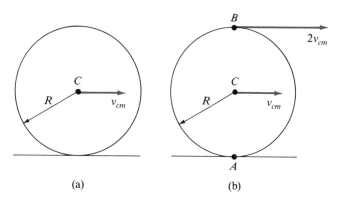

Figure 7.12
A cylinder of radius R rolling without slipping on a flat surface.

(a)   (b)

and because, from Equation (7-44), $v_{cm} = R\omega$, we may write

$$K = \frac{1}{2}I_{cm}\omega^2 + \frac{1}{2}Mv^2_{cm} \tag{7-50}$$

$$K = K_{rot} + K_{trans} \tag{7-51}$$

in which $\frac{1}{2}I_{cm}\omega^2$ is the rotational kinetic energy K_{rot} of the cylinder calculated relative to an axis through the center of mass and $\frac{1}{2}Mv^2_{cm}$ is the kinetic energy associated with the translational motion of the cylinder relative to an inertial frame.

Although we have merely illustrated the theorem, $K = K_{rot} + K_{trans}$, for a specific situation, that theorem is valid generally for bodies that are simultaneously rotating and translating.

E 7.19   A uniform solid sphere rolls without slipping. What fraction of its kinetic energy is associated with its rotation?   Answer: 2/7

We have seen how, even though the center of mass of a body may be accelerating, we may calculate the torques about this center of mass and relate the resultant of those torques to the angular acceleration α with Equation (7-46), $\Sigma\tau_{cm} = I_{cm}\alpha$. The angular acceleration α is related to the translational acceleration a_{cm} of the body's center of mass by $a_{cm} = r\alpha$. Finally, we have shown how the total kinetic energy of the system is shared by purely rotational energy and purely translational energy in Equation (7-49), $K = \frac{1}{2}I_{cm}\omega^2 + \frac{1}{2}Mv^2_{cm}$. In this case, the angular velocity ω and the velocity V_{cm} of the center of mass are related by $v_{cm} = r\omega$. Let us see, then, how these considerations may be applied to appropriate physical situations.

A sphere of mass M and radius R starts from rest and rolls without slipping down an inclined plane from a vertical height h to the bottom of the plane. What is the speed of the center of mass of the sphere when it reaches the bottom?

Figure 7.13 depicts the physical situation. We let the plane be inclined at an angle θ and call d the distance that the body moves along the plane. Is this problem one of dynamics or energetics? It involves forces (gravitational and frictional, in this case) and the motion resulting from those forces, so we could conclude that this is a dynamics problem. On the other hand, the sphere moves through a vertical distance as it descends the plane, thereby converting its gravitational potential energy to kinetic energy. We could, therefore, conclude that the problem should be treated by use of the conservation of energy principle. Then let us treat it by both approaches.

First, considering the problem as one of dynamics, we examine the forces acting on the sphere. These forces are shown in Figure 7.13(b), and the appropriate force diagram is drawn in Figure 7.13(c). The resultant force is in the direction of the assumed acceleration of the center of mass of the sphere. Newton's second law gives

$$\sum \mathbf{F} = M\mathbf{a}$$

$$Mg \sin\theta - f = Ma_{cm} \tag{7-52}$$

where $\mathbf{f}$ is the force of friction on the sphere that prevents slipping, and $\mathbf{a}_{cm}$ is the acceleration of the center of mass of the sphere. The forces on the sphere act at

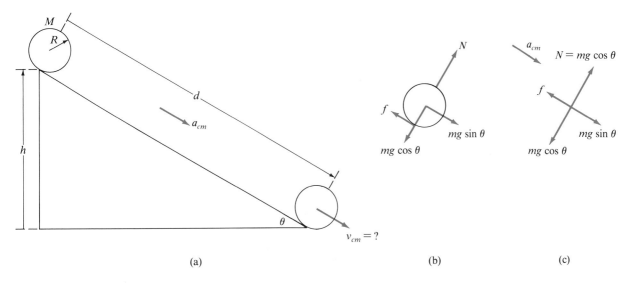

Figure 7.13
(a) A sphere of mass M and radius R rolling without slipping down a plane of length d, inclined at an angle θ with the horizontal.
(b) The forces acting on the sphere. (c) Force diagram for the sphere.

different points, as seen in Figure 7.13(b), and this fact suggests that we consider the torques on the sphere. The appropriate equation of dynamics for rotation is Equation (7-46),

$$\sum \tau_{cm} = I_{cm}\alpha$$

$$fR = \left(\frac{2}{5}MR^2\right)\left(\frac{a_{cm}}{R}\right)$$

$$f = \frac{2}{5}Ma_{cm} \tag{7-53}$$

We have used $\alpha = a_{cm}/R$ because the sphere rolls without slipping. Substitution of Equation (7-53) into Equation (7-52) gives

$$Mg\sin\theta - \frac{2}{5}Ma_{cm} = Ma_{cm}$$

$$a_{cm} = \frac{5}{7}g\sin\theta \tag{7-54}$$

Having obtained the constant acceleration of the sphere down the plane, we may use an appropriate kinematic relationship to find the speed v_{cm} of the sphere when it has moved a distance d:

$$v_{cm}^2 = v_0^2 + 2ad$$

$$v_{cm}^2 = 0 + 2\left(\frac{5}{7}g\sin\theta\right)d$$

Since $h = d\sin\theta$, we have

$$v_{cm}^2 = \frac{10}{7}gh$$

or

$$v_{cm} = \sqrt{\frac{10\,gh}{7}} \qquad (7\text{-}55)$$

Equation (7-55) gives the desired speed.

Now consider the same problem from the point of view of the conservation of energy. At the top of the inclined plane, the sphere is at rest; its total mechanical energy is Mgh, if we choose the bottom to be at the zero of gravitational potential energy. When the sphere reaches the bottom, its potential energy is zero, its translational kinetic energy is $\frac{1}{2}Mv_{cm}^2$, and its rotational kinetic energy is $\frac{1}{2}I_{cm}\omega^2$. Mechanical energy is conserved as the sphere rolls without slipping. Of course, there is a frictional force on the sphere (otherwise it would slide, not roll), but there is no relative motion between the sphere and plane at the point of contact and no energy loss because of friction. Then the conservation of energy requires that

$$Mgh = \frac{1}{2}Mv_{cm}^2 + \frac{1}{2}I_{cm}\omega^2$$

$$Mgh = \frac{1}{2}Mv_{cm}^2 + \frac{1}{2}\left(\frac{2}{5}MR^2\right)\left(\frac{v_{cm}}{R}\right)^2$$

$$Mgh = \frac{1}{2}Mv_{cm}^2 + \frac{1}{5}Mv_{cm}^2$$

$$v_{cm}^2 = \frac{10}{7}gh$$

$$v_{cm} = \sqrt{\frac{10\,gh}{7}} \qquad (7\text{-}56)$$

which is the same result obtained in Equation (7-55) using dynamics.

Another type of problem that involves simultaneous translation and rotation is the "yo-yo" situation:

A uniform disk of mass M and radius R has a string wrapped around its periphery and tied to a ceiling. What is the acceleration of the center of mass of the disk after it is released?

Figure 7.14 shows this situation along with a diagram of the forces acting on the disk. We assume that the center of mass C will accelerate downward while the disk undergoes both translation and rotation. Because forces on the disk cause translational motion and rotational motion, we may expect to use both dynamic relationships, $\Sigma\mathbf{F} = m\mathbf{a}$ and $\Sigma\tau_{cm} = I_{cm}\alpha$. First, let us use

$$\sum F = Ma_{cm}$$

$$Mg - T = Ma_{cm} \qquad (7\text{-}57)$$

Here T is the tension in the string. When the disk unwinds (and we assume the string does not slip), the length x of string that unwinds is equal to $R\theta$, where θ is the angle through which the disk turns about an axis through the center of mass. It follows, then, that $v_{cm} = R\omega$ and $a_{cm} = R\alpha$. Continuing, we use

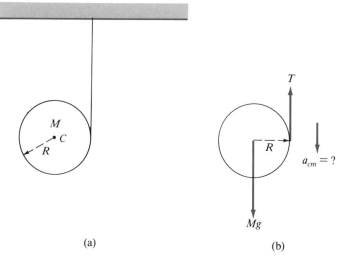

Figure 7.14
(a) A uniform disk of mass M and radius R unwinding on a string wrapped around the periphery of the disk and attached to a ceiling. (b) The forces acting on the disk.

(a)   (b)

$$\sum \tau_{cm} = I_{cm}\alpha$$

$$TR = \left(\frac{1}{2}MR^2\right)\left(\frac{a_{cm}}{R}\right)$$

$$T = \frac{1}{2}Ma_{cm} \qquad\qquad (7\text{-}58)$$

Substitution of Equation (7-58) into Equation (7-57) gives

$$Mg - \frac{1}{2}Ma_{cm} = Ma_{cm}$$

$$a_{cm} = \frac{2}{3}g \qquad\qquad (7\text{-}59)$$

The disk, translating and rotating simultaneously, accelerates downward two-thirds as rapidly as it would if it were falling freely.

E 7.20   Calculate the speed of descent of the center of mass after the disk in Figure 7.15 has fallen a distance of 3.8 m below its initial position of rest.

Answer: 7.0 m/s

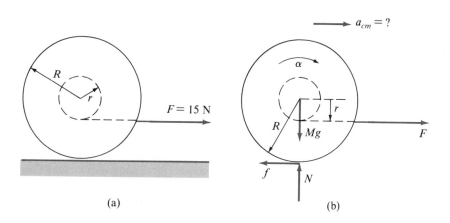

Figure 7.15   Example 7.4   (a)   (b)

E 7.21   At any instant, what fraction of the kinetic energy of the disk in Figure 7.14 is rotational?
Answer: 1/3

Example 7.4

PROBLEM   A 20-kg yo-yo–like object is made of two uniform disks of equal radii ($R = 0.30$ m) joined by a uniform axle of radius $r = 0.10$ m. The yo-yo, which has a moment of inertia of 0.81 kg $\cdot$ m^2 about its symmetry axis, is on a rough surface. A horizontal force of 15-N magnitude is exerted on a string wrapped around the axle, as shown in Figure 7.15(a). Calculate the acceleration of the center of mass of the yo-yo.

SOLUTION   Figure 7.15(b) shows the force diagram for the yo-yo. The frictional force $\mathbf{f}$ is assumed to keep the yo-yo from slipping while rolling, and the angular acceleration α is then related to the acceleration $\mathbf{a}_{cm}$ of the center of mass by $a_{cm} = R\alpha$. Applying $\Sigma F_x = Ma_{cm}$ and $\Sigma \tau_{cm} = I_{cm}\alpha$ to this yo-yo gives

$$F - f = Ma_{cm}$$

$$-Fr + fR = I\alpha = I\left(\frac{a_{cm}}{R}\right)$$

Dividing the second equation by R and then adding the two equations gives

$$F(1 - r/R) = (M + I/R^2)a_{cm}$$

or

$$a_{cm} = \frac{F(1 - r/R)}{M + I/R^2}$$

$$a_{cm} = \frac{15(1 - 0.1/0.3)}{20 + 0.81/0.09} = 0.35 \text{ m/s}^2 \quad \blacksquare$$

E 7.22   Calculate the magnitude of the frictional force acting on the yo-yo of Example 7.4.
Answer: 8.1 N

7.3 Conservation of Angular Momentum

To this point we have been discussing the rotation of particles and of rigid bodies. Sometimes, however, a physical system may change its configuration, that is, the distribution of its mass; and, in doing so, the system may change its state of rotational motion. Let us now consider physical systems that can change their mass distributions as they rotate about a fixed axis. Look at Equation (7-30), $\Sigma \tau = dL/dt$. Suppose now that the net torque $\Sigma \tau$ on a system of particles is zero. In that case, we see that

$$\frac{dL}{dt} = 0 \quad (L = \text{constant}) \tag{7-60}$$

This equation says that if the net torque $\Sigma \tau$ about the fixed axis of rotation is zero, the angular momentum L of the system about this axis is conserved. In other words, the angular momentum of a system remains constant unless a net torque is applied to the system. Of course, both external forces and internal forces may produce torques on the system, but internal forces are produced by

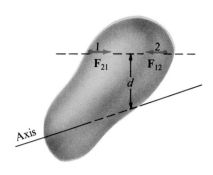

particles within the system on other particles within the system. Newton's third law ensures that for every internal force $\mathbf{F}_{12}$ on particle 2 by particle 1, there is a force $\mathbf{F}_{21}$ on particle 1 by particle 2 that is equal in magnitude and opposite in direction to $\mathbf{F}_{12}$. If these forces, $\mathbf{F}_{12}$ and $\mathbf{F}_{21}$, lie along the same line, as shown in Figure 7.16, they produce torques on the system of equal magnitudes but in opposite directions (because each force has the same lever arm l). Thus there is no net torque on the system resulting from internal forces. Then we may state the *principle of conservation of angular momentum:*

When no net external torque is applied to a system, the angular momentum of the system remains constant.

Figure 7.16
Diagram to illustrate that the net torque on an object due to internal forces $\mathbf{F}_{12}$ and $\mathbf{F}_{21}$ is zero. The distance d is the lever arm of each of the forces $\mathbf{F}_{12}$ and $\mathbf{F}_{21}$ about the axis shown.

The conservation of angular momentum becomes useful, therefore, in analyzing physical systems on which there is no net external torque. By using Equation (7-34), $L = I\omega$, to express the angular momentum of a system, we may write

$$L_i = L_f \tag{7-61}$$

$$I_i\omega_i = I_f\omega_f \tag{7-62}$$

where i and f refer to the initial and final states of the system. Equation (7-62) is a mathematical statement of the principle of conservation of angular momentum. This fundamental principle of physics is of equal stature to the principle of conservation of linear momentum.

The following examples illustrate how the principle of conservation of angular momentum applies to familiar physical situations.

A 130-lb ice skater squats and spins slowly at an angular velocity of 2.0 rad/s with one leg and both arms extended, as shown in Figure 7.17(a). In this configuration, with a massive skate at a considerable distance from the axis of rotation, her moment of inertia about the vertical axis is 4.0 slug·ft². Figure 7.17(b) shows her after she has pulled her arms and leg inward so that her moment of inertia is reduced to 1.0 slug·ft². What is her angular speed in this final configuration?

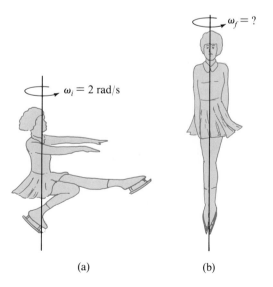

Figure 7.17
An ice skater in two positions spinning about a vertical axis.

(a)   (b)

The rotating system in this problem is the skater, and she changes her configuration by internal forces. Then no external torques are applied to the system, and its angular momentum is conserved, or $L = I\omega$ remains constant:

$$I_i\omega_i = I_f\omega_f$$

$$(4)(2) = (1)\omega_f$$

$$\omega_f = 8.0 \text{ rad/s} \qquad (7\text{-}63)$$

This problem should provide insight into why, when you spring from a diving board rotating slowly about an axis through your center of mass and contract yourself into a "cannonball," you begin to spin rapidly.

Another example of the conservation of angular momentum has some interesting aspects concerning the energy of a rotating system.

A boy of mass $m = 60$ kg stands at the center of a playground merry-go-round rotating about a frictionless axle at 1.0 rad/s. Treat this apparatus as a uniform disk of mass $M = 100$ kg with a radius of $R = 2.0$ m. If the boy jumps to a position on the merry-go-round 1.0 m from the center, what will the angular velocity of the boy–merry-go-round system be after he has landed? Is the kinetic energy of the system conserved?

Figure 7.18(a) shows the system while the boy is at the center. If we treat the boy as a point mass located at the axis of rotation, his mass makes no contribution to the moment of inertia of the system. So the initial moment of inertia I_i of the system is the moment of inertia of the disk:

$$I_i = \frac{1}{2}MR^2 = \frac{1}{2}(100)(2)^2 = 200 \text{ kg} \cdot \text{m}^2 \qquad (7\text{-}64)$$

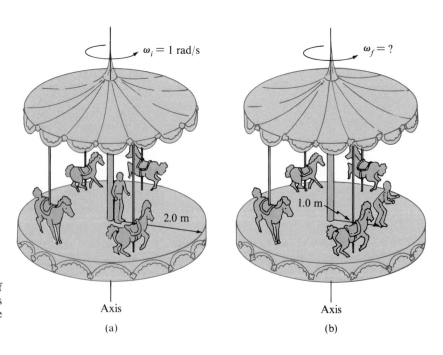

Figure 7.18
A boy on a spinning merry-go-round of radius 2.0 m before and after he leaps from the center to a point 1 m from the center.

After the boy has leapt and landed, the situation shown in Figure 7.18(b), the final moment of inertia I_f of the disk and the point mass m at a distance $r = 1.0$ m from the axis is

$$I_f = \frac{1}{2}MR^2 + mr^2$$

$$I_f = \frac{1}{2}(200)(2)^2 + (60)(1)^2 = 260 \text{ kg} \cdot \text{m}^2 \qquad (7\text{-}65)$$

No external torques were applied to the system during the transition from its initial to its final state. Therefore, the angular momentum of the system is conserved, and we may write

$$I_i\omega_i = I_f\omega_f$$

$$(200)(1) = 260\omega_f$$

$$\omega_f = 0.77 \text{ rad/s} \qquad (7\text{-}66)$$

The system is rotating more slowly after the boy has landed.

Now let us consider the kinetic energy of the system before and after the boy leaps outward. Because the motion of the system is purely rotational, we have

$$K_i = \frac{1}{2}I_i\omega_i^2 = \frac{1}{2}(200)(1)^2 = 100 \text{ J} \qquad (7\text{-}67)$$

$$K_f = \frac{1}{2}I_f\omega_f^2 = \frac{1}{2}(260)(0.77)^2 = 77 \text{ J} \qquad (7\text{-}68)$$

Obviously the system lost kinetic energy during the change in configuration. What happened to the missing energy? An inelastic collision has occurred. As the boy sprang outward from the center, his motion was entirely radial. When he touched the rotating platform, the point of contact on the disk was moving tangentially, but the boy was not. If we think of the boy as a rigid body, his shoes would have skidded on the platform as his motion was accelerated to change from a radially directed velocity to a tangentially directed velocity. Then the lost energy was dissipated as a frictional energy loss during the skid. In the case of a real, nonrigid boy, the boy's impact with the platform was an inelastic collision in which his body was momentarily deformed. We have seen in Chapter 6 how energy is lost in the deformation of bodies that collide inelastically.

If the boy jumped back to the center of the merry-go-round, the system would again be in a state identical with the initial state. The energy of the system would once again be 100 J. Where would the additional energy come from?

7.4 Problem-Solving Summary

Several questions may be asked about any problem that involves a rotating body:

1. Does the rotation take place about a fixed axis?
2. Is the rotating body a rigid body?
3. Do external forces exert a net torque on the rotating system?

4. If the body is rolling (or winding or unwinding), does it roll without slipping?

5. Is the problem one of kinematics, dynamics, or energetics?

These or similar questions serve to focus our attention on those aspects of a problem that can characterize it and simplify the problem-solving process. For example, if the rotation of a body in a problem takes place about a fixed axis, the problem involves pure rotation; we may then ignore any translational aspects of the rotating body. The kinetic energy of the body is then given by $K = \frac{1}{2}I\omega^2$. If the system under consideration is a rigid body rotating about a fixed axis, any particle located at a distance r from the axis of rotation has translational displacement, tangential speed, and tangential acceleration related to its angular variables by $s = r\theta$, $v_t = r\omega$, and $a_t = r\alpha$.

If a rotating body in a problem is not rigid, that is, if the distances between particles that comprise the body change, the problem is probably one that utilizes the principle of conservation of angular momentum. Before we conclude that the angular momentum of a system is conserved, we must be certain that external forces produce no net torque on the system (the necessary condition for conservation of angular momentum). When angular momentum of a system is conserved, the equation $I_i\omega_i = I_f\omega_f$ relates the angular velocities and moments of inertia of the system before (initially) and after (finally) the configuration has been changed to give the system a different moment of inertia.

Rolling (or unwinding) bodies translate and rotate simultaneously. If they roll without slipping, the equations $s = r\theta$, $v_t = r\omega$, and $a_t = r\alpha$ relate the translational and rotational variables of points on or within the rolling body. Special properties of the center of mass of a rigid body permit us to treat a body rolling without slipping in a simple fashion:

1. We may calculate the net torque $\Sigma\tau_{cm}$ about the center of mass of the body, use its moment of inertia I_{cm} about an axis through its center of mass, and treat the problem dynamically using $\Sigma\tau_{cm} = I_{cm}\alpha$ and $\Sigma\mathbf{F} = M\mathbf{a}_{cm}$, in which $a_{cm} = r\alpha$.

2. We may write the kinetic energy K of the body as $K = \frac{1}{2}I_{cm}\omega^2 + \frac{1}{2}Mv_{cm}^2$, in which $v_{cm} = r\omega$.

How do we distinguish between rotational problems that are appropriately treated by kinematics or dynamics or energetics? The issue is kinematic if a given problem requires us to relate any of the quantities θ, ω, α, and t among themselves. If the angular acceleration of a system is constant, the equations that relate the rotational kinematic quantities are

$$\theta = \theta_o + \omega_o t + \frac{1}{2}\alpha t^2$$

$$\omega = \omega_o + \alpha t$$

$$\omega^2 = \omega_o^2 + 2\alpha(\theta - \theta_o)$$

Rotational dynamics problems relate the net torque on a body, its moment of inertia, and its angular acceleration according to $\Sigma\tau = I\alpha$. Problems in which we seek torque or angular acceleration (or their corresponding forces or translational acceleration) are usually handled conveniently using rotational dynamics, that is, using $\Sigma\tau = I\alpha$. On the other hand, rotational problems that specifically ask for the speed of a body (or its corresponding angular velocity) when the body moves through a vertical distance, exchanging gravitational potential energy and kinetic energy, we should consider using the conservation of energy principle.

The correspondence between rotational variables and translational variables permits us to construct appropriate rotational relationships from familiar translational relationships. The following tabulation recalls the correspondence between the variables of translation and rotation:

$$x \rightarrow \theta$$
$$v \rightarrow \omega$$
$$a \rightarrow \alpha$$
$$F \rightarrow \tau$$
$$m \rightarrow I$$
$$p \rightarrow L$$

From these correspondences we may construct numerous composites that are useful in solving problems that involve rotation:

$$\sum F_x = ma_x \rightarrow \sum \tau = I\alpha$$

$$p_x = mv_x \rightarrow L = I\omega$$

$$W = \int F_x dx \rightarrow W = \int \tau d\theta$$

$$K = \frac{1}{2}mv^2 \rightarrow K = \frac{1}{2}I\omega^2$$

$$P = F_x v_x \rightarrow P = \tau\omega$$

$$x = x_0 + v_o t + \frac{1}{2}at^2 \rightarrow \theta = \theta_o + \omega_o t + \frac{1}{2}\alpha t^2$$

$$x(t) = \int v_x(t)dt + C \rightarrow \theta(t) = \int \omega(t)dt + C$$

and so on.

Two problem-solving hazards are associated with problems of rotational motion. The statements of many rotational problems use familiar units, such as degrees, revolutions, or revolutions per minute. Many of the relationships in this chapter are valid only if angles are expressed in radians and time is expressed in seconds. Therefore, the safest problem-solving procedure is to convert all angular measures to radians and all measures of time to seconds. Finally, any time that we calculate torques on a system or the moment of inertia of a system, it is important that we recognize about which axis any rotation in the problem is taking place and that we make those calculations accordingly.

Problems

GROUP **A**

7.1 What constant angular acceleration will bring a wheel (initially turning at 10 rad/s CCW) to rest

(a) in 20 s?

(b) after turning through 8 revolutions?

7.2 A wheel is initially turning at 100 rpm. If a constant torque causes the wheel to slow at an angular acceleration of 2.0 rad/s^2,

(a) how long will it take for the wheel to come to rest?

(b) through what angle does it turn in coming to rest?

7.3 A wheel that has a diameter of 4.0 ft is on a truck moving 60 mi/h (88 ft/s). What is the speed (relative to the earth) of a point

(a) at the top of the wheel?

(b) midway between the ground and the center of the wheel?

7.4 What is the moment of inertia of a cylindrical (diameter = 24 cm, length = 3.0 m) log about its axis of symmetry if the log has a uniform density of 0.80 g/cm^3?

7.5 A single particle of mass m is moving around an axis at a fixed distance r from the axis. By what multiplicative factor is the moment of inertia of the particle about the given axis changed if

(a) the mass is doubled and the distance to the axis is tripled?

(b) the mass is tripled and the distance to the axis is doubled?

7.6 The net work done accelerating a propeller from rest to an angular velocity of 200 rad/s is 3000 J. What is the moment of inertia of the propeller?

7.7 A string is wrapped around the periphery of a wheel 10 cm in diameter. As the wheel turns about a fixed axis, the string unwinds along a straight line tangential to the wheel. When the wheel has an angular speed of 3.0 rad/s and an angular acceleration of 5.0 rad/s^2,

(a) what are the linear speed and acceleration of a point on the string?

(b) what is the magnitude of the linear acceleration of a point on the periphery of the wheel?

7.8 A 16-lb hoop with a radius of 1.5 ft rolls horizontally without slipping at a translational speed of 10 ft/s.

(a) With what angular speed is the hoop rotating about its center of mass?

(b) Calculate the kinetic energy of rotation.

(c) What fraction of the total kinetic energy is rotational?

7.9 A flywheel starting from rest turns with a constant angular acceleration of 1.5 rad/s^2.

(a) What is the angular speed of the flywheel 6.0 s after starting?

(b) Through what angle does the flywheel turn during the first 10 s of its motion?

(c) What is the angular speed of the flywheel when it has turned through 10 rev?

7.10 A flywheel has a moment of inertia of 80 kg·m^2. It is turning at 9.0 rad/s CCW when a constant torque is applied to the wheel to bring it to rest.

(a) How much work must be done to stop the flywheel?

(b) If the flywheel is brought to rest in 1.5 min, what is the torque?

7.11 Suppose the angular speed of a turbine engine is given by $\omega = (2t^2 - t + 4)$ rad/s with t measured in seconds and the angular position of the engine is zero at $t = 0$.

(a) When $t = 3.0$ s, what is the angular acceleration of the turbine?

(b) What is the angular position of the well at $t = 3.0$ s?

7.12 Assume that the earth today is a uniform sphere having a density of 5.5 g/cm^3. What was the length of a day when the earth was cooling and shrinking and had a density of 2.3 g/cm^3?

7.13 A uniform meter stick of mass 0.10 kg is free to rotate about a horizontal axis through one end of the stick. The stick is released from a horizontal position.

(a) What is the angular acceleration of the stick immediately after it is released? Is this acceleration constant?

(b) What is the linear acceleration of the moving end of the stick immediately after it is released? Compare your answer to g.

(c) What are the linear speed and the linear acceleration of the moving end of the stick as that end moves through its lowest point?

7.14 A uniform disk of mass 2.5 kg has a radius of 0.50 m. What constant tangential force must be applied to the periphery of the disk if the disk is to stop 6.0 s after it is turning at 100 rpm CW?

7.15 An automobile wheel (weight = 40 lb, diameter = 28 in.) has a moment of inertia about its axle of 1.1 slug·ft^2.

(a) If the automobile is moving 55 mi/h, what is the kinetic energy of the wheel?

(b) What fraction of the kinetic energy calculated in part (a) is rotational?

7.16

(a) What is the kinetic energy of a cylindrical shell (weight = 20 lb, radius = 0.80 ft) that rolls with a (center-of-mass) speed of 4.0 ft/s on a horizontal surface? The shell rolls without slipping.

(b) What fraction of the kinetic energy calculated in part (a) is translational?

7.17 The system shown in Figure 7.19 is released from rest.

(a) What is the angular speed of the uniform disk when the 8.0-kg block has fallen a distance of 2.0 m?

(b) What fraction of the kinetic energy of the system is associated with the disk when the block has fallen a distance of 2.0 m?

(c) When the block has fallen 3.0 m, what fraction of the total kinetic energy is associated with rotation?

7.18 Initially a uniform thin rod ($M = 2.0$ kg, $L = 1.2$ m) is rotating freely with an angular speed of 10 rad/s about a vertical axis perpendicular to the rod and passing through a point 0.40 m

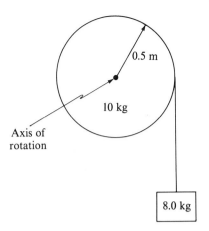

Figure 7.19 Problem 7.17

from one end. Periodically, a 20-gm bead of negligible size is released from the axis and slides to one end of the rod, where it stops and remains fixed.

(a) What are the initial moment of inertia, angular momentum, and kinetic energy of the system?

(b) After five beads are fixed on the long end and eight beads are fixed on the short end, what are the moment of inertia, angular momentum, angular velocity, and kinetic energy of the system?

7.19 If a propeller is accelerated from rest to an angular speed of 3000 rev/min for a period of 5.0 seconds by a constant torque of 1200 ft · lb,

(a) calculate the moment of inertia of the propeller.

(b) what average power is being provided to the propeller?

(c) what instantaneous power is being provided to the propeller at 4.0 s after starting?

7.20

(a) What torque is required to accelerate a flywheel (radius = 0.80 m, I = 150 kg · m^2) at a rate of 0.50 rad/s^2?

(b) If the force producing this torque is applied tangentially to the periphery of the wheel, what is the magnitude of this force?

(c) How much work is done by this force if it accelerates the flywheel (from rest) for 30 s?

7.21 Material having a density of 5.8 g/cm^3 (5800 kg/m^3) is used for making a uniform cylindrical grinding wheel (radius = 0.20 m, thickness = 0.10 m). This wheel rotates about its axis of symmetry at 3.0 rev/s.

(a) What is the kinetic energy of the wheel?

(b) What is the angular momentum of the wheel?

7.22 Two blocks (m_1 = 4.0 kg, m_2 = 6.0 kg) are connected by a string and suspended as shown in Figure 7.20 over a uniform disk (M = 4.0 kg, R = 0.50 m). The disk is free to turn about an axis through the center of the disk and perpendicular to the face of the disk. Assume that the string does not slip on the disk.

(a) What is the magnitude of the accelerations of the blocks?

(b) What is the tension in the part of the string just above m_1? m_2?

(c) What is the tension in the string supporting the disk?

(d) What is the acceleration of the center of mass of the system consisting of the disk and the two blocks?

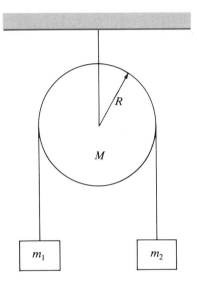

Figure 7.20 Problem 7.22

7.23 At a certain time (t = 0), a uniform sphere (M = 3.0 kg, r = 0.12 m) is rolling without slipping with a center-of-mass speed of 5.0 m/s up a surface inclined 30° with the horizontal.

(a) Calculate the initial (t = 0) kinetic energy of the sphere.

(b) How far (parallel to the plane) will the sphere roll before stopping?

(c) Determine the acceleration of the center of mass of the sphere as the sphere rolls up the inclined surface.

(d) When will the sphere come to rest (momentarily)?

7.24 The uniform sphere (R = 0.80 ft) rolls without slipping across a horizontal surface (*see* Fig. 7.21). At an instant when the center of the sphere is moving with a velocity and an acceleration given by 4.0$\hat{\mathbf{i}}$ ft/s and 10$\hat{\mathbf{i}}$ ft/s^2, calculate the (linear) velocity and acceleration with respect to the surface of

(a) point A.

(b) point B.

(c) point C.

(d) point D.

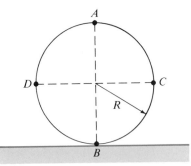

Figure 7.21   Problem 7.24

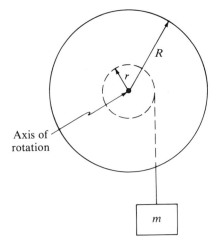

Figure 7.22   Problem 7.27

7.25 A uniform sphere (m = 4.0 kg, r = 0.12 m) is fixed so that it will rotate freely about a horizontal axis. The perpendicular distance from this axis to the center of the sphere is 0.10 m. Initially (t = 0) the sphere is released from rest with its center 0.060 m higher than the axis.

(a) What is the initial angular acceleration of the sphere?

(b) What is the kinetic energy of the sphere when its center is at the same height as the axis?

(c) What are the linear speed and acceleration of the point on the sphere farthest from the axis when this point is at its lowest position?

7.26 A uniform cylinder (weight = 10 lb, radius = 6.0 in.) is fixed so that it will rotate freely about an axis that is parallel to the axis of the cylinder. The perpendicular distance between these two axes is 5.0 in. At t = 0, the cylinder is in its lowest position, and its center of mass is moving with a speed v_o in the horizontal direction.

(a) Determine the minimum value of v_o such that the cylinder will reach its maximum possible height.

(b) If v_o = 10 ft/s, calculate the force on the cylinder by the axis at the lowest position of the cylinder.

(c) Repeat part (b) for the cylinder at its highest position.

(d) Repeat part (b) when the center of mass of the cylinder is at the same height as the axis of rotation.

7.27 A flywheel, which is made from two disks of equal radii (R = 0.40 m) joined by a cylindrical axle of radius r = 0.10 m, is mounted so that it rotates freely about its axis of symmetry. The moment of inertia of the flywheel about this axis is 2.0 kg·m². A mass (m = 5.0 kg) hangs from the end of a string that is wrapped around the axle, as shown in Figure 7.22. This system is released from rest.

(a) How fast is the mass moving after it falls 4.0 m?

(b) Account for the gravitational potential energy lost by the 5.0-kg mass as it falls 4.0 m from the rest position.

7.28 A uniform solid sphere (m = 2.0 kg, r = 0.10 m) is fixed to one end of a 0.10-m rod of negligible mass. This pendulumlike apparatus is suspended so that it swings freely from

the other end of the rod (*see* Fig. 7.23). Initially the apparatus is released from rest with the rod in a horizontal position.

(a) What is the moment of inertia of this apparatus about the axis of rotation?

(b) Calculate the initial angular acceleration of this apparatus.

(c) What is the angular speed of the apparatus as it swings through its lowest position?

(d) What is the tension in the rod at the lowest point of the swing?

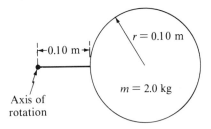

Figure 7.23   Problem 7.28

7.29 The wheel (M = 20 kg, I_{cm} = 0.36 kg·m², R = 0.40 m) shown in Figure 7.24 rolls on its axle (r = 0.050 m) without slipping on a surface inclined 30° with the horizontal. The system starts from rest.

(a) What is the acceleration of the center of mass of the flywheel?

(b) How much time is required for the wheel to move 2.0 m starting from rest?

(c) What is the translational speed of the wheel after it has moved 2.0 m from the rest position?

(d) What fraction of the total kinetic energy is rotational?

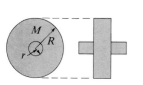

Figure 7.24 Problem 7.29

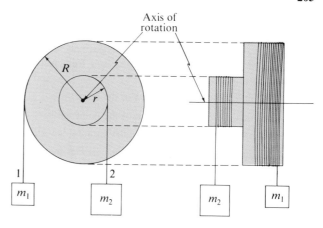

7.30

(a) Determine the speed of the center of mass of a uniform sphere (radius = 0.20 m) after it rolls from rest without slipping for a distance of 2.0 m down a surface inclined 30° with the horizontal.

(b) How much time is required for the sphere of part (a) to roll a distance of 2.0 m if it starts from rest?

(c) Repeat parts (a) and (b) for a uniform cylinder having the same 0.20-m radius.

(d) Repeat parts (a) and (b) for a uniform hoop having the same 0.20-m radius.

(e) Compare the results calculated above to the speed and time required for a block to slide (without friction) 2.0 m from a rest position.

7.31 A mass m (200 g) of negligible dimensions is fixed at the 0.75-m mark on a uniform thin meter stick of mass M (300 g). This apparatus, which is free to rotate about a horizontal axis that is perpendicular to the stick and passes through the 0.20-m mark, is released at $t = 0$ from rest in the horizontal position.

(a) What is the initial linear acceleration of m?

(b) Calculate the linear acceleration of m at the lowest point of its swing.

7.32 Masses m_1 (1.0 kg) and m_2 (6.0 kg) are suspended by strings 1 and 2 from the differential pulley ($I = 0.10$ kg·m², $R = 0.20$ m, $r = 0.080$ m), as shown in Figure 7.25. The pulley rotates freely about its axis.

(a) Calculate the accelerations of m_1 and m_2.

(b) What are the tensions in strings 1 and 2?

(c) What is the resultant torque on the pulley?

7.33 Two uniform hemispheres, each having a radius of 8.0 in. and a weight of 4.0 lb, are connected by a cylindrical axle of negligible weight and 1.0-in. radius. This yo-yo–like object, which is suspended from the ceiling by a string wrapped around the axle (*see* Fig. 7.26), is released from rest with the string vertical.

(a) What is the moment of inertia of the yo-yo about its axis of symmetry?

(b) What is the angular acceleration of the object about its symmetry axis?

Figure 7.25 Problem 7.32

(c) How far does the center of mass fall during the first 2.0 s of motion?

(d) What fraction of the kinetic energy of the yo-yo is rotational?

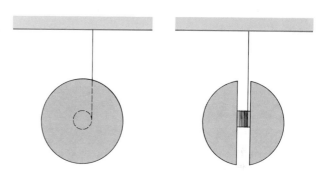

Figure 7.26 Problem 7.33

7.34 A uniform thin rod (length 0.20 m) is suspended so that it swings freely about one end. The rod is released from rest in a horizontal position.

(a) What is the initial angular acceleration of the rod?

(b) Calculate the magnitude of the initial (linear) acceleration of the moving end of the rod.

(c) What are the angular speed and angular acceleration of the rod after it has turned 45° from its initial position?

(d) Calculate the magnitude of the (linear) acceleration of the moving end of the rod after the rod has turned 45° from its initial position.

(e) What are the angular speed, the angular acceleration of the rod, and the magnitude of the (linear) acceleration of the moving end of the rod at the instant the rod is vertical?

7.35   A 0.50-kg point mass is positioned at the center of a uniform thin rod (mass = 1.0 kg, length = 0.60 m). This composite body is suspended so that it swings freely about one end of the rod. Initially the body is released from rest in a horizontal position. What is

   (a)   the initial angular acceleration of the body?

   (b)   the initial (linear) acceleration of the point mass?

   (c)   the angular speed of the body at the instant it is vertical?

   (d)   the (linear) acceleration of the point mass at the instant the rod is vertical?

   (e)   the resultant force on the body at the instant it is horizontal? vertical?

   (f)   the force on the body at the point of suspension at the instant the rod is horizontal? vertical?

7.36   The irregular 4.0-kg body (*see* Fig. 7.27) is suspended so that is swings freely about an axis that is perpendicular to the paper and through point A. The perpendicular distance from this axis to the center of mass of the body is equal to 0.20 m. The moment of inertia of the body about an axis that is perpendicular to the paper and through the center of mass is 0.040 kg·m². If the body is released from rest in the position shown, that is, when the line connecting point A and the center of mass is horizontal, what is

   (a)   the initial force on the body by the support at A?

   (b)   the force on the body by the support at A at the instant the center of mass is directly below A?

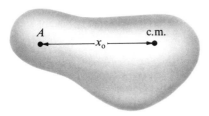

Figure 7.27   Problem 7.36

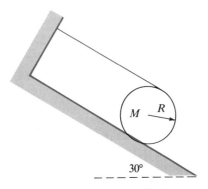

Figure 7.28   Problem 7.37

7.37   A light string is wrapped around a uniform cylindrical disk (M = 4.0 kg, R = 0.10 m) and attached to a wall as shown in Figure 7.28. The inclined surface is frictionless.

   (a)   What is the acceleration of the center of mass of the disk?

   (b)   Calculate the tension in the string.

   (c)   Determine the translational and rotational kinetic energies of the disk after its center of mass has traveled 2.0 m down the plane from an initial position of rest.

GROUP **B**

7.38   Using $\int r^2 dm$, derive the expression $I = \frac{1}{2}MR^2$, which gives the moment of inertia about the symmetry axis of a uniform cylinder of mass M and radius R.

7.39   Using $\int r^2 dm$, derive the expression $I = \frac{2}{5}MR^2$, which gives the moment of inertia about an axis through the center of a uniform sphere of mass M and radius R.

7.40   Consider a spherical shell of radius R, which has a mass M distributed uniformly over the surface. Assume the thickness of this layer of mass is negligible.

   (a)   Show that the moment of inertia about any axis through the center of this shell is given by $I = 2MR^2/3$.

   (b)   If this shell rolls without slipping, what fraction of its kinetic energy is rotational?

7.41   Figure 7.29 shows a block (w = 10 lb) suspended from a string that is wrapped around a uniform cylinder (W = 48 lb, R = 0.80 ft). The cylinder rolls without slipping on the horizontal surface. Disregard any mass or friction of the pulley.

   (a)   Determine the magnitude of the acceleration of the block.

   (b)   What is the magnitude of the acceleration of the center of mass of the cylinder?

   (c)   What is the angular acceleration of the cylinder about its center of mass?

   (d)   What is the tension in the connecting string?

   (e)   What is the frictional force on the cylinder?

   (f)   After the block has fallen 2.0 ft from an initial position of rest, calculate the kinetic energy of the block, the translational kinetic energy of the cylinder, and the rotational kinetic energy of the cylinder.

7.42   Figure 7.29 shows a block (w = 10 lb) suspended from a string that is wrapped around a uniform cylinder (W = 48 lb, R = 0.80 ft). The horizontal surface on which the cylinder moves is frictionless. Disregard any mass or friction of the pulley.

   (a)   What is the magnitude of the acceleration of the block?

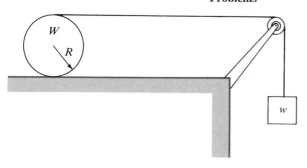

Figure 7.29   Problems 7.41 and 7.42

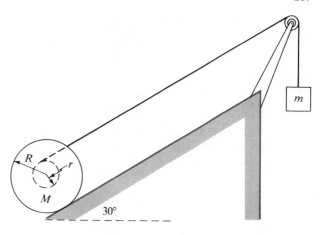

Figure 7.30   Problems 7.44 and 7.45

(b) Calculate the magnitude of the acceleration of the center of mass of the cylinder.

(c) What is the angular acceleration of the cylinder about its center of mass?

(d) What is the tension in the connecting string?

(e) After the block has fallen 2.0 ft from an initial position of rest, calculate the kinetic energy of the block, the translational kinetic energy of the cylinder, and the rotational kinetic energy of the cylinder.

7.43 A 75-kg person stands on the perimeter of a horizontal circular platform (mass = 250 kg, radius = 4.0 m) which is rotating freely about a vertical axis through its center. The platform is a uniform disk. At $t = 0$ when the platform is rotating with an angular speed of 0.10 rev/s, the person starts to walk at a constant speed of 1.0 m/s toward the center along a radial line painted on the platform.

(a) Determine the angular speed of the platform as a function of time.

(b) Calculate the initial and final values of the kinetic energy of the system consisting of the platform and the person.

(c) At what average rate is this person doing work during the walk?

7.44 A yo-yo–like apparatus ($M = 2.0$ kg, $I_{cm} = 0.16$ kg·m²) is made from two identical disks of radii $R = 0.30$ m joined by a cylindrical axle of radius $r = 0.10$ m. The yo-yo rolls without slipping on a surface inclined 30° with the horizontal. A string wrapped around the axle has its other end attached to a block ($m = 0.90$ kg), as shown in Figure 7.30. Disregard any mass and friction of the pulley. The system is released from rest. Calculate

(a) the acceleration of the block.

(b) the acceleration of the center of mass of the yo-yo.

(c) the angular acceleration of the yo-yo about its center of mass.

(d) the tension in the string.

(e) the kinetic energy of the yo-yo after the block has moved 2.0 m.

7.45 Consider the "yo-yo" of Problem 7.44 with $M = 2.0$ kg, $r = 0.10$ m, and $R = 0.30$ m. Calculate

(a) the value for m if the system does not move when released from rest.

(b) the frictional force on the yo-yo by the surface if the system remains at rest.

7.46 Repeat Problem 7.44 using Figure 7.31 with $M = 2.0$ kg, $I_{cm} = 0.16$ kg·m², $m = 0.90$ kg, $R = 0.30$ m, and $r = 0.10$ m.

7.47 Mass is distributed throughout a cylindrical volume (length = L, radius = R) with a nonuniform density given by $\rho = \alpha r (0 < r < R)$, where r is the perpendicular distance from the axis of the cylinder and α is a constant.

(a) Determine the mass M of the cylinder in terms of α, R, and L.

(b) Determine the moment of inertia I of the cylinder about its axis of symmetry. Express your result in terms of M and R.

(c) Can you explain why your result for I obeys the inequality $\frac{1}{2}MR^2 < I < MR^2$?

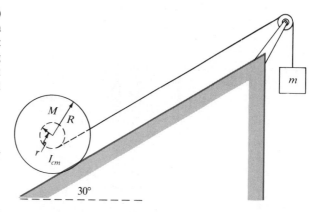

Figure 7.31   Problem 7.46

7.48 A mass M is uniformly distributed on a flat area that is rectangular in shape. The lengths of adjacent sides of the rectangle are L_1 and L_2. Calculate the moment of inertia of this distribution about an axis perpendicular to the flat surface and passing through

(a) the geometric center of the rectangle.

(b) one corner of the rectangle.

7.49 We have seen that rotational motion in a plane, like translational motion along a line, can be analyzed by using (signed) real number quantities like torque, angular velocity, and the like. But, just as with translational motion, the extension to three-dimensional motion is most easily accomplished with vector quantities. For example, the torque $\boldsymbol{\tau}$ about the origin caused by a force $\mathbf{F}$ acting at the position $\mathbf{r}$, as shown in Figure 7.32(a), is defined by

$$\boldsymbol{\tau} = \mathbf{r} \times \mathbf{F}$$

(a) Show that $\boldsymbol{\tau}$ is perpendicular to the plane containing $\mathbf{r}$ and $\mathbf{F}$ and has a magnitude given by $rF \sin\theta$, where θ is the smaller angle between $\mathbf{r}$ and $\mathbf{F}$.

(b) Show that the magnitude of the torque is still given by

$$\tau = rF_\perp$$

where $F_\perp$ is the component of $\mathbf{F}$ perpendicular to $\mathbf{r}$.

(c) Show that for a force $\mathbf{F} = F_x\hat{\mathbf{i}} + F_y\hat{\mathbf{j}}$ acting at the point $\mathbf{r} = x\hat{\mathbf{i}} + y\hat{\mathbf{j}}$, we have

$$\boldsymbol{\tau} = \hat{\mathbf{k}}(xF_y - yF_x) = \hat{\mathbf{k}}\,rF \sin(\beta - \alpha)$$

where β and α are shown in Figure 7.32(b).

(d) Show that CCW (CW) torque for forces in the $x-y$ plane is a vector in the positive (negative) $\hat{\mathbf{k}}$ direction.

7.50 The angular momentum $\mathbf{L}$ about the origin of a particle moving with momentum $\mathbf{p}$ through a position $\mathbf{r}$, as shown in Figure 7.33(a), is defined to be

$$\mathbf{L} = \mathbf{r} \times \mathbf{p}$$

(a) Show that $\mathbf{L}$ is perpendicular to the plane containing $\mathbf{r}$ and $\mathbf{p}$ and has a magnitude given by $rp \sin\phi$, where ϕ is the smaller angle between $\mathbf{r}$ and $\mathbf{p}$.

(b) Show that the magnitude of the angular momentum is still given by

$$L = rp_\perp$$

where $p_\perp$ is the component of $\mathbf{p}$ perpendicular to $\mathbf{r}$.

(c) Show that for a momentum $\mathbf{p} = p_x\hat{\mathbf{i}} + p_y\hat{\mathbf{j}}$ at the position $\mathbf{r} = x\hat{\mathbf{i}} + y\hat{\mathbf{j}}$, we have

$$\mathbf{L} = \hat{\mathbf{k}}(xp_y - yp_x) = \hat{\mathbf{k}}\,rp \sin(\beta - \alpha)$$

where β and α are shown in Figure 7.33(b).

(d) Show that the angular momentum for CCW (CW) motion in the $x-y$ plane is a vector in the positive (negative) $\hat{\mathbf{k}}$ direction.

(e) Show that for any particle

$$\frac{d\mathbf{L}}{dt} = \sum \boldsymbol{\tau}$$

where $\sum \boldsymbol{\tau}$ is the vector sum of all torques on the particle.

(f) Show that if the only force on a particle is radial, that is, the force always points toward or away from the origin, then the angular momentum (about the origin) of the particle is constant.

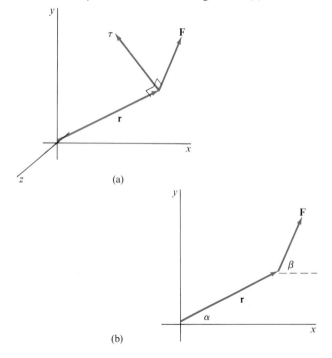

(a)

(b)

Figure 7.32   Problem 7.49

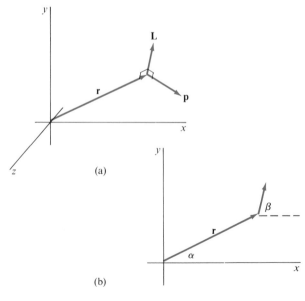

(a)

(b)

Figure 7.33   Problem 7.50

7.51 The angular momentum for a number of particles is defined to be the vector sum of the angular momenta of the particles. For a two-particle system, for example, we have

$$\mathbf{L} = \mathbf{L}_1 + \mathbf{L}_2 = \mathbf{r}_1 \times \mathbf{p}_1 + \mathbf{r}_2 \times \mathbf{p}_2$$

(a) Show that for a two-particle system with central internal forces, that is, the force on particle *A* by particle *B* acts along the line connecting the two particles, the angular momentum **L** of the two-particle system obeys

$$\frac{d\mathbf{L}}{dt} = \sum \boldsymbol{\tau}(\text{ext})$$

where $\sum \boldsymbol{\tau}(\text{ext})$ is the vector sum of torques on the system caused by agents external to the system.

(b) Suppose two particles, which interact via a central force, move in an external field that is radial (*see* Problem 7.50). Show that the angular momentum of the system is conserved.

7.52 A toy gyroscope is essentially a composite body consisting of a disk and an axle (*see* Fig. 7.34[a]). If the disk is mounted so that it rotates freely about the axle with an angular speed ω, as shown in Figure 7.34(b), the gyro has an angular momentum **L** in the direction indicated. If *I* is the moment of inertia of the gyro about its symmetry axis, then $L = I\omega$. Now, if (i) one end of the axle is mounted so that the gyro can rotate freely about a point *P*, as shown in Figure 7.35(a), (ii) the gyro is oriented so that its axle (and, therefore, **L**) is horizontal, and (iii) the free end of the axle of the gyro is started moving CCW as viewed from above, the gyro does not fall, as might be expected, but rotates about point *P* with its axle remaining horizontal as shown in Figure 7.35(b). The gyro is said to *precess* about the vertical axis through point *P*. This motion is explained by using $\sum \boldsymbol{\tau} = d\mathbf{L}/dt$. The only torque on the gyro about point *P* is caused by the weight *mg* acting with a lever arm *l*. This torque $\boldsymbol{\tau}$ ($\tau = mgl$) is shown in Figure 7.36(a). Notice that $\boldsymbol{\tau}$ is horizontal and perpendicular to the angular momentum **L**.

(a) Show that if $\boldsymbol{\tau}$ is perpendicular to **L**, that, even though **L** is changing ($d\mathbf{L}/dt = \boldsymbol{\tau}$), its magnitude remains constant. Thus, **L** is rotating such that its change d**L** in a time d*t* is given by τdt, as shown in Figure 7.36(b).

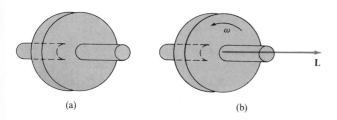

(a)   (b)

Figure 7.34   Problem 7.52

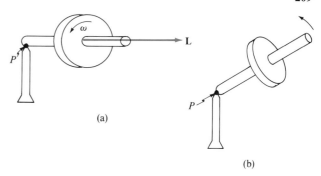

(a)

(b)

Figure 7.35   Problem 7.52

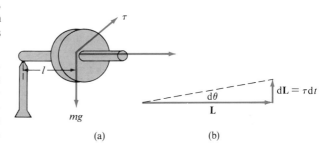

(a)   (b)

Figure 7.36   Problem 7.52

(b) Using Figure 7.36(b), show that the angle dθ through which **L** turns in a time d*t* is given by

$$d\theta = \frac{mgl}{I\omega}dt$$

or

$$\frac{d\theta}{dt} = \frac{mgl}{I\omega}.$$

Thus, **L** (and, therefore, the gyro) precesses about point *P* with a constant angular rate given by dθ/d*t*.

(c) Show that the time *T* for one cycle of precession is given by

$$T = \frac{2\pi I\omega}{mgl}$$

(d) If the gyro disk is uniform with a radius *R*, show that

$$T = \frac{\pi R^2 \omega}{gl}$$

(e) Calculate the precessional period for a gyro, which has $R = 3.0$ cm and $l = 2.0$ cm, that is spinning with a frequency of 20 Hz.

8 Oscillations

An *oscillation* is a back-and-forth motion over the same path. If an oscillating object executes each *cycle*, or repetition of its motion, in equal intervals of time, the motion is said to be *periodic*. The motion of a pendulum of a clock is an example of periodic oscillation. In this chapter we will concentrate on the simplest form of periodic oscillation, called simple harmonic motion, in which oscillatory motion takes place along a straight line. Simple harmonic motion plays a fundamental role in several areas of physics, and it is associated with several topics that are covered in subsequent chapters of this text.

Idealized systems that execute simple harmonic motion are the primary considerations of this chapter. A discussion of quantities that characterize simple harmonic motion will be followed by a detailed kinematic analysis of that kind of motion. Then we will consider the dynamics of ideal oscillators, relating the forces that cause simple harmonic motion to the motion itself. Energy considerations will complete our analysis of idealized oscillators.

Having described idealized simple harmonic motion in terms of kinematics, dynamics, and energetics, we will turn our attention to more realistic mechanical oscillators. Damped oscillations, back and forth motions with decreasing energies, will be given a brief treatment. Finally, oscillators that are driven, or forced, by an external mechanism will be discussed qualitatively. These forced oscillations demonstrate resonance, an important physical phenomenon observed in many kinds of systems.

8.1 Simple Harmonic Motion

Our description of simple harmonic motion will necessarily be essentially mathematical. You may wish to create a mental image of the kind of motion we are describing by imagining a long spring hanging from a ceiling with a block attached to the lower end of the spring. When the block has been pulled downward and released, it oscillates, vertically in this case. First we will describe an idealization, that is, a frictionless version, of this kind of motion.

Let us consider the rectilinear (straight-line) motion of a particle with displacement from $x = 0$ that is described by

$$x(t) = A \cos (\omega t + \phi) \tag{8-1}$$

where A, ω, and ϕ are constants. Motion of this type is said to be "harmonic" because it can be described by sinusoidal (sine or cosine) functions. The motion is said to be "simple" in that only one sine or cosine function is required to describe it. Equation (8-1), therefore, describes the displacement from $x = 0$ of a particle executing *simple harmonic motion*. The argument $(\omega t + \phi)$ of the cosine factor in Equation (8-1) is called the *phase* of the oscillation, and ϕ is called the *phase constant*. Since A is a constant, the displacement x of the particle is periodic because cosine is a periodic function that repeats itself each time the phase increases by 2π radians. Furthermore, x ranges from $x = A$, when the cosine factor is equal to 1, to $x = -A$, when the cosine factor is equal to -1, and back again to $x = A$, and so on, as time proceeds from $t = 0$. Thus, the motion is oscillatory as well as periodic.

In Equation (8-1), when t is such that $\cos(\omega t + \phi)$ has the value 1, the displacement x of the particle achieves its greatest value, $x = A$. The *amplitude A* of the oscillation is defined to be the maximum distance the particle moves from its central position $x = 0$. The *period T* of the oscillation is defined to be the time interval, usually measured in seconds, for the particle to complete one cycle, that is, for ωt to increase by 2π radians, or

$$T = \frac{2\pi}{\omega} \tag{8-2}$$

The *frequency ν* of the oscillation is the number of cycles the particle completes in one second. Therefore, the frequency is the reciprocal of the period, or

$$\nu = \frac{1}{T} = \frac{\omega}{2\pi} \tag{8-3}$$

The unit of frequency is understandably cycles per second, which a committee has agreed to call a hertz (Hz). Because "cycle" is not a dimensional unit, the hertz is equal to the unit $s^{-1} \equiv 1/s$.

E 8.1   An oscillator completes four cycles every 5 s. The amplitude of the motion is 3.0 cm. Calculate
(a) the frequency of the motion.   Answer: 0.80 Hz
(b) the period of the motion.   Answer: 1.3 s
(c) the total distance traveled by the oscillator during a 5.0-s time interval.
                                                                                       Answer: 0.48 m

E 8.2   The motion of a particle is described by

$$x = 0.1 \cos(6t + \pi) \text{ m}$$

where t is measured in seconds. For this motion calculate the
(a) amplitude.   Answer: 0.10 m
(b) frequency.   Answer: 1.0 Hz
(c) period.   Answer: 1.1 s

E 8.3
(a) Show that simple harmonic motion of frequency ν is described by the expression $x(t) = A \cos(2\pi\nu t + \phi)$.
(b) Show that simple harmonic motion of period T is described by the expression $x(t) = A \cos[(2\pi t)/T + \phi]$.

It is particularly instructive to relate the simple harmonic motion of a particle to circular motion. A particle moving in a circular path at constant speed is in periodic motion (though, of course, it is not in oscillation). Figure 8.1 shows a point (or a particle) P_c moving in a circular path of radius A at a constant angular velocity ω. Beneath the circle is a straight line, the x axis, along which a particle, or point, P_s is executing simple harmonic motion. In Figure 8.1, we have, for simplicity, chosen the phase constant ϕ to be zero. Further, when P_s is at $x = A$ we have chosen $t = 0$ so that Equation (8-1), with $\phi = 0$, describes the position of P_s at any time t:

$$x(t) = A \cos\omega t \tag{8-4}$$

In Figure 8.1, the projection of P_c onto the x axis is, at all times, at a position $x = A \cos\omega t$, which is Equation (8-4), the displacement of the particle in simple harmonic motion. A circle associated in this way with a particle executing simple harmonic motion is the *reference circle* for that motion. Then ω, which is the angular velocity of P_c on the reference circle, is defined to be the *angular frequency* of the simple harmonic motion. When P_c has turned through one revolution, or 2π radians, P_s has executed one cycle of its oscillation. Therefore, the angular frequency ω is 2π times the frequency ν of the simple harmonic motion, which is consistent with Equation (8-3).

The phase constant ϕ permits us to describe the motion of an oscillating particle when the particle is at an arbitrary position in its cycle at $t = 0$. Figure 8.2 shows a situation in which a phase constant $\phi > 0$ has been added to the phase of Equation (8-4). The diagram indicates that ωt is measured from the radius of the reference circle passing through P_{co}, the position on the reference circle corresponding to $t = 0$. Then the projection of P_c onto the x axis is located by

$$x(t) = A \cos(\omega t + \phi)$$

When $t = 0$, the initial position of P_s, the particle executing simple harmonic motion, is

$$x(0) = x_o = A \cos\phi \tag{8-5}$$

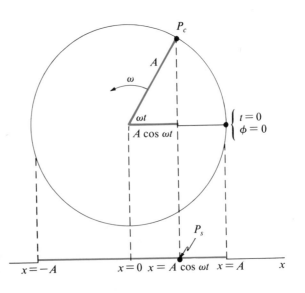

Figure 8.1
A reference circle for a particle executing simple harmonic motion with phase constant $\phi = 0$.

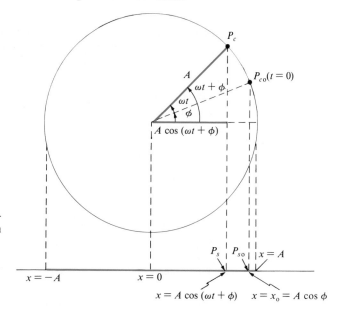

Figure 8.2
A reference circle for a particle executing simple harmonic motion with phase constant $\phi > 0$.

To illustrate the use of a phase constant, let us choose a negative value of ϕ, say $\phi = -\pi/2$ radians. Figure 8.3 shows that the projection of P_c onto the x axis is

$$x(t) = A \cos\left(\omega t - \frac{\pi}{2}\right)$$

which, because of the trigonometric identity, $\cos(\theta - \pi/2) = \sin\theta$, becomes

$$x(t) = A \sin\omega t \qquad \text{(8-6)}$$

Thus, because $\sin\omega t$ is zero when $t = 0$, the position $x(t)$ in Equation (8-6) describes a particle in simple harmonic motion that is at $x = 0$ when $t = 0$. Recall

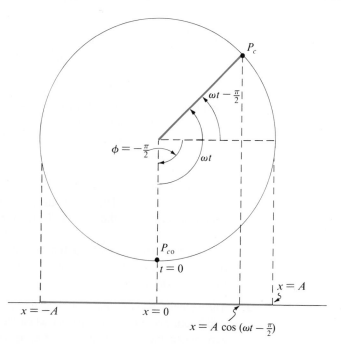

Figure 8.3
A reference circle for a particle executing simple harmonic motion with phase constant $\phi = -\pi/2$.

that Equation (8-4), $x(t) = A \cos\omega t$, describes a particle in simple harmonic motion that is at $x = A$ at $t = 0$. Therefore, the phase constant ϕ is a convenience that permits the description of a particle executing simple harmonic motion that starts at any position in the cycle.

E 8.4   While the cosine function has been used here to describe simple harmonic motion, the sine function can be used just as well. To see this, use the trigonometric identity $\cos\theta = \sin(\theta + \pi/2)$ to show that

$$x = A \sin(\omega t + \alpha)$$

is completely equivalent to

$$x = A \cos(\omega t + \phi)$$

if α is chosen as $\phi + \pi/2$.

E 8.5

(a) Write the equation $x = 0.4 \cos(6t + \pi/3)$ using the sine function.

Answer: $x = 0.4 \sin(6t + 5\pi/6)$

(b) Write the equation $x \doteq 0.6 \sin(3t - \pi)$ using the cosine function.

Answer: $x = 0.6 \cos(3t - 3\pi/2)$

Kinematics of Simple Harmonic Motion

We have already described the displacement of a particle in simple harmonic motion by Equation (8-1). The remaining kinematic quantities associated with this motion, velocity and acceleration, may be obtained in terms of the same parameters, A, ω, and ϕ, by taking the time derivatives of Equation (8-1):

$$x(t) = A \cos(\omega t + \phi) \tag{8-1}$$

$$v_x(t) = \frac{dx}{dt} = -A\omega \sin(\omega t + \phi) \tag{8-7}$$

$$a_x(t) = \frac{dv_x}{dt} = -A\omega^2 \cos(\omega t + \phi) \tag{8-8}$$

$$a_x(t) = -\omega^2 x \tag{8-9}$$

Equations (8-7) and (8-8) describe the velocity and acceleration of the particle in terms of time; Equation (8-9), however, relates the acceleration of the particle to its displacement. Through these relationships we may examine the physical character of simple harmonic motion. Look at the curves in Figure 8.4. These are graphs of Equations (8-1), (8-7), and (8-8), the displacement, velocity, and acceleration of a particle executing simple harmonic motion when the phase constant ϕ is zero. The following physical facts about the simple harmonic motion described by the graphs of Figure 8.4 are notable:

1. The displacement x is a cosine function; the displacement has its maximum value A at $t = 0$. The displacement is minimum at $x = 0$ (when $\omega t = \pi/2$), the center of the oscillatory path.

2. The velocity v_x is a sine function; its value is zero at $t = 0$. Velocity has its maximum magnitude (speed) ωA when the displacement is zero; the velocity is zero when the displacement is $\pm A$. In other words, speed is greatest when

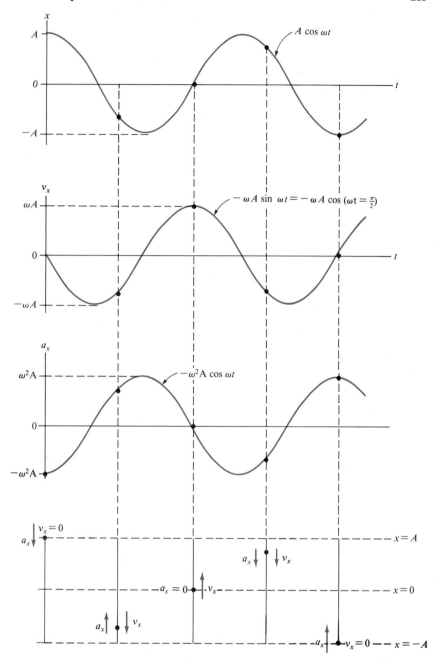

Figure 8.4
Graphs of the displacement x, velocity v_x, and acceleration a_x for a particle executing simple harmonic motion with phase constant $\phi = 0$.

displacement is zero, and the speed is zero when the particle is farthest from its equilibrium position.

3. The acceleration a_x is, like displacement, a cosine function, but its direction is opposite to that of the displacement. Therefore, the acceleration is always toward the center of the oscillatory path.

4. Acceleration has its greatest magnitude when the particle is farthest from $x = 0$ (and, therefore, when the velocity is zero). The acceleration is zero when the displacement is zero (and, therefore, when the speed is greatest).

E 8.6   The position of a particle is given by $x(t) = (0.7 \cos 3t)$ m where t is measured in seconds.
(a) Determine $v_x(t)$.   Answer: $v_x(t) = (-2.1 \sin 3t)$ m/s
(b) Determine $a_x(t)$.   Answer: $a_x(t) = (-6.3 \cos 3t)$ m/s^2
(c) For what value(s) of x is the speed of the particle a maximum?
                                                    Answer: $x = 0$
(d) When, during the first cycle after $t = 0$, is the particle's speed a maximum?
                                            Answer: $\pi/6$ s, $\pi/2$ s
(e) For what value(s) of x does the acceleration of the particle have a maximum magnitude?   Answer: ± 0.70 m

E 8.7   If the position of a particle is given by $x(t) = 0.2 \sin(\pi t + \pi/6)$ m,
(a) determine $v_x(t)$.   Answer: $v_x(t) = 0.2\pi \cos(\pi t + \pi/6)$ m/s
(b) determine $a_x(t)$.   Answer: $a_x(t) = -0.2\pi^2 \sin(\pi t + \pi/6)$ m/s^2
(c) when during the first cycle after $t = 0$ does the particle achieve its maximum speed?   Answer: 5/6 s, 11/6 s

Example 8.1   **PROBLEM**   If a particle executing simple harmonic motion about $x = 0$ with a frequency ν has a position x_0 and a velocity v_0 at $t = 0$, determine the amplitude A and phase constant ϕ (in terms of ν, x_0, and v_0) so that the position is given by $x(t) = A \cos(\omega t + \phi)$.

SOLUTION   Any simple harmonic motion is described by the expression $x(t) = A \cos(\omega t + \phi)$ with the angular frequency ω and the frequency ν related by Equation (8-3), $\omega = 2\pi\nu$. The velocity at any time is given by

$$v_x(t) = \frac{dx}{dt} = -\omega A \sin(\omega t + \phi)$$

Thus, for $t = 0$ we get, letting $x(0) = x_0$ and $v(0) = v_0$,

$$x_0 = A \cos\phi$$

$$v_0 = -\omega A \sin\phi$$

These two equations may be rewritten as

$$A \cos\phi = x_0$$

$$A \sin\phi = -\frac{v_0}{\omega}$$

The amplitude A is determined by squaring each side of these two equations and then adding to get

$$A^2 \cos^2\phi + A^2 \sin^2\phi = x_0^2 + (v_0/\omega)^2$$

Because $\cos^2\phi + \sin^2\phi = 1$, this simplifies to

$$A^2 = x_0^2 + (v_0/\omega)^2$$

or

$$A = [x_0^2 + (v_0/\omega)^2]^{1/2}$$

Once A is determined, the phase constant ϕ is determined from

$$\cos\phi = x_0/A$$

or

$$\phi = \pm\cos^{-1}(x_o/A)$$

The sign for ϕ is chosen so that

$$A \sin\phi = -\frac{v_o}{\omega}$$

is satisfied.

E 8.8   A particle is executing simple harmonic motion about $x = 0$ with a frequency of 2.0 Hz. Its $t = 0$ position and velocity are $x_o = -0.40$ m and $v_o = 0$.
(a) Determine the amplitude A.   Answer: $A = 0.40$ m
(b) Determine the phase constant ϕ.   Answer: $\phi = \pi$
(c) Write an expression for $x(t)$.   Answer: $x(t) = 0.4 \cos(4\pi t + \pi)$ m

E 8.9   Repeat Exercise 8.8 for $x_o = 0$ and $v_o = 2.0$ m/s.
Answer: (a) $A = 1/(2\pi) = 0.16$ m; (b) $\phi = -\pi/2$;
(c) $x(t) = 1/(2\pi) \cos(4\pi t - \pi/2)$ m

Dynamics of Simple Harmonic Motion

Equation (8-9), $a_x = -\omega^2 x$, was obtained by taking successive time derivatives of Equation (8-1), $x = A \cos(\omega t + \phi)$, which we used to define simple harmonic motion. The sequence could have been reversed: We could have defined simple harmonic motion by Equation (8-9) and, by successive integrations, obtained Equation (8-1). Thus, either equation serves to define simple harmonic motion.

Let us consider the acceleration of a mass m executing simple harmonic motion. According to Newton's second law, the acceleration of the mass is equal to the net force on the mass divided by the mass: $a_x = \Sigma F_x/m$. According to Equation (8-9), the acceleration of the mass is $a_x = -\omega^2 x$, so we have

$$\frac{\Sigma F_x}{m} = -\omega^2 x$$

$$\Sigma F_x = -m\omega^2 x \qquad (8\text{-}10)$$

Because m and ω are constants, Equation (8-10) requires that the net force on a mass m executing simple harmonic motion be proportional to its displacement x and be oppositely directed from that displacement. We have encountered such a force in our study of idealized springs that obey Hooke's law, $F_x = -kx$, where k is the spring constant. Therefore, if we consider a block of mass m on a frictionless surface attached to a Hooke's law spring, as shown in Figure 8.5, the net force ΣF_x on the block is $-kx$. Then Equation (8-10) becomes

$$-kx = -m\omega^2 x$$

$$\omega^2 = \frac{k}{m}$$

$$\omega = \sqrt{\frac{k}{m}} \qquad (8\text{-}11)$$

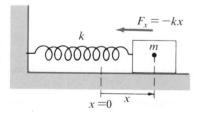

Figure 8.5
The force F_x acting on a block attached to a Hooke's law spring and free to move along a horizontal frictionless surface. The force F_x is oppositely directed from the displacement x of the block.

Thus, the block executes simple harmonic motion with angular frequency given by $\omega = \sqrt{k/m}$, and the frequency ν of the oscillation is

$$\nu = \frac{\omega}{2\pi} = \frac{1}{2\pi}\sqrt{\frac{k}{m}} \tag{8-12}$$

From Equation (8-11) and Equation (8-12) we can see that the angular frequency ω and the frequency ν depend only on the mass m of the oscillating particle and the spring constant k. In particular, *the frequency (and, therefore, the period) of the oscillation does not depend on the amplitude A of the oscillation*. This important characteristic of simple harmonic oscillators means that if we start a mass oscillating by extending the spring to which it is attached to *any* length before releasing it, the system will oscillate at the same frequency. Further, the independence of the frequency (or period) and the amplitude means that identical oscillators, extended different distances from equilibrium and released simultaneously, will reach the equilibrium position simultaneously.

We may combine the kinematics and dynamics of simple harmonic motion in a problem like the following:

The long weightless spring, shown in Figure 8.6 hanging from a ceiling, is extended 0.40 m when a 0.20-kg mass is attached to it and slowly lowered to rest. From this equilibrium position, the spring is pulled downward 0.50 m and then released from rest to oscillate. Calculate: (a) the frequency of the oscillation; (b) the maximum speed the mass achieves; (c) the magnitude of the maximum acceleration of the mass.

Figure 8.6(a) shows the spring in its unstretched position. When the mass is attached, the spring rests at its equilibrium position, $x = 0$, shown in Figure 8.6(b). The mass is then pulled downward to $x = 0.50$ m, as seen in Figure 8.6(c), where it is released from rest to oscillate vertically. We seek the frequency ν, which is related to the given mass m and the spring constant k by Equation (8-12). Therefore, we need first to find the spring constant. The force that extends the spring a length of 0.40 m is the weight $W = mg = (0.2)(9.8) = 1.96$ N. Then the spring constant k is

$$k = \frac{1.96 \text{ N}}{0.4 \text{ m}} = 4.9 \frac{\text{N}}{\text{m}} \tag{8-13}$$

from which we obtain

$$\nu = \frac{1}{2\pi}\sqrt{\frac{k}{m}} = \frac{1}{2\pi}\sqrt{\frac{4.9}{0.2}} = 0.79 \text{ Hz} \tag{8-14}$$

the required frequency of oscillation. The initial displacement x of the mass from its equilibrium position, $x = 0$, is equal to the amplitude of the oscillation if the mass is released from rest. By choosing the displacement x to be positive when the mass is below $x = 0$ and setting $\phi = 0$, we may describe the displacement x of the mass by

$$x = A \cos\omega t \tag{8-15}$$

in which the maximum displacement, the amplitude A, is equal to 0.5 m and ω is equal to $2\pi\nu = 2\pi(0.79) = 5.0 \text{ s}^{-1}$. Then Equation (8-15) becomes

$$x = 0.5 \cos(5.0 \ t) \text{ m} \tag{8-16}$$

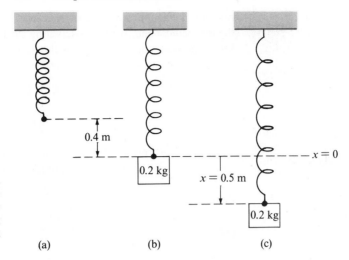

Figure 8.6
A long weightless spring hanging from a ceiling. The spring is (a) in its unstretched position; (b) at a new equilibrium position, $x = 0$, with a 0.20-kg mass attached; and (c) extended 0.50 m below $x = 0$.

The velocity of the mass is expressed by the time derivative of Equation (8-16):

$$v_x = \frac{dx}{dt} = -(0.5)(5.0)\ \sin(5.0\ t)$$

$$v_x = -2.5\ \sin(5.0\ t)\ \text{m/s} \tag{8-17}$$

The maximum speed occurs when v_x has its maximum value $(v_x)_{max}$, which happens when the sine factor of Equation (8-17) is equal to -1. At that time, then, the maximum speed is

$$\left|(v_x)_{max}\right| = 2.5\ \text{m/s} \tag{8-18}$$

The acceleration is obtained from the time derivative of Equation (8-17),

$$a_x = \frac{dv_x}{dt} = (-2.5)(5.0)\ \cos(5.0\ t)$$

$$a_x = -12\ \cos(5.0\ t)\ \text{m/s}^2 \tag{8-19}$$

which achieves its maximum value when the cosine factor is equal to -1. Then the magnitude of the maximum acceleration is

$$\left|(a_x)_{max}\right| = 12\ \text{m/s}^2 \tag{8-20}$$

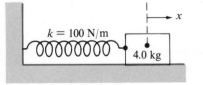

Exercise 8.10

E 8.10   A 4.0-kg block, which slides on a frictionless horizontal surface, is attached to an ideal spring ($k = 100$ N/m) as shown in the diagram.
(a) What is the frequency of this oscillator?   Answer: $5/(2\pi)$ Hz $= 0.80$ Hz
(b) What is the period of this oscillator?   Answer: $(2\pi)/5$ s $= 1.3$ s

E 8.11   The block of Exercise 8.10 is moved 0.12 m in the positive x direction from the ($x = 0$) equilibrium position and then released from rest at $t = 0$. Write an expression for $x(t)$.   Answer: $x(t) = 0.12\ \cos(5t)$ m

Energetics of Simple Harmonic Motion

A simple harmonic oscillator, such as a mass on a spring, is moving most rapidly when its displacement from equilibrium is equal to zero. It is not moving at all when the spring is fully extended or compressed. Thus, the kinetic energy is at a maximum when the elastic potential energy is minimum; the potential energy is

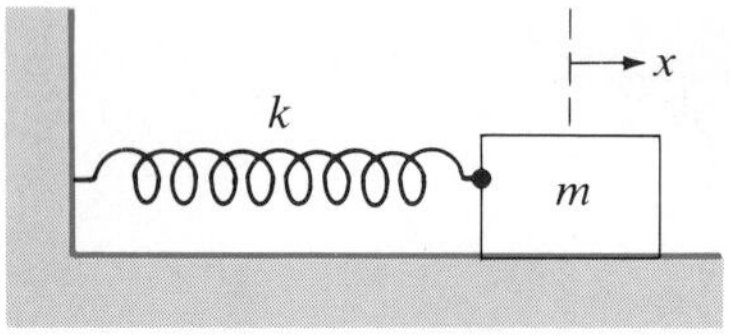

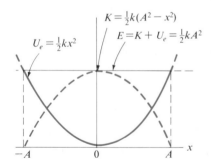

Figure 8.7
A mass m in simple harmonic motion on a horizontal frictionless surface. The mass is attached to a spring with spring constant k.

maximum when the kinetic energy is zero. For a mass attached to an ideal spring, the mechanical energy of the oscillating system is conserved. With no dissipative forces present, then, the sum of the kinetic energy K and the elastic potential energy U_e of the system is constant:

$$K + U_e = \text{constant}$$

$$\frac{1}{2}mv_x^2 + \frac{1}{2}kx^2 = \text{constant} \tag{8-21}$$

To consider further the oscillator energy, suppose a mass m is executing simple harmonic motion with amplitude A on a horizontal frictionless surface. The mass is attached to one end of an ideal spring, with spring constant k, with the other end of the spring fixed, as shown in Figure 8.7. When the mass is at $x = A$, it is not moving, and its kinetic energy is zero. The elastic potential energy stored in the spring at all times is

$$U_e = \frac{1}{2}kx^2 \tag{8-22}$$

so when $x = A$ the total mechanical energy $E = K + U_e$ of the oscillating system is equal to $U_e = \frac{1}{2}kA^2$, or

$$E = \frac{1}{2}kA^2 \tag{8-23}$$

Because no dissipative forces act on the system, the total mechanical energy of the system is conserved, and its value is $\frac{1}{2}kA^2$:

$$K + U_e = \frac{1}{2}kA^2 \tag{8-24}$$

$$\frac{1}{2}mv_x^2 + \frac{1}{2}kx^2 = \frac{1}{2}kA^2 \tag{8-25}$$

As the mass goes through a cycle of its oscillation, its energy is continuously exchanged between its kinetic and potential forms. At $x = \pm A$, the endpoints at which the mass is at rest, all the energy of the system is stored as potential energy in the spring. At $x = 0$, the equilibrium position of the spring where the potential energy of the spring is zero, all the energy of the system is kinetic energy. The energy sharing between kinetic and potential energies for a harmonic oscillator may be seen graphically by plotting Equation (8-22), $U_e = \frac{1}{2}kx^2$, which is the parabola shown for $-A \le x \le A$ in Figure 8.8. The kinetic energy K may be written in terms of displacement x from Equation (8-24),

$$K = \frac{1}{2}kA^2 - U_e$$

$$K = \frac{1}{2}kA^2 - \frac{1}{2}kx^2$$

$$K = \frac{1}{2}k(A^2 - x^2) \tag{8-26}$$

$K = \frac{1}{2}k(A^2 - x^2)$

$E = K + U_e = \frac{1}{2}kA^2$

$U_e = \frac{1}{2}kx^2$

$-A \quad\quad 0 \quad\quad A$

x

Figure 8.8
Potential energy U_e and kinetic energy K of a harmonic oscillator. The energies, U_e and K, are plotted as functions of x, the displacement of the oscillator from equilibrium.

which is also the equation of a parabola that is plotted in Figure 8.8. The maximum value of K, namely $\frac{1}{2}kA^2$, occurs when $x = 0$; the minimum value of K is zero when $x = \pm A$. The sum of the two curves of U_e and K in Figure 8.8 is equal to $E = \frac{1}{2}kA^2$ at every position, in agreement with Equation (8-23).

The velocity of the oscillating mass at any displacement x may be obtained from the expression of the conservation of energy, Equation (8-25):

$$\frac{1}{2}mv_x^2 + \frac{1}{2}kx^2 = kA^2$$

$$v_x^2 = \frac{k}{m}(A^2 - x^2) = \omega^2(A^2 - x^2)$$

$$v_x = \pm\omega\sqrt{A^2 - x^2} \qquad (8\text{-}27)$$

The two signs for v_x reflect the physical fact that for a given position x, the particle may be moving in either the positive or negative x direction. Here the velocity is expressed in terms of the displacement x. Heretofore we have seen v_x expressed only in terms of t.

E 8.12   A 0.80-kg block sliding on a frictionless horizontal surface is attached to one end of a horizontal ideal spring ($k = 500$ N/m), which has its other end held fixed. The block is released from rest when the spring has been extended 20 cm beyond its equilibrium length.
(a) Calculate the total energy of this mass-spring system.   Answer: 10 J
(b) Calculate the maximum kinetic energy of the block as it oscillates.
                                                                 Answer: 10 J
(c) What is the maximum speed of the block during a cycle of its motion?
                                                              Answer: 5.0 m/s
(d) Calculate the kinetic energy of the block when the spring is compressed
    10 cm from its equilibrium length.   Answer: 7.5 J

E 8.13   Consider the mass-spring system of Exercise 8.12. Suppose the mass has a maximum speed of 7.5 m/s as it oscillates.
(a) What is the total energy of the system?   Answer: 23 J
(b) Calculate the maximum elongation of the spring during this oscillation.
                                                       Answer: 0.30 m

Example 8.2   PROBLEM   A *pendulum* is a device that executes periodic motion by oscillating about a horizontal axis. A device like this is illustrated in Figure 8.9. As shown in Figure 8.9(a), the center of mass of the pendulum is a distance L below the axis of rotation (which is perpendicular to the paper) when the pendulum is in its equilibrium position. If the device is rotated from its equilibrium position and then released, periodic motion results. Show that
(a) this motion is simple harmonic motion for small amplitude oscillations.
(b) the frequency ν for these small amplitude oscillations is given by

$$\nu = \frac{1}{2\pi}\sqrt{\frac{mgL}{I}}$$

where m is the mass of the pendulum and I is the moment of inertia of the pendulum about the axis of rotation.

SOLUTION   As the pendulum oscillates, its center of mass moves back and forth along a circular arc with a radius L. The position of the center of mass along this arc may be specified by the variable x, the arc length measured from the equilibrium position, or by θ, the angle between the vertical direction and the line connecting the axis and the center of mass. Both x and θ are shown in Figure 8.9(b); these two quantities are related by $x = L\theta$. Because the pendulum rotates

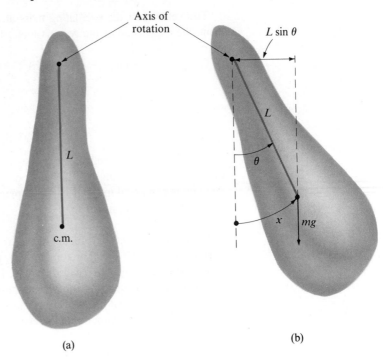

Figure 8.9   Example 8.2

about a fixed axis, its motion obeys $\Sigma\tau = I\alpha$. The only torque about the axis, neglecting any dissipative torques, is that caused by the weight of the pendulum. From Figure 8.9(b) the lever arm of the weight is $L \sin\theta$, and the torque τ is then given by

$$\tau = -mgL \sin\theta = -mgL \sin\left(\frac{x}{L}\right)$$

where the negative sign indicates a restoring torque. Using $I\alpha = \Sigma\tau$, we have

$$I\alpha = -mgL \sin\left(\frac{x}{L}\right)$$

or

$$\alpha = -\frac{mgL}{I} \sin\left(\frac{x}{L}\right)$$

Since the tangential acceleration a_t of the center of mass is equal to $L\alpha$, we now have

$$a_t = L\alpha = -\frac{mgL^2}{I} \sin\left(\frac{x}{L}\right)$$

For any angle less than 0.25 rad (approximately 14°), the sine of that angle and the angle (measured in radians) are equal to within 1 percent. (You may like to investigate this using your calculator.) Thus, for $x/L < 0.25$, we find that $\sin(x/L) \approx x/L$; and the acceleration of the center of mass is

$$a_t = -\left(\frac{mgL}{I}\right)x$$

But this equation is just the simple harmonic motion equation (Equation [8-9]) with

$$\omega^2 = \frac{mgL}{I}$$

Thus, the frequency ν for small amplitude oscillations is

$$\nu = \frac{\omega}{2\pi} = \frac{1}{2\pi} \sqrt{\frac{mgL}{I}}$$

E 8.14   A *simple pendulum*, an idealized model of a pendulum, consists of a point mass m fixed to one end of a massless rod of length L. The simple pendulum rotates freely about a horizontal axis perpendicular to the rod and through the end of the rod opposite m. Show that the frequency ν of the simple pendulum is given by

$$\nu = \frac{1}{2\pi} \sqrt{\frac{g}{L}}$$

E 8.15   What is the length of a simple pendulum that oscillates with a period of 1 s?   Answer: 0.25 m

8.2 Damped and Forced Oscillations

Two physically significant situations involving oscillators are considered in this section: damped oscillations and forced oscillations. The mathematics necessary to treat these systems in detail, differential equations, is beyond the scope of this text. Nevertheless, physical descriptions of these systems are included here because they introduce two important concepts of physics that help us to understand—in an intuitive way, at least—some physical situations that are common to everyday experience. First, we shall describe *damped oscillations*, back-and-forth motions that lose energy as time passes, which correspond to actual oscillating systems. Then we shall consider *forced oscillations*, those driven by an external agent. These oscillators are considered in order to introduce resonance, a phenomenon seen in many physical systems.

Damped Oscillations

Oscillators that we can actually observe do not move forever with undiminished amplitude. Unless energy is supplied from a source outside the oscillator, it loses energy, and its amplitude decreases as time goes on. Swinging pendula, for example, eventually come to rest. A real mass oscillating on a spring finally comes to rest. How are these facts consistent with our descriptions of simple harmonic oscillators? Real oscillators are not, of course, the ideal models we have been analyzing. We have not accounted for the losses of energy that actual oscillators experience. To do so, it is necessary to recognize that damping forces—forces that reduce, or damp, the amplitude of oscillation—are present on actual oscillators. Damped oscillations, therefore, result from air resistance;

from friction in various forms; or from dampers, devices that are designed expressly to reduce the amplitude of an oscillatory motion.

In an ideal oscillatory system, the only force acting on the moving particle is the restoring force that is always directed toward the center of the oscillatory path. In oscillators that experience damping, an additional force acts on the particle in a direction that is opposite to the direction of the velocity of that particle at every instant. The mathematical analysis of a damped oscillator is consequently more complicated than that of the ideal single harmonic oscillator. We shall not consider a mathematical analysis of the damped oscillator here (Problems in Group B at the end of this chapter will, however, treat the damped oscillator), but simply offer the analytical expression that describes damped oscillation:

$$x = Ae^{-\delta t} \cos(\omega't + \phi) \tag{8-28}$$

In Equation (8-28), A is the initial ($t = 0$) amplitude of the oscillation. The amplitude diminishes exponentially with time so that the amplitude at any time $t > 0$ is given by $Ae^{-\delta t}$. The constant δ, called the *damping constant*, is a measure of the effectiveness of the damping forces that act on the oscillating particle. The angular frequency ω' of the oscillation measures, in effect, how often the particle passes through the equilibrium position. If the damping is not excessively large, the angular frequency ω' of a damped oscillation is very nearly equal to the natural angular frequency ω for the ideal oscillator.

Figure 8.10 shows the displacement as a function of time for damped oscillators in three different damping situations, that is, oscillators with different values of the damping constant δ. In Figure 8.10(a) the oscillator with displacement that is graphed as a function of time is slightly damped (δ has a relatively small value). In Figure 8.10(b), δ is relatively large, and the damping is more pronounced. Figure 8.10(c) shows a transition point between two kinds

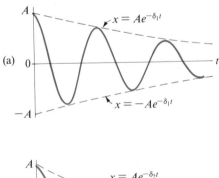

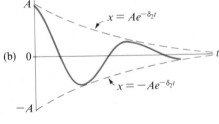

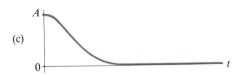

Figure 8.10
The displacement from equilibrium as a function of time of an oscillator that is (a) slightly damped, (b) more strongly damped, and (c) critically damped.

of damped motion, called *critical damping*, in which δ has a sufficiently large value to prevent the particle from reaching the equilibrium position $x = 0$. In this case, the particle approaches $x = 0$ exponentially as time passes. In both Figure 8.10(a) and Figure 8.10(b) the displacement of the particle is confined by the curves $x = Ae^{-\delta t}$ and $x = -Ae^{-\delta t}$.

Forced Oscillations: Resonance

Forced oscillations occur when an oscillating system is driven by a periodic force that is external to the oscillating system. In such a case, the oscillator is compelled to move at the frequency $\nu_D = \omega_D/2\pi$ of the driving force. The physically interesting aspect of a forced oscillator is its response—how much it moves—to the imposed driving force. Let us, therefore, examine qualitatively the response of an oscillator to a driving force.

Consider a boy seated in a playground swing. He and the swing constitute (very nearly) a simple pendulum, which has a natural frequency ν_N of oscillation. By a single push against the ground, the boy can start a gentle oscillation of the system at its natural frequency. The period T_N of his oscillation is equal to $1/\nu_N$. Now suppose a girl stands behind the swing and periodically pushes on the oscillating boy but with a driving period $T_D = 2\pi/\omega_D$ that is much shorter, i.e., very different, from the natural period of the system. How does the swing respond? It accelerates every time she pushes, of course, moving at the driving frequency, but it does not move very far from its equilibrium position. The boy is rattled around near the vertical position of the swing. Technically, we could say that the amplitude of his oscillation is small. Suppose, then, that the girl, sensing the ineffectiveness of her efforts, begins to push on the boy only when he has reached the utmost extent of his backward swing. She could then observe two facts: (a) The boy still oscillates at the frequency at which she pushes, and (b) he smoothly (relatively speaking) moves through longer and longer arcs.

By injecting energy into the oscillator at an appropriate frequency, the girl is able to achieve a large response (amplitude) in the oscillating system. We may generalize the results of this experience. When a driving force is applied to an oscillating system at a frequency near the natural frequency of the system, the amplitude of the oscillation becomes large. This relatively large response of an oscillator to being driven at a frequency that is near its natural frequency is called *resonance*. An oscillator is, therefore, said to be *resonant* at its natural frequency.

Some further insight into the phenomenon of resonance may be gotten from considering a few simple points. In saying that a driving force is applied to an oscillator at its natural frequency, we have assumed that the oscillator is driven in phase, that is, that the oscillator is driven in the same direction as its motion at every instant. This condition need not be the case when the driving force is initially applied to an oscillating system. If the driver is applied out of phase, there will be a *transient state* of motion during which the oscillator changes its motion to coincide with that of the driver, which the oscillator will eventually follow. Once the oscillator is moving at the frequency of the driver (whether or not the system is near resonance), the oscillations are said to have reached the *steady-state* condition.

Once the steady-state condition has been achieved, applying an external force to the idealized (undamped) oscillator at its natural frequency means that the applied force is in the direction of the motion of the oscillating particle at every instant. Thus the oscillating particle is always accelerated in the direction of its motion, and the extent of its motion (its amplitude) will increase with each cycle. On the other hand, if the driving force is applied at any frequency other than the natural frequency of the oscillator, part of the time the driving force will oppose the motion of the particle, thereby lessening the amplitude of the oscillation.

Finally, we may ask, "Why, then, does an oscillator driven at its natural frequency not acquire infinite amplitude, that is, acquire excursions that are without bound from the equilibrium position?" An ideal oscillator would, in principle, have infinite amplitude at resonance. But, as we have seen, real oscillators are damped by resistive forces. The inherent damping in real oscillators makes the amplitudes of their resonant oscillations finite, although relatively large.

8.3 Problem-Solving Summary

Oscillations are easy to identify. A particle moving back and forth over the same path is in oscillation. A particle executing simple harmonic motion may be identified by its periodic oscillatory motion in a straight line that is characterized by two facts:

1. The displacement x of the particle from its equilibrium position may be expressed by $x = A \cos(\omega t + \phi)$, in which A is the amplitude of the oscillation, ω is its angular frequency, and ϕ is the phase constant of the oscillation.

2. The acceleration of the particle is proportional to its displacement from its equilibrium position and is always directed toward the equilibrium position, $x = 0$, according to $a_x = -\omega^2 x$.

The period T and the frequency ν of a simple harmonic oscillator are related to its angular frequency ω by $\nu = 1/T = \omega/2\pi$. The phase constant ϕ determines the position and velocity of the oscillating particle at $t = 0$.

A useful fact of simple harmonic motion is that the frequency and period of a given oscillation are independent of the amplitude of that oscillation.

The kinematic values, velocity and acceleration, of a simple harmonic oscillator are determined by the derivatives of the displacement, $x = A \cos(\omega t + \phi)$. Thus, $v_x = -A\omega \sin(\omega t + \phi)$ and $a_x = -A\omega^2 \cos(\omega t + \phi) = -\omega^2 x$ relate the velocity and acceleration of the particle to time and position. The maximum value of v_x or a_x occurs when the factors, $\sin(\omega t + \phi)$ and $\cos(\omega t + \phi)$, have appropriate values (± 1), giving $v_x(\text{max}) = A\omega$ and $a_x(\text{max}) = A\omega^2$.

An oscillating particle of mass m attached to an ideal spring, fixed at one end and having a spring constant k, executes simple harmonic motion with an angular frequency $\omega = \sqrt{k/m}$. The restoring force F_x on such an oscillator is just the Hooke's law force of the spring, $F_x = -kx$, a force on the oscillating mass that is always directed toward the center of oscillation, $x = 0$, where the spring exerts no force on the mass.

The total mechanical energy of a simple harmonic oscillator is $\frac{1}{2}kA^2$. At the center of the oscillation, the energy of the oscillator is totally kinetic with a value of $\frac{1}{2}mv^2 = \frac{1}{2}kA^2$. At the extreme positions of the oscillation, the energy is wholly potential with a value of $\frac{1}{2}kx^2 = \frac{1}{2}kA^2$, or $x = \pm A$. These energy relationships are particularly useful in solving oscillator problems that require finding the velocities or speeds of oscillating masses.

Damped harmonic oscillations are those with amplitudes that decrease as time passes. The damping results from the dissipation of the oscillatory energy because frictional forces act on the system.

Forced oscillations occur in oscillator systems that are driven by periodic forces exerted by agents external to the oscillator system. When the frequency of the driving force is at or near the natural frequency of the oscillator, a relatively large amplitude of oscillation results; this increase in amplitude is called resonance.

The general four-step problem-solving strategy developed in Section 1.4 is directly applicable to problems concerned with simple harmonic motion. In visualizing the problem, the reference circle is often a helpful device, especially in problems in which the phase constant is a significant aspect of the problem.

Problems

GROUP **A**

8.1 The position of a particle executing simple harmonic motion is given by $x(t) = 0.2 \sin(4\pi t + \pi/3)$, where t is measured in seconds.

  (a) Determine the velocity and acceleration of the particle at $t = 0.125$ s.

  (b) Through how many cycles will the oscillator move during a 5-s time interval?

8.2 The position of a particle as it moves along the x axis is given by $x(t) = 0.4 \cos(5t + \pi/4)$ m, where t is measured in seconds.

  (a) Determine the velocity and acceleration of the particle as functions of time.

  (b) Calculate the frequency and period of this oscillator.

8.3 A simple harmonic oscillator has a frequency of 2.0 Hz and an amplitude of 0.10 m. At $t = 0$ the oscillator is at a position of maximum positive displacement, that is, $x = 0.10$ m. Write an expression for its position as a function of time.

8.4 An object executing simple harmonic motion along the x axis about $x = 0$ has a period of 0.20 s and an amplitude of 8.0 cm. At $t = 0$ it is passing through $x = 0$ moving in the positive x direction. Write an expression for its position as a function of time.

8.5 The amplitude and maximum speed of a simple harmonic oscillator are 15 cm and 180 cm/s. The midpoint of the oscillation is $x = 0$. At $t = 0$ the position is $x = -15$ cm.

  (a) What is the frequency of this oscillation?

  (b) Write an equation for the velocity of the oscillator as a function of time.

8.6 The period and maximum speed of a particle executing simple harmonic motion about $x = 0$ are 3.0 s and 2.0 m/s. At $t = 0$ the velocity of the particle is $v_x = -2.0$ m/s.

  (a) What is the amplitude of this oscillation?

  (b) Write an equation for the position of the particle as a function of time.

8.7 The maximum velocity and acceleration of a particle executing simple harmonic motion about $x = 0$ are equal to 3.0 m/s and 27 m/s². At $t = 0$ the acceleration of the particle is given by $a_x = -27$m/s². Write an expression for the velocity of the particle as a function of time.

8.8 A 2.0-kg object executes simple harmonic motion when attached to an ideal spring ($k = 50$ N/m). The maximum speed of the object during a cycle is 2.0 m/s.

  (a) Calculate the amplitude of the motion.

  (b) What is the speed of the object when it is 0.20 m from the midpoint of the oscillation?

8.9 When attached to an ideal spring ($k = 40$ lb/ft), a 6.0-lb block executes simple harmonic motion with an amplitude equal to 8 in.

  (a) What is the speed of the block as it passes through the midpoint of the oscillation?

  (b) How far is the block from the midpoint of the oscillation when its speed is 5.0 ft/s?

8.10 If a simple harmonic oscillator has a maximum speed of 2.4 m/s and its amplitude is 0.60 m, what is

  (a) the frequency of the oscillator?

  (b) the maximum acceleration of the oscillator?

8.11 The oscillatory motion of a 0.20-kg body is described by $x(t) = 0.3 \cos(\pi t)$ m where t is measured in seconds.

(a) What is the maximum speed of this body as it oscillates?

(b) When after $t = 0$ does the body first achieve its maximum speed?

(c) What is the magnitude of the maximum resultant force on the body as it oscillates?

8.12 A 0.80-kg body is suspended at the end of an ideal spring ($k = 80$ N/m) so that the body oscillates vertically.

(a) How much time elapses during six cycles of the oscillator?

(b) If the maximum speed of the body during one cycle is 2.6 m/s, what is the amplitude of the oscillation?

8.13 If a particle executes simple harmonic motion with an amplitude of 15 cm and a frequency of 0.40 Hz,

(a) what is the maximum speed of the particle as it oscillates?

(b) what is the maximum acceleration of this oscillator?

8.14 The position of a particle is $x(t) = 0.2 \sin(4t - \pi/2)$ m where t is measured in seconds.

(a) What is the period of this motion?

(b) Calculate the maximum speed of the particle as it oscillates.

(c) Determine the acceleration of the particle when its position is $x = 0.15$ m.

8.15 A 0.30-kg object undergoes simple harmonic motion with a period of 0.40 s and an amplitude of 12 cm. At $t = 0$ the object is moving through the equilibrium ($x = 0$) position in the negative direction.

(a) Write an expression for the position of the object as a function of time.

(b) When after $t = 0$ will the object first reach its maximum distance from the equilibrium position?

(c) What is the resultant force on the object when its position is $x = 8.0$ cm?

8.16 The maximum speed and acceleration for a particle executing simple harmonic motion about the position $x = 0$ are 2.0 m/s and 6.0 m/s². At $t = 0$ the velocity of the particle has its maximum value.

(a) What is the frequency of this oscillation?

(b) What is the amplitude of this oscillation?

(c) Write an expression for the acceleration of the particle as a function of time.

8.17 A 2.0-kg body attached to an ideal spring ($k = 40$ N/m) executes simple harmonic motion with a maximum speed equal to 5.0 m/s.

(a) What is the amplitude of this motion?

(b) Calculate the total energy of this oscillation.

(c) What is the frequency of the motion?

8.18 A pendulum is a thin 1.5-m rod that swings freely about one end.

(a) Calculate the period of this pendulum for small amplitude oscillations.

(b) Compare the period of this pendulum to that of a simple pendulum of the same length.

8.19 A uniform hoop of radius R is used as a pendulum by mounting it so that it rotates freely about an axis parallel to the axis of the hoop and through the periphery of the hoop. Show that the period of small amplitude oscillations of this pendulum is given by

$$T = 2\pi \sqrt{\frac{2R}{g}}$$

8.20 Simple harmonic motion has been described by an expression of the form $x(t) = A \cos(\omega t + \phi)$, where A and ϕ are constants determined from initial conditions and ω is the angular frequency of the oscillator.

(a) Show that the expression for $x(t)$ can be written

$$x(t) = B \cos\omega t + C \sin\omega t$$

where $B = A \cos\phi$ and $C = -A \sin\phi$ are again constants to be determined from initial conditions. *Hint:* Use the identity $\cos(u + v) = \cos u \cos v - \sin u \sin v$.

(b) If the position and velocity of an oscillator are $x = x_o$ and $v_x = v_o$ at $t = 0$, show that

$$x(t) = x_o \cos\omega t + \frac{v_o}{\omega}\sin\omega t$$

8.21 A pendulum is constructed using two small objects, each of mass m, arranged on a light rod of length L, as shown in Figure 8.11. The pendulum swings freely about the upper end of the rod. Let θ measure the angular displacement of the pendulum from the vertical.

(a) Show that the restoring torque for this pendulum is given by

$$\tau = -\frac{3}{2}mgL \sin\theta$$

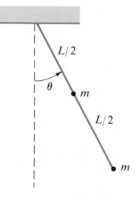

Figure 8.11   Problem 8.21

(b) Show that the tangential acceleration of the lower end of the pendulum is given by

$$a_t = -\frac{6}{5}g \, \sin\left(\frac{x}{L}\right)$$

where x measures the position of the lower end along its circular arc of radius L, that is, $\theta = x/L$.

(c) Show that for small oscillations, the motion is simple harmonic with a period given by

$$T = 2\pi \sqrt{\frac{5L}{6g}}$$

8.22 A 6.0-lb body oscillates vertically with a frequency equal to 2.0 Hz and an amplitude of 0.80 ft while it is suspended on a vertical ideal spring.

(a) Calculate the spring constant k.

(b) What maximum force does the spring exert on the body during this oscillation?

8.23 A pendulum is constructed using a uniform thin rod ($M = 2.0$ kg, $L = 0.90$ m) with a uniform sphere ($m = 0.50$ kg, $r = 0.10$ m) attached to one end of the rod. This apparatus swings freely about an axis perpendicular to the rod and through the end of the rod opposite to the sphere (*see* Fig. 8.12). Calculate the period for small-amplitude oscillations of this pendulum.

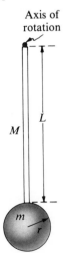

Axis of rotation

M L

m r

Figure 8.12   Problem 8.23

8.24 A uniform sphere ($W = 20$ lb, $R = 0.50$ ft) is suspended from the ceiling by a string of length L, as shown in Figure 8.13. For small-amplitude oscillations the frequency of this pendulum is 1.0 Hz.

(a) Calculate L.

(b) If the maximum angle the string makes with the vertical during such an oscillation is 12°, calculate the tension in the string at the lowest point of the swing.

8.25 A simple pendulum oscillates with a period of 1.5 s and an angular amplitude of 0.20 rad.

(a) Determine the length of the pendulum.

(b) What is the maximum angular speed of the pendulum?

(c) What is the maximum angular acceleration of the pendulum?

(d) Let θ measure the angular position (from the vertical) of the pendulum. Write an expression for $\theta(t)$ assuming that $\theta(0) = 0.20$ rad, its maximum value.

8.26 A 0.02-kg particle executes simple harmonic motion about $x = 0$ when attached to a spring. At $t = 0$ the position, velocity, and acceleration of the particle are observed to be $x = 0.15$ m, $v_x = -2.3$ m/s, and $a_x = -15$ m/s².

(a) What is the frequency of this oscillator?

(b) Calculate the amplitude of this oscillation.

(c) What is the total energy of this oscillator?

(d) Write an expression for $x(t)$, the position of the particle as a function of time.

8.27 At a particular time the position, velocity, and acceleration of a 0.04-kg particle executing simple harmonic motion about $x = 0$ are $x = 0.20$ m, $v_x = 4.0$ m/s, and $a_x = -5.0$ m/s².

(a) What is the frequency of this motion?

(b) What is the amplitude of this oscillation?

(c) What is the maximum speed of the particle as it oscillates?

(d) Calculate the total energy of the oscillator (assume the potential energy to be zero at $x = 0$).

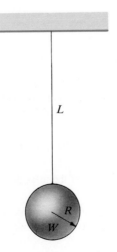

L

R W

Figure 8.13   Problem 8.24

GROUP B

8.28 An ideal spring (k = 50 lb/ft) is positioned vertically, as shown in Figure 8.14(a). At t = 0 a block (W = 5.0 lb), which is attached to the spring, is released from rest with the spring in its uncompressed position (*see* Fig. 8.14[b]).

  (a) Calculate the frequency of the ($t > 0$) oscillation.

  (b) Where is the midpoint of this oscillation?

  (c) What is the amplitude of this oscillation?

  (d) What is the maximum kinetic energy of this oscillation?

  (e) What is the maximum energy stored in the spring during this oscillation?

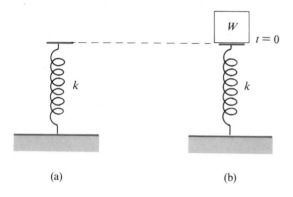

(a)   (b)

Figure 8.14   Problem 8.28

8.29 An ideal spring (k = 180 N/m) of negligible mass hangs as shown in Figure 8.15(a). A mass (m = 5.0 kg) is attached to the spring and set in vertical oscillation. At t = 0 the spring is at its initial equilibrium length with the mass moving upward with a speed v_0 = 2.0 m/s, as shown in Figure 8.15(b).

  (a) Calculate the period of the oscillation.

  (b) Where is the midpoint of this oscillation?

  (c) Determine the amplitude of the oscillation.

  (d) Determine the distances of maximum compression and maximum extension of the spring, as measured from its initial equilibrium position.

  (e) Calculate the maximum energy stored in the spring during a cycle.

  (f) Calculate the maximum kinetic energy of the mass during a cycle.

8.30 The mass shown in Figure 8.16 slides on a frictionless surface. In the equilibrium configuration (x = 0), each spring is at its natural length.

  (a) Show that the net force on the mass is given by

$$F_x = -(k_1 + k_2)x$$

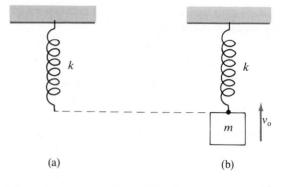

(a)   (b)

Figure 8.15   Problem 8.29

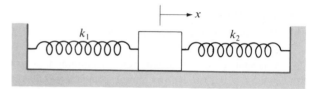

Figure 8.16   Problem 8.30

  (b) Show that the mass executes simple harmonic motion with a frequency

$$\nu = \frac{1}{2\pi} \sqrt{\frac{k_1 + k_2}{m}}$$

8.31 A uniform cylindrical disk of radius R is used to make a pendulum by mounting the disk so that it swings freely about an axis parallel to and at a perpendicular distance $x < R$ from the axis of the disk. Show that

  (a) the frequency of this pendulum for small amplitude oscillations is given by

$$\nu = \frac{1}{2\pi} \sqrt{\frac{2\,gx}{2x^2 + R^2}}$$

  (b) this frequency is a maximum for $x = R/\sqrt{2}$.

  (c) the maximum frequency is given by

$$\nu_{max} = \frac{1}{2\pi} \sqrt{\frac{g}{\sqrt{2}\,R}}$$

8.32 The t = 0 position and velocity of a 0.14-kg particle executing simple harmonic motion (period 0.20 s) about x = 0 are x_0 = 0.30 m and v_0 = 8.0 m/s.

  (a) What is the amplitude of this oscillation?

  (b) What is the maximum speed of the particle as it oscillates?

  (c) Calculate the maximum resultant force on the particle as it oscillates.

(d) Write an expression for the resultant force on the particle as a function of time.

8.33 The block ($W = 4.0$ lb) depicted in Figure 8.17 slides on a frictionless surface and is attached to a spring ($k = 20$ lb/ft). If the block is released at $t = 0$ from rest with the spring stretched 0.40 ft,

(a) when will the block first pass through the equilibrium position?

(b) what is the position of the block at $t = 0.70$ s?

(c) how fast is the block moving at $t = 0.20$ s?

(d) what total distance does the block move during the time interval $0 \leq t \leq 3.0$ s?

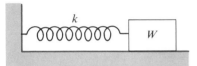

Figure 8.17   Problem 8.33

8.34 An ideal spring ($k = 120$ N/m) has a mass ($m = 8.0$ kg) attached to it (*see* Fig. 8.18). The inclined surface is frictionless. The mass is released from rest with the spring at its natural unstressed length.

(a) What is the frequency of the subsequent oscillation?

(b) What is the maximum kinetic energy of the mass during a cycle of the pendulum?

(c) Calculate the maximum energy stored in the spring during a cycle of the oscillation.

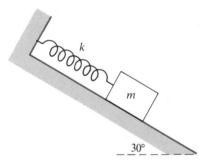

Figure 8.18   Problem 8.34

8.35 The mass shown in Figure 8.19 slides on a frictionless surface. Assume that the mass of each of the two ideal springs is negligible. Let x measure the position of the mass m relative to its equilibrium ($x = 0$) position.

(a) Show that the force on m is given by

$$F_x = -\frac{k_1 k_2}{k_1 + k_2} x$$

To do so, you may find it helpful to note that the net force on spring 2 is equal to zero at any instant.

(b) Show that the mass executes simple harmonic motion with a period given by

$$T = 2\pi \sqrt{\left(\frac{1}{k_1} + \frac{1}{k_2}\right) m}$$

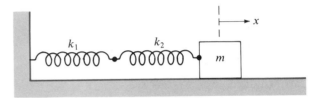

Figure 8.19   Problem 8.35

8.36 One model for a damped harmonic oscillator is a mass m (attached to an ideal spring having an elastic constant k) moving in a medium that exerts a "resistive" force R_x on the mass. If we assume that this resistive force is proportional and oppositely directed to the velocity v_x of the mass, then we may write

$$R_x = -2m\omega_1 v_x$$

where ω_1 is a constant that measures the strength of the damping. If x measures the position of the mass relative to that position with the spring unstressed, then the spring exerts a force given by $F_x = -kx$ on the mass; and the resultant force on the mass is

$$\sum F_x = -kx - 2m\omega_1 v_x = ma_x$$

or

$$a_x = -\omega_0^2 x - 2\omega_1 v_x$$

where $\omega_0 = \sqrt{k/m}$ is the angular frequency of the undamped oscillator ($\omega = 0$). The motion of this damped oscillator depends upon the relative strengths of the damping force and the spring force, that is, upon how ω_1 compares to ω_0. The position $x(t)$ for all cases is given by:

$\omega_1 = 0$ (undamped)
   $x(t) = A \cos\omega_0 t + B \sin\omega_0 t$
$0 < \omega_1 < \omega_0$ (underdamped)
   $x(t) = Ae^{-\omega_1 t} \cos\omega t + Be^{-\omega_1 t} \sin\omega t, \; \omega = \sqrt{\omega_0^2 - \omega_1^2})$
$\omega_0 = \omega_1$ (critically damped)
   $x(t) = (A + Bt)e^{-\omega_0 t}$
$\omega_0 < \omega_1$ (overdamped)
   $x(t) = Ae^{-\omega_- t} + Be^{-\omega_+ t}, \; \omega_\pm = \omega_1 \pm \sqrt{\omega_1^2 - \omega_0^2}$

In each expression, A and B are arbitrary constants. Verify that $x(t)$, as given above, satisfies $a_x = -\omega_0^2 x - 2\omega_1 v_x$ for

(a) the undamped (simple harmonic motion) case.

(b) the underdamped case.

(c)   the critically damped case.

(d)   the overdamped case.

8.37   Problem 8.36 gives the solution for $x(t)$, the position of a damped harmonic oscillator as a function of time, for each of four cases. In each case, the expression for $x(t)$ contains two constants, A and B, which are determined from initial conditions. If the position and velocity at $t = 0$ are given by $x = x_0$ and $v_x = v_0$, use the results of Problem 8.36 to show that for

(a) $\omega_1 = 0$, $A = x_0$ and $B = v_0/\omega_0$

(b) $0 < \omega_1 < \omega_0$, $A = x_0$ and
$B = (v_0 + \omega_1 x_0)/\omega$ $(\omega = \sqrt{\omega_0^2 - \omega_1^2})$

(c) $\omega_1 = \omega_0$, $A = x_0$ and $B = v_0 + \omega_0 x_0$

(d) $\omega_0 < \omega_1$, $A = \dfrac{\omega_+ x_0 + v_0}{\omega_+ - \omega_-}$ and

$B = -\dfrac{\omega_- x_0 + v_0}{\omega_+ - \omega_-}$ $(\omega_\pm = \omega_1 \pm \sqrt{\omega_1^2 - \omega_0^2})$

8.38   Consider the damped harmonic oscillator of Problems 8.36 and 8.37. Let $\omega_0 = 1.0 \text{ s}^{-1}$, $x_0 = 0.10 \text{ m}$, and $v_0 = 0$. Write an expression for $x(t)$, and sketch $x(t)$ for $0 \le t \le 10$ s if

(a) $\omega_1 = 0$ (undamped).

(b) $\omega_1 = 0.60 \text{ s}^{-1}$ (underdamped).

(c) $\omega_1 = 1.0 \text{ s}^{-1}$ (critically damped).

(d) $\omega_1 = 1.25 \text{ s}^{-1}$ (overdamped).

8.39   The response of a simple harmonic oscillator to being driven at its natural frequency is an interesting demonstration of resonance. To see this, consider a mass (m)–spring (k) oscillator that is acted upon by a force $F_x = F_0 \cos\omega_0 t$, where F_0 is a constant and $\omega_0 = \sqrt{k/m}$ is the natural (angular) frequency of the oscillator.

(a)   If the spring force is given by $-kx$ show that the acceleration of the mass is

$$a_x = -\omega_0^2 x + a_0 \cos\omega_0 t$$

where the constant a_0 is defined by $a_0 = F_0/m$.

(b)   The position of the mass is given by

$$x = A \cos\omega_0 t + \left(B + \frac{a_0 t}{2\omega_0}\right) \sin\omega_0 t$$

where A and B are arbitrary constants to be determined from initial conditions. Verify that this expression for x yields an expression for a_x that satisfies the result obtained in part (a).

(c)   If $x = x_0$ and $v_x = v_0$ at $t = 0$, show that $A = x_0$ and that $B = v_0/\omega_0$, that is,

$$x(t) = x_0 \cos\omega_0 t + \left(\frac{v_0}{\omega_0} + \frac{a_0 t}{2\omega_0}\right) \sin\omega_0 t$$

(d)   Now consider the particular initial state with the mass starting at rest $(v_0 = 0)$ from the origin $(x_0 = 0)$. For this case, show that the motion is oscillatory with an amplitude that grows linearly with time.

(e)   The essential point here is that the response of a simple harmonic oscillator to a driving force (having the same

frequency as the oscillator) is oscillatory with an amplitude that grows without bound. Comment on the inevitable incorrectness of this analysis for any real mass-spring oscillator.

8.40   Consider the driven simple harmonic oscillator of Problem 8.39 with $m = 2.0$ kg, $k = 50$ N/m, $F_0 = 0.50$ N, $x_0 = 0$, and $v_0 = 0$.

(a)   Write an expression for $x(t)$.

(b)   Sketch a graph of $x(t)$ for the first three cycles after $t = 0$.

8.41   Problems 8.39 and 8.40 consider the response of a simple harmonic oscillator to being driven at its natural frequency. In this problem we treat the case where the frequency of the driving force is not equal to that of the oscillator. Let the natural frequency of the oscillator be ω_0 and suppose the driving force is given by $F_x = F_0 \cos\omega t$, where F_0 is a constant and $\omega \ne \omega_0$.

(a)   Show that the acceleration of the oscillator is given by

$$a_x = -\omega_0^2 x + a_0 \cos\omega t$$

where the constant a_0 is equal to F_0 divided by the mass m of the oscillator, that is, $a_0 = F_0/m$.

(b)   The position $x(t)$ of the mass is given by

$$x(t) = A \cos\omega_0 t + B \sin\omega_0 t + \frac{a_0}{\omega_0^2 - \omega^2} \cos\omega t$$

Verify that this expression satisfies the equation of part (a).

(c)   If $x = x_0$ and $v_x = v_0$ at $t = 0$, show that

$$A = x_0 - \frac{a_0}{\omega_0^2 - \omega^2} \text{ and } B = \frac{v_0}{\omega_0}$$

(d)   Thus for arbitrary initial conditions, the position of the oscillator is described by the sum of terms corresponding to two different frequencies, namely ω_0 (the natural frequency of the oscillator) and ω (the frequency of the driving force). Determine the initial conditions for which $x(t)$ has no terms corresponding to ω_0. For these conditions, the position of the oscillator is then

$$x(t) = \frac{a_0}{\omega_0^2 - \omega^2} \cos\omega t$$

For $\omega < \omega_0$, the oscillator is in phase with the driving force, and for $\omega > \omega_0$, the oscillator and the driving force are out of phase by π radians. Verify this last statement. In either case the amplitude A of this steady-state response of the system to the driving force is given by

$$A(\omega) = \left| \frac{A_0}{\omega_0^2 - \omega^2} \right| = \left| \frac{a_0/\omega_0^2}{1 - (\omega/\omega_0)^2} \right|$$

(e)   Graph the amplitude of the steady-state response versus ω/ω_0 for $0 < \omega/\omega_0 < 2$. Note that as $|\omega - \omega_0|$ approaches zero, the amplitude becomes arbitrarily large. Discuss this graph in terms of resonance.

8.42 The mass-spring system shown in Figure 8.20 is called, for obvious reasons, a coupled oscillator system. The figure shows two blocks, each of mass m, sliding on a frictionless surface. The springs, each having an elastic constant k, connect the blocks to each other and to the confining walls. In the equilibrium configuration ($x_1 = x_2 = 0$), each spring is unstressed.

(a) Show that the accelerations of the blocks are given by

$$a_{1x} = -2\omega_0^2 x_1 + \omega_0^2 x_2$$
$$a_{2x} = \omega_0^2 x_1 - 2\omega_0^2 x_2$$

where $\omega_0^2 = k/m$.

(b) Show that $x_1 = A_1 \cos\omega t$ and $x_2 = A_2 \cos\omega t$ satisfy the equations of part (a) if either (i) $\omega = \omega_0$ and $A_1 = A_2$ or (ii) $\omega = \sqrt{3}\omega_0$ and $A_2 = -A_1$.

(c) Consider the first case of part (b). Let $\omega_1 = \omega_0$ and let $A = A_1 = A_2$. Then the blocks move according to

$$x_1 = A \cos\omega_1 t$$
$$x_2 = A \cos\omega_1 t$$

Thus, the blocks each execute simple harmonic motion of amplitude A and frequency ω_1. Furthermore, the two oscillators are in phase. Sketch graphs of x_1 and x_2 versus t for a few cycles.

(d) Now consider the second case of part (b). Suppose $\omega_2 = \sqrt{3}\omega_0$ and $B = A_1 = -A_2$. The motions of the blocks are then described by

$$x_1 = B \cos\omega_2 t$$
$$x_2 = -B \cos\omega_2 t = B \cos(\omega_2 t - \pi)$$

Again each block executes simple harmonic motion of amplitude B and frequency ω_2. But for this higher frequency, the two oscillators differ in phase by π radians. Sketch graphs of x_1 and x_2 versus t for a few cycles.

Comments: The two different modes of oscillation examined in parts (c) and (d) are referred to as *normal modes* of the coupled-oscillator system. The frequencies, ω_1 and ω_2, are called the *natural frequencies* of oscillation. Normal modes and natural frequencies are characteristics of most coupled-oscillator systems. Generally speaking, as the number of particles in the system is increased, the number of normal modes and natural frequencies increase proportionately. Most important, in any one normal mode, each of the oscillators has the same frequency but not necessarily the same amplitude.

8.43 Figure 8.21 shows a different coupled-oscillator system from the one studied in Problem 8.42. The spring constant for the middle spring is twice that of each of the outer springs.

(a) Show that the accelerations of the blocks are given by

$$a_{1x} = -3\omega_0^2 x_1 + 2\omega_0^2 x_2$$
$$a_{2x} = 2\omega_0^2 x_1 - 3\omega_0^2 x_2$$

where $\omega_0^2 = k/m$.

(b) Calculate the two natural frequencies of oscillation for this system, that is, determine the values of ω for which $x_1 = A_1 \cos\omega t$ and $x_2 = A_2 \cos\omega t$ satisfy the equations of part (a).

(c) Determine the normal mode corresponding to each of the two natural frequencies of part (b). Show that these modes have the general properties discussed in the comment at the conclusion of Problem 8.42.

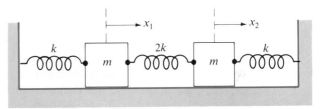

Figure 8.21   Problem 8.43

8.44 The coupled-oscillator system of Figure 8.22 has three blocks, each of mass m, coupled by four identical springs, each of elastic constant k. In the equilibrium configuration ($x_1 = x_2 = x_3 = 0$), each of the springs is unstressed.

(a) Show that the accelerations of the blocks are given by

$$a_{1x} = -2\omega_0^2 x_1 + \omega_0^2 x_2$$
$$a_{2x} = \omega_0^2 x_1 - 2\omega_0^2 x_2 + \omega_0^2 x_3$$
$$a_{3x} = \omega_0^2 x_2 - 2\omega_0^2 x_3$$

where $\omega_0^2 = k/m$.

(b) Show that the three natural frequencies of this system are $\omega_1 = \sqrt{2 - \sqrt{2}}\,\omega_0$, $\omega_2 = \sqrt{2}\,\omega_0$, and $\omega_3 = \sqrt{2 + \sqrt{2}}\,\omega_0$. To do so, find the values of ω for which $x_1 = A_1 \cos\omega t$, $x_2 = A_2 \cos\omega t$, and $x_3 = A_3 \cos\omega t$ are solutions to the equations of part (a).

(c) Determine the normal mode corresponding to each of the natural frequencies of part (b.)

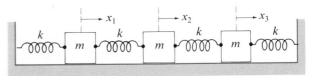

Figure 8.22   Problem 8.44

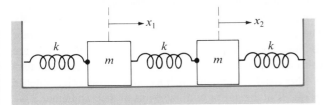

Figure 8.20   Problem 8.42

9 Mechanics of Fluids

Fluids are either liquids or gases. They are characterized by physical properties that distinguish them from solids. In this chapter we will first describe the properties of fluids that permit us to analyze their mechanics and then introduce pressure, a quantity that is useful in considering how forces may affect fluids and how fluids exert forces. Pressure will be useful in studying *fluid statics*, that is, fluids at rest, as contrasted with *fluid dynamics*, the study of fluids in motion. Fluid statics entails the interaction of fluids with their containers, with other fluids, and with solids. Included in the study of statics is the physics of buoyancy, a practical aspect of fluid studies. Another practical application of fluid statics is the use of Pascal's law, which treats the transmission of pressure in confined fluids (hydraulic systems). The fundamental relationships of fluid dynamics and characteristics of fluid flow will be described. Some idealized (and, therefore, simplified) descriptions and analyses of fluid flow will follow. Finally, our brief study of fluid motion will conclude with a few practical applications of fluid mechanics.

9.1 The Fluid State

Our concern in this chapter is primarily with the macroscopic, or gross, nature of fluids, and so we will treat fluids as if they were continuous media, even though we know that all matter is composed of molecular, atomic, or subatomic particles. We must first, however, determine the fundamental properties, that is, the microscopic character, of fluids that distinguish them from solids. Therefore, we now appeal to the particulate nature of matter to establish the properties of fluids that we will use to describe and analyze them macroscopically.

At the microscopic level, all matter is composed of particles that are constantly in random motion. The particles in a solid are strongly bound together; their random motion is confined to jiggling about equilibrium positions that are virtually fixed relative to each other. Thus solids are relatively rigid. The particles in a gas, on the other hand, essentially interact only during brief, intense collisions with other particles of the gas or with the walls of their container. (We say "essentially" because there are, in fact, other forces like gravitational attraction that are present but are negligible compared to the collision forces.) Thus the particles in a gas are not bound together at all, and a gas will fill any container in which it is confined. Gas particles are in chaotic random motion: They mix thoroughly. Particles within a liquid are bound to each other weakly. They are bound to neighboring particles more than are those in gases but less than are those in solids. The motion of particles in a liquid is random but not as chaotic as the motion of particles of a gas. The attractive forces between particles in a liquid are sufficiently strong on the average to make the liquid cohesive (it hangs together, so to speak) yet are weak enough to permit the liquid to flow. In flowing, the particles of the liquid pass one another easily without macroscopic separation of the mass of liquid.

From these qualitative microscopic descriptions of fluids, we may identify some inherent properties of fluids:

1. Fluids can flow. The particles can glide easily past one another.

2. Fluids are *isotropic*, which means that a physical property of a fluid at a given point is independent of direction within the medium. This isotropy is a consequence of the nonlocalized random motion of the fluid particles. Motion cannot be random and be directionally preferential.

3. Fluids vary from being highly *compressible* to being virtually *incompressible*. Gases are compressible in that the volume of a gas may be reduced readily. Compressing a gas changes its *density*, that is, its mass per unit volume. Liquids are almost incompressible; it is virtually impossible to alter the density of most liquids significantly.

9.2 Fluid Statics

We have established that fluids are capable of flowing. A fluid in static equilibrium, that is, at rest, must not be flowing. Therefore, a fluid at rest cannot sustain a force tangential to any surface within or at the boundary of the fluid; otherwise the tangential force would cause the fluid to flow and it would not be at rest. Thus static fluids can sustain (and, therefore, can exert) only normal forces on any surface on or within the fluid. The ability of a fluid to sustain or exert normal forces on surfaces is conveniently described by a measure called pressure.

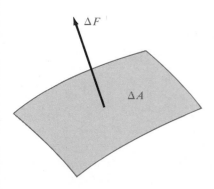

Fluid Pressure

Figure 9.1
A small surface area ΔA within a fluid. The force $\Delta \mathbf{F}$ exerted by the fluid on the surface is normal to the surface.

A small area ΔA within a fluid is depicted in Figure 9.1. The fluid exerts a force of magnitude ΔF normal to the surface. The *average pressure* $\overline{P}$ on the surface of the small area we have chosen is defined to be the ratio of the magnitude ΔF of the normal force acting on that surface to the area ΔA of that surface, or

$$\overline{P} = \frac{\Delta F}{\Delta A} \qquad (9\text{-}1)$$

Thus pressure is a scalar quantity. *Pressure* at a point within the area ΔA is defined to be

$$P = \lim_{\Delta A \to 0} \frac{\Delta F}{\Delta A} \qquad (9\text{-}2)$$

We have seen that fluids are isotropic, that the physical properties of the fluid at a point in the fluid cannot depend on direction. Pressure is a physical property of a fluid at a point and must be independent of direction. Thus *the magnitude of the force on an infinitesimal surface at a point in a fluid has the same value regardless of the orientation of that surface.*

Pressure has units of force per unit area. The metric unit of pressure then is $N/m^2 \equiv Pa$ (the pascal); the English unit is lb/ft^2. Many units of pressure are used in the sciences and engineering. Pressure units will be considered further after we have discussed the measurement of pressure.

At points within a fluid at rest, pressure varies with vertical position. The pressure at the bottom of a swimming pool is greater than at the top. The force on each unit area of the pool bottom is greater than the force on each unit area of its surface because the bottom is supporting the additional weight of the water in the pool. We say "additional" because there is a pressure at the surface of the pool that results from the weight of the atmosphere pushing downward on each unit area of the surface. Thus the pressure at points on the surface of the earth is called *atmospheric pressure*. The standard value of atmospheric pressure is 1.013×10^5 Pa or about 2120 $lb/ft^2 = 14.7$ lb/in^2 (atmospheric pressure varies somewhat depending on elevation and weather conditions).

Let us now consider the vertical variation of pressure in a fluid at rest. More specifically, let us consider a liquid. Figure 9.2 represents a region within a liquid in which we have isolated an element of volume, a slab of area A oriented horizontally and having an infinitesimal thickness dy in the vertical direction. The lower face of the slab is at a height y above an arbitrary origin and d is the depth of the slab below the surface of the liquid. Let P be the pressure along the lower face of the slab. Then $P + dP$ is the pressure along the upper face. If the density of the liquid is ρ, the mass dm of the element is equal to ρdV, where dV is its volume. The weight dW of the slab is $dW = (dm)g = g\rho dV = g\rho A dy$. Because the slab is in equilibrium, the vertical forces on the slab must sum to zero:

$$PA - (P + dP)A - g\rho A dy = 0$$

$$dP = -g\rho dy \qquad (9\text{-}3)$$

The pressure difference between any two vertical positions, y and y_0, at which the pressures are P and P_0, may be obtained by integrating both sides of Equation (9-3):

$$\int_P^{P_0} dP = -g \int_y^{y_0} \rho \, dy \qquad (9\text{-}4)$$

If we were dealing with a gas, in which the density could change appreciably with vertical distance, we would have to know how the density ρ varies with y in order to integrate Equation (9-4). For a liquid, the density is very nearly a

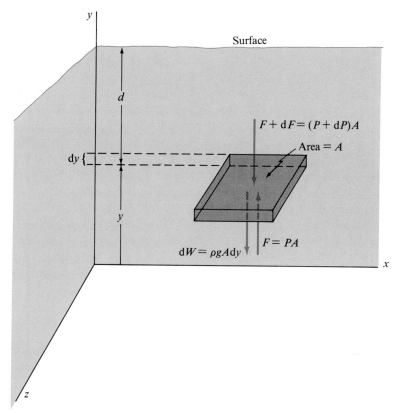

Figure 9.2
An element of volume within a liquid
at a depth d below the surface of the
liquid.

constant value, and we may assume that ρ is a constant. Then for a liquid,
Equation (9-4) becomes

$$P_o - P = -g\rho(y_o - y) \qquad (9\text{-}5)$$

If we now let the surface of the liquid be at $y = y_o$, as shown in Figure 9.3, and if
we designate P_o to be the pressure at the surface (which may be at atmospheric
pressure, for example), then the pressure P at the depth $d = y_o - y$ is

$$P = P_o + g\rho d \qquad (9\text{-}6)$$

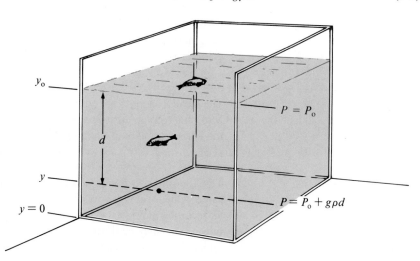

Figure 9.3
The variation of pressure with depth
below the surface of a liquid.

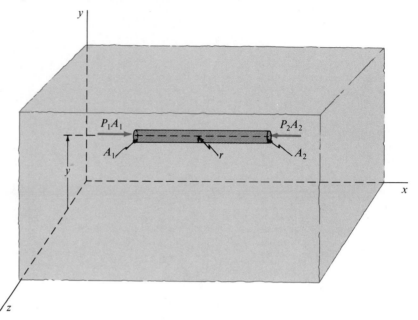

Figure 9.4
A thin cylinder oriented with its axis
lying horizontally inside a fluid.

E 9.1   A cylindrical tank (diameter = 2.0 m, height = 15 m) oriented
with its axis vertical is filled with water. The top of the tank is open to the
atmosphere.
(a) Calculate the weight of the water in the tank.   Answer: 4.6×10^5 N
(b) What force does the water exert on the bottom of the tank?
                                                          Answer: 7.8×10^5 N
(c) What is the pressure at the bottom of the tank?   Answer: 2.5×10^5 Pa

Now that we are able to determine the variation in pressure with vertical distance
in a liquid (or a gas if we know how ρ varies with height), let us consider how
pressure in a liquid varies horizontally.

Imagine a thin cylinder of radius r, like that of Figure 9.4, within a fluid at
rest. The axis of the cylinder lies horizontally, parallel to the x axis. The areas A_1
and A_2 at the ends of the cylinder are equal and lie in vertical planes. The average
pressures are $\bar{P}_1$ and $\bar{P}_2$ on the ends of the cylinders. Because the fluid within the
imaginary cylinder is at rest, the net horizontal force on the fluid within the
cylinder is zero:

$$\bar{P}_1 A_1 - \bar{P}_2 A_2 = 0 \qquad (9\text{-}7)$$

The areas A_1 and A_2 are equal, and if we let the radius r of the cylinder approach
zero, the average pressures, $\bar{P}_1$ and $\bar{P}_2$, become the pressures at two points along
the horizontal axis of the cylinder, so that

$$P_1 = P_2 \qquad (9\text{-}8)$$

Thus, at any two points along a horizontal line in a fluid, the pressure has the
same value. We may now conclude that every point at the same depth below the
surface of a liquid is at the same pressure, regardless of the shape of the container. This may be illustrated using a strange container like that of Figure 9.5 in
which differently shaped vessels communicate. The surface level in each part of

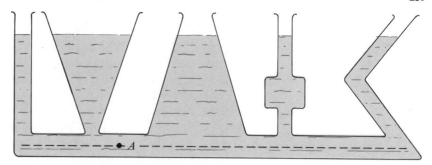

Figure 9.5
A container with communicating chambers. A liquid in the container has its surface at the same level in each chamber.

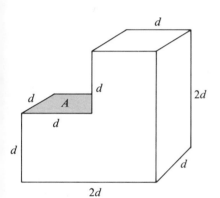

Exercise 9.2

the container must be the same because a point like point A is at a pressure determined only by its depth below any surface connected to point A by the liquid. Before the advent of fluid mechanics, the constant surface level of communicating chambers of arbitrary shapes was called the "hydrostatic paradox," which, of course, is no paradox at all.

E 9.2   An L-shaped tank, constructed as shown in the diagram, is filled with water. The tank is open at the top. If $d = 4.0$ m, calculate
(a) the pressure on the lowest surface of the tank,
(b) the pressure on the face A shown,
(c) the force on face A by the water.
      Answers: (a) 1.8×10^5 Pa; (b) 1.4×10^5 Pa; (c) 2.3×10^6 N, up

Because the pressure in a fluid varies with depth below the fluid surface, it is necessary to use an integrating process to find the total force on a wall retaining a fluid. The following problem illustrates this procedure.

What is the total force exerted by the water on the upstream vertical face of a dam 1000 ft across and 110 ft high if the water behind the dam has a depth of 100 ft?

The dam is visualized in Figure 9.6(a), and Figure 9.6(b) shows a diagram of the upstream face of the dam. We seek the total force on the face of the dam, and we can relate the force exerted by the water to the pressure caused by the water by using Equation (9-6). The pressure is constant in a liquid along a horizontal level, so we select on the face of the dam a horizontal strip of infinitesimal width dy. Then every point along the strip is at the same pressure P. The infinitesimal area of the strip is $dA = L dy$, where $L = 1000$ ft is the length of the dam. The depth of the strip is $D - y$, where D is the depth of the water at the dam and y is the vertical distance from the bottom of the dam to the strip of area dA. The magnitude of the force $d\mathbf{F}$ on the strip that results from water pressure is

$$dF = P dA = PL\, dy \tag{9-9}$$

The pressure P may be found using $d = D - y$ in Equation (9-6), that is, $P = P_o + g\rho(D - y)$. The force on the infinitesimal strip, then, has magnitude

$$dF = P_o L\, dy + g\rho(D - y)L\, dy \tag{9-10}$$

Because each of these infinitesimal forces is in the same direction, the total force on the upstream face of the dam has magnitude $F = \int dF$. The integration of the

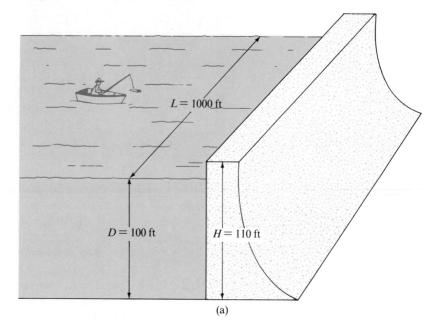

(a)

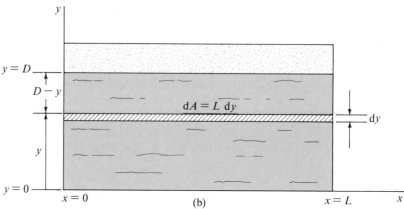

(b)

Figure 9.6
(a) A dam, 110 ft high and 1000 ft across, with water at a depth of 100 ft behind it. **(b)** The upstream face of the dam. An infinitesimal strip at a depth $D - y$ has area $dA = L\ dy$.

right side of Equation (9-10) must include that part of the face of the dam that is below the surface of the water. Integrating both sides of Equation (9-10) gives

$$F = \int_0^D P_o L\ dy + \int_0^D g\rho(D - y)L\ dy$$

$$F = P_o LD + g\rho L\left[Dy - \frac{y^2}{2}\right]_0^D$$

$$F = P_o LD + \frac{1}{2}g\rho LD^2 \tag{9-11}$$

The product $g\rho$ is called the *weight density* of a substance that has density ρ as its mass per unit volume. Weight density is measured in units of lb/ft^3 or N/m^3. Water has a weight density of 62.4 lb/ft^3 = 9800 N/m^3. If we use atmospheric pressure P_o to be 2120 lb/ft^2 and substitute the other given values, $D = 100$ ft and $L = 1000$ ft, into Equation (9-11), we obtain the total force on the dam:

$$F = (2.12 \times 10^3)(10^3)(10^2) + \frac{1}{2}(62.4)(10^3)(10^4)$$

$$F = 2.12 \times 10^8 + 3.12 \times 10^8 = 5.2 \times 10^8 \text{ lb} \qquad (9\text{-}12)$$

The water exerts a force of over half a billion pounds on the upstream face of the dam. About 40 percent of this force is exerted on the dam by the atmosphere.

Example 9.1   **PROBLEM**   Water exerts a force on the dam of Figure 9.6. Calculate the torque this force produces about the horizontal axis shown in Figure 9.7.

   SOLUTION   Figure 9.7, which includes a cross section of the dam, shows the force $d\mathbf{F}$ acting at a height y above the horizontal axis. The (clockwise) torque $d\tau$ produced by this force is

$$d\tau = y \, dF = yP(y)L \, dy$$

where y is the lever arm of $d\mathbf{F}$ and $P(y)$ is the pressure at the height y above the axis. But, as before, this pressure is given by

$$P(y) = P_o + g\rho(D - y)$$

and so we have

$$d\tau = y[P_o + g\rho(D - y)]L \, dy$$

or

$$\tau = \int_0^D [P_o Ly + g\rho L(Dy - y^2)]dy$$

$$\tau = \left[\frac{1}{2}P_o Ly^2 + g\rho L\left(\frac{1}{2}Dy^2 - \frac{1}{3}y^3\right)\right]_0^D$$

$$\tau = \frac{1}{2}P_o LD^2 + \frac{1}{6}g\rho LD^3$$

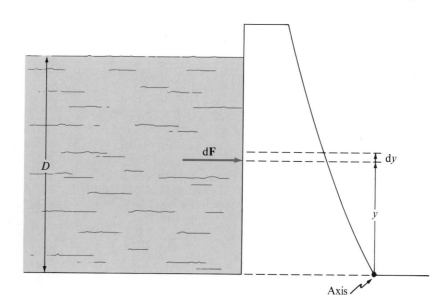

Figure 9.7   Example 9.1

Substituting $P_o = 2120$ lb/ft^2, $g\rho = 62.4$ lb/ft^3, $L = 1000$ ft, and $D = 100$ ft gives

$$\tau = \frac{1}{2}(2120)(1000)(100)^2 + \frac{1}{6}(62.4)(1000)(100)^3$$

$$\tau = 1.06 \times 10^{10} + 1.04 \times 10^{10} = 2.1 \times 10^{10} \text{ ft-lb}$$

Notice that 50 percent of the torque produced by the water is attributable to atmospheric pressure. Because the atmosphere acts on both faces of the dam, only that 50 percent of the torque caused by the weight of the water tends to topple the dam. Consequently, it is this torque that must be counteracted by torque caused by the weight of the dam or by other constraining forces on the dam. ■

The measurement of pressure in a fluid is most easily accomplished in principle by using a U-shaped tube containing a liquid, as illustrated in Figure 9.8. Such a device, called a *manometer*, has one end of the U-tube open to the atmosphere, where the pressure is P_o. The other end of the tube is open to a region (of liquid or gas) at the pressure P, which is to be measured. Using first the left side of the tube, the pressure at the lowest point inside the tube, using Equation (9-6), is $P + g\rho y$, where y is the height of the column of liquid in the left side of the tube. The pressure at the same lowest point inside the tube, using the right side of the tube, is $P_o + g\rho y_o$, where y_o is the height of the liquid column on the right side. We may, therefore, conclude that

$$P + g\rho y = P_o + g\rho y_o$$

$$P - P_o = g\rho(y_o - y) \tag{9-13}$$

or

$$P = P_o + g\rho h \tag{9-14}$$

in which h is the difference, $y_o - y$, in the heights of the liquid surfaces in the two columns. Of course, we could have obtained this result from Equation (9-6)

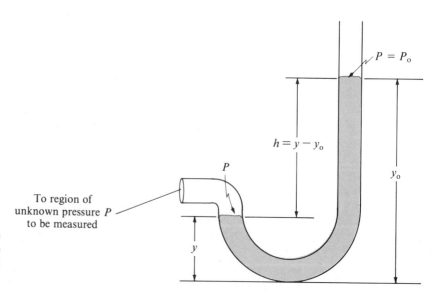

Figure 9.8
A manometer open at one end to the atmosphere and open at the other end to a region of unknown pressure, which is to be measured.

directly. Equation (9-14) gives the absolute pressure P in the region of the fluid that has the pressure being measured. Equation (9-13) gives the difference $P - P_o$ between the absolute pressure and atmospheric pressure, and this difference is called *gauge pressure*. Thus, the height difference h between surfaces of the liquid in a manometer is proportional to the gauge pressure in the region being measured.

E 9.3   Suppose the liquid in the manometer of Figure 9.8 is mercury ($\rho = 1.36 \times 10^4$ kg/m³) and the height h is measured to be 12.8 cm. Calculate the corresponding absolute and gauge pressures.

Answer: 1.2×10^5 Pa, 1.7×10^4 Pa

E 9.4   An open container holds oil of density 850 kg/m³. The depth of the oil in the container is 1.5 m. Calculate the absolute and gauge pressures at the lowest level in the oil.   Answer: 1.1×10^5 Pa, 1.2×10^4 Pa

The *barometer*, another device for measuring pressure, is usually used to measure atmospheric pressure. A barometer, shown in Figure 9.9, is a long, transparent tube, closed at one end, that has been filled with a liquid (usually mercury) before being inverted and placed in an open vessel containing the same liquid. When the tube is in place, a space is formed above the column of liquid that contains only the vapor of the liquid in the column. The vapor pressure of mercury at normal room temperatures is negligibly small compared to atmospheric pressure. In a mercury barometer, then, we may assume that the pressure at the upper liquid surface is zero, and the atmospheric pressure P_o at the surface of the mercury in the open vessel is given by

$$P_o = g\rho h \qquad (9\text{-}15)$$

where h is the difference in the vertical heights of the liquid surfaces. In mercury, ρ is equal to 13.6 g/cm³.

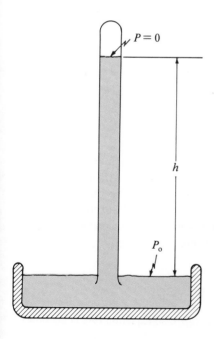

E 9.5   Calculate the height h of the mercury column of the barometer shown in Figure 9.9. Assume that $P_o = 1.013 \times 10^5$ Pa.   Answer: 0.76 m

Manometers and barometers are common laboratory instruments for the measurement of pressure, and a number of the common units of pressure reflect this usage. The wide variety of units of pressure include millimeters of mercury (mm-Hg), which is now less descriptively called the *torr*; inches of mercury (in-Hg); centimeters of mercury (cm-Hg); inches of water (in-H_2O); and so on. The fundamental units of pressure, as noted earlier, are lb/ft² and N/m². The unit N/m² is named the pascal (Pa). A pressure of 1.013×10^5 Pa = 14.7 lb/in.² is named the *atmosphere* (atm). A *bar* is defined to be exactly 10^5 Pa. The *millibar* (mbar), one thousandth of a bar, is the unit chosen by meteorologists to report atmospheric pressures (except on television, where barometric readings are frequently quoted in "inches," referring, of course, to in-Hg).

Figure 9.9
A barometer, consisting of a long tube that has been filled with a liquid, inverted, and placed in an open vessel containing the same liquid.

E 9.6   Express one atmosphere in bar, mbar, torr, in-Hg, and in-H_2O.

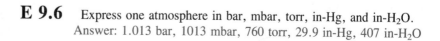
Answer: 1.013 bar, 1013 mbar, 760 torr, 29.9 in-Hg, 407 in-H_2O

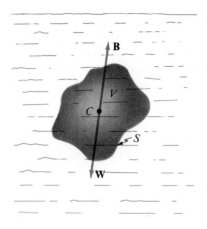

Figure 9.10
An arbitrary volume V, bounded by a surface S within a fluid. The weight of the fluid within S is W, and the buoyant force on the fluid within S is B.

Archimedes' Principle

A useful fact associated with bodies that are floating on or submerged in a fluid is expressed by *Archimedes' principle: An object completely or partially immersed in a fluid experiences a vertically upward buoyant force equal in magnitude to the weight of the fluid displaced by that object.* This principle incorporates no new physics but is an application of Newton's laws and the concept of hydrostatic pressure.

The proof of Archimedes' principle may be seen by considering Figure 9.10. Imagine an arbitrary volume V, within a static fluid, bounded by the surface S. The fluid is at rest, so the net force on the fluid within S is equal to zero. The fluid outside of S must be exerting an upward force B on the fluid inside of S that is equal in magnitude to the weight W of the fluid inside S. Further, the buoyant force B must pass through the center of mass C of the fluid inside S, but the forces exerted on the surface S cannot depend on what is, in fact, inside of S. Therefore, the net force B that the fluid outside of S exerts on any object occupying the volume V will be equal in magnitude to the weight W of the fluid of volume V and will be directed upward. Not only is it true that $B = W$, but it is also true that B passes through the point where the center of mass of the displaced liquid would be if the liquid had not been displaced.

An instructive application of Archimedes' principle is the determination of the density of an irregularly shaped solid object:

An object weighing 6.0 oz is alleged to be pure gold ($\rho = 19.3$ g/cm^3). The object is suspended by a thread and completely submerged in water ($\rho = 1.00$ g/cm^3) while hanging from a scale, which indicates that the "weight in water" of the object is equal to 5.47 oz. Is the object pure gold?

Figure 9.11(a) shows the object suspended from the scale and immersed in water; Figure 9.11(b) indicates the forces acting on the object. The weight W_a of the object in air is the pull of the earth on the object, so

$$W_a = g\rho_s V \qquad (9\text{-}16)$$

where V is the volume of the solid and ρ_s is its density, which we are seeking. The upward buoyant force B on the object is, according to Archimedes' principle, equal in magnitude to the weight of water displaced by the object. Thus we use $B = g\rho_w V$, where ρ_w is the density of water. The thread exerts an upward force on the object equal to W_w, the "weight in water" indicated on the scale. The forces on the suspended object must sum to zero:

$$B + W_w - W_a = 0$$

$$g\rho_w V = W_a - W_w \qquad (9\text{-}17)$$

We know from Equation (9-16) that $V = W_a/(g\rho_s)$, so Equation (9-17) becomes

$$\frac{g\rho_w W_a}{g\rho_s} = W_a - W_w$$

$$\rho_s = \frac{\rho_w W_a}{W_a - W_w} \qquad (9\text{-}18)$$

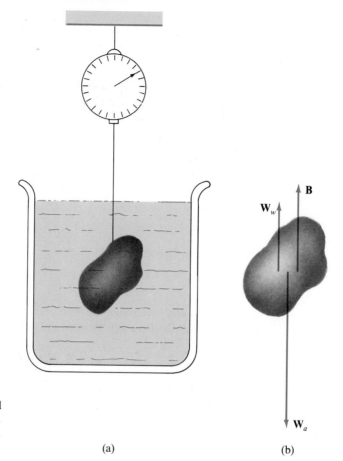

Figure 9.11
(a) An irregular solid object suspended from a scale and immersed in water.
(b) A force diagram of the solid object.

(a)   (b)

which relates the desired density ρ_s of the object to known quantities so that

$$\rho_s = \frac{(1.00 \text{ g/cm}^3)(6.00 \text{ oz})}{(6.00 \text{ oz}) - (5.47 \text{ oz})} = 11.3 \text{ g/cm}^3 \qquad (9\text{-}19)$$

The calculated density of the object, 11.3 g/cm³, is considerably less than that of gold and, in fact, suspiciously coincides with the density of lead.

E 9.7   A typical woman will float in water with very little, if any, of her body above the surface of the water. Calculate the volume of a woman weighing 550 N (124 lb). Estimate your own volume.   Answer: 0.056 m³ = 2.0 ft³

Example 9.2   **PROBLEM**   A thin uniform wooden beam, which has a length of 10 m and a density of 640 kg/m³, is pinned at its lower end and "floats," as shown in Figure 9.12(a), in water that is 6.0 m deep. Calculate the angle θ that the beam makes with the horizontal.

SOLUTION   Figure 9.12(b) shows the forces on the beam. Because the beam is uniform, the buoyant force **B** acts at a distance $x/2$ from the pin, where x measures the length of the submerged part of the beam. Similarly, the weight **W** acts at the center of the uniform beam. The force **P** is the downward force on the beam exerted by the pin. The magnitudes of **B** and **W** are given by $B = \rho_0 g A x$ and $W = \rho g A L$ where ρ_0 and ρ are the densities of water and the beam, A is the

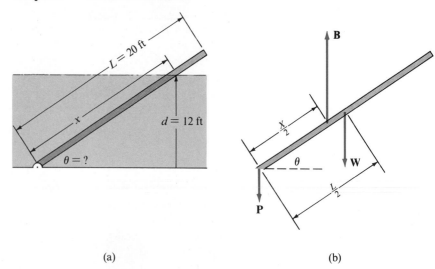

Figure 9.12   Example 9.2   (a)   (b)

area of a cross section of the beam, and L is the length of the beam. Equating the net torque about an axis through the pin to zero gives

$$\sum \tau = 0 = B\frac{x}{2}\cos\theta - W\frac{L}{2}\cos\theta$$

or

$$Bx = WL$$

$$\rho_0 gAx^2 = \rho gAL^2$$

$$x^2 = \frac{\rho}{\rho_0}L^2$$

$$x = \sqrt{\frac{\rho}{\rho_0}}L = \sqrt{\frac{640}{1000}}(10) = 8.0 \text{ m}$$

From Figure 9.12(a) we see that

$$\sin\theta = \frac{d}{x} = \frac{6}{8} = 0.75$$

$$\theta = 49° \quad \blacksquare$$

E 9.8   If the beam of Example 9.2 has a cross-sectional area of 0.040 m², determine each of the three forces acting on the beam.
Answer: $\mathbf{W} = 2.5 \times 10^3$ N, down; $\mathbf{B} = 3.1 \times 10^3$ N, up;
$\mathbf{P} = 6.3 \times 10^2$ N, down

Pascal's Law

Forces can be amplified in a closed fluid system, called a hydraulic system, by the judicious use of the pressure of a confined fluid. In an automotive brake system, for example, a few pounds of force on a brake pedal become considerably greater forces on the brake shoes that press against surfaces on the wheels of the auto. The principle of operation of such hydraulic systems may be understood from *Pascal's law: Pressure applied at any point to a confined fluid is transmitted undiminished to every point within the fluid and, therefore, to the walls of the container of the fluid.*

Pascal's law is not an independent physical principle but follows directly from the laws of mechanics we have already encountered. In Equation (9-6), where $P = P_o + g\rho d$, and in Figure 9.3, we can see that if additional pressure is applied to the top surface of the liquid to increase P_o this additional pressure increases P by the same amount at every point within the liquid. Although Pascal's law applies to gases as well as liquids, large displacements of a piston are necessary to transmit significant pressure in gases because of the compressibility of gases. Consequently, we consider only those hydraulic systems that use confined liquids.

The amplification of forces using hydraulics is illustrated by the system of Figure 9.13, which is composed of two pistons of cross-sectional areas A_1 and A_2 at the ends of a closed system of liquid. If we press with a force $\mathbf{F}_1$ on the piston of area A_1, the additional pressure transmitted to every point in the liquid of the closed system is $\Delta P = F_1/A_1$. Because this same additional pressure ΔP appears at the piston with area A_2, and $\Delta P = F_2/A_2$, we may write

$$\frac{F_1}{A_1} = \frac{F_2}{A_2}$$

or

$$F_2 = \frac{A_2}{A_1} F_1 \tag{9-20}$$

Thus the magnitude of the force $\mathbf{F}_2$ on the second piston is equal to the magnitude of $\mathbf{F}_1$ amplified by a factor equal to the ratio A_2/A_1 of the cross-sectional areas of the pistons. Nevertheless, by amplifying the applied force by a factor of A_2/A_1 we have not "gotten something for nothing." The work done by $\mathbf{F}_1$ in moving the piston of area A_1 is equal to the work done by $\mathbf{F}_2$ in moving the piston of area A_2. In other words, energy is conserved even when we achieve an amplification of the applied force.

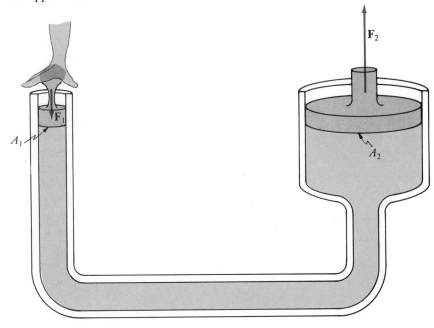

Figure 9.13
Schematic representation of a hydraulic system.

E 9.9   By exerting a force **F** on a small piston of diameter d (2.0 cm), a hydraulic system lifts a load of mass M (1200 kg) supported by a piston of diameter D (30 cm). What is the magnitude of the force **F**?   Answer: 52 N

9.3 Fluid Dynamics

The study of fluids in motion is, in general, complex, both physically and mathematically. The level of this text, therefore, restricts our considerations to a few basic concepts associated with fluid dynamics. These fundamentals are not new physical principles. They are applications of the principles of mechanics that we have discussed in earlier chapters. We will attempt to show how these simple applications of fluid dynamics permit an understanding of the physical foundations of a number of commonly observed phenomena. Consistent with our uncomplicated approach to the study of fluid motion, then, we will deal with an idealized version of a real fluid, that is, with an *ideal fluid* flowing in an uncomplicated way. Several technical terms are used to characterize an ideal fluid and its flow: incompressible, nonviscous, and steady flow. Let us consider these characteristics individually.

We have already introduced incompressibility in describing static liquids, in which the density of the liquid is constant regardless of the pressure within the liquid. Gases, when moving at speeds well below the speed of sound, flow as if they were essentially incompressible. The speed of sound in a fluid is referred to as Mach 1, and gases moving at speeds less than Mach 0.5, or half the speed of sound in the fluid, do not change density appreciably. At these subsonic speeds, the gas can flow over and around a fixed object before the gas can be compressed significantly. At speeds near Mach 1 (about 330 m/s $\cong$ 750 mi/h in air), a moving gas bunches up against an object, like the leading edge of a wing, and produces what is known as a *shock wave*. We will restrict our considerations to speed ranges in which fluids flow as if they were incompressible.

A *nonviscous fluid* experiences no forces tangential to the direction of motion of a fluid when layers of the fluid move relative to one another. Viscosity, therefore, is a sort of fluid friction. Molasses is very viscous and tends to flow slowly. Water has a much lower viscosity than molasses and more nearly approximates the "frictionless" flow of a nonviscous fluid.

Incompressibility and nonviscosity are idealized properties of fluids; steady flow is a characterization of the motion, or flow, of a fluid. A fluid is said to be in *steady flow* if the velocity of the fluid at an arbitrary point within the fluid does not change with time. Different points within the fluid may have different velocities, but the velocity at each point remains constant. The stream of water from a gently flowing faucet is an example of steady flow. When a fluid is passing over the wing of a moving airplane, an observer on the ground "sees" the position and velocity of the air changing with time. An observer on the airplane, however, sees the air in steady flow over the wing. The velocity of the fluid flow at each point in the reference frame of the airplane does not change with time; the fluid is in steady flow in the reference frame of the airplane. In our studies of fluid flow, we may eliminate time considerations by placing ourselves in the

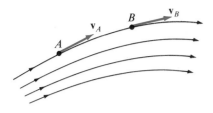

Figure 9.14
Streamlines of a moving fluid. The velocity at every point within the fluid is tangential to a streamline.

reference frame in which the fluid flow is steady. Wind tunnels are experimental facilities that take advantage of the steady-flow frame of reference.

Assuming an ideal fluid, one that is incompressible and nonviscous, moving in steady flow may seem as if we have reduced fluids to an unusable, impractical idealization. In fact, the use of this idealized approximation provides surprisingly accurate results in many simple applications of fluid mechanics.

Equation of Continuity and Bernoulli's Equation

Let us now introduce a concept that makes the study of fluids in steady flow both graphic and convenient. A *streamline* is an imaginary line, like those of Figure 9.14, within a moving fluid that has the following properties:

1. The velocity of any particle in the fluid is tangential to a streamline at all times.

2. At a fixed point, like point A, on the streamline, the velocity of every particle that reaches that point is the same, both in magnitude and direction.

Every particle of the fluid that reaches point A must proceed to point B, and so on. Thus the streamlines are the trajectories of particles in the fluid. In steady flow, streamlines cannot intersect (otherwise a particle reaching the intersection could follow either of two paths and the flow would not be steady). Therefore, in steady flow, streamlines illustrate a fixed pattern of the flow. A *stream tube* is a region in a fluid bounded by streamlines, as seen in Figure 9.15(a). In steady flow, a particle within a stream tube cannot pass outside the tube (because its streamline would then have to intersect a streamline bounding the stream tube).

Suppose in Figure 9.15(a) the stream tube is sufficiently small that the speeds of particles within the tube at a given cross section are equal. At the point P_1, where the cross-sectional area of the tube perpendicular to the streamlines is A_1, let v_1 be the speed of the particles. At another section of the tube, where A_2 is the cross-sectional area, suppose the particles at P_2 have a speed v_2. Figure 9.15(b) shows the incremental volume ΔV of fluid passing through A_1 in a small time Δt. Because the speed of every particle at A_1 is v_1, the thickness $\Delta \ell$ of the cylindrical volume is equal to $v_1 \Delta t$. The volume ΔV is then

$$\Delta V = A_1 \Delta \ell = A_1 v_1 \Delta t \tag{9-21}$$

Figure 9.15
(a) A stream tube. Particles at P_1, where the cross-sectional area of the tube is A_1, have velocity v_1. Particles at P_2, where the cross-sectional area is A_2, have velocity v_2. **(b)** The incremental volume ΔV of fluid passing through the cross-sectional area A_1 of the stream tube in a time interval Δt.

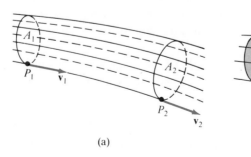

(a)   (b)

and the mass Δm_1 of the fluid within ΔV is given by

$$\Delta m_1 = \rho_1 \Delta V = \rho_1 A_1 v_1 \Delta t$$

or

$$\frac{\Delta m_1}{\Delta t} = \rho_1 A_1 v_1 \qquad (9\text{-}22)$$

where ρ_1 is the density of the liquid at the cross section A_1. In the limit as $\Delta t \to 0$, Equation (9-22) becomes

$$\lim_{\Delta t \to 0} \frac{\Delta m_1}{\Delta t} = \frac{dm}{dt} = \rho_1 A_1 v_1 \qquad (9\text{-}23)$$

The quantity dm/dt is the *mass flow rate* at A_1, and dm/dt represents the mass per unit time passing through the cross-sectional area A_1. At point P_2, where the speed of particles in the tube is v_2, a similar expression gives the mass flow rate at P_2:

$$\frac{dm_2}{dt} = \rho_2 A_2 v_2 \qquad (9\text{-}24)$$

E 9.10   At what speed must a liquid of density 1200 kg/m³ flow in a pipe that has a diameter of 4.0 cm if the pipe is to deliver 5.0 kg of the liquid per second?   Answer: 3.2 m/s

In steady flow, particles of the fluid cannot leave the tube. Therefore, the mass passing through A_1 must be equal to the mass passing through A_2 in the same time interval, or

$$\rho_1 v_1 A_1 = \rho_2 v_2 A_2 \qquad (9\text{-}25)$$

Equation (9-25) is an expression of the conservation of mass for any fluid in steady flow. For an incompressible fluid, in which $\rho_1 = \rho_2$, Equation (9-25) becomes

$$v_1 A_1 = v_2 A_2$$

or

$$vA = \text{constant} \qquad (9\text{-}26)$$

Equation (9-25), or Equation (9-26) in the case of incompressible fluids, is called the *equation of continuity* in fluid dynamics. In an incompressible fluid, the product vA is constant in a given stream tube. Where streamlines are close together (A is small), the fluid is flowing more swiftly than where the streamlines are farther apart (A is larger). The product vA is called the *volume flow rate* of a fluid and has units of m³/s or ft³/s.

E 9.11   A liquid flows in a tube (diameter = 2.0 cm) at a speed equal to 0.50 m/s. How many cubic centimeters of this liquid does the tube deliver each second?   Answer: 160 cm³

Let us now consider an important consequence of the equation of continuity applied to an incompressible fluid, that is, a fluid for which ρ is constant and, therefore, the product vA of the equation of continuity is constant at every point along a stream tube. Figure 9.16 shows an incompressible fluid moving toward the right through a stream tube. The cross-sectional area of the tube is A_1 at x_1 and A_2 at x_2. Consider an infinitesimal volume of the fluid at (x_1, y_1). Every point within this element has speed v_1 and is at a pressure P_1. In an infinitesimal time interval dt, the fluid at x_1 moves through a horizontal displacement dx_1. During this same time interval, an element of fluid at (x_2, y_2), where all points in the element have speed v_2 and are at a pressure P_2, moves through a displacement dx_2. Because these two elements have equal volumes, the displacements dx_1 and dx_2 are related by $A_1dx_1 = A_2dx_2$. Figure 9.16 shows the displacements of these elements during the time interval dt. The intervening fluid between (x_1, y_1) and (x_2, y_2) is physically identical before and after the displacements occur. To relate the speeds v_1 and v_2 to the pressures P_1 and P_2, let us use the work-energy principle: The work done by the net force on a system is equal to the change in kinetic energy of the system.

The work dW_P done by the pressure forces on the fluid that has been moved in the time interval dt is

$$dW_P = P_1A_1dx_1 - P_2A_2dx_2 \tag{9-27}$$

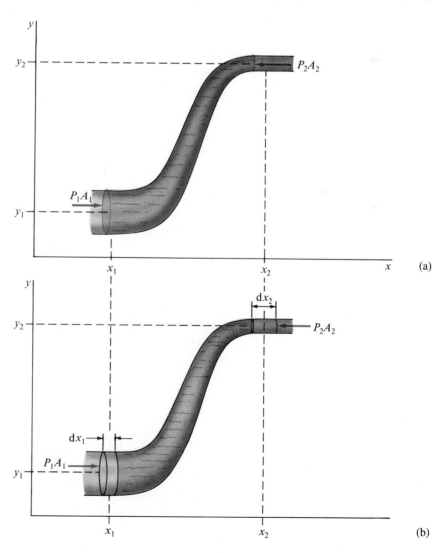

Figure 9.16
Illustrating the displacements of infinitesimal elements of fluid through a stream tube (a) at the beginning and (b) at the end of a time interval dt.

The force P_2A_2 is oppositely directed from the displacement of the fluid, which accounts for the negative work done on the element at (x_2, y_2). Because the fluid is incompressible, the equation of continuity requires that $A_1v_1 = A_2v_2$, or $A_1dx_1/dt = A_2dx_2/dt$, which means that $A_1dx_1 = A_2dx_2 = dV$, where dV is the volume of each of the elements of fluid. The work dW_P done by the pressure forces is then

$$dW_P = P_1dV - P_2dV = (P_1 - P_2)dV \qquad (9\text{-}28)$$

The effect of the flow that has taken place (displacements dx_1 of the element at x_1 and dx_2 of the element at x_2) has been to raise an element of mass $dm = \rho dV$ through a vertical height $y_2 - y_1$. Therefore, the work dW_g done on that mass by the gravitational force is

$$dW_g = -(dm)g\,(y_2 - y_1) = -\rho g(y_2 - y_1)dV \qquad (9\text{-}29)$$

Why is this work negative?

The total work $dW_P + dW_g$ done on the element is, according to the work-energy principle, equal to the change ΔK in the kinetic energy of the element:

$$dW_P + dW_g = \Delta K$$

$$(P_1 - P_2)dV - \rho g(y_2 - y_1)dV = \frac{1}{2}(\rho dV)v_2^2 - \frac{1}{2}(\rho dV)v_1^2$$

or, after some algebra,

$$P_1 + g\rho y_1 + \frac{1}{2}\rho v_1^2 = P_2 + g\rho y_2 + \frac{1}{2}\rho v_2^2 \qquad (9\text{-}30)$$

Because the levels y_1 and y_2 used here are arbitrary, we may ignore the subscripts and write

$$P + g\rho y + \frac{1}{2}\rho v^2 = \text{constant} \qquad (9\text{-}31)$$

Equation (9-30) or Equation (9-31) is called *Bernoulli's equation*. It is an expression of the conservation of energy in fluids. Like Archimedes' principle and Pascal's law, Bernoulli's equation is an application of the basic principles of mechanics.

An important and useful result is obtained in a special case of Bernoulli's equation. Figure 9.17 shows an incompressible fluid moving in the positive x direction in a horizontal stream tube. For this case the average vertical position of the fluid in the tube remains constant, or $y_1 = y_2$. The application of Equation (9-30) at x_1, where the pressure and speed are P_1 and v_1, and x_2, where the pressure and speed are P_2 and v_2, gives

$$P_1 + \frac{1}{2}\rho v_1^2 = P_2 + \frac{1}{2}\rho v_2^2 \qquad (9\text{-}32)$$

Figure 9.17
A horizontal stream tube in which an incompressible fluid is moving in the positive x direction.

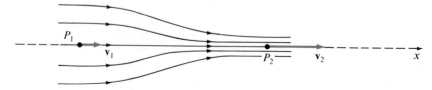

Because the streamlines are closer together at x_2 than at x_1, the speed v_2 is greater than v_1. Equation (9-32) then demands that the pressure P_1 be greater than P_2. This result may be summarized as follows: *For an incompressible fluid in horizontal steady state flow, the pressure in the fluid is less in a region where the speed of flow is greater.*

E 9.12   Use Bernoulli's equation to obtain Equation (9-6), that is, $P = P_o + g\rho d$, an expression that relates the pressure P at a depth d below the surface of a static liquid of density ρ to the pressure P_o at the surface of the liquid.

Applications of Fluid Dynamics

Bernoulli's equation, accompanied by the equation of continuity, is the fundamental relationship of fluid mechanics. In this section we shall see how fluid mechanics may be applied to explain and analyze a variety of familiar physical situations.

Several physical situations of common experience are explained by a result we have obtained from consideration of a special case of Bernoulli's equation: Pressure is greater in a horizontally flowing fluid where the speed of the fluid is less.

A spinning ball moving through air is often described as a "curve ball." Such a ball experiences a force that is transverse to its velocity; this force causes it to curve, or deviate, from the trajectory it would follow if it were not spinning. Figure 9.18(a) suggests a nonspinning ball moving to the left through air. The streamlines indicate that we are depicting the situation seen from a reference frame moving along with the ball. Figure 9.18(b) shows the situation with the ball spinning about an axis out of the page. Because air is viscous, the spinning ball drags air near its surface around with the ball. The spinning air near the ball causes an increase in the speed of air past the ball in the region of point A. This increase is indicated by the close spacing between the streamlines in the region of A. In a similar way, the speed of the air past the ball in the region of point B is decreased. Thus, the pressure at A is less than at B and a net force $\mathbf{F}$ acts on the ball in a direction from B toward A. Is it now understandable why many curve-ball pitchers "rough up" a baseball with their hands before delivering a pitch?

Figure 9.18
A ball moving toward the left through still air. The diagrams are shown in the reference frame of the ball. **(a)** A nonspinning ball. **(b)** A ball spinning about an axis out of the page.

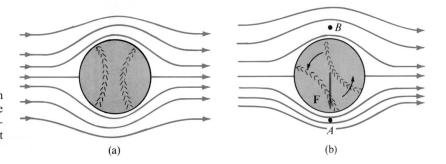

(a)   (b)

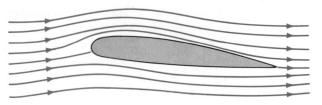

Figure 9.19
An airfoil passing through air and depicted in the reference frame of the airfoil.

A similar fluid dynamics situation is the flow of air over the surface of an airfoil, the wing of an airplane (or the rotor of a helicopter). Figure 9.19 shows the cross section of an airfoil with the streamlines of air indicated. Above the wing, where the streamlines are closer together and the speed of flow is greater than below the wing, the pressure is less than below the wing. The result of this pressure difference is a net upward force, called *lift*, on the wing.

Example 9.3

PROBLEM Figure 9.20 depicts a *Venturi meter*, a device used to measure the speed of flow of a liquid in a pipe. A liquid of density ρ flows at a speed v in a pipe of cross-sectional area A. A manometer using a liquid of density $\rho_o > \rho$ is connected with one opening at a point (1) where the pressure and flow speed are P and v. The second manometer opening is at a point (2) where the pipe is constricted to an area $A_o < A$ and where the liquid pressure and flow speed are P_o and v_o. The height difference h of the manometer liquid indicates a pressure difference, $P - P_o$. Express the flow speed v in terms of A, A_o, ρ, ρ_o, and h.

SOLUTION Bernoulli's equation relates the pressures and flow rates to give

$$P + \frac{1}{2}\rho v^2 = P_o + \frac{1}{2}\rho v_o^2$$

The equation of continuity requires that $vA = v_o A_o$, or $v_o = v(A/A_o)$. If this result is employed in Bernoulli's equation, the pressure difference, $P - P_o$, is given by

$$P - P_o = \frac{1}{2}\left(\frac{A^2 - A_o^2}{A_o^2}\right)v^2$$

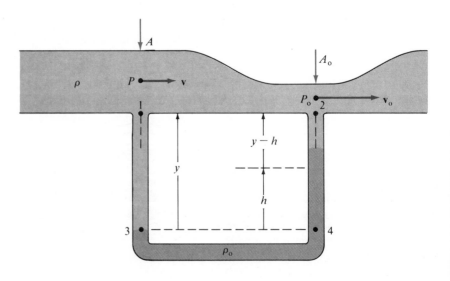

Figure 9.20   Example 9.3

Because the liquid in the manometer is at rest, the pressures at point 3 and 4 are equal. Thus, we have

$$P_3 = P_4$$

or

$$P + \rho gy = P_o + \rho g(y - h) + \rho_o gh$$

$$P - P_o = (\rho_o - \rho)gh$$

If this result is combined with the previous result obtained using Bernoulli's equation, we get

$$v = A_o\sqrt{\frac{2(\rho_o-\rho)gh}{\rho(A^2-A_o^2)}} = \sqrt{\frac{2gh[(\rho_o/\rho)-1]}{(A/A_o)^2 - 1}} \quad \blacksquare$$

E 9.13   Calculate the flow speed that corresponds to a Venturi-meter reading of $h = 12$ cm if $\rho_o/\rho = 13.6$ and $A/A_o = 3.0$.   Answer: 1.9 m/s

Venturi tubes, which are constrictions or "throats" in fluid conduits, are regions of reduced pressure that are used in a number of devices. A perfume *atomizer*, for example, operates by having air forced from a squeeze-bulb through a Venturi throat that reduces the pressure above an open tube dipped into liquid perfume. Atmospheric pressure then forces the perfume up the tube into the airstream, where the liquid is "atomized" into small droplets that are carried forward in a spray by the rapid flow of the air. An *aspirator pump* is a device designed to evacuate air from a closed space by passing water through a Venturi throat to which the airspace is connected. The reduced pressure in the throat permits the air to be forced into the water stream and carried away.

A *Prandtl tube*, more commonly known as a *Pitot tube*, is a device (*see* Fig. 9.21) used to measure the speed of a fluid. The tube is self-contained and can be inserted into a moving fluid, as shown in the figure. The fluid of density ρ is shown flowing with a speed v past the device. The pressure in the fluid is P; at the point S, called the *stagnation point* (where the fluid speed is zero), the pressure is P_S, the *stagnation pressure*. The stagnation point is at the upstream opening of the device; the device is also open to the fluid in a region where the pressure is P.

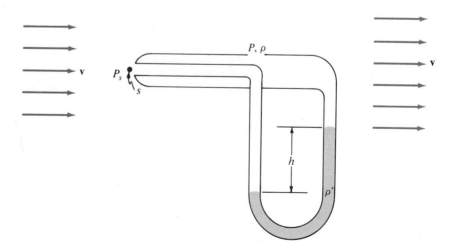

Figure 9.21
A *Prandtl* (or Pitot) tube.

Bernoulli's equation may be applied to the stagnation point and another point in the fluid where the pressure is P to give

$$P_S = P + \frac{1}{2}\rho v^2 \tag{9-33}$$

The difference, $P_S - P$, in pressure is determined by the manometric difference in height h of the fluid of density ρ' in the U-tube:

$$P_S - P = g\rho'h \tag{9-34}$$

The speed v of the fluid is found by combining Equations (9-33) and (9-34):

$$v = \sqrt{2gh\rho/\rho'} \tag{9-35}$$

A device like this can be calibrated to read speed directly. When mounted on an aircraft, a Pitot tube measures airspeed of the aircraft. On a ship or boat it measures the speed of the vessel through the water.

Let us consider one further illustration of the use of Bernoulli's equation. Consider an open tank filled with a liquid to a height h above a small orifice, or opening in the side of the tank, as shown in Figure 9.22. We apply Bernoulli's equation, Equation (9-30), at a point on the surface of the liquid, where the pressure is P_a, atmospheric pressure, and at a point just outside the orifice, where the liquid is also at atmospheric pressure but has a speed v. Letting $y = 0$ at the level of the orifice and $y = h$ be the height of the surface above the orifice, we obtain

$$P_a + g\rho h + 0 = P_a + 0 + \frac{1}{2}\rho v^2 \tag{9-36}$$

Both sides of Equation (9-36) include atmospheric pressure because the pressures appearing in Bernoulli's equation are absolute pressures. Solving Equation (9-36) for v gives

$$v = \sqrt{2gh} \tag{9-37}$$

The speed v is called the *speed of efflux* of the liquid from the orifice. The result of Equation (9-37) indicates that the speed of efflux from the orifice is equal to the speed of an object that has fallen from rest through a vertical distance equal to the depth of the orifice below the surface of the liquid.

In some problems concerning the efflux of fluid through an orifice, it becomes important to know the *effective* area of the orifice through which the fluid exits. Experiment shows that the effective area of a circular opening is about 65 percent of the actual area. This narrowing of the efflux stream results from the convergence of the streamlines as they reach the orifice, thereby constricting the stream tube to a smaller cross section as it exits the orifice.

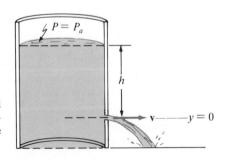

Figure 9.22
A tank, open at the top, with liquid emerging from an orifice located a vertical height h below the surface of the liquid.

E 9.14   Water drains with a speed of efflux of 8.0 m/s through an opening having a diameter of 2.0 cm. The water that drains during a 20-s time interval is collected in a container.

(a) Disregarding any constriction, what mass of water would be collected?

Answer: 50 kg

(b) If the container actually collects 33 kg of water during this time, calculate the effective area and the effective diameter of the opening.

Answer: 2.1 cm^2, 1.6 cm

9.4 Problem-Solving Summary

Problems that are concerned with fluid mechanics are generally easy to identify. In those problems, liquids or gases are functionally involved in the problem situation. Classifying problems of fluid mechanics as static, when the fluids are at rest, or as dynamic, when the fluids are in motion, is equally simple. Deciding to what extent the fluids involved in a problem may be considered ideal fluids, if, indeed, that consideration is necessary, requires more thought. The assumption of an ideal gas is not necessary in most static situations. For example, Archimedes' principle and Pascal's law apply to real liquids and gases.

From a problem-solving point of view, liquids differ from gases most significantly in that liquids may be considered incompressible; the density of a liquid may be treated as a constant value. Gases are highly compressible; we must take care in deciding if the density of a gas is essentially constant in each physical situation.

Pressure is an important analytical quantity for treating both liquids and gases. Defined to be the force per unit area at any point in a fluid, pressure is a scalar quantity. Pressure at a point in a fluid specifies how much force per unit area acts normally at any surface within or at the boundary of a fluid, regardless of the direction in which the surface is oriented. In a liquid of density ρ, the pressure P varies with depth d below the surface of the liquid, where the pressure has a value P_o, according to $P = P_o + g\rho d$. The pressure at a given depth is independent of the horizontal position within the fluid. The gauge pressure at a point within a fluid is the difference between absolute pressure P and atmospheric pressure.

Problems involving flotation or buoyancy usually require the use of Archimedes' principle: The buoyant force on an object on or within a fluid is always directed vertically upward and has magnitude equal to the weight $mg = g\rho V$ of the displaced fluid. It is important in solving problems using a buoyant force to recognize that the buoyant force on an object displacing fluid passes through the point where the center of mass of the fluid displaced would have been—not necessarily through the center of mass of the floating or submerged object.

Pascal's law, which states that an increase in pressure in a confined fluid is transmitted undiminished to every part of the fluid and to the walls containing the fluid, is mostly useful in problems dealing with pistons in hydraulic systems. Remember, it is the pressure increase, not force, that is transmitted throughout a confined fluid. Hydraulic systems are typically used, however, to amplify forces.

In problems of fluid dynamics, situations in which fluids are flowing, we have established relationships that apply to the uncomplicated flow of idealized fluids that are identified by the following properties of the fluid and characteristic of the flow:

1. The fluid may be considered incompressible. All liquids qualify in this respect. Gases flowing at speeds below about half the speed of sound in that fluid may be considered incompressible.

2. The fluid is nonviscous, which means the fluid is assumed to be "frictionless." Care must be taken in assuming a fluid to be nonviscous; no real fluids are actually nonviscous, but gases are excellent approximations to an ideal fluid in this respect. In many flow problems using water or other "thin" liquids, the nonviscous approximation is usually adequate. One should use careful technical judgment in deciding on the appropriateness of this approximation.

3. The fluid is in steady flow. Particles within the fluid follow streamlines that are fixed in some frame of reference.

Bernoulli's equation, $P + g\rho y + \frac{1}{2}\rho v^2 =$ constant, accompanied by the equation of continuity, $vA =$ constant (for constant density), is the fundamental relationship of fluid dynamics. Any problem that is concerned with the flow of an ideal fluid suggests the use of Bernoulli's equation and the equation of continuity. These equations may be applied to any two points along a streamline of a flowing ideal fluid. A word of caution: The pressures that appear in Bernoulli's equation are absolute pressures.

An important analytical consequence of Bernoulli's equation is that the pressure is reduced in a stream tube at reduced cross-sectional areas of the tube where the speed of the fluid is increased. This phenomenon, called the Venturi effect, occurs in pipes or tubes carrying moving fluids through a constriction.

Probably the most common error in solving fluid mechanics problems arises from the failure to account appropriately for atmospheric pressure. Anytime a value for pressure is included in a calculation or relationship, care should be taken to determine if that pressure is properly an absolute pressure or a gauge pressure.

Problems

GROUP A

9.1 A pipe (diameter = 0.60 m) delivers 1500 kg of oil (density = 900 kg/m³) each second. What is the flow speed of the oil in the pipe?

9.2 What force does water exert on a ball (diameter = 10 cm) of steel (density = 7800 kg/m³) that is completely submerged in a pool?

9.3 What force must be exerted on a cube (10 cm along an edge) of balsa wood (density = 130 kg/m³) to keep it submerged in water?

9.4 Mercury (density = 1.36 × 10⁴ kg/m³) fills a cylindrical container that has a height (length) of 20 cm and a diameter of 8.0 cm. The container is open at the top.

(a) What is the pressure at the lowest level in the mercury?

(b) Calculate the weight of the mercury.

(c) What force does the mercury exert on the bottom of the container?

9.5 Repeat Problem 9.4 if the cylindrical container is placed in an evacuated (zero pressure) chamber.

9.6 What fraction of the volume of an iceberg, which has a density = 920 kg/m³, is submerged when the iceberg floats in seawater (density = 1030 kg/m³)?

9.7 A hollow cube (10 cm on a side) has a mass of 0.45 kg. What volume of mercury (density = 1.36 × 10⁴ kg/m³) must be placed inside the cube if the cube is to be in equilibrium when submerged in water?

9.8 Gas is trapped in the closed part of a tube containing mercury ($\rho = 1.36 \times 10^4$ kg/m³), as shown in Figure 9.23. If $h = 1.2$ m, calculate the pressure of the gas.

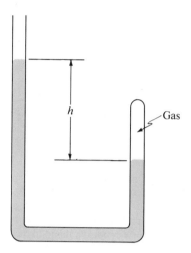

Figure 9.23   Problem 9.8

9.9 What is the tension in a cable supporting a 50-kg sphere (radius = 0.20 m) completely submerged in water?

9.10 Because a real liquid interacts with the wall of the pipe in which it flows, some energy is dissipated, and Bernoulli's equation is no longer applicable. For the situation depicted in Figure 9.24, calculate the pressure difference, $P_2 - P_1$, between points 1 and 2 if the density of the liquid is 900 kg/m³.

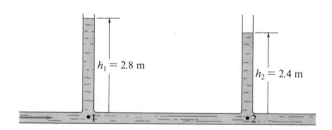

Figure 9.24   Problem 9.10

9.11 Water flows in a horizontal pipe of variable diameter at a mass flow rate of 10 kg/s. If the gauge pressure is 0.80 atm at a point where the diameter of the pipe is 10 cm, what is the gauge pressure of the water in a section of the pipe having a diameter of 6.0 cm?

9.12 Water escapes from a large tank through a hole that has a diameter of 0.10 cm and is 8.0 m below the level of water in the tank. How many kilograms of water escape per hour?

9.13 A nuclear submarine (volume = 5.0×10^4 ft³) weighs 3.5×10^6 lb in dry dock. What volume of water must the submarine take in if it is to remain at equilibrium when completely submerged? What percentage of the weight of the dry-docked submarine does this quantity of water represent?

9.14 The container shown in Figure 9.25 has horizontal cross sections that are circular, and it is filled to a depth d with mercury ($\rho = 1.36 \times 10^4$ kg/m³). If $a = 4.0$ cm, $b = 10$ cm, and $d = 30$ cm, what force does the mercury exert on the bottom of the container?

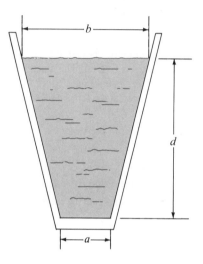

Figure 9.25   Problem 9.14

9.15 Water from reservoir A is to be pumped to reservoir B, which is 30 m above A. If the water is to be delivered at a rate of 5.0 liters/s (1 liter = 1000 cm³), what is the minimum power required for the pump?

9.16 What maximum amount of mass may be floated above a water surface while on a block of cork (density = 250 kg/m³) that measures 1.0 m × 2.0 m × 3.0 m?

9.17 If the hot air in a balloon has a density equal to three-fourths of the density of air, what minimum balloon volume is required to lift a total load of 500 kg? The density of air at 0° C and atmospheric pressure is 1.29 kg/m³.

9.18 The blimp *Sausage*, which is cylindrically shaped with a length of 80 m and a diameter of 20 m, is filled with helium (density = 0.18 kg/m³). Using the density of air as 1.29 kg/m³, calculate the total mass, excluding the mass of the helium, this blimp can lift.

9.19 If the "weight in water" of an object is 42 N and its "weight in alcohol" is 48 N, what is the weight of the object? The density of alcohol is 790 kg/m³.

9.20 The density of air is approximately 1.2 kg/m³. Estimate the buoyant force of the atmosphere on a person having a mass of 70 kg. What fraction of the person's weight is this buoyant force? (Remember that the density of a person is approximately that of water.)

9.21 Water flows at a speed of 8.0 m/s in a pipe having a 5.0-cm diameter. Calculate

  (a)   the volume flow rate for the water.

  (b)   the mass flow rate for the water.

  (c)   the kinetic energy flow rate for the water.

9.22 A person weighing 500 N wishes to float (out of the water) on a block made of plastic foam having a density equal to 110 kg/m^3.

  (a)   What is the minimum quantity (mass) of the plastic foam required?

  (b)   If this minimum block is a cube, calculate the length of any edge.

9.23 A tank is made of two coaxial cylinders (*see* Fig. 9.26). The common axis of the two cylinders is vertical. The diameters and heights shown are $D_1 = 2.0$ m, $D_2 = 0.40$ m, $H_1 = 5.0$ m, and $H_2 = 10$ m. The tank, which is open at the top, is filled with a liquid having a density of 1150 kg/m^3.

  (a)   What is the weight of the liquid in the tank?

  (b)   Calculate the force on the bottom of the tank by the liquid.

  (c)   Calculate the pressures at points A and B.

  (d)   What is the upward force on the annular ring (at the level of point B) by the fluid?

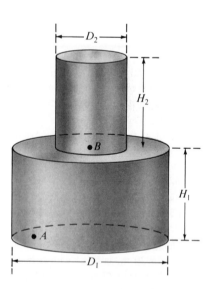

Figure 9.26   Problem 9.23

9.24 How many grams of lead (density = 11.6 g/cm^3) must be placed in one end of a hollow cylinder (radius = 2.0 cm, length = 40 cm) of negligible mass if the cylinder is to float with its axis vertical and with 5.0 cm of its length above the surface of the water?

9.25 Figure 9.27 depicts the addition of water (density $\rho_1 = 1000$ kg/m^3) to the left side of a U-tube, which initially contains only mercury ($\rho_o = 1.36 \times 10^4$ kg/m^3). As a result, the level of the mercury on the right side is observed to rise a distance d of 8.5 cm. If the inner diameter of the tube is 2.0 cm, calculate the mass of the water added to the tube.

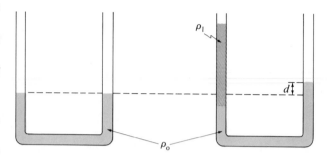

Figure 9.27   Problem 9.25

9.26 If the diameters shown in Figure 9.28 are $D = 12$ cm and $d = 9.0$ cm, calculate the volume flow rate for the liquid in the pipe if $h = 0.90$ m.

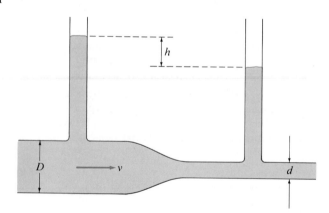

Figure 9.28   Problem 9.26

9.27 A Venturi meter (*see* Example 9.3) is used to measure the flow speed v of oil (density = 850 kg/m^3) moving in a pipe having a diameter of 30 cm. If the flow speed is to be measured over a range of $0 < v < 10$ m/s and if the maximum height difference in the manometer fluid (density = 1.36×10^4 kg/m^3) is to be 20 cm, calculate the minimum diameter of the constricted part of the pipe.

9.28 A body is suspended from a string. The tension in the string is 60 N with the body in air and 38 N with the body immersed in a liquid of density 950 kg/m^3. What is the density of the body?

9.29 Water flows from a large tank, as shown in Figure 9.29. If $h_1 = 15$ m, $h_2 = 10$ m, $d_1 = 0.60$ cm, and $d_2 = 2.5$ cm, find

(a) the flow speeds v_1 and v_2.

(b) the pressure in the lower pipe.

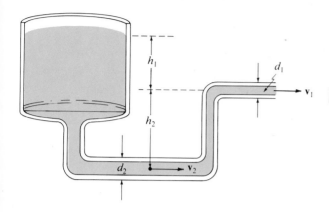

Figure 9.29   Problem 9.29

9.30   A sphere (radius = 2.0 in., weight = 0.90 lb) is dropped into a large pool of water. What volume of water will the sphere displace when it comes to rest?

9.31   A downward force of 120 N is required to keep a body (volume = 0.10 m³) submerged in water. To keep the same body submerged in a second liquid requires a downward force equal to 470 N. What is the density of the second liquid?

9.32   A composite cube is constructed of two different materials (ρ_1 = 1200 kg/m³, ρ_2 = 500 kg/m³), as shown in Figure 9.30. If a = 10 cm, how far will the top of the cube be above the surface of the liquid if the cube is floated in water?

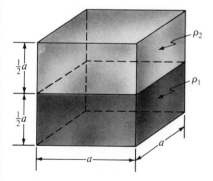

Figure 9.30   Problem 9.32

9.33   Water flows from a 0.40-cm diameter hole in the side of a large tank (*see* Fig. 9.31). If h = 1.0 m and b = 3.0 m, calculate the height H of the water surface above the hole.

9.34   An orifice (area = 1.5 cm²) in the side of a tank is 5.0 m below the surface of the water in the tank. What is the mass flow rate of water from the orifice?

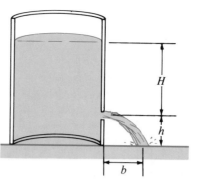

Figure 9.31   Problem 9.33

9.35   A pump P forces water through the pipe (*see* Fig. 9.32). Each second 5.0 kg of water flows into the atmosphere at a speed of 10 m/s from end C. The diameter of the pipe at point A is equal to 3.0 cm, and at point B the diameter is 5.0 cm.

(a) What is the diameter of the pipe at point C?

(b) Calculate the pressure at point B.

(c) How fast is the water flowing at point A?

(d) Assuming that the water taken in by the pump is initially at rest, at what rate is the pump doing work on the water?

Figure 9.32   Problem 9.35

9.36   Water fills a hemispherical pool having a radius of 5.0 m. Calculate

(a) the pressure at the lowest point in the pool.

(b) the total downward force on the pool by the water.

9.37   A sphere (radius = 8.0 cm, mass = 1.8 kg) is submerged in water and released from rest. What is its initial acceleration?

9.38   An irregular object weighs 1700 N and has a volume of 0.25 m³.

(a) What volume of water will this object displace when it is resting in a large tank of water?

(b) If the object is held in a submerged position by a cable, what is the tension in the cable?

(c) If the cable pulling down on the object has a tension of 300 N, what volume of water is the object displacing?

9.39   The composite beam of length L and uniform cross-sectional area A is constructed of two materials of densities ρ_1 and ρ_2 (*see* Fig. 9.33). If the beam is suspended while immersed in water, as shown in the figure, calculate the tension in each of the two ropes from which it hangs. Let L = 4.0 m, A = 0.25 m², ρ_1 = 2500 kg/m³, and ρ_2 = 1200 kg/m³.

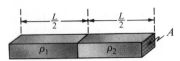

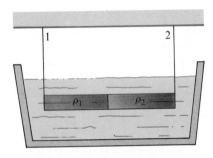

Figure 9.33   Problem 9.39

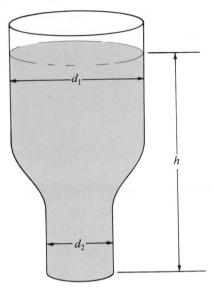

Figure 9.35   Problem 9.42

9.40   Figure 9.34 shows a pump P that takes water from rest in a pool and delivers a stream of water to the atmosphere at point A with a speed of 25 m/s. Point A, where the diameter of the pipe is 2.0 cm, is displaced vertically a distance h of 20 m above the pump. At point B the diameter of the pipe is 5.0 cm.

(a)   Calculate the gauge pressure at point B.

(b)   At what rate is the pump doing work on the water?

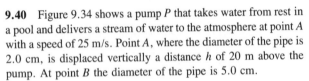

Figure 9.34   Problem 9.40

9.41   Liquid of a density $\rho = 1500$ kg/m^3 partially fills a tank having a horizontal cross-sectional area of 1.5 m^2. An irregularly shaped object is held in a submerged position by a cable attached to the bottom of the tank. In this position the tension in the cable is 570 N. With the object submerged, the surface of the liquid is 12 cm higher than before the object is placed in the liquid. Calculate the density of the submerged object.

9.42   Water flows from the lower end of the vertical pipe shown in Figure 9.35. The diameters have the values $d_1 = 2.5$ cm and $d_2 = 1.5$ cm. At an instant when $h = 2.0$ m, calculate

(a)   the speed with which water flows from the lower end.

(b)   the rate at which the top surface of the water is falling.

9.43   Water flows through a 5.0-m pipe inclined 40° with the horizontal (*see* Fig. 9.36). The water rises a height of 15 m above the top end of the pipe. The diameter of the pipe at the lower end is 3.0 cm; at the upper end the diameter is 1.0 cm.

(a)   What is the gauge pressure at the lower end of the pipe?

(b)   What volume of water passes through the pipe in 5.0 minutes?

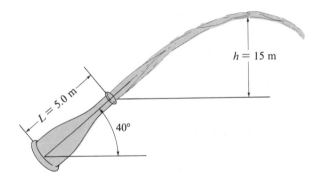

Figure 9.36   Problem 9.43

GROUP **B**

9.44   An airplane being designed is to weigh 5000 N and cruise at 50 m/s. The streamline along the top of the wing is 15 percent longer than a streamline along the lower surface of the wing. Estimate the wing area required. Use 1.3 kg/m^3 as the density of air.

9.45 The cross section of a proposed low-speed "waterfoil" is shown in Figure 9.37. The top surface is an arc of a circle of radius R. Let $x = 2.0$ m and $t = 0.20$ m.

(a) Show that $R = (x^2 + 4t^2)/8t$.

(b) Calculate the path length along the upper surface.

(c) Estimate the lift per meter of length of the waterfoil if the waterfoil is to move through the water at a speed v_o of 2.0 m/s.

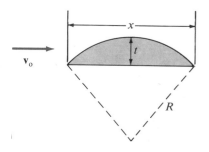

Figure 9.37   Problem 9.45

9.46 A uniform sphere of density $\rho = 800$ kg/m³ floats in water. How far is the highest point on the sphere above the surface of the water?

9.47 Cork ($\rho = 250$ kg/m³) is used to construct a buoyant cone ($r = 0.25$ m, $h = 1.0$ m) that is connected by a 4.0-m cable to the bottom of a lake of depth d (see Fig. 9.38).

(a) What is the minimum value of d for which the cable (disregard its weight) is taut?

(b) If $d = 4.8$ m, what is the tension in the cable?

(c) If $d = 6.0$ m, what is the tension in the cable?

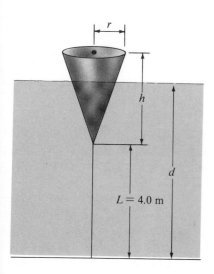

Figure 9.38   Problem 9.47

9.48 A hollow cylinder is weighted on one end so that it floats vertically in water with its lowest point a distance $L = 0.80$ m below the surface of the water (see Fig. 9.39). If the cylinder is raised a vertical distance $x < L$ and released, it will oscillate.

(a) Show that this oscillation is simple harmonic motion if damping forces are negligible.

(b) Calculate the frequency of this oscillation.

(c) If the same cylinder is floated in a liquid that has a density of 1200 kg/m³, calculate the frequency of the oscillation.

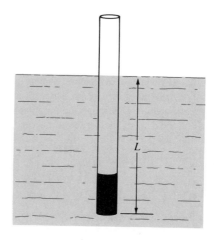

Figure 9.39   Problem 9.48

9.49 Two kilograms of water are placed in a U-tube with a cross-sectional area of 10 cm² (see Fig. 9.40). If one of the two water surfaces is pushed down and then released, the surface oscillates vertically. Disregarding any dissipative forces, calculate the period of this oscillation.

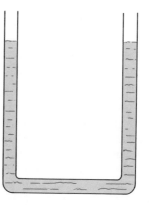

Figure 9.40   Problem 9.49

9.50 One vertical side (5.0 m × 8.0 m) of a tank has an opening (2.0 m × 4.0 m) cut in it as shown in Figure 9.41. This opening is covered by a "door" hinged along the top of the door. The door is held in a vertical position by a horizontal force **F** that is exerted at the center of the door.

(a) What is the magnitude of the force the water exerts on the door?

(b) What is the magnitude of the force **F**?

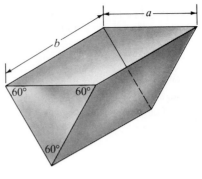

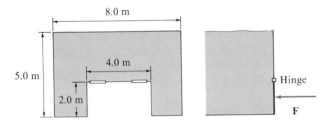

Figure 9.41   Problem 9.50

Figure 9.43   Problem 9.52

9.51 A "dam" having the rectangular cross section shown in Figure 9.42 is constructed of material of density ρ_0. The dam is held in place by a footing of height h. Determine the minimum value for a, the thickness of the dam, if the dam is to have no tendency to topple when the water level reaches the top of the dam. Express your result in terms of h, ρ_0, ρ (density of water), and b (height of the dam).

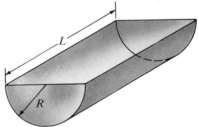

Figure 9.44   Problem 9.53

9.53 The open container shown in Figure 9.44 is filled with water. If $R = 5.0$ m and $L = 8.0$ m, calculate the magnitude of the force exerted by the water on

(a) either of the semicircular ends.

(b) the curved bottom of the container.

9.54 A cylindrical (diameter = 4.0 ft, length = 16 ft) water tank with its axis vertical is emptied by letting water flow through a circular opening (diameter = 1.0 in.) in the bottom of the tank. If the tank is full at $t = 0$, calculate

(a) the height y of the water remaining in the tank at the time t.

(b) the time required for the tank to empty.

(c) the volume flow rate of water from the opening as a function of time.

9.55 An open hemispherical tank (radius = 0.80 m) initially filled with water is drained through a hole (radius = 1.0 cm) at the lowest point in the tank. How much time is required to drain the tank completely?

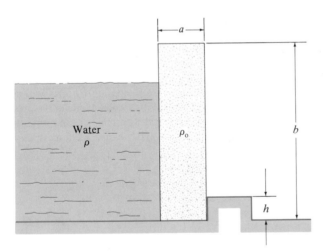

Figure 9.42   Problem 9.51

9.52 An open container (*see* Fig. 9.43) is filled with mercury (density = 1.36×10^4 kg/m³). If $a = b = 1.0$ m, calculate the magnitude of the force exerted by the mercury on

(a) one of the triangular ends.

(b) one of the rectangular sides.

10 Heat and Thermodynamics

The physics of thermal phenomena is a treatment of certain relationships among macroscopic, measurable quantities that are fundamentally microscopic in nature. The basic principles of thermal physics are known as the laws of thermodynamics. We will approach the study of these principles by first defining thermal equilibrium, temperature, and temperature scales and then considering several practical applications of thermal physics: the thermal expansion of substances; heat, the flow of thermal energy; and some techniques for the measurement (called calorimetry) of some thermal properties of materials.

We will define a thermodynamic system and the variables by which such a system is characterized and discuss the first law of thermodynamics in terms of these variables. The second law of thermodynamics will be discussed in terms of heat engines. An idealized heat engine, called a Carnot engine, provides a standard by which the performance of any engine may be judged. Further, the Carnot engine will be seen to provide a conceptual basis for establishing an absolute temperature scale.

Finally, the physical quantity, entropy, will be considered briefly.

10.1 Thermal Equilibrium and Temperature

Matter is composed of microscopic particles that are in continuous random motion. This chaotic movement, which is called *thermal motion*, is exemplified by a gas. Relatively high speeds (hundreds of meters per second at room temperature) and extremely high collision rates (billions per second for an air molecule at room temperature and pressure) characterize this highly chaotic molecular motion. The molecules of gas in the balloon of Figure 10.1 are in rapid random motion, which is to be distinguished from the more orderly movement of the macroscopic body, the air-filled balloon, as it moves through space. We refer to the kinetic energy and potential energy associated with the random thermal motion of microscopic particles in a substance as the *thermal energy* of that substance. In this chapter we will focus attention on some macroscopic quantities used to define the physical state of a substance, that is, a macroscopic state that exists because of the microscopic conditions within that substance. Some of the *state variables*, those quantities that characterize the macroscopic state of a

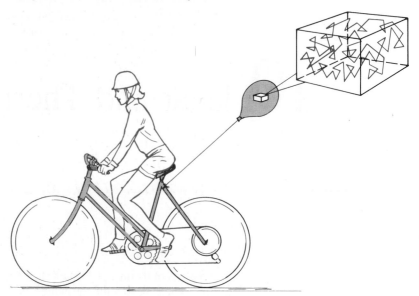

Figure 10.1
Contrasting the ordered motion of the balloon with the chaotic motion of air molecules within the balloon.

substance, are already familiar. Volume and pressure, for example, are variables that are descriptive of the physical state of a particular body of matter. Other quantities that may be used to characterize a state, like temperature, remain to be defined. The relationships between the macroscopic variables that reflect what is taking place at the microscopic level are collectively called *thermodynamics*.

Before we define temperature, it is convenient to define a number of terms used in describing thermodynamic systems and their interactions. First, a *simple thermodynamic system* refers to a well-defined collection of microscopic particles. Such a simple system is said to be in *thermal equilibrium* if it is macroscopically homogeneous and has the same average thermal energy in every equal nonmicroscopic volume within that system. A simple thermodynamic system at thermal equilibrium may be characterized uniquely by a set of state variables such as pressure, volume, and temperature.

A thermodynamic system may, in general, interact with other systems or with any part of the remainder of the universe that is external to that system. An *isolated system* is a well-defined set of particles that does not exchange energy with the universe external to that system. We say that a system is *thermally insulated* if it cannot exchange thermal energy with its environment. Two simple thermodynamic systems may interact in one or more of three ways: (1) They may exchange thermal energy; (2) they may do work on each other; and (3) they may exchange particles, or mix. In this section we will consider the exchange of thermal energy between simple systems, and in Section 10.3, the work done on or by systems. The mixing processes are beyond the scope of this book.

Let us propose a means of bringing into *thermal contact* two systems that do not mix and in which neither does any work. We may imagine two systems separated by a rigid partition, called a *diathermal wall*, which permits the free exchange of thermal energy between the systems yet is impermeable to the particles of either system. Then, when two otherwise isolated systems are brought into contact through a diathermal wall, they may exchange thermal energy until no net energy is passing through the wall. The systems are then said to be in *thermal equilibrium with each other*.

Although we have not yet defined temperature, we may now recognize that temperature is one of the variables that can be used to characterize the physical state of a thermodynamic system. (For the moment, let us think of temperature as that which is measured by a common thermometer.) Pressure, volume, temperature, and many other measurable quantities may be used as state variables. Let us now consider again two thermodynamic systems in thermal contact. If neither system undergoes a change in *any* state variable when these systems are brought into thermal contact, we say the systems are in thermal equilibrium. In fact, we say that the two systems have the same temperature. Therefore, we define *temperature* as that property of a system that determines whether or not the system will be in thermal equilibrium with other systems with which it is brought into thermal contact. What is more, we may conclude that if two systems in thermal contact are at different temperatures, they are certainly *not* in thermal equilibrium.

It is an experimental fact that two otherwise thermally isolated systems, each in thermal equilibrium with a third system, are in thermal equilibrium with each other. This fact, called (cutely but logically) the *zeroth law of thermodynamics*, permits us to assign a number to the temperature of a system with the assurance that any other system having that value of temperature will be in thermal equilibrium with the first system.

It remains, then, to adopt a temperature scale. But, first, let us digress briefly to characterize temperature further.

We associate temperature with how hot something is. But what makes an object hot? Statistical mechanics, an advanced branch of physics that treats systems of large numbers of particles from a probabilistic, mechanical point of view, relates temperature to microscopic motion. The temperature of a substance is related directly to the average thermal energy of the particles of that substance. Thus, at the microscopic level, the temperature of a body is a measure of how fast its particles are moving about in disorganized motion, independent of any mechanical energies the object as a whole may have.

At a more familiar level, temperature is, of course, a measure of how hot something is. The human sense of touch includes a temperature sense. Over a narrow range of hotness, we can distinguish roughly by touch which of two objects is the hotter, but we cannot do so with much precision or with repeatability. Obviously, for scientific purposes we need a *thermometer*, a device for measuring temperature.

Temperature Scales

Any object that undergoes an observable, reproducible change with a corresponding change in its temperature can be used as a thermometer. The familiar mercury-in-glass thermometer, shown in Figure 10.2, will illustrate how this may be done. This thermometer is made with a bulb containing a considerable volume of mercury, and the bulb communicates with a uniform capillary tube inside a transparent material. The region in the tube above the mercury column is evacuated. The bulb serves as the temperature-sensing element; it is placed in the substance the temperature of which is to be measured and is allowed to come to thermal equilibrium with that substance. Mercury, like most materials, expands with increasing temperature and contracts with decreasing temperature. As the

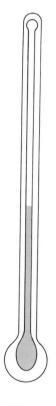

Figure 10.2
A mercury-in-glass thermometer.

temperature of the bulb increases, the volume of the enclosed mercury increases, causing the surface of the mercury to move farther from the bulb. This device becomes a practical thermometer if we establish a scale by first marking the tube to coincide with the mercury surface at two easily reproducible temperatures. As Anders Celsius first suggested in the eighteenth century, let us choose as reference temperatures the *ice point* and the *steam point* of water. These reference points are the temperatures of equilibrium mixtures of ice and water and of steam and water at standard pressure (1.0 atm). The level of the mercury surface is marked on the tube at each of these points and labeled 0° C at the ice-point mark and 100° C at the steam-point mark. Then if we place 99 equally spaced marks between the 0° C and the 100° C marks, we have constructed a thermometer calibrated on the *Celsius scale*, the temperature scale used most commonly throughout the world and in most scientific laboratory measurements of temperature. The symbol ° C is read "degrees Celsius."

We have made several assumptions in devising the mercury-in-glass thermometer. We assumed the capillary tube to have a constant cross-sectional area between the 0° C mark and the 100° C mark. We also assumed that, throughout the scale, equal volumes of mercury move into the capillary tube for equal changes in temperature of the bulb, that is, that the volumetric expansion of mercury is linear with change in temperature. Actually, two different mercury-in-glass thermometers would not necessarily agree on exactly the same temperature reading except at 0° C and 100° C. Thermometers like this one are further limited by their abilities to be made operative only within the range of temperatures between the melting point and boiling point of mercury. Because of the limitations of any thermometer that depends on the properties of a specific material, a thermodynamic scale, called the *absolute scale*, has been adopted as a standard. The absolute scale eliminates any dependence of temperature measurement on the properties of any material. We will see how this scale is defined in a later section of this chapter. For now, let us assume that a mercury-in-glass thermometer is sufficiently accurate. Nevertheless, in anticipation of the standard scale, let us introduce several facts about the absolute temperature scale that will be useful now:

1. The *triple-point of water*, the temperature and pressure at which water can exist simultaneously as solid, liquid, and gas, is assigned the exact value of 273.16 K, where K is the symbol for *kelvin*, a "degree" on the absolute scale.

2. A change in temperature of 1 K is equal to 1° C.

3. The ice point of water (0° C) is, using three significant figures, equal to 273 K; the steam point of water (100° C) is equal to 373 K. Therefore, a temperature t_C on the Celsius scale may be converted to absolute temperature T by

$$T = t_C + 273 \qquad (10\text{-}1)$$

The *Fahrenheit temperature scale* is still in use in the United States and Great Britain. On this scale, the ice and steam points of water are 32° F and 212° F, respectively, a difference of 180 Fahrenheit degrees. Thus, temperatures on the Celsius and Fahrenheit scales are related by

$$t_F = \frac{9}{5} t_C + 32° \text{ F} \qquad (10\text{-}2)$$

$$t_C = \frac{5}{9} (t_F - 32° \text{ F}) \qquad (10\text{-}3)$$

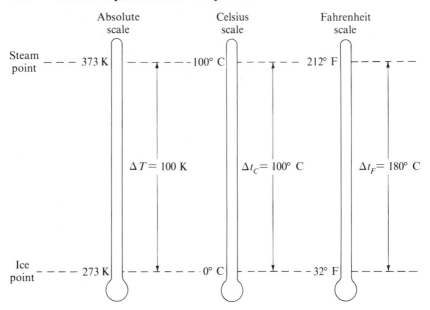

Figure 10.3
A comparison of the absolute, Celsius, and Fahrenheit temperature scales.

Figure 10.3 illustrates the three temperature scales discussed here.

E 10.1   Derive an expression relating absolute temperature T to Fahrenheit temperature t_F.   Answer: $T = 5t_F/9 + 255$

E 10.2   Express a typical room temperature of 68° F on the Celsius and absolute scales.   Answer: 20° C, 293 K

E 10.3   Express a temperature change of 55° C on the absolute and Fahrenheit scales.   Answer: 55 K, 99° F

Thermal Expansion

Most substances expand with increasing temperature. (A notable exception is water between 0° C and 4° C, which contracts with increasing temperature.) The increase in the linear dimension of a solid object is known as *linear expansion*. An increase that takes place because of an increase in temperature of the object is called *temperature expansion* or *thermal expansion*. Let L be the length of any dimension of a solid, like the length of a rod, for example. Let ΔL be the change in L that takes place as a result of a temperature increase ΔT of the rod, as shown in Figure 10.4. Experiment shows that, for most materials, ΔL is very nearly

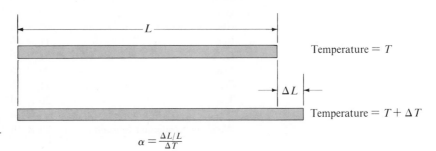

Figure 10.4
A rod, which changes from a length L to a length $L + \Delta L$ when the temperature changes from T to $T + \Delta T$.

$$\alpha = \frac{\Delta L/L}{\Delta T}$$

directly proportional to both L and ΔT, or $\Delta L \propto L\Delta T$. Thus, for a given material, we may write

$$\Delta L = \alpha L \Delta T \qquad (10\text{-}4)$$

where the constant of proportionality α is called the *coefficient of linear expansion* for that material. Equation (10-4) can be rearranged as

$$\alpha = \frac{(\Delta L/L)}{\Delta T} \qquad (10\text{-}5)$$

from which it may be seen that α represents the fractional change in length per unit change in temperature. The unit for α is therefore K^{-1}. Equations (10-4) and (10-5) define an average value for α over the temperature range ΔT. At a specific temperature T, $\alpha(T)$ is defined by

$$\alpha(T) = \lim_{\Delta T \to 0} \frac{(\Delta L/L)}{\Delta T} = \frac{1}{L}\frac{dL}{dT} \qquad (10\text{-}6)$$

Because the "shape" of a fluid is almost always determined by the container holding the fluid, the volume V, and not the length, of a fluid is usually the significant geometric measure associated with a fluid. Consequently, a *coefficient of volume expansion* β is defined by

$$\beta(T) = \frac{1}{V}\frac{dV}{dT} \qquad (10\text{-}7)$$

Thus, β is a measure of the fractional change in volume per unit change in temperature.

Table 10.1 lists some representative values of α and β for various materials near room temperature (~ 300 K). Because α is so small compared to unity, changes in length ΔL are small compared to L, and it makes no appreciable difference whether L is taken to be the length at the lower or higher temperature.

E 10.4   Determine the change in length of a 20-m railroad track made of steel if the temperature of the track is changed from $-15°$ C to $35°$ C.

Answer: 1.1 cm

TABLE 10.1   **Coefficients of Thermal Expansion for Some Materials Near 300 K**

Material	$\alpha(K^{-1})$	Material	$\beta(K^{-1})$
Aluminum	24×10^{-6}	Benzene	12×10^{-4}
Copper	17×10^{-6}	Gasoline	9.5×10^{-4}
Diamond	1.2×10^{-6}	Mercury	1.8×10^{-4}
Glass, soft	$\sim 9 \times 10^{-6}$	Water	2.1×10^{-4}
Glass, ovenware	$\sim 3 \times 10^{-6}$		
Iron	11×10^{-6}		
Ice	51×10^{-6}		
Lead	29×10^{-6}		
Steel	11×10^{-6}		
Tin	20×10^{-6}		
Wood, parallel to grain	$\sim 5 \times 10^{-6}$		
Wood, perpendicular to grain	$\sim 40 \times 10^{-6}$		

E 10.5   Calculate the change in volume of 15 gal of gasoline, where 1 gal is equal to 231 in.^{3}, if the temperature of the gasoline is increased by 45° F.

Answer: 82 in.^{3}

E 10.6   Consider a cube made of a solid isotropic material. Show, using $V = L^3$, that the coefficients of volume and linear expansion are related by

$$\beta(T) = 3\alpha(T)$$

E 10.7   Calculate the change in volume of a copper cube, 15 cm along an edge, if the temperature of the copper is decreased by 25° C.

Answer: -4.3 cm^3

10.2 Heat and Calorimetry

Experience and experiment tell us that when two bodies at different temperatures are placed in thermal contact, the hotter body decreases in temperature while the colder body undergoes an increase in temperature. Microscopically, the more energetic particles of the hotter body collide with the less energetic particles of the colder body, thereby transferring thermal energy to the colder body. When the two bodies reach thermal equilibrium, their final common temperature is intermediate between their original temperatures. We call the energy that is transferred between bodies because of their temperature difference *heat energy*, or simply *heat*.

Heat, thermal energy in transit, may be measured using any unit associated with energy, but the standard unit of heat is the *calorie* (cal). The calorie is defined to be that quantity of heat that must be transferred to a gram of water at 14.5° C to raise the temperature of that water to 15.5° C. In practical usage, the calorie may be considered the quantity of heat that, when absorbed by 1 g of water, will raise the temperature of that water by 1 K. The calorie is related to the English unit of heat, the *British thermal unit* (Btu), and to the usual metric unit of energy, the joule, by

$$252 \text{ cal} = 1 \text{ Btu} \tag{10-8}$$

$$1 \text{ cal} = 4.184 \text{ J} \tag{10-9}$$

The relationship of Equation (10-9) is called the *mechanical equivalent of heat*.

E 10.8   Determine the energy equivalent (in J and ft·lb) to the British thermal unit.   Answer: 1.0 Btu = 1.1×10^3 J = 7.8×10^2 ft·lb

The quantity of heat that must be absorbed by an object to raise the temperature of that object by 1 K is called the *heat capacity C* of that object. If the temperature of an object changes by an amount ΔT as a result of that object having absorbed a quantity of heat Q, the heat capacity C of that body is given by

$$C = \frac{Q}{\Delta T} \tag{10-10}$$

The unit of heat capacity is cal/K. If C is measured by transferring heat to a substance at constant pressure (usually such measurements are made at atmospheric pressure), that condition is denoted by labeling the heat capacity C_p. In some cases, when the measurement is made at constant volume, the heat capacity is labeled C_v.

The *specific heat c* of a substance is defined to be the ratio of the heat capacity C of an object composed of that substance to the mass m of the object:

$$c = \frac{C}{m} \tag{10-11}$$

The unit of specific heat is cal/(kg·K), or more commonly, cal/(g·K).

Notice that heat capacity is a property of a given object, whereas specific heat is a property of a particular material. A block of lead and a block of wood may have equal heat capacities, although equal masses of lead and wood have unequal heat capacities because their specific heats are different.

Table 10.2 lists the specific heats of a number of materials near room temperature. Although specific heats vary markedly over wide ranges of temperature, in this book we will treat the specific heat of a given material as if it were a constant value within the temperature ranges we use.

Using Equations (10-10) and (10-11), we may relate the heat Q that must be absorbed by a system of mass m to effect a change ΔT in the temperature of the system by

$$Q = C\Delta T = cm\Delta T \tag{10-12}$$

This equation establishes an important sign convention that we will adopt and adhere to throughout this chapter. Because a positive change ΔT (an increase) in the temperature of a system requires the input of heat Q into the system, we agree that *Q is positive when heat is added to a system and Q is negative when heat is removed from a system.*

TABLE 10.2

Specific Heats of Some Common Materials Near 300 K

Material	Specific Heat $\frac{cal}{g \cdot K}$
Solids	
Aluminum	0.21
Brass	~0.09
Copper	0.092
Gold	0.03
Ice	0.48
Iron	0.11
Lead	0.031
Silver	0.056
Liquids	
Benzene	0.41
Ethanol	0.58
Mercury	0.033
Water	1.0

E 10.9   How much heat must be absorbed by 2.5 kg of iron if the temperature of the iron is to change from 20° to 35° C?
Answer: 4.1×10^3 cal = 4.1 kcal

E 10.10   How many calories of heat are required to raise the temperature of 1.0 lb of water by 1.0° F? (Remember that 0.45 kg weighs 1.0 lb.)
Answer: 2.5×10^2 cal

E 10.11   What is the heat capacity of 200 g of aluminum?
Answer: 42 cal/K

The heat capacity or specific heat of a material is conveniently measured by simple calorimetric (heat measuring) methods. The basic problem-solving process in calorimetry is the *method of mixtures*, based on the fact that for an isolated system, the net heat flow to or from such a system is zero. Practically, this means that the heat gained by any part of an isolated system is equal to the heat lost by other parts of the system.

E 10.12   A 120-g aluminum block is immersed in 80 g of water. If both the block and the water are initially at 15° C, how much heat must be added to this system if its temperature is to be raised to 65° C?   Answer: 5.3 kcal

E 10.13 A solid object (mass = m_1, specific heat = c_1), initially at a temperature T_1, is immersed in a liquid (mass = m_2, specific heat = c_2), initially at a temperature $T_2 > T_1$. This two-component system, which is thermally insulated from its environment, achieves an equilibrium temperature T_3, where $T_1 < T_3 < T_2$. What is Q_S for the solid during this process? Is Q_S positive or negative?
Answer: $Q_S = m_1c_1(T_3 - T_1)$, positive

E 10.14 For Exercise 10.13, express Q_L, the heat added to the liquid, in terms of m_2, c_2, T_2, and T_3. Is Q_L positive or negative?
Answer: $Q_L = m_2c_2(T_3 - T_2)$, negative

E 10.15 The system of Exercise 10.13 is thermally insulated, so it follows that $Q_S + Q_L = 0$. Determine the equilibrium temperature in terms of m_1, m_2, c_1, c_2, T_1, and T_2. Answer: $T_3 = (m_1c_1T_1 + m_2c_2T_2)/(m_1c_1 + m_2c_2)$

When a substance changes from one of its possible forms, solid, liquid, or gas, to another of those forms, the substance undergoes a *phase change*. During a phase change, heat may be absorbed or removed from a substance *without changing the temperature of the substance*. While a solid is melting, for example, energy must be supplied to the solid to do work on its atoms against the forces binding them into the solid phase. This added thermal energy does not increase the temperature of the solid material. The heat transferred to or from a substance undergoing a phase change is called *latent heat*. As a substance changes from a solid phase to a liquid phase, the heat required to effect this change for a unit mass of the substance is called its *latent heat of fusion L_F*. Thus, the heat Q required to cause a mass m to change from solid to liquid is given by

$$Q = mL_F \quad \text{(fusion)} \tag{10-13}$$

As the substance changes from the liquid to the solid phase, it releases an equal quantity of heat, that is, $Q = -mL_F$. Similarly, the heat per unit mass transferred to a substance as it changes from the liquid phase to the gaseous phase is called its *latent heat of vaporization L_V*. For a mass m undergoing this phase change from liquid to gas, the heat Q required to effect this change is

$$Q = mL_V \quad \text{(vaporization)} \tag{10-14}$$

When this substance changes from the gaseous phase to the liquid phase, it releases an equal amount of heat, that is, $Q = -mL_V$. The most common unit of latent heat is cal/g. For water at atmospheric pressure, L_F has the value 80 cal/g, and L_V has the value 540 cal/g.

E 10.16 If $L_F = 5.9$ cal/g for lead, how much heat is required to change 2.0 kg of lead at its melting temperature (600 K) from the solid to the liquid phase?
Answer: 1.2×10^4 cal = 12 kcal

E 10.17 What is the value of Q for a process in which 1.0 kg of water at 0° C is frozen?
Answer: -8.0×10^4 cal

E 10.18 How much heat must be added to 2.0 g of ice at 0° C to create 2.0 g of steam at 100° C?
Answer: 1.4 kcal

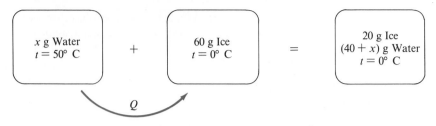

Figure 10.5   Example 10.1

Example 10.1

PROBLEM   How many grams of water at 50° C must be mixed with 60 g of ice at 0° C to produce an equilibrium mixture of water and ice having 20 g of ice?

SOLUTION   Figure 10.5, which depicts the problem symbolically, shows 60 g of ice at 0° C receiving Q cal of energy from x grams of water at 50° C to produce a final ($t = 0°$ C) mixture containing 20 g of ice and $(40 + x)$ g of water. The heat Q absorbed by the ice must melt 40 g of ice, that is,

$$Q = (40 \text{ g}) \cdot \left(80 \frac{\text{cal}}{\text{g}}\right) = 3200 \text{ cal}$$

This heat must be thermal energy released by cooling the x grams of water from 50° C to 0° C, that is,

$$Q = (x \text{ g}) \cdot \left(1.0 \frac{\text{cal}}{\text{g} \cdot {}^\circ\text{C}}\right) \cdot (50° \text{ C}) = 50x \text{ cal}$$

Equating these two expressions for Q yields

$$50x = 3200$$

$$x = 64$$

Thus, 64 g of water at 50° C will produce an equilibrium mixture containing 20 g of ice and 104 g of water. ∎

Example 10.2

PROBLEM   If 50 g of ice at 0° C is combined with 20 g of steam at 100° C in a thermally insulated container, determine the equilibrium state of the system.

SOLUTION   Figure 10.6 depicts the problem and shows the three distinctly different possible final states. For state I to prevail, the 20 g of steam must condense and cool to 0° C and, in doing so, release enough thermal energy to melt x grams of ice. State II results if the 50 g of ice melts, the 20 g of steam condenses, and the 70 g of water achieves an equilibrium temperature t_f in the range 0° C $\le t_f \le$ 100° C. Finally, state III results if the 50 g of ice melts, is heated to 100° C, and, in doing so, absorbs enough thermal energy from the stream to condense x g of steam. Which one of these corresponds to the final state? We can decide by selecting either state II or III and determining if that situation is possible. Suppose, for example, we assume state II to be the final state. The problem, then, is to determine the final temperature t_f. This is done by equating the thermal energy absorbed by the ice to the thermal energy released by the steam:

$$(50 \text{ g}) \cdot \left(80 \frac{\text{cal}}{\text{g}}\right) + (50 \text{ g}) \cdot \left(1.0 \frac{\text{cal}}{\text{g} \, °\text{C}}\right) \cdot (t_f) = (20 \text{ g}) \cdot \left(540 \frac{\text{cal}}{\text{g}}\right) + (20 \text{ g}) \cdot \left(1.0 \frac{\text{cal}}{\text{g} \, °\text{C}}\right) \cdot (100 - t_f)$$

(thermal energy absorbed by ice = thermal energy released by steam)

$$4000 + 50 \, t_f = 10800 + 2000 - 20 \, t_f$$

$$t_f = 126° \text{ C}$$

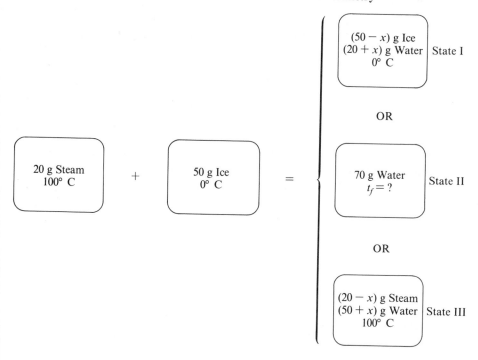

Figure 10.6   Example 10.2

This result, of course, represents an impossible state: water at 126° C. A similar analysis shows that state I cannot exist. The high temperature suggests that state III is the appropriate final state. For this state, we must determine the number of grams of steam that must condense in order to release enough thermal energy to melt 50 g of ice and then increase the temperature of the 50 g of water from 0° C to 100° C. An energy balance, then, gives

$$(50 \text{ g}) \cdot \left(80\frac{\text{cal}}{\text{g}}\right) + (50 \text{ g}) \cdot \left(1.0\frac{\text{cal}}{\text{g °C}}\right) \cdot (100° \text{ C}) = (x \text{ g}) \cdot \left(540\frac{\text{cal}}{\text{g}}\right)$$

(thermal energy absorbed by ice) = (thermal energy released by steam)

$$9000 \text{ cal} = 540x \text{ cal}$$

$$x = 16.7 \text{ g}$$

Thus, the final state consists of a mixture of 66.7 g of water and 3.3 g of steam at a temperature of 100° C. ∎

Energy in the form of heat may be transferred from place to place by three processes:

1. *Convection* is the transfer of energy by the mechanical movement of a substance. Air heated by a furnace and then blown about a house by a fan is an example of convection.

2. *Conduction* is the flow of energy through matter without any change in position of the matter itself. When a metallic object conducts heat, for example, the thermal energy is transferred by collisions between the electrons and atoms within the object. The handle of a silver spoon becomes hot when the other end is in hot coffee because heat is readily conducted through the silver.

3.   *Radiation* is the transfer of energy by electromagnetic waves (examples of which are light, radio waves, and x rays). Anyone who has taken a sunbath has felt the effects of thermal energy radiated from the sun. Radiation will be considered further in a later chapter.

An understanding of the conduction of heat has considerable practical value. In particular, knowing how to impede heat conduction effectively permits us to insulate our buildings, beer, feet, and food better. A simple analysis of heat conduction may be developed from consideration of the situation depicted in Figure 10.7. A slab of material of thickness d and area A has a constant temperature difference ΔT maintained across its thickness. The heat that passes through the slab in a time interval Δt is directly proportional to A, ΔT, and Δt and is inversely proportional to d:

$$Q = k\frac{\Delta T \, A \, \Delta t}{d} \tag{10-15}$$

The constant of proportionality k in Equation (10-15) is called the *thermal conductivity* of the slab material. The units of k may be written cal/(m·s·K), J/(m·s·K) = W/(m·K), or Btu/(ft·h·°F). Some representative values of thermal conductivities for a variety of materials are listed in Table 10.3. The table suggests some generalities about thermal conductivity. Metals are relatively good thermal conductors (k is relatively large), and gases are poor thermal conductors (or, therefore, good thermal insulators). These facts suggest why Styrofoam (filled with air pockets) rather than copper is the more appropriate material used to make ice chests.

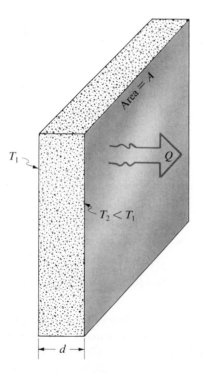

Figure 10.7
A slab of material with area A and thickness d. One face of the slab is maintained at temperature T_1; and the other face is maintained at temperature T_2, where $T_1 > T_2$ and $\Delta T = T_1 - T_2$. Heat Q flows through the slab in the direction indicated.

TABLE 10.3   **Thermal Conductivities of Some Common Materials Near 300 K**

Material	Thermal Conductivity $\dfrac{\text{cal}}{\text{s} \cdot \text{cm} \cdot \text{K}}$
Metals	
Aluminum	0.48
Copper	0.92
Gold	0.76
Iron	0.19
Mercury	0.02
Silver	1.0
Nonmetals	
Air	0.56×10^{-4}
Brick	17×10^{-4}
Concrete	41×10^{-4}
Glass	16×10^{-4}
Ice	53×10^{-4}
Water	14×10^{-4}
Wood	$\sim 7 \times 10^{-4}$

E 10.19   How much energy is conducted during a time interval of 1.0 h through a plate glass window (area = 2.0 m², thickness = 2.5 mm) if a temperature difference of 5.0° C exists across the thickness of the window?

Answer: 2.3×10^6 cal = 9.6×10^6 J

E 10.20   Express the energy determined in Exercise 10.19 in kWh (kilowatt-hour) and estimate the cost of replacing this energy with electrical energy costing 10¢/kWh.

Answer: 2.7 kWh, 27¢

10.3   Thermodynamics

Just as mechanics is based on Newton's laws and the principles of conservation of energy and momentum, the study of thermal energy and how it is related to work is expressed by the laws of thermodynamics. The laws of thermodynamics are primarily energy relationships. They are expressed and developed in terms of *thermodynamic systems*, each of which is a well-defined collection of matter that may interact with its environment in a number of ways, one of which is the exchange of heat between the system and its environment. We shall restrict our considerations for the most part to the interactions of simple systems and their environments. For example, when we speak of a gas that may be compressed or expanded in cylinders by pistons, the gas is the "simple system" under consideration.

Thermodynamic States and Processes

The *thermodynamic state* of a system is defined only when the system is in thermal equilibrium. *State variables*, such as pressure P, volume V, and temperature T, are used to specify a thermodynamic state. For a specific system, which contains a definite number of particles, a relationship between its state variables that determine any one variable in terms of the others is called an *equation of state*. For example, the equation of state of an *ideal gas*, a simple but very useful model of a gas, is

$$PV = nRT \quad \text{(ideal gas)} \tag{10-16}$$

where $R = 8.3$ J/(mol·K) is called the *universal gas constant*, and n is the number of moles (mol) of gas in the system. One mole of a substance consists of 6.0×10^{23} (Avogadro's number) molecules of that substance. Real gases have equations of state that are more complex than Equation (10-16), but for real gases at relatively low pressures, the equation of state of an ideal gas is a reasonable approximation.

E 10.21   One *mole* of a substance is defined to be a quantity of the substance having a mass (in grams) equal in value to the molecular weight of the substance. If the atomic weight of hydrogen (H), carbon (C), and oxygen (O) are 1.008, 12.011, and 15.999, respectively, determine the mass of 1.0 mol of (a) H_2O, (b) H_2, and (c) CO_2.

Answers: (a) 18.006 g; (b) 2.016 g; (c) 44.009 g

E 10.22   How many atoms of oxygen are there in 1.5 mol of CO_2?

Answer: 9.0×10^{23}

E 10.23   How many grams of hydrogen are there in 4.5 mol of H_2O?

Answer: 9.1 g

E 10.24   How many moles are there in a 50-g sample of CO_2?

Answer: 1.1

E 10.25   What is the pressure of 0.40 mol of an ideal gas if it is confined to a volume of 2.0 liters (1 L = 1.0×10^{-3} m^3) at a temperature of 300 K?

Answer: 5.0×10^5 N/m^2

E 10.26   Express the pressure of Exercise 10.25 in atmospheres, where 1.0 atm = 1.0×10^5 N/m^2.

Answer: 5.0 atm

E 10.27   Calculate the particle density (number of molecules/m^3) of an ideal gas at a pressure of 1.0 atm and a temperature of 300 K.

Answer: 2.4×10^{25} molecules/m^3

Because the equation of state of an ideal gas is known, any two of the state variables are sufficient to specify the state of an ideal-gas system. Let us use P and V. Figure 10.8 shows a point (P_1, V_1) in the P-V plane. This point represents a thermodynamic state of a given system (in this case, a gas confined within a cylinder by a piston). Another point in the plane, (P_2, V_2), represents another

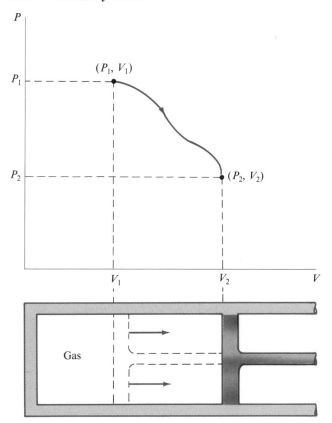

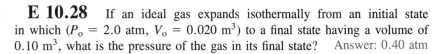

Figure 10.8
A representation of an expansion process on a *P-V* diagram.

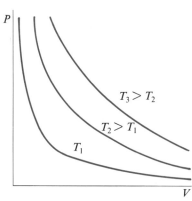

Figure 10.9
Representations of three isothermal process lines on a *P-V* diagram, where *PV* = constant appears as a family of hyperbolas. Each hyperbola corresponds to a different temperature.

state of the same system. A directed line, such as the one connecting (P_1, V_1) and (P_2, V_2), represents a *thermodynamic process*. Because $V_2 > V_1$ in this example, the process illustrated is an expansion of the gas in the cylinder. If the direction of the process line were reversed, that is, if it were directed from (P_2, V_2) to (P_1, V_1), the process would be a compression.

A few processes are of special interest in thermodynamics. For example, a process that takes place with the temperature of the system held fixed is called an *isothermal process*, and the ideal gas equation of state becomes

$$PV = nRT = \text{constant}$$

Thus for a given temperature, PV = constant is a hyperbola on a *P-V* diagram. Figure 10.9 shows three such hyperbolas. These curves, each representing an isothermal process, are called *isotherms*.

E 10.28 If an ideal gas expands isothermally from an initial state in which $(P_o = 2.0 \text{ atm}, V_o = 0.020 \text{ m}^3)$ to a final state having a volume of 0.10 m^3, what is the pressure of the gas in its final state?   Answer: 0.40 atm

Every point on a process line represents an equilibrium state. Then how can we compress a gas and have the system in equilibrium at every stage of the compression? By compressing the gas a small amount, waiting for equilibrium to be reestablished, and then compressing the gas a bit further, the system is never

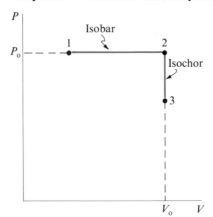

Figure 10.10
An isobaric process from state 1 to state 2 and an isochoric process from state 2 to state 3. The pressure is constant throughout the isobaric process. The volume is constant during the isochoric process.

far from equilibrium. A process that proceeds in this fashion (the system may always be considered in equilibrium) is called a *quasi-static process*. Thus a process line is an infinite succession of equilibrium states. A process line, therefore, may be imagined as the idealized limiting case of a real process carried out more and more slowly.

Two additional special processes are illustrated in Figure 10.10. The process represented by the line from state 1 to state 2 takes place at a constant pressure P_o. Such a process is called an *isobaric process*, and the line from 1 to 2 is called an *isobar*. The process from state 2 to state 3, which takes place at a constant volume V_o, is called an *isochoric process*; and a line like that from 2 to 3 is called an *isochor*. One further process that will be of interest is one for which no heat enters or leaves the system. This is an *adiabatic process*, and a line representing such a process is called an *adiabat*. For an ideal gas, adiabats are the family of curves corresponding to the relation PV^{γ} = constant. The quantity γ for the system is equal to C_p/C_v, the ratio of the heat capacities at constant pressure and constant volume. For monatomic gases, γ is ~1.7; for diatomic gases, γ is ~1.4.

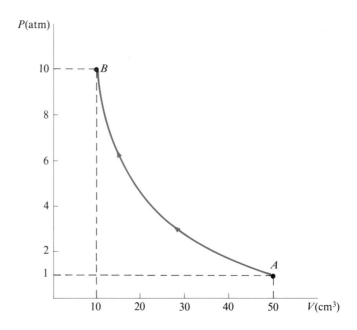

Exercise 10.30

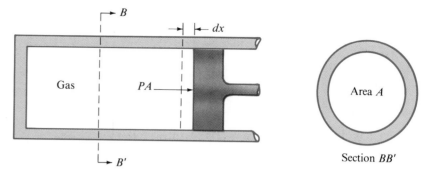

Figure 10.11
A system composed of a gas enclosed in a cylinder fitted with a frictionless piston.

E 10.29   The volume of 0.60 mol of an ideal gas is tripled by an isobaric expansion from a state having $P = 1.5 \times 10^5$ N/m^2 and $V = 7.5 \times 10^{-3}$ m^3. What are the initial and final temperatures of the gas?   Answer: 230 K, 680 K

E 10.30   An ideal gas is compressed adiabatically from state A to state B, as shown in the diagram. Determine γ for this gas. Is this gas monatomic or diatomic?   Answer: 1.4, diatomic

E 10.31   Show that for an ideal gas changing isothermally, that is, so that $PV = $ constant,

$$\frac{dP}{dV} = -\frac{P}{V}$$

but if the same gas changes adiabatically, that is, $PV^\gamma = $ constant,

$$\frac{dP}{dV} = -\gamma\frac{P}{V}$$

Thus, at any point in a P-V diagram where an adiabatic-process curve and an isothermal-process curve cross, the adiabat has a steeper (more negative) slope than the isotherm.

Consider now a gas contained within a cylindrical vessel by a tight-fitting, but frictionless and movable, piston like that illustrated in Figure 10.11. Let us see how work done on or by such a system may be expressed in terms of the thermodynamic variables of the system. The system is a gas of pressure P within the cylindrical volume of cross-sectional area A. The magnitude of the force exerted by the gas on the face of the piston is equal to PA. As the gas expands and pushes the piston through a displacement dx, the work dW done by the force on the piston is

$$dW = PA\,dx = P\,dV \qquad\qquad (10\text{-}17)$$

in which $A\,dx$ is equal to dV, the change in volume of the gas. When the piston moves so as to change the volume of the system from an initial value V_i to a final value V_f, the work done by the gas during that process is

$$W = \int_{V_i}^{V_f} P\,dV \qquad\qquad (10\text{-}18)$$

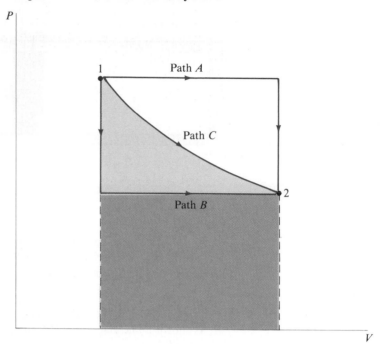

Figure 10.12
Three possible paths along which a system may pass from state 1 to state 2. The shaded region indicates the work done when the system proceeds along path C.

The pressure is, in general, a function of volume, and it is necessary to know how P varies with volume in order to evaluate the work done by the gas. Figure 10.12 shows three of the countless ways that P could change as the system passes from state 1 to state 2. The work done by the system during a given process from state 1 to state 2 is graphically represented by the area under the corresponding process line from 1 to 2. The figure indicates by a shaded area the work done if the system follows path C. Obviously, the areas under paths A and B represent other values of work. Thus, the work done by a system in changing from one state to another depends on the path between those states, that is, on the particular process by which the system changes state. Because the work done by a system is not determined by the state of the system (or even by two states, because the work depends on path), *work is not a state variable*. It is, therefore, meaningless to associate a quantity of work with a state of a system.

E 10.32   An ideal gas, starting at state A, undergoes the four-step process shown in the accompanying figure, namely, step (1), an isobaric expansion;

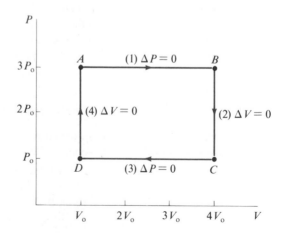

step (2), an isochoric pressure reduction; step (3), an isobaric compression; and step (4), an isochoric pressure increase. Express the work (by the gas) in terms of P_o and V_o for each of the four steps.

Answers: $W_1 = 9\ P_oV_o$, $W_2 = 0$, $W_3 = -3\ P_oV_o$, $W_4 = 0$

E 10.33   Determine the net work by the gas for the four-step process of Exercise 10.32. To what area does this work correspond?

Answer: $6\ P_oV_o$, area of the rectangle enclosed by the process lines

E 10.34   An ideal gas expands isothermally from a state (P_o, V_o) to a new state having a volume equal to $4\ V_o$. Using $PV = P_oV_o$, show that the work done by the gas is given by

$$W = P_oV_o\ \ln 4$$

Example 10.3   **PROBLEM**   An ideal gas ($\gamma = 1.4$), starting in state A, undergoes the three-step process shown in Figure 10.13, namely, step (1), an adiabatic expansion; step (2), an isothermal compression; and step (3), an isochoric pressure increase. If $P_o = 1.8$ atm and $V_o = 0.60$ L, determine the total work by the gas as it goes through the three-step process.

SOLUTION   During the adiabatic process, the pressure and volume are related by $PV^\gamma = $ constant, or

$$PV^\gamma = P_oV_o{}^\gamma$$

At state B, for which $V = 3\ V_o$, the pressure is given by

$$P_B = \frac{P_oV_o{}^\gamma}{(3V_o)^\gamma} = \frac{P_o}{3^\gamma} = \frac{1.8\ \text{atm}}{3^{1.4}} = 0.39\ \text{atm}$$

and the work by the gas during step (1) is

$$W_1 = \int_{V_o}^{3V_o} P\ dV = \int_{V_o}^{3V_o} \frac{P_oV_o{}^\gamma}{V^\gamma}\ dV = \frac{P_oV_o}{\gamma - 1}(1 - 3^{1-\gamma}) = 96\ \text{J}$$

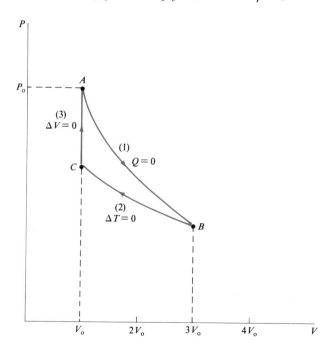

Figure 10.13   Example 10.3

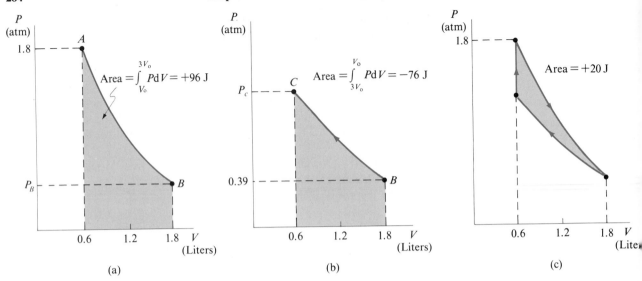

Figure 10.14   Example 10.3

Figure 10.14(a) shows the area corresponding to this work. During the iso-, thermal process, the pressure and volume are related by

$$PV = P_BV_B = \frac{P_o}{3^\gamma}(3\ V_o) = \frac{P_oV_o}{3^{\gamma-1}}$$

so that

$$P_C = \frac{P_o}{3^{\gamma-1}} = 1.3\ \text{atm}$$

Thus, the work by the gas during step (2) is shown graphically in Figure 10.14(b) and is given by

$$W_2 = \int_{3V_o}^{V_o} P\ dV = \int_{3V_o}^{V_o} \frac{P_oV_o}{3^{\gamma-1}}\frac{dV}{V} = -\frac{P_oV_o}{3^{\gamma-1}}\ln3 = -76\ \text{J}$$

The gas is compressed during this step, so the work is negative. Of course, the gas does no work during step (3), an isochoric pressure increase; and, consequently, the total work by the gas during this three-step process is

$$W_\text{total} = W_1 + W_2 + W_3 = 96 - 76 + 0 = 20\ \text{J}$$

Figure 10.14(c) shows the graphical equivalent of this total work, that is, the area enclosed by the three process lines. ■

During a thermodynamic process, heat may, in general, enter or leave the system. In an isothermal expansion, for example, heat must be added to the system in order to maintain a constant temperature. It has been found experimentally that the heat transferred to a system is quite different for processes that proceed along different process lines. This experimental fact means that heat flow into or out of a system, like the work done by the system, depends on the path between initial and final states of the system. Thus, as in the case with work, *heat is not a state variable*. It is, therefore, meaningless to speak of a system in a given state as "containing" a quantity of heat. *Both heat and work are quantities associated with processes—not with states.*

The First Law of Thermodynamics

Heat absorbed by a system may increase the thermal energy associated with the chaotic motion of the microscopic particles that constitute that system. The total thermal energy of a system is called the *internal energy U* of the system. We see that three different forms of energy are associated with thermodynamic systems and processes: heat Q, corresponding to the flow of thermal energy into or out of the system; work W done by the system; and a change ΔU in the internal energy of the system. The following sign conventions will be used for Q and W:

1.   A quantity of heat Q is positive if it represents thermal energy added to the system; Q is negative if thermal energy is removed from the system.

2.   The work W associated with a process is positive if energy is removed from the system in the performance of this work; W is negative if energy is added to the system (by external agents performing the work).

The results of countless experiments show that the quantity $Q - W$ has the same value for all processes that take a system from any specific initial equilibrium state to any specific final equilibrium state. This is true even though Q and W each will generally have values that change with each different process linking the initial and final states. Because Q is the energy added to the system as heat and W is the energy removed from the system as work, the quantity $Q - W$ represents the net energy added to the system as it is changed from the initial state to the final state. Consequently, this energy, $Q - W$, is equal to the change ΔU in the internal energy of the system as the system evolves from an initial state with internal energy U_i to a final state with internal energy U_f, that is,

$$Q - W = \Delta U = U_f - U_i \qquad (10\text{-}19)$$

or

$$Q = \Delta U + W \qquad (10\text{-}20)$$

Equation (10-20) is our statement of the *first law of thermodynamics*. The first law is illustrated schematically in Figure 10.15, which indicates that any heat Q added to a system is exactly accounted for by the sum of the change ΔU of the internal energy of the system and the work W performed by the system.

Because only *changes* in the internal energy of a system are measurable, the value of the internal energy for any particular equilibrium state has no absolute significance. Once the internal energy of any one equilibrium state has been assigned an arbitrary value, the first law of thermodynamics assures that the internal energy of all other equilibrium states is uniquely specified.

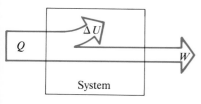

Figure 10.15
A schematic illustration of the first law of thermodynamics. Some of the heat Q entering the system is used to cause a change ΔU in the internal energy of the system; the remainder is used in the performance of work W by the system.

E 10.35   If 45 cal of heat are added to a system during a process for which the system performs 95 J of work on its environment, calculate the corresponding change in the internal energy of the system.   Answer: 93 J

E 10.36   If the internal energy of a system is equal to 25 J in state A and 70 J in state B and if the work by the system as it is changed from state A to state B during a certain process is equal to -60 J, calculate Q for this process.
Answer: -3.6 cal

Previously we have seen that an ideal gas obeys the equation $PV = nRT$. This equation of state is predicted when the methods of statistical mechanics are applied to a system of noninteracting molecules. A second important result predicted by statistical mechanics asserts that the internal energy U of an ideal gas is proportional to the absolute temperature T according to

$$U = C_v T \quad \text{(ideal gas)} \qquad (10\text{-}21)$$

where C_v is the heat capacity at constant volume, that is, the heat that must be added to the gas to increase its temperature by 1 K while the volume of the gas is held constant.

The first law of thermodynamics, when applied to specific processes, takes on specialized forms:

1. In an isobaric (constant pressure) process, the work done by the system becomes $W = \int_{V_i}^{V_f} P \, dV = P(V_f - V_i)$. In this case, Equation (10-20) may be written $Q = \Delta U + P(V_f - V_i)$.

2. In an isochoric (constant volume) process, no work is done on or by the system, and the first law becomes $Q = U_f - U_i$. In this case, all the heat added to the system is used to increase the internal energy of the system.

3. An adiabatic process is one in which no heat enters or leaves the system, so that $Q = 0$. Then the first law becomes $U_f - U_i = -W$, and any (positive) work done by the system results in a decrease in the internal energy of the system.

4. In an isothermal (constant temperature) process, none of the quantities of the first law, Q, W, and U, are, in general, constant.

E 10.37   During the isobaric process shown, 550 cal of heat are ejected from the system. Determine the change in internal energy of the system.

Answer: -8.0×10^2 J

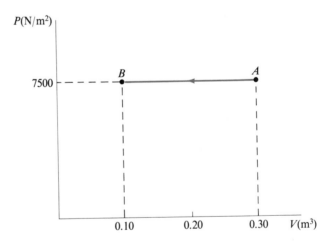

Exercise 10.37

Example 10.4   PROBLEM   Show that for an ideal gas, $C_p = C_v + nR$, where C_p and C_v are the heat capacities of n moles of the gas at constant pressure and constant volume, respectively.

SOLUTION   The heat dQ that must be added to a system during an isobaric process, if the temperature of the system changes by an amount dT, is given by

$$dQ = C_p \, dT$$

According to the first law of thermodynamics, however, this heat dQ is also given by

$$dQ = dU + P\,dV$$

where dU is the change in internal energy of the system and $P\,dV$ is the work done by the system. Because $U = C_v T$ and $PV = nRT$ for an ideal gas, we get

$$dU = C_v\,dT$$

and for an isobaric process,

$$P\,dV = nR\,dT$$

Thus, we have

$$C_p\,dT = C_v\,dT + nR\,dT$$

and dividing by dT gives

$$C_p = C_v + nR \quad \blacksquare$$

E 10.38   For an ideal monatomic gas, the ratio of C_p to C_v is equal to 5/3, that is, $\gamma = 5/3$. Determine C_v and C_p for 1.0 mol of an ideal monatomic gas.
                                                             Answer: $3R/2$, $5R/2$

Example 10.5   **PROBLEM**   Show that for a quasi-static adiabatic process, the pressure and volume of an ideal gas obey $PV^\gamma = $ constant, where $\gamma = C_p/C_v$.

SOLUTION   Because $dQ = 0$ for an infinitesimal adiabatic process, the first law requires for a gas undergoing a quasi-static adiabatic process that

$$0 = dU + P\,dV$$

But $dU = C_v\,dT$ for an ideal gas, and so we get

$$0 = C_v\,dT + P\,dV$$

Because $PV = nRT$ for an ideal gas, we have

$$d(PV) = d(nRT)$$

$$P\,dV + V\,dP = nR\,dT$$

or

$$dT = \frac{P\,dV + V\,dP}{nR}$$

Substituting this result for dT into $C_v\,dT + P\,dV = 0$ gives

$$\frac{C_v}{nR}(P\,dV + V\,dP) + P\,dV = 0$$

$$(C_v + nR)P\,dV + C_v V\,dP = 0$$

But, as we saw in Example 10.4, C_p and C_v for an ideal gas are related according to $C_p = C_v + nR$, so we now have

$$C_p P\,dV + C_v V\,dP = 0$$

or

$$\frac{C_p}{C_v}\frac{dV}{V} + \frac{dP}{P} = 0$$

Letting $\gamma = C_p/C_v$ and integrating gives

$$\gamma \ln V + \ln P = K$$

where K is an arbitrary integration constant. This simplifies to

$$PV^\gamma = e^K = \text{constant} \quad \blacksquare$$

Heat Engines and the Second Law of Thermodynamics

A *heat engine* is a device that accepts energy from a high temperature source of heat, called a reservoir; converts heat energy to work; and rejects, or exhausts, heat to a reservoir at a lower temperature. A heat engine may be an actual device, like a steam engine, a gasoline engine, or a diesel engine. Here we will consider heat engines, both actual and idealized, that operate in a *cycle*, that is, they function in a sequence of thermodynamic processes (or may be so approximated) that periodically returns the system to its original state.

A *heat reservoir* is an infinite source or sink (acceptor) of heat and is characterized by a single fixed temperature. Therefore, no finite quantity of heat added to or extracted from a given reservoir can alter its temperature. Then all heat exchanges with a given reservoir take place at a well-defined temperature.

Because a cyclic heat engine returns to its original state after each cycle, the internal energy of its "working substance," usually a gas, assumes its original value. Then ΔU is zero for a complete cycle, and the first law relates the net heat Q entering the system each cycle to the net work W that the engine performs each cycle: $Q = W$. Thus, the net heat absorbed by a cyclic heat engine in one cycle is equal to the net work done by the engine during that cycle. Let Q_1 be the heat absorbed by an engine from high-temperature reservoirs (HTR). Let Q_2 be the heat exhausted by the engine to low-temperature reservoirs (LTR). Notice that Q_2, according to our sign convention, is a negative quantity. Figure 10.16 is a

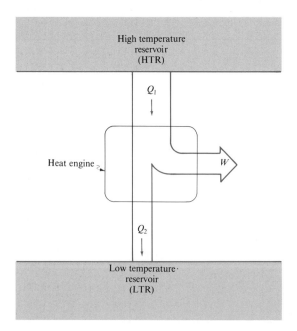

Figure 10.16
A schematic representation of a heat engine. Part of the heat Q_1 absorbed by the engine from the high-temperature reservoir is used by the engine to perform work W. The remainder is ejected as heat Q_2 to a low-temperature reservoir.

schematic representation of this heat engine. Part of the heat Q_1 absorbed from the HTR is converted to work W by the engine. The unused heat Q_2 is exhausted to the LTR.

The object of an engine is to provide as much external work as possible per unit of heat energy supplied to it, and we are interested in how efficiently heat engines do so. The *thermal efficiency ϵ of a cyclic heat engine* is defined to be the ratio of the net work W done by the engine each cycle to the heat Q_1 absorbed each cycle, or

$$\epsilon = \frac{W}{Q_1} = \frac{Q_1 - |Q_2|}{Q_1} = 1 - \left|\frac{Q_2}{Q_1}\right| \qquad (10\text{-}22)$$

where $|Q_2|$ is the heat rejected each cycle. Equation (10-22) indicates that a heat engine would be 100 percent efficient if $Q_2 = 0$, that is, if no heat were exhausted by the engine so that all the absorbed heat were converted to work. It is an experimental fact that no cyclic engine can achieve an efficiency of 100 percent, and this fact is incorporated into the *second law of thermodynamics*:

No heat engine, operating continuously in a cycle, can extract heat from a reservoir and convert all of that heat into work.

The second law denies the possibility of using energy—no matter how much energy may be available—in particular ways. The first law expresses the conservation of energy but does not forbid, for example, our converting the enormous internal energy of the atmosphere into work to fly an aircraft indefinitely. The second law ensures that we may not enjoy such an atmospheric bonanza. It has been said that the first law ensures that one cannot win, but the second law makes sure that one cannot break even. In what thermodynamic sense is this casino proverb true? For a cyclic process, the first law forbids getting more work out of an engine than the heat put into it. The second law, however, asserts that the work performed will be less than the heat absorbed.

The operations of a real engine, a diesel engine, may be approximated by a thermodynamic cycle like that shown in Figure 10.17. The cycle begins at A,

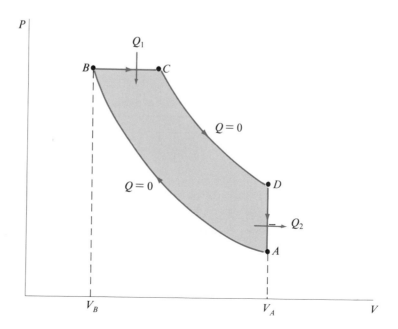

Figure 10.17
An approximation of the diesel cycle.

which corresponds to the real engine with air in a cylinder at atmospheric pressure. The compression stroke of the engine is approximated in the diagram by the adiabat AB. An increase in temperature accompanies the compression. The volume ratio V_A/V_B is called the compression ratio of the engine; typically, V_A/V_B is about 15 for diesel engines. At the end of the compression stroke of an actual engine, fuel is injected into the cylinder, where it ignites in the high-temperature air and burns. In the thermodynamic model, this part of the "power stroke," in which the engine performs external work, is approximated by an isobaric expansion. The remainder of the power stroke of the engine is approximated by the adiabatic expansion CD. In the engine, the valves of the cylinder open, exhausting heat and returning the contents of the cylinder to atmospheric pressure. This exhaust portion of the cycle is represented by the isochoric process DA in the diagram.

Example 10.6   **PROBLEM**   An ideal gas ($\gamma = 1.4$) is operated in the diesel cycle of Figure 10.18 having a compression ratio, $r = V_A/V_B$, equal to 18 and an expansion ratio, $\mathcal{R} = V_C/V_B$, equal to 2.0. Determine the thermal efficiency of this heat engine.

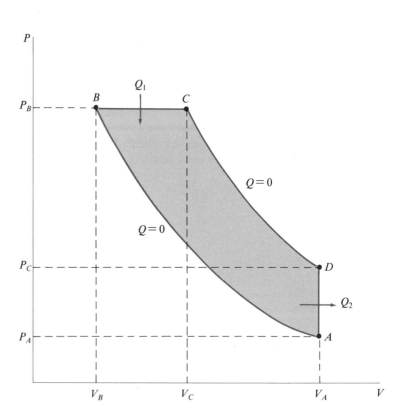

Figure 10.18   Example 10.6

SOLUTION   First, we solve for the state variables P, V, and T for states B, C, and D in terms of those for state A, that is, P_A, V_A, and T_A. We can then determine Q_1 and Q_2 using $Q_1 = C_P(T_C - T_B)$ and $Q_2 = C_v(T_A - T_D)$. Then, the thermal efficiency ϵ of the engine is given by $\epsilon = 1 - |Q_2/Q_1|$.

For state B, because $V_A = rV_B$ and $PV^\gamma = $ constant for process AB, we have:

$$P_B = \frac{P_A V_A{}^\gamma}{V_B{}^\gamma} = r^\gamma P_A$$

$$T_B = \frac{P_B V_B}{nR} = \frac{r^\gamma P_A}{nR}\frac{V_A}{r} = r^{\gamma-1}\frac{P_A V_A}{nR} = r^{\gamma-1}T_A$$

For state C, we get:

$$P_C = P_B = r^\gamma P_A$$

$$V_C = \mathscr{R}V_B = \frac{\mathscr{R}V_A}{r}$$

$$T_C = \frac{P_C V_C}{nR} = \frac{r^\gamma P_A}{nR}\frac{\mathscr{R}V_A}{r} = \mathscr{R}r^{\gamma-1}\frac{P_A V_A}{nR} = \mathscr{R}r^{\gamma-1}\,T_A$$

Because $V_D = V_A$ and $PV^\gamma = $ constant for process CD, we get for state D:

$$P_D = \frac{P_C V_C{}^\gamma}{V_D{}^\gamma} = \frac{r^\gamma P_A}{V_A{}^\gamma}\left(\frac{\mathscr{R}V_A}{r}\right)^\gamma = \mathscr{R}^\gamma P_A$$

$$T_D = \frac{P_D V_D}{nR} = \mathscr{R}^\gamma \frac{P_A V_A}{nR} = \mathscr{R}^\gamma T_A$$

Thus, the heats, Q_1 and Q_2, are given by:

$$Q_1 = C_p(T_C - T_B) = C_p r^{\gamma-1}\,(\mathscr{R} - 1)T_A$$

$$Q_2 = C_v(T_A - T_C) = C_v\,(1 - \mathscr{R}^\gamma)T_A$$

Finally, the thermal efficiency ϵ is

$$\epsilon = 1 - \left|\frac{Q_2}{Q_1}\right| = 1 - \frac{\mathscr{R}^\gamma - 1}{\gamma r^{\gamma-1}(\mathscr{R} - 1)} = 1 - \frac{2^{1.4} - 1}{1.4(18^{0.4})(2 - 1)} \approx 0.63$$

Thus, 63 percent of the energy Q_1 available during each cycle is converted to "useful" work, and 37 percent of the available energy is "wasted" as heat Q_2 rejected during the cycle. ■

E 10.39   For the diesel cycle of Example 10.6, determine the ratio of T_C (the maximum temperature) to T_A (the minimum temperature).   Answer: 6.4

A *refrigerator* is a heat engine in which the energy flows are reversed from the usual engine designed to perform work. In the refrigeration cycle represented schematically in Figure 10.19, work is done *on* the thermodynamic system by an external agent; heat is extracted from an LTR and exhausted to an HTR. Thermal energy is caused to flow from a cold region to a hot region because an external agent does work on the refrigerant. The most effective refrigerator is one that removes the most heat Q_2 from the LTR for the least expenditure of work. Therefore, refrigerators are characterized by a *performance coefficient* η, defined to be the ratio of the heat Q_2 the refrigerator extracts each cycle to the external work $(-W)$ done on it each cycle, or

$$\eta = \frac{Q_2}{|W|} = \frac{Q_2}{|Q_1| - Q_2} \tag{10-23}$$

where $|Q_1|$ is the heat rejected each cycle.

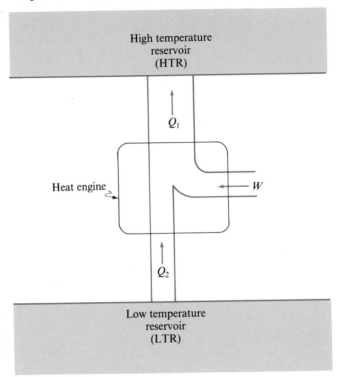

Figure 10.19
A schematic representation of a refrigerator. Work W by an external agent
is done on the heat engine, which extracts heat Q_2 from a low-temperature
reservoir and ejects heat Q_1 to a high-
temperature reservoir.

Example 10.7

PROBLEM An ideal gas ($\gamma = 1.4$) is operated in the inverse diesel cycle of
Figure 10.20 as a refrigerator, having a compression ratio, $\mathcal{R} = V_C/V_B$, equal to
2.0 and an expansion ratio, $r = V_A/V_B$, equal to 18. Calculate the performance
coefficient for this refrigeration cycle.

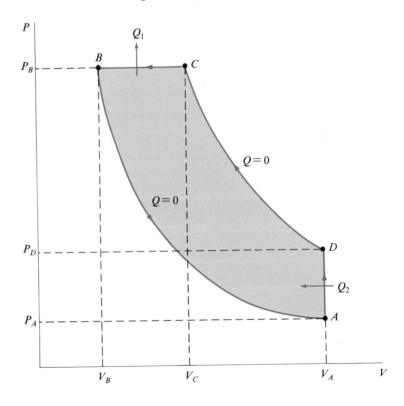

Figure 10.20   Example 10.7

SOLUTION   The expressions obtained in Example 10.6 are appropriate here if the signs of Q_1 and Q_2 are changed. (Why?) Thus, we have:

$$Q_1 = -C_p r^{\gamma-1}(\mathfrak{R} - 1)T_A$$

$$Q_2 = C_v(\mathfrak{R}^{\gamma-1})T_A$$

The performance coefficient η for this refrigeration cycle is given by

$$\eta = \frac{Q_2}{|Q_1| - Q_2} = \frac{\mathfrak{R}^\gamma - 1}{\gamma r^{\gamma-1}(\mathfrak{R} - 1) - (\mathfrak{R}^\gamma - 1)} \simeq 0.58$$

Because $\eta = Q_2/|W|$, where $|W|$, the work that must be done by an external agent acting on the working substance, is given by

$$|W| = \frac{Q_2}{\eta} \simeq 1.7 \, Q_2$$

Thus, for each calorie of heat removed by the refrigeration cycle, 1.7 calories (7.1 J) of work must be performed by the external agent, and 2.7 calories of heat are ejected to the environment. ∎

The Carnot Cycle and the Absolute Temperature Scale

The second law of thermodynamics forbids the existence of a heat engine with 100 percent efficiency. Then, what is the most efficient engine possible? In 1824 Sadi Carnot proposed an idealized heat engine cycle that has the maximum possible efficiency that can be obtained by any heat engine operating in a cycle between two given heat reservoirs. Called the Carnot cycle, the operating cycle of this hypothetical engine consists of isotherms and adiabats.

Consider an engine operating quasi-statically between the fixed temperatures of two given reservoirs. If this engine accepts heat Q_1 only from an HTR (a process that proceeds at a temperature T_H along an isotherm) and rejects heat $|Q_2|$ only to an LTR (a process that proceeds along an isotherm at a given temperature $T_L < T_H$), the paths connecting the isotherms must be adiabats. The Carnot cycle operating as an engine (if the arrows were reversed in direction, it would be operating as a refrigerator) is shown in Figure 10.21. The Carnot cycle consists of the following steps:

A→B: Isothermal expansion absorbing heat Q_1 at temperature T_H.
B→C: Adiabatic expansion cooling the gas from T_H to T_L.
C→D: Isothermal compression rejecting heat $|Q_2|$ at T_L.
D→A: Adiabatic compression heating the gas from T_L to T_H.

The area ABCDA enclosed by these process paths represents the net work performed by the engine in one cycle. Then, using Equation (10-22), we find the efficiency of the Carnot cycle to be

$$\epsilon_C = 1 - \left|\frac{Q_2}{Q_1}\right| \qquad (10\text{-}24)$$

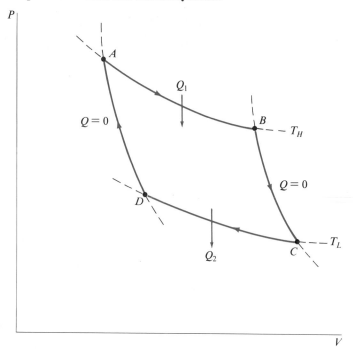

Figure 10.21
A Carnot cycle operating as a heat engine.

An important proposition, proposed and proved by Carnot, is equivalent to our statement of the second law of thermodynamics: *A Carnot engine is the most efficient heat engine that can operate between two given temperatures.* Any engine that absorbs heat from an HTR, rejects heat to an LTR, and exchanges no heat in the intervening processes necessarily operates only along isotherms and adiabats, that is, such an engine is a Carnot engine. Furthermore, it follows from Carnot's proposition that the efficiencies of all Carnot engines operating between two given temperatures are identical (otherwise some would be more efficient than others, thereby violating the proposition). Thus the efficiency of a Carnot engine is independent of the working substance of the engine. In other words, any Carnot engine operating between fixed temperatures, T_H and T_L, has the same ratio of heat rejected to heat absorbed, $|Q_2/Q_1|$, regardless of the working substance of the engine.

Because Q_2/Q_1 is the same for any working substance of a Carnot engine, we can define an absolute temperature scale (it is absolute because it does not depend on the properties of any substance). Let us define the ratio of absolute (Kelvin) temperatures T_H and T_L to be equal to the ratio of the magnitudes of Q_1 and Q_2, or

$$\frac{T_H}{T_L} = \frac{|Q_1|}{|Q_2|} = -\frac{Q_1}{Q_2} \tag{10-25}$$

Equation (10-25) defines only the ratio of absolute temperatures. A fixed point is also needed to define a temperature scale completely. Recall that the triple point of water has been assigned the absolute temperature of 273.16 K. Then, the absolute temperature scale is defined in terms of the Carnot heats Q_1 and Q_2. Thus, the efficiency of any Carnot engine may be written in terms of the reservoir temperatures:

$$\epsilon_C = 1 - \left|\frac{Q_2}{Q_1}\right| = 1 - \frac{T_L}{T_H} \tag{10-26}$$

Equation (10-26) indicates that a Carnot engine would be 100 percent efficient, or $\epsilon_C = 1$, only when $T_L = 0$, that is, when the LTR is at *absolute zero*. When $T_L = 0$, Q_2 is zero, and the work done by a Carnot engine is equal to Q_1. All the heat absorbed by the engine is converted into work, but an engine that converts all the heat that enters the engine into work violates the second law. Therefore, achieving a temperature of absolute zero is a violation of the second law of thermodynamics.

E 10.40   A cyclic heat engine is to be designed to operate in a temperature range between the ice point and the steam point of water. What is the maximum thermal efficiency of such an engine?   Answer: 0.27

E 10.41   If the heat engine of Exercise 10.40 is to perform 50 J of work during each cycle, what minimum quantity of heat must the engine acquire from its (higher temperature) energy source during each cycle? (*Hint:* Recall that the efficiency $\epsilon = W/Q_1 \leq \epsilon_C$.)   Answer: 45 cal

E 10.42   If the heat engine of Exercises 10.40 and 10.41 is to deliver work at the rate of 1.0 hp (approximately 750 watts), how many cycles per minute must the engine execute?   Answer: 900

E 10.43   Calculate the thermal efficiency of a Carnot heat engine operating between the maximum and minimum temperatures of the (ideal gas) diesel heat engine of Example 10.6 and Exercise 10.39. Compare this value to the thermal efficiency of the diesel engine.   Answer: $\epsilon_C = 0.84 > 0.63 = \epsilon_{diesel}$

Entropy

Entropy is introduced here because it is associated with a very important physical principle, but a detailed consideration appropriately belongs in a more advanced text. In order to present the principle with which it is connected in a meaningful way, we will define entropy and consider three of its aspects, each of which is intended to provide insight into the physical significance of entropy:

1. Entropy as a measure of how much of a given quantity of heat cannot be converted into work by a heat engine.

2. Entropy as a state variable in thermodynamics.

3. Entropy as a measure of randomness.

In the preceding section, we saw the relationship, Equation (10-25), between the heats transferred and the reservoir temperatures associated with a Carnot cycle. Let us rewrite Equation (10-25) as

$$|Q_2| = \left(\frac{Q_1}{T_H}\right) T_L \qquad \text{(10-27)}$$

where $|Q_2|$ is the quantity of heat that is wasted in each cycle. In other words, $|Q_2|$ is energy that is unavailable for performing work. We now define the quantity Q_1/T_H to be the *entropy S* associated with the heat absorbed by the Carnot engine from the HTR. For an exhaust temperature T_L, say $T_L = 300$ K, let us calculate the wasted heat $|Q_2|$ for a few values of T_H. First, when 800 cal of heat is absorbed by the engine at $T_H = 800$ K, the entropy $Q_1/T_H = 1$ cal/K. Then Equation (10-27) gives $|Q_2| = (1 \text{ cal/K})(300 \text{ K}) = 300$ cal, so that 300 cal of heat is not available to perform work. Now suppose the same quantity, 800 cal, is absorbed at a temperature $T_H = 400$ K; then we have $S = Q_1/T_H = 2$ cal/K and $|Q_2| = (2 \text{ cal/K})(300 \text{ K}) = 600$ cal. The wasted, or unavailable, energy is twice the value of that obtained when the entropy was 1 cal/K. Indeed, if $T_H = 300$ K, all the energy absorbed is wasted. Thus, entropy may be considered a property associated with the heat entering a Carnot engine. *Low-entropy heat provides more energy available to do work than does high-entropy heat.* Thus, entropy is a characteristic of heat that measures its unavailability for the performance of work. This aspect of entropy has useful engineering applications.

Let us now consider a more physical, or thermodynamic, view of entropy. Recall that temperature is constant along an isotherm, volume is constant along an isochor, and pressure is constant along an isobar. An obvious question is: What is constant along an adiabat? Of course, heat Q absorbed during an adiabatic process is equal to zero, and it may appear that Q is a state variable that remains constant during an adiabatic process. But heat is not a state variable, as are T, V, and P. Is there a state variable that remains constant during an adiabatic process? To answer this, let us consider the Carnot cycle once again.

For each infinitesimal quantity of heat dQ that enters or leaves the system during a quasi-static process, let us define a change dS in the entropy *of the system* to be $dS = dQ/T$. Then in Figure 10.21 the change $\Delta S = S_B - S_A$ of the entropy of the system between states A and B is $\int_A^B dQ/T$, where T is equal to the constant value T_H. Similarly, as the system proceeds from state C to state D, the change in entropy, $S_D - S_C = \int_C^D dQ/T$, in which dQ is a negative quantity and T has the constant value T_L. Along the adiabats, from B to C and from D to A, no heat is transferred to or from the system. So dQ is equal to zero, and thus we have $\Delta S = \int dQ/T = 0$ along both of these paths. Now let us investigate the change in entropy around the closed cycle:

$$\oint dS = \int_A^B \frac{dQ}{T} + \int_C^D \frac{dQ}{T} = \frac{Q_1}{T_H} - \frac{|Q_2|}{T_L} \tag{10-28}$$

where the symbol $\oint$ means that the integration is performed about a closed path. From Equation (10-27), we know that $|Q_1|/T_H = |Q_2|/T_L$, so Equation (10-28) becomes

$$\oint dS = \oint \frac{dQ}{T} = 0 \tag{10-29}$$

We have proved that the change in entropy is zero for an arbitrary Carnot cycle. This proof may be generalized to apply to an arbitrary quasi-static cycle. The essential argument of this generalization is based on the fact that any quasi-static process line may be approximated arbitrarily well by a series of alternating adiabats and isotherms. Equation (10-29) implies that S is a state variable: The change in S between two states is independent of the path connecting those two states. Like internal energy U, the entropy S of a system may be assigned an arbitrary value in a given state; only changes ΔS of the entropy of the system are significant.

From the foregoing analysis, we see that not only is entropy S a state variable, but S *is that state variable that does not change as the system proceeds along an adiabat.* Because S, like P, V, T, and U, is a state variable, we can use a plot of

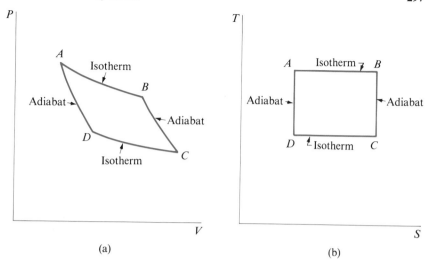

Figure 10.22
A Carnot cycle represented on a P-V diagram and on a T-S diagram.

T versus S, much as we have been using P versus V, to represent states and processes. Figure 10.22 shows a Carnot cycle on a P-V diagram and on a T-S diagram. On the P-V diagram, the work done by the system is represented by the area $\int P \, dV$ under a process line. On the T-S diagram, an area under a process line represents the heat flow Q into the system according to $dQ = T \, dS$, or $Q = \int T \, dS$. Thus entropy is a useful thermodynamic state variable by which systems undergoing quasi-static processes may be analyzed.

E 10.44   For the Carnot cycle shown, calculate (a) Q_1 and Q_2, (b) the net work W performed during one cycle, and (c) the thermal efficiency of the cycle.

Answers: (a) 1.6×10^4 cal, -4.0×10^3 cal; (b) 5.0×10^4 J; (c) 0.75

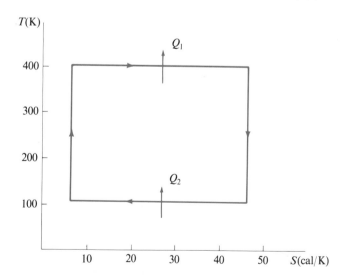

Exercise 10.44

Let us now examine entropy in terms of the randomness of a physical system. We shall continue, for the moment, to consider only one simple thermodynamic system.

Determine the change in entropy of 50 g of water that are vaporized at 100° C.

A change in entropy of a simple thermodynamic system may be calculated for any quasi-static process using

$$\Delta S = \int \frac{dQ}{T} \tag{10-30}$$

Because T is constant while the water is changing phase, from liquid to gas, we may place T outside the integral. Next we must assume that the vaporization takes place quasi-statically; such a process may be accomplished by placing the water in contact with a temperature reservoir such that their temperature difference is vanishingly small. Then we have

$$\Delta S = \frac{1}{T} \int dQ = \frac{mL_V}{T} \tag{10-31}$$

where $\int dQ$ is equal to the heat absorbed by the water in changing phase to a gas. This heat is given by Equation (10-14) to be mL_V, where m is the mass of the water and $L_V = 540$ cal/g is the latent heat of vaporization of water. Substitution of $m = 50$ g and $T = 373$ K into Equation (10-31) gives

$$\Delta S = \frac{(50 \text{ g})(540 \text{ cal/g})}{373 \text{ K}} = 72 \text{ cal/K} \tag{10-32}$$

The important point of this result is that the entropy of the system (the water) has increased as a result of adding thermal energy to the water. By adding heat to the water, the water changes from the relatively ordered liquid phase, where it is bound together by molecular forces, to the more chaotic gaseous phase. The entropy of the system has increased as the system becomes more chaotic, or more random. This example illustrates the final aspect of entropy that we shall consider: *An increase in the entropy of a system corresponds to a greater randomization of that system.*

Example 10.8

PROBLEM   If 2.0 mol of an ideal gas are compressed isothermally from an initial volume V_1 to a final volume $V_2 = 0.20\,V_1$, calculate the change in entropy of the gas.

SOLUTION   According to the first law of thermodynamics, that is, $dQ = dU + P\,dV$, we get, because $dS = dQ/T$ and $dU = C_v\,dT = 0$ (why?):

$$dS = \frac{dQ}{T} = \frac{P}{T}\,dV$$

But for an ideal gas, the equation of state gives $P = nRT/V$. Thus we have

$$dS = nR\frac{dV}{V}$$

Integrating this expression, we get

$$\Delta S = nR \int_{V_1}^{V_2} \frac{dV}{V} = nR \, \ln\frac{V_2}{V_1}$$

Using $n = 2.0$ mol, $R = 8.3$ cal/(mol·K), and $\ln(V_2/V_1) = \ln(0.20)$, we get

$$\Delta S = -27\frac{\text{cal}}{\text{K}}$$

Thus, as the gas is compressed, it is forced into a more ordered, that is, more localized, state; and, as we expect, the entropy of the gas decreases.   ■

The simple thermodynamic systems we have been considering do, in fact, interact with the remainder of the universe. A body of gas interacts, for example, with its container and with the piston that moves to permit expansion or compression. We have seen that in the course of a cyclic process in which the system undergoes only quasi-static processes, the net change in the entropy of the system is zero. We now define a quasi-static process in which there are no attendant dissipative effects in the system *or* in its environment to be a *reversible process*. Thus the piston has no friction, and there are no viscous forces between the system and its container in a reversible process.

It is generally true that, if a process is reversible, the change in entropy of the universe (the system and its environment) is zero. But for all real processes—no actual physical process is reversible—the entropy of the universe increases, or

$$\Delta S_{universe} > 0 \qquad\qquad (10\text{-}33)$$

This inequality is the fundamental principle we have anticipated, the principle of increasing entropy:

Every physical process proceeds in such a way that the entropy of the universe increases.

It can be shown that this statement is another form of the second law of thermodynamics. This form, however, indicates the direction in which natural processes proceed. For example, we know that bourbon and water will mix, but a mixture of bourbon and water never unmixes. When we let a confined gas escape into an evacuated region, it never reconfines itself. All natural processes proceed along one-way streets, and the direction of each street is toward increasing entropy.

As natural processes proceed so as to increase the entropy of the universe, all energy is continually being degraded, being converted into even more disorganized, random motion that is less available for performing work. Thermal energy at one temperature can be converted to work only if a lower temperature is available. Eventually, all energy will become unavailable for performing work when no lower temperature exists. All the universe will be at the same temperature, and no work of any kind can be done. All ordered motion will cease. Thus, the principle of increasing entropy *predicts* the so-called heat death of the universe.

10.4 Problem-Solving Summary

Problems that concern thermometry, heat flow, thermal expansion, and calorimetry may be identified by characteristic words: temperature, expansion, thermal conductivity, heat, heat capacity, specific heat, and latent heat. Most problems in these areas can be successfully solved merely by knowing and using the definitions of these words. Thermodynamics problems are similarly identified by key words: isothermal, isobaric, isochoric, adiabatic, quasi-static, heat engine, refrigerator, reversible, and entropy. Again, knowing precise definitions is a helpful (and necessary) head start on any problem.

It is usually assumed in problems that the student is completely familiar with conversions between the Celsius, Fahrenheit, and absolute (or Kelvin) scales. Remember that *changes* in temperature on the Celsius and absolute scales have the same numerical value, that is, a change of one degree Celsius (°C) is equal to a change of one kelvin (K).

Problems about thermal expansion usually yield to direct substitution of values into the definition of the coefficient of linear expansion α, which has the unit of K^{-1}. In volumetric expansion, the appropriate quantity is the coefficient of volume expansion β. For isotropic solids, α and β are related by $\beta = 3\alpha$.

Almost all calorimetric problems may be solved using the "method of mixtures": For an isolated system, the heat lost by the initially hotter portion of the system is equal to the heat gained by the initially colder portion. By writing the heat lost and the heat gained in terms of the specific heats of the materials, their masses, and the appropriate temperature changes, according to $Q = cm\Delta T$, the "heat lost = heat gained" expression may be used to find any unknown quantity. It is particularly helpful to remember that the common unit of specific heat is cal/(g · K). When a material undergoes a phase change (either it melts or freezes or it vaporizes or condenses) the heats required to change the phases without any change in temperature are expressed in terms of the latent heats (of fusion or of vaporization), $Q = \pm mL_F$ or $Q = \pm mL_V$. The positive sign applies for melting or vaporizing; the negative sign applies for freezing or condensing.

The sign convention for heat and work, both in heat and thermodynamics, is focused on whether energy is entering or leaving the system under consideration:

1. Heat is positive for thermal energy entering a system, negative for thermal energy leaving.

2. Work is positive when energy is lost from the system in performing work, negative when energy is added to the system as a result of external agents doing work on the system.

A simple thermodynamic system is a definite collection of particles that is macroscopically homogeneous when the system is in thermal equilibrium. The thermodynamic state of a system is uniquely specified by state variables: $P, V, T, U,$ or S. A line between states, on a P-V diagram, for example, represents a quasi-static process, a succession of equilibrium states. An equation of state relates state variables for a given substance. Work and heat are *not* state variables. The work or heat associated with a process depends on the path of the process line: $W = \int P \, dV$, the area under a process line on a P-V diagram and $Q = \int T \, dS$, the area under a process line on a T-S diagram.

Special thermodynamic processes should be known by name:

1. An isothermal process takes place at constant temperature.

2. An isobaric process takes place at constant pressure.

3. An isochoric process takes place at constant volume.

4. An adiabatic process takes place at constant entropy, or equivalently, it takes place with no heat entering or leaving the system.

Because an ideal gas is so frequently used as the working substance in thermodynamic problems, we emphasize the following relationships for ideal gases:

1. The equation of state for an ideal gas is $PV = nRT$.

2. The internal energy of an ideal gas is given by $U = C_v T$, where C_v is the heat capacity of the gas at constant volume.

3. The equation PV^γ = constant applies to an ideal gas undergoing an adiabatic process. The quantity γ is a constant defined by $\gamma = C_p/C_v$.

4. The heat capacities, C_p and C_v, for an ideal gas are related by $C_p = C_v + nR$.

The first law of thermodynamics may be thought of as a statement of the conservation of energy. This statement relates the heat absorbed or ejected by a system, the work done on or by a system, and the change in the internal energy of the system: $Q = W + \Delta U$. The sign convention for heat and work is important in problems that use the first law. The first law takes special forms for special processes. For example, in an adiabatic process, where $Q = 0$, the first law becomes $\Delta U = -W$.

Heat engines, cyclic devices that convert heat energy to external work, are usually considered in terms of their efficiencies. For heat Q_1 absorbed from a high-temperature reservoir (HTR) and heat Q_2 ejected to a low-temperature reservoir (LTR), the efficiency of a heat engine is defined to be the ratio of work W per unit of heat Q_1 absorbed: $\epsilon = W/Q_1 = 1 - |Q_2/Q_1|$. Notice that the sign of Q_2 is negative (thermal energy flowing to the LTR from the system), so the efficiency is less than 1. Refrigerators, heat devices for which work is done on the system (W is negative) by an external agent, extract heat from the LTR (Q_2 is positive) and eject heat to the HTR (Q_1 is negative). The performance coefficient η for a refrigerator is $\eta = Q_2/(|Q_1| - Q_2)$.

The most efficient engine operating cyclically between two given fixed temperatures is the hypothetical Carnot engine, which extracts and ejects heats along isotherms and changes the temperature of its working substance along adiabats. The efficiency of a Carnot engine is $\epsilon_C = 1 - |Q_2/Q_1|$. Because the absolute temperature scale is defined in terms of the heats Q_1 and Q_2 of a Carnot engine, the Carnot efficiency ϵ_C may be written as $1 - T_L/T_H$. Thus the absolute zero of temperature is the temperature of the LTR for which a Carnot engine would have 100 percent efficiency, $\epsilon_C = 1$.

Entropy S is a measure of the unavailability of energy for the performance of work, a measure of randomness, and a thermodynamic state variable. We can measure changes in the entropy of a simple system using $\Delta S = \int [dQ/T]$ over any quasi-static path. For any actual process, which is necessarily irreversible, the change in entropy of the universe, that is, system-plus-its-environment, is greater than zero.

Problems

GROUP A

10.1 Determine the volume (in liters) of 1.0 mol of an ideal gas at STP (standard temperature and pressure), $T = 0°$ C and $P = 1.0$ atm.

10.2 The "dietary calorie" used in determining the energy content of food is equal to 1000 calories. If a person consumes approximately 2500 "dietary calories" per day, estimate (in watts and horsepower) this person's rate of expending energy.

10.3 If 2.6×10^4 J of energy must be added to 2.0 kg of a metal to raise its temperature 15° C, what is the specific heat of the metal?

10.4   An aluminum measuring tape is calibrated to be 10 m long at 20° C. What correction must be applied to a measurement of 1.0 m made with this tape when the temperature is 40° C?

10.5   An oil tanker made of steel is 1000 ft long when steaming in the Arctic Ocean where the water temperature is 32° F. How much does the length of the tanker increase when it is in the Persian Gulf, where the water temperature is 68° F?

10.6   (a) At what Celsius temperature are the Celsius and Fahrenheit temperatures numerically equal? (b) At what Fahrenheit temperature are the Kelvin and Fahrenheit temperatures numerically equal?

10.7   Calculate the change in entropy that takes place as 1.0 kg of ice at 0° C is converted to water at the same temperature.

10.8   How much heat is required to increase the temperature of 2.0 lb of water by 90° F?

10.9   Show that, for a quasi-static adiabatic process, the temperature T and volume V of an ideal gas are related by

$$T V^{\gamma-1} = \text{constant}$$

where $\gamma = C_p/C_v$.

10.10   If the thermal efficiency of an engine being designed is to be 0.35 and if the engine is to eject heat at a temperature equal to 50° C, what is the minimum value for the highest temperature of the cycle?

10.11   If the heat capacity at constant pressure (C_p) for a 2.0-g sample of helium gas (atomic weight = 4.0 g/mol) is measured to have a value of 2.5 cal/K, calculate (a) C_v, the heat capacity at constant volume; (b) c_p, the specific heat of helium gas at constant pressure; and (c) c_v, the specific heat at constant volume.

10.12   If 25 g of ice at 0° C and 95 g of an unknown liquid at 110° C are mixed in a thermally insulated container and the final product is a liquid mixture at 15° C, calculate the specific heat of the liquid.

10.13   If an LTR (T_L = 250 K) is placed in thermal contact with an HTR (T_H = 450 K) and 8.0 kcal of heat is transferred to the LTR, calculate the change in entropy of (a) the LTR; (b) the HTR; and (c) the universe. (d) Is this a reversible process?

10.14   How many molecules of a gas are contained in a 10-L vessel at 30° C and at 2.0 atm pressure?

10.15   A square hole (8.0 cm along each side) is cut in a sheet of copper. Calculate the change in the area of this hole if the temperature of the sheet is increased by 50 K.

10.16   A sample of ideal gas with a molecular weight of 4.0 is in a thermodynamic state having P = 1.2 atm, V = 8.8 L, and T = 85° C. Determine the mass of this sample of gas.

10.17   The latent heat of fusion L_F for water is 80 cal/g, and for lead, L_F is equal to 5.9 cal/g. How much heat is required to melt 1.0 cubic centimeter (cc) of (a) water, and (b) lead? (The density of lead is equal to 11.3 g/cm³.)

10.18   Two moles of an ideal gas (γ = 5/3) are compressed quasi-statically and adiabatically from an initial state, P = 1.5 atm and T = 300 K, to a final state having a volume equal to one-fifth of the initial volume. Determine P, V, and T for the final state.

10.19   A heat engine operates using the cycle shown in Figure 10.23. If Q_{AB} = 50 cal, Q_{BC} = 150 cal, and Q_{CA} = −90 cal,

(a)   what is the thermal efficiency of the engine?

(b)   calculate the enclosed area $ABCA$.

(c)   determine the performance coefficient of the refrigerator obtained by reversing the cycle.

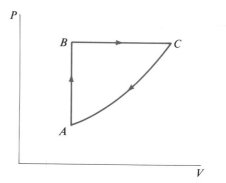

Figure 10.23   Problem 10.19

10.20   Two moles of an ideal gas are in a cylinder fitted with a movable piston. The gas is initially at a pressure of 1.0 atm at a temperature of 300 K.

(a)   What is the volume of the gas initially?

(b)   If the temperature of the gas is increased to 400 K at constant pressure, what will be its change in volume?

(c)   If the gas is returned to its original volume while being maintained at a constant temperature of 400 K, what will be the final pressure of the gas?

10.21   On a cold winter day when the outside temperature is constant at 0° C, the temperature inside a house is maintained at 20° C. During a 24-h period how much heat passes through a glass window with an area of 4.0 m² and a thickness of 0.40 cm?

10.22   A wall 15 cm thick has an area of 20 m². When a 20°-C difference in temperature is maintained across the wall, it is found that 3.5×10^6 J of thermal energy passes through the wall each hour. What is the thermal conductivity of the wall?

10.23

(a)   What thickness d of brick is equivalent as a thermal insulator to a 1.0-cm-thick layer of air?

(b)   Repeat part (a) for copper rather than brick.

10.24   Determine the percentage change in the volume of a steel ball if its temperature is increased from 30° C to 130° C.

10.25   A heat engine with a thermal efficiency of 0.35 develops 95 hp when executing 4.0×10^3 cycles per minute. Calculate

(a)   the work done per cycle.

(b)   the heat absorbed per cycle.

(c)   the heat rejected per cycle.

(d)   the lowest possible maximum temperature of the cycle if the heat is ejected at a temperature of 150° C.

10.26   If 50 g of copper BB's at a temperature of 200° C are added to 100 g of water at 20° C, at what temperature will they reach thermal equilibrium?

10.27   A hollow sphere (inner radius = 5.0 cm and an outer radius = 6.0 cm) made of lead (density = 11.3 g/cm³) is filled with water at 50° C. Calculate the heat capacity of this apparatus.

10.28   The temperature reservoirs of a Carnot engine are 300 K and 600 K. If the engine does 2000 J of work, (a) how much heat is exhausted to the lower-temperature reservoir, and (b) how much heat is taken from the higher-temperature reservoir?

10.29   Determine the change in the internal energy of a system that

(a)   absorbs 500 cal of heat energy while doing 800 J of external work

(b)   absorbs 500 cal of heat energy while 500 J of external work is done on the system

(c)   is maintained at a constant volume while 1000 cal is removed from the system

10.30   A 1.0-kg ball of metal at 100° C is placed in a calorimeter containing 0.50 L of water at 20° C. When the ball and water reach equilibrium, the temperature is 34° C. What is the specific heat of the metal?

10.31   An ideal gas is compressed to half its original volume while its temperature is held constant.

(a)   If 1000 J of energy is removed from the gas during the compression, how much work is done on the gas?

(b)   What is the change in the internal energy of the gas during the compression?

10.32

(a)   What is the heat capacity of 4.0 L of a liquid that has a density of 800 kg/m³ and a specific heat of 1.1 cal/(g·K)?

(b)   How much work must be done to raise the temperature of this particular volume of fluid from 30° C to 50° C?

10.33   If 0.80 mol of an ideal gas is compressed quasi-statically from a volume of 16 L to a volume of 2.0 L while the gas is in thermal contact with a heat reservoir (T = 300 K), how much heat is absorbed by the gas?

10.34   If the density of a sample of an isotropic material decreases by 1.2 percent when the temperature of the sample is increased by 200° F, calculate the (thermal) coefficient of linear expansion.

10.35   A spherical glass ($\alpha = 9.0 \times 10^{-6}$) bulb having an inside radius of 8.0 cm is filled with mercury ($\beta = 1.8 \times 10^{-4}$) at a temperature of 20° C. If the bulb has a small hole at its highest point, how many cubic centimeters of mercury will flow out of the bulb if the temperature is increased to 50° C?

10.36   A Carnot engine with a high-temperature reservoir at 100° C exhausts into a reservoir of ice at 0° C. When the engine has done 10^4 J of work, how much ice will it have melted with its exhaust?

10.37   A Carnot engine with an efficiency of 40 percent is used as a refrigerator.

(a)   What is the coefficient of performance of the refrigerator?

(b)   How much work must be done on the refrigerator to remove 1000 cal from the cold temperature reservoir?

10.38   How long will it take a 1.0-kW electric heater to change 1000 cm³ of ice at 0° C to water vapor at 100° C if no heat is lost?

10.39   How much heat must be added to 10 g of ice at −20° C to change it to water vapor at 100° C?

10.40   Estimate the extra energy that must be given to one water molecule to "free" it from the liquid state (at 100° C) to the gaseous (vapor) state.

10.41   A closed steel ($\alpha = 11 \times 10^{-6}$ K^{-1}) tank (volume of 1.0 m³) at 20° C is filled with alcohol ($\beta = 1.0 \times 10^{-3}$ K^{-1}) at 20° C. A vertical glass ($\alpha = 9.0 \times 10^{-6}$ K^{-1}) overflow tube (inner radius = 1.0 cm) is attached to the top of the tank. How high will the alcohol rise in the tube for a 1.0 K change in temperature of the entire apparatus?

10.42   If 100 g of lead at 200° C, 250 g of aluminum at 250° C, and 75 g of water at 50° C are mixed and no heat is exchanged with the environment, determine the final temperature of the mixture.

10.43   When 400 g of metal shot is heated to 100° C and placed in a 100 g aluminum cup containing 200 g of water at 20° C, the mixture is stirred until it reaches thermal equilibrium at 24° C. Calculate the specific heat of this sample.

10.44   A cylindrically shaped container with its (vertical) curved sides and top made of a good thermal insulator has a flat lower surface made of copper (thickness d = 1.5 cm), as shown in Figure 10.24. The inner diameter D of the container is equal to

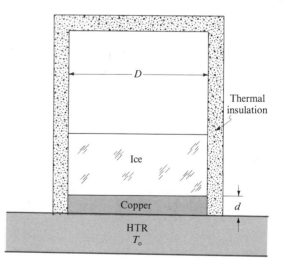

Figure 10.24   Problem 10.44

4.0 cm. If 0.45 kg of ice at 0° C is placed in the container and the device is then set upon an HTR ($T_o = 50°$ C), how much time will be required to melt the ice?

10.45 If an ideal gas ($C_v = 3.0$ cal/K, $C_p = 5.0$ cal/K) undergoes a quasi-static adiabatic process and, as a result, the temperature of the gas changes from 30° C to 80° C, what is the percentage change in the volume of the gas?

10.46 A Carnot engine extracts heat at 200° C.

(a) If the efficiency of the engine is 30 percent, what is its exhaust temperature?

(b) If this heat engine exhausts 100 J, how much work does it perform?

(c) When this engine operates as a refrigerator, what is its performance coefficient?

(d) With the high temperature reservoir fixed at 200° C, what must be the temperature of the reservoir from which it extracts heat as a refrigerator if its performance coefficient is to be equal to 4.0?

10.47 An ideal gas ($n = 0.40$ mol, $C_v = 1.2$ cal/K) undergoes the quasi-static process shown in Figure 10.25.

(a) Determine Q for this process.

(b) Calculate the change in the internal energy of the gas.

(c) How much work does the gas perform during this process?

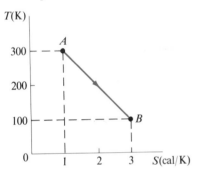

Figure 10.25   Problem 10.47

10.48 An ideal gas ($C_v = 6.0$ cal/K) is the working substance for a heat engine following the cycle shown in Figure 10.26.

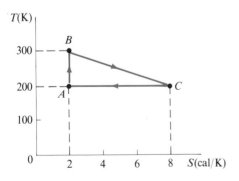

Figure 10.26   Problem 10.48

(a) Determine the thermal efficiency for this heat engine.

(b) If the operational cycle is reversed, calculate the performance coefficient for the resulting refrigerator.

10.49 If the pressure of 2.5 mol of an ideal gas for which $C_v = 7.5$ cal/K, is increased from 1.0 atm to 3.0 atm by an isochoric ($V = 0.040$ m³) process, calculate the

(a) heat Q for this process.

(b) change ΔS in the entropy of the gas.

10.50 If 2.0 mol of an ideal gas ($C_v = 6.0$ cal/K) are expanded at a constant pressure of 1.5 atm from an initial volume equal to 0.030 m³ to a final volume of 0.070 m³, calculate the

(a) heat Q for this isobaric process.

(b) change ΔS in the entropy of the gas.

10.51 Aboard an orbiting space vehicle, a solar collector with an area of 10 m² is oriented to absorb maximum radiation from the sun, which supplies energy to the upper atmosphere of the earth at a rate of 330 cal/(m²·s). If this energy is supplied to a heat engine having a Carnot efficiency of 25 percent, what is the maximum power that may be expected to be developed by the solar engine?

10.52 An ideal gas is expanded isothermally from an initial state ($P_1 = 4.0$ atm, $V_1 = 0.20$ L) to a final state with a volume $V_2 = 0.60$ L. During this (quasi-static) process, how much

(a) work is done by the gas?

(b) heat must be added to the gas?

10.53 An ideal gas ($C_v = 0.10$ cal/K) undergoes an isobaric expansion from a state ($P_1 = 3.0$ atm, $V_1 = 2.5 \times 10^{-4}$ m³, $T_1 = 250$ K) to a final state ($V_2 = 4.0 \times 10^{-4}$ m³). Determine Q for this (quasi-static) process.

10.54 If 10 g of steam at 100° C is added to 50 g of ice at 0° C (without any heat loss),

(a) will all the ice melt?

(b) what is the final state and temperature of the combination when it reaches thermal equilibrium?

10.55 If 150 g of ice at 200 K is mixed in a thermally insulated container with 20 g of steam at 100° C, determine the final state of the mixture.

10.56 A solid object ($C_p = 40$ cal/K) is heated quasi-statically at atmospheric pressure from an initial temperature of 250 K to a final temperature of 450 K.

(a) Calculate the change in entropy of the solid object.

(b) If the object is heated by placing it (with its initial temperature equal to 250 K) in direct thermal contact with a heat reservoir ($T = 450$ K), determine the change in entropy of the universe.

(c) Discuss part (b) as it relates to the law of increasing entropy. In particular, is the process reversible?

10.57 If 50 g of ice at 0° C is mixed with 150 g of hot water at 90° C in a thermally insulated container, calculate the change in entropy of the universe. Is this process reversible?

10.58 Starting from a state (P_1 = 4.0 atm, V_1 = 0.20 L), an ideal gas (C_v = 0.15 cal/K, γ = 1.4) expands adiabatically to a final state with V_2 = 0.60 L. For this (quasi-static) process,

(a) how much work is done by the gas?

(b) calculate the change in the internal energy of the gas.

(c) determine the change in the temperature of the gas.

(d) what is the final temperature of the gas?

10.59 An ideal gas (n = 1.5 mol, C_v = 4.5 cal/K) is the working substance for a heat engine that executes the cycle shown in Figure 10.27. Calculate the thermal efficiency for

(a) this heat engine.

(b) a Carnot engine operating in the same temperature range.

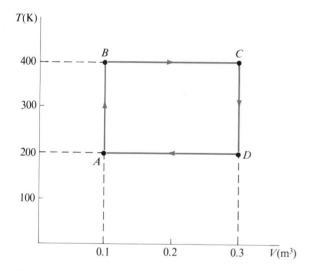

Figure 10.27   Problem 10.59

10.60 Three cylindrical metal conductors, each with a length d = 15 cm and cross-sectional area A = 3.0 cm^2, are arranged, as shown in Figure 10.28, between an HTR (T_H = 100° C) and an LTR (T_L = 0° C).

(a) Determine the temperatures, T_1 and T_2, of the two interfaces.

(b) Calculate the rate at which heat is conducted into the LTR.

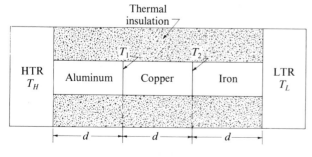

Figure 10.28   Problem 10.60

10.61 Two slabs of thermal conductors are placed betweeen an LTR and an HTR (*see* Fig. 10.29). Derive an expression for

(a) the temperature of the interface between the conductors.

(b) the heat per unit area per unit time that is conducted to the LTR.

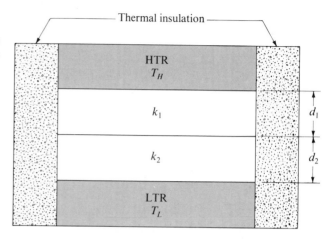

Figure 10.29   Problem 10.61

10.62 Figure 10.30 shows (a) parallel and (b) series combinations of equal volumes of two thermal conductors (k_1, k_2) placed between an HTR (temperature = T_H) and LTR (temperature = T_L). In the series combination the cross-sectional area of each slab is denoted by A and the thickness of each slab by d.

(a) Derive an expression for the rate R_p at which heat is conducted into the LTR in the parallel configuration.

(b) Repeat part (a) for R_s, the series rate.

(c) If medium 1 is air, medium 2 is concrete, d = 2.0 cm, A = 25 cm^2, and $T_H - T_L$ = 30 K, evaluate R_p and R_s. Which is the more effective insulating geometry?

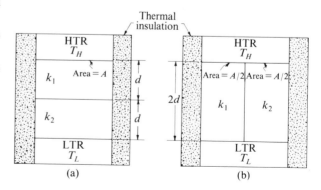

Figure 10.30   Problem 10.62

10.63 An ideal gas (n = 1.6 mol, C_v = 4.8 cal/K) is the working substance in a heat engine using the cycle in Figure 10.31. Calculate the thermal efficiency of this heat engine.

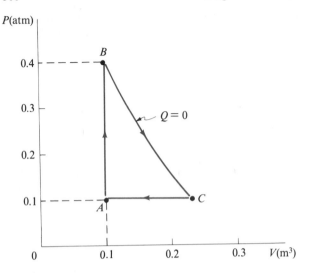

Figure 10.31   Problem 10.63

10.64   If the heat engine of Problem 10.63 is operated as a refrigerator, determine the performance coefficient.

10.65   An ideal gas ($n = 6.0$ mol, $\gamma = 1.4$) is the working substance for a heat engine executing the cycle shown in Figure 10.32.

(a)   Calculate the thermal efficiency for this engine.

(b)   Determine the thermal efficiency for a Carnot engine operating over the same temperature range.

(c)   If the cycle is reversed, determine the performance coefficient of the resulting refrigerator.

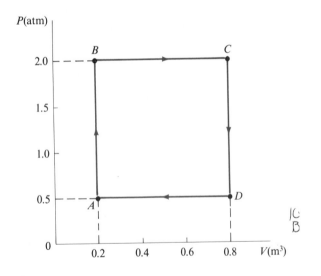

Figure 10.32   Problem 10.65

10.66   An ideal gas ($C_v = 1.5\ nR$) is the working substance for a heat engine operating in the three-process cycle:

Process AB: Isochoric pressure increase, $(P_1, V_1, T_1) \rightarrow (4P_1, V_1, T_2)$

Process BC: Adiabatic expansion, $(4P_1, V_1, T_2) \rightarrow (P_2, V_2, T_1)$

Process CA: Isothermal compression, $(P_2, V_2, T_1) \rightarrow (P_1, V_1, T_1)$

(a)   Determine the thermal efficiency of this engine.

(b)   Calculate the thermal efficiency of a Carnot engine operating between T_1 and T_2.

10.67   The Otto cycle (*see* Fig. 10.33) is an idealized cycle that is a model for a gasoline engine. Briefly, the Otto cycle, and its real counterpart, are:

Process AB: Adiabatic compression (compression stroke)

Process BC: Isochoric heat injection (ignition of gasoline vapor)

Process CD: Adiabatic expansion (power stroke)

Process DA: Isochoric heat ejection (exhaust)

(a)   If the working substance for this cycle is an ideal gas, show that the thermal efficiency for this heat engine is given by

$$\epsilon = 1 - \left(\frac{1}{r}\right)^{\gamma-1}$$

where the compression ratio $r = V_2/V_1$ and $\gamma = C_p/C_v$.

(b)   Evaluate ϵ for $\gamma = 1.4$ and $r = 8.5$.

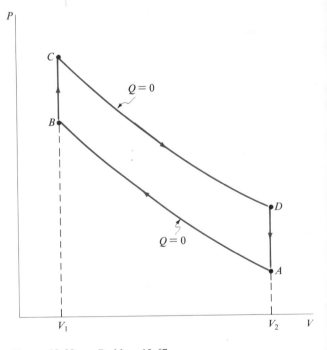

Figure 10.33   Problem 10.67

10.68   An ideal gas ($n = 1.2$ mol, $C_v = 3.6$ cal/K) is expanded quasi-statically such that its pressure and volume are related by $P = \alpha V$, where $\alpha = 0.20$ atm/m³. If the initial and final volumes of the gas are $V_i = 0.10$ m³ and $V_f = 0.30$ m³,

(a) sketch the process line in the P-V plane.

(b) determine the work by the gas during this expansion.

(c) calculate the heat added to the gas.

(d) what is the change of the entropy of the gas?

GROUP **B**

10.69 In the ideal-gas model for a low density gas, interactions between the particles (molecules) of the gas are neglected. Consequently, a gas molecule moves with a constant velocity except when it "bounces" off a wall of the containing vessel. We may approximate this bounce by assuming that the velocity component normal to the surface is reversed while the tangential component is not changed.

(a) If molecule i ($i = 1, 2, \ldots, N$) having a mass m and a velocity $\mathbf{v}_i = v_{xi}\hat{\mathbf{i}} + v_{yi}\hat{\mathbf{j}} + v_{zi}\hat{\mathbf{k}}$ collides with a wall parallel to the y-z plane, show that the change in momentum of the molecule is given by

$$\Delta \mathbf{p}_i = -2\, m\, v_{xi}\hat{\mathbf{i}}$$

(b) If the containing vessel is a cube (each edge = L), show that the time Δt between successive collisions of molecule i with the wall of part (a) is given by

$$\Delta t = \frac{2L}{|v_{xi}|}$$

(c) Show that the average force on the wall by molecule i has a magnitude given by

$$\langle F_i \rangle = \frac{mv_{xi}^2}{L}$$

(d) Show that the pressure is given by

$$P = \frac{N}{V} \cdot m \, \langle v_x^2 \rangle$$

where V is the volume of the container and $\langle v_x^2 \rangle$ is the average value of v_x^2, i.e.,

$$\langle v_x^2 \rangle = \frac{1}{N} \sum_{i=1}^{N} v_{xi}^2$$

(e) Assuming the velocities of the molecules to be randomly distributed, show that

$$PV = \frac{2}{3}K$$

where K is the total translational kinetic energy of the gas.

(f) Show that the total translational kinetic energy K of an ideal gas is related to the temperature T of the gas by

$$K = \frac{3}{2}nRT$$

where n, the number of moles of the gas, is given by

$$n = \frac{N}{N_A}$$

in which Avogadro's number, N_A, is equal to 6.0×10^{23}.

10.70

(a) Show that the average translational kinetic energy per molecule of an ideal gas is equal to 3 $kT/2$, where the *Boltzmann constant k* is given by

$$k = \frac{R}{N_A} = \frac{8.3}{6.0 \times 10^{23}} = 1.4 \times 10^{-23} \frac{\text{J}}{\text{molecule} \cdot \text{K}}$$

(b) Calculate the average kinetic energy per ideal gas molecule at a temperature of 300 K.

(c) The *root-mean-square-speed* v_{rms} of a particle is defined by $v_{rms} = \sqrt{\langle v^2 \rangle}$. Determine this speed for a helium (gas) molecule at $T = 300$ K.

10.71 The internal energy U of a monatomic ideal gas is equal to the translational kinetic energy of the gas molecules (atoms). For n moles of an ideal gas, demonstrate that (a) $U = 3nRT/2$; (b) $C_v = 3nR/2$; (c) $C_p = 5nR/2$; and (d) $\gamma = 5/3$.

10.72 The *equipartition of energy theorem* asserts that the internal energy U of a simple thermodynamic system is given by

$$U = \frac{N}{2}nRT$$

where N is the number (per molecule) of independent modes of energy absorption by the system and n is the number of moles of the substance constituting the system. For example, the kinetic energy of a monatomic gas atom is purely translational and given by

$$K_{tran.} = \frac{1}{2}mv_x^2 + \frac{1}{2}mv_y^2 + \frac{1}{2}mv_z^2$$

And this atom has three independent modes of absorbing energy, that is, motion in the x direction, motion in the y direction, and motion in the z direction. Thus, for a monatomic gas, we have $N = 3$.

(a) Show that the equipartition of energy theorem predicts results that agree with those of Problem 10.71.

(b) A molecule of a diatomic gas has two additional modes of absorbing energy, namely, rotational kinetic energy about each of the two mutually perpendicular axes that are through the center of mass of the molecule and perpendicular to the axis of the molecule. That is, if a set of coordinate axes are constructed with the origin at the center of mass and with the x axis through the two atoms, the rotational kinetic energy of such a molecule is given by

$$K_{rot} = \frac{1}{2}I_y\,\omega_y^2 + \frac{1}{2}I_z\,\omega_z^2$$

where I_y and I_z are the moments of inertia of the molecule about the y- and z-axes. Determine N, $U(T)$, C_v, C_p and γ for a diatomic gas.

(c) Experiments show that $\gamma \approx 4/3$ for a gas consisting of molecules having three or more atoms not all lying along the same axis. Can you explain this result?

10.73 The *Dulong-Petit rule* (an empirical result of the early nineteenth century) states that the heat capacity at constant volume of most solids is given by $C_v = 3\,n\,R$, where n is the number of moles of the material and R is the universal gas constant.

(a) Can you offer an explanation for this empirical result? (*Hint:* Use the equipartition of energy theorem of Problem 10.72 and a model for a solid which (i) has the atoms located at the vertices of a cubic lattice and (ii) approximates the interaction of each atom with its six nearest neighbors as though each pair were connected by an ideal spring.)

(b) Use the Dulong-Petit rule to estimate the specific heat of aluminum (molecular weight = 27 g/mol) and compare this result to the measured value of 0.21 cal/(g·K).

10.74 The ideal-gas model neglects any interactions of the constituent molecules of a gas. Even inert-gas atoms, because they have a small, but finite, volume, will suffer elastic "billiard-ball" collisions. Suppose an atom (diameter = d) suffers such a collision at $t = 0$ (*see* Fig. 10.34). Until the next collision this atom will move with a constant velocity **v**. This next collision will occur when the center of another atom lies within the shaded volume in the figure, namely, the volume of a cylinder having a diameter equal to $2d$ and a length equal to the distance vt the atom has traveled since the last collision. On the average, we expect this should happen when the swept-out volume, $vt\pi d^2$, is equal to the volume V/N (V = container volume, N = number of atoms) available per atom, i.e.,

$$vt\pi d^2 = \frac{V}{N}$$

Thus, the *average time between successive collisions* is estimated to be

$$\langle t \rangle = \frac{V}{vN\pi d^2}$$

and the average distance between successive collisions, a distance defined to be the *mean free path* $\langle \ell \rangle$, is approximately

$$\langle \ell \rangle = v\langle t \rangle = \frac{V}{N\pi d^2}$$

Estimate the mean free path and the number of collisions per second for an atom of a helium gas at STP ($P = 1.0$ atm and $T = 273$ K). Let d equal 1.0×10^{-10} m for a helium atom.

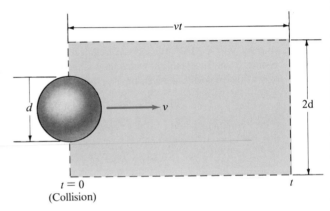

Figure 10.34   Problem 10.74

11 Electric Charge and Electric Fields

This chapter begins our study of electric and magnetic phenomena, collectively called *electromagnetism*. Electromagnetic forces, like gravitational forces, are fundamental interactions that occur between particles. Just as mass is that property of matter that engenders gravitational attraction, *electric charge* is a property of certain fundamental particles, like electrons and protons, that causes electromagnetic interaction. Our investigation of *electrostatics*, the analysis of interactions between charges at rest, is based on a force law for charges and on the concept of electric fields. The force law for charges, called Coulomb's law, is the electrical analogue of the law of universal gravitation. The electric field is an abstract concept used to characterize how the presence of charge alters the properties of the space around that charge.

11.1 Electric Charge and Coulomb's Law

Electric charges exist in two forms, called *positive charge* and *negative charge*. The natural unit of charge is the positive charge of a proton or the negative charge of an electron. The SI unit of charge is the coulomb (C), which will be defined later in terms of electric current. This unusual definition is used because of the difficulty of making reproducible measurements on quantities of charge. For now, we will use as a measure of charge the experimentally obtained value for e, the magnitude of the *charge of an electron or proton*:

$$e = 1.60 \times 10^{-19} \text{ C} \tag{11-1}$$

Equivalently, a coulomb is the quantity of charge carried by 6.25×10^{18} electrons.

An electron (or proton) bears the smallest unit of charge found in nature, so any quantity of charge q occurs as an integral number n of the basic charge e, or $q = ne$. Thus, as with any physical quantity made up of a number of "minimum parcels," we say that electric charge is *quantized;* the least quantity that occurs, the *quantum* of charge, is the electronic charge e. Because the quantum of charge is so small, we frequently treat charge as if it were available in continuous quantities.

E 11.1   A typical charge on a macroscopic object may be of the order of 10^{-9} C. Calculate the number of excess electronic charges that corresponds to a charge of 1.0×10^{-9} C.   Answer: 6.25×10^9

E 11.2   How many electronic charges must be transferred to or from an object having a charge of 2.0×10^{-9} C if this charge is to be changed by 0.1 percent?   Answer: 1.25×10^7

E 11.3   If only one electronic charge is transferred to or from an object having a charge of 2.0×10^{-9} C, what is the fractional change in the charge on the object?   Answer: 8.0×10^{-11}

Matter is said to be *neutral* if it bears no *net* charge. Of course, a material is composed of atoms containing many electrons and protons, but a net charge occurs only if there is an imbalance in the numbers of positive and negative charges. In general, electrons are the more mobile particles. Therefore, when negative charges (electrons) are added to a neutral object, we say that object is negatively charged; when electrons are removed from a neutral object, we say that object is positively charged. When a transfer of charge takes place between two otherwise isolated objects, the net negative charge added to one object is equal in magnitude to the net positive charge remaining on the other object. Thus, the total charge on the two objects remains constant; in other words, the total charge of the two objects is conserved.

Contact between different materials may cause a transfer of electrons between the materials, resulting in a net charge on each material when they are separated. Rubbing materials together serves to increase the microscopic regions that make contact and thereby increases the net charge transferred. The time-honored technique for producing small quantities of charge that have a known sign is rubbing a rubber rod with fur to produce a net negative charge on the rod. Similarly, rubbing a glass rod with silk produces a net positive charge on the rod.

Electric charges (electrons) are relatively free to move about in materials called (electrical) *conductors*. In general, metals are good conductors. In other materials, called (electrical) *insulators*, charges cannot move about freely. Nonmetals are, in general, good insulators (or poor conductors). Frequently it will be convenient to idealize these properties and refer to "perfect" insulators or conductors.

A traditional device used for the demonstration and examination of the effects of static charges is the pith ball, a small sphere of very light material, usually coated with a thin layer of metal to provide a conducting surface. Using pith balls, we can deduce from direct observation a number of qualitative facts about the nature of electric charge. For instance, Figure 11.1(a) shows two neutral pith balls, each suspended by a thread. If each ball is charged by placing it in contact with a positively charged rod, the balls exert forces on each other that push the balls apart, as shown in Figure 11.1(b). If both balls are charged negatively, as shown in Figure 11.1(c), they again repel each other. If one ball is charged positively and the other is charged negatively, as in Figure 11.1(d), however, they attract each other. We may conclude: Like charges repel; unlike charges attract.

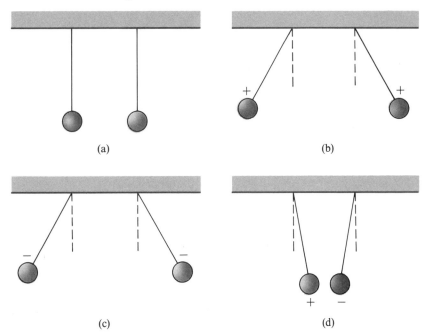

Figure 11.1
(a) Two neutral pith balls. **(b)** Positively charged pith balls. **(c)** Two negatively charged pith balls. **(d)** Pith balls with charges of unlike signs.

Induction

We can produce a net charge on an object without actually touching it with another charged object. Consider a neutral metal-coated sphere that rests on an insulating stand, as indicated in Figure 11.2(a). Suppose a positively charged rod is brought near the sphere, but is not permitted to touch it. Negative charges on the sphere are attracted toward the positively charged rod; positive charges are repelled, as shown in Figure 11.2(b). In Figure 11.2(c) the side of the sphere away from the rod is connected through a conducting wire to the earth. The earth acts as a reservoir of charge, which means that it can supply or receive an arbitrary quantity of charge without being measurably changed. We say that an object connected to the earth through a conducting material is *grounded*. The symbol labeled "ground" in Figure 11.2(b) is used conventionally to indicate a connection to the ground. In this case, the ground supplies negative charges to neutralize the positive charges on the right side of the sphere in Figure 11.2(c), and in Figure 11.2(d) the ground connection has been removed. When the positively charged rod is removed, as in Figure 11.2(e), the sphere has a net negative charge. The sphere is then said to have been charged by *induction*. The same charged rod could be used to induce a similar charge on an arbitrary number of conducting objects, because no charge has been transferred from the rod.

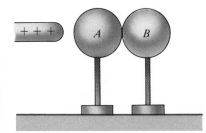

Exercise 11.4

E 11.4   Two identical conducting spheres on insulating stands are initially uncharged and are touching. A positively charged rod is placed near one of the spheres, as shown in the figure. While the charged rod is in place, the spheres are separated. If the charged rod is then removed from the vicinity, what is known about the charge on each sphere?   Answer: $q_A = -q_B > 0$

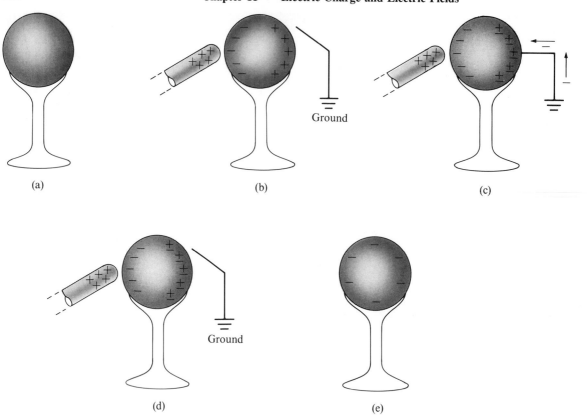

Figure 11.2
(a) A metal-coated sphere resting on an insulating stand. (b) Negative charges on the sphere are attracted to a positively charged rod; positive charges are repelled. (c) Positive charges on the sphere are neutralized by negative charges that flow onto the sphere through a connection from ground. (d) The ground is disconnected from the sphere. (e) The sphere is negatively charged by induction.

E 11.5   You are given two identical spheres on insulating stands. One sphere carries a positive charge, and the other is uncharged. How can you manipulate the two spheres so that when you have finished, they carry equal positive charges?

A quantitative relationship for the interaction forces between electric charges was deduced in the eighteenth century by Charles Coulomb:

The force on each of two point charges of measure q_1 and q_2 lies along the straight line passing through the charges. The magnitude of the force on each charge is proportional to the product $q_1 q_2$ and is inversely proportional to the square of the distance r between them.

Called Coulomb's law, this principle may be expressed mathematically by

$$F = k \frac{|q_1 q_2|}{r^2} \tag{11-2}$$

where the constant of proportionality k has been measured to be

$$k = 9.0 \times 10^9 \frac{N \cdot m^2}{C^2} \tag{11-3}$$

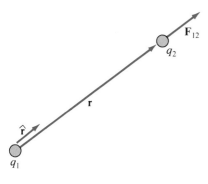

Figure 11.3
Illustrating the force $\mathbf{F}_{12}$ exerted on charge q_2 by the charge q_1.

when force is measured in newtons (N), charge is measured in columbs (C), and distance is measured in meters (m). If we let the force exerted by q_1 on q_2 be $\mathbf{F}_{12}$ and the force exerted by q_2 on q_1 be $\mathbf{F}_{21}$, Newton's third law requires that

$$\mathbf{F}_{12} = -\mathbf{F}_{21} \tag{11-4}$$

The directions of the forces, whether attractive or repulsive, are determined by the signs of the two charges.

Because it will be convenient in later developments of electromagnetism, we relate the constant k to another constant ϵ_o:

$$k = \frac{1}{4\pi\epsilon_o} \tag{11-5}$$

The constant ϵ_o is called the *permittivity of free space* and has the value

$$\epsilon_o = 8.85 \times 10^{-12} \frac{C^2}{N \cdot m^2} \tag{11-6}$$

E 11.6   Any realistic electrostatic charge is always many orders of magnitude smaller than 1.0 C. To see that 1.0 C is an unrealistically large unit for static charge, calculate the magnitude of the force on a 1.0-C charge by an identical charge 1.0 m away.   Answer: 9.0×10^9N

E 11.7   The magnitude of the force on either of two identical charged particles separated by 10 cm is 1.0 N. What is the magnitude of either charge?
                                                                            Answer: 1.1 μC

Coulomb's law may be written using vector notation as

$$\mathbf{F}_{12} = \frac{kq_1q_2}{r^2}\frac{\mathbf{r}}{r} = \frac{kq_1q_2}{r^2}\hat{\mathbf{r}} \tag{11-7}$$

where $\mathbf{F}_{12}$ is the force exerted by q_1 on q_2, $\mathbf{r}$ is the position vector from q_1 (the charge exerting the force) to q_2 (the charge on which the force acts), and $\hat{\mathbf{r}}$ is a unit vector having the same direction as $\mathbf{r}$, that is, $\hat{\mathbf{r}} = \mathbf{r}/r$. Each of these vectors is shown in Figure 11.3.

Example 11.1   **PROBLEM**   Charges q_1 (500 μC) and q_2 (100 μC) are located on the xy plane at the positions $\mathbf{r}_1 = 3.0\hat{\mathbf{j}}$ m and $\mathbf{r}_2 = 4.0\hat{\mathbf{i}}$ m. Use Equation 11-7 to calculate the electrostatic force exerted on q_2.

SOLUTION   The position vector from q_1 to q_2 is, as shown in Figure 11.4,

$$\mathbf{r} = (4\hat{\mathbf{i}} - 3\hat{\mathbf{j}}) \text{ m}$$

and the unit vector $\hat{\mathbf{r}}$ having the same direction as $\mathbf{r}$ and pointing from q_1 to q_2, is

$$\hat{\mathbf{r}} = \frac{\mathbf{r}}{r} = \frac{4\hat{\mathbf{i}} - 3\hat{\mathbf{j}}}{5} = 0.8\hat{\mathbf{i}} - 0.6\hat{\mathbf{j}}$$

Thus, the force $\mathbf{F}_{12}$ exerted on q_2 by q_1 is

$$\mathbf{F}_{12} = \frac{kq_1q_2}{r^2}\hat{\mathbf{r}}$$

$$\mathbf{F}_{12} = \frac{9 \times 10^9 \times 5 \times 10^{-4} \times 1 \times 10^{-4}}{25}(0.8\hat{\mathbf{i}} - 0.6\hat{\mathbf{j}})$$

$$\mathbf{F}_{12} = 18(0.8\hat{\mathbf{i}} - 0.6\hat{\mathbf{j}})\text{N} = (14\hat{\mathbf{i}} - 11\hat{\mathbf{j}})\text{N} = 18 \text{ N} \underline{/-37°} \quad \blacksquare$$

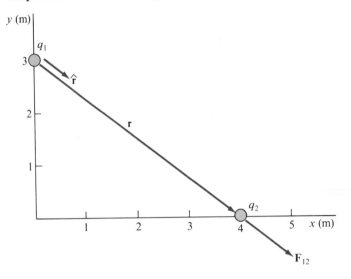

Figure 11.4   Example 11.1

Both electrical and gravitational forces may act on charged objects. Let us consider the relative magnitudes of electrical and gravitational forces on a charged particle.

Compare the magnitudes of the Coulomb force F_e and the gravitational force F_g exerted on either of two protons ($m_p = 1.67 \times 10^{-27}$ kg) that are separated by a distance of 1.0 cm.

The magnitude of the repulsive electrical force acting on either proton ($q = e = 1.6 \times 10^{-19}$ C) is given by

$$F_e = k\frac{e^2}{r^2} = \frac{9.0 \times 10^9 \times (1.6 \times 10^{-19})^2}{(10^{-2})^2} = 2.30 \times 10^{-24} \text{ N}$$

The magnitude of the attractive gravitational force is

$$F_q = G\frac{m_p^2}{r^2} = \frac{6.67 \times 10^{-11} \times (1.67 \times 10^{-27})^2}{(10^{-2})^2} = 1.86 \times 10^{-60} \text{ N}$$

The ratio F_e/F_g is 1.2×10^{36}, a value that indicates that, for atomic particles, the gravitational force is negligible compared to the electrical force.

E 11.8   Compare the magnitudes of the electrical and gravitational forces exerted on an object (mass = 10 g, charge = 2.0 μC) by an identical object that is placed 10 cm from the first.   Answer: $F_e/F_g = 5.4 \times 10^{12}$

It is frequently necessary to calculate the net force on a charge caused by a number of other point charges fixed in space. Let $\mathbf{F}_1$, $\mathbf{F}_2$, ... $\mathbf{F}_n$ be the forces exerted on a charge q by each of n fixed point charges $Q_1, Q_2, \ldots Q_n$. It has been verified experimentally that the *net* force $\mathbf{F}$ on q may be obtained by calculating the force on q caused by each fixed charge (using Coulomb's law) as if it were the only charge exerting a force on q and summing all of these forces *vectorially*. Thus,

$$\mathbf{F} = \mathbf{F}_1 + \mathbf{F}_2 + \ldots + \mathbf{F}_n = \sum_{i=1}^{n} \mathbf{F}_i \tag{11-8}$$

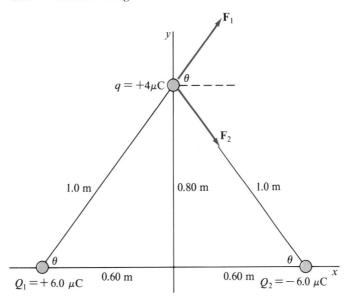

Figure 11.5
An arrangement of fixed charges. The forces $\mathbf{F}_1$ and $\mathbf{F}_2$ are the forces exerted on q by Q_1 and Q_2.

is an expression of the *principle of superposition* for electrostatic forces. In other words, the forces on a charge q caused by a number of individual charges may be superposed (placed over each other, that is, they act at the same point) and added vectorially to find the total force.

Figure 11.5 shows two fixed charges, $Q_1 = 6.0\mu C$ and $Q_2 = -6.0\mu C$, and a third charge $q = 4.0\mu C$. Calculate the net electrostatic force on the charge q.

The electrostatic forces $\mathbf{F}_1$ and $\mathbf{F}_2$ exerted on q by Q_1 and Q_2 are shown in Figure 11.5. The directions of $\mathbf{F}_1$ and $\mathbf{F}_2$ are in accordance with the rules for attraction and repulsion of positive and negative charges. Because of the symmetry of the charge distribution, the magnitudes of the forces $\mathbf{F}_1$ and $\mathbf{F}_2$ are equal, and their values are specified by Coulomb's law:

$$F_2 = F_1 = k\frac{Q_1 q}{r_1^2} = \frac{(9 \times 10^9)(6 \times 10^{-6})(4 \times 10^{-6})}{(1)^2} = 0.216 \text{ N}$$

The net force $\mathbf{F}$ on q is the sum of the vectors $\mathbf{F}_1$ and $\mathbf{F}_2$. The y components of $\mathbf{F}_1$ and $\mathbf{F}_2$ sum to zero, and their x components add to give

$$F_x = F_{1x} + F_{2x} = F_1 \cos\theta + F_2 \cos\theta = 2(0.216)(0.6) = 0.259 \text{ N}$$

Thus the net force $\mathbf{F}$ on the charge q is

$$\mathbf{F} = 0.26\hat{\mathbf{i}} \text{ N} \qquad\qquad (11\text{-}9)$$

E 11.9   Calculate the net electrostatic force on the charge q of Figure 11.5 if $Q_1 = Q_2 = 6.0 \ \mu C$.   Answer: $0.35\hat{\mathbf{j}}$ N

Example 11.2   **PROBLEM**   Charges Q_1, Q_2, and q are positioned as shown in Figure 11.6. Express the total electrostatic force acting on q in terms of Q_1, Q_2, q, r, d, and constants as needed. Show that this result reduces appropriately for the distance r much greater than the distance d, that is, for $r \gg d$.

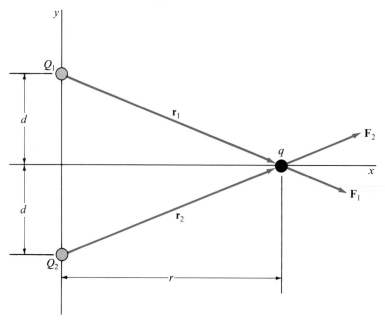

Figure 11.6   Example 11.2

SOLUTION   The force $\mathbf{F}$ acting on q is the vector sum of the forces exerted on q by Q_1 and Q_2. These two forces are given by

$$\mathbf{F}_1 = \frac{kqQ_1}{r_1^2}\frac{\mathbf{r}_1}{r_1} = \frac{kqQ_1}{(r^2 + d^2)^{3/2}}\,(r\hat{\mathbf{i}} - d\hat{\mathbf{j}})$$

$$\mathbf{F}_2 = \frac{kqQ_2}{r_2^2}\frac{\mathbf{r}_2}{r_2} = \frac{kqQ_2}{(r^2 + d^2)^{3/2}}\,(r\hat{\mathbf{i}} + d\hat{\mathbf{j}})$$

where $\mathbf{r}_1(= r\hat{\mathbf{i}} - d\hat{\mathbf{j}})$ and $\mathbf{r}_2(= r\hat{\mathbf{i}} + d\hat{\mathbf{j}})$ are the position vectors shown in Figure 11.6. Adding these two forces to obtain $\mathbf{F}$ gives

$$\mathbf{F} = \mathbf{F}_1 + \mathbf{F}_2 = \frac{kq}{(r^2 + d^2)^{3/2}}\,[(Q_1 + Q_2)\,r\hat{\mathbf{i}} + (Q_2 - Q_1)d\hat{\mathbf{j}}]$$

If $Q_1 + Q_2 \neq 0$, the limiting $(r \gg d)$ behavior of $\mathbf{F}$ is

$$\mathbf{F} \xrightarrow[r \gg d]{} \frac{kq(Q_1 + Q_2)}{r^2}\hat{\mathbf{i}}$$

Thus if the distance r is large compared to the distance $2d$ separating Q_1 and Q_2, the force exerted on q by the (two-charge) distribution is essentially the same as that of a point charge of the same net charge $(Q_1 + Q_2)$. If $Q_1 + Q_2 = 0$, these conclusions are not appropriate. The following example considers this important case. ■

Example 11.3

PROBLEM   An *electric dipole* is a charge distribution consisting of a positive charge Q and a negative charge $-Q$. In most applications the distance separating the two charges is usually small compared to the distances from the dipole to other charges. Figure 11.7 shows such a dipole. Calculate the force on a third charge q positioned as shown. Determine the limiting $(r \gg d)$ behavior of this force.

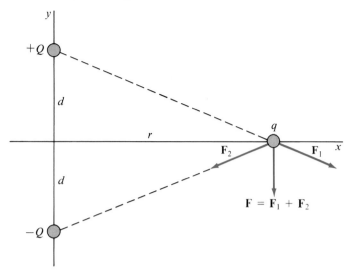

Figure 11.7   Example 11.3

SOLUTION   If we set $Q_1 = Q$ and $Q_2 = -Q$, the result of Example 11.2 for the force on q simplifies to

$$\mathbf{F} = -\frac{2kqQd}{(r^2 + d^2)^{3/2}}\hat{\mathbf{j}}$$

For $r \gg d$ this reduces to

$$\mathbf{F} \xrightarrow[r \gg d]{} -\frac{kqp}{r^3}\hat{\mathbf{j}}$$

where the *electric dipole moment* $\mathbf{p}$ of the distribution is defined as the product of the charge Q and the displacement $2d\hat{\mathbf{j}}$ separating the two charges, or $p = 2Qd$. Interestingly, the force on q is *not* in the direction from the dipole to the charge, but rather is perpendicular to this direction. Furthermore, the magnitude of the force decreases like the cube, rather than the square, of the distance separating the dipole and the charge. In summary, even though a localized charge distribution has a net charge of zero, this distribution may still exert a force on another charge placed at a relatively large distance away. This force, however, decreases much more rapidly with increasing distance than does the usual coulomb force between two point charges.   ■

Calculation of the force exerted on a point charge q by charge distributed continuously throughout a region of space involves a special problem-solving technique. Because the quantum of charge is small, we can usually treat charge distributions as if they were continuous. For example, if a quantity of charge Q is uniformly distributed along a line of length L, we may treat the line as if it has a uniform *linear charge density* $\lambda = Q/L$. Similarly, a surface area A over which a charge Q is uniformly distributed may be treated as if the surface has a uniform *surface charge density* $\sigma = Q/A$. And if a charge Q is distributed uniformly throughout a volume V, we may treat that volume as if it has a uniform volume charge density $\rho = Q/V$.

Figure 11.8 represents a positive charge q that experiences a force because of the presence of a continuous positive charge distribution labeled Q. Each infinitesimal element of charge dQ within Q may be treated as if it were a point

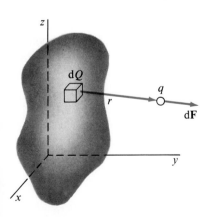

Figure 11.8
The force $d\mathbf{F}$ exerted on a positive charge q by the infinitesimal element of charge dQ within a distribution having a total positive charge Q.

charge that exerts an infinitesimal force dF on q. Then, according to Coulomb's law, the force dF is given by

$$dF = k\frac{qdQ}{r^2}\hat{r} \tag{11-10}$$

where r is the distance between the element dQ and the point charge q and $\hat{r}$ is a unit vector in the direction from dQ toward q. The principle of superposition assures that integration, which is a summation process, will provide the total force $\mathbf{F}$ acting on q, or

$$\mathbf{F} = \int_Q d\mathbf{F} \tag{11-11}$$

in which the subscript Q indicates that the integration is performed over the entire charge distribution. The actual integration process is perhaps best illustrated by specific examples.

A semi-infinite line of positive charge extending from $x = 0$ along the x axis has a uniform linear charge density λ. Determine the force that this line charge exerts on a point charge q located at the point $(0, D)$ in the x-y plane, as shown in Figure 11.9.

Figure 11.9 shows the force dF on q exerted by the infinitesimal element of charge dQ, which lies a distance x from the origin. If r is the distance from dQ to q, Coulomb's law expresses the magnitude of the force dF:

$$dF = k\frac{qdQ}{r^2} \tag{11-12}$$

Of course, the force on q is away from dQ because q is positive. Because λ is the charge per unit length along the line, or $\lambda = dQ/dx$, we may express dQ as $\lambda\,dx$ and write the components dF_x and dF_y of the force dF as

$$dF_x = -dF\,\sin\theta = -\frac{kq\lambda\,dx\,\sin\theta}{r^2} \tag{11-13}$$

$$dF_y = dF\,\cos\theta = \frac{kq\lambda\,dx\,\cos\theta}{r^2} \tag{11-14}$$

where $dF = kq\lambda\,dx/r^2$ and θ is the angle shown in Figure 11.9. The x component of the total force on q is obtained by integrating Equation (11-13) over the entire charge distribution:

$$F_x = -kq\lambda \int_0^\infty \frac{\sin\theta\,dx}{r^2} \tag{11-15}$$

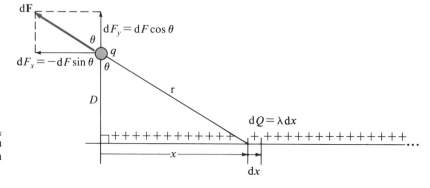

Figure 11.9
The force dF, and its components dF_x and dF_y, exerted on a charge q by an infinitesimal charge element dQ within a semi-infinite line charge distribution.

The integrand of Equation (11-15) contains three variables, which must be related so that we may express the integrand in terms of a single variable. Using the geometry of Figure 11.9, we may express dx and r in terms of the single variable θ and the constant D by

$$x = D \tan\theta$$

$$dx = D \sec^2\theta \, d\theta$$

$$r = D \sec\theta$$

$$r^2 = D^2 \sec^2\theta$$

which may be substituted into Equation (11-15) to give

$$F_x = -kq\lambda \int_0^{\pi/2} \frac{\sin\theta \, D \, \sec^2\theta \, d\theta}{D^2 \sec^2\theta} = -\frac{kq\lambda}{D} \int_0^{\pi/2} \sin\theta \, d\theta$$

$$F_x = -\frac{kq\lambda}{D} \qquad \qquad \text{(11-16)}$$

Notice that the limits on the integral from $x = 0$ to $x = \infty$ correspond to values of θ from $\theta = 0$ to $\theta = \pi/2$. A similar calculation for F_y gives

$$F_y = \frac{kq\lambda}{D} \int_0^{\pi/2} \cos\theta \, d\theta = \frac{kq\lambda}{D} \qquad \qquad \text{(11-17)}$$

The resultant force $\mathbf{F}$ on the charge q is equal to $F_x \hat{\mathbf{i}} + F_y \hat{\mathbf{j}}$, or

$$\mathbf{F} = \frac{kq\lambda}{D}(-\hat{\mathbf{i}} + \hat{\mathbf{j}}) = \sqrt{2}\,\frac{kq\lambda}{D} \; \underline{/135°} \qquad \qquad \text{(11-18)}$$

Example 11.4

PROBLEM   Charge of uniform linear density λ is distributed along the x axis from $x = 0$ to $x = L$, as shown in Figure 11.10. Calculate the force this line charge exerts on a point charge q positioned on the x axis at the point $x = R$, where $R > L$.

SOLUTION   The force $d\mathbf{F}$ exerted on q by dQ, an infinitesimal element of charge positioned a distance x from the origin, has a magnitude

$$dF = \frac{kq\,dQ}{r^2}$$

where r is the distance from dQ to q. Because $dQ = \lambda \, dx$, dF may be expressed as

$$dF = \frac{kq\lambda\,dx}{r^2}$$

In this case $d\mathbf{F}$ has no y component, and the x component of $d\mathbf{F}$ is then

$$dF_x = dF = \frac{kq\lambda\,dx}{r^2}$$

Because $r = R - x$ (*see* Fig. 11.10), it follows that $dx = -dr$. With this substitution, the expression for F_x becomes

$$F_x = kq\lambda \int_{x=0}^{x=L} \frac{dx}{r^2} = kq\lambda \int_{r=R}^{r=R-L} \frac{-dr}{r^2} = kq\lambda \left[\frac{1}{r}\right]_{r=R}^{r=R-L}$$

$$F_x = \frac{kq\lambda L}{R(R - L)} \qquad \blacksquare$$

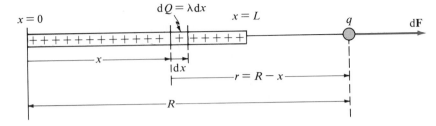

Figure 11.10   Example 11.4

Finally, the force exerted on q is

$$\mathbf{F} = \frac{kq\lambda L}{R(R - L)}\hat{\mathbf{i}}$$ ∎

E 11.10   Realizing that the total distributed charge Q in Example 11.4 is given by $Q = \lambda L$, show that the expression (obtained in Example 11.4) for the force **F** exerted on q by Q behaves appropriately for the limiting case in which $R \gg L$.   Answer: $\mathbf{F} \xrightarrow[R \gg L]{} (kqQ/R^2)\hat{\mathbf{i}}$

The following problem illustrates how this general problem-solving procedure is employed for another geometry.

A positive charge Q is distributed uniformly along a circular ring of radius R. What force does Q exert on a positive point charge q located on the axis of the ring at a distance d from the plane of the ring?

Figure 11.11 shows the geometry of this problem. The center of the ring of charge is placed at the origin, and q lies on the x axis. An infinitesimal element of arc length ds on the ring bears an element of charge $dQ = \lambda\, ds$, where the linear charge density of the ring is $\lambda = (Q/2\pi R)$. Thus the element of charge dQ on the element ds is

$$dQ = \frac{Q\, ds}{2\pi R} \tag{11-19}$$

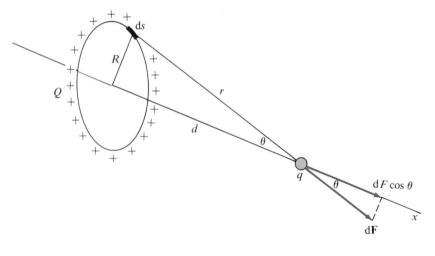

Figure 11.11
The force d**F** exerted on a charge q by an element of charge, which extends over an infinitesimal arc length ds.

The magnitude dF of the force exerted on q by the charge element dQ is given by Coulomb's law,

$$dF = k\frac{q\,dQ}{r^2} \tag{11-20}$$

where r is the distance from dQ to q. Because of the symmetry of the ring, the resultant force on q is in the x direction and has magnitude $F = \int dF\cos\theta$. The components of $d\mathbf{F}$ perpendicular to the x axis sum to zero because elements of equal charge on opposite sides of the ring produce component vectors of $d\mathbf{F}$ perpendicular to the x axis that sum to zero. Then the net force on q is

$$F = kq\int\frac{dQ\,\cos\theta}{r^2} = \frac{kqQ}{2\pi R}\int\frac{ds\,\cos\theta}{r^2} \tag{11-21}$$

where θ is shown in Figure 11.11. We see that $\cos\theta = d/r$ and $r^2 = R^2 + d^2$, so that Equation (11-21) becomes

$$F = \frac{kqQd}{2\pi R(R^2 + d^2)^{3/2}}\int ds \tag{11-22}$$

The integral of ds must extend over the entire charge distribution, the circumference of the ring, so that $\int ds = 2\pi R$. Then Equation (11-22) becomes

$$F = \frac{kqQd}{(R^2 + d^2)^{3/2}} \tag{11-23}$$

or

$$\mathbf{F} = \frac{kqQd}{(R^2 + d^2)^{3/2}}\,\hat{\mathbf{i}} \tag{11-24}$$

In the limiting case in which d becomes large compared to R, that is, when q is a great distance from the ring of charge, Equation (11-23) approaches

$$F = \frac{kqQ}{d^2} \tag{11-25}$$

Thus, at a great distance from q, the ring carrying a charge Q has the same effect as if it were a point charge Q exerting a force on q given by Coulomb's law for the force between point charges Q and q separated by a distance d.

E 11.11   Show that the force expression of Equation (11-23) behaves appropriately for the special case $d = 0$.

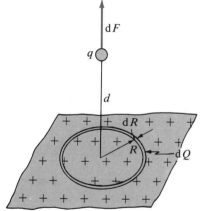

Figure 11.12   Example 11.5

Example 11.5

PROBLEM   Charge of uniform density $\sigma > 0$ is distributed over the entire x-y plane. Calculate the magnitude of the force exerted on a positive charge q located a distance d from this charged plane.

SOLUTION   Figure 11.12 shows the charge q and a charged ring of inner radius R and outer radius $R + dR$. If dQ is the charge on the ring, then according to Equation (11-23) the infinitesimal force exerted on q by this ring has a magnitude given by

$$dF = \frac{kqd\,dQ}{(R^2 + d^2)^{3/2}}$$

Because the force from any such ring is perpendicular to and away from the plane, the magnitude of the total force exerted on q is just the sum of the magnitudes from all rings, that is,

$$F = \int dF = \int \frac{kqd \; dQ}{(R^2 + d^2)^{3/2}}$$

Because the infinitesimal area dS of the ring is equal to $2\pi R \; dR$ (circumference times width), the charge dQ on the ring is given by

$$dQ = \sigma \; dS = 2\pi\sigma R \; dR$$

Thus the force exerted on q has a magnitude of

$$F = \int_0^\infty \frac{kqd \; 2\pi\sigma R \; dR}{(R^2 + d^2)^{3/2}} = \pi k\sigma qd \int_0^\infty \frac{2R \; dR}{(R^2 + d^2)^{3/2}}$$

The integral is evaluated by the substitution $u = R^2 + d^2$, for which $du = 2R \; dR$. Thus we have

$$\int_0^\infty \frac{2R \; dR}{(R^2 + d^2)^{3/2}} = \int_{d^2}^\infty \frac{du}{u^{3/2}} = \frac{u^{-1/2}}{-1/2}\Big|_{u = d^2}^{u = \infty} = \frac{2}{d}$$

Consequently, the magnitude of the force exerted on q is

$$F = \pi k\sigma qd \left(\frac{2}{d}\right) = 2\pi k\sigma q = \frac{\sigma q}{2\epsilon_o} \quad \blacksquare$$

E 11.12 How does the force exerted on q by the uniformly charged plane change if the distance from the plane to the charge is doubled?

11.2 Electric Field

A *field* is an abstraction that we use to describe a particular property of space. An *electric field* exists at a point in space if an electrical force acts on a charge placed at that point. We consider the electric field a property of space, and the electric field exists at points in space whether or not a charge is actually placed at those points to experience a force. Quantitatively, we define the electric field $\mathbf{E}$ at a point to be

$$\mathbf{E} = \frac{\mathbf{F}}{q} \tag{11-26}$$

where $\mathbf{F}$ is the electrical force that would be exerted on a charge q placed at that point. Thus, for a given electric field $\mathbf{E}$ at a point, the force $\mathbf{F}$ that is exerted on a particle with charge q is given by $\mathbf{F} = q\mathbf{E}$. The force on a charged particle at a point where the field is $\mathbf{E}$ is in the same direction as $\mathbf{E}$ if the charge is positive; the force is in a direction opposite to $\mathbf{E}$ if the charge is negative. The unit of electric field is that of force per unit charge, or N/C.

The force that appears in Equation (11-26) is a force of electrical origin. In static situations, only charges can exert forces on other charges, and therefore, an electric field exists at a point in space if there is charge somewhere that causes this field. Thus the presence of charge alters the electrical properties of the space surrounding that charge.

The "test charge" q used to measure **E** must be vanishingly small so that its presence does not alter the charge distribution that causes the field **E**. Therefore, we refine the definition of electric field in Equation (11-26):

$$\mathbf{E} = \lim_{q \to 0} \frac{\mathbf{F}}{q} \tag{11-27}$$

We may now observe that it is the presence of a charge that produces an electric field, which, in turn, can affect another charge. Then why bother with the concept of a field at all when Coulomb's law provides a description of the effect of one charge on another without any intermediary field? For one thing, electric fields are a convenience in many kinds of calculations that we shall encounter. The compelling reason for the concept of electric field, however, is that Coulomb's law tacitly assumes that the force on one charge by another is transmitted instantaneously across the intervening space between them. This assumption (which, in fairness to M. Coulomb, is not specifically addressed by Coulomb's law) is, in fact, incorrect. The effect of a slight movement of one charge is not transmitted instantaneously to another charge some distance from the first. We now know that the effect is transmitted at the "speed of light," which is, in fact, the speed of electromagnetic waves. The concept of fields will permit us to describe the transmission of the effects of the motion of a charge in terms of electromagnetic waves, a topic that will be considered in later chapters.

Let us now examine the electric field produced by a positive fixed point charge of magnitude Q. If another particle bearing a positive charge q is placed a distance r from Q, we know from Coulomb's law that the force exerted on the particle has magnitude

$$F = k \frac{Qq}{r^2} \tag{11-28}$$

Then the magnitude of the electric field at every point a distance r from Q is

$$E = \frac{F}{q} = k \frac{Qq}{qr^2} = k \frac{Q}{r^2} \tag{11-29}$$

Because the force **F** exerted on q by Q is repulsive, the direction of the electric field is radially *outward* from the *positive* charge Q. If the charge Q were *negative*, the force on the positive test charge q would be attractive, and the electric field would be directed radially *inward*—toward the negative charge—at every point around Q. Figure 11.13 shows a graphical representation of the

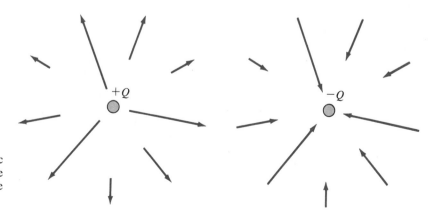

Figure 11.13
Graphical representations of the electric fields in the space around a positive charge $+Q$ and around a negative charge $-Q$.

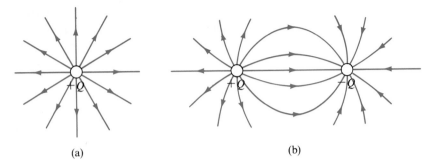

Figure 11.14
Conventional representation of electric fields (a) around a positive point charge and (b) around a pair of charges, one positive and one negative.

electric field in the space about a positive charge $+Q$ and about a negative charge $-Q$. Figure 11.14(a) shows a conventional representation of the electric field about a positive point charge. In such a representation, the radial lines are called *lines of force*. The arrows indicate the direction of the electric field, which is tangent to the lines at each point. In regions where the lines of force are closely spaced, the magnitude of the electric field is greater. Figure 11.14(b) shows the lines of force in the space surrounding two equal charges, one positive and one negative.

The electric field of a point charge Q may be expressed in vector notation. At a point a distance r from Q, the electric field is given by

$$\mathbf{E} = k\frac{Q}{r^2}\hat{\mathbf{r}} = \frac{1}{4\pi\epsilon_o}\frac{Q}{r^2}\hat{\mathbf{r}} \tag{11-30}$$

where $\hat{\mathbf{r}}$ is a unit vector in a direction radially outward from Q. The direction of $\mathbf{E}$ is radially outward when Q is positive and radially inward when Q is negative.

E 11.13   A point charge $Q = 4.0 \times 10^{-9}$ C is placed at the origin. Calculate the electric field at the position $\mathbf{r} = 3.0\hat{\mathbf{i}}$ m.   Answer: $4.0\hat{\mathbf{i}}$ N/C

E 11.14   A point charge $Q = -8.0 \times 10^{-8}$ C is placed at the origin. Calculate the electric field at a point 2.0 m from the origin on the positive z axis.
Answer: $-180\hat{\mathbf{k}}$ N/C

E 11.15   Determine the electric field at the position $\mathbf{r} = (3\hat{\mathbf{i}} + 4\hat{\mathbf{j}})$ m caused by a point charge $Q = 5.0 \times 10^{-8}$ C placed at the origin.
Answer: $(11\hat{\mathbf{i}} + 14\hat{\mathbf{j}})$ N/C

11.3 Motion of a Charged Particle in an Electric Field

A particle that has a charge q and is at a point where the electric field is $\mathbf{E}$ experiences a force $\mathbf{F}$ given by

$$\mathbf{F} = q\mathbf{E} \tag{11-31}$$

If that particle has mass m, it will experience an acceleration, which, according to Newton's second law, is

$$\mathbf{a} = \frac{\mathbf{F}}{m} = \frac{q\mathbf{E}}{m} \qquad (11\text{-}32)$$

E 11.16   A proton ($q = 1.6 \times 10^{-19}$ C, $m = 1.7 \times 10^{-27}$ kg) moves in a region where the electric field is given by $\mathbf{E} = 500\hat{\mathbf{k}}$ N/C. What are the force and acceleration of the proton in this region?

Answer: $8.0 \times 10^{-17}\hat{\mathbf{k}}$ N, $4.8 \times 10^{10}\hat{\mathbf{k}}$ m/s^2

Suppose a charged particle is in a region of *uniform electric field*, a region in which the magnitude and direction of $\mathbf{E}$ is the same everywhere. A region of uniform electric field may be approximated in practice by maintaining equal and opposite charges on two parallel conducting plates. (As we will see in later chapters, this arrangement can be achieved by connecting the terminals of a battery to the conducting plates.)

Equation (11-32) asserts that a charged particle in a uniform electric field experiences a constant acceleration. Then all the kinematic, dynamic, and energetic relationships associated with particles undergoing constant acceleration apply to a charged particle in a uniform electric field. For example, if we assume a constant electric field of magnitude $\mathbf{E}$ in the y direction, a particle of mass m bearing a charge q in that field has constant acceleration $a_y = qE/m$, a situation essentially identical to the projectile problem of Section 2.2. Then the kinematic equations for constant acceleration apply:

$$y = y_o + v_{oy}t + \frac{qEt^2}{2m} \qquad (11\text{-}33)$$

$$v_y = v_{oy} + \frac{qEt}{m} \qquad (11\text{-}34)$$

$$v_y^2 = v_{oy}^2 + \frac{2qE}{m}(y - y_o) \qquad (11\text{-}35)$$

An electron is projected at a speed of 10^4 m/s in a direction of $45°$ above the horizontal into a region where a uniform electric field is directed vertically upward. If the electron travels a distance $d = 4$ cm horizontally before returning to its initial vertical height, what is the magnitude of the electric field?

Let us choose the x direction as horizontal and the y direction as vertical and let the initial position of the electron be at the coordinate origin, as shown in Figure 11.15. The initial velocity $\mathbf{v}_o$ of the electron has components $v_{ox} = v_{oy} = v_o/\sqrt{2}$. Because $\mathbf{E}$ is in the positive y direction, the constant acceleration of the negatively charged electron is in the negative y direction. Because the electron experiences no acceleration in the x direction, we may adapt Equation (11-33) for both the x and the y positions of the electron at any time t:

$$x = v_{ox}t \qquad (11\text{-}36)$$

$$y = v_{oy}t - \frac{eE}{2m}t^2 \qquad (11\text{-}37)$$

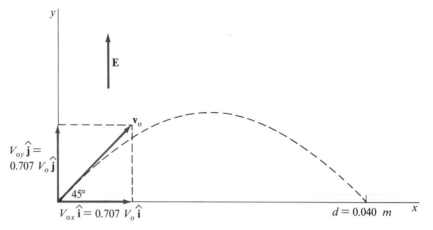

Figure 11.15
The trajectory of an electron launched at an angle of 45° above the horizontal. The uniform electric field **E** is directed vertically upward.

When the electron has returned to its original height, $y = 0$, Equations (11-36) and (11-37) may be written

$$d = v_{ox}t \tag{11-38}$$

$$v_{oy} = \frac{eE}{2m}t \tag{11-39}$$

Eliminating t from Equations (11-38) and (11-39) and using $v_{ox} = v_{oy} = v_o/\sqrt{2}$ gives

$$E = \frac{2m\, v_{ox}v_{oy}}{ed} = \frac{mv^2_o}{ed}$$

$$E = \frac{(9.1 \times 10^{-31})(10^4)^2}{(1.6 \times 10^{-19})(4 \times 10^{-2})} = 1.4 \times 10^{-6} \text{ N/C} \tag{11-40}$$

Finally, consider the energetics of a charged particle in a uniform electric field. Let us examine the kinetic energy of a charged particle after it has been released from rest in a uniform electric field that is in the positive y direction, that is, $\mathbf{E} = E\hat{\mathbf{j}}$. Suppose the particle has mass m and charge q. When the particle has moved from the origin to a position y, the particle will have acquired kinetic energy $K = \frac{1}{2}mv_y^2$. Equation (11-35) provides that $v_y^2 = 2qEy/m$, so that at the position y the particle has kinetic energy

$$K = \frac{1}{2}m\left(\frac{2qEy}{m}\right) = qEy \tag{11-41}$$

The same value of the kinetic energy of the particle may be obtained using the work-energy principle: The work done by the resultant force on a particle is equal to the change in the kinetic energy of the particle. When a particle with charge q moves from the origin to a position y, the work done on that particle by the constant force $qE\hat{\mathbf{j}}$ is qEy. Thus the change in the kinetic energy of the particle, and therefore its kinetic energy at the position y, is $K = qEy$, a value identical with that of Equation (11-41).

E 11.17   A charged particle ($q = 4.0\ \mu C$) moves in a constant electric field given by $\mathbf{E} = 800\hat{\mathbf{j}}$ N/C. If the particle is released from rest at the origin, what is its kinetic energy when its y coordinate is 0.25 m?

Answer: 8.0×10^{-4} J

11.4 Problem-Solving Summary

This chapter and its Group A and Group B Problems are concerned with electrical forces on charged particles. Electrical forces on a charge are caused by other charges, which, in a given problem, may be specifically present or may have an existence that is inferred from the presence of an electric field.

In those problems that expressly specify the charge or charges that produce a force on a given charge, Coulomb's law is used to determine the force on that charge. Three types of static charge distributions may exert electrical forces on a point charge: (a) a point charge, (b) a set of individual point charges, and (c) a continuous distribution of charge.

For a point charge Q that exerts a force $\mathbf{F}$ on another point charge q located a distance r from Q, Coulomb's law specifies the magnitude of the force on q to be $F = kQq/r^2$. The direction of the force on q is toward Q if the signs of Q and q are different; the force on q is away from Q if the signs of Q and q are the same.

The net electrical force on a point charge q exerted by an array of charges Q_1, Q_2, . . . Q_n is found by

1. Calculating the force on q by each point charge of the array (using Coulomb's law) just as if each charge in the array were the only charge exerting a force on q.

2. Summing *vectorially* the forces caused by the individual charges.

Determination of the total electrical force on a charge q exerted by a continuous charge distribution deserves special problem-solving attention. The total electrical force on the charge q is found by first treating each infinitesimal charge element dQ within the distribution as if it were a point charge. Then the magnitude of the force $d\mathbf{F}$ exerted on q by the element dQ is given by $dF = kq\,dQ/r^2$, where r is the distance from dQ to q. The net force on q is found by integrating over the entire continuous charge distribution. The integration of a vector quantity like $d\mathbf{F}$ is most readily accomplished by integrating its components, say dF_x and dF_y, separately. At the same time, it is important to take advantage of any symmetry of the charge distribution to simplify the force components. Performing the integration for each component of the force requires that we express both r and dQ in terms of a common variable, usually an angle or one of the coordinate variables. Expressing r in terms of a chosen variable may be accomplished directly from the geometry of the charges q and dQ (*see* Examples 11.4 and 11.5). The charge element dQ can be expressed in terms of coordinate variables by using appropriate charge densities:

1. For a linear (one-dimensional) charge distribution, we use $dQ = \lambda\,dx$, where λ is the charge per unit length.

2. For an areal (two-dimensional) charge distribution, we use $dQ = \sigma\,dA$, where σ is the charge per unit area.

3. For a volumetric (three-dimensional) charge distribution, we use $dQ = \rho\,dV$, where ρ is the charge per unit volume.

Solutions to problems that involve a charge distribution with a finite charge can often be supported by checking limiting cases. For example, when a charge distribution having a total charge Q exerts a force on a point charge q, the solution, in the limiting case of a large distance d separating the charge distribution and the point charge q, can be expected to reduce to the solution for point charges: $F = kqQ/d^2$. In other words, when q is sufficiently far from a charge distribution with total charge Q, the distribution will have the effect on q of a point charge Q.

An electric field $\mathbf{E}$ at a point in space is defined by $\mathbf{E} = \mathbf{F}/q$, where $\mathbf{F}$ is the force exerted on a test charge q placed at that point. Because the force $\mathbf{F}$ on the test charge q by another point charge Q is $\mathbf{F} = \hat{\mathbf{r}} \, kqQ/r^2$, the electric field of a point charge Q has magnitude kQ/r^2. The direction of the field of a point charge Q is radially outward (away from Q) if Q is positive and is radially inward (toward Q) if Q is negative.

A particle with mass m and charge q in an electric field $\mathbf{E}$ experiences a force of magnitude $F = qE$ in the direction of $\mathbf{E}$ if q is positive and opposite to the direction of $\mathbf{E}$ if q is negative. Then, according to Newton's second law, the acceleration experienced by the particle is $\mathbf{a} = \mathbf{F}/m$, or $\mathbf{a} = q\mathbf{E}/m$. If $\mathbf{E}$ is uniform, the particle will have constant acceleration, and certain special kinematic relationships (*see* Chapter 2) are applicable.

Most problems that involve forces on charges may be treated, for the most part, in component form. The directions of the forces are determined by applying the basic sign rule of charges: Charges of like sign repel; charges of unlike sign attract. It is important to remember that the answers to problems requiring a vector quantity, like force, electric field, or acceleration, must include a direction as well as a magnitude with appropriate units.

Problems

GROUP A

11.1 What is the acceleration of an electron when it is 2.0 cm from a proton? What is the acceleration of the proton at the same instant?

11.2 The magnitude of the electric field at a point 5.0 cm from a negative point charge is 80 N/C. Determine the charge.

11.3 What is the electric field at a point where a charged particle ($m = 3.0 \times 10^{-5}$ kg, $q = -5.0$ μC) experiences an acceleration of 1.8×10^3 $\hat{\mathbf{k}}$ m/s^2?

11.4 When one particle ($m_1 = 2.0 \times 10^{-4}$ kg, $q_1 = 5.0$ μC) and a second particle ($m_2 = 5.0 \times 10^{-4}$ kg, $q_2 = -10$ μC) are 0.30 m apart, what is the magnitude of the acceleration of each particle?

11.5 Calculate the magnitude of the electric field at a point that is a distance of 5.0×10^{-11} m from a proton. This is the field experienced by an electron in a hydrogen atom.

11.6 Calculate the electric field at the point (4, 2) m resulting from an electric charge of 5.0×10^{-8} C placed at (0, 5) m.

11.7 What is the distance between two protons at an instant when each experiences an acceleration having a magnitude of 5.0×10^4 m/s^2?

11.8 Charges $Q_1 = 40$ μC and $Q_2 = -50$ μC are positioned on the x-y plane at $\mathbf{r}_1 = (8.0\hat{\mathbf{i}} + 16\hat{\mathbf{j}})$ cm and $\mathbf{r}_2 = 20\hat{\mathbf{i}}$ cm. Calculate the force exerted on Q_1 by Q_2.

11.9 A particle (mass = m, charge = $q > 0$) is released from rest in a region where the electric field is uniform. Its speed after moving a distance d is v. Determine the magnitude of the electric field.

11.10 The three charges in Figure 11.16 are located on the vertices of an equilateral triangle. Determine the net force on q.

11.11 A positive charge of 8.0 μC is fixed at the origin. At $t = 0$ a particle ($m = 1.5 \times 10^{-4}$ kg, $q = -4.5$ μC) has a position and velocity given by $\mathbf{r} = 0.25\hat{\mathbf{i}}$ m and $\mathbf{v} = v_o \, \hat{\mathbf{j}}$.

(a) Calculate the value of v_o required if the particle is to execute circular motion centered on the origin.

(b) What is the period of this circular motion?

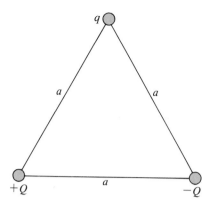

Figure 11.16   Problem 11.10

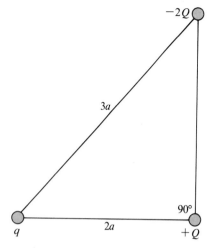

Figure 11.19   Problem 11.15

11.12   Three point charges $q_1 = 4.0\ \mu C$, $q_2 = -5.0\ \mu C$, and $q_3 = 8.0\ \mu C$ are fixed at $\mathbf{r}_1 = \mathbf{0}$, $\mathbf{r}_2 = 3.0\hat{\mathbf{i}}$ m, and $\mathbf{r}_3 = 4.0\hat{\mathbf{j}}$ m. Calculate the force exerted on q_3.

11.13   Four point charges are placed on the vertices of a rectangle (*see* Fig. 11.17). If $a = 50$ cm, $b = 25$ cm, and $Q = 20\ \mu C$, calculate the force on the $4Q$ charge.

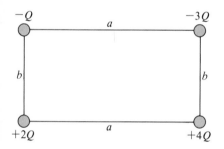

Figure 11.17   Problem 11.13

11.14   Four charges are positioned on the vertices of the parallelogram in Figure 11.18. If $q = 25\ \mu C$, $Q = 64\ \mu C$, $\alpha = 60°$, $b = 0.80$ m, and $a = 0.50$ m, calculate the net force on q.

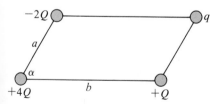

Figure 11.18   Problem 11.14

11.15   Three charges are placed at the vertices of the right triangle in Figure 11.19. Calculate the angle between the direction of the force on q and the line connecting q and Q.

11.16   Point charges Q_1 and Q_2 are placed on the y axis at the points $y = \pm d$ (*see* Fig. 11.20).

(a)   Determine the force exerted on a point charge q placed on the y axis at $y = r > d$.

(b)   Show that the expression obtained in part (a) reduces appropriately for $r \gg d$.

(c)   Suppose that Q_1 and Q_2 form an electric dipole, that is, $Q_1 = Q$ and $Q_2 = -Q$. Show that the limiting ($r \gg d$) behavior for the force exerted on q is

$$\mathbf{F} \xrightarrow[r \gg d]{} \frac{2kqp}{r^3}\hat{\mathbf{j}}$$

where $p = 2Qd$ is the magnitude of the dipole moment of the two charges.

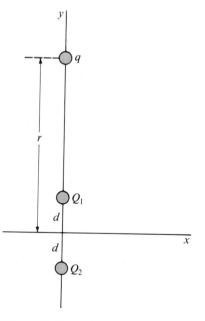

Figure 11.20   Problem 11.16

11.17 A negative charge $Q = -35 \ \mu C$ is fixed. If $q = 65 \ \mu C$ is released from rest when it is a large distance from Q, determine the kinetic energy of q when it is 1.5 m from Q.

11.18 A particle ($m = 2.0 \times 10^{-4}$ kg, $q = 50 \ \mu C$) moves in a region where the electric field is $\mathbf{E} = 85\hat{\mathbf{j}}$ N/C. At $t = 0$ the position and velocity of the particle are $\mathbf{r} = \mathbf{0}$ and $\mathbf{v_o} = 20\hat{\mathbf{i}}$ m/s. Disregard any gravitational force.

    (a) What are the position and velocity of the particle at time $t = 2.0$ s?

    (b) How much work does the electric force do on the particle during this 2-s time interval?

11.19 Point charges, $Q_1 = 80 \ \mu C$ and $Q_2 = -40 \ \mu C$, are located on the x axis at $x_1 = 0$ and $x_2 = 2.0$ m. Where must a third charge $q = 25 \ \mu C$ be placed so that the force on it is zero?

11.20 Charges ($q_1 = 3Q$ and $q_2 = Q$) are held in position at a distance d apart. Where must a third charge be placed so that the force on it is zero?

11.21 Charges ($q_1 = 8.0 \ \mu C$ and $q_2 = 2.0 \ \mu C$) are located on the x axis at $x_1 = 0$ and $x_2 = 3.0$ m. A third charge q_3 is positioned so that the net force on each of the three charges is zero. Determine q_3 and its location.

11.22 A particle ($m = 3.5 \times 10^{-5}$ kg, $q = -1.2 \ \mu C$) is suspended from the ceiling by a thread. The electric field in the vicinity of the particle is uniform and has a magnitude equal to 170 N/C. Calculate the tension in the thread (once the ball achieves static equilibrium) if the electric field is directed (a) vertically upward, (b) vertically downward, (c) horizontally.

11.23 Two identical particles, each having a mass m and a charge q, are each suspended from a common point by threads of length L. The static equilibrium positions of the particles are shown in Figure 11.21. If $L = 1.2$ m, $m = 3.2 \times 10^{-3}$ kg, and $\theta = 25°$, calculate q and the tension in each of the strings.

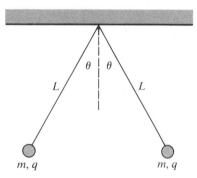

Figure 11.21   Problem 11.23

11.24 If a charge Q is fixed at the origin and a second charge moves along the x axis from $x = a$ to $x = b$, how much work is done by the electric force exerted on q by Q?

11.25 If a particle ($q = -7.5 \ \mu C$) is released from rest when it is 75 cm from a fixed charge $Q = 12 \ \mu C$, what is the kinetic energy of the particle when it is 15 cm from Q?

11.26 Charge of uniform density $\lambda = 2.5 \ \mu C/m$ is distributed along the entire negative x axis. Calculate the force this distributed charge exerts on a 45-μC charge on the x axis at $x = 2.0$ m.

11.27 A 24-μC charge is distributed uniformly along a thin nonconducting rod that is 1.2 m long. If the charged rod is bent to form a semicircle, calculate the magnitude of the force on a 15-μC charge at the center of the circle.

11.28 The electric field in a region of space is given by

$$\mathbf{E} = E_o(1 + x/d)\hat{\mathbf{i}}$$

where E_o and d are positive constants. A positive charge q is released from rest at the origin. Determine its kinetic energy when it has moved a distance of $6d$.

11.29 The electric field in a region of space is given by

$$\mathbf{E} = E_o(1 - 8x^3/d^3)\hat{\mathbf{i}}$$

where E_o and d are positive constants. If a particle of mass m and charge $q > 0$ is released from rest at the origin,

    (a) what is its initial acceleration?

    (b) how far does it move in the x direction?

    (c) what is its maximum speed?

11.30 Charge of uniform density $\lambda > 0$ is distributed along an infinite straight line. Calculate the magnitude of the force this distributed charge exerts on a point charge $q > 0$ that is a (perpendicular) distance r from the line.

11.31 A charge of 3.5 μC is fixed at the origin. If a particle ($q = 16 \ \mu C$) is released from rest at $\mathbf{r} = 2.3\hat{\mathbf{i}}$ m, what is the kinetic energy of the particle when it is very far from the origin?

11.32 If a charge $Q = 75 \ \mu C$ is fixed at the origin and a particle ($m = 2.2 \times 10^{-3}$ kg, $q = -25 \ \mu C$) has a velocity equal to $70\hat{\mathbf{i}}$ m/s at an instant when its position is $2.5\hat{\mathbf{i}}$ m, what maximum distance from Q will the particle subsequently achieve?

11.33 Charge of uniform linear density $\lambda = 1.8 \ \mu C/m$ is distributed along the entire x axis. Determine the force this distributed charge exerts on a 15-μC charge placed on the y axis at $y = 3.0$ m.

11.34 If a charge of a uniform density $\lambda_1 > 0$ is distributed along the entire x axis and charge of uniform density $\lambda_2 > 0$ is distributed along the entire y axis, determine the magnitude of the force on a charge $q > 0$ located at

    (a) $\mathbf{r} = a\hat{\mathbf{i}} + b\hat{\mathbf{j}}$, where a and b are positive quantities.

    (b) $\mathbf{r} = a\hat{\mathbf{k}}$, where a is a positive quantity.

11.35 The net force $\mathbf{F}$ exerted on the charge q by the charges Q_1 and Q_2, arranged as shown in Figure 11.22, is 5.5 N $\underline{/25°}$. If $q = 16 \ \mu C$, $a = 0.20$ m, and $b = 0.40$ m, determine Q_1 and Q_2.

11.36 A particle (mass $= m$, charge $= q > 0$) moves toward a fixed charge $Q > 0$. When the distance separating the two is r_o, the speed of the moving particle is v_o. How close will q come to Q?

11.37 Two positive charges separated by 2.0 cm have a total charge of 0.50 μC.

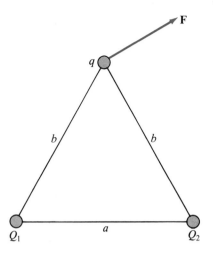

Figure 11.22   Problem 11.35

(a)   What is the magnitude of the maximum force either charge can exert on the other?

(b)   If the force on one charge by another has a magnitude of 0.90 N, calculate the charge on the particle having the greater charge.

11.38   A positive charge Q is distributed uniformly along a circular arc. The length of the arc equals the radius R of the circle. Calculate the magnitude of the force exerted by Q on a positive point charge q placed at the center of the circle.

11.39   A positive charge Q is distributed uniformly along a circular arc of radius R. The arc is subtended by an angle β.

(a)   Calculate the magnitude of the force exerted by Q on a positive point charge q placed at the center of the circle.

(b)   Show that the answer obtained in part (a) reduces appropriately for the limiting cases $\beta \to 0$ and $\beta \to 2\pi$.

11.40   Two equal charges, each of magnitude Q, are fixed on the y axis at $y = \pm d$ (*see* Fig. 11.23). A third positive charge q moves along the x axis.

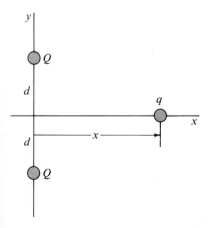

Figure 11.23   Problem 11.40

(a)   Write an expression for the net electrical force exerted on q.

(b)   Calculate the work done by the electrical force on q as the charge moves from $x = 0$ to $x = 5d$.

(c)   For what value of x does the electrical force on q have its greatest magnitude? What is the value of this maximum magnitude?

11.41   A charge of 400 μC is uniformly distributed along the y axis from $y = -1.0$ m to $y = 3.0$ m. Calculate the force this continuous charge exerts on a point charge of 50 μC positioned on the x axis at $x = 4.0$ m.

11.42   A continuous charge of uniform linear density λ is distributed along the x axis, the y axis, and the circular arc (*see* Fig. 11.24). Determine the magnitude of the force on a point charge q placed at the origin.

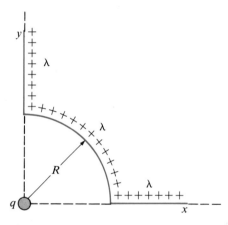

Figure 11.24   Problem 11.42

11.43   Two continuous charges are arranged on the x-y plane, as shown in Figure 11.25. One is a charge of uniform density 3λ distributed along the entire x axis, and the second of uniform density λ is distributed along the semicircle. A point charge q is placed at the center of the circle. Determine the angle between the net force exerted on q and the unit vector $\hat{\mathbf{i}}$.

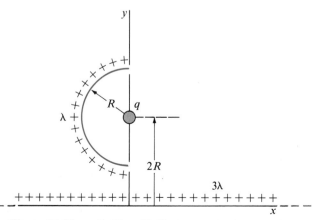

Figure 11.25   Problem 11.43

11.44 A continuous charge of uniform density σ is distributed over a flat circular surface of radius R. A point charge q is placed on the axis of this charged disk at a distance r from the disk.

(a) Show that the magnitude of the force exerted on q by the distributed charge is given by

$$F = 2\pi kq\sigma\left(\frac{1}{r} - \frac{1}{\sqrt{r^2 + R^2}}\right)$$

(b) Show that the expression obtained in part (a) reduces appropriately for $r \gg R$.

GROUP **B**

11.45 A nonuniform charge is distributed along the x axis with a linear charge density given by $\lambda(x) = \alpha x$ for $-L \le x \le L$ where α is a positive constant. A positive point charge q is placed on the y axis at $y = R$.

(a) Calculate the total charge Q distributed along the x axis.

(b) Calculate the magnitude of the force exerted on q by Q. Express your answer in terms of q, Q, L, and R.

(c) Show that the answer for part (b) reduces appropriately for $R \gg L$.

11.46 Charge of uniform density λ_1 is distributed along the x axis from $x = 0$ to $x = a$. Calculate the force this charge exerts on a second charge of uniform density λ_2 distributed along the x axis from $x = b$ to $x = c$. Assume that $0 < a < b < c$. (*Hint:* First determine the force dF_{12} exerted on an element of charge $dQ_2 = \lambda_2\, dx_2$ by the charge of density λ_1. Then integrate this expression over the charge of density λ_2.)

11.47 Charge of uniform density $\lambda_1 > 0$ is distributed along the entire x axis. A second charge of uniform density $\lambda_2 > 0$ is distributed along the y axis on the interval $a \le y \le b$ where a is positive. Calculate the magnitude of the force exerted on the second charge by the first.

11.48 A charge Q is distributed uniformly along a circle of radius a. A point charge q is positioned in the plane of the circle at a distance $r > a$ from the center of the circle (*see* Fig. 11.26).

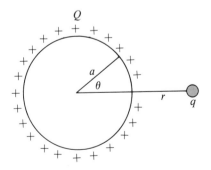

Figure 11.26   Problem 11.48

(a) Set up, but do not evaluate, an integral for the force exerted on q by Q. Use θ, the angle shown, as the variable of integration.

(b) Show that the integral of part (a) reduces appropriately for $r \gg a$.

11.49 Charge of uniform density λ is distributed along the spiral path $r = ae^{\alpha\theta}$, $0 \le \theta \le \pi$, where a and α are constants. This path is shown in Figure 11.27. A point charge q is placed at the origin.

(a) Calculate the total charge distributed along the spiral path.

(b) Determine F_x and F_y, the x and y components of the force exerted on q by the distributed charge. (Note that $\int_0^\pi e^{-\alpha\theta}\cos\theta\, d\theta = \alpha(1 + e^{-\alpha\pi})/(1 + \alpha^2)$ and that $\int_0^\pi e^{-\alpha\theta}\sin\theta\, d\theta = (1 + e^{-\alpha\pi})/(1 + \alpha^2)$.)

(c) Show that the expressions obtained in parts (a) and (b) yield appropriate results for $\alpha = 0$.

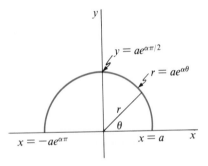

Figure 11.27   Problem 11.49

11.50 The force exerted on a point charge by a continuous charge may be estimated to any desired accuracy by replacing the continuous charge by an appropriate number of point charges. The force $\mathbf{F}$ exerted on a point charge q by a continuous charge Q is given by

$$\mathbf{F} = \int_Q \frac{kq\,\mathbf{r}\,dQ}{r^3}$$

where $\mathbf{r}$ is the displacement vector from dQ to q, as shown in Figure 11.28(a). Figure 11.28(b) shows a set of point charges $(Q_1, Q_2, \ldots, Q_N)$ that sum to Q. The net force $\mathbf{F}_N$ exerted on q by these point charges is given by

$$\mathbf{F}_N = \sum_{n=1}^N \frac{kq\,Q_n\mathbf{r}_n}{r_n^3}$$

where $\mathbf{r}_n$ is the displacement vector from Q_n to q. As an example, suppose 600 μC is distributed uniformly along the x axis from $x = -1.0$ m to $x = 1.0$ m and a point charge $q = 20$ μC is placed on the x axis at $x = 3.0$ m.

(a) Estimate the force exerted on q by approximating the continuous charge by one point charge Q at the center $(x = 0)$ of the continuous charge.

(b) Improve the estimate in part (a) by placing three 200-μC charges at $x = -1/2$ m, $x = 0$, and $x = 1/2$ m.

(c) Calculate the exact answer for the force exerted on q and compare the results of parts (a) and (b) to this result.

(d) Improving the estimate for the force on q by using more and more point charges to approximate Q is, of course, tedious and time consuming. But this is just the type of repetitive calculation for which programmable calculators and computers are designed. If you have access to either, you might enjoy trying a higher order calculation. The table below summarizes some results of such a calculation.

N	F_N	F_N/F_{exact}
1	12.0000	0.88889
2	12.4538	0.92250
5	12.9555	0.95967
10	13.1982	0.97764
50	13.4340	0.95111
100	13.4666	0.99753
500	13.4933	0.99950
1000	13.4967	0.99976
5000	13.4997	0.99998

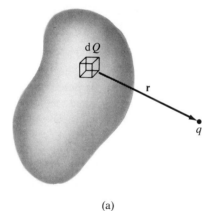

(a)

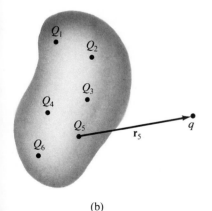

(b)

Figure 11.28 Problem 11.50

11.51 A charge $Q = 600$ μC is uniformly distributed along the x axis from $x = -1.0$ m to $x = 1.0$ m. A point charge $q = 20$ μC is placed on the y axis at $y = 3.0$ m.

(a) Use the technique of Problem 11.50 to estimate the force exerted on q by Q. Do this for N = 1, 2, 3.

(b) Calculate the exact value for the force exerted on q and compare it with the results obtained in part (a).

11.52 Charge is distributed along the x axis from $x = -1.0$ m to $x = 1.0$ m according to

$$\lambda(x) = \lambda_o \cos(\pi x/2)$$

where $\lambda_o = 200$ μC/m.

(a) Calculate the total charge Q of this continuous charge distribution.

(b) Estimate the force that Q exerts on a charge $q = 40$ μC placed on the x axis at $x = 3.0$ m. Do this by replacing the continuous charge by a point charge Q at $x = 0$.

(c) Improve the estimate of part (b) by using two point charges at $x = -1/3$ m and $x = 1/3$ m to approximate the continuous charge.

(d) If three charges are used to approximate the distribution, these charges should not be chosen to be equal. Why? Do this calculation. (*Hint:* $Q_1 = Q_3 = \int_{1/3}^{1} \lambda(x)\, dx$ and $Q_2 = \int_{-1/3}^{1/3} \lambda(x)\, dx = Q - Q_1 - Q_3$.)

11.53

(a) Repeat Problem 11.52 with $\lambda(x) = \lambda_o \sin(\pi x)$ for the interval -1 m $\leq x \leq 1$ m and $\lambda_o = 200$ μC/m.

(b) Estimate the dipole moment of this charge distribution using the results of part (a) and Problem 11.16.

12 Calculation of Electric Fields

Electric fields play a key role in electromagnetic physics. It is, therefore, important that we be able to determine the electric fields associated with a wide variety of charge distributions. This chapter introduces several techniques appropriate to the calculation of electric fields that result from relatively simple distributions of charge.

Our first consideration is the calculation of electric fields at points in space caused by collections of fixed point charges. This procedure will then be extended to include the calculation of electric fields caused by continuous charge distributions.

Certain symmetrical arrangements of charge produce electric fields that may be determined by a simple, powerful, problem-solving procedure. This process is based on a relationship called Gauss's law, which, in turn, is based on the concept of electric flux. Therefore, we will illustrate electric flux and see how it is used in Gauss's law to calculate the electric fields of symmetrical charge distributions.

Electric fields and the way that electric fields are related to Gauss's law permit us to deduce how static charges position themselves in and on conducting materials. Therefore, our final consideration is how static charges and the electric fields associated with those charges distribute themselves in and around electrical conductors.

12.1 Electric Fields of Point Charges

In the preceding chapter we saw that a point charge q located a distance r from a fixed point charge Q experiences a force $\mathbf{F}$ given by Coulomb's law,

$$\mathbf{F} = k \frac{qQ}{r^2} \, \hat{\mathbf{r}} \tag{12-1}$$

where $\hat{\mathbf{r}}$ is a unit vector in the direction from Q toward q. We saw also that the electric field $\mathbf{E}$ at the position of q is defined to be

$$\mathbf{E} = \frac{\mathbf{F}}{q} \tag{12-2}$$

Combining Equations (12-1) and (12-2), we obtain the electric field of the point charge Q at a distance r from Q:

$$\mathbf{E} = k\frac{Q}{r^2}\,\hat{\mathbf{r}} \tag{12-3}$$

A point in space at which we calculate a field is referred to as a *field point*. Thus, the electric field $\mathbf{E}$ at a field point P located at a distance r from a point charge Q is in the same direction as would be the force $\mathbf{F}$ on a positive test charge q if it were placed at P, and the magnitude of $\mathbf{E}$ is equal to $F/q = kQ/r^2$. It follows, then, that the principle of superposition applies not only to electrostatic forces but also to electric fields. In other words, the total electric field (at a field point) caused by n fixed charges $Q_1, Q_2, \ldots Q_n$ may be calculated by finding the electric field at P produced by each of those n charges as if each were the only charge present and summing those fields *vectorially*. Mathematically we may write

$$\mathbf{E} = \mathbf{E}_1 + \mathbf{E}_2 + \ldots + \mathbf{E}_n = \sum_{i=1}^{n} \mathbf{E}_i \tag{12-4}$$

where $\mathbf{E}_i$ is the field at P produced by the charge Q_i.

The relationship between the electric field $\mathbf{E}$ at a field point produced by point charges and the force on a test charge q placed at that field point is graphically demonstrated in Figure 12.1. In Figure 12.1(a) a positive test charge q is located a distance r_1 from a positive charge Q_1 and a distance r_2 from a negative charge $-Q_2$. The forces $\mathbf{F}_1$ and $\mathbf{F}_2$ that the charges Q_1 and $-Q_2$ exert on q are indicated in their appropriate directions. These forces may be calculated from Equation (12-1). According to the principle of superposition, the net force on q is $\mathbf{F}$, the vector sum of $\mathbf{F}_1$ and $\mathbf{F}_2$. Figure 12.1(b) shows the same arrangement of Q_1 and $-Q_2$ but with q removed. At the point P, where q had been located, the electric field $\mathbf{E}$ is the vector sum of $\mathbf{E}_1$ and $\mathbf{E}_2$, the fields produced by Q_1 and $-Q_2$, respectively. Notice that the direction of $\mathbf{E}$ is the same as that of $\mathbf{F}$ in Figure 12.1(a). Further, the magnitude of $\mathbf{E}$ is, according to Equation (12-2), equal to F/q. Thus, if $\mathbf{F}$ has been determined to be the force exerted on a charge q by a point charge distribution, it becomes trivial to determine $\mathbf{E}$: It is only necessary to divide F by q.

E 12.1   Charges Q_1 and Q_2 exert the forces $\mathbf{F}_1 = (6.0\hat{\mathbf{i}} + 8.0\hat{\mathbf{j}})$ N and $\mathbf{F}_2 = -16\hat{\mathbf{j}}$ N on a charge $q = 2.0 \times 10^{-4}$ C when q is placed at the point P. Calculate the electric field at point P.

Answer: $(3.0\hat{\mathbf{i}} - 4.0\hat{\mathbf{j}}) \times 10^4$ N/C $= 5.0 \times 10^4$ N/C $\underline{/-53°}$

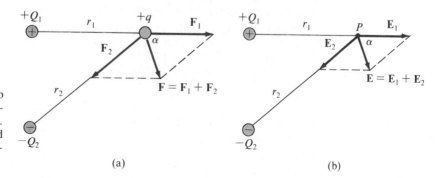

Figure 12.1
(a) The total force $\mathbf{F}$ exerted by two fixed charges $+Q_1$ and $-Q_2$ on a positive test charge q, located by r_1 and r_2.
(b) The total electric field $\mathbf{E}$ at a field point P at the same location. The vectors $\mathbf{E}$ and $\mathbf{F}$ have the same direction.

E 12.2   A total force of $0.08°$ N$\underline{/60°}$ is exerted on a charge $q = -400\ \mu C$ placed at a point. Determine the electric field at that point.

Answer: 200 N/C $\underline{/240°}$

The following problem (1) illustrates the calculation of the field of more than one point charge, and (2) shows how mathematical approximations permit simplifications of the description of a physical situation.

An electric dipole consists of two charges, Q and $-Q$, separated by a fixed distance a. Calculate the electric field of a dipole (a) at a point far from the dipole along the line on which the charges lie, and (b) at a point far from the dipole on the perpendicular bisector of the line segment joining the charges.

The electric dipole shown in Figure 12.2(a) is oriented along the y axis. The electric field $\mathbf{E}$ at a point $(0, y)$ on the y axis is $\mathbf{E} = \mathbf{E}_+ + \mathbf{E}_-$, where $\mathbf{E}_+$ is the field of $+Q$ and $\mathbf{E}_-$ is the field of $-Q$. The direction of $\mathbf{E}$ is in the positive y direction (because the field point is nearer to $+Q$ than to $-Q$), and $\mathbf{E}$ has the magnitude

$$E = k\ \frac{Q}{\left(y - \dfrac{a}{2}\right)^2} + k\ \frac{(-Q)}{\left(y + \dfrac{a}{2}\right)^2}$$

$$E = kQ\left[\frac{y^2 + ay + \dfrac{a^2}{4} - y^2 + ay - \dfrac{a^2}{4}}{\left(y - \dfrac{a}{2}\right)^2\left(y + \dfrac{a}{2}\right)^2}\right] =$$

$$kQ\left[\frac{2ay}{\left(y - \dfrac{a}{2}\right)^2\left(y + \dfrac{a}{2}\right)^2}\right] \tag{12-5}$$

Far from the dipole, where $y \gg a$, we may disregard $a/2$ compared to y. The denominator of Equation (12-5) becomes y^4, and the magnitude of the electric field becomes

$$E = \frac{kQ2ay}{y^4} = \frac{2kQa}{y^3} = \frac{Qa}{2\pi\epsilon_o y^3} \tag{12-6}$$

The electric field $\mathbf{E}$ at a point $(x, 0)$ on the perpendicular bisector of the line segment joining the charges of the dipole is seen in Figure 12.2(b) to be in the negative y direction. Because $\mathbf{E}_+$ and $\mathbf{E}_-$ have equal magnitudes ($E_+ = E_-$) and because the components of $\mathbf{E}_+$ and $\mathbf{E}_-$ in the x direction, $E_+ \sin\theta$ and $-E_- \sin\theta$, sum to zero, the magnitude of $\mathbf{E}$ is equal to $E_+ \cos\theta + E_- \cos\theta = 2E_+ \cos\theta$, or

$$E = 2E_+ \cos\theta = 2k\ \frac{Q}{r^2}\ \cos\theta \tag{12-7}$$

from Figure 12.2(b), we see that

$$r^2 = x^2 + \left(\frac{a}{2}\right)^2$$

and

$$\cos\theta = \frac{a/2}{r} = \frac{a/2}{\sqrt{x^2 + (a/2)^2}}$$

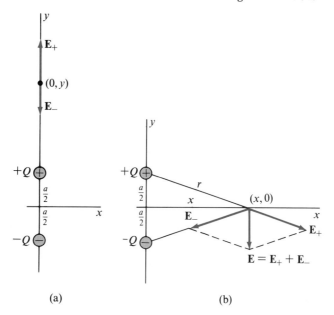

Figure 12.2
(a) The electric fields, $\mathbf{E}_+$ and $\mathbf{E}_-$, at the point $(0, y)$. These fields are caused by a dipole oriented along the y axis. **(b)** The electric field $\mathbf{E}$ and its component vectors, $\mathbf{E}_+$ and $\mathbf{E}_-$, at the point $(x, 0)$. These fields are caused by a dipole oriented along the y axis.

Then E becomes

$$E = \frac{2kQ\left(\dfrac{a}{2}\right)}{\left[x^2 + \left(\dfrac{a}{2}\right)^2\right]^{3/2}} = \frac{kQa}{\left[x^2 + \left(\dfrac{a}{2}\right)^2\right]^{3/2}} \qquad (12\text{-}8)$$

Far from the dipole, where $x \gg a/2$, we may again neglect $a/2$ compared to x. The magnitude of the electric field becomes

$$E = \frac{kQa}{x^3} = \frac{Qa}{4\pi\epsilon_o x^3} \qquad (12\text{-}9)$$

Examination of Equations (12-6) and (12-9) yields several facts about the field of an electric dipole. First, at a distance far from a dipole, the electric field decreases with distance r as $1/r^3$ (which may be compared to the electric field of a single point charge, a field that falls off as $1/r^2$). Thus, at a point far from a dipole, the field of the two charges, one positive and one negative, is less than the field of a single charge—but the field is not eliminated. Further, a comparison of Equations (12-6) and (12-9) shows that (at distances far from a dipole) the electric field of the dipole is twice as large along the line passing through its charges as along the perpendicular bisector of the dipole. The electric field in a plane of a dipole is illustrated in Figure 12.3.

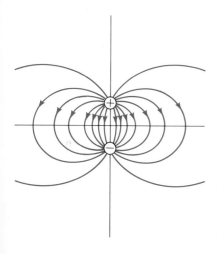

Figure 12.3
The electric field in a plane of a dipole.

12.2   Electric Fields of Continuous Charge Distributions

Electric fields of continuous charge distributions may be calculated using a procedure similar to that described in Chapter 11 for determining the force on a charge by a continuous distribution of charge. Indeed, the electric field $\mathbf{E}$ produced at a point by a continuous charge distribution may be determined by first finding the force **F** on a charge q at that point and then using the relationship, **E** = **F**/q, to specify the electric field at that point. Here, however, we will see how the electric fields of continuous charge distributions may be calculated directly.

A continuous charge distribution may be considered a collection of infinitesimal elements of charge dQ, each of which causes an electric field $d\mathbf{E}$ at a field point located a distance r from the charge element. In other words, each element of charge within the continuous distribution of charge is treated as if it were a point charge causing an electric field. The total electric field **E** at the field point, then, is the vector sum of the individual electric fields $d\mathbf{E}$ caused by all the charge elements. The summation of infinitesimal elements is, of course, an integration process. The total electric field is given by

$$\mathbf{E} = \int d\mathbf{E} \qquad (12\text{-}10)$$

The electric field $d\mathbf{E}$ produced by each element of charge is

$$d\mathbf{E} = k\frac{dQ}{r^2}\hat{\mathbf{r}} = \frac{1}{4\pi\epsilon_o}\frac{dQ}{r^2}\hat{\mathbf{r}} \qquad (12\text{-}11)$$

where r is the distance from the element dQ to the field point and $\hat{\mathbf{r}}$ is a unit vector in the direction from dQ to the field point. Because **E** is a vector quantity, it is usually most easily calculated by first expressing $d\mathbf{E}$ in terms of its components, integrating the components of $d\mathbf{E}$, and summing the resulting component vectors to give the total electric field vector. Because Equation (12-11) contains more than one variable, it is necessary to express both dQ and r^2 in terms of a single variable (at least in the problems we will consider) in order to perform the required integration. The following problem demonstrates the direct calculation of the electric field produced by a continuous charge distribution.

A charge Q is distributed uniformly along a line segment bent into a semicircle of radius R. Calculate the electric field at the center of the semicircle.

Let the semicircle be oriented so that each end of the line charge lies on the x axis, as shown in Figure 12.4. An arbitrary element of charge dQ on the line of charge may be expressed in terms of $\lambda = Q/(\pi R)$, the linear charge density (charge per unit length of arc) of the line:

$$dQ = \lambda\, ds = \frac{Q}{\pi R} \cdot R\, d\theta = \frac{Q}{\pi}\, d\theta \qquad (12\text{-}12)$$

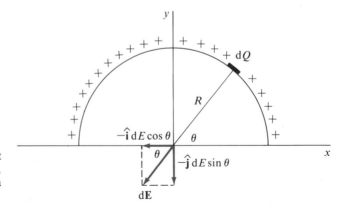

Figure 12.4
The electric field $d\mathbf{E}$ and its component vectors, $-dE\,\cos\theta\hat{\mathbf{i}}$ and $-dE\,\sin\theta\hat{\mathbf{j}}$, caused by an element of charge dQ on a semicircular line of charge.

The infinitesimal arc length ds along the line of charge is equal to $R\,d\theta$, where R is the constant value of the radius of the semicircle and θ is the angle that locates the arbitrary element of charge relative to the x axis.

The electric field d$\mathbf{E}$ produced by dQ at the field point has the components $dE_x = -dE\cos\theta$ and $dE_y = -dE\sin\theta$. Thus, Equation (12-12) provides that E_x is given by

$$E_x = -k\int \frac{dQ}{r^2}\cos\theta = -\frac{kQ}{\pi R^2}\int_0^\pi \cos\theta\,d\theta = 0 \qquad (12\text{-}13)$$

where r has the constant value R for every element dQ in the charge distribution. The integral in Equation (12-13) has the value zero, so the x component of the electric field at the field point is zero. The y component, on the other hand, is

$$E_y = -k\int \frac{dQ}{r^2}\sin\theta = -\frac{kQ}{\pi R^2}\int_0^\pi \sin\theta\,d\theta \qquad (12\text{-}14)$$

The integral in Equation (12-14) has the value 2, and the y component of the total electric field is

$$E_y = -\frac{2kQ}{\pi R^2} = -\frac{Q}{2\pi^2\epsilon_o R^2}$$

Therefore, the total electric field $\mathbf{E} = E_x\hat{\mathbf{i}} + E_y\hat{\mathbf{j}}$ at the center of the semicircle is in the negative y direction and is given by

$$\mathbf{E} = -\frac{2kQ}{\pi R^2}\hat{\mathbf{j}} = -\frac{Q}{2\pi^2\epsilon_o R^2}\hat{\mathbf{j}} \qquad (12\text{-}15)$$

If the foregoing problem had stated the charge distribution in terms of the linear charge density λ instead of the total charge Q, the answer would appropriately have been expressed in terms of $\lambda = Q/(\pi R)$:

$$\mathbf{E} = -\frac{2k\lambda}{R}\hat{\mathbf{j}} = -\frac{\lambda}{2\pi\epsilon_o R}\hat{\mathbf{j}} \qquad (12\text{-}16)$$

Example 12.1

PROBLEM   A linear charge of uniform density λ is spread along the entire x axis. Determine the electric field at a point P on the y axis at $y = d > 0$.

SOLUTION   The magnitude dE of the electric field (at point P) caused by the charge $dQ = \lambda dx$ on the x axis is given by

$$dE = \frac{k\,dQ}{r^2}$$

where $r = d\sec\theta$ is the distance from dQ to point P, as shown in Figure 12.5. We see, using the figure, that $x = d\tan\theta$ and, consequently, that $dx = d\sec^2\theta\,d\theta$. Thus, we have

$$dE = \frac{k\,dQ}{r^2} = \frac{k\lambda\,dx}{r^2} = \frac{k\lambda d\sec^2\theta\,d\theta}{d^2\sec^2\theta} = \frac{k\lambda}{d}\,d\theta$$

The rectangular components of d$\mathbf{E}$ are

$$dE_x = -dE\sin\theta = -\frac{k\lambda}{d}\sin\theta\,d\theta$$

$$dE_y = \quad dE\cos\theta = \frac{k\lambda}{d}\cos\theta\,d\theta$$

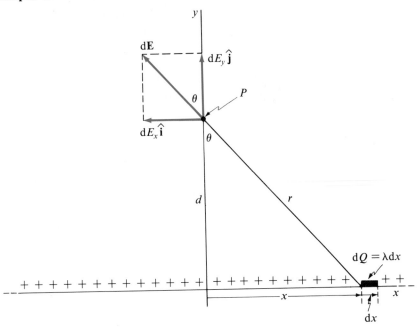

Figure 12.5   Example 12.1

Integrating from $\theta = -\pi/2$ ($x \to -\infty$) to $\theta = \pi/2$ ($x \to \infty$) gives

$$E_x = -\frac{k\lambda}{d} \int_{-\pi/2}^{\pi/2} \sin\theta \; d\theta = 0$$

$$E_y = \frac{k\lambda}{d} \int_{-\pi/2}^{\pi/2} \cos\theta \; d\theta = \frac{2k\lambda}{d} = \frac{\lambda}{2\pi\epsilon_o d}$$

The zero value for E_x is, of course, a result of the symmetry of the charge distribution about the y axis. Thus, the electric field at point P is

$$\mathbf{E} = \frac{2k\lambda}{d} \, \hat{\mathbf{j}} = \frac{\lambda}{2\pi\epsilon_o d} \, \hat{\mathbf{j}} \quad \blacksquare$$

Example 12.2   PROBLEM   The circular arc (radius $= R$) shown in Figure 12.6 carries a continuous charge described by the linear density

$$\lambda(\theta) = \lambda_o \sin\theta$$

where λ_o is a positive constant and θ is the angle shown. Determine the electric field at point P, the center of the semicircle.

SOLUTION   The magnitude dE of the electric field caused by the infinitesimal charge dQ shown is

$$dE = \frac{k \, dQ}{R^2}$$

The charge dQ is given by

$$dQ = \lambda R \; d\theta = \lambda_o R \sin\theta \; d\theta$$

so that

$$dE = \frac{k\lambda_o}{R} \sin\theta \; d\theta$$

Resolving dE into rectangular components, we get

$$dE_x = -dE \cos\theta = -\frac{k\lambda_o}{R} \sin\theta \cos\theta \, d\theta$$

$$dE_y = -dE \sin\theta = -\frac{k\lambda_o}{R} \sin^2\theta \, d\theta$$

Integrating over the interval $-\pi/2 \le \theta \le \pi/2$ gives

$$E_x = -\frac{k\lambda_o}{R} \int_{-\pi/2}^{\pi/2} \sin\theta \cos\theta \, d\theta = 0$$

$$E_y = -\frac{k\lambda_o}{R} \int_{-\pi/2}^{\pi/2} \sin^2\theta \, d\theta = -\frac{\pi k\lambda_o}{2R} = -\frac{\lambda_o}{8\epsilon_o R} \quad \blacksquare$$

Can you explain, using symmetry considerations, why E_x is necessarily zero?

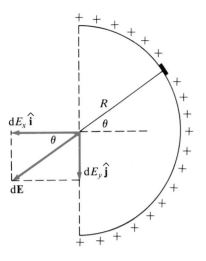

Figure 12.6   Example 12.2

E 12.3   In Example 11.5 we found that the force $\mathbf{F}$ exerted on a point charge q by an infinite plane of charge having a uniform density σ has a magnitude given by $F = q\sigma/(2\epsilon_o)$ (for $\sigma > 0$). Determine the electric field caused by a uniformly charged plane.

Answer: $\mathbf{E} = \sigma/(2\epsilon_o)$, away from the plane if $\sigma > 0$

E 12.4   Suppose the y-z plane is uniformly charged with a density that is given by $\sigma = 4.0 \times 10^{-10}$ C/m². Calculate the electric field on the x axis at (a) $x = 2.0$ m and (b) $x = -3.0$ m.   Answer: (a) $23\hat{\mathbf{i}}$ N/C; (b) $-23\hat{\mathbf{i}}$ N/C

E 12.5   Repeat Exercise 12.4 for $\sigma = -3.0 \times 10^{-10}$ C/m².
Answer: (a) $-17\hat{\mathbf{i}}$ N/C; (b) $17\hat{\mathbf{i}}$ N/C

12.3 Electric Flux and Gauss's Law

It is sometimes convenient to characterize a region of space in terms of the electric field on the boundary of that region. Such a characterization may be accomplished using the concept of electric flux. The notion of flux will again be useful when we consider magnetic fields, but for now we will use flux in the calculation of electric fields produced by charge distributions that have special symmetry properties.

Electric flux is always associated with a surface. A small area on a surface may be represented by a vector quantity $\Delta\mathbf{S}$, which has magnitude equal to the area and a direction normal to the surface. Conventionally, we designate the direction of an area to be *outward* from a surface that encloses a region of space. If the surface is not closed, the sense of the (perpendicular) direction of $\Delta\mathbf{S}$ must be specified. Figure 12.7 shows a closed surface on which a small area is represented by the vector $\Delta\mathbf{S}$. The electric flux $\Delta\Phi_E$ evaluated for a "small" element of area $\Delta\mathbf{S}$ at which an electric field $\mathbf{E}$ exists is approximated by

$$\Delta\Phi_E \cong \mathbf{E} \cdot \Delta\mathbf{S} = E \, \Delta S \cos\theta \qquad (12\text{-}17)$$

where θ is the angle between $\mathbf{E}$ and $\Delta\mathbf{S}$. By small, we mean the element of area is sufficiently small that the electric field is essentially constant over the area ΔS. If,

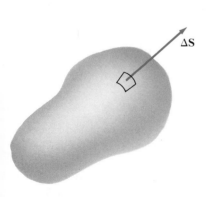

Figure 12.7
A closed surface on which a small area is represented by the vector $\Delta\mathbf{S}$.

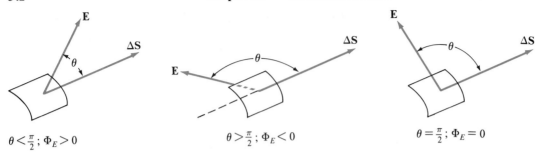

$$\theta < \frac{\pi}{2}; \Phi_E > 0 \qquad\qquad \theta > \frac{\pi}{2}; \Phi_E < 0 \qquad\qquad \theta = \frac{\pi}{2}; \Phi_E = 0$$

Figure 12.8
Three orientations of **E** and **ΔS** for which the electric flux Φ_E is positive, negative, and zero, respectively.

in fact, **E** is constant over a large flat area **ΔS**, Equation (12-17) is then an exact equality. Figure 12.8 shows three orientations of **E** and **ΔS** for which the electric flux $\Delta\Phi_E$ is positive, negative, and zero.

E 12.6   At each point on the flat rectangular surface shown in the diagram, the electric field has a magnitude of 350 N/C and makes an angle of 50° with **ΔS**. Calculate the electric flux for this surface.   Answer: 6.7 N·m²/C

E 12.7   The electric field is constant at each point on a flat circular surface having a 5.0-cm radius. Calculate the electric flux for a surface in which **E** has a magnitude of 3.2×10^3 N/C and is directed 120° from **ΔS** (*see* the accompanying diagram).   Answer: −13 N·m²/C

E 12.8   The electric field in a region of space is constant and given by $\mathbf{E} = (35\hat{\mathbf{i}} - 18\hat{\mathbf{j}} + 45\hat{\mathbf{k}})$ N/C. Calculate the magnitude of the electric flux for a flat surface (area = 0.18 m²) that is parallel to the *x-y* plane.   Answer: 8.1 N·m²/C

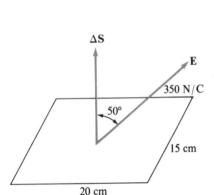

Exercise 12.6

If a large area *S* is composed of a number of small areas **ΔS** (over each of which the electric field is essentially constant) to which directions have been assigned, the electric flux for *S* is approximated by

$$\Phi_E \cong \sum_S \mathbf{E} \cdot \Delta\mathbf{S} = \sum_S E\,\Delta S \cos\theta \qquad (12\text{-}18)$$

in which the summation takes place over the entire surface *S*. If now each small area **ΔS** becomes arbitrarily small, we may precisely define the *electric flux* Φ_E for a surface to be

$$\Phi_E = \int_S \mathbf{E} \cdot d\mathbf{S} \qquad (12\text{-}19)$$

in which the integration takes place over the surface *S*. Notice that in Equation (12-19), $\mathbf{E} \cdot d\mathbf{S} = E\,dS\cos\theta$ is the product of the component of **E** normal to the surface at each point on the surface and the infinitesimal element of area *dS*. If the surface *S* is closed and *d***S** is directed outward (from the enclosed volume) at every point on the surface, we define the *net electric flux for a closed surface* as

$$\Phi_E = \oint \mathbf{E} \cdot d\mathbf{S} = \oint E\cos\theta\,dS \qquad (12\text{-}20)$$

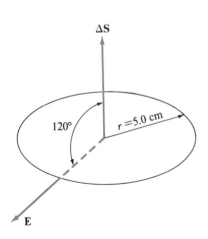

Exercise 12.7

A surface integral, like that of Equation (12-20), is difficult to perform for arbitrary electric fields and surfaces. We will use this relationship only in situations for which the integration is extremely simple and is quite useful.

For example, using Figure 12.9 let us calculate the electric flux for the surface of each of two cubes located in two different positions relative to a very large plane sheet of positive charge. Recall from Exercise 12.3 that such a charged sheet, on which there is a uniform positive charge density σ, has a uniform electric field on each side of the sheet. The magnitude of the field is $E = \sigma/(2\epsilon_o)$, and the field is normal to and away from the charged plane. Figure 12.9(a) shows such a plane of charge and a cubic region of space that lies entirely outside the plane of charge. Each face of the cube has area A, and the cube is oriented so that two of its faces are parallel to the charged plane. The electric flux for the surface of the cube is given by Equation (12-20). The evaluation of the integral is simple if we execute the integration piecemeal. At every point on the cube face labeled A_1, for example, $\mathbf{E}$ is constant and is in the same direction as $d\mathbf{S}$, that is, $\theta = 0$, so that

$$\int_{A_1} \mathbf{E} \cdot d\mathbf{S} = E \int_{A_1} dS = EA \qquad (12\text{-}21)$$

At all points on the face labeled A_2, however, $\mathbf{E}$ and $d\mathbf{S}$ are oppositely directed ($\theta = \pi$), and the flux for A_2 is

$$\int_{A_2} \mathbf{E} \cdot d\mathbf{S} = -E \int_{A_2} dS = -EA \qquad (12\text{-}22)$$

Everywhere on the remaining four faces, $\mathbf{E}$ is perpendicular to $d\mathbf{S}$ ($\theta = \pi/2$), and the flux for each of these faces is equal to zero. Then the net flux for the closed surface of the cube is equal to the sum of Equations (12-21) and (12-22):

$$\Phi_E = EA - EA = 0 \qquad (12\text{-}23)$$

Equation (12-23) says the flux for the closed cubic surface is equal to zero when the surface is in a region of uniform electric field. Although we will not prove it here, we assert that the electric flux for *any* closed surface is zero if the surface lies entirely within a region of constant electric field.

Now suppose an identical cubical surface is positioned, as shown in Figure 12.9(b), so that the charged plane passes through the cube parallel to two faces of the cube. The electric flux for A_1 is equal to EA, and the flux for A_2 is likewise

Figure 12.9
A cubic region of space positioned so that two of its faces are parallel to a plane of charge. **(a)** The cubic region lies entirely outside the plane of charge. **(b)** The cubic region includes a portion of the plane of charge.

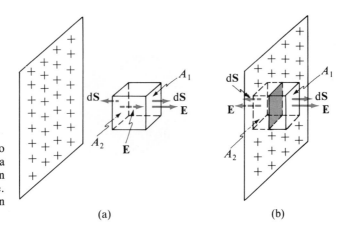

(a)   (b)

equal to *EA*. Because the flux for each of the remaining four faces of the cube is once again equal to zero, the net flux for the closed surface of the cube is

$$\Phi_E = 2EA \tag{12-24}$$

But because $E = \sigma/2(\epsilon_o)$, Equation (12-24) becomes

$$\Phi_E = 2\left(\frac{\sigma}{2\epsilon_o}\right)A = \frac{\sigma A}{\epsilon_o} \tag{12-25}$$

The product σA is equal to the net change Q_{net} lying within the closed surface of the cube. We may, therefore, write Equation (12-25) as

$$\Phi_E = \frac{Q_{net}}{\epsilon_o} \tag{12-26}$$

The result given by Equation (12-26) was obtained for a special—and an easily calculable—situation. Nevertheless, the conclusions obtained in this special case are quite generally applicable to *any* closed surface in *any* electric field:

The electric flux for any closed surface is proportional to the *net* charge enclosed by that surface.

This relationship between electric flux for a surface and the net charge enclosed by that surface is called *Gauss's law*, which may be stated mathematically as

$$\oint_S \mathbf{E} \cdot d\mathbf{S} = \frac{Q_{net}}{\epsilon_o} \tag{12-27}$$

in which Q_{net} is the algebraic sum of the charges within the closed surface S, referred to as a gaussian surface. Notice that a gaussian surface need not correspond to a physical surface; it may be any convenient imaginary closed surface. As we will see, an important class of electrostatic problems may be treated using Gauss's law, but before we explore the utility of Gauss's law, let us examine one further simple situation that relates that law to Coulomb's law.

Consider a closed spherical surface S having a radius R and a positive point charge Q at its center, as shown in Figure 12.10. Every element of area $d\mathbf{S}$ is

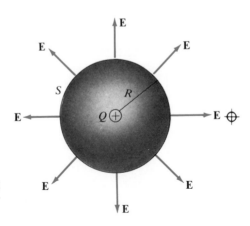

Figure 12.10
A closed spherical surface S having radius R and enclosing a charge Q located at the center of the sphere.

radially outward and has a magnitude given by

$$E = k\frac{Q}{R^2} = \frac{1}{4\pi\epsilon_{o}}\frac{Q}{R^2} \qquad (12\text{-}28)$$

Equation (12-28) was obtained, we should recall, from Coulomb's law, $\mathbf{F} = (kQq/r^2)\hat{\mathbf{r}}$, and the definition of electric field, $\mathbf{E} = \mathbf{F}/q$. The electric flux for the closed spherical surface is

$$\Phi_E = \oint_S \mathbf{E}\cdot d\mathbf{S} = \frac{Q}{4\pi\epsilon_{o}R^2}\oint_S dS = \frac{Q}{4\pi\epsilon_{o}R^2}\cdot 4\pi R^2 \qquad (12\text{-}29)$$

in which we have used the fact that the surface area of a sphere of radius R is $4\pi R^2$. Then Equation (12-29) becomes

$$\oint_S \mathbf{E}\cdot d\mathbf{S} = \frac{Q_{\text{net}}}{\epsilon_{o}}$$

which is precisely Gauss's law, Equation (12-27).

E 12.9   Charge is uniformly distributed throughout a region of space with a density $\rho = 1.5 \times 10^{-8}$ C/m^3. Calculate the electric flux for a spherical surface (radius = 0.20 m) in this region.   Answer: 57 N·m^2/C

E 12.10   At each point on the surface of the cube shown in the accompanying figure, the electric field is in the positive x direction. On face A, $\mathbf{E}$ is constant and given by $\mathbf{E}_A = 95\hat{\mathbf{i}}$ N/C; and on face B, $\mathbf{E}_B = 40\hat{\mathbf{i}}$ N/C. Calculate the electric flux for the entire cubical surface.   Answer: 14 N·m^2/C

E 12.11   Calculate the net charge contained in the cube of Exercise 12.10.   Answer: 1.2×10^{-10} C

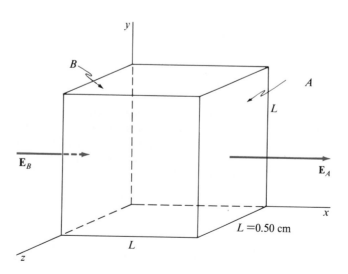

Exercise 12.10

E 12.12   Charge having a uniform density of 8.0×10^{-10} C/m³ is distributed throughout a spherical volume with a radius of 0.40 m. Determine the electric flux for a concentric spherical surface having a radius of 0.20 m.

Answer: 3.0 N·m²/C

E 12.13   Repeat Exercise 12.12 for a concentric spherical surface having a radius of 0.50 m.   Answer: 24 N·m²/C

Gauss's law is useful in electrostatics because it permits us to determine the electric fields in and around certain charge distributions that are highly symmetrical. We have just seen, for example, how the electric field of a point charge is related to the flux for a closed spherical surface centered on that charge. Our result was dependent on the spherical symmetry associated with a point charge. Let us now see how we may use Gauss's law to determine the electric field of a charge distribution with a different symmetry:

A long cylinder of radius R contains a positive charge of uniform density ρ throughout its volume. Determine the electric field outside and inside the cylindrical charge distribution.

First let us consider the consequences of the cylindrical symmetry of the charge distribution. Figure 12.11 shows a side view and a cross section of the cylindrical charge distribution. An arbitrary charge element dQ_1 produces an electric field $d\mathbf{E}_1$, which is directed away from dQ_1, at a field point P located a radial distance r from the axis of the cylinder. Because the charge distribution is a "long" cylinder, we are assured that there is a charge element dQ_2 that is symmetrically located with respect to the point A (on the radial line from the axis to the field point P) and produces the field $d\mathbf{E}_2$ at P. Because of the symmetric locations of the elements dQ_1 and dQ_2, all the nonradial components of $d\mathbf{E}_1$ and $d\mathbf{E}_2$ sum to zero. The radial components of $d\mathbf{E}_1$ and $d\mathbf{E}_2$ add to give a resultant electric field $d\mathbf{E}$ that is radially outward from the axis of the cylinder. Because the charge element dQ_1 used here is at an arbitrary location, we may conclude that the electric field at every field point is directed radially outward. Furthermore, we may observe from the symmetry of the cross section in Figure 12.11 that the electric fields at all points a distance r from the axis have the same magnitude.

We may now use these symmetry properties of the charge distribution to evaluate the electric field at points outside the cylinder of charge. Keeping in mind the symmetry properties of the electric field, we construct a cylindrical surface of radius r and length L, as shown in Figure 12.12. At every point on this surface, the electric field $\mathbf{E}$ is radially outward and has a constant magnitude E. If we include the flat "caps," or ends, of this cylindrical surface, we have constructed a closed surface. We now use this closed surface as a gaussian surface. Using Gauss's law,

$$\oint_S \mathbf{E} \cdot d\mathbf{S} = \frac{Q_{net}}{\epsilon_0} \qquad (12\text{-}30)$$

we must evaluate the integral over the chosen gaussian surface S and express Q_{net}, the charge enclosed within that surface, in terms of the given quantities ρ

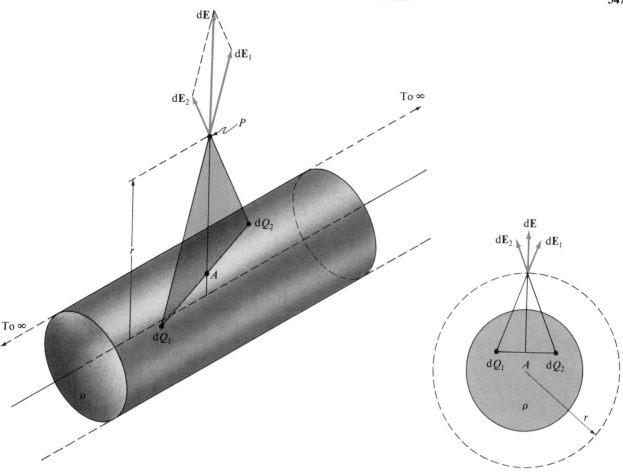

Figure 12.11
(a) A perspective view and (b) a cross-sectional view of a solid cylinder with volume charge density ρ. The electric field d**E** at a perpendicular distance r from the axis of the cylinder has component vectors d**E**$_1$ and d**E**$_2$ caused by charge elements dQ_1 and dQ_2, which are located symmetrically with respect to point A.

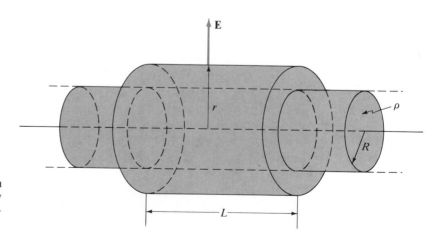

Figure 12.12
A cylindrical gaussian surface of length L and radius r constructed coaxially with and outside of a long charged cylinder of radius R.

and R. The surface integral of Equation (12-30) may be expressed as the sum of two integrals, one over the curved surface (c.s.) and the other over the caps of the surface:

$$\oint_S \mathbf{E} \cdot d\mathbf{S} = \int_{\text{c.s.}} \mathbf{E} \cdot d\mathbf{S} + \int_{\text{caps}} \mathbf{E} \cdot d\mathbf{S} \qquad (12\text{-}31)$$

The integral over the caps is equal to zero because $\mathbf{E}$ is perpendicular to $d\mathbf{S}$ at every point on the caps ($\mathbf{E}$ is radially directed and $d\mathbf{S}$ is parallel to the axis at every point). Because $\mathbf{E}$ is normal to the gaussian surface (parallel to $d\mathbf{S}$) at every point on the curved surface and because the magnitude of the electric field is constant at every point on the curved surface, we may write Equation (12-31) as

$$\oint_S \mathbf{E} \cdot d\mathbf{S} = E \int_{\text{c.s.}} dS + 0 \qquad (12\text{-}32)$$

And because the area of the curved surface is the product of the circular distance $2\pi r$ around the surface and its length L, Equation (12-32) becomes

$$\oint_S \mathbf{E} \cdot d\mathbf{S} = E(2\pi rL) \qquad (12\text{-}33)$$

The charge Q_{net} enclosed within the gaussian surface is that charge within the volume V of the cylinder of radius R and length L:

$$Q_{\text{net}} = \rho V = \rho(\pi R^2 L) \qquad (12\text{-}34)$$

Substituting Equations (12-33) and (12-34) into Equation (12-30) gives

$$E(2\pi rL) = \frac{\rho \pi R^2 L}{\epsilon_o}$$

$$E = \frac{\rho R^2}{2\epsilon_o r} \qquad (12\text{-}35)$$

Thus, the electric field outside the charge distribution is directed radially outward from the axis of the cylinder and has a magnitude that decreases with distance r from the axis in proportion to $1/r$.

The electric field inside the charge distribution is similarly obtained using Gauss's law, except that we now construct the cylindrical gaussian surface at a radius $r < R$, as shown in Figure 12.13. The same symmetry arguments apply to the electric field at points on this surface, and Equation (12-33) applies to this gaussian surface. On the other hand, we must reevaluate Q_{net} for a surface within the charge distribution. If we let V_{in} represent the volume within the gaussian surface, the net charge within the surface is

$$Q_{\text{net}} = \rho V_{\text{in}} = \rho \pi r^2 L \qquad (12\text{-}36)$$

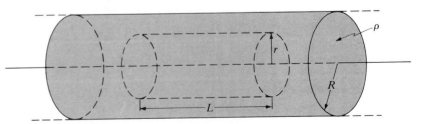

Figure 12.13
A cylindrical gaussian surface of length L and radius r constructed coaxially with and inside of a long charged cylinder of radius R.

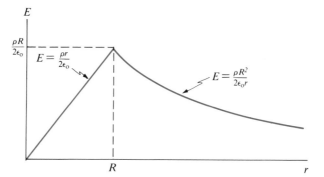

Figure 12.14
A graph of the magnitude of the radial electric field of a long cylinder carrying a uniform volume charge density ρ, plotted as a function of r, the distance from the axis of the cylinder of radius R.

Substituting Equations (12-33) and (12-36) into Gauss's law, Equation (12-30), gives

$$E \cdot 2\pi rL = \frac{\rho \pi r^2 L}{\epsilon_o}$$

$$E = \frac{\rho r}{2\epsilon_o} \qquad\qquad (12\text{-}37)$$

which indicates that the magnitude of the radial electric field within the cylindrical charge distribution increases in direct proportion to the distance r from the axis of the cylinder.

Thus the solution of this problem may be stated

$$\mathbf{E} = \frac{\rho R^2}{2\epsilon_o r}\hat{\mathbf{r}}, \qquad r \geq R \qquad\qquad (12\text{-}38)$$

$$\mathbf{E} = \frac{\rho r}{2\epsilon_o}\hat{\mathbf{r}}, \qquad r \leq R \qquad\qquad (12\text{-}39)$$

and a check on our solution is afforded by evaluating the solution at $r = R$, the radius of the cylinder of charge. In that special case, the value of E is obtained by equating r and R in Equations (12-38) and (12-39):

$$E = \frac{\rho R}{2\epsilon_o}, \qquad r = R \qquad\qquad (12\text{-}40)$$

Figure 12.14 shows the magnitude of **E**, which is directed radially outward from the axis of the cylinder, as a function of r, the distance from the axis.

Example 12.3

PROBLEM   A continuous charge of uniform surface density $\sigma > 0$ is distributed over an infinite plane. Use Gauss's law to determine the magnitude of the electric field that results from this distribution.

SOLUTION   The planar symmetry of the charge distribution ensures that: (1) the electric field **E** is directed perpendicularly from the charged plane; (2) the magnitude of **E** is constant on a flat surface that is parallel to the charged plane; and (3) the magnitude of **E** is the same for any two points on opposite sides of, but equidistant from, the charged plane.

With these symmetry considerations in mind, construct a cylindrical gaussian surface, oriented as shown in Figure 12.15, with its axis oriented perpendicularly to the charged plane and with its flat (circular) ends, each of area A, on opposite sides of, and equidistant from, the charged plane. Because **E** is parallel to the

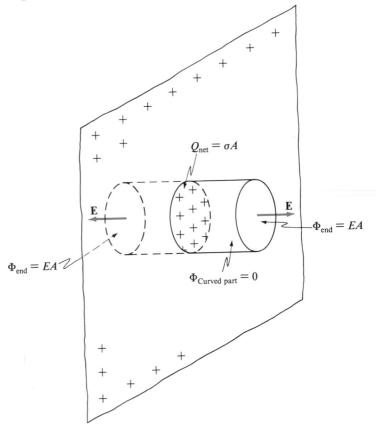

Figure 12.15   Example 12.3

axis of the cylinder, the electric flux for the curved part of the cylinder is equal to zero. The electric flux for each end is equal to EA, where E is the magnitude of the electric field on either end. Thus the total electric flux for the gaussian surface is equal to $2EA$. Finally, because the net charge enclosed by the surface is equal to σA, Gauss's law requires that

$$2EA = \frac{1}{\epsilon_o} \sigma A$$

or

$$E = \frac{\sigma}{2\epsilon_o}$$

in agreement with the result of Exercise 12.3.   ■

Example 12.4   PROBLEM   A positive charge Q is distributed uniformly throughout a spherical shell of inner radius a and outer radius b, as shown in Figure 12.16(a). Use Gauss's law to determine the magnitude of the electric field as a function of distance from the center of the spherical volume.

SOLUTION   The spherical symmetry of the charge distribution ensures that the electric field $\mathbf{E}$ exhibits spherical symmetry, that is, $\mathbf{E}$ is radially outward and its magnitude depends only upon the radial distance r from the center. An appropriate gaussian surface for this charge distribution is a sphere that is concentric with the charge center. The electric flux for the gaussian sphere having a radius r is then

$$\phi_E = EA = 4\pi r^2 E$$

where A is the surface area of the sphere. The net charge enclosed by this gaussian surface depends upon the value of r according to (*see* Fig. 12.16(b))

$$Q_{net} = 0 \qquad\qquad\qquad\qquad\qquad\qquad\qquad 0 < r < a$$

$$Q_{net} = \rho\frac{4}{3}\pi(r^3 - a^3) = \frac{Q}{\frac{4}{3}\pi(b^3 - a^3)} \cdot \frac{4}{3}\pi(r^3 - a^3)$$

$$= Q\frac{r^3 - a^3}{b^3 - a^3} \qquad\qquad\qquad\qquad a < r < b$$

$$Q_{net} = Q \qquad\qquad\qquad\qquad\qquad\qquad\qquad b < r$$

Equating the electric flux to Q_{net}/ϵ_o gives for each of these intervals

$$4\pi r^2 E = 0 \qquad\qquad\qquad\qquad 0 < r < a$$

$$4\pi r^2 E = \frac{Q}{\epsilon_o}\frac{r^3 - a^3}{b^3 - a^3} \qquad\quad a < r < b$$

$$4\pi r^2 E = \frac{Q}{\epsilon_o} \qquad\qquad\qquad\quad b < r$$

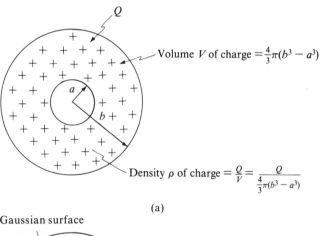

Volume V of charge $= \frac{4}{3}\pi(b^3 - a^3)$

Density ρ of charge $= \frac{Q}{V} = \dfrac{Q}{\frac{4}{3}\pi(b^3 - a^3)}$

(a)

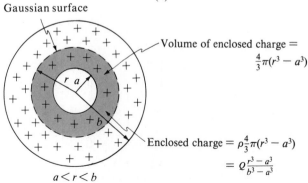

Gaussian surface

Volume of enclosed charge $=$
$$\frac{4}{3}\pi(r^3 - a^3)$$

Enclosed charge $= \rho\frac{4}{3}\pi(r^3 - a^3)$
$$= Q\frac{r^3 - a^3}{b^3 - a^3}$$

$a < r < b$

(b)

Figure 12.16   Example 12.4

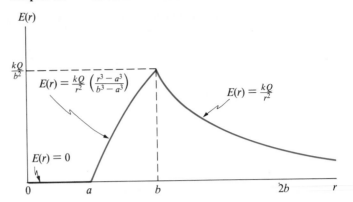

Figure 12.17   Example 12.4

Solving for E gives

$$E = 0 \qquad\qquad\qquad 0 < r < a$$

$$E = \frac{kQ}{r^2}\,\frac{r^3 - a^3}{b^3 - a^3} \qquad a < r < b$$

$$E = \frac{kQ}{r^2} \qquad\qquad\quad b < r$$

This dependence of E on r is shown graphically in Figure 12.17. Notice that for points inside the hollow part of the charge, E is identically zero; and for points outside the charged volume, E is the same as if the total charge were a point charge Q at the center of the charged volume. Satisfy yourself that the expressions for E behave appropriately at $r = a$ and $r = b$. ∎

Example 12.5

PROBLEM   A continuous charge of density $\rho > 0$ fills the infinite "slab" of thickness $2a$ defined by $-a < x < a$, as shown in Figure 12.18(a). Determine the electric field caused by this charge.

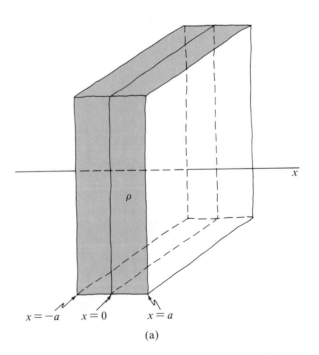

Figure 12.18(a)   Example 12.5   (a)

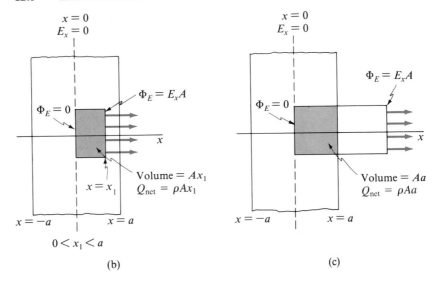

Figures 12.18(b) and (c)
Example 12.5

(b)   (c)

SOLUTION   The planar symmetry of this charge distribution ensures that the electric field $\mathbf{E}$ has an x component E_x only and that E_x depends upon x only. Consequently, an appropriate gaussian surface is the surface of a cylinder constructed with its axis parallel to the x axis and its two flat ends located at $x = 0$ and $x = x_1$, as shown in Figures 12.18(b) and (c). The electric flux is equal to zero for (i) the curved part of the cylindrical surface, because $\mathbf{E}$ is parallel to this part of the surface, and (ii) the end at $x = 0$ because $\mathbf{E} = \mathbf{0}$ in this symmetry plane. (Why?) Thus the total electric flux for this gaussian surface is equal to $E_x A$ where E_x is equal to the magnitude of $\mathbf{E}$ for $x = x_1$ and A is the cross-sectional area of the cylinder. For $0 \leq x_1 \leq a$, as in Figure 12.18(b), the net charge enclosed by this gaussian surface is given by

$$Q_{net} = \rho V = \rho A x_1 \qquad 0 \leq x_1 \leq a$$

where the volume V of enclosed charge is equal to Ax_1. For $a < x_1$, as in Figure 12.18(c), the net charge is given by

$$Q_{net} = \rho V = \rho A a \qquad a < x_1$$

because the enclosed charge is in the volume $V = Aa$. Using Gauss's law gives

$$E_x A = \frac{\rho A x_1}{\epsilon_o} \qquad 0 \leq x_1 \leq a$$

$$E_x A = \frac{\rho A a}{\epsilon_o} \qquad a < x_1$$

Solving for E_x and dropping the subscript from x_1 (because x_1 was arbitrary) gives

$$E_x(x) = \frac{\rho x}{\epsilon_o} \qquad 0 \leq x \leq a$$

$$E_x(x) = \frac{\rho a}{\epsilon_o} \qquad a < x$$

Figure 12.19 shows this dependence of E_x upon x. ∎

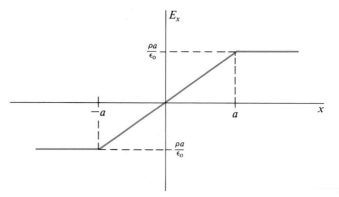

Figure 12.19   Example 12.5

Example 12.6

PROBLEM   Two uniformly charged, parallel sheets are arranged as shown in Figure 12.20(a). The sheet on the left has a uniform charge density $\sigma > 0$, and the sheet on the right has a uniform charge density equal to -2σ. Determine the electric field at the points A, B, and C shown.

SOLUTION   Figure 12.20(b) shows the fields $\mathbf{E}_+$ and $\mathbf{E}_-$ (caused by the positive and negative charges) at each of the three points. Because $|\mathbf{E}_+| = \sigma/(2\epsilon_0)$ and $|\mathbf{E}_-| = \sigma/\epsilon_0$, we see, referring to the figure, that

$$\mathbf{E}_A = \mathbf{E}_+ + \mathbf{E}_- = -\frac{\sigma}{2\epsilon_0}\hat{\mathbf{i}} + \frac{\sigma}{\epsilon_0}\hat{\mathbf{i}} = \frac{\sigma}{2\epsilon_0}\hat{\mathbf{i}}$$

$$\mathbf{E}_B = \mathbf{E}_+ + \mathbf{E}_- = \frac{\sigma}{2\epsilon_0}\hat{\mathbf{i}} + \frac{\sigma}{\epsilon_0}\hat{\mathbf{i}} = \frac{3\sigma}{2\epsilon_0}\hat{\mathbf{i}}$$

$$\mathbf{E}_C = \mathbf{E}_+ + \mathbf{E}_- = \frac{\sigma}{2\epsilon_0}\hat{\mathbf{i}} - \frac{\sigma}{\epsilon_0}\hat{\mathbf{i}} = -\frac{\sigma}{2\epsilon_0}\hat{\mathbf{i}}$$

where $\hat{\mathbf{i}}$ is a unit vector having the usual left to right direction. ∎

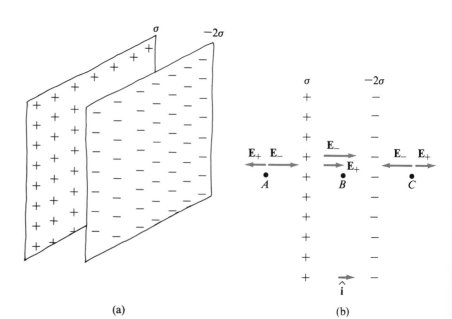

Figure 12.20   Example 12.6   (a)   (b)

12.4   Electrostatic Properties of Conductors

We now have all the tools necessary to establish the basic properties of conducting materials in the electrostatic situation. The key word here is *electrostatic*, the condition in which any charges that may be present are at rest.

First, the electrostatic condition requires that the electric field be zero at all points within a conducting medium. Recall that a conductor is a material containing charges (electrons) that are free to move about. If an electric field existed within a conductor, the free charges would experience a net force and would accelerate, thereby violating the condition of electrostatic equilibrium. Thus, we may conclude that $\mathbf{E} = \mathbf{0}$ everywhere within a conducting medium in an electrostatic situation.

Now suppose a net charge is placed on a solid conductor insulated from its surroundings. Where does the electrostatic charge reside? Figure 12.21 represents a conducting material inside of which we have constructed a gaussian surface *immediately* beneath the surface. Because the electric field is equal to zero inside the conducting medium, the net electric flux for this gaussian surface must be equal to zero. Then, according to Gauss's law, no net charge resides within the gaussian surface. We may conclude, therefore, that no net charge may reside inside a solid conductor. It follows that any net charge on a solid conductor must reside on the surface of that conductor.

What is the direction of the electric field at the surface of a charged conductor in the electrostatic situation? There can be no component of $\mathbf{E}$ parallel to the surface; otherwise, free charges on the surface would accelerate along the surface in response to the force caused by the electric field, thereby violating the assumption of static equilibrium. Therefore, the electric field at the surface of a charged conductor is normal to the surface at every point on the surface.

Finally, we may determine the magnitude of the electric field immediately outside the surface of a charged conductor. Suppose a conductor has an electrostatic charge density σ on its surface. Let us construct a gaussian surface in the shape of a cylindrical pillbox, as shown in Figure 12.22, that has ends, each with a small area ΔS and encloses a net charge $\sigma \Delta S$ on the surface of the conductor. Because the electric field is equal to zero inside the conductor and is normal to the conducting surface (and, therefore, is normal to the end of the pillbox), the net electric flux for the ends of the pillbox is $E \Delta S$. The flux for the curved side of the pillbox is equal to zero, because $\mathbf{E}$ and $\Delta \mathbf{S}$ are perpendicular to each other at every point on that surface. Then using Gauss's law, we obtain

$$E \, \Delta S = \frac{\sigma \Delta S}{\epsilon_0}$$

$$E = \frac{\sigma}{\epsilon_0} \tag{12-41}$$

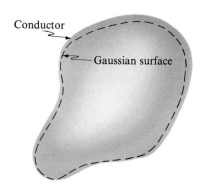

Figure 12.21
A gaussian surface constructed immediately beneath the surface of a conducting material.

Figure 12.22
A small cylindrical gaussian surface including a region on the surface of a conductor. The caps of the cylinder are parallel to the surface of the conductor; therefore the electric field caused by charge on the surface is normal to both the conducting surface and the cap of the cylinder.

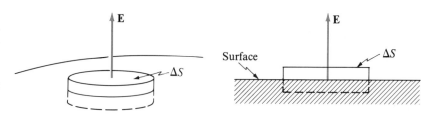

which relates at each point the electric field at the surface of a conductor to the charge density on that surface.

We have now established four properties of a conductor in an electrostatic situation:

1. $\mathbf{E} = \mathbf{0}$ everywhere inside a conducting medium.

2. No net charge resides inside a solid conductor.

3. $\mathbf{E}$ is everywhere normal to the surface of a conductor.

4. The magnitude of $\mathbf{E}$ immediately outside the surface of the conductor is equal to σ/ϵ_o.

A nonconducting sphere of radius a has a uniform positive charge density ρ throughout its volume. The charged sphere is surrounded concentrically by a neutral conducting spherical shell with inner radius b and outer radius c. Determine the surface charge density σ on the outer surface of the conducting shell.

We have just obtained four properties of a conducting material in an electrostatic situation. Let us solve the present problem in two ways. First we will use properties 1 and 2.

The physical situation of the problem is pictured in Figure 12.23. Let us construct a spherical gaussian surface, concentric with the sphere and shell and having a radius r that is greater than b and less than c. Thus, the imaginary surface lies within the conducting shell, where, according to property 1, the electric field is equal to zero. The total flux for the gaussian surface is, therefore, equal to zero, and Gauss's law demands that the net charge Q_{net} within that surface must be equal to zero. But the net charge within the spherical region with radius r is equal to the algebraic sum of the charge Q_s within the sphere of radius a and the induced charge Q_i on the inside surface of the conducting shell (property 2 permits no net charge within the conductor):

$$Q_{\text{net}} = Q_s + Q_i$$

$$0 = \rho\left(\frac{4}{3}\pi a^3\right) + Q_i$$

$$Q_i = -\frac{4\pi\rho a^3}{3} \tag{12-42}$$

Thus the induced charge on the inner surface of the conducting shell is negative. Because the shell has no net charge, there must be a surface charge Q_o on the outer surface of the shell that is equal in magnitude but opposite in sign to the charge Q_i on the inner surface, or

$$Q_o = +\frac{4\pi\rho a^3}{3} \tag{12-43}$$

The surface charge density on the outer surface of the shell is, therefore, the induced charge Q_o on the outer surface divided by the area $4\pi c^2$ of the outer surface:

$$\sigma = \frac{4\pi\rho a^3}{3(4\pi c^2)} = \frac{\rho a^3}{3c^2} \tag{12-44}$$

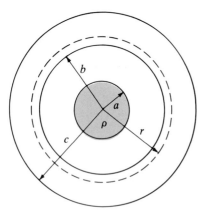

Figure 12.23
A nonconducting sphere having radius a and having a uniform volume charge density ρ. The sphere is surrounded by a concentric conducting shell with inner radius b and outer radius c. A spherical gaussian surface of radius r has been constructed within the conducting material.

Now let us solve the same problem using properties 3 and 4. Consider a spherical gaussian surface immediately outside the surface of radius c. Property 3 assures us that the electric field here is normal to this surface, so we may apply Gauss's law to this surface, giving

$$E(4\pi c^2) = \frac{Q_{net}}{\epsilon_o} \tag{12-45}$$

The net charge within the gaussian surface is equal to the charge $4\pi\rho a^3/3$ inside the sphere of radius a. Thus Equation (12-45) becomes

$$E(4\pi c^2) = \frac{4\pi\rho a^3}{3\epsilon_o}$$

$$E = \frac{\rho a^3}{3\epsilon_o c^2} \tag{12-46}$$

But property 4 equates E to σ/ϵ_o, or:

$$\frac{\sigma}{\epsilon_o} = \frac{\rho a^3}{3\epsilon_o c^2}$$

$$\sigma = \frac{\rho a^3}{3c^2} \tag{12-47}$$

which is identical to the result obtained in Equation (12-44).

E 12.14   The electric field at the surface of a spherical conductor has a magnitude of 85 N/C and points radially inward. Determine the density of charge on the surface of the conductor.   Answer: -7.5×10^{-10} C/m²

Example 12.7

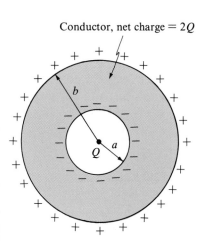

Conductor, net charge $= 2Q$

Figure 12.24   Example 12.7

PROBLEM   A point charge $Q > 0$ is placed at the center of a hollow spherical conductor carrying a net charge of $2Q$, as shown in Figure 12.24. Determine the magnitude of the electric field that results from this charge distribution.

SOLUTION   First, because $\mathbf{E} = 0$ for all points inside the conducting medium, there must be an induced charge equal to $-Q$ distributed uniformly on the inner ($r = a$) surface of the hollow conductor. Consequently, the outer ($r = b$) surface must have a uniformly distributed charge equal to $3Q$. The spherical symmetry of this charge distribution ensures that the electric field $\mathbf{E}$ is spherically symmetric and that the appropriate gaussian surfaces are spheres with radius r and are centered on Q. For any such gaussian sphere, the electric flux is equal to $4\pi r^2 E$, where E is the magnitude of $\mathbf{E}$ on the surface. For $0 < r < a$, the enclosed charge is equal to Q, and Gauss's law gives

$$4\pi r^2 E = \frac{Q}{\epsilon_o}$$

or

$$E = \frac{kQ}{r^2} \qquad 0 < r < a$$

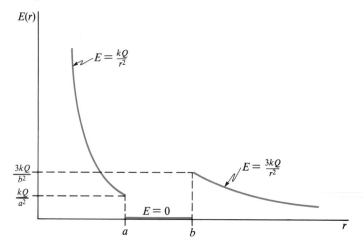

Figure 12.25   Example 12.7

For $b < r$, the enclosed charge is equal to $3Q$, and we get

$$4\pi r^2 E = \frac{3Q}{\epsilon_0}$$

or

$$E = \frac{3kQ}{r^2} \qquad b < r$$

Of course, in the conducting medium, that is, $a < r < b$, the electric field is equal to zero. Figure 12.25 depicts graphically the dependence of E upon r. ∎

One final point may help to avoid confusion. We have seen that, just outside a conductor having a surface charge density σ, the electric field has a magnitude equal to σ/ϵ_0. In Example 12.3 we saw that the constant electric field outside an infinite plane of charge density σ is $\sigma/(2\epsilon_0)$. A brief consideration of Gauss's law makes it clear why the electric field outside the plane of charge is half that just outside the conductor. Figure 12.26(a) shows a gaussian surface in the shape of a

Figure 12.26
A cylindrical gaussian surface in the shape of a pillbox enclosing an area A containing charge on (a) a plane having a surface charge density σ, and (b) the surface of a conductor having a surface charge density σ.

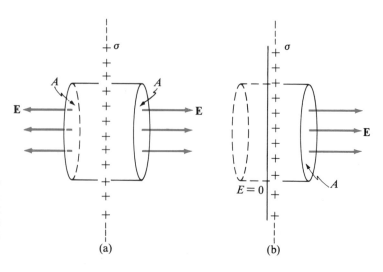

(a)   (b)

cylindrical pillbox that encloses an area A on a large plane having surface charge density σ. Electric flux for *each* cap of the pillbox is equal to EA. Because $\mathbf{E}$ is parallel to the curved surface of the pillbox at every point, the flux for that surface is equal to zero; the total flux for the gaussian surface is $2EA$. Then Gauss's law requires that $2EA = \sigma A/\epsilon_o$, or $E = \sigma/(2\epsilon_o)$. In Figure 12.26(b), however, a similar pillbox encloses a surface area A on a conductor bearing a surface charge density σ. The flux for the cap of the pillbox outside the conductor is again equal to EA; inside the conductor, where $\mathbf{E} = \mathbf{0}$, the flux for the inner cap is equal to zero. Then because the total flux for this gaussian surface is equal to EA, Gauss's law gives $EA = \sigma A/\epsilon_o$, or $E = \sigma/\epsilon_o$.

12.5 Problem-Solving Summary

The electric field $\mathbf{E}$ at a point in space is a vector quantity that describes the force $\mathbf{F}$ that a test charge q would experience if it were placed at that point, according to $\mathbf{E} = \mathbf{F}/q$. Because electrical forces on charges are caused by other charges, we often need to describe the electric fields in space around charges—a point charge, groups of point charges, and continuous distributions of charge.

The electric field at a distance r from a point charge Q has a magnitude given by $E = kQ/r^2$. The direction of $\mathbf{E}$ is radially outward, or away from, a positive charge and radially inward, or toward, a negative charge. Both the magnitude and direction of an electric field can be expressed by $\mathbf{E} = \hat{\mathbf{r}}kQ/r^2$, in which Q may be positive or negative and $\hat{\mathbf{r}}$ is a unit vector radially outward from the point charge.

The electric field caused by a group of n point charges is determined by first calculating the field (a vector) caused by each individual point charge and then summing (vectorially) the fields of these charges: $\mathbf{E} = \mathbf{E}_1 + \mathbf{E}_2 + \ldots + \mathbf{E}_n$.

Finding the electric field of a continuous charge distribution requires, in general, that we perform one or more integrations. To determine the electric field $\mathbf{E}$ caused by a continuous charge distribution, we choose an arbitrary charge element dQ within the distribution and treat that element as if it were a point charge. That is, dQ produces an electric field $d\mathbf{E}$ with a magnitude at a distance r given by $dE = k\,dQ/r^2$. If the electric fields $d\mathbf{E}$ caused by the different charge elements of the distribution are in different directions, the field of each element may be broken down into convenient component parts, like $dE_x\hat{\mathbf{i}}$ and $dE_y\hat{\mathbf{j}}$, for example. Then each component of $\mathbf{E}$ can be calculated separately by integrating the appropriate component of $k\,dQ/r^2$ throughout the entire region containing charge. In order to perform the integration, dQ and r can usually be expressed in terms of a single variable. Expressing the factors of the integral in terms of one variable is usually most easily done by carefully considering and using the geometry of the problem. Thus the visualization of the problem—depiction is especially important in these problems—should be carefully indicated and labeled. When the integration of the components has been carried out, the total electric field at the field point is determined by vectorially adding the component vectors.

Gauss's law is a practical shortcut to the solution of problems that ask us to find the electric fields caused by charge distributions that have spherical, cylindrical, or planar symmetry. Spherical charge distributions, for example, permit us to construct spherical gaussian surfaces over which the electric field $\mathbf{E}$ is both normal to the surface at all points on the surface and constant in magnitude at those points. If the radius of the gaussian surface is r, the electric flux for the surface is $E(4\pi r^2)$, which, according to Gauss's law is equal to Q_{net}/ϵ_o, where Q_{net} is the net charge within the gaussian sphere. Sometimes the charge within the gaussian surface is specified in terms of a constant volume charge density ρ. In that case, Q_{net} is equal to ρV, where V is the volume of the charge enclosed by the gaussian sphere. Of course, $\mathbf{E}$ is outward from the gaussian surface if Q_{net} is positive and inward if Q_{net} is negative.

For problem situations having cylindrically symmetric charge distributions (for example, a *long* straight line with linear charge density λ), we may choose a closed cylindrical gaussian surface that is coaxial with the charge distribution. Circular caps normal to the axis close the cylindrical surface. The electric flux for the caps of the closed cylinder is equal to zero because the electric field is parallel to the surface everywhere on the caps. Because $\mathbf{E}$ has a constant magnitude over the curved portion of the gaussian surface and $\mathbf{E}$ is normal to that surface at every point, the electric flux for the gaussian surface is $E(2\pi rL)$, where L is the length of the cylindrical gaussian surface. Then, using Gauss's law, $E(2\pi rL)$ is equal to Q_{net}/ϵ_o. Again Q_{net} must be determined in terms of r, L, and the appropriate charge density, λ, σ, or ρ.

Charge distributions with planar symmetry have a volume charge density ρ or a surface charge density σ (maybe both). In these cases, we may choose closed cylindrical gaussian surfaces with the circular caps parallel to the plane (or planes) of charge. Then the total flux for the gaussian surface is only that for the caps, because $\mathbf{E}$ is everywhere normal to the charged surface. Gauss's law again provides that the electric flux for the chosen surface is equal to Q_{net}/ϵ_o, where Q_{net} is the net charge inside the gaussian surface.

The properties of a conductor in the electrostatic situation are extremely useful in solving problems. The fact that $\mathbf{E} = \mathbf{0}$ everywhere inside a conducting medium, for example, often provides a head start in solving electrostatic problems. Similarly, it is useful to remember that any net charge on a conductor must reside on its surface. Finally, it is often helpful to recognize that immediately outside a conducting surface, $\mathbf{E}$ has a direction normal to that surface and has a magnitude related to the surface charge density on the conductor by $E = \sigma/\epsilon_o$. Electrostatic problems are frequently simplified by mentally ticking off these four basic facts about charged conductors: $\mathbf{E} = \mathbf{0}$ inside the conductor; Q_{net} is on the surface of the conductor; $\mathbf{E}$ is normal to the conducting surface; and E is equal to σ/ϵ_o just outside the conductor.

Problems

GROUP A

12.1 A closed surface having an area equal to 0.035 m^2 has an electric flux of 75 N·m^2/C. Calculate the net charge contained within this surface.

12.2 Two particles, each having a charge of 5.0×10^{-9} C, are placed on the x axis at $x = -3.0$ m and $x = 3.0$ m. Calculate the electric field at the points (a) $\mathbf{r} = 4.0\hat{\mathbf{j}}$ m and (b) $\mathbf{r} = 2.0\hat{\mathbf{i}}$ m.

12.3 Two particles, each having a charge Q, are positioned on the y axis at $y = -d$ and $y = d$.

(a) Calculate the electric field at the point $(x, 0)$ on the x axis.

(b) Show that the result of part (a) behaves appropriately for $x = 0$, $x \gg d$, and $x \ll -d$.

12.4 The electric flux for a cubical surface 12 cm along an edge is -85 N·m²/C.

(a) Determine the net charge within this cube.

(b) Calculate the average charge density within the volume enclosed by this surface.

12.5 Charge of uniform density $\rho = 4.0 \times 10^{-8}$ C/m³ fills a right cylinder having a length of 2.5 cm and a radius of 1.4 cm. Calculate the electric flux for the surface of this cylinder.

12.6 A cube, 15 cm along an edge, has a 7.0×10^{-8} C charge distributed uniformly throughout its volume.

(a) Calculate the charge density ρ inside the cube.

(b) Calculate the electric flux for a spherical surface (radius = 5.0 cm) concentric with the geometric center of the cube.

(c) Repeat part (b) for a radius of 15 cm.

12.7 The point charges $Q_1 = -5.0 \ \mu C$ and $Q_2 = 3.0 \ \mu C$ are positioned on the y axis at $y_1 = 2.0$ m and $y_2 = 8.0$ m, respectively. Calculate the electric field at the points (a) $(0, 4.0)$ m, (b) $(0, 9.0)$ m, (c) $(0, 0)$, and (d) $(4.0, 5.0)$ m.

12.8 Two point charges, $Q_1 = 4.0 \ \mu C$ and $Q_2 = -7.0 \ \mu C$, are located on the x axis at $x_1 = 0$ and $x_2 = 2.0$ m, respectively. At which points, if any, in the x-y plane is the electric field equal to zero?

12.9 Three point charges are positioned as shown in Figure 12.27 at the vertices of an isosceles triangle. Determine the electric field at point P, the midpoint of the base of the triangle, if $Q = 1.2 \times 10^{-10}$ C.

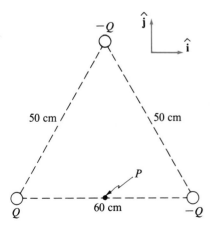

Figure 12.27 Problem 12.9

12.10 Four point charges are arranged on the vertices of a rectangle (*see* Fig. 12.28). Calculate the electric field at the point P shown if $a = 0.20$ m, $b = 0.48$ m, and $Q = 1.5 \times 10^{-9}$ C.

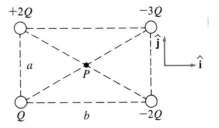

Figure 12.28 Problem 12.10

12.11 Three charges are positioned as shown in Figure 12.29. If $Q = 8.0 \ \mu C$ and $d = 0.50$ m, determine q if the electric field at point P is zero.

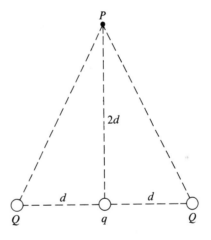

Figure 12.29 Problem 12.11

12.12 A charge of 4.0×10^{-9} C is distributed uniformly along the x axis from $x = 0$ to $x = 2.0$ m. Calculate the electric field at the point $\mathbf{r} = 5.0\hat{\mathbf{i}}$ m.

12.13 Three point charges are positioned as shown in Figure 12.30.

(a) Determine the electric field at the point P shown.

(b) Determine the behavior of the electric field for $r \gg d$. How does this behavior compare to the behavior of the electric field of a point charge? a dipole? (This charge distribution, which has a net charge of zero and a dipole moment of zero, is an example of an *electric quadrupole*.)

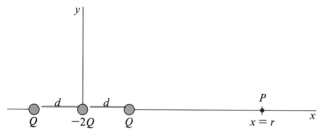

Figure 12.30   Problem 12.13

12.14 Repeat Problem 12.13 for a field point P on the y axis at $y = r$.

12.15 A linear charge of uniform density $\lambda > 0$ is distributed along the entire x axis. Use Gauss's law to show that the magnitude of the electric field at a (perpendicular) distance r from this charged line is given by

$$E = \frac{2k\lambda}{r} = \frac{\lambda}{2\pi\epsilon_o r}$$

12.16 A continuous charge of uniform density $\lambda > 0$ is distributed along the x axis from $x = -L$ to $x = L$.

(a) Determine the electric field at the point $\mathbf{r} = r\hat{\mathbf{j}}$ on the positive y axis.

(b) Show that the result of part (a) behaves appropriately for $r \gg L$.

(c) Show that the result of part (a) reproduces the result of Problem 12.15 for $L \gg r$.

12.17 A point charge $Q = -8.0 \times 10^{-13}$ C is placed at the center of a hollow spherical conductor that has an outer radius of 2.0 cm. The electric field just outside the (outer) surface of the conductor is 6.0 N/C, radially outward.

(a) What is the surface charge density on the outer surface of the conductor?

(b) Calculate the net charge on the conductor.

12.18 Two parallel charged planes carry uniform charge densities σ and $-\sigma$, where $\sigma = 6.0 \times 10^{-8}$ C/m². Calculate the electric field at a point (a) between the planes, and (b) neither between the planes nor on either of the planes.

12.19 A flat disk (radius = 2.0 cm, thickness = 0.10 mm), cut from a copper sheet, has a net charge of 4.0×10^{-10} C on it. Estimate the magnitude of the electric field

(a) at a distance of 2.0 m from the center of the disk.

(b) just outside, and near the center of, a face of the disk.

12.20 The plane $z = 0$ carries a uniform surface charge density σ_1, and the plane $z = L$ has the uniform surface charge density σ_2. The electric field at the point $\mathbf{r} = \frac{1}{2}L\hat{\mathbf{k}}$ is equal to $85\hat{\mathbf{k}}$ N/C, while the value of $\mathbf{E}$ at $\mathbf{r} = 2L\hat{\mathbf{k}}$ is equal to $-45\hat{\mathbf{k}}$ N/C. Determine σ_1 and σ_2.

12.21 The electric field in a particular region of space is given by $\mathbf{E} = \alpha/(b + y)\hat{\mathbf{j}}$, where $\alpha = 150$ N·m/C and $b = 2.0$ m. Calculate the net charge contained within the volume (*see* Fig. 12.31) if $a = 3.0$ m, $b = 2.0$ m, and $c = 1.0$ m.

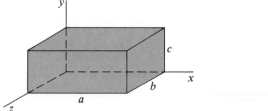

Figure 12.31   Problem 12.21

12.22 The plane $x = 0$ is covered with a uniform surface charge density $\sigma_1 = 5.0 \times 10^{-8}$ C/m², and the plane $y = 0$ has a uniform charge density given by $\sigma_2 = -3.0 \times 10^{-8}$ C/m². Calculate the magnitude of the electric field at any point not on either plane.

12.23 The electric field at any point P in a region of space is given by

$$\mathbf{E} = \alpha\mathbf{r}$$

where $\alpha = 750$ NC^{-1} m^{-1} and $\mathbf{r}$ is the position vector from the origin to the point P. Calculate the net charge contained inside a spherical shell (inner radius = 1.0 m, outer radius = 2.0 m) centered on the origin.

12.24 Repeat Problem 12.23 if $\mathbf{E}$ is given by

$$\mathbf{E} = \frac{\alpha}{r^4}\mathbf{r}$$

where $\alpha = 800$ N·m³/C.

12.25 The electric field in a region of space has a direction that is radially inward toward the origin and has a magnitude that is given by $E = \alpha/r^2$ where $\alpha = 270$ N·m²/C and r is the distance from the origin to the point where $\mathbf{E}$ is to be determined. Calculate the net charge enclosed in a sphere having a radius of (a) 2.0 m and (b) 4.0 m. Explain these two results.

12.26 A charge Q is distributed uniformly along the circular arc shown in Figure 12.32.

(a) Determine the magnitude of the electric field at the center of the arc.

(b) Show that the result of part (a) behaves appropriately for $\alpha \to 0$ and $\alpha \to 2\pi$.

12.27 Four point charges are arranged as shown in Figure 12.33.

(a) Show that the electric field at a point on the x axis is given by

$$\mathbf{E} = \frac{2kQ}{x^2}\left[\left(1 + \frac{d^2}{x^2}\right)\left(1 - \frac{d^2}{x^2}\right)^{-2} - \left(1 + \frac{d^2}{x^2}\right)^{-3/2}\right]\hat{\mathbf{i}}$$

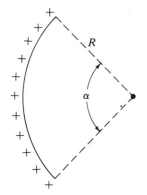

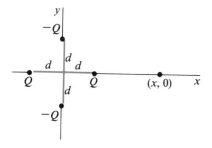

Figure 12.32   Problem 12.26

Figure 12.33   Problem 12.27

(b) Use the first two terms of the binomial expansion, that is, $(1 + \epsilon)^p \cong 1 + p\epsilon$, to show that for $x \gg d$,

$$\mathbf{E} \cong \frac{9kQd^2}{x^4}\hat{\mathbf{i}}$$

12.28 The infinite slab defined by $4.0 \text{ cm} \leq x \leq 8.0 \text{ cm}$ is filled with a charge of a uniform density $\rho = 1.8 \times 10^{-8} \text{ C/m}^3$. Calculate the electric field at (a) $x = 2.0$ cm, (b) $x = 7.0$ cm, and (c) $x = 10.0$ cm.

12.29 A long cylindrical volume having an inner radius of 2.0 mm and an outer radius of 5.0 mm is filled with a charge of uniform density $\rho = 7.0 \ \mu\text{C/m}^3$. Calculate the magnitude of the electric field at a distance (from the symmetry axis) of (a) 1.0 mm, (b) 3.0 mm, and (c) 8.0 mm.

12.30 A charge Q is distributed uniformly along a circle that is located in the y-z plane and centered on the origin. The radius of the circle is equal to R.

(a) Determine the electric field at a point $(x, 0, 0)$ on the x axis.

(b) Show that the result of part (a) behaves appropriately for $x \to 0$ and $x \gg a$.

12.31 Charge of a uniform density ρ fills a nonconducting cylindrical region (radius $= 2a$), which has a metallic coating of thickness a. The net charge on this coating is equal to zero.

(a) Calculate the electric field at a distance $r = 4a$ from the axis of the cylinder.

(b) Calculate the surface charge density on the inner surface of the metallic coating.

(c) Repeat part (b) for the outer surface of the metal.

12.32 Charge of a uniform density $\rho_1 = -5.0 \times 10^{-8} \text{ C/m}^3$ fills a spherical volume having a radius of $a = 2.0$ cm. Charge of a uniform density $\rho_2 = 5.0 \times 10^{-9} \text{ C/m}^3$ fills the concentric spherical volume having an inner radius of $b = 4.0$ cm and an outer radius of $c = 6.0$ cm. If r measures distance from the center of this charge distribution, calculate the electric field for (a) $r = 3.0$ cm, (b) $r = 5.0$ cm, and (c) $r = 7.0$ cm.

12.33 For the charge distribution given in Problem 12.32 determine the value of r for which $\mathbf{E} = \mathbf{0}$.

12.34 A circular disk has a radius R, is located in the y-z plane, is centered on the origin, and carries a uniform surface charge density σ.

(a) Determine the electric field at a point $(x, 0, 0)$ on the x axis.

(b) Show that the result of part (a) behaves appropriately for $x \gg R$.

(c) Repeat part (b) for $|x| \ll R$.

12.35 Charge of uniform surface density σ is distributed on the x-y plane on the "long ribbon" surface, $-b \leq x \leq b$. Calculate the electric field at a point $(d, 0)$ on the positive x axis. Assume that $d > b$.

12.36 Charge of uniform surface density σ is distributed on the x-y plane along the "long ribbon" surface, $-b \leq x \leq b$. Calculate the electric field at a point $(0, 0, d)$ on the positive z axis.

12.37 Charge of uniform linear density $\lambda > 0$ is distributed along the entire y axis. Charge of uniform linear density $-\lambda$ is distributed along the line that is parallel to the y axis and passes through the point $(R, 0, 0)$ where $R > 0$. Calculate the electric field at the points (a) $\mathbf{r} = \frac{1}{2}R\hat{\mathbf{i}}$ and (b) $r = R\hat{\mathbf{k}}$.

12.38 Charge is distributed uniformly along the entire z axis with a linear density $\lambda > 0$. The coaxial cylindrical volume, which has a radius R, is filled with charge of a constant volume density $\rho > 0$. If r measures the (perpendicular) distance of a point from the z axis, determine $E(r)$, the magnitude of the electric field as a function of r.

12.39 A charge $Q > 0$ is distributed uniformly along the x axis from $x = 0$ to $x = L$, and a charge $-Q$ is distributed uniformly along the x axis from $x = -L$ to $x = 0$.

(a) Calculate the magnitude of the electric field at the point $(x, 0, 0)$, on the x axis, for $x > L$.

(b) Investigate the behavior of the result obtained in part (a) for $x \gg L$. Can you explain this behavior?

12.40 A sphere (radius $= 2.0$ cm) is filled with a continuous charge of volume density $\rho(r) = \alpha r^2 \ (0 < r < 2.0 \text{ cm})$, where $\alpha = 3.0 \times 10^{-4} \text{ C/m}^5$ and r measures distance from the center of the sphere. Calculate the magnitude of the electric field for (a) $r = 1.0$ cm and (b) $r = 3.0$ cm.

12.41 A sphere of radius R is filled with a continuous charge described by the volume density $\rho(r) = \alpha r^3$ $(0 < r < R)$, where α is a constant and r measures distance from the center of the sphere. Determine the magnitude of the electric field as a function of r.

12.42 Charge of uniform linear density λ is distributed along the perimeter of a square that has a length $2b$ along each side.

(a) Calculate the magnitude of the electric field at a distance r (from the plane of the square) measured along a line that passes through the geometric center of the square and is perpendicular to the plane of the square.

(b) Show that the result obtained in part (a) behaves appropriately for $r = 0$ and $r \gg d$.

GROUP **B**

12.43 Charge is distributed along a circular arc with a linear charge density given by $\lambda(\theta) = \lambda_o \theta$ for $0 < \theta < 2\pi$, where λ_o is a positive constant and θ is the angle shown in Figure 12.34. Calculate the electric field at the origin.

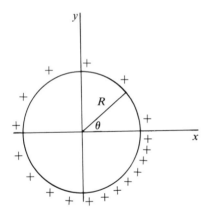

Figure 12.34   Problem 12.43

12.44 A continuous charge that is described by the linear density $\lambda(x) = \alpha x$ ($\alpha = 3.0 \times 10^{-9}$ C/m^2) is distributed along the x axis from $x = -3.0$ m to $x = 3.0$ m. Determine the electric field at the point $\mathbf{r} = 4.0\hat{\mathbf{j}}$ m on the y axis.

12.45 A continuous charge that is described by the linear density $\lambda(x) = \alpha x$ ($\alpha = 3.0 \times 10^{-9}$ C/m^2) is distributed along the x axis from $x = -3.0$ m to $x = 3.0$ m. Determine the electric field at the point $\mathbf{r} = 4.0\hat{\mathbf{i}}$ m on the x axis.

12.46 The (spherically symmetric) electric field in a region of space is given by

$$\mathbf{E}(r) = \frac{\alpha - \beta r}{r^3}\mathbf{r}$$

where α and β are positive constants and $\mathbf{r}$ is the position vector from the origin to the field point. Determine the charge distribution in this region of space.

12.47 The (cylindrically symmetric) electric field in a region of space is given by

$$\mathbf{E}(r) = \alpha r^2 \mathbf{r}$$

where α is a constant and $\mathbf{r}$ is the position vector drawn perpendicularly from the symmetry axis to the field point. Determine the charge distribution in this region of space.

12.48 The (planar-symmetric) electric field in a region of space is given by

$$\mathbf{E}(r) = \left(\frac{\alpha}{r} + \beta r\right)\mathbf{r}$$

where α and β are positive constants and $\mathbf{r}$ is the position vector drawn perpendicularly from the symmetry plane to the field point. Determine the charge distribution in this region of space.

12.49 The electric field in a region of space is given by

$$\mathbf{E}(r) = \begin{cases} \dfrac{\alpha}{r^3}\mathbf{r} & 0 < r < R \\[2ex] -\dfrac{\alpha}{r^3}\mathbf{r} & R < r \end{cases}$$

where α is a positive constant and $\mathbf{r}$ is the position vector from the origin to the field point. What charge distribution will produce this field?

12.50 A linear charge of uniform density λ is distributed along the perimeter of a circle of radius R. The circle is centered on the origin and lies in the x-y plane.

(a) Set up, but do not evaluate, an integral for the electric field at the point $\mathbf{r} = r\hat{\mathbf{i}}$ $(r > R)$. Use θ, the angle shown in Figure 12.35, as the variable of integration.

(b) Show that the integral of part (a) reduces appropriately for $r \gg R$.

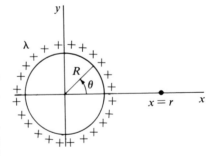

Figure 12.35   Problem 12.50

12.51 The electric field caused by a continuous charge may be estimated to any desired accuracy by approximating that continuous charge by an appropriate number of point charges, Q_1, $Q_2, \ldots, Q_n$, i.e.,

$$\mathbf{E} = \int \frac{k\mathbf{r}\,dQ}{r^3} \cong \sum_{n=1}^{N} \frac{k\,\mathbf{r}_n Q_n}{r_n^3}$$

where $\mathbf{r}_n$ is the position vector from Q_n to the field point where $\mathbf{E}$ is to be determined (see Fig. 12.36). Use this technique to estimate the electric field (at the point $\mathbf{r} = 2.0\hat{\mathbf{i}}$ m) caused by a charge of $Q = 8.0 \times 10^{-10}$ C, which is distributed uniformly along the x axis from $x = -1.0$ m to $x = 1.0$ m. Do this estimate for (a) $N = 1$, (b) $N = 2$, and (c) $N = 4$, as shown in Figure 12.36. (d) Calculate the exact answer for $\mathbf{E}$ at $\mathbf{r} = 2.0\mathbf{i}$ m and compare this result to the estimates of parts (a), (b), and (c).

12.52 If you have access to either a programmable calculator or a computer, you may wish to extend the calculation of Problem 12.51. The results of such a calculation are summarized at right.

N	E_N
2	2.1760 N/C
4	2.3333 N/C
8	2.3823 N/C
16	2.3955 N/C
32	2.3989 N/C
64	2.3997 N/C
128	2.3999 N/C
256	2.4000 N/C
512	2.4000 N/C

12.53 Use the technique of Problem 12.51 and the point charge distributions of Figure 12.37 to estimate the electric field of Problem 12.50 at the field point $\mathbf{r} = 2R\hat{\mathbf{i}}$.

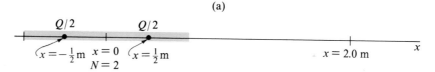

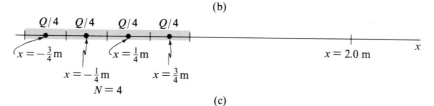

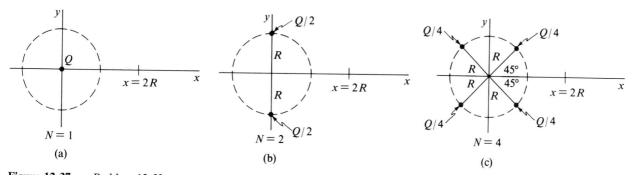

Figure 12.36   Problem 12.51

Figure 12.37   Problem 12.53

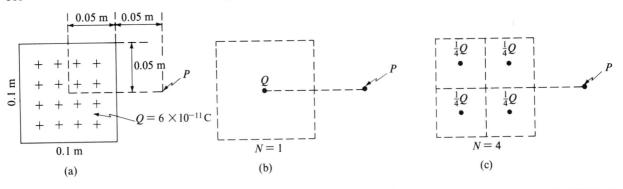

Figure 12.38   Problem 12.55

12.54 If you have access to either a programmable calculator or a computer, you may wish to extend the calculation of Problem 12.53. The results of such a calculation are summarized below.

N	$\dfrac{E_N}{kQ/R^2}$
2	0.1789
4	0.2638
8	0.3073
12	0.3111
16	0.3114
20	0.3114

12.55 A charge of $Q = 6.0 \times 10^{-11}$ C is distributed uniformly over the surface of a square, which is 0.10 m along each side. Estimate the magnitude of the electric field at the point P shown in Figure 12.38(a). Do this for $N = 1$ (Fig. 12.38[b]) and for $N = 4$ (Fig. 12.38[c]).

12.56 If you have access to either a programmable calculator or a computer, you may wish to extend the calculation of Problem 12.55. The results of such a calculation are summarized below.

N	E_N
1	54.00 N/C
4	57.28 N/C
16	59.11 N/C
100	59.70 N/C
400	59.79 N/C
2500	59.82 N/C
10000	59.82 N/C
40000	59.82 N/C

13 Electric Potential

We have stressed that electric fields are a means of characterizing a physical property of space caused by the presence of electrostatic charge. Electric fields are vector fields that have both magnitude and direction at every point at which the field is defined. The electrical properties of space can also be described by electric potential, which is, in some respects, a simpler and more practical concept than the electric field. Electric potential is simpler than electric fields because electric potential is a scalar quantity and, therefore, has no direction associated with it. Electric potential is more practical than the electric field because differences in potential, at least on conductors, are more readily measured directly.

Electric potentials and electric fields in a given region are related to each other, and either can be used to describe the electrostatic properties of space. In a given problem situation, either one of these descriptions may be more convenient to calculate. We will see how each of these descriptions may be determined if the other is known or has been calculated.

13.1 Electric Potential and Electric Fields

We have already encountered the notion of potential energy. In Chapter 5 the gravitational potential energy function was introduced to relate the position of an object that has mass to the work done by gravitational forces on that object as it moves from one point to another. The following aspects of gravitational potential energy proved to be particularly important and useful:

1. *Differences* in potential energy are the measurable and physically significant quantities associated with gravitational potential energy.

2. The gravitational force is a *conservative* force. Equivalently, the work done on an object by the gravitational force as the object moves from one point to another (and, hence, the corresponding change in gravitational potential energy) is independent of the path connecting the two points.

3. The gravitational potential energy of an object at a point is meaningful only in terms of the difference in potential energy between that point and the potential energy specified at some reference point.

Electric potential should have characteristics similar to those of gravitational potential energy if we expect electric potential to be equally useful. In the case of electric potential, however, we will consider the charge, rather than the mass, of an object and the electrostatic coulomb force, rather than the gravitational force. We will also consider both the difference in electric potential between any two points and the specification of electric potential at a point measured relative to some reference point. First, however, we need to establish that the electric force is conservative, so that we are assured that any difference in potential between two points depends only on the positions of those points and not on the path connecting those points. We have established that the force **F** on a charge q at a given point is related to the electric field **E** at that point by $\mathbf{F} = q\mathbf{E}$, and we saw in Chapter 5 that a conservative force is one for which $\oint \mathbf{F} \cdot d\mathbf{s} = 0$. Similarly, showing that the electric field is conservative is equivalent to showing that $\oint \mathbf{E} \cdot d\mathbf{s} = 0$, where the integration takes place over any closed path. The following example and exercise establish the conservative nature of the electric field.

Example 13.1

PROBLEM   Show that the electric field of a point charge is conservative.

SOLUTION   Figure 13.1(a) shows a point charge Q, a path connecting two field points, A and B, and an infinitesimal displacement d**s** along this path. Because the electric field **E** associated with the charge Q is $\mathbf{E} = kQ\hat{\mathbf{r}}/r^2$, where $\hat{\mathbf{r}}$ is a unit vector in the direction from Q to the field point, the quantity $\mathbf{E} \cdot d\mathbf{s}$ may be written

$$\mathbf{E} \cdot d\mathbf{s} = \frac{kQ}{r^2}\, \hat{\mathbf{r}} \cdot d\mathbf{s}$$

Because $\hat{\mathbf{r}} \cdot d\mathbf{s} = ds\, \cos\theta$, where θ is the angle between $\hat{\mathbf{r}}$ and d**s**, we have

$$\mathbf{E} \cdot d\mathbf{s} = \frac{kQ}{r^2}\, ds\, \cos\theta$$

Using Figure 13.1(b) we see that $ds\, \cos\theta$ is equal to the projection of d**s** onto **r**. In other words, the displacement d**s** produces a change dr in the magnitude of **r**

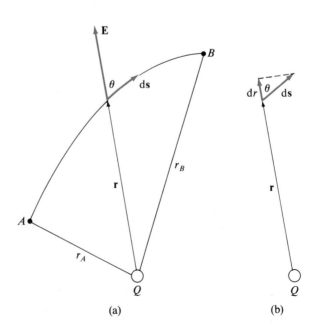

(a)   (b)

Figure 13.1   Example 13.1

that is equal to $ds \cos\theta$. Substituting $ds \cos\theta = dr$ and summing (integrating) from point A to point B gives

$$\int_A^B \mathbf{E} \cdot d\mathbf{s} = \int_{r_A}^{r_B} \frac{kQ\,dr}{r^2} = kQ \left(\frac{1}{r_A} - \frac{1}{r_B} \right)$$

where r_A and r_B are the distances shown in Figure 13.1(a). If the path of integration is closed, points A and B coincide, i.e., $r_A = r_B$, and we have

$$\oint \mathbf{E} \cdot d\mathbf{s} = 0 \quad \blacksquare$$

E 13.1   Use the superposition principle for electric fields ($\mathbf{E} = \Sigma_n \mathbf{E}_n$) and the result of Example 13.1 to show that the electric field resulting from N point charges is conservative.

Suppose a particle with a charge q_o is moved from point A to point B along an arbitrary path, as suggested by Figure 13.2. Let $(W_{A \to B})_{ext}$ be the work done on the particle by an external force provided that (a) the only forces on the particle are the electrical force and the external force and, (b) there is no net change in the kinetic energy K of the particle, that is, $K_A = K_B$. The *difference in electric potential*, $V_B - V_A$, between the points A and B is then defined by

$$V_B - V_A = \frac{(W_{A \to B})_{ext}}{q_o} \tag{13-1}$$

Thus the change, $V_B - V_A$, in electric potential is equal to the external work done per unit charge in moving a charged particle from point A to point B without causing a net change in the kinetic energy of the particle. The work $(W_{A \to B})_{ext}$ is equal to the negative of the work $(W_{A \to B})_E$ done by the electric field $\mathbf{E}$. In other words, according to the work-energy theorem, the work done on the charge by the resultant force must be equal to the change in the kinetic energy of the particle; because the change in kinetic energy is equal to zero, we see that $(W_{A \to B})_{ext} + (W_{A \to B})_E = 0$, or $(W_{A \to B})_{ext} = -(W_{A \to B})_E$. Then the difference in electric potential, $V_B - V_A$, may be expressed as

$$V_B - V_A = -\frac{(W_{A \to B})_E}{q_o}$$

and because $(W_{A \to B})_E = \int_A^B q_o \mathbf{E} \cdot d\mathbf{s}$, we have

$$V_B - V_A = -\frac{1}{q_o} \int_A^B q_o \mathbf{E} \cdot d\mathbf{s} = -\int_B^A \mathbf{E} \cdot d\mathbf{s} \tag{13-2}$$

or

$$V_B - V_A = \int_B^A \mathbf{E} \cdot d\mathbf{s} \tag{13-3}$$

Equation (13-1) defines a difference in electric potential to be a quantity of work (J) per unit charge (C). Electric potential, therefore, has the unit joule/coulomb (J/C), which is defined to be the *volt* (V):

$$1 \text{ V} = 1 \text{ J/C} \tag{13-4}$$

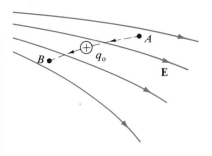

Figure 13.2
A particle having charge q_o moving in an electric field from point A to point B.

Notice that the roman V represents the unit volt, and the italic V represents the electric potential. When V indicates the unit volt, it is usually preceded by a number, so the symbolism is less confusing than it may appear.

From Equation (13-3) we can see that the unit (V) of potential is equal to the product of the unit (N/C) of electric field and the unit (m) of displacement. Thus, electric field may be expressed in terms of the unit V/m, which is equal to the N/C:

$$1 \text{ N/C} = 1 \text{ V/m} \qquad (13\text{-}5)$$

E 13.2   If an external agent does 3.0 μJ of work while moving a charge $q_o = 0.05$ μC from point A to point B at a constant speed, calculate (a) the potential difference, $V_B - V_A$, (b) the work by the electrical force acting on q_o as q_o moves from A to B, and (c) the potential difference, $V_A - V_B$.

Answers: (a) 60 V; (b) -3.0 μJ; (c) -60V

E 13.3   As a charge $q_o = -6.0$ μC moves from point A to point B, the work done by the electrical force acting on q_o is equal to 72 μJ. Calculate (a) the work an external agent must perform in order to move q_o from point A to point B without changing the kinetic energy of q_o, and (b) the potential difference, $V_B - V_A$.

Answers: (a) -72 μJ; (b) 12 V

E 13.4   If the change, $V_B - V_A$, in electric potential between points A and B is equal to 42 V, calculate (a) the work required of an external agent as it moves a charge $q_o = 5.0 \times 10^{-5}$ C from point A to point B at a constant speed, and (b) the work by the electrical force acting on q_o as the charge moves from A to B.

Answers: (a) 2.1 mJ; (b) -2.1 mJ

Example 13.2

PROBLEM   If the electric field in a region of space is uniform and given by $\mathbf{E} = E_o\hat{\mathbf{i}}$, where $E_o = 35$ V/m, calculate the potential difference, $V_B - V_A$, for the points $\mathbf{r}_A = 4\hat{\mathbf{i}}$ m and $\mathbf{r}_B = 5\hat{\mathbf{j}}$ m shown in Figure 13.3.

SOLUTION   In Equation (13-3), the potential difference, $V_B - V_A$, is equal to $\int_B^A \mathbf{E} \cdot d\mathbf{s}$. An arbitrary infinitesimal displacement $d\mathbf{s}$ may be written

$$d\mathbf{s} = \hat{\mathbf{i}}dx + \hat{\mathbf{j}}dy + \hat{\mathbf{k}}dz$$

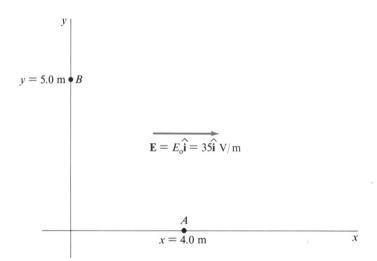

Figure 13.3   Example 13.2

so that, in this case, we obtain

$$\mathbf{E} \cdot d\mathbf{s} = E_o \hat{\mathbf{i}} \cdot (\hat{\mathbf{i}}dx + \hat{\mathbf{j}}dy + \hat{\mathbf{k}}dz) = E_o dx$$

Thus the desired potential difference is given by

$$V_B - V_A = \int_{x_B}^{x_A} E_o dx = E_o \int_0^4 dx = \left(35 \frac{\text{V}}{\text{m}}\right)(4\text{m}) = 140 \text{ V} \quad \blacksquare$$

E 13.5   Using the data of Example 13.2, calculate the potential differ-
ence, $V_C - V_A$, if $\mathbf{r}_C = (10\hat{\mathbf{i}} + 8\hat{\mathbf{j}})$ m.   Answer: -210 V

A "feeling" for the relationship between a change of position within a region
of electric field and the corresponding change in electric potential may be
strengthened by considering first that Equation (13-2) can be written in its
differential form,

$$dV = -\mathbf{E} \cdot d\mathbf{s} \tag{13-6}$$

Let us interpret Equation (13-6) in a few simple cases, illustrated in Figure 13.4.
A change in position d**s** in a direction parallel to a given electric field **E**, as
indicated in Figure 13.4(a), provides a change dV in electric potential that is
negative; *the electric potential decreases as a result of a change in position in the
direction of the electric field.* The decrease in potential occurs because the scalar
product **E** · d**s** has a positive value when **E** and d**s** are in the same direction. In
Figure 13.4(b) **E** and d**s** are antiparallel, in opposite directions, and **E** · d**s** is
negative. In this case, Equation (13-6) provides that dV is positive: *The potential
increases as a result of a change in position opposite to the electric field.* Finally,
in Figure 13.4(c) we see that a change d**s** in position perpendicular to the
direction of **E** means that the scalar product **E** · d**s** is equal to zero. In this case, a
change of position produces, according to Equation (13-6), no change in electric
potential.

When a displacement d**s** is in the same direction as a given electric field **E**,
E · d**s** has its maximum value, and the change dV in potential has its greatest
negative value. In other words, an electric field is always in the direction toward
which the electric potential is decreasing most rapidly.

Because changes of position along lines that are at every point perpendicular to
the electric field produce no change in electric potential (in other words, because
lines perpendicular to the electric field are lines of constant potential), any
surface on which the electric field is perpendicular to that surface at every point is
called an *equipotential surface.*

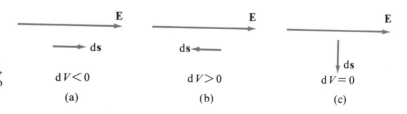

Figure 13.4
Displacements d**s** that are (a) parallel,
(b) antiparallel, and (c) perpendicular to
an electric field **E**.

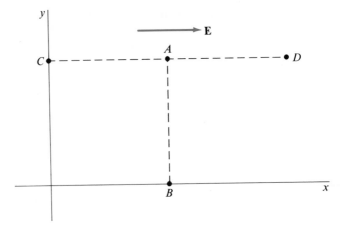

Exercise 13.6

E 13.6   The electric field in a region of space is in the positive x direction. With reference to the points shown in the figure, determine the appropriate sign $(+, -, 0)$ for the potential difference (a) $V_C - V_A$, (b) $V_D - V_A$, and (c) $V_B - V_A$
                                              Answers: (a) $+$; (b) $-$; (c) 0

We have been discussing *differences* in electric potential between two points. These differences are physically meaningful and often easily measurable. (An instrument used to measure differences in electric potential in conducting circuits is called a *voltmeter*.) Often, however, it is convenient to specify the value of electric potential at a point. In this way, we identify an important electrical property associated with a point in space, much as we do when we specify the electric field at a point.

To specify the electric potential at a point, we must choose a reference point at which the electric potential is taken to be a specific value, say zero. Thus, in Equations (13-2) and (13-3), if we let the point A be the reference point (where $V_A = 0$), the *electric potential* at the point B is

$$V_B = \int_B^{ref} \mathbf{E} \cdot d\mathbf{s} \qquad (13\text{-}7)$$

in which the limits on the integral refer to the reference point, where the electric potential is zero, and the point B, where the electrical potential has the value V_B. Because the electric field is conservative, the electric potential V_B at B is determined by starting at the point B and integrating $\mathbf{E} \cdot d\mathbf{s}$ along *any* path to the reference point. In order to evaluate the integral of Equation (13-7), we must know the value of the electric field $\mathbf{E}$ at every point along the chosen path of integration. Knowing the electric field at a single point is not sufficient to determine the electrical potential at that point.

Unless otherwise specified, the reference point at which the potential is zero is conventionally taken to be an infinite distance from the charge distribution that causes this potential. In such cases, Equation (13-7) becomes

$$V_B = \int_B^{\infty} \mathbf{E} \cdot d\mathbf{s} \qquad (13\text{-}8)$$

In the following sections we will see how Equations (13-7) and (13-8) may be used to determine electric potential in and around various charge distributions.

13.2 Electric Potential of Point Charges

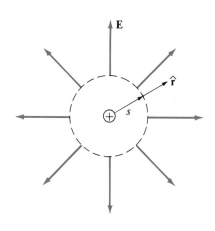

Figure 13.5
The electric field of a single point charge Q at a distance s from the charge. The unit vector $\hat{\mathbf{r}}$ is directed radially outward for a positive charge.

The electric field $\mathbf{E}$ of a single point charge Q at a distance s from the charge is expressed by

$$\mathbf{E} = \frac{kQ}{s^2}\hat{\mathbf{r}} = \frac{Q}{4\pi\epsilon_0 s^2}\hat{\mathbf{r}} \qquad (13\text{-}9)$$

where $\hat{\mathbf{r}}$ is a unit vector directed radially outward from the charge, as shown in Figure 13.5. The electric potential V at an arbitrary distance r from the point charge may be found using Equation (13-8) in the form

$$V = \int_r^\infty \mathbf{E} \cdot d\mathbf{s} \qquad (13\text{-}10)$$

If we choose a radially outward path of integration, then $d\mathbf{s} = ds\,\hat{\mathbf{r}}$ is an infinitesimal displacement in the radial direction. Because $\mathbf{E}$ and $d\mathbf{s}$ are in the same direction, the substitution of Equation (13-9) into Equation (13-10) gives

$$V = kQ \int_r^\infty \frac{ds}{s^2} = kQ\left[-\frac{1}{s}\right]_r^\infty = \frac{kQ}{r} = \frac{Q}{4\pi\epsilon_0 r} \qquad (13\text{-}11)$$

The electric potential V is positive if Q is positive or negative if Q is negative.

E 13.7 A point charge $Q = 8.0 \times 10^{-8}$ C is located at the origin of a coordinate system. Calculate the value of the electric potential at the points (a) $\mathbf{r} = 4.0\hat{\mathbf{i}}$ m, (b) $\mathbf{r} = -4.0\hat{\mathbf{i}}$ m, (c) $\mathbf{r} = 3.0\hat{\mathbf{j}}$ m, (d) $\mathbf{r} = (-4.0\hat{\mathbf{i}} + 3.0\hat{\mathbf{j}})$ m
Answers: (a) 180 V; (b) 180 V; (c) 240 V; (d) 140 V

E 13.8 A point charge of 6.0 μC is located at the origin of a coordinate system. Describe the equipotential surface corresponding to a potential of 30,000 V.   Answer: Sphere (radius = 1.8 m) centered on the origin

The principle of superposition for electric fields implies a principle of superposition for electric potential. The electric potential V_P that results at a point P from the presence of a group of n point charges is calculated by (a) finding the potential V_i at P caused by each charge Q_i, just as if that charge were the only one present; and (b) algebraically summing the potentials V_i. The resulting electric potential at P is then

$$V_P = k\left(\frac{Q_1}{r_1} + \frac{Q_2}{r_2} + \ldots + \frac{Q_n}{r_n}\right) = k\sum_{i=1}^{n}\frac{Q_i}{r_i} = \frac{1}{4\pi\epsilon_0}\sum_{i=1}^{n}\frac{Q_i}{r_i} \qquad (13\text{-}12)$$

in which r_i is the distance from Q_i to the point P. Recall that each charge Q_i may be positive or negative, so each Q_i may yield a corresponding positive or negative term in the sum of Equation (13-12).

E 13.9 Use the superposition principle for electric fields, Equation (13-8), and Equation (13-11) to obtain Equation (13-12).

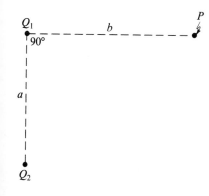

Exercise 13.10

E 13.10 Two point charges, Q_1 and Q_2, are positioned as shown in the figure. If $Q_1 = 8.0\ \mu$C, $Q_2 = -5.0\ \mu$C, $a = 3.0$ m, and $b = 4.0$ m, calculate the electric potential at point P.   Answer: 9.0×10^3 V = 9.0 kV

E 13.11   Two particles, each having a positive charge Q, are located on the x axis at $x = \pm d$. Determine the electric potential at a point P on the x axis at $x = r > d$.

Answer: $V_P = \dfrac{2kQr}{r^2 - d^2}$

E 13.12   Show that the result of Exercise 13.11 behaves appropriately for $r \gg d$.

Equation (13-11) says that the electric potential of a point charge changes with distance r from the charge in proportion to $1/r$. The greater the distance a point is from the charge, the less is the magnitude of the potential. Equation (13-12) says that when more than one charge is present, the effect of each charge on the electric potential at a point is algebraically additive. It is useful perhaps to acquire a physical "feeling" for the way that electric potential varies with position. Look, for example, at Figure 13.6, a sketch of the electric potential V as function of position x at points on a straight line through two charges, one positive and one negative, of equal magnitude. Some of the more important physical facts associated with the variation of the electric potential as a function of position may be summarized as follows:

> The closer a point is to a positive charge, in the vicinity of the positive charge with no negative charge intervening, the more positive becomes the electric potential at that point. The closer a point is to a negative charge, in the vicinity of the negative charge with no positive charge intervening, the more negative becomes the potential at that point. The effects at any point on the electric potential of more than one charge are algebraically additive.

Figure 13.6 confirms that the electric potentials caused by the positive and negative charges of equal magnitude sum to zero at the midpoint between them. To the right of the negative charge, the effect of the negative charge dominates,

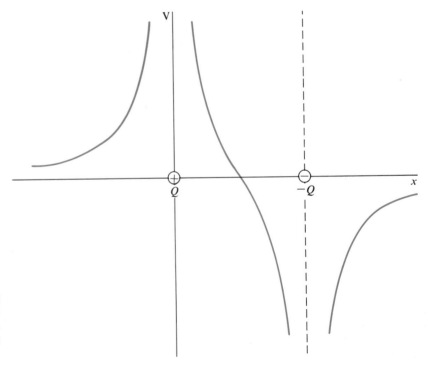

Figure 13.6
Electric potential V as a function of position x. The potential is caused by two charges, one positive and one negative, of equal magnitude. Both charges lie on the x axis.

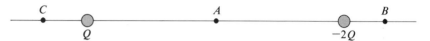

Exercise 13.13

and the net electric potential is negative. A similar effect causes the net electric potential to be positive to the left of the positive charge.

> **E 13.13**   A positive point charge Q and a negative point charge $-2Q$ are positioned as shown in the accompanying figure. Points A, B, and C lie on the line passing through the two charges. Arrange the electric potentials, V_A, V_B, and V_C, in order of decreasing electric potential.   Answer: V_C, V_A, V_B

13.3   Electric Potential of Continuous Charge Distributions

Suppose that a charge distribution of *finite* spatial extent is spread continuously throughout a definite region of space, as indicated in Figure 13.7(a). Each infinitesimal element of charge dQ within the distribution contributes an electric potential dV at the point P given by

$$dV = k\frac{dQ}{r} = \frac{1}{4\pi\epsilon_o}\frac{dQ}{r} \tag{13-13}$$

where r is the distance from the charge element dQ to the point P. The charge element dQ may be positive or negative. The electric potential V at P resulting from the entire charge distribution is determined by integrating Equation (13-13) so that the integration includes all the charge within the distribution:

$$V = k\int \frac{dQ}{r} = \frac{1}{4\pi\epsilon_o}\int \frac{dQ}{r} \tag{13-14}$$

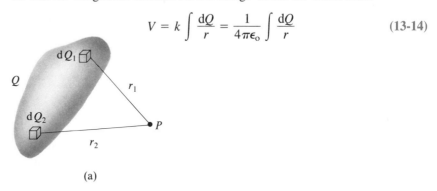

(a)

Figure 13.7
(a) A field point P near a charge distribution of finite extent. (b) A field point P far from a charge distribution of finite extent.

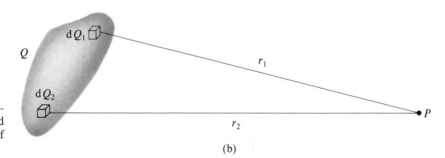

(b)

Equation (13-14) applies only to charge distributions of finite extent, distributions with electric potential that has a limiting value at an infinite distance from the charge (and that value has been chosen to be zero). Figure 13.7(b) illustrates that as a point P becomes farther removed from a finite charge distribution with a total charge of Q, the distances r_1 and r_2 from any different charge elements dQ_1 and dQ_2 to the point P become more nearly identical. In other words, as P becomes arbitrarily far from the finite charge distribution, the distribution produces at P the same electrical effect as a point charge of magnitude Q, that is, V is very nearly equal to kQ/r. Thus, for charge distributions of finite extent, the electric potential approaches zero as the distance r becomes arbitrarily large.

The following problems illustrate the calculation of electric potential in the space around charge distributions of both finite and infinite extent.

A straight line segment of length L has a linear charge density λ. Calculate the electric potential at points lying along an extension of the charged line segment.

Let us orient the line of charge along the x axis so that its ends are at $x = 0$ and $x = L$, as shown in Figure 13.8. Let the field point at which we calculate the electric potential be at $x = d > L$. An arbitrary element of charge $dQ = \lambda\,dx$ is located at a distance x from the origin. Then the distance from the charge element to the field point is $r = d - x$. According to Equation (13-14), the electric potential V at $x = d$ is

$$V = k \int \frac{dQ}{r} = k\lambda \int_0^L \frac{dx}{d - x} \tag{13-15}$$

in which the integration includes the entire distribution of charge, from $x = 0$ to $x = L$. Executing the integration of Equation (13-15) gives

$$V = k\lambda \left[-\ln|d - x| \right]_0^L = k\lambda \ln\left(\frac{d}{d - L}\right) \tag{13-16}$$

which specifies the electric potential at $x = d$ outside the charged line. Then, at points along the x axis for which $x > L$, we may write Equation (13-16) in terms of x:

$$V = k\lambda \ln\left(\frac{x}{x - L}\right) \quad ; \quad x > L \tag{13-17}$$

This result reduces to what is expected in the limiting case in which the field point is arbitrarily far from the line of charge. As the distance $x \to \infty$, the fraction $x/(x - L) \to 1$ and Equation (13-17) provides that $V \to k\lambda(\ln 1) = 0$, the

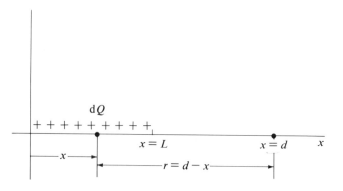

Figure 13.8
A linear distribution of charge oriented along the x axis from $x = 0$ to $x = L$.

expected zero value of electric potential resulting from a finite charge at a sufficiently great distance from the charge.

E 13.14   Use the approximation $\ln(1 + \epsilon) \simeq \epsilon$ for $|\epsilon| \ll 1$ to show that the result given in Equation (13-17) behaves appropriately for $x \gg L$. *Hint:* $\ln(x/(x - L)) = -\ln(1 - L/x)$.   Answer: $V \simeq k\lambda L/x$

E 13.15   A charge Q is distributed uniformly along the perimeter of a circle of radius a. Consider a point P on the axis of the ring at a distance r from the plane of the ring. Determine the electric potential at point P, realizing that each point on the ring is the same distance from point P.
Answer: $V_P = kQ/\sqrt{r^2 + a^2}$

E 13.16   Show that the result of Exercise 13.15 behaves appropriately for (a) $r \gg a$ and (b) $r = 0$.

Determine the electric potential at a perpendicular distance r from an infinitely long straight line that has a linear charge density λ.

This problem involves a charge distribution of infinite spatial extent. In general, the calculation of electric potential from Equation (13-14), $V = k\int dQ/r$, is a tedious and sometimes intractable task. Therefore, when a distribution of charge has an infinite extent, we may inquire whether this condition provides the necessary symmetry, either cylindrical or planar, that permits the use of Gauss's law to determine the electric field $\mathbf{E}$ as a function of position. Knowing the electric field function, we may use Equation (13-3), $V_B - V_A = \int_B^A \mathbf{E} \cdot d\mathbf{s}$, to determine the desired potential difference.

In this problem, the infinite line of charge provides cylindrical symmetry, and the simpler technique, which uses Gauss's law, is applicable. To determine the magnitude E_r of the radial electric field, we construct a cylindrical gaussian surface coaxial with the line of charge, as shown in Figure 13.9. Then Gauss's law gives

$$E_r \cdot 2\pi rL = \frac{Q_{\text{net}}}{\epsilon_o} = \frac{\lambda L}{\epsilon_o}$$

$$E_r = \frac{\lambda}{2\pi\epsilon_o r} \tag{13-18}$$

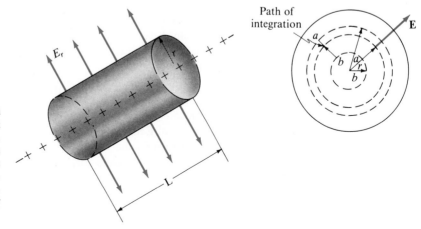

Figure 13.9
Two views of a cylindrical gaussian surface constructed coaxially with an infinite line of linear charge density λ. The cross-sectional view shows the radial path of integration used in the text to calculate the difference in electric potential between points at distances $r = a$ and $r = b$ from the line of charge.

We may now substitute this electric field into Equation (13-3) to determine the difference in potential, $V_b - V_a$, between two arbitrary points on a radial line at the distances b and a from the line of charge:

$$V_b - V_a = \int_b^a \mathbf{E} \cdot d\mathbf{s} = \int_b^a E_r dr \qquad (13\text{-}19)$$

Here the path of integration is chosen to be along a radial line so that $\mathbf{E} = E_r \hat{\mathbf{r}}$ and $d\mathbf{s} = dr\hat{\mathbf{r}}$. The cross-sectional view in Figure 13.9 shows the radial path of integration from $r = b$ to $r = a$. Substituting Equation (13-18) into Equation (13-19) gives

$$V_b - V_a = \frac{\lambda}{2\pi\epsilon_0} \int_b^a \frac{dr}{r} = \frac{\lambda}{2\pi\epsilon_0} \ln\frac{a}{b} \qquad (13\text{-}20)$$

If, now, we wish to specify the electric potential V at an arbitrary radial distance r from the line of charge, we may try specifying the potential V_∞ at an infinite radial distance from the line of charge to be zero. In that case, we may set $V_a = V_\infty = 0$ and $V_b = V(r)$ in Equation (13-20) gives

$$V(r) = \frac{\lambda}{2\pi\epsilon_0} \ln\frac{\infty}{r}$$

which is an undefined quantity. In order to define an electric potential at an arbitrary radial distance r for this infinite charge distribution, we first choose a reference point at a radial distance $r = r_0$ where V is assigned the value zero. Then, letting b be an arbitrary distance r and $a = r_0$ in Equation (13-20), we obtain

$$V(r) = \frac{\lambda}{2\pi\epsilon_0} \ln\frac{r_0}{r} \qquad (13\text{-}21)$$

We may verify that Equation (13-21) satisfies our choice that $V = 0$ at $r = r_0$. That substitution, $r = r_0$, into Equation (13-21) gives $V = (\lambda/2\pi\epsilon_0)\ln 1 = 0$.

We have seen how electric potential may be calculated directly from a given charge distribution and how either electric potential or difference in electric potential may be calculated from a known electric field. Sometimes, when the electric potential has been calculated or has been given, we can determine the electric field $\mathbf{E}$ from the electric potential function V. Equation (13-2),

$$'V_B - V_A = -\int_A^B \mathbf{E} \cdot d\mathbf{s}$$

implies, as we have seen in Equation (13-6), that there exists a differential relationship between an infinitesimal displacement $d\mathbf{s}$ from a point where the electric field is $\mathbf{E}$ and a corresponding infinitesimal change dV in electric potential:

$$dV = -\mathbf{E} \cdot d\mathbf{s} \qquad (13\text{-}22)$$

Because $\mathbf{E} \cdot d\mathbf{s}$ is a scalar product the value of dV depends not only on the magnitudes of $\mathbf{E}$ and $d\mathbf{s}$ but on the relative directions of $\mathbf{E}$ and $d\mathbf{s}$. For example, if $d\mathbf{s}$ were perpendicular to $\mathbf{E}$, there would be no change in electric potential. On the other hand, if $d\mathbf{s}$ were parallel to $\mathbf{E}$ and in the same direction, the change dV would have the magnitude

$$dV = -E \, ds \quad (ds \text{ in the same direction as } \mathbf{E})$$

which may be written in the form of a derivative,

$$E = -\frac{dV}{ds} \tag{13-23}$$

In a more specific situation, suppose that **E** is directed radially outward, so that $\mathbf{E} = E_r \hat{\mathbf{r}}$, and that $\mathbf{ds} = dr\,\hat{\mathbf{r}}$ is a radial displacement in the same direction as **E**. Then Equation (13-22) becomes $dV = -E_r\,dr$, or

$$E_r = -\frac{dV}{dr} \tag{13-24}$$

In a similar manner, for $\mathbf{E} = E_x \hat{\mathbf{i}}$ and a displacement $\mathbf{ds} = dx\,\hat{\mathbf{i}}$, Equation (13-22) is, in this case, $dV = -E_x\,dx$, or

$$E_x = -\frac{dV}{dx} \tag{13-25}$$

Thus, if the electric potential V in a region of space is a known function of radial distance or of a rectangular coordinate, we can determine the electric field in that region as a function of that particular variable using Equation (13-24) or Equation (13-25). It is interesting to note that in the general case, V is a function of all the spatial coordinates of a given coordinate system, like $V(x, y, z)$ in a rectangular system, for example. In that case, the rectangular components of the electric field are specified by partial derivatives:

$$E_x = -\frac{\partial V}{\partial x} \quad ; \quad E_y = -\frac{\partial V}{\partial y} \quad ; \quad E_z = -\frac{\partial V}{\partial z}$$

A partial derivative, like $\partial V/\partial x$, is the derivative of V with respect to x with y and z held constant.

The following problem illustrates how the electric field may be calculated from a known electric potential function.

Equation (13-17), $V = k\lambda \ln[x/(x - L)]$, specifies the electric potential at points $x > L$ along the x axis, where a charge of uniform density λ has one end at $x = 0$ and the other at $x = L$. Determine the electric field **E** as a function of x for points $x > L$ lying on the x axis.

We may determine the electric field component E_x from the given potential V using Equation (13-25):

$$E_x = -\frac{dV}{dx} = -\frac{d}{dx}\left[k\lambda \ln\left(\frac{x}{x - L}\right) \right]$$

$$E_x = \frac{k\lambda L}{x(x - L)}$$

Then the required electric field is

$$\mathbf{E} = \frac{k\lambda L}{x(x - L)}\hat{\mathbf{i}} \tag{13-26}$$

13.4 Equipotential Surfaces and Charged Conductors

Any surface along which the electric potential has a constant value is called an *equipotential surface*. Equation (13-6), $dV = -\mathbf{E} \cdot d\mathbf{s}$, requires that for a displacement $d\mathbf{s}$ along an equipotential surface, where $dV = 0$, the electric field $\mathbf{E}$ is perpendicular to the displacement, so that $\mathbf{E} \cdot d\mathbf{s} = 0$. Therefore, the electric field at any point is perpendicular to the equipotential surface passing through that point.

The relationship between an electric field and its corresponding equipotential surfaces is conventionally pictured in diagrams like those of Figure 13.10, a point charge and an electric dipole. In this figure the electric field (solid lines) and equipotential surfaces (dashed lines) are displayed in a cross-sectional plane passing through all points on an axis of symmetry. For charge distributions with axes of symmetry, the equipotential surfaces are those surfaces of revolution generated by rotating the dashed lines about an axis of symmetry. Notice that the equipotential surfaces are perpendicular to the electric field lines at every point.

The surface of a charged conductor is an equipotential surface, because the electric field immediately outside the surface is perpendicular to that surface. Inside a conducting material, where the electric field $\mathbf{E}$ is identically equal to zero, the change ($dV = -\mathbf{E} \cdot d\mathbf{s}$) in potential is equal to zero. Thus the electric potential inside the material of a conductor has the same value as the potential on the surface of the conductor.

The following problem illustrates how these relationships between electric fields, electric potentials, and conductors permit the analysis of electrostatically charged conductors.

An insulated solid conducting sphere of radius R has a net positive charge Q. Determine the electric field and the electric potential within and outside this sphere.

We have already seen in Chapter 12 that by construction of a spherical gaussian surface (radius $r > R$) concentric with the conducting sphere, we can find the magnitude of the radially outward electric field to be

$$E_r = \frac{kQ}{r^2} \quad ; \quad r > R \qquad (13\text{-}27)$$

Figure 13.10
Representations of electric fields (solid lines) and equipotential surfaces (dashed lines) for a positive point charge and an electric dipole. The equipotential surfaces are perpendicular to the lines of electric field at every point.

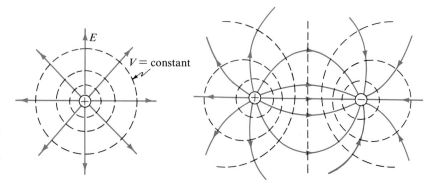

And because the electric field inside a conductor must be equal to zero in an electrostatic situation, we know that

$$E_r = 0 \quad ; \quad r < R \tag{13-28}$$

Because we are dealing here with a charge distribution of finite extent, we may obtain the electric potential (relative to a zero value of potential at an infinite distance from the sphere) using Equation (13-8):

$$V = \int_r^\infty \mathbf{E} \cdot d\mathbf{s} = kQ \int_r^\infty \frac{ds}{s^2} = \frac{kQ}{r} \quad ; \quad r \geq R \tag{13-29}$$

Here we have integrated $\mathbf{E} \cdot d\mathbf{s}$ along a radial path that extends from $r = R$ to infinity, expressing the radial electric field of Equation (13-29) as $E = kQ/s^2$, where s is the distance from the center of the sphere. Inside the surface of the sphere, where the electric field is equal to zero, the electric potential cannot change from its value $V = kQ/R$ at the surface. Therefore, at every point within the sphere, the electric potential is

$$V = \frac{kQ}{R} \quad ; \quad r \leq R \tag{13-30}$$

Figure 13.11 depicts the charged sphere and the curves of E_r and V as functions of radial position r from the center of the sphere. The values of E_r and V at points

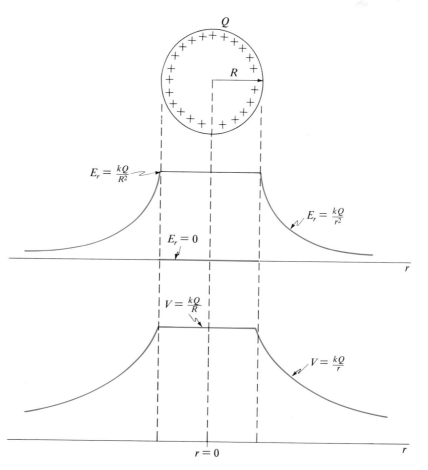

Figure 13.11
An insulated solid conducting sphere of radius R having a positive charge Q. The curves below the sphere show the magnitude of the radial electric field E_r and the electric potential V as functions of radial distance r from the center of the sphere.

lying outside the sphere are the same as they would be if the sphere were replaced by a point charge Q at the center of the sphere.

It is important to recognize and to remember that in problems involving electric potential inside and around conductors, even though the electric field is required to be zero inside a conducting material, the electric potential need not be equal to zero. Indeed, the electric potential has a constant value inside and on the surface of a conducting medium.

Example 13.3

PROBLEM   A large, flat sheet of a conductor carries a uniform surface charge density of $\sigma = 0.20\ \mu\text{C/m}^2$ on each of its surfaces. If the conductor is assigned an electric potential of zero, calculate the electric potential at a (perpendicular) distance of 3.0 cm from either surface.

SOLUTION   Choose the positive x direction perpendicular to and outward from either surface. We have seen, using Gauss's law, that the electric field outside a conductor is given by $E_x = \sigma/\epsilon_o = 4\pi k\sigma$. Because $E_x = -dV/dx$ or $dV = -E_x\,dx$, we get

$$V = -\int_0^d E_x\,dx = -4\pi k\sigma d$$

where x has been assigned the value $x = 0$ on the surface of the conductor. Substituting the specified values for σ and d, we get

$$V = -4\pi(9 \times 10^9)(2 \times 10^{-7})(3 \times 10^{-2}) = -680\ \text{V}$$

That is, a point 3.0 cm outside the surface of the conductor has a potential that is 680 V less than that of the potential of the conductor.   ∎

E 13.17   The equipotential surfaces associated with the charged conductor of Example 13.3 are flat planes that are parallel to the surface of the conductor. Calculate the distance between two such equipotential surfaces that differ in potential by 1000 volts.   Answer: 4.4 mm

Example 13.4

PROBLEM   A long, solid, cylindrical conductor has a radius of $R = 0.8$ cm and carries a uniform surface charge density $\sigma = 5.0 \times 10^{-8}$ C/m^2. If the potential of the conductor is specified to be 100 V, calculate the electric potential of a coaxial cylindrical surface having a radius of 4.0 cm.

SOLUTION   Applying Gauss's law to a coaxial cylindrical surface of radius $r \geq R$ and length L gives

$$E_r(2\pi rL) = \frac{1}{\epsilon_o}\sigma(2\pi RL)$$

$$E_r = \frac{\sigma R}{\epsilon_o r}$$

Because $E_r = -dV/dr$, or $dV = -E_r\,dr$, we get

$$\int_{V_1}^{V_2} dV = -\int_{r_1}^{r_2} \frac{\sigma R}{\epsilon_o r}\,dr$$

or

$$V_2 - V_1 = -\frac{\sigma R}{\epsilon_o}\ln(r_2/r_1)$$

$$V_2 = V_1 - \frac{\sigma R}{\epsilon_o}\ln(r_2/r_1)$$

Substituting the values $V_1 = 100$ V, $\sigma = 5.0 \times 10^{-8}$ C/m^2, $r_2 = 0.040$ m, and $r_1 = R = 0.0080$ m gives the desired potential

$$V_2 = 100 - \frac{5 \times 10^{-8} \times 8 \times 10^{-3}}{8.85 \times 10^{-12}} \ln\left(\frac{0.04}{0.008}\right) = 100 - 73 = 27 \text{ V} \quad \blacksquare$$

13.5 Electrostatic Potential Energy of Charge Collections

Electrostatic potential energy U_{el}, like all energy, is a measure of the ability of a system to do work. In the case of U_{el}, the system is composed of charged particles that interact through electrostatic forces. Like all potential energies, the value of U_{el} is measured relative to a reference configuration for which the potential energy is chosen to be equal to zero. Usually we specify electrostatic potential energy to be equal to zero when the charges constituting the system are separated by an infinite distance. Then U_{el} for a system of charges in a given configuration is equal to the work that must be done against the electrical forces in order to assemble that configuration from an initial state of infinite separation of the charges. This work, and, therefore, the energy of the configuration, is also equal to the work that the electrical forces among the charges would perform as the charges are caused to be separated by an infinite distance.

To find the electrostatic potential energy of a configuration of point charges, we may imagine that the individual charges are brought—one at a time—from an infinite distance away to their final positions in the configuration. Consider first the simplest configuration: two point charges, Q_1 and Q_2, separated by a distance r_{12}. The electric potential at a distance r_{12} from Q_1 alone is

$$V = \frac{kQ_1}{r_{12}} = \frac{Q_1}{4\pi\epsilon_o r_{12}} \tag{13-31}$$

It follows from our definition of the difference in electric potential, Equation (13-1), that the work necessary to bring the second charge Q_2 from an infinite distance away from Q_1 to a distance r_{12} from Q_1, is equal to $Q_2 V$. Then the potential energy $U_{el,2}$ of this two-charge system is

$$U_{el,2} = \frac{kQ_1 Q_2}{r_{12}} \tag{13-32}$$

Notice that if both charges have the same sign, the electrostatic potential energy is positive, indicating that (positive) work must be done *on* the system by an external agent against the repulsive electrostatic forces between the charges in order to assemble the given configuration. If the charges have unlike signs, U_{el} is negative, indicating that (negative) work must be done *by* an external agent against the attractive forces between the charges.

If now a third charge Q_3 is brought from an infinite distance away from the first two charges to the point P, as suggested by Figure 13.12, additional work is required. If P is located a distance r_{13} from Q_1 and r_{23} from Q_2, the electric potential V (resulting from Q_1 and Q_2) at the point P is

$$V = k\left[\frac{Q_1}{r_{13}} + \frac{Q_2}{r_{23}}\right]$$

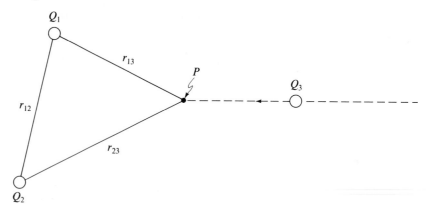

Figure 13.12
A charge Q_3 being brought to a point P
near two fixed charges Q_1 and Q_2.

The work necessary to bring Q_3 from infinity to P is Q_3V. If this energy is added to the electrostatic potential energy, given by Equation (13-32), of the first two charges, then the potential energy $U_{el,3}$ of the three-charge configuration is

$$U_{el,3} = k\left[\frac{Q_1Q_2}{r_{12}} + \frac{Q_1Q_3}{r_{13}} + \frac{Q_2Q_3}{r_{23}}\right] \qquad (13\text{-}33)$$

A similar procedure may be continued to include an arbitrary number of charges, $Q_1, Q_2, \ldots, Q_n$. The electrostatic potential energy of a configuration of n charges is

$$U_{el,n} = k\sum_{\text{pairs}} \frac{Q_iQ_j}{r_{ij}} \qquad (13\text{-}34)$$

where the summation is carried out over all pairs, Q_i and Q_j, among n charges; each pair, Q_i and Q_j, is separated by a distance r_{ij}. The sign of each term in Equation (13-34) is determined by the signs of the charge pair. The term is positive if the signs of Q_i and Q_j are the same; the term is negative if the signs of Q_i and Q_j are different.

E 13.18   Calculate the work required to bring two particles, each having a charge $Q = 20 \ \mu C$, from a very large separation to a separation of 5.0 cm.

Answer: 72 J

E 13.19   Three identical particles, each having a charge of 15 μC, are located at the vertices of an equilateral triangle that is 9.0 cm along each side. Calculate the electrostatic potential energy of this charge configuration.

Answer: 68 J

Example 13.5   PROBLEM   Two identical particles, of a charge $q = 4.0 \times 10^{-7}$ C and mass $m = 3.0 \times 10^{-4}$ kg, are released from rest when they are separated by a distance of 2.0 cm. Calculate the speed of either particle at the instant when the two are 5.0 cm apart.

SOLUTION   Figure 13.13(a) shows the two particles initially at rest and separated by a distance r_o. Figure 13-13(b) shows the particles later when they are separated by a distance r_1 and are each moving with a speed v_1. Because no external forces are acting on the system, the total energy is conserved, i.e.,

$$\frac{kq^2}{r_o} = \frac{kq^2}{r_1} + 2\left(\frac{1}{2}mv_1^2\right)$$

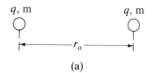

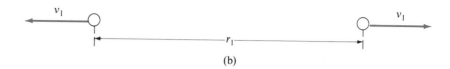

Figure 13.13   Example 13.5

(b)

Solving for v_1 gives

$$v_1 = \sqrt{\frac{kq^2}{m}\left(\frac{r_1 - r_0}{r_0 r_1}\right)}$$

$$v_1 = \left[\frac{9 \times 10^9 \times 16 \times 10^{-14}}{3 \times 10^{-4}}\left(\frac{0.03}{0.02 \times 0.05}\right)\right]^{1/2} = 12 \text{ m/s} \quad \blacksquare$$

E 13.20   The solution of Example 13.5 disregarded any consideration of gravitational potential energy $(U_g = -Gm^2/r)$. Was this reasonable?

Answer: Yes, $|U_g/U_e| \simeq 4.0 \times 10^{-15}$

A continuous distribution of charge also has associated with it an electrostatic potential energy. The following problem illustrates the process by which U_{el} may be calculated for a given continuous charge.

What is the electrostatic potential energy of a conducting sphere (of radius R) that has a net charge Q?

Figure 13.14 shows the sphere of radius R at a time when an infinitesimal quantity of charge dq is being brought to the surface of the sphere. At the instant shown in the figure, the sphere has a charge q already present on its surface. The surface of the sphere, therefore, is at the potential $V = kq/R$. The element of charge dq is being brought to the surface of the sphere from an infinite distance away. The work dW required to bring the element of charge dq to the surface with potential V is

$$dW = V\,dq = \frac{kq}{R}dq \qquad\qquad (13\text{-}35)$$

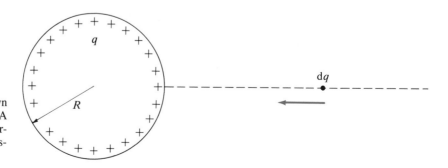

Figure 13.14
A conducting sphere of radius R, shown at a time when it has a charge q. A charge dq is being brought to the surface of the sphere from an infinite distance away.

Because the electrostatic potential energy U_{el} of the sphere is equal to the work required to accumulate the total charge Q, we may write

$$U_{el} = W = \frac{k}{R} \int_0^Q q \, dq = \frac{kQ^2}{2R} \tag{13-36}$$

the required value of electrostatic potential energy of the charged sphere.

E 13.21   Assume that the electrostatic potential energy of an electron ($q = -1.6 \times 10^{-19}$ C, $m = 9.1 \times 10^{-31}$ kg) manifests itself as the famous Einsteinian rest-mass energy, $E = mc^2$. Calculate the corresponding value for the radius of the electron.   Answer: 1.4×10^{-15} m

13.6 Problem-Solving Summary

The basic definition of electric potential is expressed in terms of a difference in potential, $V_B - V_A$, between two points A and B: $V_B - V_A$ is equal to the work $W_{A \to B}$ per unit charge necessary to move a particle with charge q from A to B without changing the kinetic energy of the particle. Because the electrostatic field is conservative, the difference in electric potential between A and B is independent of the path along which $W_{A \to B}$ is calculated. Therefore, in solving problems, we may choose the most convenient path for purposes of computing potential differences.

The definition of potential difference, $V_B - V_A = W_{A \to B}/q$, permits us to assign a value to the electric potential at a single point if we have chosen a reference point at which the potential is assigned the value zero. Usually the potential is chosen to be equal to zero at an infinite distance from a given charge. This choice of a zero of potential is appropriate if the charge distribution being considered has a finite extent; otherwise the position at which the potential is chosen to be equal to zero must be specified.

The electric potential V of a point charge Q at a distance r from the charge is given by $V = kQ/r$. The potential at point P caused by a collection of point charges is found by first calculating the potential at P caused by each individual charge. Then the electric potential at P is the algebraic sum of each of the potentials caused by the charges in the entire collection. The algebraic sign of the potential computed for each charge Q is positive if Q is positive and negative if Q is negative.

The electric potential at a point P caused by a continuous distribution of charge of finite extent is calculated by treating each infinitesimal element of charge dQ as if it were a point charge at a distance r from the point P. Then the potential at P is found by integrating the contribution of each charge element so that the integration spans the entire distribution of charge:

$$V = k \int \frac{dQ}{r} = \frac{1}{4\pi\epsilon_o} \int \frac{dQ}{r}$$

In order to carry out the integration in most problems, dQ and r must be expressed in terms of a single variable. Setting up an appropriate integral requires careful consideration of the geometry of a given problem. The illustrations and examples in the text should be examined carefully and considered as guides for the solution of problems in this category.

In electrostatics problems, the effects of the presence of charges may be described in terms of either electric fields or electric potentials. In a given problem situation, one or the other description may be the more convenient to calculate. But it is often necessary to calculate electric potential from a known (given or calculated) electric field or an electric field from an electric potential. Both processes are summarized here:

1. Suppose $\mathbf{E}$ is given or can be easily obtained (using Gauss's law, for example) in a given problem situation. Then the definition of difference in potential leads to $V_B - V_A = \int_B^A \mathbf{E} \cdot d\mathbf{s}$, which may be evaluated if $\mathbf{E}$ is known as a function of spatial coordinates throughout the length of *any* path connecting A and B. The potential at any point, say the point B, may be determined by letting A be the reference point at which the potential is chosen to be zero. Then the potential at point B becomes

$$V_B = \int_B^{\text{ref}} \mathbf{E} \cdot d\mathbf{s}$$

2. Suppose we are given (or have found) the electric potential function V for a change distribution. Then we may find the corresponding electric field by calculating the negative derivative of the potential function with respect to the appropriate spatial coordinate. We may use, $E_x = -dV/dx$ or $E_r = -dV/dr$, for example, as appropriate.

An equipotential surface is any surface that has the same value of electric potential at every point. The electrostatic relationship between electric potential V and electric field $\mathbf{E}$, $dV = -\mathbf{E} \cdot d\mathbf{s}$, requires that:

1. The electric field is always perpendicular to equipotential surfaces.

2. The electric field always points toward the direction in which the electric potential decreases most rapidly.

3. The surface of every conductor is an equipotential surface.

Because the electric field $\mathbf{E}$ inside a conductor is equal to zero in electrostatic situations, the electric potential V inside a conducting material cannot change from its value on the surface of the conductor. Therefore, the value of V inside a conducting material is not, in general, equal to zero, even though $\mathbf{E}$ is identically equal to zero.

The electric potential energy U_{el} of a given configuration of charges is equal to the work (by an external agent) required to assemble the configuration. Because the work that must be done against electrical forces to bring a point charge Q from an infinite separation to a point where the electric potential is V is equal to QV, the electric potential energy U_{el} of a pair of point charges Q_1 and Q_2, separated by a distance r_{12}, is given by kQ_1Q_2/r_{12}. For a collection of more than two charges, U_{el} is found by summing (algebraically, noting that each charge has a signed value) the energies of all pairs of charges, Q_i and Q_j, which are separated by a distance r_{ij} or

$$U_{\text{el}} = k \sum_{\text{pairs}} \frac{Q_i Q_j}{r_{ij}}$$

Problems

GROUP A

13.1   If the kinetic energy of a charged particle ($q = -5.0\ \mu$C) changes from $K_A = 12$ mJ to $K_B = 28$ mJ as it moves from point A to point B under the influence of electrical forces only, determine the potential difference, $V_B - V_A$.

13.2   If the potential difference between points A and B is equal to $V_B - V_A = 1.5 \times 10^4$ V, how much work must be done to move a charged particle ($q = 2.0\ \mu$C) from point B to point A without changing the kinetic energy of the particle?

13.3   A point charge $Q = 8.0 \times 10^{-8}$ C is placed at the origin of a coordinate system. Calculate the electric potential difference, $V_B - V_A$, if points A and B have the rectangular coordinates (2.0, 0, 0) m and (0, 3.0, 4.0) m, respectively.

13.4   Two points charges, $Q_1 = 7.0$ nC and $Q_2 = -4.0$ nC, are positioned as shown in Figure 13.15. Calculate the electric potential difference, $V_A - V_B$, for the points A and B.

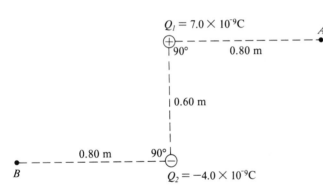

Figure 13.15   Problem 13.4

13.5   Calculate the electrostatic potential energy of each of the charge distributions shown in Figure 13.16.

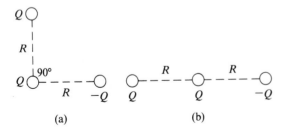

(a)   (b)

Figure 13.16   Problem 13.5

13.6

(a) Calculate the electric potential at the point P shown in Figure 13.17.

(b) Determine the (dipole) behavior of the potential at point P for $r \gg a$.

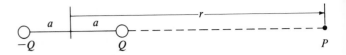

Figure 13.17   Problem 13.6

13.7   Three collinear point charges are positioned as shown in Figure 13.18.

(a) Calculate the electric potential at point P.

(b) Determine the (quadrupole) behavior of the potential at point P for $r \gg a$.

Figure 13.18   Problem 13.7

13.8   A charge of uniform linear density λ is distributed along a semicircle of radius R. Determine the electric potential at the center of the semicircle.

13.9   Two point charges are at the vertices of a rectangle (*see* Fig. 13.19). If $Q_1 = 6.0 \times 10^{-8}$ C, $Q_2 = -2.0 \times 10^{-8}$ C, $a = 0.30$ m, and $b = 0.40$ m, calculate the potential difference, $V_A - V_B$.

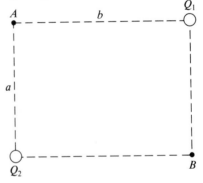

Figure 13.19   Problem 13.9

13.10   Show that a constant electric field is a conservative field.

13.11   Two particles, each having a charge Q, are placed on the y axis at $y = +a$ and $y = -a$.

(a) Find an expression for the electric potential at any point $(x, 0, 0)$ on the x axis.

(b) Use the expression obtained in part (a) to determine E_x, the x component of the electric field, at a point $(x, 0, 0)$ on the positive axis.

13.12   An electron is accelerated from rest through a potential difference of 1000 V.

(a) Calculate the speed of the electron after this acceleration.

(b) Compare (that is, find the ratio of) this speed to the speed of light, $c = 3.0 \times 10^8$ m/s.

13.13 A solid spherical conductor (radius = 0.20 cm) has a uniform surface charge density of 3.0×10^{-7} C/m². Calculate the electric potential

(a) at the center of the sphere.

(b) at the surface of the sphere.

(c) at a point 3.0 cm from the center of the sphere.

13.14 Four identical point charges ($Q = 8.0 \ \mu$C) are placed at the vertices of a rectangle (see Fig. 13.20). Calculate the electrostatic potential energy of this charge configuration.

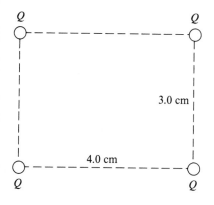

Figure 13.20   Problem 13.14

13.15 Equipotential planes associated with a uniform plane charge of surface density $\sigma > 0$ differ in potential by 100 V and are separated by 2.5 cm. Calculate σ.

13.16 Charges $Q_1 = 5.0 \ \mu$C and $Q_2 = -9.0 \ \mu$C are positioned as shown in Figure 13.21. Calculate

(a) the potential difference, $V_A - V_B$, for the points A and B.

(b) the work an external agent will perform in moving a 20-μC charge from A to B without changing its kinetic energy.

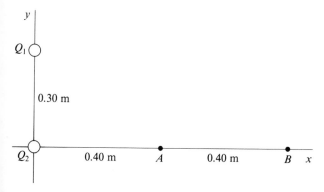

Figure 13.21   Problem 13.16

13.17 Charge of uniform linear density $\lambda = 5.0 \times 10^{-8}$ C/m is distributed along the entire z axis. Point A is 7.0 cm from the z axis, and point B is 2.0 cm from the z axis. Calculate the potential difference, $V_B - V_A$.

13.18 Charge of uniform linear density λ is distributed along the perimeter of a circle of radius R.

(a) If the point P is on the line that is perpendicular to the plane of the circle and through the center of the circle and if P is a distance r from the plane of the circle, determine the electric potential at P.

(b) Show that the expression obtained in part (a) behaves appropriately for $r = 0$ and $r \gg R$.

(c) Use the expression obtained in part (a) to determine E_n, the component of the electric field normal to the plane of the circle, as a function of r.

13.19 Two identical point charges ($Q = 2.0 \ \mu$C) are positioned as shown in Figure 13.22. Calculate the work an external agent must do to move a particle having a charge $q = 15 \ \mu$C from point A to point B without changing the kinetic energy of the particle.

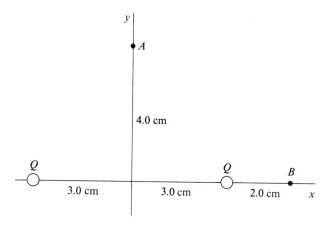

Figure 13.22   Problem 13.19

13.20 As a particle ($m = 3.0$ mg, $q = 5.0 \ \mu$C) moves from point A to point B, its speed changes from 120 m/s to 190 m/s. Only electrical forces act on the particle. Calculate

(a) the work done by the electrical forces acting on the particle as it moves from A to B.

(b) the potential difference, $V_B - V_A$.

13.21 Two protons ($m_p = 1.7 \times 10^{-27}$ kg, $q_p = 1.6 \times 10^{-19}$ C) are at rest when the distance between them is 2.5×10^{-11} m. What is the speed of either proton at the instant when the distance separating them is 1.0 cm?

13.22 Suppose two protons and an electron are positioned at the vertices of an equilateral triangle ($r_o = 1.0 \times 10^{-10}$ m) and released from rest (see Fig. 13.23). Disregarding any motion of the considerably more massive protons ($m_p = 1.7 \times 10^{-27}$ kg, $m_e = 9.1 \times 10^{-31}$ kg),

(a) what is the speed of the electron as it passes the midpoint between the protons?

(b) estimate the time that elapses between the initial state and the state of part (a).

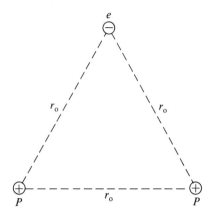

Figure 13.23   Problem 13.22

13.23 Charges Q_1 and Q_2 are placed at two vertices of a rectangle, as shown in Figure 13.24. If the electric potential difference between points A and B is $V_A - V_B = 270$ V, determine Q_2 if $Q_1 = 1.4 \times 10^{-8}$ C.

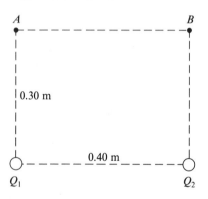

Figure 13.24   Problem 13.23

13.24 Two point charges, $Q_1 = Q$ and $Q_2 = -3Q$, are located on the x axis at $x_1 = 0$ and $x_2 = L$. Determine the point(s) on the x axis for which the electric potential is zero.

13.25 Four identical charged particles ($Q = 2.5\ \mu$C) are placed at the vertices of a square (length of a side $= 2.0$ cm) and released simultaneously from rest. What is the kinetic energy of any one of the particles after it has moved 10 cm?

13.26 A long, solid, cylindrical conductor (radius $= 0.50$ cm) carries a uniform surface charge density of 6.0×10^{-8} C/m². Calculate the magnitude of the potential difference between a point on the axis of the cylinder and a point 5.0 cm from the axis.

13.27 Two particles, each having a charge $Q > 0$, are held fixed a distance L apart, as shown in Figure 13.25. A third

particle ($q > 0$) is released from rest with $\alpha = 30°$. Determine the kinetic energy of the third particle

(a) at the instant when $\alpha = 60°$.

(b) as α approaches $90°$.

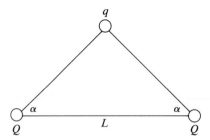

Figure 13.25   Problem 13.27

13.28 Two particles ($m_1 = m_2 = M$, $q_1 = -q_2 = Q > 0$) are released simultaneously from rest when the distance between the two is r_o. Determine the speed of either particle when the distance separating them is $r_o/5$.

13.29 Two parallel charged planes that have surface densities $\sigma_1 = 5.0 \times 10^{-9}$ C/m² and $\sigma_2 = 2.0 \times 10^{-9}$ C/m² and are separated by 7.0 cm (*see* Fig. 13.26). Calculate the potential difference, $V_B - V_A$, for the points A and B.

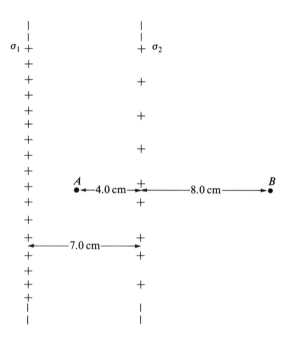

Figure 13.26   Problem 13.29

13.30 The electric field in a region of space is given by

$$\mathbf{E} = -\frac{E_o x}{L}\hat{\mathbf{i}}$$

where E_o and L are positive constants. If points A, B, and C have the locations $\mathbf{r}_A = 0$, $\mathbf{r}_B = 2L\hat{\mathbf{i}}$, and $\mathbf{r}_C = (3L\hat{\mathbf{i}} + 2L\hat{\mathbf{j}})$, calculate the potential differences for

(a) $V_B - V_A$.

(b) $V_C - V_B$.

(c) $V_A - V_C$.

13.31 Charge of uniform linear density $\lambda = 4.0 \times 10^{-8}$ C/m is distributed along the entire y axis. If points A and B are located on the x axis at $x_A = 12$ cm and $x_B = 32$ cm, calculate the potential difference, $V_A - V_B$.

13.32 A charge, $Q = 5.0 \times 10^{-8}$ C, is uniformly distributed along the x axis from $x = -2.0$ m to $x = 0$. Calculate the electric potential at the point on the x axis, $x = 3.0$ m.

13.33 A point charge, $Q = 7.0 \times 10^{-9}$ C, is placed at the center of a hollow spherical conductor having an inner radius of 2.0 cm and an outer radius of 4.0 cm. If point A is 1.0 cm from Q and point B is 5.0 cm from Q, calculate the potential difference, $V_A - V_B$.

13.34 Repeat Problem 13.33 if the hollow spherical conductor has a net charge of -9.0×10^{-9} C.

13.35 Two identical charges ($Q = 8.0$ μC) are fixed on the y axis at $y = \pm 0.30$ m (*see* Fig. 13.27). A particle that has $m = 4.0 \times 10^{-4}$ kg and $q = 5.0$ μC has a velocity $\mathbf{v} = v\hat{\mathbf{i}}$ when its position is $\mathbf{r} = -0.40\hat{\mathbf{i}}$ m.

(a) Calculate the minimum value of v for which the particle will reach $x = 0$.

(b) If $v = 40$ m/s, how close will the particle get to $x = 0$?

(c) If $v = 100$ m/s, how fast will the particle be moving as it passes through $x = 0$?

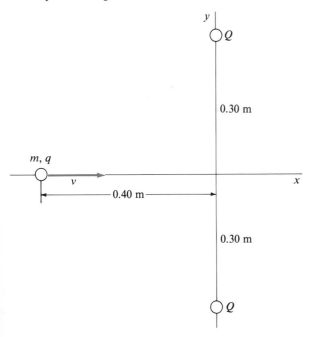

Figure 13.27 Problem 13.35

13.36 A charge Q is uniformly distributed throughout a spherical volume of radius R. Determine the electric potential difference, $V_1 - V_2$, if V_1 is the electric potential at the center of the sphere and V_2 is the potential on the outer surface of the sphere.

13.37 Charge having a nonuniform linear density $\lambda(x) = \alpha x$, where α is a positive constant, is distributed along the x axis from $x = 0$ to $x = L$.

(a) Determine the total distributed charge Q in terms of α and L.

(b) Determine the electric potential at the point, $\mathbf{r} = y\hat{\mathbf{j}}$, on the positive y axis.

(c) Show that the result of part (b) behaves appropriately for $r \gg L$. (*Hint:* for $|\epsilon| \ll 1$, $\sqrt{1 + \epsilon} \simeq 1 + \frac{1}{2}\epsilon$.)

(d) Use the expression obtained in part (b) to determine E_y, the y component of the electric field, at the point $\mathbf{r} = y\hat{\mathbf{j}}$ on the positive y axis.

13.38 Three identical ($m = 4.0 \times 10^{-4}$ kg, $q = 3.0$ μC) particles are released (from rest) simultaneously from the vertices of an equilateral triangle (length of a side $= 0.20$ m). Calculate the speed of any one of the particles after it has traveled 0.10 m.

13.39 Two particles ($m_1 = 25$ μg, $m_2 = 60$ μg, $q_1 = 0.60$ μC, $q_2 = 0.45$ μC) are released simultaneously from rest when the distance between them is 0.25 m. Calculate the speed of the particle of mass m_1 when the distance separating the two particles is 1.5 m.

13.40 Charge of uniform surface density σ covers a flat circular disk of radius R. Point P is on the line that is perpendicular to, and through the center of, the disk. The distance from center of the disk to point P is r.

(a) Determine the electric potential at point P.

(b) Show that the result of part (a) behaves appropriately for $r \gg R$. (*Hint:* $(1 + \epsilon)^{1/2} \simeq 1 + \frac{1}{2}\epsilon$ for $|\epsilon| \ll 1$.)

(c) Use the expression obtained in part (a) to determine E_n, the component of the electric field normal to the plane of the disk, as a function of r.

13.41 Two large, flat, parallel, conducting plates, each having a thickness of 0.50 cm, are separated by 1.5 cm, as shown in Figure 13.28. The uniform surface charge densities are such that $\sigma_1 + \sigma_2 = 3\sigma$ and $\sigma_3 + \sigma_4 = \sigma$, where $\sigma = 4 \times 10^{-10}$ C/m². Calculate the charge densities, σ_1, σ_2, σ_3, and σ_4 and the magnitude of the potential difference between the two conductors.

13.42 A charge Q is distributed uniformly throughout the volume of a spherical shell of inner radius R and outer radius $2R$. If A is the point at the center of the sphere and B is a point at a distance of $3R$ from A, calculate the electric potential difference, $V_B - V_A$.

13.43 A spherical region, having a radius R, is filled with a nonuniform, spherically symmetric charge distribution having a volume charge density given by $\rho(r) = \alpha r^2$, where α is a constant.

(a) Calculate the total distributed charge Q in terms of α and R.

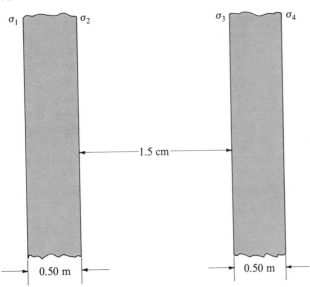

Figure 13.28   Problem 13.41

(b) If point A is a distance of $R/2$ from the center of the sphere and point B is a distance of $2R$ from the center, determine the potential difference, $V_A - V_B$, in terms of Q and R.

GROUP **B**

13.44 A charge of 3.0×10^{-8} C is distributed uniformly along the perimeter of a square that is 0.20 m along each side. Point P is on the line perpendicular to the plane of the square and passes through the geometrical center of the square. If the distance to P from the geometrical center of the square is 0.50 m, calculate the electric potential at point P.

13.45 An electric dipole consists of two charges, $+Q$ and $-Q$, positioned on the x axis at $x = a$ and $x = -a$, respectively.

(a) Determine the electric potential for any point (x, y) in the x-y plane.

(b) Defining the dipole moment to be $\mathbf{p} = 2Qa\hat{\mathbf{i}}$, show that for $r = \sqrt{x^2 + y^2} \gg a$, the electric potential is given by

$$V = \frac{k\mathbf{p} \cdot \hat{\mathbf{r}}}{r^2}$$

13.46 Continuous charge having a nonuniform linear density given by $\lambda(x) = \alpha x$, where α is a constant, is distributed along the x axis on the interval $-L \le x \le L$.

(a) Calculate the total distributed charge.

(b) Determine the electrical potential at $\mathbf{r} = r\hat{\mathbf{i}}$ $(r > L)$ on the positive x axis.

(c) Determine the behavior for $r \gg L$ of the potential expression obtained in part (b)? Why is it not $1/r$?

(d) Use the results of part (c) and Problem 13.45 to determine the magnitude of the electric dipole moment of this charge distribution.

13.47 A uniform linear charge of density λ is distributed along a segment of the spiral $r = ae^{\alpha\theta}$, $0 \le \theta < 2\pi$, where a and α are constants and r and θ are as shown in Figure 13.29.

(a) Calculate the total charge on the segment.

(b) Calculate the electric potential at the origin.

(c) Show that the expressions obtained in parts (a) and (b) behave appropriately as α approaches zero.

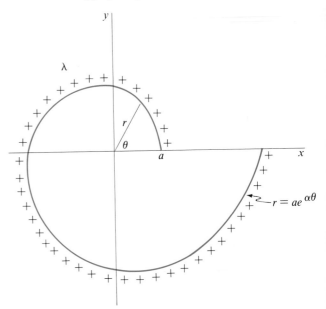

Figure 13.29   Problem 13.47

13.48 For the spiral charge distribution of Problem 13.47, show that the electric potential at any point in the x-y plane (not on the spiral) is given by

$$V(x, y) = [\sqrt{\alpha^2 + 1}\, k\lambda a] \cdot$$

$$\left[\int_0^{2\pi} \frac{e^{\alpha\theta}}{[x^2 + y^2 + a^2 e^{2\alpha\theta} - 2ae^{\alpha\theta}(x\cos\theta + y\sin\theta)]^{3/2}}\, d\theta \right]$$

13.49 Two identical, collinear, charged rods, each having a length L and a uniform linear charge density λ, are arranged as shown in Figure 13.30. Show that

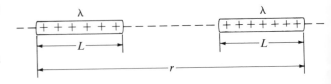

Figure 13.30   Problem 13.49

(a) the interaction potential energy of the two rods is given by

$$U_{el} = k\lambda^2 \int_0^L \ln\left(1 + \frac{L}{r - x}\right)dx$$

(b) the result of part (a) reduces appropriately for $r \gg L$. Recall that $\ln(1 + \epsilon) \approx \epsilon$ for $|\epsilon| \ll 1$.

13.50 The electric potential of any continuous charge distribution, which is of finite spatial extent, may be estimated to any desired accuracy by approximating the distributed charge by an appropriate set of point charges, i.e.,

$$V = \int \frac{kdQ}{r} \approx k \sum_{n=1}^N \frac{Q_n}{r_n}$$

Suppose a charge $Q = 4.0 \times 10^{-9}$ C is distributed uniformly along the x axis from $x = -1.0$ m to $x = 1.0$ m (see Fig. 13.31). Point P is on the x axis at $x = 3.0$ m.

(a) Show that the electric potential at the point P is given by $V_P = 18 \ln 2$ V ≈ 12.477 V.

(b) Use the $N = 1$ and $N = 2$ point charge distributions in the figure to estimate the electric potential at point P.

13.51 If you have access to a computer or a programmable calculator, you may enjoy continuing the sequence of electric potential estimates begun in Problem 13.50. The results of such a calculation are summarized below.

N	$V_{P,N}$ (volts)
1	12.000
2	12.343
4	12.442
8	12.468
16	12.475
32	12.476
64	12.477
128	12.477
256	12.477
Exact (N → ∞)	18 ln2 ≈ 12.477

13.52 A charge of $Q = 4.0 \times 10^{-9}$ C is distributed uniformly over the surface of a square that is 2.0 m along each side, as shown in Figure 13.32. Use the $N = 1$ and $N = 4$ point charge distributions to estimate the electric potential at the point P shown.

13.53 If you have access to a computer or a programmable calculator, you may wish to continue the sequence of electric potential estimates begun in Problem 13.52. The results of such a calculation are summarized below.

N	$V_{P,N}$ (volts)
1	12.000
4	12.151
16	12.197
64	12.209
256	12.212
1024	12.213
4096	12.213

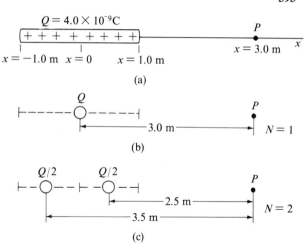

(a)

(b)

(c)

Figure 13.31   Problem 13.50

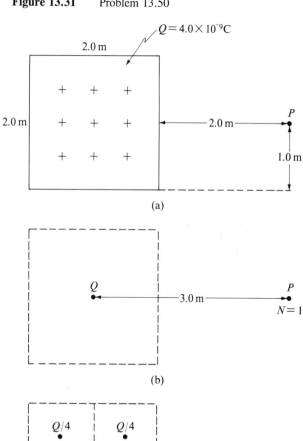

(a)

(b)

(c)

Figure 13.32   Problem 13.52

14 Capacitance, Current, and Resistance

This chapter examines the fundamental characteristics of two components of electrical circuitry, capacitors and resistors. We will treat these circuit elements as if they were idealized components, called *lumped passive elements*. They are said to be "lumped" because each is treated as though its electrical characteristics are localized to a single specific position in a given circuit. They are said to be "passive" because their functions are to respond electrically to charge in a circuit in ways that depend only on their configuration and on the material of which they are composed.

Our study of electromagnetism proceeds from the electrostatic case to a consideration of charges in motion. Electric current, which will be a primary consideration in electrical circuitry, will be briefly examined in this chapter.

14.1 Capacitance

Suppose two electrical conductors, labeled A and B in Figure 14.1, are separated in space. A positive charge Q has been transferred from B to A, leaving B with a negative charge $-Q$. Such a two-conductor arrangement in which the conductors have charges of equal magnitudes but opposite signs is called a *capacitor*. The electric field in the space around the capacitor is indicated by the lines of force in

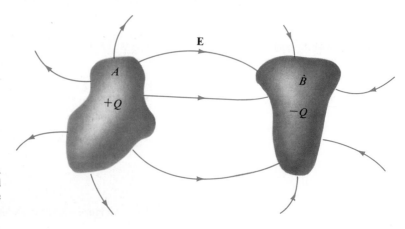

Figure 14.1
Two electrical conductors, A and B. A positive charge Q has been transferred from B to A, leaving B with a negative charge $-Q$.

the figure; the lines originate on the positively charged conductor and terminate on the negatively charged conductor. The electric potential V_A on the positively charged conductor A is greater than the potential V_B on B. The difference in electric potential ΔV between the separated conductors of a capacitor is defined to be the positive value, $V_A - V_B$. If Q is the magnitude of the charge on either conductor of a capacitor and ΔV is the potential difference between the conductors, the *capacitance* C of a two-conductor system is defined by

$$C = \frac{Q}{\Delta V} \tag{14-1}$$

which is an inherently positive quantity because both Q and ΔV are positive. As we will see, the ratio of charge Q to potential difference ΔV is a constant value for a fixed-configuration capacitor. Thus, the capacitance of a given capacitor with fixed conductors is a constant value; if the transferred charge Q of a given capacitor is changed, the potential difference ΔV changes proportionately.

The unit of capacitance, as indicated by Equation (14-1), is a coulomb (C) per volt (V), which is defined to be a farad (F), or

$$1 \text{ F} \equiv 1 \frac{\text{C}}{\text{V}} \tag{14-2}$$

Convenient values of capacitance are considerably smaller than a farad, and the most frequently encountered measures of capacitance are the microfarad ($1 \ \mu\text{F} = 10^{-6}$ F) and the picofarad ($1 \text{ pF} = 10^{-12}$ F).

Capacitance of Symmetrical Capacitors

The most common capacitor is the parallel-plate capacitor, illustrated in Figure 14.2. In this arrangement, the separation d between the parallel conducting plates is usually small compared to the linear dimensions of the plates. Then the electric field is constant and is perpendicular to each plate, except near the edges of the plates where the field "fringes" slightly. In practical capacitors (and in our considerations) the fringing effects may be disregarded. We may assume, therefore, that the net charge on each plate is uniformly distributed over its interior surface.

Useful relationships among the variables associated with parallel-plate capacitors may be obtained by again considering Figure 14.2. Suppose that the electric field between the plates is in the x direction so that $\mathbf{E} = E_x \hat{\mathbf{i}}$ and suppose that ΔV is the potential difference between the plates, which are at $x = 0$ and $x = d$. Then we have

$$\Delta V = \int \mathbf{E} \cdot \mathbf{ds} = \int_0^d E_x \, dx = E_x d \tag{14-3}$$

which relates the constant electric field, difference in potential, and separation between the plates of a parallel-plate capacitor. Further, the electric field immediately outside a conductor, which has a surface charge density σ, is equal to σ/ϵ_0. And because $\sigma = Q/A$, where Q is the charge on the positive plate and A is the area of the face of each plate, we may express the magnitude of the electric field as

$$E_x = \frac{\sigma}{\epsilon_0} = \frac{Q}{\epsilon_0 A} \tag{14-4}$$

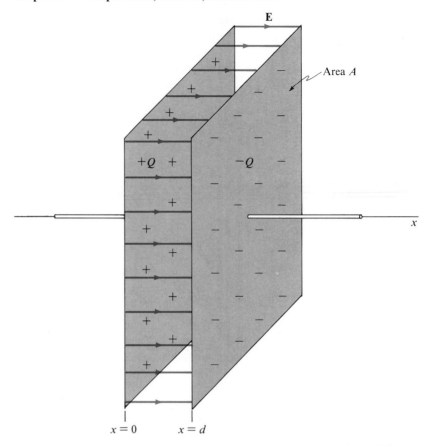

Figure 14.2
A parallel-plate capacitor.

$x = 0$ $x = d$

Then the capacitance C of the parallel plate capacitor may be written, using Equations (14-1), (14-3), and (14-4), as

$$C = \frac{Q}{\Delta V} = \frac{Q}{E_x d} = \frac{Q}{\dfrac{Q}{\epsilon_o A} d}$$

$$C = \frac{\epsilon_o A}{d} \tag{14-5}$$

Equation (14-5) confirms the anticipated fact that the capacitance of a fixed pair of conductors is dependent only on the geometry of the conductors. This equation further indicates that the capacitance of a parallel-plate capacitor may be increased by either enlarging the area A of the plates or decreasing the separation d between them.

E 14.1   A parallel-plate capacitor consists of two square plates, 2.0 cm along a side, separated by a distance of 1.0 mm. Determine the capacitance of this device.   Answer: 3.5 pF

E 14.2   Is it feasible to construct a parallel-plate capacitor that has its two plates separated by 0.10 mm and has a capacitance of 1.0 F?
                Answer: No, each plate would have to have an area of 1.1×10^7 m^2!

Some capacitors have spherical or cylindrical symmetry. The following problems illustrate the process by which their capacitances can be determined.

Express the capacitance of two spherically concentric conductors in terms of their geometry. Use the result to explain what is meant by "the capacitance of an isolated conducting sphere."

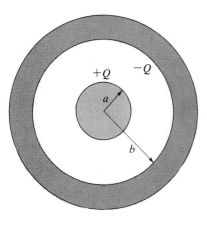

Figure 14.3
A cross section of two spherically symmetric conductors. A positive charge Q has been transferred from the outer conductor to the inner conductor.

The two concentric conductors are shown in Figure 14.3. The radius of the inner sphere is a, and the inner radius of the outer conductor is b. Recalling that capacitance C is defined by Equation (14-1), $C = Q/\Delta V$, we may imagine that a positive charge of magnitude Q has been transferred from the outer conductor to the inner one. Then a difference in potential, $\Delta V = V_a - V_b$, exists between the two conductors. To determine ΔV in terms of the geometry of this system, we first recognize that the electric field outside the inner sphere with net charge Q is the same as that of a point charge Q located at the center of that sphere, namely, a radially outward electric field with magnitude

$$E_r = \frac{kQ}{r^2}; \quad a < r < b \tag{14-6}$$

We might have obtained Equation (14-6) by constructing a spherical gaussian surface concentric with and enclosing the inner sphere and applying Gauss's law. In either case, we can determine ΔV using Equation (14-6) in the relationship between $\mathbf{E}$ and ΔV:

$$\Delta V = V_a - V_b = \int_a^b E_r \, dr \tag{14-7}$$

Substituting Equation (14-6) into Equation (14-7) gives

$$\Delta V = kQ \int_a^b \frac{dr}{r^2} = kQ \left(\frac{1}{a} - \frac{1}{b} \right)$$

$$\Delta V = \frac{kQ(b - a)}{ab} \tag{14-8}$$

Then the capacitance of the two-sphere capacitor is

$$C = \frac{Q}{\Delta V} = \frac{ab}{k(b - a)} = \frac{4\pi\epsilon_0 ab}{(b - a)} \tag{14-9}$$

If we suppose that the radius b of the outer shell becomes arbitrarily large, that is, if we suppose that $b \to \infty$, then $b - a \to b$ and $C \to 4\pi\epsilon_0 a$, which is the capacitance of an isolated sphere of radius a.

E 14.3   Show that the capacitance of a spherical capacitor may be written

$$C = \frac{\epsilon_0 A}{d} \left(1 + \frac{d}{a} \right)$$

where A is the area of the inner sphere (of radius a) and d is the distance between the inner and outer conductors, or $d = b - a$.

E 14.4   Show that if the two conductors of a spherical capacitor are sufficiently close together, the capacitance of the device is given by $C = \epsilon_0 A/d$, (where A is the area of either plate and d is the distance separating the plates), an expression identical to that for the capacitance of a parallel-plate capacitor.

Example 14.1

PROBLEM   A long cylindrical capacitor consists of an inner conductor that is a cylinder of radius a and a coaxial outer conductor with an inner radius b. If the length of either plate is L ($\gg b$), find an expression for the capacitance of the device.

SOLUTION   Figure 14.4(a) shows the capacitor in perspective. If the capacitor is assumed to have a charge Q on the inner cylinder, then the capacitance of this device may be obtained from $C = Q/\Delta V$, where ΔV is the potential difference between the plates. This potential difference may be calculated using

$$\Delta V = \int_a^b E(r)\, dr$$

where $E(r)$ is the magnitude of the (radial) electric field at a distance r from the symmetry axis of the cylinder. The quantity $E(r)$ may be determined by applying Gauss's law to a coaxial surface of radius r, as shown in cross section in Figure 14.4(b). If the fringing effects near the ends are disregarded, the electric field exhibits cylindrical symmetry, and Gauss's law gives

$$E(2\pi rL) = \frac{Q}{\epsilon_o}$$

where $2\pi rL$ is equal to the area of the curved part of the gaussian surface. Solving for $E(r)$ gives

$$E(r) = \frac{Q}{2\pi\epsilon_o Lr}$$

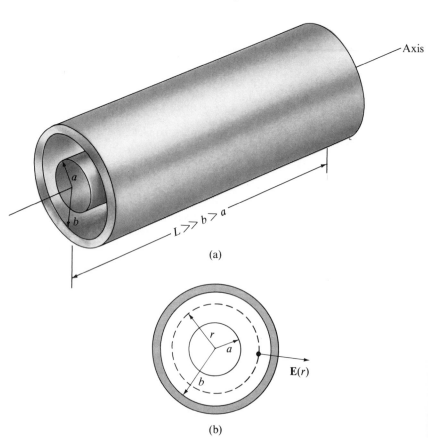

(a)

(b)

Figure 14.4   Example 14.1

and the potential difference between the plates is given by

$$\Delta V = \int_a^b \frac{Q}{2\pi\epsilon_o L r}\, dr = \frac{Q}{2\pi\epsilon_o L}\ln(b/a)$$

Thus, the capacitance of the device is

$$C = \frac{Q}{\Delta V} = \frac{Q}{\dfrac{Q}{2\pi\epsilon_o L}\ln(b/a)}$$

or

$$C = \frac{2\pi\epsilon_o L}{\ln(b/a)}$$

As a check for this result, we note that as the distance between the plates approaches zero, that is, as $b \to a$, or $(b/a) \to 1$, or $\ln(b/a) \to 0$, then $C \to \infty$, just as for the parallel-plate capacitor. ■

Capacitors in Series and in Parallel

Because capacitors are common circuit components, they are often connected by conducting wires to other capacitors and to other circuit components. Figure 14.5(a) illustrates a parallel-plate capacitor connected to wires; Figure 14.5(b) shows the conventional symbol used to represent a capacitor in a circuit diagram. The horizontal straight lines in Figure 14.5(b) indicate connecting wires.

Figure 14.6(a) symbolically shows a combination of two capacitors that are connected *in parallel*. Both of the plates that are positively charged are at the same potential V_A, because they and the wire between them form an equipotential surface. Similarly, both of the negatively charged plates are at the same potential V_B. The two capacitors, labeled C_1 and C_2, have the same potential difference, $\Delta V = V_A - V_B$. If the charge on C_1 is Q_1 and the charge on C_2 is Q_2, we may apply Equation (14-1) to each capacitor:

$$Q_1 = \Delta V\, C_1 \qquad \text{and} \qquad Q_2 = \Delta V\, C_2 \tag{14-10}$$

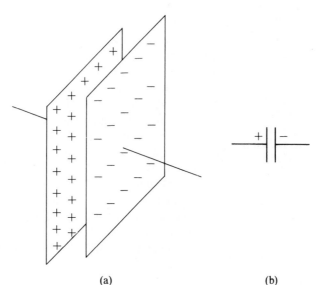

Figure 14.5
(a) A parallel-plate capacitor connected to conducting wires.
(b) The symbol used to represent a capacitor in a circuit diagram.

(a)   (b)

Figure 14.6
(a) A diagram showing two capacitors connected in parallel.
(b) Two capacitors connected in series.

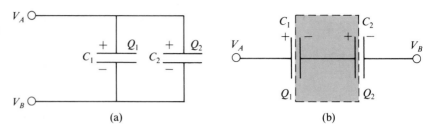

(a) (b)

The total charge Q stored on both capacitors is then

$$Q = Q_1 + Q_2 = \Delta V\, C_1 + \Delta V\, C_2 = (C_1 + C_2)\Delta V \qquad (14\text{-}11)$$

The equivalent capacitance C_{eq} of the two capacitors connected in parallel is defined to be the total charge Q stored on both capacitors divided by their common difference in potential ΔV, or

$$C_{eq} = \frac{Q}{\Delta V} = \frac{(C_1 + C_2)\Delta V}{\Delta V} = C_1 + C_2 \qquad (14\text{-}12)$$

We call C_{eq} the equivalent capacitance because a single capacitor with capacitance C_{eq} has the same, or equivalent, electrical effect as the parallel combination of the two capacitors with capacitances C_1 and C_2.

E 14.5 Two capacitors, $C_1 = 40$ pF and $C_2 = 120$ pF, are arranged in parallel, as shown in the accompanying figure. Calculate the equivalent capacitance C_{eq} of this combination.
Answer: 160 pF

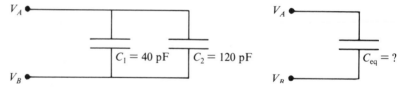

Exercise 14.5

The foregoing analysis may be extended to any number n of capacitors connected in parallel:

$$C_{eq} = C_1 + C_2 + \ldots + C_n \qquad \text{(Capacitors in parallel)} \qquad (14\text{-}13)$$

The essential physical effect of connecting one capacitor to another in parallel is to increase the original plate area, thereby increasing the original capacitance. Thus the connection of additional capacitors in parallel permits the combination to store a greater charge for a given potential difference across the combination.

The two capacitors, C_1 and C_2, shown in Figure 14.6(b), are connected *in series*. The potential difference, $\Delta V = V_A - V_B$, across this combination is the difference in potential between the left plate of C_1 and the right plate of C_2. Placing a charge $+Q$ on the left plate of C_1 induces a charge $-Q$ on the right plate of C_2. We can see that this must happen from Figure 14.6(b): The net charge on the inner plates of the combination and the wire connecting them (the shaded region of the figure) cannot change, because no net charge can enter or leave that region. According to Equation (14-1), the potential difference ΔV across C_1 is equal to Q/C_1, and ΔV_2 across C_2 is equal to Q/C_2. Then the total difference in potential ΔV across the series combination is

$$\Delta V = \Delta V_1 + \Delta V_2 = \frac{Q}{C_1} + \frac{Q}{C_2} = Q\left(\frac{1}{C_1} + \frac{1}{C_2}\right) \qquad (14\text{-}14)$$

The equivalent capacitance C_{eq} of the two capacitors connected in series is equal to $Q/\Delta V$. Using Equation (14-14), we obtain

$$\frac{1}{C_{eq}} = \frac{\Delta V}{Q} = \frac{Q\left(\frac{1}{C_1} + \frac{1}{C_2}\right)}{Q}$$

$$\frac{1}{C_{eq}} = \frac{1}{C_1} + \frac{1}{C_2} \tag{14-15}$$

For two capacitors in series, their equivalent capacitance C_{eq} may be expressed by rearranging Equation (14-15):

$$C_{eq} = \frac{C_1 C_2}{C_1 + C_2} \tag{14-16}$$

For n capacitors connected in series, the analysis above may be extended to give

$$\frac{1}{C_{eq}} = \frac{1}{C_1} + \frac{1}{C_2} + \ldots + \frac{1}{C_n} \qquad \text{(Capacitors in series)} \tag{14-17}$$

E 14.6   Two capacitors, $C_1 = 40$ pF and $C_2 = 120$ pF, are arranged in series, as shown in the accompanying figure. Calculate the equivalent capacitance C_{eq} of this combination.   Answer: 30 pF

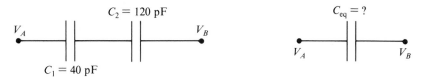

Exercise 14.6

In solving problems that involve networks of capacitors, three basic facts are generally helpful in sorting out details:

1.   Capacitors in parallel have equal potential differences across them.

2.   Capacitors in series have equal charges on each of them.

3.   Capacitors in series or in parallel may be replaced, for purposes of analysis, by a single equivalent capacitor.

The following problem illustrates how these facts may be used to analyze networks of capacitors.

A potential difference of 50 V is maintained between points A and B in the network of capacitors shown in Figure 14.7.
(a) What is the equivalent capacitance of the entire network? (b) What is the charge on the 60-μF capacitor? (c) What is the potential difference across the 20-μF capacitor?

Figure 14.8 indicates how the given network of capacitors may be reduced to simpler systems for purposes of analysis. In the figure, the series combination of C_1 and C_2 is replaced by C_{12}, the equivalent capacitance of the combination; C_{12} is evaluated using Equation (14-16):

$$C_{12} = \frac{C_1 C_2}{C_1 + C_2} = \frac{(20 \times 10^{-6})(30 \times 10^{-6})}{(20 \times 10^{-6}) + (30 \times 10^{-6})} = 12 \times 10^{-6} = 12 \ \mu\text{F} \tag{14-18}$$

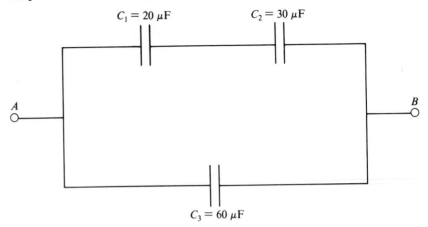

Figure 14.7
A network of capacitors.

In the second stage of Figure 14.7, C_{12} and C_3 are connected in parallel, and that combination is replaced by a single equivalent capacitance C_{eq}, found from Equation (14-12):

$$C_{eq} = C_{12} + C_3 = 12 \ \mu F + 60 \ \mu F = 72 \ \mu F \qquad (14\text{-}19)$$

This value is, therefore, the equivalent capacitance of the entire network.

Because the total difference in potential (50 V) across the given network appears across the 60-μF capacitor, the charge Q_3 on C_3 may be obtained by applying the definition of capacitance to the 60-μF capacitor:

$$Q_3 = C_3 \Delta V_3 = (60 \times 10^{-6})(50) = 3.0 \times 10^{-3} \ C \qquad (14\text{-}20)$$

Finally, in order to determine the potential difference ΔV_1 across C_1, the 20-μF capacitor, using $C_1 = Q_1/\Delta V_1$, we need to know its charge Q_1.

First we recognize that Q_1 is equal to Q_2, the charge on C_2, because C_1 and C_2 are in series. These charges are each equal to Q_{12}, the charge that would be stored on C_{12} (a replacement capacitance for C_1 and C_2). Because the potential difference ΔV_{12} across C_{12} has the value 50 V, the charge Q_{12} is given by

$$Q_{12} = C_{12} \Delta V_{12} = (12 \times 10^{-6})(50) = 6.0 \times 10^{-4} \ C \qquad (14\text{-}21)$$

Thus, we may write

$$Q_1 = Q_2 = Q_{12} = 6.0 \times 10^{-4} \ C \qquad (14\text{-}22)$$

and

$$\Delta V_1 = \frac{Q_1}{C_1} = \frac{6.0 \times 10^{-4}}{20 \times 10^{-6}} = 30 \ V \qquad (14\text{-}23)$$

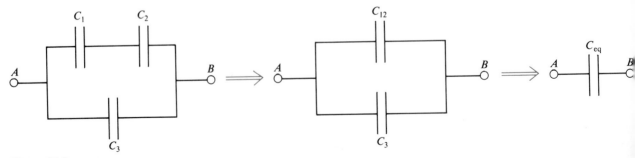

Figure 14.8
Simplifying a network of capacitors.

TABLE 14.1

Typical Dielectric Constants of Common Materials at Room Temperature

Substance	κ
Air	1.0006
Beeswax	2.9
Glass (pyrex)	4.0
Mica	5.0–7.5
Paraffin	2.3
Porcelain	~7
Rubber	~3
Water (distilled)	80

We may check this solution by first using Equation (14-22) to find the potential difference ΔV_2 across C_2, the 30-μF capacitor:

$$\Delta V_2 = \frac{Q_2}{C_2} = \frac{6.0 \times 10^{-4}}{30 \times 10^{-6}} = 20 \text{ V} \qquad (14\text{-}24)$$

This value, when added to the 30-V potential difference across C_1, must sum to 50 V, the potential difference across the series combination of C_1 and C_2.

Effects of Dielectric Materials

A nonconducting material is called a *dielectric material*, or simply a *dielectric*. When a dielectric material completely fills the space between the conductors of a capacitor, the capacitance of the capacitor is increased by a factor κ, called the *dielectric constant* of the nonconducting material. Suppose that the capacitance of a capacitor with a vacuum between its conductors is C_o. When the space between the capacitors is filled with a material of dielectric constant κ, the capacitance C of the dielectric-filled capacitor is

$$C = \kappa C_o \qquad (14\text{-}25)$$

The dielectric constant κ, which is a unitless number, is characteristic of a given material. We will assume that dielectric materials are isotropic and homogeneous. (Some dielectrics are anisotropic; their dielectric properties depend on the orientation of crystalline axes. Inhomogeneous materials may cause further complications. We will consider only simple dielectric materials.) Equation (14-25) indicates that the dielectric constant of free space (a vacuum) is equal to 1. Air ($\kappa = 1.0006$) is usually taken to have $\kappa = 1$. Typical values of κ are listed in Table 14.1 for some common materials used in capacitors.

The following experimental situations illustrate some of the important physical effects of filling the space between the plates of a capacitor with dielectric material.

Suppose the isolated capacitor of Figure 14.9(a) has a vacuum between its plates. Its capacitance is C_o, and it has a fixed charge Q. The potential difference ΔV_o across its plates is given by $\Delta V_o = Q/C_o$. Now suppose a material with dielectric constant κ is inserted between the plates, as shown in Figure 14.9(b). According to Equation (14-25), the capacitance increases to a value $C = \kappa C_o$. Because the capacitor is isolated, the charge Q is not changed by the presence of the dielectric. The potential difference, however, decreases to the new value $\Delta V = Q/(\kappa C_o) = \Delta V_o/\kappa$. Thus, the potential difference across a capacitor with a fixed charge is decreased by a factor κ because of the presence of a material with

Figure 14.9
(a) A charged capacitor C_o with a vacuum between its plates.
(b) The same capacitor with a dielectric material (dielectric constant $= \kappa$) filling the space between its plates.

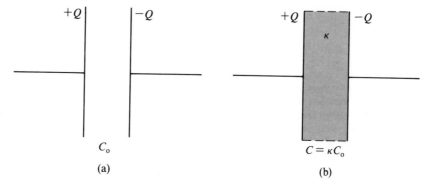

C_o

(a)

$C = \kappa C_o$

(b)

Figure 14.10
The polarization electric field $\mathbf{E}_p$ caused by the application of an electric field $\mathbf{E}_o$ to the molecules of a dielectric material. The resultant electric field $\mathbf{E}$ inside the dielectric material is the sum of $\mathbf{E}_o$ and $\mathbf{E}_p$.

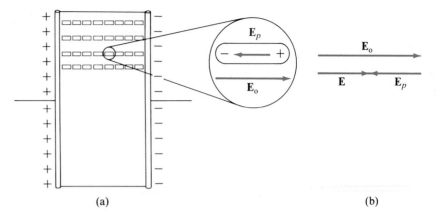

(a)   (b)

dielectric constant κ filling the space between the plates. Further, because the magnitude of the electric field $\mathbf{E}_o$ within the space-filled (evacuated) capacitor is equal to V_o/d, the presence of a dielectric in a capacitor having a fixed charge reduces the magnitude of the electric field to a value E given by

$$E = \frac{E_o}{\kappa} \qquad (14\text{-}26)$$

The reduction of the magnitude of the electric field between the plates of a space-filled capacitor because of the presence of a dielectric material may be further understood on the microscopic level. Each molecule within the dielectric becomes *polarized* by the electric field $\mathbf{E}_o$, which results from the charges on the capacitor plates. A molecule is polarized when its positively charged part is pushed in the direction of $\mathbf{E}_o$ and the negatively charged part is pushed in the opposite direction. This situation is suggested by Figure 14.10(a), in which the charge separation of each molecule of the dielectric material is in a direction parallel to $\mathbf{E}_o$. But each polarized molecule produces a polarization electric field $\mathbf{E}_p$ that is in a direction opposite to the direction of $\mathbf{E}_o$. The average value $\overline{\mathbf{E}}_p$ of the polarization field in the dielectric adds to the electric field $\mathbf{E}_o$, as shown in Figure 14.10(b), to give a resultant electric field $\mathbf{E} = \mathbf{E}_o + \overline{\mathbf{E}}_p$, which has a magnitude equal to E_o/κ.

Now suppose the space-filled capacitor C_o in Figure 14.11(a) has a fixed potential difference ΔV across its plates. The charge Q_o on the plates is given by $Q_o = C_o \Delta V$. If a material with dielectric constant κ is inserted between the plates of the capacitor, as shown in Figure 14.11(b), the capacitance is $C = \kappa C_o$.

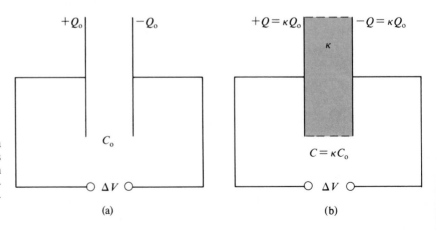

Figure 14.11
(a) A space-filled capacitor C_o with a fixed difference in potential ΔV across its plates. (b) The same capacitor when filled with a material of dielectric constant κ. The capacitance of the capacitor has increased from C_o to $C = \kappa C_o$.

(a)   (b)

Then the charge Q on the plates increases to $Q = C\Delta V = \kappa C_o \Delta V = \kappa Q_o$. Thus the charge on a capacitor with a fixed potential difference between its plates is increased by a factor κ because of the presence of a material with dielectric constant κ.

E 14.7   An isolated, space-filled capacitor has a capacitance of 5.0 μF and carries a charge of 40 μC. If a dielectric material ($\kappa = 1.6$) is inserted so as to fill the region between the plates, calculate the resulting capacitance and potential difference between the plates.   Answer: 8.0 μF, 5.0 V

E 14.8   The space-filled capacitor ($C_o = 5.0$ μF) of Exercise 14.7 is arranged so that the potential difference of 8.0 V across the capacitor remains constant while a dielectric ($\kappa = 1.6$) is inserted so as to fill the region between the plates. Calculate the resulting capacitance and charge on the capacitor.
Answer: 8.0 μF, 6.4 $\times$ 10^{-5} C

14.2   Current and Resistance

When a net charge moves from one place to another, the flow of charge is called an *electric current*, or simply *current*. The flow of charge is most often encountered through a wire or some other fixed conductor; but a beam of charged particles, like a stream of electrons moving through the evacuated space within a television tube, for example, also constitutes a current. For now, let us concentrate on the flow of charge within a fixed conducting wire.

Microscopically, a conducting wire is composed of a latticework of atoms. Charged particles (usually electrons) moving through the conductor are impeded in their progress by continual collisions with the atoms of the lattice. The effect of the collisions on the charged particles is much like a frictional effect: A force must be exerted on the charged particles to sustain their motion. The electrical force that impels the charged particles through the conductor results from an electric field inside the conductor. Thus, contrary to an electrostatic situation, an electric field is present inside a conductor in which charges are flowing.

In order to see the relationship between current and electric field in a conductor, let us consider the segment of conducting wire with cross-sectional area A shown in Figure 14.12. Suppose an electric field $\mathbf{E}$ within the wire is directed toward the right. Then the unbound electrons within the conductor each experience a force $\mathbf{F} = -e\mathbf{E}$ toward the left. The electrons accelerate under the

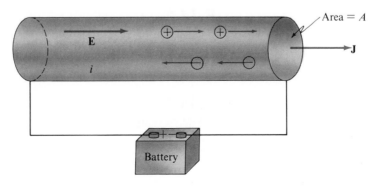

Figure 14.12
A segment of conducting wire with cross-sectional area A. When an electric field $\mathbf{E}$ exists within the wire, the wire carries a current $\mathbf{i}$. The current density $\mathbf{J}$ in the wire is in the direction of $\mathbf{E}$.

influence of the electric field until the resistive forces of collisions with atoms of the conductor are equal in magnitude to the impelling electrical forces. Then the electrons move with a constant average speed, called the *drift speed*, through the wire. It can be shown that the magnitude of the drift speed of electrons in a typical conductor is about 1 mm/s. This is a very slow speed compared to the speed at which a pulse of current propagates along a wire, about 3×10^8 m/s.

Let us now define the *average current $\bar{i}$* in the wire of Figure 14.12, in terms of the net charge Δq that passes through the cross-sectional area A of the wire during a time interval Δt, by

$$\bar{i} = \frac{\Delta q}{\Delta t} \tag{14-27}$$

Then the *instantaneous current i* in the wire is

$$i = \lim_{\Delta t \to 0} \frac{\Delta q}{\Delta t} = \frac{dq}{dt} \tag{14-28}$$

From Equation (14-28) we may conclude that the unit of current must be equal to the unit of charge (C) per unit of time (s). The SI unit of current is the ampere (A), which will be operationally defined in a later chapter. For the present, we will use

$$1 \text{ A} \equiv 1 \, \frac{C}{s} \tag{14-29}$$

Although the ampere is not an unusually large quantity of current, we often encounter smaller quantities, like the milliampere (1 mA = 10^{-3} A) and the microampere (1 μA = 10^{-6} A). An instrument used to measure current is called an ammeter.

Current is, by definition, a scalar quantity. Although current has no direction, we often hear (from an urge toward brevity rather than toward imprecision, it is hoped) of the "direction of a current" in a wire or in a circuit. What is intended by this loose phrasing is the direction of the *current density $\mathbf{J}$*, a vector quantity with a direction that is usually in the direction of the electric field within a conductor. The magnitude of $\mathbf{J}$ is equal to the current per unit cross-sectional area, and the unit of current density is A/m^2. Thus, if the wire of Figure 14.12 has a current i, $\mathbf{J}$ has a magnitude given by

$$J = \frac{i}{A} \tag{14-30}$$

Because the direction of $\mathbf{J}$ is in the direction of $\mathbf{E}$, the current density is in a direction *opposite* to the flow of electrons.

E 14.9   A wire (diameter = 1.0 mm) carries a current of 2.5 A. Calculate the magnitude of the current density in the wire, assuming that the current is uniformly distributed over a cross section of the wire.

Answer: 3.2×10^6 A/m^2

E 14.10   The magnitude of the current density is 2.8×10^4 A/m^2 in a wire carrying a current of 3.0 mA. Calculate the cross-sectional area of the wire.

Answer: 1.1×10^{-7} m^2

TABLE 14.2

Typical Resistivities of Common Metals at Room Temperature

Substance	$\rho(\Omega \cdot m)$
Aluminum	2.8×10^{-8}
Copper	1.7×10^{-8}
Iron	1.0×10^{-7}
Nickel	7.1×10^{-8}
Platinum	1.1×10^{-7}
Silver	1.6×10^{-8}
Tungsten	5.6×10^{-8}

Resistivity and Ohm's Law

If certain materials are maintained at a constant temperature, it is found experimentally that the electric field **E** within those materials is directly proportional to the current density **J** in the material, or

$$\mathbf{E} = \rho\mathbf{J} \qquad (14\text{-}31)$$

where ρ is a constant called the *resistivity* of the material. Equation (14-31) is called *Ohm's law for a substance*, and materials to which it applies are called ohmic substances. Materials in which **E** is not proportional to **J** are known as nonlinear substances. Equation (14-31) reflects the fact that **E** and **J** are in the same direction and indicates that the resistivity ρ has the value

$$\rho = \frac{E}{J} \qquad (14\text{-}32)$$

Because the unit of **E** may be expressed as V/m and **J** has the unit A/m², the unit of resistivity ρ is the V $\cdot$ m/A. Table 14.2 lists the resistivities of some common metals at room temperature. It is convenient to define the unit *ohm* (Ω) to be one volt per ampere, or

$$1 \; \Omega \equiv 1 \, \frac{V}{A} \qquad (14\text{-}33)$$

Then the unit of resistivity ρ becomes the ohm-meter ($\Omega \cdot$ m).

Suppose, instead of considering the resistive character of a substance, we focus on a specific object. Let us consider the length L of wire with cross-sectional area A shown in Figure 14.13. If we apply a difference in potential ΔV across the length and observe that the wire carries a current i, we define the resistance R of this particular length of wire to be

$$R = \frac{\Delta V}{i} \qquad (14\text{-}34)$$

Figure 14.13
A length L of wire with cross-sectional area A. The resistivity of the material of the wire is ρ. When a difference in potential ΔV is across the length of wire, it carries a current i, and the current density in the wire is **J**.

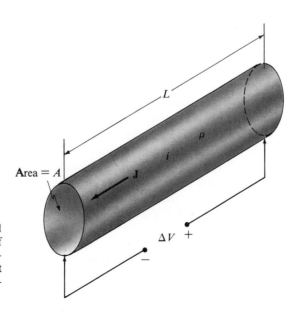

in which ΔV is measured in volts (V), i is measured in amperes (A), and the resistance R is measured in ohms (Ω). Equation (14-34) is called *Ohm's law for a specific object*. Thus, while resistivity ρ is a characteristic of a material, resistance R characterizes a particular specimen. We can relate the resistance of a given specimen to the resistivity of the material that constitutes that specimen. Consider again the wire with uniform cylindrical cross section, shown in Figure 14.13. Because the difference in potential ΔV is across its length L, the material has within it an electric field with magnitude $E = \Delta V/L$; if the wire is carrying a current i, the current density has magnitude $J = i/A$. Putting these values for E and J into Equation (14-32) and substituting R for $\Delta V/i$ from Equation (14-34) gives

$$\rho = \frac{\Delta V/L}{i/A} = R\frac{A}{L}$$

or

$$R = \rho\frac{L}{A} \qquad (14\text{-}35)$$

From Equation (14-35) we can see that the greater the length of a wire of a given material, the greater is its resistance; and the greater the cross-sectional area of the wire, the less is its resistance.

E 14.11   Calculate the resistance of a 15-m length of copper wire that has a diameter of 0.50 mm.   Answer: 1.3 Ω

E 14.12   The potential difference between the ends of a wire of length 2.0 m and of diameter 1.0 mm carrying a current of 100 A is 5.6 V. Calculate (a) the resistance of the wire, and (b) the resistivity of the substance of which the wire is made.   Answers: (a) 0.056 Ω; (b) 2.2 $\times$ 10^{-8} $\Omega \cdot$ m

Resistors and Combinations of Resistors

In electrical circuits we frequently use circuit elements specifically designed to impede the flow of charge. Such elements are called *resistors*. Figure 14.14(a) illustrates a common form of resistor, a cylinder of pressed carbon encased in plastic and attached to connector wires. Figure 14.14(b) shows the symbol used in circuit diagrams to represent a resistor; the wires are represented by straight lines. We usually assume, both in circuit diagrams and in actual circuits, that the wires connecting the components of the circuit have negligible resistance compared to that of the resistors in the circuit.

Figure 14.15 illustrates several conventions used in circuit diagrams. The resistor in the diagram is labeled R, and we speak of "the resistor R," by which we mean "the resistor that has resistance of R." The resistor carries a current i, which is a scalar quantity, and the arrow associated with the current on the

Figure 14.14
(a) A common form of a resistor.
(b) The symbol used in circuit diagrams to represent a resistor.

(a)   (b)

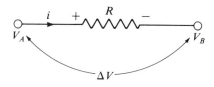

Figure 14.15
Some conventions associated with a resistor R carrying a current i. Here the potential V_A is greater than V_B.

diagram indicates only the sense of the current (toward the right, as opposed to toward the left, in this diagram). The arrow suggests that a free positive charge in the resistor would proceed from left to right, from a higher to a lower potential. Because point A is at a higher potential than point B, the difference in potential, $\Delta V = V_A - V_B$, has a positive value, and the $+$ and $-$ signs on either side of the resistor symbol indicate that the left side of the resistor is at a higher potential than the right side. Sometimes we refer to the potential difference $\Delta V = iR$ across a resistor as a "potential drop" or "drop in potential." These phrases are intended to mean that the electric potential decreases, or drops, in proceeding from the high-potential to the low-potential side of the resistor, that is, the potential decreases in proceeding through the resistor in the sense indicated by the current arrow.

E 14.13   A 12-Ω resistor carries a current of 15 A, as shown in the diagram. Calculate the potential difference, $V_2 - V_1$. Answer: -1.8×10^2 V

Exercise 14.13

$i = 15$ A $R = 12\ \Omega$

1   2

E 14.14   A potential difference of 60 V across a 15-Ω resistor has the polarity (placement of signs) shown in the accompanying figure. Determine the current in the resistor and the sense of this current.

Answer: 4.0 A, right to left

$\Delta V = 60$ V

Exercise 14.14

$R = 15\ \Omega$

Resistors are frequently used in combinations in electrical circuits. When resistors are connected so that the same charge must pass through each of them, they are said to be connected *in series*. Figure 14.16(a) shows two resistors, R_1 and R_2, connected in series. Each of these resistors carries a current i when there is a potential difference, $\Delta V = V_A - V_C$, across the combination. The potential difference ΔV is equal to the sum of the potential differences ΔV_{AB} and ΔV_{BC} across the individual resistors, or

$$\Delta V = \Delta V_{AB} + \Delta V_{BC} = iR_1 + iR_2 = i(R_1 + R_2) \qquad (14\text{-}36)$$

in which we have applied Ohm's law for a resistor, $\Delta V = iR$, to each resistor. Figure 14.16(b) shows a single resistor R_{eq}, which is equivalent to the combination of R_1 and R_2 in that when it carries the same current i, the same difference in

Figure 14.16
Illustrating how (a) two resistors R_1 and R_2, connected in series are equivalent to (b) an equivalent resistor R_{eq}.

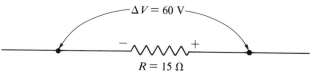

(a)   (b)

potential ΔV exists across R_{eq}. Applying Ohm's law to R_{eq} and using Equation (14-36) gives

$$R_{eq} = \frac{\Delta V}{i} = \frac{i(R_1 + R_2)}{i} = R_1 + R_2 \qquad (14\text{-}37)$$

If the foregoing analysis is extended to include n resistors in series, Equation (14-37) generalizes to

$$R_{eq} = R_1 + R_2 + \ldots + R_n \qquad \text{(Resistors in series)} \qquad (14\text{-}38)$$

Thus the equivalent resistance for resistors in series is the sum of their individual resistances.

E 14.15   Calculate R_{eq} for the series combination shown in the diagram.
Answer: 45 Ω

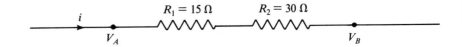

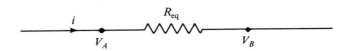

Exercise 14.15

In Figure 14.17(a), the resistors, R_1 and R_2, are connected *in parallel*. Both the resistors have the same potential difference, $\Delta V = V_A - V_B$, across them. Because charge does not accumulate at the point where the wire divides, the current i carried by the combination is equal to the sum of the currents, i_1 and i_2, in R_1 and R_2. Figure 14.17(b) shows a single equivalent resistor R_{eq} that carries the same total current, $i = i_1 + i_2$, as the combination when the potential difference ΔV is across R_{eq}, or

$$R_{eq} = \frac{\Delta V}{i} = \frac{\Delta V}{i_1 + i_2} \qquad (14\text{-}39)$$

Ohm's law applied to each of the resistors, R_1 and R_2, requires that $i_1 = \Delta V / R_1$ and $i_2 = \Delta V / R_2$, so that Equation (14-39) becomes

$$R_{eq} = \frac{\Delta V}{\dfrac{\Delta V}{R_1} + \dfrac{\Delta V}{R_2}} = \frac{1}{\dfrac{1}{R_1} + \dfrac{1}{R_2}}$$

or

$$\frac{1}{R_{eq}} = \frac{1}{R_1} + \frac{1}{R_2} \qquad (14\text{-}40)$$

For two resistors in parallel, the equivalent resistance R_{eq} is, therefore,

$$R_{eq} = \frac{R_1 R_2}{R_1 + R_2} \qquad (14\text{-}41)$$

Notice that the equivalent resistance for two resistors in parallel is always less

Figure 14.17
Illustrating how (a) two resistors R_1 and R_2, connected in parallel, are equivalent to (b) an equivalent resistor R_{eq}.

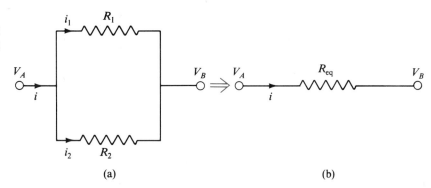

(a)   (b)

than that of either resistor in the parallel combination. For n resistors in parallel, the foregoing analysis may be extended to give

$$\frac{1}{R_{eq}} = \frac{1}{R_1} + \frac{1}{R_2} + \dots + \frac{1}{R_n} \qquad \text{(Resistors in parallel)} \qquad \text{(14-42)}$$

and, again, the equivalent resistance for resistors in parallel is always less than that of any resistor in the parallel combination.

E 14.16   Calculate R_{eq} for the parallel combination of resistors shown in the diagram.   Answer: $10\ \Omega$

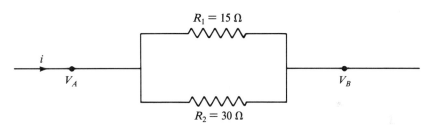

Exercise 14.16

E 14.17   Calculate R_{eq} for the combination of resistors shown in the diagram.   Answer: $20\ \Omega$

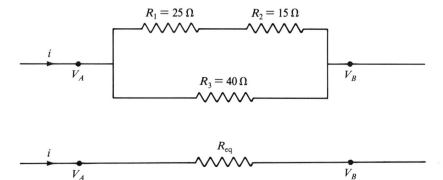

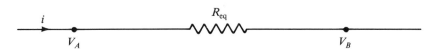

Exercise 14.17

14.3 Energetics of Resistors and Capacitors

The circuit elements introduced in this chapter, capacitors and resistors, function quite differently in a circuit. Capacitors store energy by separating charge; resistors impede the flow of charge. We now consider the energy changes that take place in resistors and capacitors as they function.

Electric Power Loss in Resistors

When there is current in a resistor, the electrons moving through the lattice of the conductor suffer continual collisions with the ions of the lattice. These collisions convert some of the kinetic energy of the electrons into thermal energy, thereby heating the resistor. Let us consider that conversion of energy.

Suppose a positive quantity of charge ΔQ flows from A to B in a time interval Δt in the conducting material of the resistor in Figure 14.18. And suppose the electric potential at A is V_A and at B is V_B.

The charge ΔQ has work done on it by the electrical forces that impel the charge through the resistor. As the charge moves through the resistor from A to B, this work by the impelling electrical forces is given by

$$\Delta W = \Delta Q(V_A - V_B) = \Delta Q \Delta V \qquad (14\text{-}43)$$

where $\Delta V = V_A - V_B$ is a positive quantity (A is at a higher potential than B).

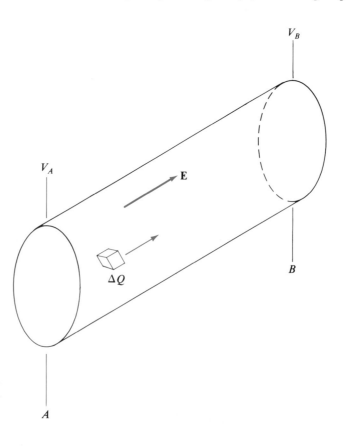

Figure 14.18
A charge ΔQ flowing in the direction of the electric field **E** inside the material of a resistor.

This work ΔW on ΔQ replaces the kinetic energy that the moving charge loses in microscopic collisions with the lattice. Thus the rate at which thermal energy is added to the lattice is found by dividing Equation (14-43) by the time interval Δt, or

$$\frac{\Delta W}{\Delta t} = \frac{\Delta Q}{\Delta t} \Delta V = i \, \Delta V \qquad (14\text{-}44)$$

where $i = \Delta Q / \Delta t$ is the current in the resistor. The rate at which energy is lost by the charges in passing through the resistor is called the *electric power loss P* within the resistor, or

$$P = i \, \Delta V \qquad (14\text{-}45)$$

By using the definition of resistance, $R = \Delta V / i$, we may write Equation (14-45) in terms of the resistance R of a particular resistor:

$$P = i^2 R \qquad (14\text{-}46)$$

or

$$P = \frac{(\Delta V)^2}{R} \qquad (14\text{-}47)$$

The unit of electric power loss P is indicated by Equation (14-45). Because the unit of current is $A = C/s$ and the unit of potential is $V = J/C$, the unit of power loss is $J/s = W$.

E 14.18   A 15-Ω resistor carries a constant current of 2.0 A.
(a) At what rate is electrical energy being converted into thermal energy in the resistor?   Answer: 60 W
(b) How much heat (1 cal = 4.19 J) is generated in the resistor during a 1-minute time interval?   Answer: 8.6×10^2 cal

Energy Stored in Capacitors

Charging a capacitor, transferring charge between the two conductors of the capacitor, requires an expenditure of energy. The energy required to charge a capacitor is stored in the charged capacitor, and that stored energy may be recovered when the capacitor is discharged. Let us calculate the energy stored in a capacitor C when a charge Q is stored on its plates.

Imagine that a capacitor C is being charged, starting from a completely uncharged ($Q = 0$) condition. At some stage of the charging process, when a charge q has been transferred from one conductor to the other, the potential difference ΔV across the capacitor is

$$\Delta V = \frac{q}{C} \qquad (14\text{-}48)$$

The work dW necessary to transfer an additional element of charge dq is

$$dW = \Delta V \, dq = \frac{q}{C} \, dq \qquad (14\text{-}49)$$

The total work that must be done to increase the transferred charge from $q = 0$ to $q = Q$ is equal to the energy U_C stored in the capacitor when it has charge Q:

$$U_C = W = \int dW = \frac{1}{C} \int_0^Q q \, dq$$

$$U_C = \frac{1}{2} \frac{Q^2}{C} \tag{14-50}$$

Using the definition of capacitance, $C = Q/\Delta V$, we may express the energy stored by the capacitor in terms of the potential difference ΔV across its plates:

$$U_C = \frac{1}{2} C(\Delta V)^2 \tag{14-51}$$

$$U_C = \frac{1}{2} Q\Delta V \tag{14-52}$$

It is often convenient to consider that the energy stored in a capacitor resides in the electric field $\mathbf{E}$ between its conductors. As Q or ΔV increases, E increases; as Q or ΔV becomes equal to zero, E becomes equal to zero.

E 14.19   Verify the units in each of Equations (14-50), (14-51), and (14-52).

E 14.20   A 0.50-μF capacitor has a potential difference of 15 V across it. Calculate the charge and energy stored by the capacitor.

Answer: 7.5 μC, 56 μJ

Example 14.2

PROBLEM   Two capacitors, $C_1 = 15\ \mu$F and $C_2 = 30\ \mu$F, connected in series have a potential difference of 120 V across the combination. Calculate the energy stored by the combination and the energy stored in each capacitor.

SOLUTION   Figure 14.19(a) shows the two capacitors, C_1 and C_2, with a potential difference, $\Delta V = 120$ V, across the pair and a charge Q on each capacitor. The equivalent capacitor C_{eq}, shown in Figure 14.19(b), has the same 120-V potential difference across it and stores the same charge Q as each of the series pair. The value of C_{eq} is obtained from

$$C_{eq} = \frac{C_1 C_2}{C_1 + C_2} = \frac{(15\ \mu\text{F})(30\ \mu\text{F})}{45\ \mu\text{F}} = 10\ \mu\text{F}$$

The energy stored in a 10-μF capacitor, charged to a 120-V potential difference, is

$$U_C = \frac{1}{2} C_{eq}(\Delta V)^2 = \frac{1}{2}(10\ \mu\text{F})(120\ \text{V})^2 = 72\ \text{mJ}$$

The charge Q on C_{eq} and on each of the capacitors, C_1 and C_2, is

$$Q = C_{eq}\ \Delta V = (10\ \mu\text{F})(120\ \text{V}) = 1.2 \times 10^{-3}\ \text{C}$$

Consequently, the energy stored in each of the capacitors is

$$U_{C_1} = \frac{1}{2} \frac{Q^2}{C_1} = \frac{1}{2} \frac{(1.2 \times 10^{-3}\ \text{C})^2}{15 \times 10^{-6}\ \text{F}} = 48\ \text{mJ}$$

$$U_{C_2} = \frac{1}{2} \frac{Q^2}{C_2} = \frac{1}{2} \frac{(1.2 \times 10^{-3}\ \text{C})^2}{30 \times 10^{-6}\ \text{F}} = 24\ \text{mJ}$$

A simple check on these results is effected by noting that the two energies, U_{C_1} and U_{C_2}, sum to the total energy in the combination, $U_C = 72$ mJ. ∎

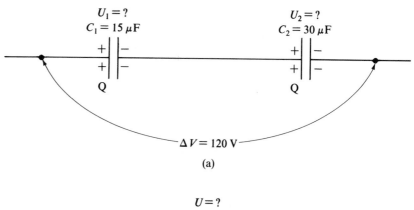

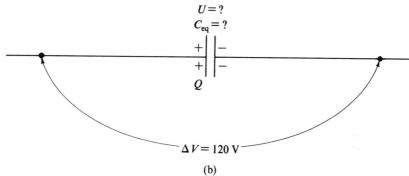

Figure 14.19 Example 14.2

The rate at which electrical energy is stored in a capacitor is the rate at which work is done on the charges being stored on its plates. The rate at which energy is stored on a capacitor may be found by using Equation (14-49) to get

$$\frac{dU_C}{dt} = \frac{dW}{dt} = \Delta V \frac{dq}{dt} = \Delta V\, i \qquad (14\text{-}53)$$

Equation (14-53) is a specific case of the general relationship between the rate P (power) at which energy is being transformed from one form to another in a circuit component that carries a current i and has a potential difference ΔV across it:

$$P = i\,\Delta V \qquad (14\text{-}54)$$

If the component is a resistor, then ΔV is equal to iR, and $P = i^2R$ is the rate at which thermal energy is being generated in the resistor. For a capacitor, ΔV is equal to Q/C, and the rate at which energy is being stored (or released) by the capacitor is given by $P = iQ/C$.

E 14.21 Determine the energy stored in a 6.0-μF capacitor and the rate at which this energy is being stored at an instant when the capacitor has a potential difference of 50 V across it and a current of 0.10 mA charging it, as shown in the figure. Answer: 7.5 mJ, 5.0 mW

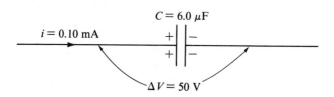

Exercise 14.21

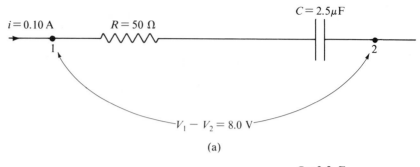

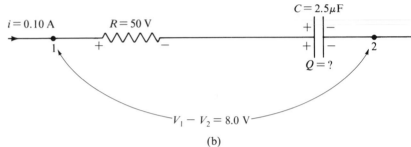

Figure 14.20   Example 14.3

(b)

Example 14.3   **PROBLEM**   Figure 14.20(a) shows a series combination of a 50-Ω resistor and a 2.5-μF capacitor. Determine the charge on the capacitor at an instant when the current i is equal to 0.10 A and the potential difference, $V_1 - V_2$, is equal to 8.0 V.

SOLUTION   Figure 14.20(b) shows the given situation with the appropriate signs for the potential changes across the resistor R and the capacitor C. We see that the potential difference, $V_1 - V_2$, is equal to the sum of the potential changes across the resistor, $V_R = iR$, and across the capacitor, $V_C = Q/C$. Thus we have

$$V_1 - V_2 = iR + Q/C$$

or

$$Q = C(V_1 - V_2 - iR) = (2.5 \ \mu F) \ [(8.0 - 0.10 \times 50)V] = 7.5 \ \mu C$$

A simple check to ensure that no calculational blunders have occurred may be accomplished by calculating V_R and V_C,

$$V_R = iR = (0.10 \ A)(50 \ \Omega) = 5.0 \ V$$

$$V_C = Q/C = (7.5 \ \mu C)/(2.5 \ \mu F) = 3.0 \ V$$

and verifying that the sum of these two potential differences is indeed equal to the potential difference, $V_1 - V_2 = 8.0$ V. ■

E 14.22   Determine the rate at which the energy stored in the capacitor is changing at the instant described in Example 14.3.   Answer: 0.30 W

14.4 Problem-Solving Summary

Problems that involve capacitors, current, and resistors are generally self-identifying. In such problems, the basic tools to be used are the defining relationships for capacitance, current, and resistance. Let us review these relationships separately.

A capacitor is composed of two conductors, and its capacitance C is defined to be $C = Q/\Delta V$, in which ΔV is the (positive) difference in potential between the conductors when a net charge Q has been transferred from one conductor to the other. In most problems a given capacitor has a fixed capacitance that depends on the geometry of the conductors. A parallel-plate capacitor with plates of area A separated by a distance d, for example, has capacitance $C = \epsilon_o A/d$. The capacitance of any symmetrical capacitor (spherical, cylindrical, or planar) can be calculated using the following procedure:

1. Assume that a charge Q has been transferred from one conductor to the other.

2. Use Gauss's law to find an expression for the electric field in the region between the conductors.

3. Use $\Delta V = V_b - V_a = \int_b^a \mathbf{E} \cdot d\mathbf{s}$ to calculate the potential difference across the capacitor.

4. Use the definition of capacitance, $C = Q/\Delta V$, to find the capacitance in terms of the geometry of the conductors.

When capacitors are connected in series or in parallel combinations, two physical characteristics are important in the analysis of networks of capacitors:

1. Each capacitor in a parallel combination of capacitors has the same potential difference ΔV across it.

2. Each capacitor in a series combination of capacitors has the same charge Q stored on it.

It follows that two capacitors, C_1 and C_2, connected in parallel have the same electrical effect as a single capacitor with capacitance, $C_{eq} = C_1 + C_2$. It further follows that two capacitors, C_1 and C_2, connected in series have the effect of a single equivalent capacitor with capacitance, $C_{eq} = C_1 C_2/(C_1 + C_2)$.

If the space between the conductors of a capacitor C_o is filled with a dielectric, a nonconducting material characterized by a unitless number $\kappa > 1$ called the dielectric constant of the material, two principal physical effects result: (a) the capacitance C_o is increased to a value equal to κC_o, and (b) the magnitude of the electric field at any point between the plates of the capacitor is reduced from E to a value equal to E/κ if the charge on the capacitor remains constant.

Electric current i in a conductor (or even in a region of free space) is a scalar quantity defined to be the rate at which charge passes a cross section of the conductor. Current density $\mathbf{J}$ is a vector in the direction of the electric field $\mathbf{E}$ inside a conductor. The magnitude of $\mathbf{J}$ is equal to the current i passing through a cross-sectional area A divided by that area A, or $J = i/A$.

In a given material, current density $\mathbf{J}$ is proportional to the electric field $\mathbf{E}$, according to Ohm's law for a substance, $\mathbf{E} = \rho \mathbf{J}$. The resistivity ρ is a characteristic of a conducting material. A specific conducting object, like a resistor, has a resistance $R = \Delta V/i$, where ΔV is the potential difference across the ends of the object and i is the current carried by the object. If the conducting object has length L and cross-sectional area A, its resistance is related to the resistivity of the conductor by $R = \rho L/A$.

In solving problems that involve resistors carrying current, remembering one important fact can avoid crucial errors: The current density (and, therefore, the current arrow in a circuit diagram) is in the direction from the high-potential side of a resistor to the low-potential side.

Resistors connected in series carry the same current. Two resistors, R_1 and R_2, in series have the same electrical effect as a resistance, $R_{eq} = R_1 + R_2$. Resistors connected in parallel have the same potential difference across them. Two resistors, R_1 and R_2, connected in parallel have the same electrical effect as an equivalent resistance, $R_{eq} = R_1 R_2/(R_1 + R_2)$.

When a potential difference ΔV is across a resistor R carrying current i, the rate at which energy is converted to heat in the resistor is called the electric power loss P. Power loss may be expressed in terms of any two of the variables R, i, and ΔV: $P = i\Delta V = i^2 R = (\Delta V)^2/R$.

When a potential difference ΔV is across a capacitor C with charge Q stored on it, the electrical energy U_C stored in the capacitor is given by $U_C = (1/2)Q^2/C = (1/2)C(\Delta V)^2 = (1/2)Q\Delta V$. The rate at which energy is stored in the capacitor is given by $dU_C/dt = \Delta V\, i$.

Problems

GROUP A

14.1 A current of 1.5 A in a resistor R results in 4.5×10^2 J of electrical energy being converted to heat in R during a 25-s time interval. Calculate R.

14.2 What is the resistance of a section of aluminum ($\rho = 2.8 \times 10^{-8}\ \Omega \cdot m$) wire that has a length of 1.0 km and a diameter of 2.0 mm?

14.3 Determine the resistance R of a resistor that, when combined appropriately with a 20-Ω resistor, will yield an equivalent resistance of

(a) 28 Ω.

(b) 14 Ω.

14.4 Determine the capacitance C of a capacitor that, when combined appropriately with a 15-μF capacitor, will yield an equivalent capacitance of

(a) 25 μF.

(b) 5.0 μF.

14.5 An energy equal to 2.7×10^{-3} J is stored by a capacitor that has a potential difference of 85 V across it. Determine

(a) the capacitance.

(b) the charge on the capacitor.

14.6 When a potential difference of 2.5 mV is applied across the ends of a rod (length = 0.80 m, diameter = 0.75 mm) made of an unknown material, a current of 5.8 mA results. Calculate the resistivity of the unknown material.

14.7 Wire A is twice as long as wire B, and the two have the same volume. If both wires are made of the same substance, calculate the ratio of the resistance of A to that of B.

14.8 A cylindrical conductor of length L and diameter d is made of a substance having a resistivity ρ. If a potential differ-

ence V_o is maintained across the ends of the conductor, determine the magnitude of the current density within the conductor.

14.9 At an instant when $i = 0.60$ A and $Q = 15\ \mu$C for the RC combination shown in Figure 14.21, calculate the potential difference, $V_A - V_B$.

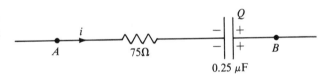

Figure 14.21   Problem 14.9

14.10 Determine the equivalent resistance of all different combinations using three 10-Ω resistors. Include only those combinations that utilize all three resistors.

14.11 Determine the equivalent capacitance of all different combinations using three 15-μF capacitors. Include only those combinations that utilize all three capacitors.

14.12 Determine the equivalent capacitance of all different combinations using two 10-μF capacitors and one 15-μF capacitor. Include only those combinations utilizing all three capacitors.

14.13 Determine the equivalent resistance of all different combinations using two 15-Ω resistors and one 30-Ω resistor. Include only those combinations that utilize all three resistors.

14.14 Determine the equivalent resistance of different combinations utilizing all three resistors having the values $R_1 = 10\ \Omega$, $R_2 = 15\ \Omega$, and $R_3 = 30\ \Omega$.

14.15 Two capacitors, $C_1 = 15$ μF and $C_2 = 30$ μF, are connected in series; and this combination has a potential difference of 5.0 V across it. Calculate

(a) the charge on the 15-μF capacitor.

(b) the energy stored in the 30-μF capacitor.

14.16 An isolated, space-filled, parallel-plate capacitor has a charge $Q = 4.5$ μC when a potential difference of 5.0×10^4 V exists across its plates (separation distance = 0.50 mm).

(a) Calculate the capacitance, the area of the charge-carrying face of either plate, the charge density on the positive plate, the energy stored, and the magnitude of the electric field between the plates.

(b) Assume the capacitor remains isolated and the space between the plates is filled with a dielectric ($\kappa = 1.4$) material, and repeat part (a).

14.17 If the region between the conductors of a spherical capacitor ($a = 0.80$ cm, $b = 0.90$ cm) is filled with a dielectric material ($\kappa = 1.4$) and the charge on the capacitor is 0.65 μC, calculate the energy stored by this device.

14.18 If $V_A - V_B = 64$ V for the combination of resistors shown in Figure 14.22, determine

(a) the equivalent resistance for the combination.

(b) the current in the 40-Ω resistor.

(c) the rate at which heat is generated in the 20-Ω resistor.

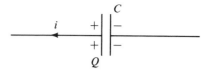

Figure 14.23   Problem 14.20

14.21 Capacitors $C_1 = 2.0$ μF, $C_2 = 3.0$ μF, and $C_3 = 6.0$ μF, are connected in series; and a potential difference of 400 V is applied across this combination. Determine the charge on each capacitor and the energy stored in the combination.

14.22 Capacitors C_1 (20 μF) and C_2 (40 μF) are connected in parallel. The potential difference across the combination is held constant at 80 V while the region between the plates of C_1 is filled with a dielectric material ($\kappa = 1.5$). Calculate the final charge on each capacitor and the energy stored by the combination.

14.23 Repeat Problem 14.22 with C_1 and C_2 connected in series.

14.24 If $V_A - V_B = 60$ V for a combination of capacitors (see Fig. 14.24), find

(a) the equivalent capacitance of the combination.

(b) the energy stored by the combination.

(c) the charge on the 60-μF capacitor.

(d) the energy stored on the 10-μF capacitor.

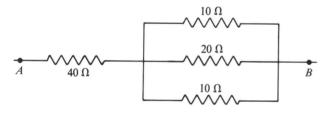

Figure 14.22   Problem 14.18

14.19 A current-carrying, copper ($\rho = 1.7 \times 10^{-8}$ $\Omega \cdot$m) cylindrical conductor (length = 2.0 m, diameter = 0.10 mm) has within it a uniform electric field parallel to the axis of the conductor. If $E = 0.54$ V/m, what is the

(a) magnitude of the current density in the conductor?

(b) magnitude of the potential change across the conductor?

(c) current in the conductor?

14.20 The charge Q and current i for a 750-μF capacitor are shown in Figure 14.23. At an instant when $i = 0.40$ mA and $Q = 5.5 \times 10^{-4}$ C, calculate the

(a) rate at which Q is changing.

(b) energy U_C stored by the capacitor.

(c) rate at which U_C is changing.

(d) potential difference ΔV across the capacitor.

(e) rate at which ΔV is changing.

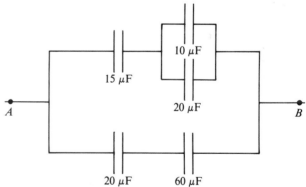

Figure 14.24   Problem 14.24

14.25 For the combination of resistors shown in Figure 14.25, the potential difference, $V_B - V_C$, is equal to 12 V. Find

(a) the current i shown.

(b) the potential difference $V_A - V_C$.

(c) the rate at which heat is generated in the 20-Ω resistor.

14.26 A resistor network (*see* Fig. 14.26) has a current of 30 mA. Determine

(a) the equivalent resistance of the network.

(b) the current in the 40-Ω resistor.

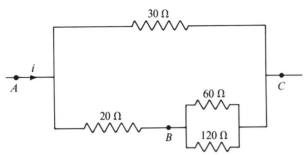

Figure 14.25   Problem 14.25

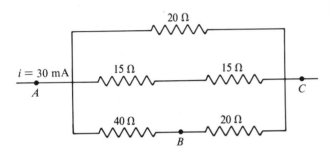

Figure 14.26   Problem 14.26

(c)   the potential difference, $V_A - V_C$.

(d)   the potential difference, $V_C - V_B$.

(e)   the rate at which heat is being generated in either of the 15-Ω resistors.

14.27   At an instant when $i = 25$ mA and $Q = 0.12$ μC for the resistor-capacitor combination shown in Figure 14.27, calculate

(a)   the potential difference, $V_A - V_B$.

(b)   the energy U_C stored by the capacitor.

(c)   the rate at which U_C is changing.

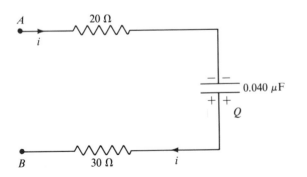

Figure 14.27   Problem 14.27

14.28   At an instant when $i = 2.0$ A and $V_1 - V_2 = 100$ V for the resistor-capacitor combination shown in Figure 14.28, calculate

(a)   the potential difference across the 9.0-μF capacitor.

(b)   the energy stored by the 4.5-μF capacitor.

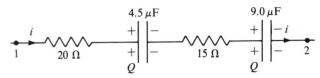

Figure 14.28   Problem 14.28

14.29   An isolated parallel-plate capacitor, which has a capacitance of 30 μF when the region between its plates is filled with a dielectric ($\kappa = 1.5$), is charged to a potential difference of 80 V.

(a)   What minimum amount of work is required to remove the dielectric from the capacitor completely?

(b)   Calculate the potential difference across the capacitor after the dielectric is removed.

14.30   If an isolated parallel-plate capacitor ($C = 4.5$ μF) has an initial potential difference of 40 V across it, determine the minimum work required to triple the separation between the plates.

14.31   For a combination of capacitors (*see* Fig. 14.29), the potential difference, $V_A - V_D$, is equal to 60 V. Calculate

(a)   the equivalent capacitance of the combination.

(b)   the potential difference, $V_B - V_D$.

(c)   the charge on the 15-μF capacitor.

(d)   the energy stored in the 10-μF capacitor.

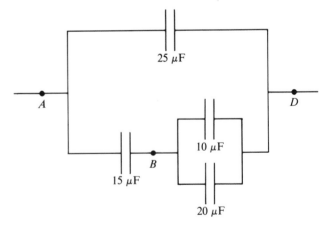

Figure 14.29   Problem 14.31

14.32   If the current through the 45-Ω resistor is 0.20 A as shown in Figure 14.30, determine $V_A - V_B$.

14.33   If a capacitor ($C_1 = 8.0 \times 10^{-10}$ F), initially charged to a potential difference of 5.0×10^2 V, is connected as shown in Figure 14.31 to a second capacitor ($C_2 = 5.0 \times 10^{-10}$ F), initially uncharged, calculate the final charge on each capacitor.

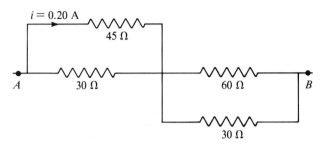

Figure 14.30   Problem 14.32

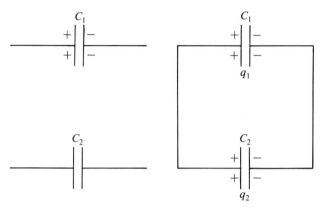

Figure 14.31   Problem 14.33

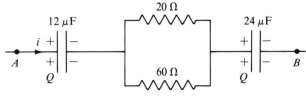

Figure 14.32   Problem 14.38

GROUP **B**

14.39   If $i = 12$ A and $R = 14$ Ω, calculate

(a)   the equivalent resistance of the resistor combination shown in Figure 14.33.

(b)   the potential difference, $V_A - V_B$.

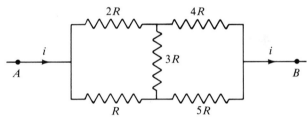

Figure 14.33   Problem 14.39

14.34   Two capacitors ($C_1 = 20$ μF and $C_2 = 10$ μF) are each charged to a potential difference of 50 V. The two charged capacitors are then connected by wires, positive plate to positive plate and negative to negative. Calculate the final charge on each capacitor.

14.35   Repeat Problem 14.34 if the charged capacitors are connected by joining plates of different polarity.

14.36   A 10-Ω resistor and a 3.0-μF capacitor are connected in series. At an instant when the charge on the capacitor is equal to 60 μC, the potential difference across this series combination is equal to 15 V. Determine the current in the resistor at this instant.

14.37   A 48-Ω resistor and a 12-μF capacitor are connected in series. Calculate the energy stored by the capacitor and the rate at which this energy is changing at an instant when the current through the resistor is 0.50 A and the potential difference across this series combination is 66 V.

14.38   At an instant when $i = 4.0 \cdot$A and $V_A - V_B = 100$ V for the combination of resistors and capacitors shown in Figure 14.32, calculate

(a)   the charge Q on either capacitor.

(b)   the current in the 20-Ω resistor.

(c)   the rate at which the energy stored by the 24-μF capacitor is changing.

14.40   The resistor depicted in Figure 14.34 consists of a cylinder (radius $= a$, length $= L$) made of material having a resistivity ρ_1 and a coaxial cylindrical shell (inner radius $= a$, outer radius $= b$, length $= L$) made of material having a resistivity ρ_2. Determine the resistance across the length of this device and show that your result reduces appropriately for $\rho_1 = \rho_2$.

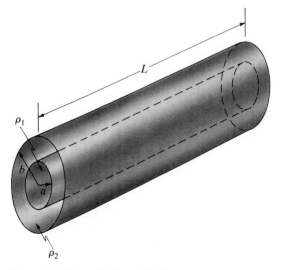

Figure 14.34   Problem 14.40

14.41   A cylindrical resistor of length L and radius b has a resistivity ρ that varies with distance r from the axis of the cylinder according to

$$\rho(r) = \rho_0(1 + \alpha r)$$

for $0 \le r \le b$ where ρ_0 and α are constants. Calculate the resistance of this device and show that your result reduces appropriately for $\alpha b \ll 1$.

14.42   A resistor of length L has a circular cross section with a radius that increases linearly as shown in Figure 14.35. If the resistor is made of material having a resistivity ρ,

(a)   calculate the resistance R of this resistor in terms of ρ, L, a, and b.

(b)   show that the expression obtained in part (a) reduces appropriately for $a = b$.

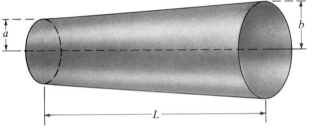

Figure 14.35   Problem 14.42

14.43   The region between the plates of a parallel-plate capacitor is "half filled" with a dielectric ($\kappa > 1$) material (*see* Fig. 14.36). If the capacitance of the device is equal to C_0 before the dielectric is inserted, determine the final capacitance. Does your result reduce appropriately for $\kappa = 1$?

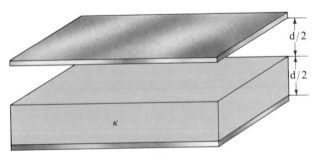

Figure 14.36   Problem 14.43

14.44   The region between the plates of a parallel-plate capacitor is filled with two dielectric slabs, as shown in Figure 14.37. If the capacitance of the device is equal to C_0 before the dielectrics are inserted, determine the final capacitance.

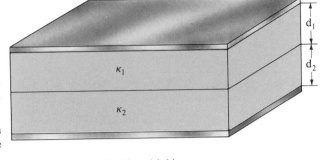

Figure 14.37   Problem 14.44

14.45   The region between the plates of a parallel-plate capacitor is "half filled" with a dielectric ($\kappa > 1$) material (*see* Fig. 14.38). If the capacitance of the device is equal to C_0 before the dielectric is inserted, determine its final capacitance.

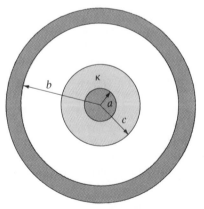

Figure 14.38   Problem 14.45

14.46   The region between the conductors of a spherical capacitor is partially filled with a concentric dielectric ($\kappa > 1$) shell (*see* Fig. 14.39). If C_0 is the capacitance of the device before the dielectric is inserted, determine its final capacitance.

Figure 14.39   Problem 14.46

14.47 The region between the conductors of a spherical capacitor is "half filled" with a dielectric (*see* Fig. 14.40). If C_o is the capacitance of the device before the dielectric is inserted, determine its final capacitance.

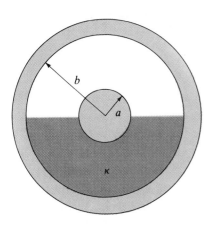

Figure 14.40 Problem 14.47

14.48 When a dielectric material is inserted between the plates of an isolated parallel-plate capacitor having a charge Q_o, the magnitude of the electric field between the plates is reduced from a value E_o ($= Q_o/\epsilon_o A$) to a value $E = E_o/\kappa$, where κ is the dielectric constant of the material and A is the area of the positively charged surface.

(a) Use Gauss's law to show that the total charge on the interface between the positive plate and the dielectric is Q_o/κ, which is less than Q_o.

(b) The reduction of the net charge from Q_o to Q_o/κ is understood by realizing that the initial electric field $\mathbf{E}_o$ induces a negative "polarization charge" $-Q_p$ on the dielectric surface adjoining the positive plate. Show that $Q_p = (\kappa - 1)Q_o/\kappa$.

(c) After a parallel-plate capacitor ($A = 2.0$ cm^2, $d = 0.10$ mm) is charged to a potential difference of 50 V, it is isolated, and a dielectric ($\kappa = 1.4$) slab is inserted so as to fill the region between the plates. Calculate Q_o, E_o, and Q_p.

14.49 Consider a parallel-plate capacitor.

(a) Show that the energy stored by this capacitor may be written $U_C = (1/2)\,\epsilon_o E^2 V$, where E is the magnitude of the (constant) electric field between the plates and V is the volume between the plates.

(b) If the energy U_C is divided by the volume V, show that the resulting energy density is equal to $(1/2)\epsilon_o E^2$.

(c) The energy density obtained in part (b) is defined as the *energy density* u_E associated with the electric field $\mathbf{E}$, i.e.,

$$u_E = \tfrac{1}{2}\,\epsilon_o E^2.$$

Show that if this energy density is integrated over the volume between the conductors of a spherical capacitor, the resulting energy is equal to that stored by the spherical capacitor.

(d) Repeat part (c) for a cylindrical capacitor.

(e) Calculate the energy density in J/m^3 and J/cm^3 at a point in space where $E = 1.0 \times 10^6$ V/m.

15 Direct-Current Circuits

An *electric circuit* is a collection of components through which electric charges may flow. Two components of circuits, resistors and capacitors, were introduced in Chapter 14. In this chapter we will introduce one more component and begin consideration of circuits that carry *steady currents*, that is, currents with a constant magnitude. Circuits in which charges flow in only one sense are called *direct-current circuits*. Finally, we analyze some circuits in which capacitors are charged and discharged through resistors. Such circuits are characterized by *transient currents*, currents with magnitudes that change before reaching steady-state values.

15.1 Energy Reservoirs in DC Circuits

When current is present in a circuit, energy changes continuously take place within the components of the circuit. Resistors convert the kinetic energy of the charged particles moving through them into thermal energy. Capacitors convert the kinetic energy of the moving charged particles of a circuit into electrical potential energy as the charges are accumulated on the plates of the capacitors. When a capacitor is discharged, it converts the electrical potential energy stored in the electric field between its plates into kinetic energy of the moving charged particles in the circuit. Every circuit that carries a steady current must include a component that provides energy to maintain the flow of charges through that circuit.

A circuit component that serves as a reservoir of energy in a current-carrying circuit is called an *emf* (pronounced ee-em-eff). The term *emf* is an abbreviation for the phrase "electromotive force," a misleading phrase that lingers although an emf is not, by any means, a force. We will refer to an energy reservoir in a circuit as an emf, which will be symbolized by $\mathscr{E}$ in circuit diagrams and in equations.

An *emf* has two physical characteristics that define its role as a circuit component:

1. It maintains a constant difference in potential $\mathscr{E}$, measured in volts, across itself (we usually say "across its *terminals*," which are its points of connection to the remainder of the circuit).

2. It functions as a reservoir of energy for the circuit.

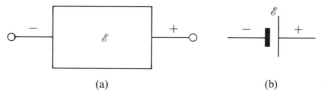

Figure 15.1
(a) An emf. (b) Conventional symbol for an emf.

(a)   (b)

When we write, "an emf $\mathscr{E}$ of 12 volts," we mean an emf, labeled $\mathscr{E}$, as in Figure 15.1(a), that maintains a constant difference of potential of 12 volts across its terminals. The $+$ sign indicates the terminal of higher potential. Figure 15.1(b) illustrates the conventional symbol for an emf. Notice that the longer bar of the symbol is the higher potential side of the emf.

In physics a reservoir is an "endless" source or sink for providing or absorbing a physical quantity. Recall, for example, that the earth, acting as an electrical ground (see Chapter 11, *Electric Charge and Electric Fields*), is a reservoir of charge. The earth can supply or absorb an arbitrary quantity of charge without affecting its own electric potential; this property is a measure of the ability of the earth to provide or accept charge. Similarly, a heat reservoir (see Chapter 10, *Heat and Thermodynamics*) can supply or absorb an arbitrary quantity of heat without affecting its own temperature, and this property is a measure of its ability to provide or accept heat. An emf, then, is a reservoir of energy in a circuit and can supply or absorb an arbitrary quantity of energy without affecting the difference in potential across its terminals. The constancy of this potential difference is a measure of the ability of the emf to supply or absorb energy.

Figure 15.2
A circuit consisting of an emf $\mathscr{E}$ and a resistor R.

Figure 15.2 shows a simple circuit composed of an emf $\mathscr{E}$ and a resistor R. The circuit carries a current i. Suppose dq is a quantity of (positive) charge passing any cross section of the circuit in a time interval dt. As the charge transits the emf (from $-$ to $+$), the emf must do a quantity of work dW on the charge in order to raise the potential of the charge by the difference in potential $\mathscr{E}$ across the terminals of the emf. Then the difference in potential $\mathscr{E}$ across an emf is defined in terms of the work dW done on (or the energy delivered to) a charge dq, by

$$\mathscr{E} = \frac{dW}{dq} \qquad (15\text{-}1)$$

Thus the emf $\mathscr{E}$ is equal to the work done per unit charge and, therefore, has the unit of joule per coulomb (J/C), or volt (V).

Exercise 15.1

E 15.1   At what rate is the emf shown in the figure supplying energy?
Answer: 12W

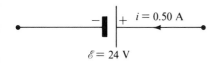

Exercise 15.2

E 15.2   Distinguish between the energetics of the emf shown in the accompanying figure with that of the emf in Exercise 15.1.
Answer: This emf is *storing* energy at a rate of 12 W.

E 15.3   Determine the potential difference, $V_B - V_A$, for each of the cases in the accompanying figure.   Answers: (a) 40 V; (b) -10V

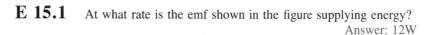

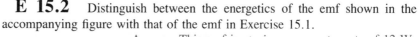

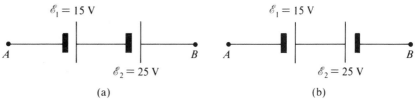

Exercise 15.3

(a)   (b)

Battery

Figure 15.3
A representation of a battery consisting
of an emf $\mathcal{E}$ and an internal resistance
R_i.

Negative
terminal

Positive
terminal

In Figure 15.2, the potential difference $\mathcal{E}$ across the emf is equal to ΔV, the potential difference across the resistor R. Then, when an emf $\mathcal{E}$ is connected across a single resistor R, using Ohm's law for a resistor, $\Delta V = iR$, we may write

$$\mathcal{E} = iR \qquad\qquad (15\text{-}2)$$

The emf, as we have defined it, is an idealized reservoir of energy in a circuit. Actual devices that function as emfs in circuits—batteries, electrolytic cells, and dc generators are typical devices—are often approximated by an emf. A battery or cell, however, has within it an internal resistance R_i, and a battery or cell may be more realistically represented by a series combination of an emf $\mathcal{E}$ and a resistor R_i, as shown in Figure 15.3. For simplicity here, we will usually restrict our considerations of energy reservoirs in circuits to idealized emfs and generally ignore any resistance that would be associated with an actual device.

E 15.4 Calculate the potential difference, $V_B - V_A$, for each of the cases depicted in the figure.

Answers: (a) 17 V; (b) 33V

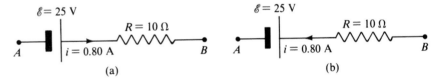

Exercise 15.4

15.2 Analysis of DC Circuits with Steady Currents

Direct-current circuits that carry a steady current can be represented by combinations of emfs and resistors. Some of the terminology of dc circuits is defined and illustrated in Figure 15.4:

1. A *junction* is a point in a circuit at which a conducting path splits into more than one path. Points a and b in Figure 15.4 are junctions of the circuit.

2. A *branch* of a circuit is any single conducting path connecting two junctions without passing through any intermediate junction. Thus, if more than one circuit component is in a branch, those components are connected in series and, therefore, carry the same current. In Figure 15.4, the four branches of the circuit are *acb*, *adb*, *aeb*, and *afb*.

3. A *loop* of a circuit is any closed path that passes through no junction more than once. In Figure 15.4 there are six loops: *acbfa*, *adbfa*, *aebfa*, *acbda*, *adbea*, and *acbea*.

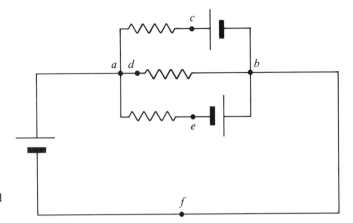

Figure 15.4
A circuit illustrating branches and loops.

Kirchhoff's Rules

Two simple rules—actually statements of conservation of charge and of conservation of energy—form the basis of circuit analysis. Called *Kirchhoff's rules*, they may be stated as:

1. The *junction rule: At any junction of a circuit, the net current entering a junction is equal to the net current leaving that junction*. If currents entering a junction are assigned positive values and currents leaving that junction are assigned negative values, the algebraic sum of currents of all branches that meet at a junction is zero, or

$$\sum_{\text{junction}} i = 0 \qquad (15\text{-}3)$$

2. The *loop rule: The sum of all the changes in potential around a loop is equal to zero*. A change in potential across a component of the circuit is either an increase (rise) in potential (+) or a decrease (drop) in potential (−). Using this sign convention, we may express the loop rule as

$$\sum_{\text{loop}} \Delta V = 0 \qquad (15\text{-}4)$$

E 15.5   The figure accompanying this exercise shows a junction in a circuit. Determine the current i.   Answer: 12 A

E 15.6   The figure shows a junction in a circuit. Determine the magnitude and sense of the current i_4.   Answer: 4.0 A toward the junction

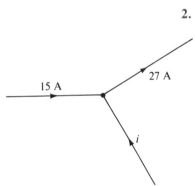

Exercise 15.5

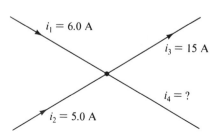

Exercise 15.6

E 15.7   The accompanying figure shows a single-loop circuit consisting of three components. Determine the potential difference (magnitude and polarity) across component C.   Answer: $\Delta V_C = 12$ V, right side is positive

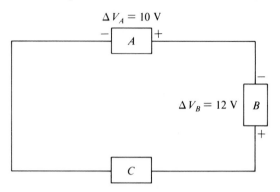

Exercise 15.7

E 15.8   The accompanying figure shows a single-loop circuit consisting of two batteries and a resistor. Determine the potential difference (magnitude and polarity) across the resistor.   Answer: 15 V, top side is positive

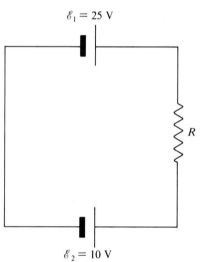

Exercise 15.8

E 15.9   Calculate the current (magnitude and sense) in the single-loop circuit shown in the figure.   Answer: 15 A, clockwise

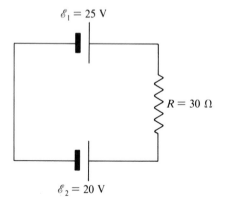

Exercise 15.9

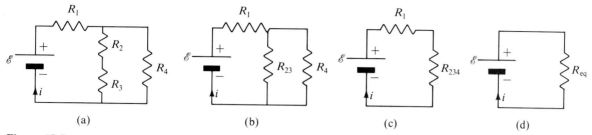

(a)   (b)   (c)   (d)

Figure 15.5
Replacing combinations of resistors within a circuit.

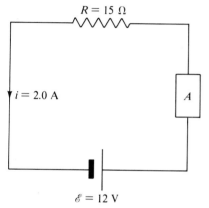

Exercise 15.10

E 15.10   Determine the potential difference (magnitude and polarity) across the component labeled A in the single-loop circuit in the figure.

Answer: 18 V, top side is positive

Although dc circuits can usually be analyzed by directly applying Kirchhoff's rules, the analysis can sometimes be simplified by replacing combinations of resistors within a circuit by the appropriate equivalent resistance (see Chapter 14, *Capacitance, Current, and Resistance*) before applying Kirchhoff's rules. In Figure 15.5(a), for example, suppose it were required to find the current i in the emf $\mathcal{E}$ of the circuit. The resistors R_2 and R_3 form a series combination that may be replaced by an equivalent resistance $R_{23} = R_2 + R_3$, as shown in Figure 15.5(b). In the same figure, the resistors R_{23} and R_4 form a parallel combination that may be replaced by $R_{234} = R_{23} R_4/(R_{23} + R_4)$. Finally, in Figure 15.5(c) the resistors, R_1 and R_{234}, constitute a series combination that may be replaced, as shown in Figure 15.5(d), by $R_{eq} = R_1 + R_{234}$. Then the current carried by the emf $\mathcal{E}$ is, according to Equation (15-2),

$$i = \frac{\mathcal{E}}{R_{eq}} \tag{15-5}$$

Thus, the multiple-loop circuit of Figure 15.5(a) has been reduced to the simple, single-loop circuit of Figure 15.5(d).

It is incorrect, however, to treat two resistors as a parallel combination if either resistor is in a branch that includes an emf. Furthermore, two resistors in different branches of a circuit *cannot* be treated as a series combination, because they do not, in general, carry the same current. Figure 15.6 shows a circuit that cannot be reduced to a simpler circuit by using the equivalent resistance of a

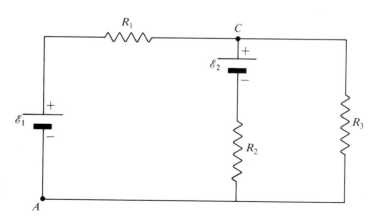

Figure 15.6
A circuit that cannot be analyzed by using the equivalent resistance of parallel or series combinations of resistors.

parallel or series combination of resistors. The resistors R_2 and R_3 are not a parallel combination because R_2 is in a branch with an emf. Resistors R_1 and R_3 are not a series combination because they are each in a different branch of the circuit. The circuit of Figure 15.6 may be analyzed by applying Kirchhoff's rules directly.

> In the circuit of Figure 15.6, let $\mathscr{E}_1$ = 6.0 V, $\mathscr{E}_2$ = 12 V, R_1 = 2.0 Ω, R_2 = 4.0 Ω, and R_3 = 6.0 Ω. (a) What is the rate at which energy is converted into heat in R_2? (b) What is the difference in potential, $V_C - V_A$, between the points labeled A and C on the circuit?

In most dc circuit problems, the values of the circuit components are given. In such cases, analysis begins by finding the current in each branch of the circuit. With that information, virtually any question that may reasonably be asked about the circuit can be answered. We begin by redrawing the circuit, as in Figure 15.7, labeling each component with its numerical value. Next we *assume* a sense for the current in each branch of the circuit and indicate those assumptions by a labeled current arrow in each branch. In Figure 15.7 we have designated the currents i_1, i_2, and i_3. Of course, the assumed sense of the current in each branch is a guess, but as we shall see, guessing incorrectly does not prevent accurate analysis of the circuit. Now label each component of the circuit with + and − signs, indicating the high- and low-potential terminals of each component. The signs on each emf are in accordance with the convention of Figure 15.1: The long bar is the high-potential terminal, labeled +, and the short bar is the low-potential terminal, labeled −. The signs on the resistors must correspond to the assumed sense of the current through each resistor: The assumed current through each resistor is from + to −. With the diagram appropriately labeled, Kirchhoff's rules may be applied.

Applying the junction rule at point C on the diagram gives $i_1 + i_2 - i_3 = 0$, or

$$i_1 + i_2 = i_3 \tag{15-6}$$

Had we used the junction at E, we would have obtained the same equation.

The loop rule may be applied to any of the three loops of this circuit. Let us use the loop $ABCEA$. Starting from point A, we sum the changes in potential around the loop. Here we proceed around the loop in the sense indicated by the arrow within the loop. (Proceeding in the opposite sense yields an equivalent result, so the choice of sense in summing the potential changes around a loop is

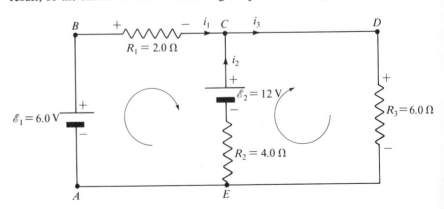

Figure 15.7
The circuit of Figure 15.6, redrawn for analysis using Kirchhoff's rules.

arbitrary.) Remember in applying the loop rule that passing through a component from $-$ to $+$ is a positive change (an increase in potential) and from $+$ to $-$ is a negative change (a decrease in potential). Then the loop rule for the path $ABCEA$ yields

$$6 - 2i_1 - 12 + 4i_2 = 0$$

or

$$2i_2 - i_1 = 3 \qquad (15\text{-}7)$$

where we have used Ohm's law for a resistor to specify (in terms of the currents) the potential change across R_1 ($\Delta V = -R_1 i_1 = -2i_1$) and the potential change across R_2 ($\Delta V = R_2 i_2 + 4i_2$).

Equations (15-6) and (15-7) provide two equations containing three unknown quantities, i_1, i_2, and i_3. We, therefore, need one further relationship between the currents. Let us apply the loop rule to the loop $CDEC$:

$$-6i_3 - 4i_2 + 12 = 0$$

or

$$3i_3 + 2i_2 = 6 \qquad (15\text{-}8)$$

We now have three equations relating i_1, i_2, and i_3. Using the loop rule for the remaining loop $ABCDEA$ would yield only redundant information. Solving Equations (15-6), (15-7), and (15-8) simultaneously gives the current in each branch of the circuit:

$$i_1 = -\frac{3}{11}\text{ A} \quad ; \quad i_2 = \frac{15}{11}\text{ A} \quad ; \quad i_3 = \frac{12}{11}\text{ A} \qquad (15\text{-}9)$$

The negative value for i_1 in Equation (15-9) signals that we guessed incorrectly the sense of the current in the branch of the circuit containing $\mathscr{E}_1$ and R_1. Our guesses for the senses of i_2 and i_3 were correct.

E 15.11   Using the results of Equation (15-9), verify that the junction rule is satisfied at point E in the circuit of Figure 15.7.

E 15.12   Using the results of Equation (15-9), verify that the loop rule is satisfied for path $ABCDEA$ of the circuit of Figure 15.7.

Now that we know the values and senses of the currents in each branch, the specific quantities required in the problem may be determined. The rate at which energy is converted into heat in a resistor R carrying a current i is the power loss, $P = i^2R$, in that resistor. Because the current i_2 is carried by R_2, the power loss in R_2 is

$$P = i_2^2 R_2 = \left(\frac{15}{11}\right)^2 (4) = 7.4 \text{ W} \qquad (15\text{-}10)$$

To determine the potential difference, $V_C - V_A$, between points C and A, we may assume that the electric potential is equal to zero at point A (because only differences in potential are physically significant). Then progressing from A to C via the path ABC, we algebraically sum the potential changes across each component between points A and C, paying scrupulous attention to the signs of the potential changes.

$$V_C - V_A = V_C - 0 = V_C = + \mathcal{E}_1 - i_1 R_1$$

$$V_C = +6 - \left(-\frac{3}{11}\right)(2) = \frac{72}{11} \text{ V} \qquad (15\text{-}11)$$

The potential change across $\mathcal{E}_1$ is an increase (positive) as we proceed along the path ABC. The potential change across R_1, when we use the signs in the circuit diagram for the *assumed* current sense, is a potential change given by Ohm's law to be equal to $-i_1 R_1$. But the assumed current i_1 has the negative value $-3/11$ A. Therefore, as indicated in Equation (15-11), $V_C - V_A$ has the positive value $72/11$ V.

A reassuring check on the solution of Equation (15-11) can be made by calculating $V_C - V_A$ along another path, say along AEC. Using that path, we obtain

$$V_C - V_A = -4i_2 + 12 = -4\left(\frac{15}{11}\right) + 12 = \frac{72}{11} \text{ V} \qquad (15\text{-}12)$$

a result that is identical to that of Equation (15-11). An additional check is available via path $AEDC$.

E 15.13   Calculate the potential difference, $V_C - V_A$, by summing the potential changes along the path $AEDC$ of Figure 15.7.   Answer: $\frac{72}{11}$ V

Example 15.1   **PROBLEM**   Show that the energy provided by the components of the circuit of Figure 15.6 (or Figure 15.7) is equal to the energy absorbed or dissipated by other components. In other words, account for all the energy in the circuit.

SOLUTION   Figure 15.8 shows the circuit of Figure 15.7 with the currents and potential polarities indicated. The emf $\mathcal{E}_2$ is supplying energy to the circuit at the rate

$$P_2 = i_2 \mathcal{E}_2 = \left(\frac{15}{11} \text{ A}\right) \cdot (12 \text{ V}) = \frac{180}{11} \text{ W}$$

while $\mathcal{E}_1$ is absorbing energy (why?) at the rate

$$P_1 = i_1 \mathcal{E}_1 = \left(\frac{3}{11} \text{ A}\right) \cdot (6 \text{ V}) = \frac{18}{11} \text{ W}$$

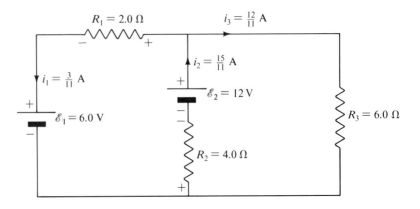

Figure 15.8   Example 15.1

Each of the resistors is dissipating energy at a rate given by i^2R. Thus heat is being generated in the circuit at the rate

$$i_1{}^2R_1 + i_2{}^2R_2 + i_3{}^2R^3 = \left(\frac{3}{11}A\right)^2 \cdot (2) + \left(\frac{15}{11}\right)^2 \cdot (4) + \left(\frac{12}{11}\right)^2 \cdot (6) = \frac{162}{11}\,W$$

Summing the rates at which energy is absorbed by $\mathcal{E}_1$ and dissipated by the resistors gives $\frac{180}{11}\,W$, a value equal to the rate $\mathcal{E}_2$ is supplying energy. ■

Ammeters and Voltmeters in DC Circuits

Modern electrical meters are frequently instruments that have digital displays or that print out measured values of such quantities as current or potential difference. In this section we will concentrate on some older types of electrical meters, specifically the galvanometer. A *galvanometer* is an electrical instrument containing a movable coil of wire that deflects an indicator needle in direct proportion to the current that passes through it. These moving coil meters are useful in investigating the essential characteristics and limitations common to instruments that measure current or potential difference. Any device that provides an output proportional to the current passing through it may be used as a "galvanometer" in the construction of an instrument for measuring current, potential difference, or related quantities.

The coil in a galvanometer through which the current flows has a resistance that is typically between 10 Ω and 100 Ω. The most common types of ammeters and voltmeters, instruments used to measure currents and potential differences, respectively, in electrical circuits, are both constructed using a galvanometer as an indicator, or read-out device.

Figure 15.9(a) shows a galvanometer G, having an internal (coil) resistance R_g, to which a low-resistance resistor R_s, called a *shunt*, has been connected *in*

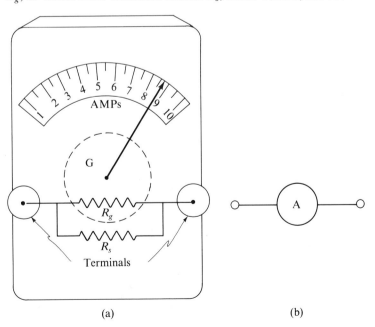

Figure 15.9
(a) An ammeter. **(b)** The conventional symbol for an ammeter used in circuit diagrams.

(a)   (b)

parallel. The resulting device, which has a very low resistance between its terminals, is an *ammeter*, indicated in circuit diagrams by the symbol shown in Figure 15.9(b). The low-resistance ammeter may be inserted into a branch of a circuit (the circuit must be broken at the point of insertion) in series with other circuit components in that branch. The scale of the indicator within the ammeter is calibrated to read directly (in amperes, milliamperes, and so on) the current in that branch. Because of the very low resistance between the terminals of most ammeters, the presence of an ammeter in a circuit does not significantly change the currents in that circuit.

Figure 15.10(a) shows a galvanometer G, which has an internal resistance R_g, to which a high-resistance resistor R_m, called a *multiplier* has been connected *in series*. The resulting device, which has a very high resistance between its terminals, is a *voltmeter*, indicated in circuit diagrams by the symbol shown in Figure 15.10(b). The high-resistance voltmeter, when connected in parallel with a resistor R, for example, in a circuit, is calibrated to read directly (in volts, millivolts, and so on) the potential difference across the circuit component R. Because the resistance between the terminals of most voltmeters is very large, the presence of a voltmeter in the circuit in parallel with R does not appreciably change the potential difference across R. The terminals of a voltmeter may be connected to any two points in a circuit to measure the difference in potential between those points.

An ammeter placed in a circuit affects the current in that circuit less when the shunt of the ammeter has a very low resistance. Similarly, a voltmeter attached to a circuit affects the current in that circuit less when the multiplier of the voltmeter has a very high resistance. But lower-resistance shunts or higher-resistance multipliers can be employed only by using galvanometers or equivalent devices that are more sensitive and, therefore, more expensive.

Figure 15.11(a) shows a simple circuit in which two resistors, R_1 and R_2, are connected in series with an emf $\mathcal{E}$, which carries a current i. In Figure 15.11(b), an ammeter and a voltmeter are shown appropriately connected into the circuit so

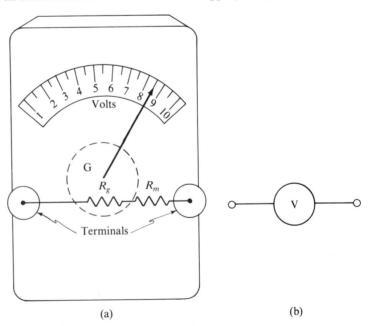

Figure 15.10
(a) A voltmeter. (b) The conventional symbol for a voltmeter used in circuit diagrams.

(a)   (b)

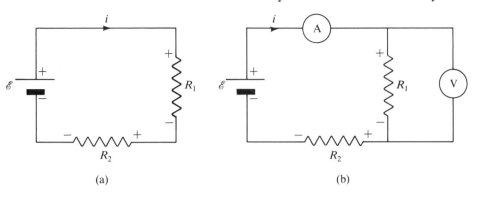

Figure 15.11
(a) A simple circuit. **(b)** An ammeter and voltmeter connected in the circuit so that the ammeter indicates the current in the emf ξ and the voltmeter indicates the potential difference across the resistor R_1.

that the ammeter A measures the current i and the voltmeter V measures the potential difference ΔV_1 across R_1.

Example 15.2   **PROBLEM**   An 80-Ω galvanometer that requires a current of 0.50 mA for a full-scale deflection is to be used to construct an ammeter that has a full-scale deflection for a current of 10 A. Determine the value of the shunt resistance in the ammeter.

SOLUTION   Figure 15.12 depicts the desired situation: Essentially all of the 10-A current into the ammeter is bypassed through the low-resistance shunt, which is chosen so that the galvanometer current is equal to 0.50 mA. Points A and B represent the terminals of the ammeter. Applying the junction rule to point A gives $i = i_g + i_s$ or $i_s = i - i_g = 10$ A $- 0.00050$ A $\simeq 10$ A. Because R_s and R_g are parallel resistors, the potential differences across the two are equal, i.e., $i_s R_s = i_g R_g$ or

$$R_s = \frac{i_g R_g}{i_s} = \left(\frac{0.00050 \text{ A}}{10 \text{ A}}\right)(80 \ \Omega) = 0.0040 \ \Omega = 4.0 \text{ m}\Omega \quad \blacksquare$$

E 15.14   Determine the resistance of the ammeter of Example 15.2 and the potential difference across this ammeter when it carries a full-scale current of 10 A.   Answer: 0.0040 Ω, 0.040 V

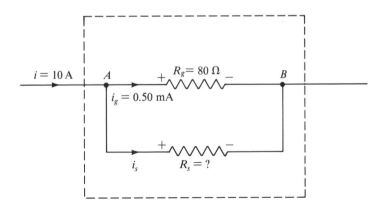

Figure 15.12   Example 15.2

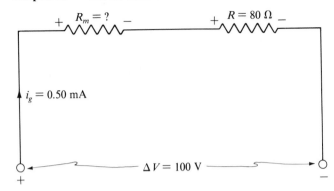

Figure 15.13   Example 15.3

Example 15.3

PROBLEM   An 80-Ω galvanometer that requires a current of 0.50 mA for a full-scale deflection is to be used to construct a voltmeter that has a full-scale deflection for a potential difference of 100 V. Determine the value of the multiplier resistance.

SOLUTION   Figure 15.13 shows the required situation: R_m is chosen so that a potential difference of 100 V across the terminals (points A and B) of the voltmeter causes a current of 0.50 mA through the meter. The sum of the potential differences across R_m and R_g is equal to 100 V, because the two resistances are in series, that is,

$$i_g R_m + i_g R_g = \Delta V = 100 \text{ V}$$

or

$$R_m = \frac{\Delta V}{i_g} - R_g$$

$$R_m = \frac{100 \text{ V}}{0.00050 \text{ A}} - 100 \text{ } \Omega = 2.0 \times 10^5 \text{ } \Omega$$

The multiplier resistance is 2500 times the resistance of the galvanometer.   ■

15.3 RC Circuits

The circuits we have considered in the previous section contained only emfs and resistors and carry steady currents. When a capacitor is included in a branch of a dc circuit, however, the current in that branch is no longer steady; the current changes with time. The analysis of dc circuits containing capacitors is, like all dc circuits, based on Kirchhoff's rules. But because the currents in dc circuits containing capacitors are changing, the analysis of these circuits is somewhat different. Let us begin our considerations of time-dependent currents in dc circuits with the discharge of a capacitor through a resistor.

Figure 15.14(a) shows a capacitor C, which has an initial charge Q_o, connected to a resistor R and a switch S. Before the switch is closed, the following conditions prevail in the open circuit:

1. The potential difference across C is $\Delta V_o = Q_o/C$.

2. The potential difference across the switch is equal to ΔV_o, the potential difference across C.

3. No current is in the circuit, so the difference in potential across R is equal to zero.

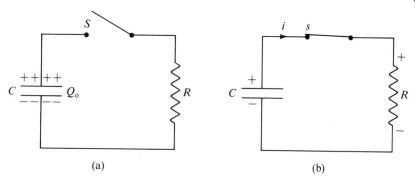

Figure 15.14
(a) A capacitor C that has been charged to a value Q_o and connected to a resistor R and a switch S. (b) The same circuit after the switch has been closed.

Suppose the switch is closed at a time $t = 0$. At that instant, the following initial ($t = 0$) conditions exist:

1. The loop rule requires that the initial value of the potential difference across R is equal to ΔV_o.

2. The initial current i_o in the circuit is, according to Ohm's law, equal to $\Delta V_o/R$.

At any time after $t = 0$, both Q, the charge on C, and i, the current in the circuit, change with time. Let us determine $i(t)$ by applying Kirchhoff's loop rule to the closed circuit of Figure 15.14(b). For any time $t > 0$, the loop rule gives

$$\frac{Q}{C} - iR = 0 \qquad (15\text{-}13)$$

Differentiating Equation (15-13) with respect to time gives

$$\frac{1}{C}\frac{dQ}{dt} - R\frac{di}{dt} = 0 \qquad (15\text{-}14)$$

But the current i in the circuit is equal to the rate at which the charge Q on C is *decreasing*, or $i = -dQ/dt$. Using this relationship between i and dQ/dt in Equation (15-14) and rearranging the terms, we may write

$$\frac{di}{dt} = -\frac{i}{RC}$$

which may be further rearranged to give

$$\frac{di}{i} = -\frac{1}{RC}\,dt \qquad (15\text{-}15)$$

If we integrate Equation (15-15) from time $t = 0$, when $i = i_o = \Delta V_o/R$, to a particular but nevertheless arbitrary time t_f, when $i = i_f$, we obtain

$$\int_{i = \Delta V_o/R}^{i = i_f} \frac{di}{i} = -\frac{1}{RC}\int_{t = 0}^{t = t_f} dt$$

$$\ln \frac{i_f}{\Delta V_o/R} = -\frac{t_f}{RC} \qquad (15\text{-}16)$$

Solving Equation (15-16) for i_f and dropping the subscript f, because i and the time t corresponding to i are arbitrary, gives the current i at a time $t \geq 0$:

$$i = \frac{\Delta V_o}{R} e^{-t/RC} = i_o\, e^{-t/RC} \qquad (15\text{-}17)$$

A graph of current as a function of time, according to Equation (15-17), is

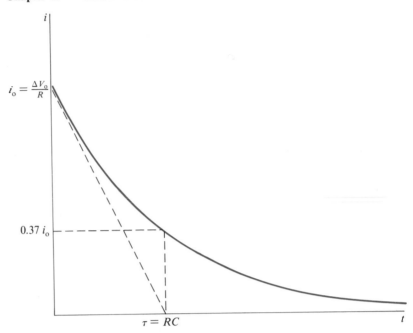

Figure 15.15
A graph of Equation (15-17).

shown in Figure 15.15. Starting at $t = 0$, when $i_o = \Delta V_o/R$, the current decays exponentially with passing time, approaching zero asymptotically (becoming arbitrarily close to zero but theoretically never achieving a zero value). An important property of a given RC combination is the measure of how rapidly the current decays as the capacitor discharges through a resistance. The slope of the i-versus-t curve at $t = 0$ is

$$\left.\frac{di}{dt}\right|_{t=0} = -\frac{i_o}{RC}\, e^{-t/RC}\bigg|_{t=0} = -\frac{i_o}{RC} \tag{15-18}$$

This value is the slope of the dashed straight line indicated in Figure 15.15. Thus, if the current continued to decrease at its initial rate of decrease, the current would become equal to zero at a time $\tau = RC$. This characteristic time τ associated with an RC combination is called the *time constant* of the combination. Setting $t = \tau = RC$ in Equation (15-17), we find the current is reduced from its initial value i_o to a value $i_o/e \cong 0.37\, i_o$. Thus, after a time interval equal to one time constant, the current is reduced to a value that is approximately equal to 37 percent of its original value.

E 15.15   Verify that the (SI) unit of the quantity RC is the second.

E 15.16   How many time constants must elapse in a discharging RC circuit before the current is reduced to a millionth of its initial value?

Answer: 14

Example 15.4   **PROBLEM**   A 20-μF capacitor is charged to an initial potential difference of 100 V. It is then discharged through a 50-kΩ resistor. How much heat is generated in the resistor?

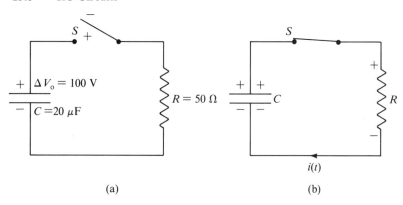

Figure 15.16   Example 15.4   (a)   (b)

SOLUTION   Figure 15-16(a) shows the initial situation before the capacitor C is discharged. After the switch S is closed, the discharging of the capacitor results in heat being generated in the resistor R, and the rate at which this heat is generated is given by Ri^2. Because the current is changing, we must use calculus to determine the total heat generated. If the current at the time t is denoted by $i(t)$, then the heat generated during a time interval of length dt is given by $Ri^2(t)\,dt$. Thus the total heat H released in the resistor is

$$H = \int_0^\infty R\,i^2(t)\,dt$$

Using Equation (15-17), $i(t) = (\Delta V_0/R)\,e^{-t/RC}$, gives

$$H = \int_0^\infty R\left(\frac{\Delta V_0}{R}\,e^{-t/RC}\right)^2 dt$$

$$H = \frac{(\Delta V_0)^2}{R}\int_0^\infty e^{-2t/RC}\,dt$$

$$H = \frac{1}{2}C(\Delta V_0)^2 = \frac{1}{2}(20 \times 10^{-6}\text{ F})\,(100\text{ V})^2 = 0.10\text{ J}$$

To check this result, we note that the heat H generated in the resistor is equal to the initial energy $[\frac{1}{2}C(\Delta V_0)^2]$ stored in the capacitor.   ∎

E 15.17   Evaluate $\int_0^\infty e^{-2t/RC}dt$ and show that $H = \frac{1}{2}C(\Delta V_0)^2$, as given above in Example 15.4.   Answer: $\int_0^\infty e^{-2t/RC}dt = \frac{1}{2}RC$

E 15.18   What is the maximum rate at which heat is released in the resistor during the discharge of the capacitor of Example 15.4?
                                                                    Answer: 0.20 W

E 15.19   If the switch S in Example 15.4 is closed at $t = 0$, determine the time t at which heat is being generated in R at a rate of 0.10 W.
                                                                    Answer: 0.35 s

Figure 15.17(a) shows a circuit in which a capacitor C, in series with a resistor R, is to be charged by an emf $\mathscr{E}$. When the switch is initially closed at time $t = 0$, no charge is on the capacitor, so no difference in potential exists across its plates; the potential difference across R is equal to $\mathscr{E}$, and the initial current in the circuit

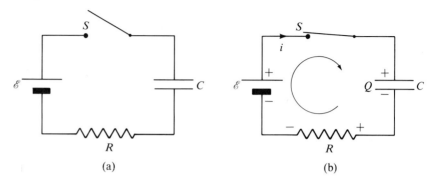

Figure 15.17
(a) A capacitor C, in series with a resistor R, an emf $\mathcal{E}$, and a switch S.
(b) The same circuit after the switch has been closed.

is $i_0 = \mathcal{E}/R$. The current at any time $t > 0$ may be found by applying the Kirchhoff loop rule to Figure 15.17(b):

$$\mathcal{E} - \frac{Q}{C} - iR = 0 \tag{15-19}$$

Using $i = dQ/dt$, the time derivative of Equation (15-19) gives,

$$-\frac{1}{C}i - R\frac{di}{dt} = 0$$

or

$$\frac{di}{i} = -\frac{1}{RC}\,dt \tag{15-20}$$

This result is identical to Equation (15-15), which was obtained for a discharging capacitor. The solutions for Equations (15-15) and (15-20) are identical, except that the initial current i_0 in this case, the charging of a capacitor, is equal to $\mathcal{E}/R$. Then the current at any time $t > 0$ is

$$i = i_0\,e^{-t/RC} = \frac{\mathcal{E}}{R}\,e^{-t/RC} \tag{15-21}$$

Figure 15.18(a) shows a graph of the current i as a function of time t.

E 15.20   Use Equations (15-19) and (15-21) to show that the charge on the capacitor at any time $t \geq 0$ is given by

$$Q(t) = C\mathcal{E}(1 - e^{-t/RC}) \tag{15-22}$$

Figure 15.18(b) shows a plot of the charge Q on the capacitor as a function of

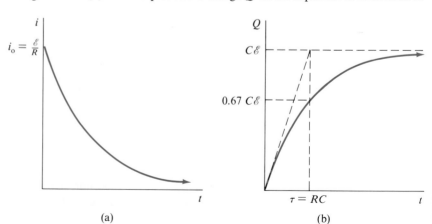

Figure 15.18
(a) A graph of Equation (15-21). **(b)** A graph of Equation (15-22).

time t. In one time constant, that is, when $t = \tau = RC$, the capacitor is charged to $(1 - 1/e) \cong 63\%$ of its final charge.

Let us examine the potential differences across the capacitor (ΔV_C) and across the resistor (ΔV_R) during the charging process. Because $\Delta V_C = Q/C$, we may use Equation (15-22) to give

$$\Delta V_C = \mathscr{E}(1 - e^{-t/RC}) \qquad (15\text{-}23)$$

And because $\Delta V_R = iR$, we may use Equation (15-21) to obtain

$$\Delta V_R = \mathscr{E}\,e^{-t/RC} \qquad (15\text{-}24)$$

These relationships, Equations (15-23) and (15-24), reconfirm that at $t = 0$, when the switch is closed, (a) the potential difference ΔV_C across C is equal to zero, and (b) the potential difference ΔV_R across R is equal to $\mathscr{E}$. Further, a very long time after the switch has been closed, when the circuit has reached a steady-state condition, (a) the potential difference ΔV_C across C is equal to $\mathscr{E}$, and (b) the potential difference ΔV_R across R is equal to zero.

These findings may lead us to some physical insights that are helpful in considering the initial and steady states of RC circuits:

1. When an initially uncharged capacitor begins to receive charge, there is no potential difference across the capacitor. The capacitor, in effect, acts like a *short*, a conductor of negligible resistance, in the circuit.

2. When an RC circuit has reached its steady-state condition, no current is present in the capacitors of the circuit. No branch containing a capacitor can then carry current. In effect, the capacitor acts like an *open*, a conductor of arbitrarily large resistance, in the circuit.

These characterizations of the initial and steady states of RC circuits are useful in problems similar to the following:

In the circuit of Figure 15.19, the capacitor C is uncharged before the switch S is closed. Find the current i in the emf $\mathscr{E}$ (a) immediately after the switch is closed, and (b) after the current has reached its steady state.

Immediately after the switch is thrown, at $t = 0$, the capacitor C is, in effect, a short. The potential difference across C is equal to zero; and because R_2 is connected in parallel with C, the potential difference across R_2 is equal to zero. Consequently R_2 carries no current, and the current i_o in $\mathscr{E}$ and R_1 is given by

$$i_o = \frac{\mathscr{E}}{R_1} \qquad (t = 0) \qquad (15\text{-}25)$$

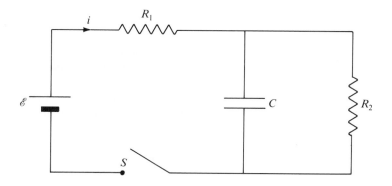

Figure 15.19
A multiple-loop circuit.

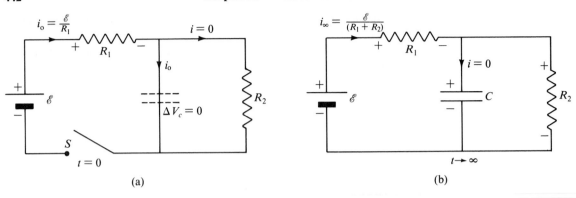

Figure 15.20
The (a) initial and (b) steady-state situations for the multiple-loop circuit of Figure 15.19.

This situation at $t = 0$ is shown in Figure 15.20(a).

When the current has reached a steady state, long after the switch has been closed, the capacitor carries no current; the capacitor is, in effect, an open in the circuit. Then R_1 and R_2 are connected in series, and the current i_∞ through $\mathscr{E}$ and R_1 and R_2 is

$$i_\infty = \frac{\mathscr{E}}{R_1 + R_2} \qquad (t \to \infty) \qquad (15\text{-}26)$$

Figure 15.20(b) shows the situation for $t \to \infty$.

E 15.21   Determine the potential difference across the capacitor C in the steady-state $(t \to \infty)$ situation.   Answer: $R_2\mathscr{E}/(R_1 + R_2)$

15.4 Problem-Solving Summary

The successful analysis of direct-current circuits begins with an understanding of the role of each component in the circuit. We have seen in an earlier chapter that a capacitor stores charge, thereby storing energy in the electric field between its plates. And we have seen that a resistor dissipates the energy of the moving charges of the circuit by converting that energy into thermal energy. In this chapter, we introduced the emf, a circuit element that functions as an energy reservoir for the moving charges of the circuit.

An emf $\mathscr{E}$, measured in volts, is an idealized circuit component that is defined by the work W (or energy) per unit charge q it supplies to or absorbs from the charges that pass through it. Thus, $\mathscr{E} = W/q$ is a measure of the constant potential difference between the terminals of an emf. When charges pass from the low- to the high-potential terminals of an emf, the emf supplies energy to the charges; when charges pass in the opposite sense, the emf absorbs energy from the charges.

Direct-current circuits can sometimes be simplified by combining similar components that are in series or parallel combinations. Remember, however, that a resistor or capacitor in the same branch with an emf cannot be treated as if it were in a parallel combination with resistors or capacitors in another branch of that circuit.

The basic tools of dc-circuit analysis are Kirchhoff's junction rule and loop rule. The junction rule requires that the net current entering a junction is equal to the net current leaving that junction. The loop rule requires that the potential changes around any complete loop of the circuit must sum to zero.

Direct-current circuits that contain only emfs and resistors are common. Such circuits may occur in both single- and multiple-loop networks. In most dc-circuit problems, the currents in the branches of the circuit are not known. Most specific questions in dc-circuit problems can be answered only after the branch currents have been determined. The following problem-solving procedure is appropriate to the analysis of most dc circuits:

1. Assume a direction for the sense of the current in each branch of the circuit. Draw and label current arrows in each branch accordingly.

2. Label the terminals of every component in the circuit with $+$ and $-$ signs, indicating the high- and low-potential terminals. The signs on the terminals of resistors must be consistent with the *assumed* current arrows (the current sense in a resistor is from $+$ to $-$).

3. Apply the junction rule to as many junctions as necessary to include every different current in the circuit in at least one algebraic relationship.

4. Apply the loop rule to as many loops as necessary in order to provide as many algebraic relationships as there are unknown currents.

5. Solve the algebraic equations simultaneously for the unknown quantities. Any assumed current value that is found to be negative means that its corresponding current arrow was chosen to have the wrong sense.

6. Reverse the sense of any current arrows that were chosen incorrectly. Reverse the $+$ and $-$ signs on the terminals of all resistors in those branches in which current arrows were changed.

7. Use the currents that have been found to answer any specific questions required by the statement of the problem.

8. Check the currents obtained by selecting two points, say A and B, in the circuit and calculating the potential difference, $V_A - V_B$, by proceeding along two different circuit paths connecting A and B.

Direct-current circuits containing capacitors involve time-varying currents. The charging by an emf and the discharging of a capacitor C connected in series with a resistor R produce currents given by $i = i_o e^{-t/RC}$, where i_o is the value of the current at time $t = 0$. The time constant $\tau = RC$ of the resistor-capacitor combination is a measure of how rapidly the current approaches its steady-state value (zero); τ is the time for the current to decrease to $1/e \cong 0.37 = 37\%$ of its initial value. The charge Q on a capacitor being discharged from an initial charge of Q_o is given by $Q = Q_o\, e^{-t/RC}$. The charge Q on a capacitor being charged (from an initially uncharged state) by an emf $\mathscr{E}$ is given by $Q = C\mathscr{E}(1 - e^{-t/RC})$.

Complex RC circuits that contain initially uncharged capacitors may be analyzed in their initial and steady states using two physical characteristics of capacitors in dc circuits:

1. A capacitor functions as a short when it is uncharged.

2. A capacitor functions as an open when the currents in the circuit have reached their steady-state values.

Problems

GROUP A

15.1 For the emf-resistor combination shown in Figure 15.21, calculate the potential difference, $V_B - V_A$.

Figure 15.21   Problem 15.1

15.2 If $i = 1.6$ A in the branch shown in Figure 15.22, calculate the potential difference, $V_B - V_A$.

Figure 15.22   Problem 15.2

15.3 Suppose $V_B - V_A = 8.0$ V for the network shown in Figure 15.23. Calculate the current in this branch.

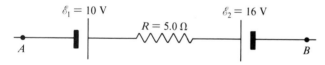

Figure 15.23   Problem 15.3

15.4 At what rate is heat being produced in the circuit (*see* Fig. 15.24)?

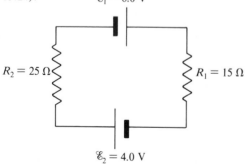

Figure 15.24   Problem 15.4

15.5 A 50-Ω resistor is rated at 100 W, which means that it is designed to dissipate no more than 100 W.

   (a)   What is the maximum current the resistor is designed to carry?

   (b)   For what maximum difference in potential across its terminals is the resistor designed?

15.6 A light bulb is rated for 100 W at 12 V.

   (a)   What current is the bulb designed to carry?

   (b)   What is the resistance of the bulb when it is operating according to its design?

15.7 For the circuit shown in Figure 15.25, calculate

   (a)   the power delivered by the emf.

   (b)   the potential difference across the 12-Ω resistor.

   (c)   the power dissipated in the 18-Ω resistor.

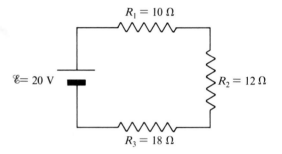

Figure 15.25   Problem 15.7

15.8 A branch of a circuit is depicted in Figure 15.26. At what rate is the remainder of the circuit, which is not shown, removing energy from or supplying energy to this branch?

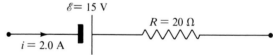

Figure 15.26   Problem 15.8

15.9 If $i = 2.0$ A in the network (*see* Fig. 15.27), calculate the potential difference, $V_B - V_A$.

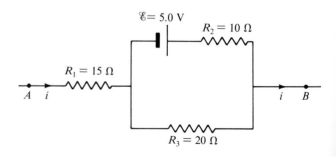

Figure 15.27   Problem 15.9

15.10 How many time constants must elapse for a capacitor in an RC circuit to become charged to within 1 percent of its final value?

15.11 At an instant when $i = 0.25$ mA in the circuit shown in Figure 15.28, determine the charge on the 15-μF capacitor. (*Hint:* This can be done using the loop rule.)

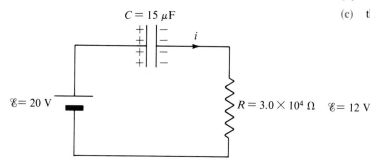

Figure 15.28   Problem 15.11

15.12 A 120-V emf is in a circuit fused for a maximum current of 30 A.

(a) How many 100-W light bulbs, connected in parallel, can be operated in this circuit?

(b) How many 100-W bulbs can be used when connected in parallel with a 1500-W toaster?

15.13 If $V_B - V_A = 90$ V for the branch in Figure 15.29, determine the charge and polarity of the capacitor.

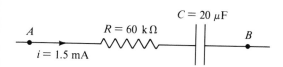

Figure 15.29   Problem 15.13

15.14 If $i = 1.4$ mA and $Q = 6.0 \times 10^{-4}$ C for the branch in Figure 15.30, determine

(a) the potential difference, $V_B - V_A$.

(b) the rate at which the remainder (not shown) of the circuit is removing energy from or supplying energy to this branch.

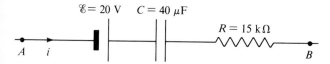

Figure 15.30   Problem 15.14

15.15 A 75-μF capacitor, initially charged to a potential difference of 80 V, is discharged through a 40-kΩ resistor.

(a) What is the initial current?

(b) How much time elapses after the start of the discharge before the current is equal to 0.80 mA?

15.16 In the circuit shown in Figure 15.31, calculate

(a) the current in the emf.

(b) the current in the 30-Ω resistor.

(c) the power dissipated in the 20-Ω resistor.

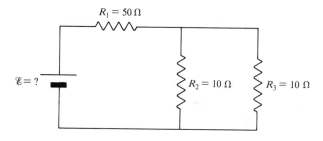

Figure 15.31   Problem 15.16

15.17 What is the potential difference across the emf $\mathscr{E}$ in Figure 15.32 if the 5.0-Ω resistor is dissipating energy at a rate of 20 W?

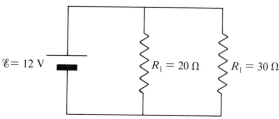

Figure 15.32   Problem 15.17

15.18 An ammeter having a full-scale reading of 10 mA is composed of a galvanometer having a resistance of 150 Ω and a shunt resistor R_s. The galvanometer deflects to full scale when it carries a current of 12 μA.

(a) What is the resistance of R_s?

(b) What is the potential difference across the ammeter when the ammeter is indicating 8.0 mA?

15.19 A voltmeter deflects to a full-scale reading of 10 V when its 150-Ω galvanometer carries a current of 10 μA.

(a) What is the resistance R_m of the multiplier resistor in the voltmeter?

(b) What is the voltmeter current when it is reading 4.0 V?

15.20 Figure 15.33 shows the internal resistors of a voltmeter that has full-scale readings of 5.0 V, 10 V, and 100 V. If the resistance of the galvanometer is 100 Ω and the galvanometer deflects to full scale when it carries a current of 0.10 mA, calculate the resistances of R_1, R_2, and R_3.

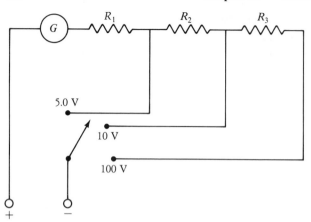

Figure 15.33   Problem 15.20

15.21 If $i = 2.0$ A in the network shown in Figure 15.34, calculate the rate at which the 10-V emf is supplying energy.

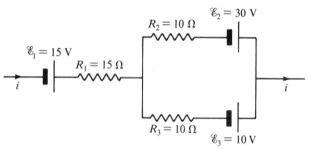

Figure 15.34   Problem 15.21

15.22 The 1.0-Ω resistor in the circuit shown in Figure 15.35 dissipates energy at a rate of 4.0 W. What power is delivered by the emfs, $\mathscr{E}_1$ and $\mathscr{E}_2$?

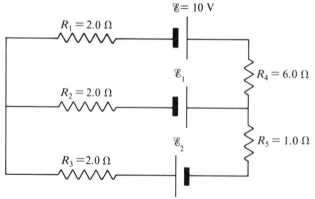

Figure 15.35   Problem 15.22

15.23 For the circuit in Figure 15.36 calculate

(a)   the current in the emf.

(b)   the potential difference across the 20-Ω resistor.

(c)   the power dissipated in the 30-Ω resistor.

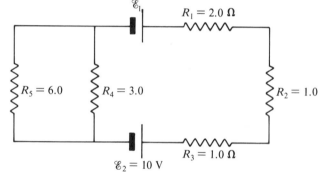

Figure 15.36   Problem 15.23

15.24 What potential difference is across the terminals of $\mathscr{E}_1$ if the 3.0-Ω resistor is dissipating 4/3 W in the circuit (*see* Fig. 15.37)?

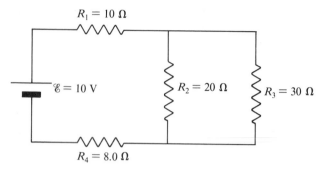

Figure 15.37   Problem 15.24

15.25 In the circuit shown in Figure 15.38, calculate

(a)   the power delivered to the resistors by the two emfs.

(b)   the potential difference across the 4.0-Ω resistor.

(c)   the power dissipated in the 6.0-Ω resistor.

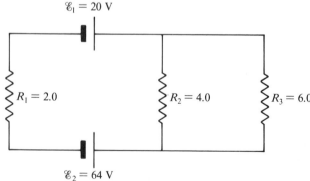

Figure 15.38   Problem 15.25

Check your solutions by comparing the power delivered by the emfs to the power dissipated in all the resistors.

15.26 What resistor R must be added in parallel with the 10-Ω resistor in the circuit (*see* Fig. 15.39) in order that the 10-V emf carry a current of 0.30 A?

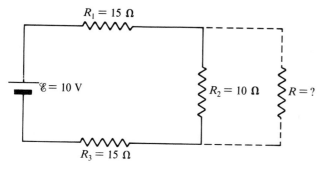

Figure 15.39   Problem 15.26

15.27 Find the power dissipated by the 1.0-Ω resistor in the circuit shown in Figure 15.40.

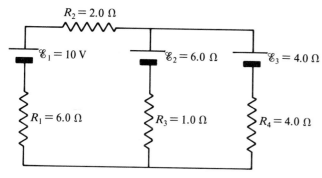

Figure 15.40   Problem 15.27

15.28 In the circuit shown in Figure 15.41, calculate
(a) the current in each resistor.
(b) the potential difference, $V_A - V_B$.

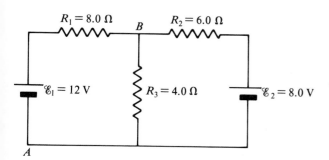

Figure 15.41   Problem 15.28

15.29 For the circuit shown in Figure 15.42, determine
(a) the current in the 15-Ω resistor.
(b) the rate at which heat is being generated in the circuit.

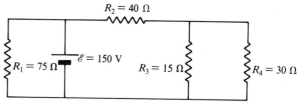

Figure 15.42   Problem 15.29

15.30 For the circuit shown in Figure 15.43, calculate
(a) the current through each resistor.
(b) the power dissipated in the 2.0-Ω resistor.
(c) the rate at which energy is delivered to the remainder of the circuit by the 8.0-V emf.

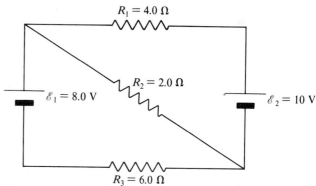

Figure 15.43   Problem 15.30

15.31 A portion of a circuit is shown in Figure 15.44. At the instant depicted, determine the rate at which the charge Q is changing.

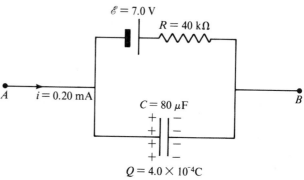

Figure 15.44   Problem 15.31

15.32 In an RC circuit like that of Figure 15.17, $R = 100\ \Omega$, $C = 10\ \mu F$, and $\mathscr{E} = 12$ V.

(a) How long after the switch is closed does the initial current in the emf reach half its initial value?

(b) What is the potential difference across the resistor when the current becomes half its initial value?

15.33 At an instant when the current through the 15-kΩ resistor (*see* Fig. 15.45) is 1.2 mA, determine

(a) the rate at which the emf is supplying energy.

(b) the charge on the 20-μF capacitor.

(c) the rate at which the energy stored in the capacitor is changing.

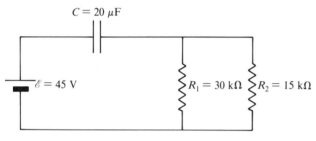

Figure 15.45   Problem 15.33

15.34 At $t = 0$ when the switch S is closed (*see* Fig. 15.46), the charge on the capacitor is equal to zero. When is energy being stored in the capacitor at a maximum rate? What is this maximum rate?

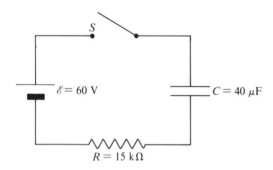

Figure 15.46   Problem 15.34

15.35 When the switch S is closed (*see* Fig. 15.47), the potential difference across the capacitor is equal to zero. What is

(a) the current in the 30-kΩ resistor immediately after S is closed?

(b) the steady-state potential difference across the capacitor?

(c) the steady-state current in the emf?

15.36 At an instant when the current i is equal to 3.0 mA (*see* Fig. 15.48), determine

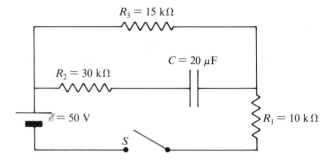

Figure 15.47   Problem 15.35

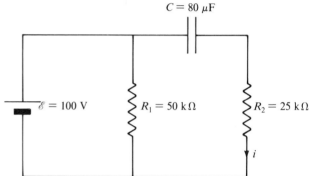

Figure 15.48   Problem 15.36

(a) the rate at which the 100-V emf is supplying energy.

(b) the energy stored by the 80-μF capacitor.

(c) the rate at which the energy stored by the capacitor is changing.

15.37 If the current $i = 0.50$ A in the network shown in Figure 15.49,

(a) calculate the current (magnitude and sense) in $\mathscr{E}_3$.

(b) calculate the potential difference, $V_B - V_A$.

(c) account for the energy being supplied, absorbed, dissipated, and so on.

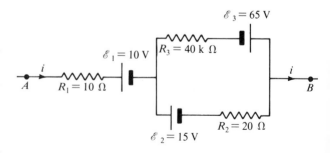

Figure 15.49   Problem 15.37

15.38 As the capacitor C in the RC circuit of Figure 15.17 is charged,

(a) how much work is done by the emf $\mathscr{E}$?

(b) what fraction of this work is stored by the capacitor?

(c) show that the energy dissipated by the resistor R accounts for the energy not stored in the capacitor.

15.39 At $t = 0$ when the switch S is closed (*see* Fig. 15.50), the capacitors are uncharged. Calculate (in terms of $\mathscr{E}$, C, and R)

(a) the initial current through $\mathscr{E}$.

(b) the steady-state current through $\mathscr{E}$.

(c) the energy stored by the capacitors in the steady-state condition.

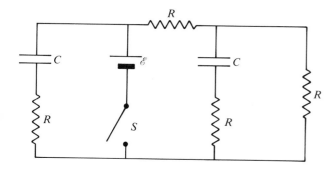

Figure 15.50 Problem 15.39

15.40 In the circuit in Figure 15.51,

(a) find the difference in potential across the open switch S.

(b) what is the charge on the 2.0-μF capacitor when the switch is open?

(c) when the switch has been closed for a long time, what is the potential difference across the 2.0-Ω resistor?

(d) what is the change in the charge on the 2.0-μF capacitor that results from closing the switch S?

(e) what change in the charge on the 4.0-μF capacitor results from closing the switch?

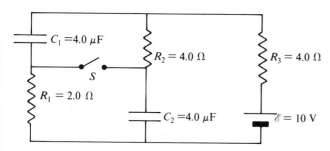

Figure 15.51 Problem 15.40

GROUP B

15.41 Determine the three currents shown in Figure 15.52 and account for the circuit energetics.

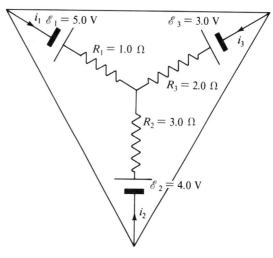

Figure 15.52 Problem 15.41

15.42 Determine the four currents shown in Figure 15.53 and account for the circuit energetics.

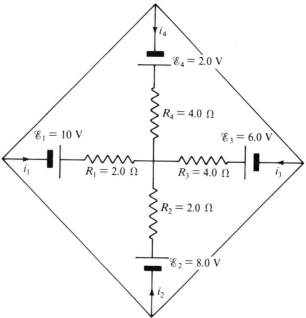

Figure 15.53 Problem 15.42

15.43 For the circuit shown in Figure 15.54, determine (in terms of $\mathcal{E}$ and R)

  (a) the current carried by R_3.

  (b) the equivalent resistance for the network of five resistors.

  (c) a value for R_4 for which the current in R_3 is equal to zero.

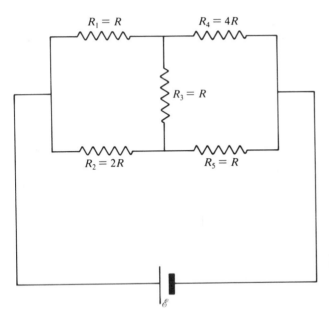

Figure 15.54   Problem 15.43

15.44 When the switch S is closed, at $t = 0$ (see Fig. 15.55), the capacitor C_2 is uncharged and the capacitor C_1 has a potential difference of 40 V across it. Calculate the initial ($t = 0$) and steady-state ($t \to \infty$) values for

  (a) the current carried by the emf.

  (b) the potential difference across C_2.

  (c) the potential difference across C_1.

  (d) the potential difference across R_1.

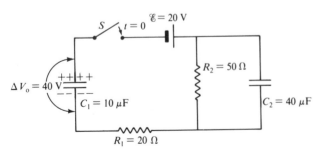

Figure 15.55   Problem 15.44

15.45 At $t = 0$ when the switch S is closed (see Fig. 15.56), the capacitor C_2 is uncharged and the potential difference across capacitor C_1 is equal to V_0.

  (a) Show that the charge on C_1 is given by

$$Q_1(t) = \left(\frac{C_1 + C_2\, e^{-t/\tau}}{C_1 + C_2} \right) C_1 V_0 \qquad (t \ge 0)$$

  where the time constant $\tau = RC_1C_2/(C_1 + C_2)$.

  (b) Show that the charge on C_2 is given by

$$Q_2(t) = \frac{C_1 C_2 V_0}{C_1 + C_2} (1 - e^{-t/\tau}) \qquad (t \ge 0)$$

  (c) If $C_1 = 15\ \mu\text{F}$, $C_2 = 30\ \mu\text{F}$, $R = 0.20\ \text{M}\Omega$, and $V_0 = 60$ V, sketch graphs of ΔV_1 and ΔV_2, the potential differences across C_1 and C_2, versus time. Be sure to show the $t = 0$, $t = \tau$, and $t \to \infty$ values.

  (d) Using the values given in part (c), account for the circuit energetics.

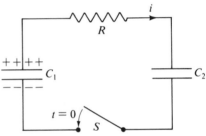

Figure 15.56   Problem 15.45

15.46 At $t = 0$ when the switch S is closed (see Fig. 15.57), the potential difference across C has the polarity shown and a magnitude given by $\Delta V_0 = 2\mathcal{E}$.

  (a) Show that the ($t \ge 0$) current i shown is given by

$$i(t) = \frac{3\mathcal{E}}{R} e^{-t/RC}$$

  (b) Account for the circuit energetics.

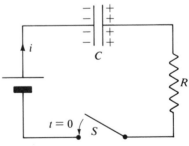

Figure 15.57   Problem 15.46

15.47 At $t = 0$ when the switch S is closed (*see* Fig. 15.58), the potential difference across the capacitor C is equal to $2\mathcal{E}$ and has the polarity shown.

(a) Show that the ($t \geq 0$) current i shown is given by

$$i(t) = \frac{2\mathcal{E}}{R} e^{-t/RC}$$

(b) Show that the potential difference ΔV_C across the capacitor is given by

$$\Delta V_C = \mathcal{E}(1 + 2e^{-t/RC}) \qquad (t \geq 0)$$

(c) Sketch a graph of ΔV_C versus time being sure to show the $t = 0$, $t = RC$, and $t \to \infty$ values.

(d) Account for the energy discharged by the capacitor.

15.48 At $t = 0$ when the switch S is closed (*see* Fig. 15.59), the capacitor C is uncharged. If i_1 and i_2 are the ($t \geq 0$) currents shown,

(a) show that $i_1 = \frac{1}{2}\mathcal{E}/R + \frac{1}{2}i_2$

(b) show that $di_2/dt = -2i_2/(RC)$

(c) show that $i_2(t) = (\mathcal{E}/R)e^{-2t/RC}$ ($t \geq 0$)

(d) show that $i_1(t) = (\mathcal{E}/2R)(1 + e^{-2t/RC})$ ($t \geq 0$)

(e) show that the potential difference ΔV_C across C is given by

$$\Delta V_C = \frac{\mathcal{E}}{2}(1 - e^{-2t/RC}) \qquad (t \geq 0)$$

(f) sketch i_1, i_2, and ΔV_C versus t, showing their values at $t = 0$, $t = RC$, and $t \to \infty$.

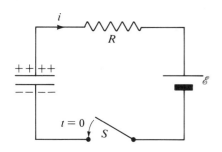

Figure 15.58   Problem 15.47

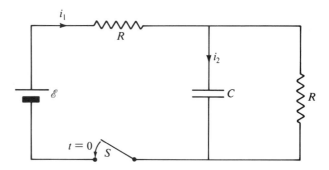

Figure 15.59   Problem 15.48

16 Magnetic Fields I

Magnetism refers to a group of phenomena that results from forces that act on charges because of the motion of those charges. These magnetic forces exerted on moving charges are usually described in terms of *magnetic fields* in a way that is similar to that of describing forces exerted on static charges in terms of electric fields. Thus, electric fields have their origins in static charges, and magnetic fields originate within moving charges.

We will describe magnetic forces in terms of fields by considering:

1. How the presence of a magnetic field causes force to be exerted on moving charges.

2. How magnetic fields are themselves produced by moving charges.

16.1 Magnetic Forces on Moving Charges

Just as the electric field **E** is defined at a point in terms of the force $\mathbf{F}_e = q\mathbf{E}$ on a static charge q at that point, we may define the magnetic field **B** at a point in terms of a force relationship between the magnetic field and a moving charge. The *magnetic field* **B** is a vector quantity defined at a point in space by

$$\mathbf{F}_m = q\mathbf{v} \times \mathbf{B} \tag{16-1}$$

where q is the charge on a particle located at that point and moving with velocity **v**. The force $\mathbf{F}_m$ on the particle does not include any forces that would act on the particle if it were at rest, for example, electrical or gravitational forces. The defining equation, Equation (16-1), for the magnetic field **B** contains a vector cross product (see Chapter 1), which is an elegant and concise way to express the following physical facts relating the vector quantities $\mathbf{F}_m$, **v**, and **B**:

1. The force $\mathbf{F}_m$ is directed along a line perpendicular to a plane containing **v** and **B**. The direction of $\mathbf{v} \times \mathbf{B}$ is determined by the right-hand rule illustrated in Figure 16.1. When **v** and **B** are represented by vectors originating at a common point, $\mathbf{v} \times \mathbf{B}$ (and, therefore, the force $\mathbf{F}_m$ on a positive charge q) is in the direction of the extended thumb of the right hand when the fingers of that hand are curled to indicate a turning sense from **v** toward **B** through θ, the smaller angle between **v** and **B**.

2. The magnitude of $\mathbf{F}_m$ has a value equal to $|q|vB \sin\theta$.

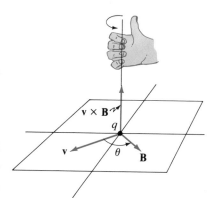

Figure 16.1
Illustrating the right-hand rule for determining the direction of the cross product $\mathbf{v} \times \mathbf{B}$.

The first of these facts means that any magnetic force acting on a moving charged particle must be at all times perpendicular to the velocity of that particle. Consequently, the magnetic force on a charged particle can do no work on that particle, and the kinetic energy of the particle does not change as a result of the interaction with the magnetic field. The second of these facts means that the magnitude of the magnetic force on a moving charged particle is directly proportional to q, v, and B: Increasing the charge or speed of the particle or increasing the magnitude of the magnetic field causes a proportionate increase in the magnitude of the magnetic force on the particle.

The simple name, magnetic field, is sometimes called by more impressive names—*magnetic induction, magnetic flux density,* and *magnetic field induction.* We shall not do so. The SI unit of magnetic field is the tesla (T), which, from Equation (16-1) may be expressed in terms of more fundamental units as

$$1\ \text{T} \equiv \frac{\text{N}}{\text{C} \cdot \text{m/s}} = \frac{\text{N}}{\text{A} \cdot \text{m}} = \frac{\text{kg}}{\text{A} \cdot \text{s}^2} \qquad (16\text{-}2)$$

Another traditional unit of magnetic field that is still frequently used is the gauss (G), which is related to the tesla by

$$1\ \text{G} \equiv 10^{-4}\ \text{T} \qquad (16\text{-}3)$$

Consider the motion of a charged particle in a region of constant magnetic field $\mathbf{B}$, that is, in a region where $\mathbf{B}$ has the same magnitude and direction at every point. The analysis of the motion is perhaps simplified by considering first a particle moving with velocity $\mathbf{v}_{\parallel}$, which is parallel to the uniform magnetic field $\mathbf{B}$. The product $\mathbf{v}_{\parallel} \times \mathbf{B}$ is equal to zero, and the magnetic field produces no force on the particle. The velocity of the particle is unaffected by the field.

Now consider a particle with a positive charge q and mass m moving with a velocity $\mathbf{v}_{\perp}$ that is perpendicular to a uniform magnetic field $\mathbf{B}$. A force equal to $q\mathbf{v}_{\perp} \times \mathbf{B}$ acts on the particle, and that force is perpendicular to both $\mathbf{v}_{\perp}$ and $\mathbf{B}$. The particle, therefore, experiences an acceleration perpendicular to $\mathbf{v}_{\perp}$, and that acceleration turns the velocity vector $\mathbf{v}_{\perp}$ (in a plane perpendicular to $\mathbf{B}$, as shown in Figure 16.2) without changing the speed of the particle. The particle moves in a circular path in a plane perpendicular to $\mathbf{B}$. The magnetic force $\mathbf{F}_m = q\mathbf{v}_{\perp} \times \mathbf{B}$ on the particle has a constant magnitude

$$\text{F}_m = q\text{v}_{\perp}\text{B} \qquad (16\text{-}4)$$

and, as Figure 16.2 indicates, the magnetic force $\mathbf{F}_m$ on the particle is perpendicular to both $\mathbf{v}_{\perp}$ and $\mathbf{B}$ at every point in the part of the particle. The acceleration of a particle moving in a circle is centripetal and has magnitude

$$a = \frac{F_m}{m} = \frac{q\text{v}_{\perp}B}{m} \qquad (16\text{-}5)$$

Because the centripetal acceleration of a particle moving in a circular path at constant speed $v_{\perp}$ is equal to $v_{\perp}^2/r$, where r is the radius of the circular path, we may write

$$a = \frac{v_{\perp}^2}{r} = \frac{q\text{v}_{\perp}B}{m}$$

$$r = \frac{m\text{v}_{\perp}}{qB} \qquad (16\text{-}6)$$

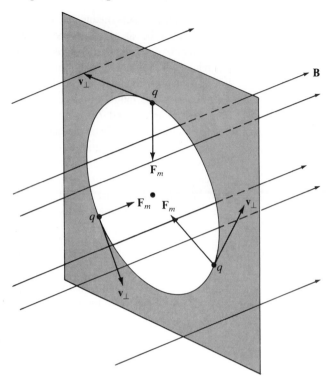

Figure 16.2
A positively charged particle moving with a velocity $\mathbf{v}_\perp$ that is perpendicular to a uniform magnetic field $\mathbf{B}$. The particle moves in a circular path.

E 16.1 Verify that the units of Equation (16-6) are correct.

E 16.2 An electron ($m = 9.1 \times 10^{-31}$ kg, $q = -1.6 \times 10^{-19}$ C) moves with a speed of 6.0×10^6 m/s in a direction that is perpendicular to a uniform magnetic field having a magnitude of 0.20 T. Calculate
(a) the magnitude of the magnetic force on the electron,
(b) the magnitude of the acceleration of the electron, and
(c) the radius of the circular trajectory of the electron.
Answers: (a) 1.9×10^{-13} N; (b) 2.1×10^{17} m/s²; (c) 0.17 mm

E 16.3 A proton ($m = 1.7 \times 10^{-27}$ kg, $q = 1.6 \times 10^{-19}$ C) moving with a speed of 4.2×10^5 m/s follows a circular path having a radius of 3.4 cm. Calculate the magnitude of the uniform magnetic field acting on the proton.
Answer: 0.13 T

A particle moving in a circular path is in periodic motion, and the period T of that motion is the time required to complete one cycle, or one revolution, of its motion. Because the distance the particle moves in one revolution is equal to $2\pi r$, where r is given by Equation (16-6), the period T of the circular motion of the particle is

$$T = \frac{2\pi r}{v_\perp} = \frac{2\pi m}{qB} \qquad (16\text{-}7)$$

The frequency $\nu = 1/T$ of the periodic motion is given by

$$\nu = \frac{qB}{2\pi m} \qquad (16\text{-}8)$$

This frequency, which is independent of the speed of the particle, is sometimes called the *cyclotron frequency*. A cyclotron is a device that makes use of the cyclic effect discussed here to accelerate charged particles along circular paths.

E 16.4   An alpha particle ($m = 6.7 \times 10^{-27}$ kg, $q = 3.2 \times 10^{-19}$ C) moves with a speed of 5.0×10^5 m/s along a circular path in a uniform magnetic field having a magnitude of 45 G. What is the cyclotron frequency?

Answer: 34 kHz

E 16.5   How is the result of Exercise 16.4 changed if the speed of the alpha particle is doubled?   Answer: No change

We can express an arbitrary velocity **v** of a charged particle moving in a uniform magnetic field in terms of the component vectors, $\mathbf{v}_\parallel$ and $\mathbf{v}_\perp$, that are parallel and perpendicular to **B**:

$$\mathbf{v} = \mathbf{v}_\parallel + \mathbf{v}_\perp \qquad (16\text{-}9)$$

We have seen that neither of the magnitudes of these velocity components is changed by the presence of a uniform magnetic field. (The velocity component vector $\mathbf{v}_\parallel$ changes direction, but its magnitude remains constant.) Because the speed, $v = \sqrt{v_\parallel^2 + v_\perp^2}$, of the particle does not change, *the presence of a uniform magnetic field cannot change the kinetic energy of a charged particle.*

E 16.6   A charged particle ($m = 7.0 \times 10^{-8}$ kg, $q = 2.8$ μC) has a velocity of $\mathbf{v} = (3.0\hat{\mathbf{i}} + 5.0\hat{\mathbf{k}}) \times 10^4$ m/s at a point where $\mathbf{B} = 0.040\,\hat{\mathbf{k}}$ T. At this instant, what
(a) is the magnetic force acting on the particle?
(b) are $\mathbf{v}_\perp$ and $\mathbf{v}_\parallel$?
Answers: (a) $-3.4\hat{\mathbf{j}} \times 10^{-3}$ N; (b) $3.0\hat{\mathbf{i}} \times 10^5$ m/s, $5.0\hat{\mathbf{k}} \times 10^5$ m/s

E 16.7   A charged particle ($q = 6.5$ μC) has a speed of 5.0×10^4 m/s in a uniform magnetic field having a magnitude of 75 G. The angle between the velocity of the particle and B is equal to 55°. Determine the magnitudes of $\mathbf{v}_\perp$, $\mathbf{v}_\parallel$, and the magnetic force acting on the particle.
Answer: 4.1×10^4 m/s, 2.9×10^4 m/s, 2.0×10^{-3} N

Consider now a charged particle moving in an arbitrary direction with velocity **v** in a region of uniform magnetic field **B**. The motion of the particle is perhaps most easily understood by considering separately the effect of the magnetic force on each component of the velocity $\mathbf{v} = \mathbf{v}_\parallel + \mathbf{v}_\perp$. This magnetic force causes the particle to circulate perpendicularly to **B** along a path of radius r, given by Equation (16-6). Because there is no magnetic force parallel to **B**, the particle proceeds in a direction parallel to **B** at a constant speed $v_\parallel$. The superposition of these two motions yields a trajectory that is a helix, a corkscrewlike path similar to the handrail of a spiral staircase. Figure 16.3 shows the helical path of a positively charged particle with velocity **v**, which has components both parallel and perpendicular to a uniform magnetic field **B**. The *pitch p* of the helix is the distance that the particle moves parallel to **B** during the time it takes to complete one turn, that is, the time equal to one period T. Then using Equation (16-7), we may write

$$p = v_\parallel T = \frac{2\pi m v_\parallel}{qB} \qquad (16\text{-}10)$$

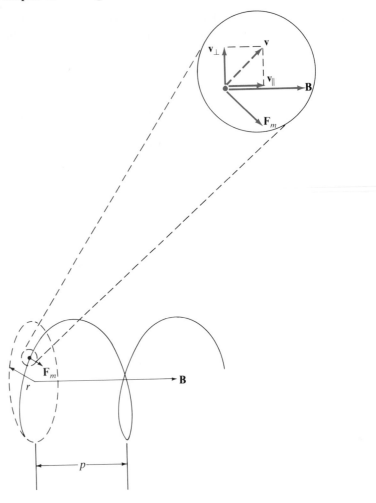

Figure 16.3
The helical path of a positively charged particle having velocity **v**, which has components parallel and perpendicular to a uniform magnetic field **B**.

E 16.8   A deuteron ($m = 3.3 \times 10^{-27}$ kg, $q = 1.6 \times 10^{-19}$ C) moves with a speed of 7.0×10^4 m/s in a uniform magnetic field having a magnitude of 85 G. If the angle between the velocity of the deuteron and the magnetic field is equal to 70°, calculate the radius and pitch of the helical path of the particle.

Answer: $r = 16$ cm, $p = 36$ cm

Example 16.1

PROBLEM   A charged particle ($m = 1.2 \times 10^{-9}$ kg, $q = 6.0$ μC) moves in a uniform magnetic field given by $\mathbf{B} = 0.040\,\hat{\mathbf{k}}$ T. At $t = 0$ the position and velocity of the particle are $\mathbf{r}_o = \mathbf{0}$ and $\mathbf{v}_o = (20\hat{\mathbf{j}} + 15\hat{\mathbf{k}})$ m/s. Determine the position of the particle at $t = (\pi/400)$ s.

SOLUTION   As we have seen, the particle follows a helical path that may be thought of as a superposition of a constant-velocity ($15\,\hat{\mathbf{k}}$ m/s) motion parallel to **B** and a constant-speed (20 m/s) circular motion in a plane (x-y) perpendicular to **B**. The $t = 0$ acceleration of the particle, namely

$$\mathbf{a}_o = \frac{q\mathbf{v}_o \times \mathbf{B}}{m} = 4000\hat{\mathbf{i}} \text{ m/s}^2$$

is shown in Figure 16.4, which depicts the motion in the x-y plane. Because $\mathbf{a}_o$ is

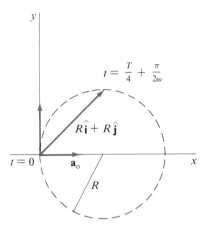

Figure 16.4   Example 16.1

a centripetal acceleration, it must point toward the center of the circle that has a radius R determined from the expression $a = v^2/R$, or

$$R = \frac{v^2}{a} = \frac{(20 \text{ m/s})^2}{4000 \text{ m/s}^2} = 0.10 \text{ m}$$

Figure 16.4 shows this circular motion. The time for one cycle around the circle, that is, the period T, is given by

$$T = \frac{2\pi R}{v} = \frac{2\pi(0.10 \text{ m})}{20 \text{ m/s}} = \frac{\pi}{100} \text{ s}$$

Thus, the time $t = (\pi/400)$ s is equal to one-fourth of this period, and the particle has, therefore, moved one-fourth of the distance around the circle, as shown in Figure 16.4. During this time the particle has also moved (in the positive z direction) a distance equal to $v_z t$ or

$$z = 15 \frac{m}{s} \cdot \frac{\pi}{400} \text{ s} = \frac{3\pi}{80}\text{m} \approx 0.12 \text{ m}$$

Consequently, the position of the particle at the time specified is

$$\mathbf{r} = R\hat{\mathbf{i}} + R\hat{\mathbf{j}} + z\hat{\mathbf{k}} = (0.10\hat{\mathbf{i}} + 0.10\hat{\mathbf{j}} + 0.12\hat{\mathbf{k}}) \text{ m} \quad \blacksquare$$

E 16.9   Consider the helical motion of Example 16.1. Determine the velocity of the particle at the time, $t = (\pi/400)$ s specified in the problem.

Answer: $(20\hat{\mathbf{i}} + 15\hat{\mathbf{k}})$ m/s

A charged particle may be in a region where both an electric field $\mathbf{E}$ and a magnetic field $\mathbf{B}$ are present simultaneously. In that case, the total electromagnetic force $\mathbf{F}$ on a particle having charge q and moving with velocity $\mathbf{v}$ is equal to the vector sum of the electrical force $\mathbf{F}_e = q\mathbf{E}$ and the magnetic force $\mathbf{F}_m = q\mathbf{v} \times \mathbf{B}$, or

$$\mathbf{F} = q(\mathbf{E} + \mathbf{v} \times \mathbf{B}) \tag{16-11}$$

Equation (16-11) is known as the *Lorentz force law*.

E 16.10   At a point where $\mathbf{E} = 2.5\hat{\mathbf{i}} \times 10^3$ V/m and $\mathbf{B} = 0.20\hat{\mathbf{j}}$ T, a charged particle ($q = 4.0$ μC) has a velocity $\mathbf{v} = 5.0\hat{\mathbf{k}} \times 10^4$ m/s. Determine the force acting on the particle.   Answer: $0.0060\hat{\mathbf{i}}$ N

The following problem illustrates a practical use of *crossed electric and magnetic fields* in a region of space.

Electrons that have a wide range of speeds are moving in the positive x direction where they pass into an evacuated region between two flat charged conducting plates, as shown in Figure 16.5. In that region is a uniform magnetic field given by $\mathbf{B} = -0.010\hat{\mathbf{k}}$ T. As the electrons exit the region through a slit S on the x axis, it is observed that only electrons having a velocity $v_s = 1.0 \times 10^5\hat{\mathbf{i}}$ m/s have negotiated the path along the x axis through the plates. If the plates of this *velocity selector* are separated by 1.0 cm in the y direction, determine the difference in potential between the plates. What is the fate of electrons with velocities other than v_s?

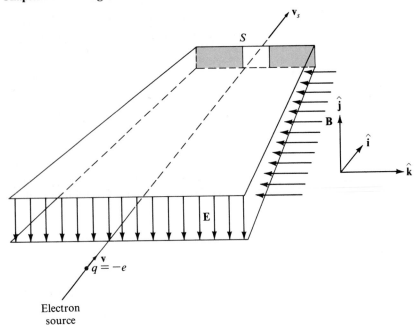

Figure 16.5
Electrons with a range of speeds entering an evacuated space between two flat
charged conducting plates. The fields
between the plates are **E** and **B**. This
arrangement of plates is called a velocity selector.

Because those electrons that pass from source to slit through the velocity
selector are undeviated in the presence of a magnetic field, there must exist a
force caused by an electric field in order that the net force on each electron with
velocity $\mathbf{v}_s$ is equal to zero:

$$\mathbf{F} = -e(\mathbf{E} + \mathbf{v}_s \times \mathbf{B}) = 0$$

$$\mathbf{E} = -\mathbf{v}_s \times \mathbf{B} \tag{16-12}$$

The vector product $\mathbf{v}_s \times \mathbf{B}$ is in the positive y direction, so the direction of **E** is in
the negative y direction. The upper plate, therefore, is at the higher potential, and
the potential difference ΔV between the plates separated by a distance d is related
to the magnitude of **E** by

$$E = \frac{\Delta V}{d} \tag{16-13}$$

From Equation (16-12), the magnitude of **E** is equal to $v_s B$, because $\mathbf{v}_s$ and **B** are
mutually perpendicular. The E is given by

$$E = v_s B = (1.0 \times 10^5 \text{ m/s})(1.0 \times 10^{-2} \text{ T}) = 1000 \text{ V/m}$$

and the difference in potential ΔV between the plates is obtained from Equation
(16-13) to be

$$\Delta V = Ed = (1.0 \times 10^3 \text{ V/m})(1.0 \times 10^{-2} \text{ m}) = 10 \text{ V} \tag{16-14}$$

Figure 16.6(a) shows a side view (looking in the negative z direction) of the
velocity selector when an electron with speed $\mathbf{v}_s$ is passing through it. The
symbol $\otimes$ indicates a vector quantity, **B** in this case, directed into the page; you
may wish to think of the symbol as the tail feathers of an arrow moving away
from you. (The symbol $\odot$ is used to indicate a vector quantity directed out of the
page—the tip of an arrow moving toward you.) The electric and magnetic
forces, $\mathbf{F}_e = -e\mathbf{E}$ and $\mathbf{F}_m = -e\mathbf{v} \times \mathbf{B}$, on the electron are indicated in

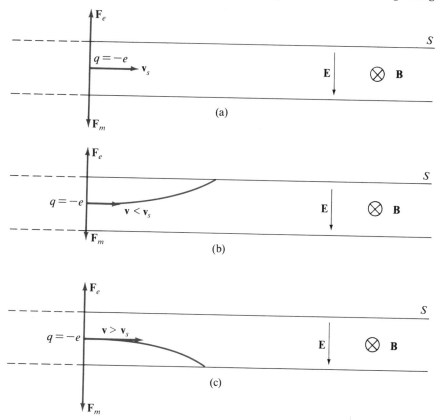

Figure 16.6
Side views of a velocity selector with crossed fields, **E** and **B**. **(a)** An electron which has speed v_s is passing undeviated through the selector when $\mathbf{F}_e = -\mathbf{F}_m$. **(b)** An electron with speed $v < v_s$ has a net force, $\mathbf{F}_e + \mathbf{F}_m$, on it that bends its path upward. **(c)** An electron with speed $v > v_s$ has a net force, $\mathbf{F}_m + \mathbf{F}_e$, on it that bends its path downward.

appropriate directions. For electrons having a speed $v_s = E/B$, the magnitudes of $\mathbf{F}_e$ and $\mathbf{F}_m$ are equal, and these electrons proceed undeviated through the velocity selector.

But what happens to those electrons having speeds other than v_s? Figure 16.6(b) shows the path of an electron having a speed $v < v_s$. The magnetic force on this electron has magnitude $F_m = evB < ev_sB = F_e$. Then the electric force $\mathbf{F}_e$, which is independent of the speed of the electron, causes this electron to strike the upper plate. Similarly, if an electron has a speed of $v > v_s$, as indicated in Figure 16.6(c), the magnitude F_m of the magnetic force is greater than the magnitude F_e of the electric force, and the electron is deflected so that it strikes the lower plate.

Just as a single charge moving in a magnetic field experiences a magnetic force, so a collection of charges moving within a conducting wire, for example, in a magnetic field, experiences a force that may be observed as a force on the wire. Suppose a thin wire of length l is carrying a current i in a region where the magnetic field is **B**. If the wire is sufficiently short and thin, we may assume that **B** is the same at every point within the wire. If we sum the magnetic forces on the moving charges, then the magnetic force $\mathbf{F}_m$ on the wire is given by

$$\mathbf{F}_m = i\mathbf{l} \times \mathbf{B} \tag{16-15}$$

where **l** is a vector that has a magnitude equal to the length l of the wire and is in the direction of the current arrow (the direction of the flow of positive charge) in the wire. Thus Equation (16-15) follows from Equation (16-1), $\mathbf{F}_m = q\mathbf{v} \times \mathbf{B}$; either equation may be used as a basis for a measurement of the magnetic field **B**.

E 16.11   A 12-cm segment of wire carries a current of 8.0 A in the positive x direction in a region where **B** is given by $0.25\hat{\mathbf{j}}$ T. Calculate the magnetic force on this wire segment.   Answer: $0.24\hat{\mathbf{k}}$ N

Each infinitesimal segment **dl** of a current-carrying wire in a magnetic field **B** experiences a magnetic force $d\mathbf{F}_m$ given by

$$d\mathbf{F}_m = i\ d\mathbf{l} \times \mathbf{B} \qquad (16\text{-}16)$$

where i is the current in the wire and **dl** is a vector representing an infinitesimal length of the wire in the direction of the current arrow. The quantity i **dl** is a *current element* of the wire. We may calculate the total force on a current-carrying wire by integrating Equation (16-16) over all the current elements that compose the wire:

$$\mathbf{F}_m = i \oint d\mathbf{l} \times \mathbf{B} \qquad (16\text{-}17)$$

E 16.12   Show that if **B** is constant, the net magnetic force acting on any closed current loop is equal to zero. (*Hint:* Here $\oint(d\mathbf{l} \times \mathbf{B}) = (\oint d\mathbf{l}) \times \mathbf{B}$, because **B** is constant.)

16.2   The Biot-Savart Law

To this point we have considered the effects of magnetic fields on moving charges. Now we will consider how moving charges themselves produce magnetic fields—how moving charges are the sources of magnetic fields. The law of Biot and Savart is the result of experimental observation: It relates the magnetic field at a point in space to the source of that field, a current element.

This important relationship plays the equivalent role in magnetism to the role of Coulomb's law in electricity. Consider Figure 16.7. The *Biot-Savart law* specifies the magnetic field **dB**, which is caused by a current element i **dl**, at a field point P located by a displacement vector **r** extending from the current element to the field point:

$$d\mathbf{B} = \frac{\mu_o}{4\pi} \frac{i\ d\mathbf{l} \times \mathbf{r}}{r^3} \qquad (16\text{-}18)$$

The quantity $\mu_o/4\pi$ is a constant having the precise value

$$\frac{\mu_o}{4\pi} = 10^{-7} \frac{\text{T} \cdot \text{m}}{\text{A}} \qquad (16\text{-}19)$$

The constant $\mu_o = 4\pi \times 10^{-7}$ T·m/A is called the *permeability of free space*.

Figure 16.7 illustrates the direction of the magnetic field **dB** at a field point P as specified by the Biot-Savart law. At every point on the dashed circle, the magnetic field is tangential to the circle, which is centered on the line that

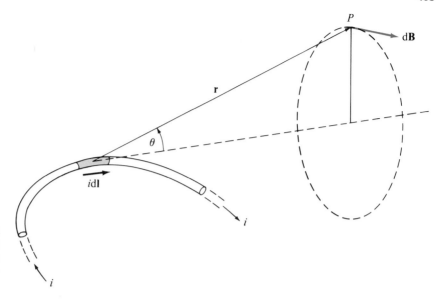

Figure 16.7
Illustrating the magnetic field d**B** at a point P, which is located relative to the current element i d**l** by the vector **r**. The angle θ is the angle between the direction of i d**l** and **r**.

contains d**l**. The sense of the direction of the field is determined by another right-hand rule, illustrated in Figure 16.8: With the right hand grasping the wire so that the thumb is extended in the direction of the current element i d**l**, the curved fingers of that hand indicate the sense of the direction of d**B** at points outside the wire. The magnitude of d**B** at any point a distance r from a current element i d**l** is obtained from Equation (16-18). Let θ be the angle between d**l** and **r** in Figure 16.7, and recall that $|d\mathbf{l} \times \mathbf{r}| = |dl| \cdot r \sin\theta$. Then the magnitude of d**B** is

$$d\mathbf{B} = \frac{\mu_o i}{4\pi} \frac{|dl| \sin\theta}{r^2} \tag{16-20}$$

Notice in Figure 16.8 and Equation (16-20) that the magnitude dB of the magnetic field caused by the current element i d**l** is equal to zero at points on the line

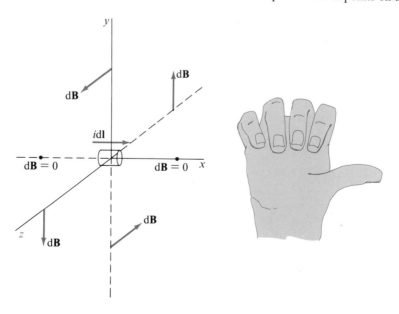

Figure 16.8
The magnetic field d**B**, caused by a current element i d**l**, at different positions around the current element.

passing through and parallel to the current element, because at all such points, $\sin\theta$ is equal to zero.

E 16.13 Consider a 1.0-cm wire segment that is centered on the origin and carries a current of 25 A in the positive x direction. Use Equation (16-18), the Biot-Savart law, with $d\mathbf{l} \approx 0.01\hat{\mathbf{i}}$ m to estimate the resulting magnetic field at (a) $\mathbf{r} = 2.0\hat{\mathbf{i}}$ m, (b) $\mathbf{r} = 2.0\hat{\mathbf{j}}$ m, and (c) $\mathbf{r} = 2.0\hat{\mathbf{k}}$ m.

Answers: (a) zero; (b) $6.3\hat{\mathbf{k}} \times 10^{-9}$ T; (c) $-6.3\hat{\mathbf{j}} \times 10^{-9}$ T

It has been verified experimentally that the resultant magnetic field at a point is equal to the vector sum of the magnetic fields at that point caused by two or more current elements. In other words, magnetic fields obey the principle of superposition, just as electric fields do. Then, just as we are able to express the total electric field **E**, at a point, caused by a distribution of charge elements by

$$\mathbf{E} = \frac{1}{4\pi\epsilon_0} \int \frac{\mathbf{r}\,dQ}{r^3}$$

we can express the total magnetic field **B**, at a point, caused by a distribution of current elements by

$$\mathbf{B} = \frac{\mu_0}{4\pi} \int \frac{i\,d\mathbf{l} \times \mathbf{r}}{r^3} \qquad (16\text{-}21)$$

Because no isolated current element exists, Equation (16-21) is the more practical form of the Biot-Savart law.

Let us see how the Biot-Savart law may be applied to find the magnetic fields caused by current-carrying wires in some simple geometries.

Calculate the magnitude of the magetic field at a perpendicular distance R from a long thin wire carrying a constant current i.

The physical situation is suggested by Figure 16.9, in which the wire carries a current i in the positive x direction along the x axis. An arbitrary current element $i\,d\mathbf{l}$ is located at a distance x from the origin and $d\mathbf{l}$ points in the positive x direction. The field point P at a distance R from the x axis is located from $d\mathbf{l}$ by

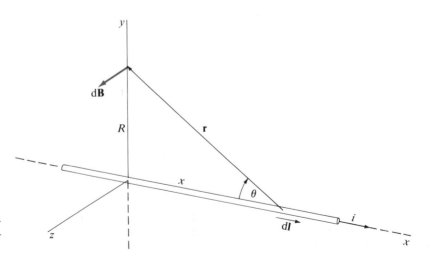

Figure 16.9
The magnetic field d**B** caused by a current element i d**l**, which is in the positive x direction.

the vector $\mathbf{r}$, which makes an angle θ with the x axis. The direction of the magnetic field $d\mathbf{B}$ at P is in the positive z direction, the direction of $d\mathbf{l} \times \mathbf{r}$. The magnitude dB is given by the Biot-Savart law,

$$dB = \frac{\mu_0}{4\pi} \frac{|i d\mathbf{l} \times \mathbf{r}|}{r^3} = \frac{\mu_0 i}{4\pi} \frac{\sin\theta \, dx}{r^2} \tag{16-22}$$

in which we have used $|d\mathbf{l}| = dx$ and $|d\mathbf{l} \times \mathbf{r}| = r \, dx \, \sin(\pi - \theta) = r \, dx \, \sin\theta$. The total magnetic field $\mathbf{B}$ at P is found by integrating Equation (16-22) throughout the entire current distribution, that is, from $x = -\infty$ to $x = \infty$:

$$B = \frac{\mu_0 i}{4\pi} \int_{x=-\infty}^{x=\infty} \frac{\sin\theta \, dx}{r^2} \tag{16-23}$$

To evaluate the integral of Equation (16-23), let us express dx and r^2 in terms of the single variable θ. From the geometry of Figure 16.9, we see that these variables are related by

$$x = R \cot\theta$$

$$dx = -R \csc^2\theta \, d\theta \tag{16-24}$$

and

$$r = R \csc\theta$$

$$r^2 = R^2 \csc^2\theta \tag{16-25}$$

Substituting Equations (16-24) and (16-25) into Equation (16-23) gives

$$B = \frac{\mu_0 i}{4\pi} \int_{\theta=\pi}^{\theta=0} \frac{\sin\theta \, (-R \csc^2\theta \, d\theta)}{R^2 \csc^2\theta} \tag{16-26}$$

The limits on θ in Equation (16-26) may be obtained by referring to Figure 16.9; as $x \to -\infty$, $\theta \to \pi$, and as $x \to \infty$, $\theta \to 0$. Simplification of Equation (16-26) yields

$$B = -\frac{\mu_0 i}{4\pi R} \int_\pi^0 \sin\theta \, d\theta$$

which may be integrated to give

$$B = \frac{\mu_0 i}{2\pi R} \tag{16-27}$$

The direction of $\mathbf{B}$ at any distance R from the straight current-carrying wire is consistent with the right-hand rule. At a point on the negative y axis in Figure 16.9, for example, the direction of $\mathbf{B}$ is in the negative z direction. Figure 16.10 indicates the *lines of magnetic field* around a wire carrying current i (out of the page). Such a representation indicates that the direction of the field at any point is tangential to the $\mathbf{B}$-field lines in the sense of the arrows shown. Further, the spacing of the lines is intended to indicate the relative magnitude of $\mathbf{B}$ at various radial distances from the wire, in accordance with Equation (16-27). The closer the spacing of the $\mathbf{B}$-field lines, the greater the magnitude of $\mathbf{B}$ in that region.

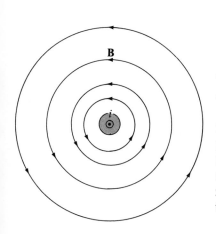

Figure 16.10
Lines of magnetic field surrounding a wire carrying current i out of the page.

E 16.14   Calculate the magnitude of the magnetic field at a distance of 1.0 cm from a long wire carrying a current of 5.0 A.

Answer: $1.0 \text{ G} = 1.0 \times 10^{-4} \text{ T}$

E 16.15   Determine the current in a long wire if the resulting magnetic field 1.0 m from the wire is to have a magnitude equal to 1.0 T. Is this a reasonable way to create a 1.0-T field?   Answer: 5.0×10^6 A

Let us use the result of the foregoing problem to (a) establish a relationship that expresses the forces acting on two long parallel wires carrying currents, and (b) provide a definition of the ampere, a definition that has been postponed until now.

Suppose two long parallel wires, labeled 1 and 2 in Figure 16.11, are separated by a distance d. The wires carry currents i_1 and i_2 in the same directional sense. At a point on wire 2, the magnitude of the magnetic field $\mathbf{B}_1$ caused by the current i_1 in wire 1 is given by Equation (16-27):

$$B_1 = \frac{\mu_0 i_1}{2\pi d} \tag{16-28}$$

The force $d\mathbf{F}_2$ on the current element $i_2 d\mathbf{l}_2$ is produced by the magnetic field $\mathbf{B}_1$. The magnitude of $d\mathbf{F}_2$ is found by using Equations (16-16) and (16-28):

$$dF_2 = i_2 |d\mathbf{l}_2 \times \mathbf{B}_1|$$

$$dF_2 = i_2 dl_2 \left(\frac{\mu_0 i_1}{2\pi d}\right) = \frac{\mu_0 i_1 i_2 dl_2}{2\pi d}$$

which, when integrated over a length L of wire 2, gives the magnitude F_2 of the force on a length L of wire 2:

$$F_2 = \frac{\mu_0 i_1 i_2 L}{2\pi d} \tag{16-29}$$

The direction of $\mathbf{F}_2$ is perpendicular to wire 2 and toward wire 1. Similarly, the force $\mathbf{F}_1$ on a length L of wire 1 has the same magnitude as $\mathbf{F}_2$ and is directed oppositely to $\mathbf{F}_2$. Thus, the wires of Figure 16.11 attract each other. A similar analysis for parallel wires carrying currents in opposite directional senses indicates that such wires repel each other.

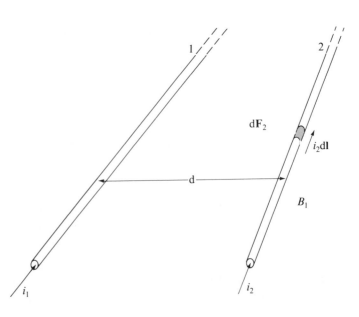

Figure 16.11
Two long parallel wires carrying currents i_1 and i_2 in the same directional sense.

E 16.16   Two parallel wires separated by a distance of 2.0 cm each carry a current of 15 A. Calculate the magnitude of the magnetic force on a 50-cm length of each wire.   Answer: 1.1×10^{-3} N

We may now define the ampere in terms of the force of Equation (16-29). An *ampere* is that quantity of current that, if carried in each of two long parallel wires separated by a distance of 1 meter, causes each meter of each wire to experience a force of exactly 2×10^{-7} N.

Because it is inconvenient to use long wires in actual measurements, the scientists at the National Bureau of Standards use parallel coils of wire carrying currents to standardize the ampere. The force on either of the coils is measured with a spring balance, and the standardization apparatus is called a *current balance*. Once the ampere has been standardized with a current balance, the coulomb is defined in terms of the ampere: The coulomb (C) is that quantity of charge that, each second, passes through a cross section of wire carrying 1 ampere.

Example 16.2   **PROBLEM**   A circular conducting loop of radius a carries a current i. Determine the resulting magnetic field on the axis of the loop at a distance x from the plane of the loop.

SOLUTION   Figure 16.12 shows the current loop and the field d**B** caused by a typical current element i d**l**. The symmetry of the current distribution assures that the nonaxial field components sum to zero. Thus, the total field is axial (in the x direction):

$$B = \int dB_x = \int \sin\theta \, dB$$

Using the Biot-Savart law to get d**B** gives

$$d\mathbf{B} = \frac{\mu_o}{4\pi} \frac{i|d\mathbf{l} \times \mathbf{r}|}{r^3}$$

Because d**l** and **r** are perpendicular, we have $|d\mathbf{l} \times \mathbf{r}| = r \, dl$, and so the expression for dB becomes

$$dB = \frac{\mu_o}{4\pi} \frac{i \, dl}{r^2}$$

From Figure 16.12 we see that $\sin\theta = a/r$ so that

$$B = \int \frac{a}{r} \frac{\mu_o}{4\pi} \frac{i \, dl}{r^2}$$

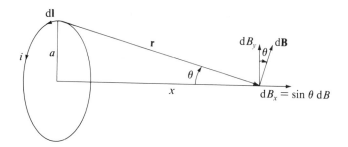

Figure 16.12   Example 16.2

Because i, a, and r are each independent of just which current element appears in the integral, the expression for B simplifies to

$$B = \frac{\mu_o}{4\pi} \frac{i\,a}{r^3} \int dl$$

Of course, $\int dl$ is equal to the circumference of the loop, i.e., $\int dl = 2\pi a$. Substituting this value gives

$$B = \frac{\mu_o i a^2}{2r^3}$$

or because $r^2 = a^2 + x^2$, the expression for B may be written

$$B = \frac{\mu_o i a^2}{2(x^2 + a^2)^{3/2}} \quad \blacksquare$$

This current loop is a magnetic equivalent to the electric dipole. In fact, the *magnetic dipole moment* $\boldsymbol{\mu}$ is perpendicular to the plane of the loop and has a magnitude given by $\mu = $ (current) $\cdot$ (area) $= i\,\pi\,a^2$. With this definition, the field of the magnetic dipole is given by

$$B = \frac{\mu_o}{2\pi} \frac{\mu}{r^3}$$

Just as with the electric dipole, the field of the magnetic dipole decreases as the cube of the distance from the dipole to the field point.

16.3 Gauss's Law for Magnetic Fields and Ampère's Law

Two integral relationships for the electrostatic field incorporate its essential characteristics: Gauss's law, $\oint_S \mathbf{E} \cdot d\mathbf{S} = q_{net}/\epsilon_o$, and the conservative nature of the electrostatic field, $\oint_C \mathbf{E} \cdot d\mathbf{s} = 0$. In a like manner, two integral relationships embody the essentials of the magnetic field: Gauss's law for magnetic fields and Ampère's law. We will now consider these fundamental relationships.

Gauss's Law for Magnetic Fields

Similarly to the way that electric flux was defined in Chapter 12, that is, $\phi_E = \int_S \mathbf{E} \cdot d\mathbf{S}$, we define the magnetic flux ϕ_m for a surface S to be

$$\phi_m = \int_S \mathbf{B} \cdot d\mathbf{S} \tag{16-30}$$

where $d\mathbf{S}$ is an element of area directed normally (and in a chosen sense) to the surface S. The integration indicated in Equation (16-30) extends over the entire surface S. Physically speaking, the magnetic flux for a surface is the product of the surface area and the average value over that surface of the component of $\mathbf{B}$ that is normal to the surface.

An experimental observation that is fundamental to electromagnetism is expressed by

$$\oint_S \mathbf{B} \cdot d\mathbf{S} = 0 \qquad (16\text{-}31)$$

which is called *Gauss's law for magnetic fields*. Equation (16-31) asserts that the magnetic flux for any closed surface is identically equal to zero. The vector $d\mathbf{S}$ is taken to be directed normally outward from S at every point on the surface.

The significance of Equation (16-31) may be emphasized by recalling that Gauss's law for electric fields is interpreted to mean that the electric flux for a closed surface is proportional to the net electric charge within that surface (see Chapter 12). Equation (16-31) suggests, therefore, that there is no magnetic analogue of an electric charge. Magnetic fields can be described in terms of magnetic "poles"; a magnet is sometimes said to have a "north pole" and a "south pole" but never only a single pole—never a monopole. In these terms, then, we may say that Gauss's law for magnetic fields, Equation (16-31), forbids the existence of a magnetic monopole. And despite extensive experimental searches, no magnetic monopole has been discovered and confirmed.

One conceptual consequence of Gauss's law for magnetic fields is that the field lines we use to represent magnetic fields are continuous. Whereas electric field lines begin and end on electric charges, there are no corresponding entities on which the lines of magnetic field can terminate. Thus the lines of magnetic fields are continuous and never end unless they end on themselves, a situation exemplified by the field lines illustrated for regions outside a current-carrying wire in Figure 16.10.

Ampère's Law

In electrostatics the use of Gauss's law simplified the analysis of electric fields produced by static charge distributions having sufficient spatial symmetry. The analysis of magnetic fields produced by steady currents is the object of the studies called *magnetostatics*. The analyses of magnetostatic fields produced by current distributions with sufficient spatial symmetry are similarly simplified by a line-integral relationship called "Ampère's law."

Ampère's law relates the components of a magnetic field $\mathbf{B}$ lying along a closed curve C to the net current through any open surface S bounded by C. Mathematically, *Ampère's law* may be written as

$$\oint_C \mathbf{B} \cdot d\mathbf{s} = \mu_o \, i_{net} \qquad (16\text{-}32)$$

where i_{net} is the net current through any surface enclosed by the closed curve C. The directional sense of the current is defined below.

Consider the conductor of Figure 16.13 carrying a (net) current $i = i_{net}$. Suppose C is a closed curve (a line not necessarily associated with any physical object) that bounds the open surface S, through which the current-carrying conductor passes. Then

1. The directional sense of i_{net} is determined by yet another right-hand rule: With the fingers of the right hand curved along a chosen sense for traversing the closed

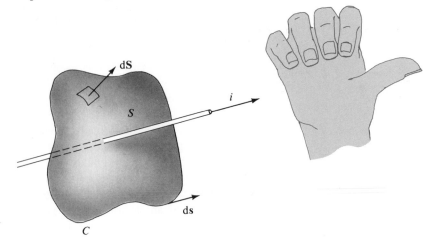

Figure 16.13
A conductor carrying a current i through a surface S, which is bounded by the curve C.

curve C, the extended thumb is in the positive sense of the net current i_{net} through S. As indicated in Figure 16.13, the positive direction of each element dS of the surface area is chosen to have the same directional sense as that of the extended thumb.

2. The line integral of Equation (16-32) is the sum of the products $\mathbf{B} \cdot d\mathbf{s}$ for every element $d\mathbf{s}$ along the closed curve C.

Ampère's law applies to any closed curve in a magnetostatic situation. Nevertheless, it is useful only if a given current distribution is sufficiently symmetrical so that we can construct a line integral in which the tangential component of $\mathbf{B}$ is constant at every point on some closed curve C. In such cases the line integral becomes trivial to evaluate. The following problem illustrates how the magnetic field caused by an appropriately symmetrical distribution may be determined using Ampère's law.

A total current I is distributed uniformly over the cross section of a long cylindrical conducting shell having an inner radius a and an outer radius b, shown in Figure 16.14. Determine the magnitude of the magnetic field caused by the current in this shell as a function of radial distance r from the axis of the shell.

The current distribution in this problem is cylindrically symmetric. We there-

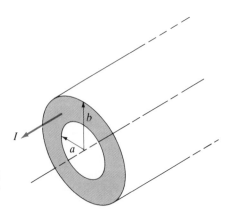

Figure 16.14
A long cylindrical conducting shell carrying a current I, which is uniformly distributed over its cross section.

fore expect the magnetic field to be tangential to any circle lying in a plane perpendicular to and concentric with the axis of the cylindrical shell. Then, for such a circle, $\mathbf{B} \cdot \mathbf{ds} = \pm B\, ds$ (the negative sign applies if $\mathbf{B}$ and $\mathbf{ds}$ are in opposite directions). Let us construct circular paths, as shown in Figure 16.15(a). Because of the symmetry of the current distribution, the magnitude of $\mathbf{B}$ is constant at every point on each circle, and we have

$$\oint_C \mathbf{B} \cdot \mathbf{ds} = \oint_C (+B\,ds) = B \oint ds = 2\pi rB \qquad (16\text{-}33)$$

Because the net current i_{net} through the circle of radius r is equal to zero for any $r \leq a$, Ampère's law yields

$$2\pi rB = \mu_0 i_{net} = 0$$

$$B = 0 \quad ; \quad r \leq a \qquad (16\text{-}34)$$

To find $\mathbf{B}$ in the region $a \leq r \leq b$, consider the circle of radius r in the region between a and b, as shown in Figure 16.15(b). Applying Ampère's law to this circle, where B is again a constant because of the symmetry, gives

$$\oint_C \mathbf{B} \cdot \mathbf{ds} = B \oint ds = 2\pi rB = \mu_0 i_{net} \qquad (16\text{-}35)$$

Here, the net current i_{net} through the circle of radius r lies within the shaded region in Figure 16.15(b). The area of the shaded region is equal to $(\pi r^2 - \pi a^2)$, and the total area of the cross section of the conductor is equal to $(\pi b^2 - \pi a^2)$, so the current within the shaded area is (because the current is distributed uniformly) equal to the total current I multiplied by the ratio of these areas:

$$i_{net} = I\left(\frac{\pi r^2 - \pi a^2}{\pi b^2 - \pi a^2}\right) = I\left(\frac{r^2 - a^2}{b^2 - a^2}\right)$$

Then Equation (16-35) becomes

$$2\pi rB = \mu_0 I\left(\frac{r^2 - a^2}{b^2 - a^2}\right)$$

$$B = \frac{\mu_0 I}{2\pi r}\left(\frac{r^2 - a^2}{b^2 - a^2}\right) \quad ; \quad a \leq r \leq b \qquad (16\text{-}36)$$

Finally, by considering a circle with radius greater than b, as shown in Figure 16.15(c), we find, using Ampère's law, that B in the region $r \geq b$ is given by

$$2\pi rB = \mu_0 I$$

Figure 16.15
Constructions of circular paths about the axis of the cylinder in Figure 16.14 at radii (a) inside the conductor, (b) within the conductor, and (c) outside the conductor.

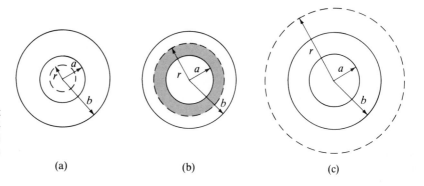

(a)   (b)   (c)

because the net current through the circle of radius r is equal to the total current I. Then B in the region outside the cylindrical conductor is

$$B = \frac{\mu_0 I}{2\pi r} \quad ; \quad r \geq b \qquad (16\text{-}37)$$

which is identical to the field of a thin wire lying along the axis of the cylindrical conductor and carrying a current I.

The solution, Equation (16-36), that applies to the region inside the conducting material may be examined in the limiting cases for which $r = a$ and for which $r = b$ to serve as a check on the reasonableness of that solution.

E 16.17   Show that Equation (16-36) reduces appropriately for $r = a$ and $r = b$.

E 16.18   If a long cylindrical wire (radius = 2.0 mm) carries a current of 100 A, calculate the magnitude of the magnetic field at a point that is (a) 1.0 mm, and (b) 3.0 mm from the axis of the wire.   Answers: (a) 50 G; (b) 67 G

16.4 Applications

A convenient and inexpensive way to produce a relatively uniform magnetic field over a considerable region of space is an arrangement of coiled current-carrying wire called a *solenoid*. Figure 16.16 shows a solenoid of length L with N turns of wire coiled into a helix having a very small pitch. Each turn of the wire is, in effect, a circular loop. Therefore, when a current i is in the wire, the magnetic field of the solenoid may be analyzed in terms of the fields produced by a "continuous" line of circular loops, each of which produces a magnetic field like that of a circle of wire carrying a current i, the situation considered in Example 16.2.

Let us determine the magnetic field **B** at an arbitrary point P on the axis of a solenoid like that of Figure 16.16. Figure 16.17 shows a schematic drawing of the solenoid oriented so that its axis coincides with the x axis. Let P be at $x = 0$, and let the radius of the solenoid be a. We treat the solenoid as if its current loops carrying current i may be divided (at least conceptually) into segments of width dx. Then each "current loop" of width dx carries the current $i(N/L)dx = i\,n\,dx$, where $n = N/L$ is the number of turns of wire per unit length along the x

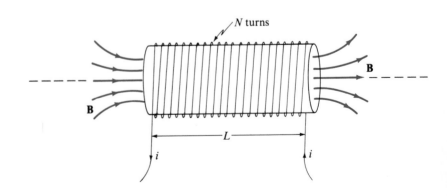

Figure 16.16
A solenoid.

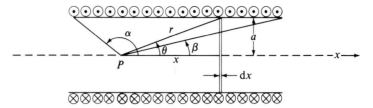

Figure 16.17
A schematic diagram of a solenoid with its axis lying along the x axis.

direction. The distance from the wire in the "loop" of width dx to the field point P on the axis is r. The center of that "loop" lies a distance x from the origin at point P. According to the result of Example 16.2, the magnetic field at P caused by the chosen "loop" of width dx is

$$d\mathbf{B} = \frac{\mu_o n i a^2 dx}{2r^3}\,\hat{\mathbf{i}} \tag{16-38}$$

We may write both r and x in terms of the radius a of the solenoid and θ, the angle between the x axis and $\mathbf{r}$:

$$r = a\,\csc\theta$$

$$x = a\,\cot\theta$$

$$dx = -a\,\csc^2\theta\,d\theta$$

Substituting these values into Equation (16-38) gives

$$d\mathbf{B} = \frac{\mu_o n i a^2 (-a\,\csc^2\theta\,d\theta)}{2a^3\,\csc^3\theta}\,\hat{\mathbf{i}} = -\frac{1}{2}\mu_o n i\,\sin\theta\,d\theta\,\hat{\mathbf{i}}$$

$$\mathbf{B} = -\frac{\mu_o n i\,\hat{\mathbf{i}}}{2}\int_{\theta=\alpha}^{\theta=\beta}\sin\theta\,d\theta = \frac{\mu_o n i}{2}(\cos\beta - \cos\alpha)\,\hat{\mathbf{i}} \tag{16-39}$$

The angles α and β in Equation (16-39) specify the spatial extent of the solenoid, as shown in Figure 16.17.

Figure 16.18 shows the magnetic field lines associated with a solenoid with a length roughly equal to twice its diameter. Equation (16-39) provides an accurate measure of the magnetic field at points on the axis; points not on the axis are

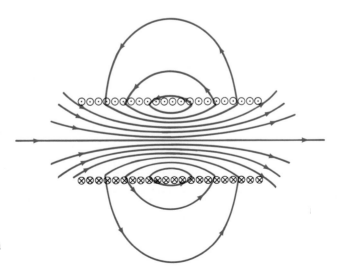

Figure 16.18
The magnetic field lines associated with a solenoid.

more difficult to calculate. The spacing of the field lines in Figure 16.18 indicates that **B** is considerably smaller outside the solenoid than inside.

A *long solenoid* is very long compared to its radius. The magnetic field on the axis of a long solenoid is found from Equation (16-39) when $\alpha \rightarrow \pi$ and $\beta \rightarrow 0$, corresponding to a linear extension of the solenoid from $x \rightarrow -\infty$ to $x \rightarrow \infty$:

$$\mathbf{B} = \mu_o n i \,\hat{\mathbf{i}} \qquad \text{(long solenoid)}$$

As a solenoid like that of Figure 16.18 becomes longer, the magnetic field lines become more nearly parallel within the solenoid and farther apart outside the solenoid. Consider the long solenoid of Figure 16.19(a), where the solenoid is depicted in cross section. The essential characteristics of the magnetic field of an ideal solenoid may be seen by using Ampère's law. We have constructed a rectangular amperean path *abcd*, which encloses a length L of the closely spaced turns of the solenoid. The path *ab* lies along the axis of the solenoid; *cd* is at an arbitrary distance from the axis but lies outside the solenoid. We apply Ampère's law, traversing the closed path in the direction indicated:

$$\oint \mathbf{B} \cdot d\mathbf{s} = \int_a^b \mathbf{B} \cdot d\mathbf{s} + \int_b^c \mathbf{B} \cdot d\mathbf{s} + \int_c^d \mathbf{B} \cdot d\mathbf{s} + \int_d^a \mathbf{B} \cdot d\mathbf{s} = \mu_o i_{net} \qquad \textbf{(16-41)}$$

The line integral of Ampère's law has been written in Equation (16-41) as the sum of the line integrals along each of the four path segments. The net current enclosed by the path *abcd* is $i_{net} = inL$. Then the sum of the four line integrals of Equation (16-41) is equal to $\mu_o \, i_{net} = \mu_o inL$, or

$$\oint \mathbf{B} \cdot d\mathbf{s} = \mu_o n i L \qquad \textbf{(16-42)}$$

But the line integral over the path *ab* alone is, using Equation (16-40),

$$\int_a^b \mathbf{B} \cdot d\mathbf{s} = \mu_o n i L \qquad \textbf{(16-43)}$$

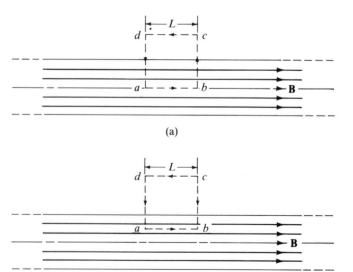

(a)

(b)

Figure 16.19
Amperean paths in a plane through the axis of a long solenoid (a) with one segment of the path coinciding with the axis of the solenoid, and (b) with one segment inside the solenoid at an arbitrary distance from the axis of the solenoid.

which is identical with the value indicated in Equation (16-42). It follows from Equations (16-41), (16-42), and (16-43) that

$$\int_b^c \mathbf{B} \cdot \mathbf{ds} + \int_c^d \mathbf{B} \cdot \mathbf{ds} + \int_d^a \mathbf{B} \cdot \mathbf{ds} = 0 \qquad (16\text{-}44)$$

The first and third integrals of Equation (16-44) are equal to zero because $\mathbf{B}$ and $\mathbf{ds}$ are perpendicular at each point on the path segments bc and da. If a field exists outside the solenoid, the symmetry of the solenoid requires that $\mathbf{B}$ in that region be parallel to the path cd and constant along that path segment. Then we have $\int_c^d \mathbf{B} \cdot \mathbf{ds} = \int_c^d B ds = BL$. Thus Equation (16-44) requires that $BL = 0$ or, because $L \neq 0$, that $B = 0$ outside the solenoid. Therefore, for a long solenoid, the magnetic field is entirely contained within the solenoid.

Now that we have established that no magnetic field exists outside of an isolated long solenoid, let us examine further the field inside. Suppose we change the amperean path $abcd$ so that ab lies within the solenoid parallel to the axis but *not* coincident with the axis; this situation is shown in Figure 16.19(b). Equation (16-43) still applies. And because the net current enclosed by $abcd$ is still that given by Equations (16-42) and (16-43), we must conclude that $\mathbf{B}$ is the same along ab as it was when ab lay along the axis of the solenoid. Therefore, because ab was chosen at an arbitrary distance from the axis, *the magnetic field is identical at every point inside a long solenoid.*

E 16.19   Consider a "half-infinite" solenoid having an axis that co-incides with the positive x axis. Use Equation (16-36) to show that the axial field at the open end ($x = 0$) is equal to one-half that of an infinitely long solenoid.

E 16.20   A long solenoid that has 100 turns per cm carries a current of 0.20 A. Calculate the magnitude of the magnetic field inside the solenoid.

Answer: 2.5×10^{-3} T = 25 G

A helical coil of wire that is curved into the shape of a doughnut is called a *toroid*, illustrated in Figure 16.20(a). We may calculate the magnetic field at points within the interior of the toroid by using Ampère's law. Figure 16.20(b) shows a cross section through the toroid where the lines of magnetic field, because of symmetry, form concentric circles within the toroid. Figure 16.20(b) shows a circular amperean path having a radius r and lying within the interior of

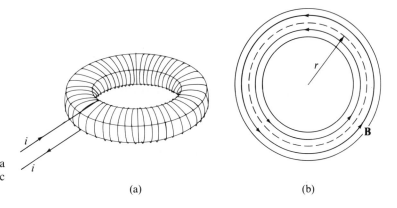

Figure 16.20
(a) A toroid. **(b)** A cross section of a toroid showing the concentric magnetic field lines within the toroid.

(a)   (b)

the toroid. This path encloses a net current $i_{net} = Ni$, where N is the number of turns on the toroid and i is the current in the wire. Then Ampère's law gives the magnitude of $\mathbf{B}$:

$$\oint \mathbf{B} \cdot d\mathbf{s} = \mu_o i_{net}$$

$$B(2\pi r) = \mu_o Ni$$

$$B = \frac{\mu_o Ni}{2\pi r} \tag{16-45}$$

Notice that the magnetic field within a toroid is not constant across the cross section but decreases in magnitude with radial distance from the center.

E 16.21   Using Ampère's law, show that, because the magnetic field lines for a toroid are circles that are centered on the axis of the toroid, the magnetic field is equal to zero at any point not enclosed by the toroid.

E 16.22   A toroid with an inner radius of 10 cm is tightly wound with one layer of wire that has a diameter of 0.20 mm. How many turns (N) are there on the toroid?   Answer: 3.1×10^3

E 16.23   If the toroid of Exercise 16.22 carries a current of 0.50 A, determine the magnitude of $\mathbf{B}$ at a distance 11 cm from the axis of the toroid, assuming that the outer radius of the toroid is greater than 11 cm.
                                    Answer: 2.9×10^{-3} T = 29 G

Example 16.3   **PROBLEM**   A long wire carries a current i_1 along the z axis in the positive z direction. A second wire segment having a length b carries a current i_2 in the positive x direction along the x axis from $x = a$ to $x = a + b$. Determine the magnetic force on the second wire.

SOLUTION   Figure 16.21 shows the physical situation described. The force $d\mathbf{F}$ shown acts on a current element that has a length dx and that is located at a distance x from the long wire. This force is given by

$$d\mathbf{F} = i_2 \, d\mathbf{l} \times \mathbf{B}$$

where $d\mathbf{l} = \hat{\mathbf{i}} \, dx$ and $\mathbf{B}$ is the magnetic field of the long wire, i.e.,

$$\mathbf{B} = \frac{\mu_o i_1}{2\pi x} \hat{\mathbf{j}}$$

Substituting for $d\mathbf{l}$ and $\mathbf{B}$ gives

$$d\mathbf{F} = \frac{\mu_o i_1 i_2}{2\pi x}(\hat{\mathbf{i}} \times \hat{\mathbf{j}})dx = \frac{\mu_o i_1 i_2 \hat{\mathbf{k}}}{2\pi} \frac{dx}{x}$$

Integrating this expression from $x = a$ to $x = a + b$ yields

$$\mathbf{F} = \frac{\mu_o i_1 i_2}{2\pi} \ln\left(1 + \frac{b}{a}\right) \hat{\mathbf{k}}$$

A simple check verifies that $\mathbf{F}$ is equal to zero if either of the currents is equal to zero. Finally, if the distance a from the long wire to the closest point of the second wire is sufficiently large, that is, as $a \to \infty$, we expect the force to approach a value of zero. Indeed, this is the case because as $a \to \infty$, $\ln(1 + b/a) \to \ln(1) = 0$. ■

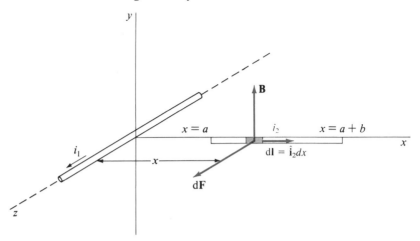

Figure 16.21   Example 16.3

16.5 Problem-Solving Summary

In problems that involve magnetic fields that do not change with time (magnetostatics problems), a magnetic field either is given or is required to be determined from a given current distribution. In general, we may treat these problems as follows:

1. When the magnetic field is given in a problem, we are usually required to relate the magnetic force $\mathbf{F}_m$ on one or more charged particles to the charge q and velocity $\mathbf{v}$ of each of those particles. In these cases we use either of two equivalent relationships: $\mathbf{F}_m = q\mathbf{v} \times \mathbf{B}$ or $d\mathbf{F}_m = i\,d\mathbf{l} \times \mathbf{B}$, in which $i\,d\mathbf{l}$ is a current element.

2. When a current distribution along a thin wire is given in a problem and we are required to find the magnetic field it produces, we may, for a general case, use the Biot-Savart law,

$$\mathbf{B} = \frac{\mu_o}{4\pi} \int \frac{i\,d\mathbf{l} \times \mathbf{r}}{r^3}$$

where $\mathbf{r}$ locates the field point from the position of a current element $i\,d\mathbf{l}$, which points in the directional sense of the current. The integration in the Biot-Savart law extends over the entire current distribution. In more special cases, when the current distribution is sufficiently symmetrical, we may use Ampère's law,

$$\oint_C \mathbf{B} \cdot d\mathbf{s} = \mu_o i_{net}$$

to relate a closed line integral of $\mathbf{B} \cdot d\mathbf{s}$ to the net current passing through a surface bounded by C, the closed amperean path of integration.

Although $\mathbf{F}_m = q\mathbf{v} \times \mathbf{B}$ specifies the magnetic force on a particle of charge q moving with velocity $\mathbf{v}$ at a point where the magnetic field is $\mathbf{B}$, this relationship is especially useful in problems in which a charged particle is moving in a region of uniform magnetic field $\mathbf{B}$. In that special case, we may simplify most problems by recognizing that the velocity $\mathbf{v}$ of a particle may be considered in terms of its component velocities, $\mathbf{v}_\parallel$ and $\mathbf{v}_\perp$, which are parallel and perpendicular to $\mathbf{B}$.

The presence of $\mathbf{B}$ does not affect $\mathbf{v}_{\parallel}$. On the other hand, $\mathbf{B}$ causes $\mathbf{v}_{\perp}$ to turn—without changing its magnitude—in a plane perpendicular to $\mathbf{B}$. The particle moves in a circular path of radius $r = mv_{\perp}/qB$ with a period given by $T = 2\pi m/qB$. A particle with arbitrary velocity $\mathbf{v} = \mathbf{v}_{\parallel} + \mathbf{v}_{\perp}$ in a uniform magnetic field B moves in a helical path with the axis of the helix parallel to $\mathbf{B}$ and with a pitch $p = 2\pi mv_{\parallel}/qB$. But regardless of the path of the particles in any static magnetic field, the field cannot produce a change in the kinetic energy of the particle. A magnetic field may change the direction of the velocity of a charged particle, but it cannot change the speed of the particle.

A particle having a charge q and moving with velocity $\mathbf{v}$ in the presence of both an electric field $\mathbf{E}$ and a magnetic field $\mathbf{B}$ experiences a total force given by $\mathbf{F} = q(\mathbf{E} + \mathbf{v} \times \mathbf{B})$.

To calculate the magnetic force $\mathbf{F}_m$ on a segment of wire carrying current i in the presence of a magnetic field $\mathbf{B}$, we may integrate $d\mathbf{F}_m = i\,d\mathbf{l} \times \mathbf{B}$ over every current element of the wire. Because the current is constant in every part of the wire, the total magnetic force on the wire is given by $\mathbf{F} = i \int d\mathbf{l} \times \mathbf{B}$.

Solving problems in which we use the Biot-Savart law to calculate $\mathbf{B}$, caused by a given current in a thin wire, at a field point is made significantly easier by careful visualization of the geometry of the physical situation. We need to have a clear understanding of the direction of $d\mathbf{l} \times \mathbf{r}$, where $d\mathbf{l}$ is a displacement along the wire in the directional sense of the current and $\mathbf{r}$ is a vector from a current element $i\,d\mathbf{l}$ to the field point. In general, before the integral for $\mathbf{B}$ can be evaluated, it is necessary to express all the quantities of the integral in terms of specified quantities and a single variable of integration.

If a given current distribution has sufficient symmetry (this usually means that it is cylindrically symmetrical), we can use Ampère's law to find the correspondingly symmetric magnetic field caused by the current distribution. For cylindrically symmetric current distributions, the appropriate amperean path around which $\mathbf{B} \cdot d\mathbf{s}$ is integrated is a circle along which $\mathbf{B}$ has a constant magnitude and is directed tangentially to the circle at every point. Then $2\pi rB$ may be set equal to $\mu_o i_{net}$, where r is the radius of the circle and i_{net} is the net current through the flat surface bounded by the circle.

Be careful when you assume that a current distribution has sufficient symmetry for Ampère's law to be useful: The symmetry must be valid for the entire circuit and not for just a segment of the circuit. Thus we may *not*, for example, use Ampère's law to calculate the magnetic field at an arbitrary point around a *finite* length of wire. We may, however, use Ampère's law to calculate the magnetic field around a finite length of wire to a fair approximation at points very near the wire.

Ampère's law is useful for calculating the magnetic field throughout the interior of a long solenoid or a toroid.

Problems

GROUP A

16.1 A duck flying due north at 15 m/s is passing over Atlanta, where the magnetic field of the earth is 5.0×10^{-5} T in a direction 60° below a horizontal line running north and south. Because the weather is dry, the duck has accumulated a positive charge of 4.0×10^{-8} C. What force (magnitude) acts on the duck as a result of moving through the magnetic field of the earth?

16.2

(a) What is the speed of electrons that pass undeviated through crossed (perpendicular) electric and magnetic fields if $E = 4.0 \times 10^4$ V/m and $B = 8.0 \times 10^{-3}$ T?

(b) What speed must protons have to negotiate the same region without being deviated?

16.3 An electron is introduced at a speed of 3.0×10^7 m/s into a region of uniform magnetic field **B**. The electron moves in a direction perpendicular to **B**. If the electron follows a circular orbit of radius 15 cm,

(a) what is the magnitude of **B**?

(b) what is the period of the orbital motion of the electron?

16.4 An electron with velocity $5.0 \times 10^5 \hat{\mathbf{j}}$ m/s is at a point where the electric field is $(3.0 \times 10^4 \hat{\mathbf{i}} - 4.0 \times 10^4 \hat{\mathbf{j}})$ V/m and the magnetic field is $2.0 \times 10^{-2} \hat{\mathbf{i}}$ T. At that instant, what is the net force on the electron?

16.5 An electron is at a point where $\mathbf{E} = (16\hat{\mathbf{i}} - 20\hat{\mathbf{j}})$ N/C and $\mathbf{B} = (2.0\hat{\mathbf{i}} - 6.0\hat{\mathbf{k}}) \times 10^{-5}$ T. At that instant, the electron has a velocity $\mathbf{v} = (3.0\hat{\mathbf{i}} + 1.0\hat{\mathbf{k}}) \times 10^6$ m/s. Calculate the force on the electron at the given instant.

16.6 Two long thin wires carrying currents of 100 A and 50 A in opposite directional senses are separated by 5.0 cm. What is the magnitude of the force exerted on each meter of length of each wire? Are the forces attractive or repulsive?

16.7 A current of 50 A is uniformly distributed over the cross section of a long wire that has a diameter of 2.0 mm.

(a) Determine $B(r)$, the magnitude of the magnetic field, as a function of the distance r from the axis of the wire.

(b) Graph $B(r)$ for $0 \leq r \leq 4.0$ mm.

16.8 A particle having a positive charge of 4.0×10^{-16} C is moving with a velocity $\mathbf{v} = (-2.0\hat{\mathbf{i}} + 4.0\hat{\mathbf{j}}) \times 10^4$ m/s at a point where the magnetic field is $\mathbf{B} = (3.0\hat{\mathbf{i}} + 2.0\hat{\mathbf{j}}) \times 10^{-3}$ T. What is the magnetic force on the particle at that instant?

16.9 A proton ($m_p = 1.7 \times 10^{-27}$ kg) moving with a speed of 4.0×10^6 m/s is introduced into a region of uniform magnetic field **B**, which has a magnitude of 3.0×10^{-2} T. The initial direction of the motion of the proton makes an angle of 60° with the direction of **B**.

(a) How far in the direction of **B** does the proton travel while making two turns of its helical path?

(b) What is the radius of its helical path?

(c) How much kinetic energy does the proton gain while making two turns of its helical path?

16.10 A toroid with inner radius 30 cm and outer radius 40 cm has 4000 turns of wire and carries a current of 10 A. What are the maximum and minimum magnitudes of the magnetic field within the interior of the toroid?

16.11 Three long parallel wires each carry 10 A in the directional senses shown in Figure 16.22. What force is exerted on each length of 1.0 m of

(a) wire C?

(b) wire B?

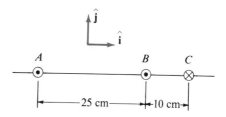

Figure 16.22   Problem 16.11

16.12 A proton is accelerated from rest through a potential difference of 2.5 kV. It then enters a region having a uniform magnetic field **B** and travels a circular path having a 9.0-cm radius. Calculate the magnitude of the magnetic field.

16.13 A current of 30 A is uniformly distributed over the cross section of a long hollow wire with an inner radius of 1.0 mm and an outer radius of 2.0 mm. Graph $B(r)$, the magnitude of the magnetic field, as a function of the distance r from the axis of the wire for $0 \leq r \leq 5.0$ mm.

16.14 Two long coaxial conductors, each of negligible thickness, have radii a and b with $a < b$. The currents in the two conductors have the same magnitude i but opposite directional senses.

(a) Determine $B(r)$, the magnitude of the magnetic field, as a function of the distance r from the axis of the cylinders.

(b) With $i = 15$ A, $a = 0.50$ cm, and $b = 1.0$ cm, graph $B(r)$ for $0 \leq r \leq 2.0$ cm.

16.15 The segment of the current-carrying wire shown in Figure 16.23 is in a uniform magnetic field of 0.080 T into the paper. If $i = 25$ A, $L_1 = 0.80$ m, and $L_2 = 0.60$ m, determine the magnitude of the magnetic force on this segment.

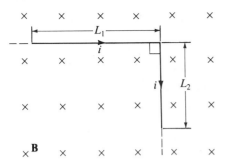

Figure 16.23   Problem 16.15

16.16   The semicircular segment of current-carrying wire shown in Figure 16.24 is in a uniform magnetic field of 0.25 T into the paper. If $i = 45$ A and $R = 5.0$ cm, determine the magnetic force on this segment.

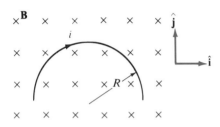

Figure 16.24   Problem 16.16

16.17   A wire segment (not necessarily straight) carries a current i from point A to point B in a region of uniform magnetic field **B**. Show that the magnetic force on this segment of wire is given by

$$\mathbf{F}_m = i\,\mathbf{L} \times \mathbf{B}$$

where **L** is the position vector from point A to point B.

16.18   A segment of wire carries a current of 12 A in the $\hat{\mathbf{i}}$-direction along the x axis from $x = 0$ to $x = 2.0$ m in a region where a magnetic field is described by $\mathbf{B} = \beta x\,\hat{\mathbf{j}}$, with $\beta = 0.050$ T/m. Determine the magnetic force on the wire segment.

16.19   Three long parallel wires carry currents perpendicular to the paper with the directional senses shown in Figure 16.25. If $i_1 = 16$ A, $i_2 = 25$ A, $i_3 = 18$ A, $a = 0.40$ cm, and $b = 0.30$ cm, determine the magnetic field at the point P shown.

16.20   A cylindrical plastic tube having a length of 50 cm and a diameter of 4.0 cm is to be used as a form for making a solenoid. Wire capable of carrying 10 A is to be used to construct the solenoid, which is to provide a magnetic field of 5.0×10^{-2} T near the center of the solenoid.

  (a) How many turns of the wire must be evenly wound about the plastic form?

  (b) What total length of wire will be needed to construct the solenoid?

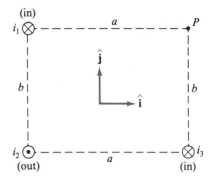

Figure 16.25   Problem 16.19

16.21   At $t = 0$ a particle with $m = 7.2 \times 10^{-9}$ kg and $q = -2.5 \times 10^{-4}$ C has a velocity given by $\mathbf{v} = (120\hat{\mathbf{i}} + 50\hat{\mathbf{j}})$ m/s. For $t \geq 0$ the particle follows a helical path having a pitch of 4.5 cm and an axis parallel to the y axis. Calculate

  (a) the radius of the helix.

  (b) the number of times per second that the particle circles the helix.

  (c) the magnitude of the magnetic field in the region.

16.22   A current of 25 A follows a circular path having a 0.50-cm radius. Calculate the magnitude of the magnetic field at the center of the circle.

16.23   An electron moves in the vicinity of a long straight wire carrying a current of 75 A. At an instant when the electron is 0.60 cm from the wire and moving toward the wire with a speed of 4.0×10^6 m/s, what is the force on the electron?

16.24   A rectangular loop carries a current of 50 A (see Fig. 16.26). Calculate the net magnetic force on the loop by the long wire, which carries 25 A.

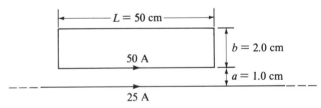

Figure 16.26   Problem 16.24

16.25   Calculate the magnetic field at the point P caused by the circuit shown in Figure 16.27.

16.26   Write an expression for the magnitude of the magnetic field at the point P caused by the current i in the circuit segment shown in Figure 16.28.

16.27   Calculate the magnitude of the magnetic field at the center of a square made of a wire carrying a current i. The length of each side of the square is d.

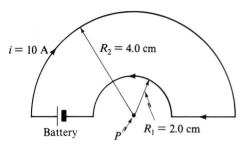

Figure 16.27   Problem 16.25

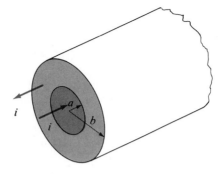

Figure 16.30   Problem 16.29

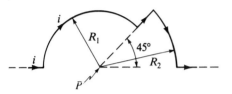

Figure 16.28   Problem 16.26

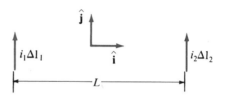

Figure 16.31   Problem 16.30

16.28 Two long wires, each having a mass per unit length of 30 g/m and each carrying a current i (see Fig. 16.29) are hanging by threads 1.0 m long attached to a bar parallel to the wires. If the angle between the threads of each pair is 10°, what is the magnitude of i?

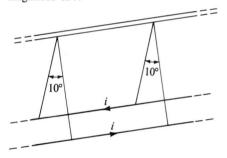

Figure 16.29   Problem 16.28

16.29 A current i is in the positive x direction parallel to the axis of the cylinder of radius a in Figure 16.30. A cylindrical shell with inner radius a and outer radius b is coaxial with the cylinder and insulated from it. The shell carries a current i in the negative x direction. Calculate the magnitude of the magnetic field caused by these currents in the regions

(a) $0 \leq r \leq a$

(b) $a \leq r \leq b$

(c) $r \geq b$

16.30 Two parallel current elements are shown in Figure 16.31. If $i_1 = i_2 = 4.0$ A, $\Delta l_1 = \Delta l_2 = 0.50$ cm, and $L = 1.2$ m, estimate the magnetic force on

(a) Δl_1

(b) Δl_2

16.31 Two current elements are shown in Figure 16.32. If $i_1 = i_2 = 4.0$ A, $\Delta l_1 = \Delta l_2 = 0.50$ cm, and $L = 1.2$ m, estimate the magnetic force acting on

(a) $i_1 \Delta l_1$

(b) $i_2 \Delta l_2$. What happened to Newton's third law?

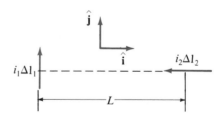

Figure 16.32   Problem 16.31

16.32 One of two long wires carries a current of 85 A along the z axis in the positive z direction. The second wire carries a current of 40 A in the positive x direction along the line in the x-y plane corresponding to $y = 3.0$ cm. Determine the magnetic field at the point $\mathbf{r} = 2.0\hat{\mathbf{j}}$ cm.

16.33 Copper ($\rho = 1.7 \times 10^{-8}$ $\Omega \cdot$m) wire having a 1.0-mm diameter is used to wind a cylindrical solenoid (radius = 1.0 cm, length = 40 cm). The current in the solenoid is to be 5.0 A.

(a) What minimum number of layers of windings will be required to produce a magnetic field (near the center of the solenoid) having a magnitude of at least 100 G?

(b) At what rate is heat generated in the solenoid of part (a)?

16.34 Two circular loops of wire each carry a 15-A current. The radii of the loops are equal to 1.0 cm and 1.5 cm, the centers of the loops are coincident, and the planes of the loops are perpendicular. Calculate the magnitude of the magnetic field at the center of the two circles.

16.35 A current of 10 A is uniformly distributed across the cross section of a long thin strip of metal that is 1.0 cm wide (*see* Fig. 16.33). What is the magnetic field at a point P that lies in the plane of the strip a distance of 5.0 cm from the centerline of the strip. Check the solution by treating the strip as if it were a thin wire.

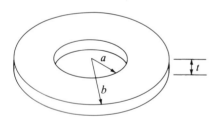

Figure 16.33   Problem 16.35

16.36 A thin, flat annular ring of constant thickness t and radii a and b (*see* Fig. 16.34) has an azimuthal (perpendicular to the radial direction) current density that varies only in the radial direction according to $J = K/r$. Write an expression for the magnitude of the magnetic field at the center of the ring.

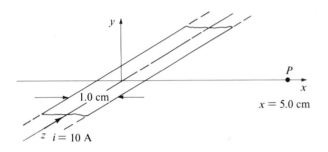

Figure 16.34   Problem 16.36

16.37 A charged particle ($m = 2.5 \times 10^{-9}$ kg, $q = -3.0$ μC) moves in a region where $\mathbf{B} = 0.40\hat{\mathbf{k}}$ T. At $t = 0$ the position and velocity of the particle are $\mathbf{r} = \mathbf{0}$ and $\mathbf{v} = 200\hat{\mathbf{i}}$ m/s. What are t, $\mathbf{r}$, and $\mathbf{v}$ when the particle has moved a distance of 0.50 m?

16.38 Two identical coils, each with N turns (radius a) carrying a current i, are positioned with a common axis. The distance between the coils is L, as shown in Figure 16.35. Show that for $L = a$, the derivatives, dB_x/dx, d^2B_x/dx^2, d^3B_x/dx^3, of B_x at point P, the axial midpoint, are each equal to zero. Thus, this pair of coils, called *Helmholtz coils*, produces a very uniform field in the region surrounding point P.

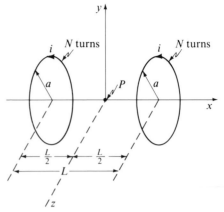

Figure 16.35   Problem 16.38

16.39 A straight wire segment of length L carries a current i (*see* Fig. 16.36). Show that the resulting magnetic field at the point P shown is given by

$$B = \frac{\mu_0 i}{4\pi R}\left[\frac{x}{\sqrt{x^2 + R^2}} - \frac{x - L}{\sqrt{(x - L)^2 + R^2}}\right]$$

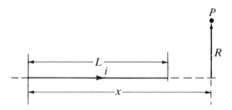

Figure 16.36   Problem 16.39

16.40

(a) Use the result of Problem 16.39 to show that the magnitude of the magnetic field at the point P shown in Figure 16.37 is given by

$$B = \frac{\mu_0 i}{2\pi R}\frac{1}{\sqrt{1 + 4R^2/L^2}}$$

(b) Use the result of part (a) to obtain the expression for the magnetic field of a long, straight, current-carrying wire.

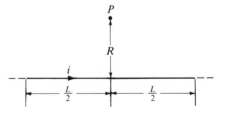

Figure 16.37   Problem 16.40

16.41

(a) Use the result of Problem 16.39 to show that the magnitude of the magnetic field at the point P shown in Figure 16.38 is given by

$$B = \frac{\mu_o i}{4\pi R} \frac{1}{\sqrt{1 + R^2/L^2}}$$

(b) Use the result of part (a) to obtain an expression for the magnetic field of "a half-infinite wire," that is, the magnetic field at point P for $L \gg R$.

Figure 16.38   Problem 16.41

16.42 Two long parallel wires, each carrying a current i_1 in the $\hat{\mathbf{k}}$-direction, pass through the x axis at $x = \pm a$. Determine the magnetic force exerted by these two wires on a segment of wire carrying a current i_2 in the $\hat{\mathbf{j}}$-direction along the y axis from $y = 0$ to $y = 3a$.

GROUP **B**

16.43 A current i follows the perimeter of an equilateral triangle. If the length of each side of the triangle is given by L, determine the magnitude of the magnetic field at the geometric center of the triangle.

16.44 Figure 16.39 shows a cross section of a long solenoid ($n = 2.0 \times 10^3$ turns/m, $i = 5.0$ A). What is the velocity of an electron at point P ($x_0 = 1.0$ cm) if the electron follows a helical path that is coaxial with the solenoid and has a pitch of 2.0 cm?

16.45 Charge of uniform density σ is distributed over an annular ring of inner radius a and outer radius b. If this ring is rotated with a frequency ν about the axis shown in Figure 16.40, determine the magnitude of the magnetic field at the center of the ring. (*Hint:* The current di for an infinitesimal ring of radius r and thickness dr is given by $2\pi\nu\sigma r dr$.)

16.46 A square path carries a current i, as shown in Figure 16.41.

(a) Show that the magnetic field at the point P shown is given by

$$\mathbf{B} = \frac{\mu_o i a^2}{2\pi\left(x^2 + \dfrac{a^2}{4}\right)\sqrt{x^2 + \dfrac{a^2}{2}}} \hat{\mathbf{k}}$$

(b) Investigate the $x \gg a$ behavior of the result obtained in part (a) and compare this result to that obtained in Example 16.2.

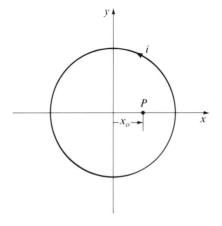

Figure 16.39   Problem 16.44

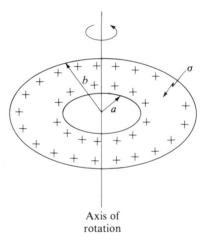

Axis of rotation

Figure 16.40   Problem 16.45

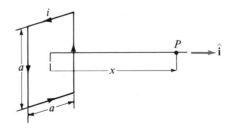

Figure 16.41   Problem 16.46

16.47 A 20-A current follows a rectangular path (*see* Fig. 16.42). Calculate the magnitude of the magnetic field at the point P shown.

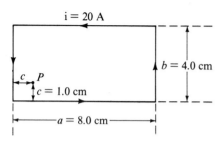

Figure 16.42   Problem 16.47

16.48 Two parallel wire segments carry currents i_1 and i_2 as shown in Figure 16.43.

(a) Show that the magnitude of the (attractive) force on either segment by the other is given by

$$F = \frac{\mu_0 i_1 i_2}{2\pi} \left[\sqrt{1 + \frac{L^2}{R^2}} - 1 \right]$$

(b) Verify that the expression obtained in part (a) behaves appropriately for $L \gg R$.

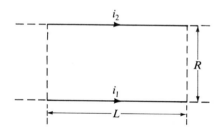

Figure 16.43   Problem 16.48

16.49 The spiral path $r = ae^{\alpha\theta}$ $(0 \le \theta \le 2\pi)$ carries a current i (*see* Fig. 16.44).

(a) Show that the magnetic field at the origin is given by

$$\mathbf{B} = \frac{\mu_0 i}{4\pi\alpha a} (1 - e^{-2\pi\alpha}) \, \hat{\mathbf{k}}$$

(b) Verify that this expression for $\mathbf{B}$ behaves appropriately as $\alpha \to 0$.

16.50 Figure 16.45 shows the square cross section of a long straight conductor carrying a current i out of the paper. If the current is uniformly distributed over the cross section, show that the magnetic field at point P $(x > a)$ is given by

$$\mathbf{B} = \left\{ \frac{\mu_0 i}{16\pi a^2} \int_{-a}^{a} \ln\left[\frac{(x+a)^2 + y^2}{(x-a)^2 + y^2} \right] dy \right\} \hat{\mathbf{j}}$$

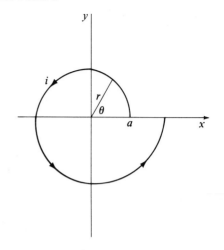

Figure 16.44   Problem 16.49

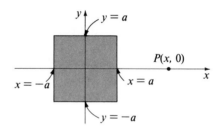

Figure 16.45   Problem 16.50

16.51 A particle (mass $= m$, charge $= q > 0$) moves in a region where both the electric and magnetic fields are uniform and given by $\mathbf{E} = E_0\hat{\mathbf{i}}$ and $\mathbf{B} = B_0\hat{\mathbf{i}}$ where E_0 and B_0 are positive quantities. At $t = 0$ the position and velocity of the particle are $\mathbf{r}_0 = \mathbf{0}$ and $\mathbf{v}_0 = v_0\hat{\mathbf{j}}$.

(a) Determine the position and velocity of the particle at the time $t = \pi m/qB_0$.

(b) Describe the motion of the particle for $t \ge 0$.

(c) Determine the kinetic energy of the particle as a function of the time t for $t \ge 0$.

16.52 An elliptical path described by

$$\mathbf{r} = \hat{\mathbf{i}} \, a \cos\theta + \hat{\mathbf{j}} \, b \sin\theta \qquad (0 \le \theta < 2\pi)$$

where $a \ge b$, is shown in Figure 16.46 carrying a current i. Show that

(a) the magnitude of the magnetic field at the origin is given by

$$B = \frac{\mu_0 i \sqrt{1 - \epsilon^2}}{\pi a} \int_0^{\pi/2} \frac{d\theta}{(1 - \epsilon^2 \sin^2\theta)^{3/2}}$$

where the eccentricity ϵ of the ellipse is defined by $\epsilon^2 = 1 - b^2/a^2$.

(b) the result obtained in part (a) behaves appropriately for
     $\epsilon = 0$.

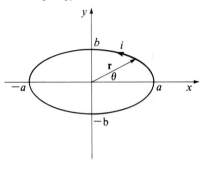

Figure 16.46   Problem 16.52

16.53 The (elliptic) integral obtained in part (a) of Problem
16.52 cannot be evaluated by any of the usual techniques of
calculus. If you have access to a computer, evaluate the function

$$f(\epsilon) = \frac{2\sqrt{1 - \epsilon^2}}{\pi} \int_0^{\pi/2} \frac{d\theta}{(1 - \epsilon^2 \sin^2\theta)^{3/2}}$$

to two-significant-figure accuracy for $\epsilon = 0.1, 0.2, \ldots, 0.9$
and plot B at the center of the ellipse as a function of the
eccentricity ϵ of the ellipse for $0 \leq \epsilon \leq 0.9$. *Hint:*

$$\int_0^{\pi/2} \frac{d\theta}{(1 - \epsilon^2 \sin^2\theta)^{3/2}} = \lim_{N \to \infty} \frac{\pi}{2N} \sum_{k=0}^{N-1} \frac{1}{(1 - \epsilon^2 \sin^2\theta_k)^{3/2}}$$

where $\theta_k = \pi(1 + 2k)/4N$.

17 Magnetic Fields II

In 1831 Michael Faraday and Joseph Henry independently discovered that a changing magnetic field induces, or generates, an emf and, therefore, can cause currents in closed circuits. This discovery proved to have considerable practical and theoretical significance. Its practical application was in providing the means, through generators, for converting mechanical energy into electrical energy. From a theoretical perspective the discovery of induced emf marked the onset of a comprehensive understanding of *electromagnetism*, a term describing those phenomena associated with the interdependence of electric and magnetic fields. This understanding led James Clerk Maxwell to formulate the theory of electromagnetic radiation, which we consider in Chapter 18.

We will study induced emf and one further circuit element, the inductor, and then investigate briefly the role of inductors in circuits with resistors. We will also discuss some of the physical effects associated with magnetic materials and the basic principles of electromagnetism in a few compact relationships, called Maxwell's equations.

17.1 Induced Emf

Faraday's Law and Lenz's Law

A statement of the results of the experiments of Faraday and Henry is called *Faraday's law*:

Any change in the total magnetic flux for a surface bounded by a closed path induces an emf along that path.

The magnetic flux $\phi_m = \int \mathbf{B} \cdot d\mathbf{S}$ evaluated for a surface is loosely spoken of as the magnetic flux through that surface. (Actually, the continuous lines of the magnetic field may pass through a surface, but the magnetic flux for a surface is a value found by summing scalar quantities evaluated at points on that surface.)

Faraday's law is expressed quantitatively by relating the induced emf $\mathcal{E}_i$ and the time rate of change of magnetic flux $d\phi_m/dt$ by

$$\mathcal{E}_i = -\frac{d\phi_m}{dt} \qquad (17\text{-}1)$$

E 17.1   Show that the SI unit for $d\phi_m/dt$ is the volt.

The negative sign in Equation (17-1) conforms to what will be our sign convention in specifying the directional sense of the induced emf. (We will consider the directional sense of induced emf in more detail below.) Because magnetic flux for a surface S is given by $\phi_m = \int_S \mathbf{B} \cdot d\mathbf{S}$, Faraday's law, Equation (17-1), may be written

$$\mathcal{E}_i = -\frac{d}{dt}\int_S \mathbf{B} \cdot d\mathbf{S} \qquad (17\text{-}2)$$

Faraday's law asserts that an emf is induced in a closed path whenever the flux through the surface enclosed by that path *changes*—regardless of how that change comes about. If the path bounding a surface through which the flux is changing is a conductor, current is induced along the path because of the induced emf. Then if we consider a closed wire loop, which may have lines of magnetic field $\mathbf{B}$ passing through it, the magnetic flux through that loop may change by any one or more of the following means:

1. $\mathbf{B}$ may change in magnitude.

2. $\mathbf{B}$ may change in direction relative to the loop.

3.   The area of the loop may change.

The physical arrangement of Figure 17.1(a) shows a magnetic field, caused by loop 2, passing through loop 1. A galvanometer G is attached to the first loop to indicate an induced emf in that loop; the field is produced by loop 2 because it carries a current determined by the emf $\mathcal{E}$ and the variable resistor R. The changes in flux enumerated above may be accomplished in the following ways:

1.   When the loops of Figure 17.1(a) are fixed in space, the magnitude of $\mathbf{B}$ passing through loop 1 may be changed by varying R, thereby changing the current in loop 2 and the field $\mathbf{B}$. Figure 17.1(a) shows $\mathbf{B}$ increasing, and the flux through loop 1 is changing.

2.   Figure 17.1(b) shows loop 2 being turned about an axis through loop 1, which is fixed. The direction of $\mathbf{B}$ at the surface bounded by loop 1 is changing; the angle between $\mathbf{B}$ and each infinitesimal surface area $d\mathbf{S}$ is changing. As a result, the magnetic flux through loop 1 is changing.

3.   In Figure 17.1(c) the total area of the surface bounded by loop 1 is being changed by squeezing the loop. The magnetic flux for the area, therefore, is changing.

Lenz's law is a verbalization of the experimentally observed facts that govern the directional sense of any induced emf:

An induced emf always opposes the external change that produces it.

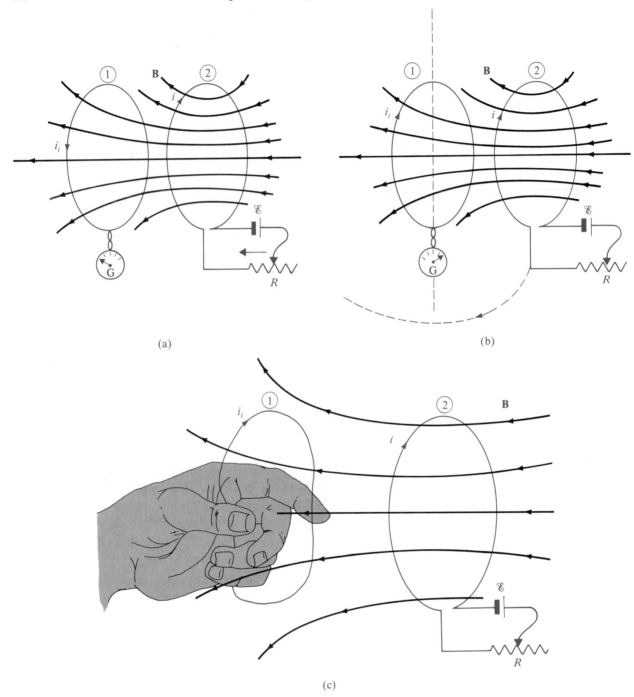

(a)

(b)

(c)

Figure 17.1
Three ways to change the magnetic flux through a coil. **(a)** A changing current in coil 2 causes a change in the magnitude of **B** in coil 1, thereby causing a change in magnetic flux through coil 1. **(b)** Swinging coil 2 about the axis shown through coil 1 causes a change in direction of **B** through coil 1, thereby causing a change in magnetic flux through coil 1. **(c)** Squeezing coil 1 causes a change in the area of coil 1, thereby causing a change in magnetic flux through coil 1.

This rule affords us a quick means for the determination of the directional sense of the induced emf (or, therefore, the induced current) in a circuit. For example, in Figure 17.1(a), where the magnitude of **B** is increasing in loop 1 and the magnetic flux through that loop is increasing, a current i_i is induced in loop 1 in the sense indicated. Using the right-hand rule, grasping the wire of loop 1 with the extended thumb along the current sense, we note that the curved fingers inside the loop point in the direction of a magnetic field that *opposes* the increase in magnetic field that is taking place. Similarly, in Figure 17.1(b), where the magnetic flux through 1 is decreasing, the induced current i_i causes a magnetic field in a direction that opposes that decrease in magnetic flux. Can you convince yourself that the directional sense of the induced current i_i in Figure 17.1(c) is appropriate?

E 17.2   A conducting rectangular loop of wire is positioned, as shown in the accompanying figure, relative to a long wire (in the plane of the loop) carrying a current i. What is the sense of the induced current in the rectangular loop if the current i is (a) increasing, (b) decreasing, (c) not changing.

Answers: (a) CCW; (b) CW; (c) no induced current

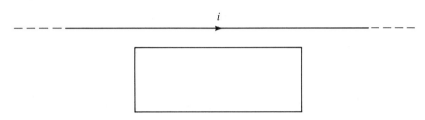

Exercise 17.2

E 17.3   The square ($L = 20$ cm) loop shown in the figure is positioned (in the x-y plane) in a uniform magnetic field ($B = 0.015$ T) into the paper. Calculate the magnitude of the magnetic flux for this surface.

Answer: 6.0×10^{-4} T·m^2

E 17.4   If the loop of Exercise 17.3 is rotated about an axis coincident with the x axis, will the sense of the initial induced emf be CW or CCW? Does the answer depend on the sense of this rotation about the x axis?

Answers: CW; no

E 17.5   If the loop of Exercise 17.3 is caused to expand, that is, L increases, what is the sense of the induced emf?   Answer: CCW

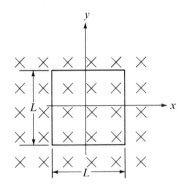

Exercise 17.3

A significant aspect of Faraday's law is that it may be formulated to describe quantitatively how a changing magnetic flux may produce an electric field. In Chapter 15 we defined an emf $\mathscr{E}$ to be a reservoir of electrical energy, capable of performing a quantity of work W on each quantity of charge q that transits the emf as

$$\mathscr{E} = \frac{W}{q} = \frac{1}{q} \int_C \mathbf{F} \cdot d\mathbf{s} \tag{17-3}$$

where $\mathbf{F}$ is the force that performs work on particles carrying the charge q as they pass through the emf along the curve C. If $\mathbf{F}$ is an electrical force, it may be expressed in terms of an electric field $\mathbf{E} = \mathbf{F}/q$, so that the emf $\mathscr{E}$ of Equation (17-3) may be represented by

$$\mathscr{E} = \int_C \mathbf{E} \cdot d\mathbf{s} \tag{17-4}$$

But according to Faraday's law, an induced emf $\mathscr{E}_i$ is developed around a *closed* path through which the magnetic flux is changing. Then Faraday's law of Equation (17-2) may be written

$$\mathscr{E}_i = \oint_C \mathbf{E}_N \cdot \mathbf{ds} = -\frac{d}{dt} \int_S \mathbf{B} \cdot \mathbf{dS} \qquad (17\text{-}5)$$

where S is any open surface bounded by the closed curve C. We have represented the induced component of the total electric field by $\mathbf{E}_N$ because this component of the electric field is *nonconservative*. Recall that the electrostatic component $\mathbf{E}_C$ of the total electric field is conservative (see Chapter 12 for a discussion of conservative electric fields), that is, $\oint \mathbf{E}_C \cdot \mathbf{ds} = 0$. It is this conservative property of $\mathbf{E}_C$ that permitted us to devise a meaningful electric potential function V. But here, in the *magnetodynamic* case, in which a changing magnetic flux generates an induced emf $\mathscr{E}_i$, the line integral of the induced electric field, $\oint_C \mathbf{E}_N \cdot \mathbf{ds}$, is obviously *not* equal to zero. It is, in fact, equal to the negative rate of change of the magnetic flux through the closed curve C. Therefore, the induced electric field $\mathbf{E}_N$ is a nonconservative electric field.

When the closed curve C along which an emf $\mathscr{E}_i$ is induced happens to be a conducting path, like a wire for example, the induced electric field $\mathbf{E}_N$ within the conductor causes an induced current i_i in the conductor. This situation is just as if the emf, which we have heretofore considered a distinct circuit element, were distributed continuously along the conducting path.

Faraday's law, written in the form of Equation (17-5), expresses the relationship between a changing magnetic flux and a nonconservative electric field. Thus Equation (17-5) is the first connection, or relationship, we have encountered between the magnetic field and the electric field. As mentioned earlier, this relationship serves as the fountainhead from which electromagnetic radiation theory may be developed (and was, by Maxwell).

Example 17.1

PROBLEM   A square conducting loop, which has sides 16 cm long and a 0.10-Ω resistance, is oriented so that a long wire carrying a current i is in the plane of the loop, as shown in Figure 17.2(a). The distance a is equal to 0.50 cm. At any instant when the current i is increasing at a rate of 15 A/s, calculate the current induced in the loop.

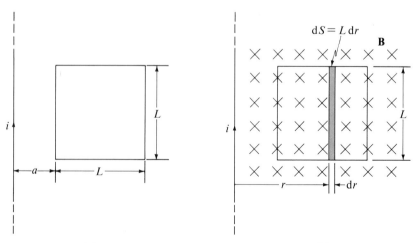

Figure 17.2   Example 17.1   (a)   (b)

SOLUTION Figure 17.2(b) shows an infinitesimal strip (length $= L$, width $= dr$) at a distance r from the long wire. The magnetic flux through this strip is

$$d\phi_m = B \, dS$$

where B is the magnitude of the (into-the-paper) magnetic field at a distance r from the wire, that is, $B = \mu_o i / 2\pi r$, and dS, the area of the strip, is given by $dS = L \, dr$. Thus, the magnetic flux for the strip is given by

$$\phi_m = \int_a^{a+L} \frac{\mu_o i}{2\pi r} \, L \, dr = \frac{\mu_o i L}{2\pi} \int_a^{a+L} \frac{dr}{r} = \frac{\mu_o i L}{2\pi} \ln\left(1 + \frac{L}{a}\right)$$

Using Faraday's law, the magnitude of the emf induced around the loop is

$$|\mathscr{E}| = \frac{\mu_o L}{2\pi} \ln\left(1 + \frac{L}{a}\right) \left|\frac{di}{dt}\right|$$

and the magnitude of the induced current I in the loop is

$$I = \frac{|\mathscr{E}|}{R} = \frac{\mu_o L}{2\pi R} \ln\left(1 + \frac{L}{a}\right) \left|\frac{di}{dt}\right|$$

$$I = \left(2 \times 10^{-7} \frac{\text{T} \cdot \text{m}}{\text{A}}\right) \cdot \left(\frac{0.16 \text{ m}}{0.10 \, \Omega}\right) \cdot \ln\left(1 + \frac{0.16}{0.0050}\right) \cdot 15 \frac{\text{A}}{\text{s}}$$

$$I = 1.7 \times 10^{-5} \text{ A} = 17 \, \mu\text{A}$$

A check of the units gives

$$\left(\frac{\text{T} \cdot \text{m}}{\text{A}}\right) \cdot \left(\frac{\text{m}}{\Omega}\right) \cdot \left(\frac{\text{A}}{\text{s}}\right) = \frac{\text{N} \cdot \text{m}}{\text{A}^2 \cdot \text{m}} \cdot \frac{\text{m} \cdot \text{A}}{\text{V}} \cdot \frac{\text{A}}{\text{s}} = \frac{\text{N} \cdot \text{m} \cdot \text{C}}{\text{J} \cdot \text{s}} = \frac{\text{C}}{\text{s}} = \text{A} \quad \blacksquare$$

E 17.6   A circular (radius $= 4.0$ cm) loop of wire is oriented with its plane perpendicular to the direction of a uniform magnetic field. At an instant when the magnitude of this field is changing at a rate of 500 G/s, calculate the magnitude of the emf induced in the loop.   Answer: 0.25 mV

Motional Emf and Faraday's Law

When a closed conducting loop of fixed shape moves without rotating relative to a uniform magnetic field, no net emf is induced in the loop because the magnetic flux through the loop does not change. Figure 17.3 shows a closed loop translating in a direction perpendicular to a uniform magnetic field **B**. The magnetic flux

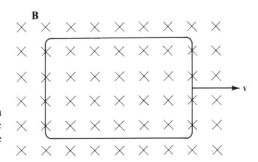

Figure 17.3
A closed loop translating in a direction perpendicular to a uniform magnetic field **B**. No net emf is induced in the loop.

through the loop is constant, and according to Faraday's law, no net emf is induced in the loop.

A different situation is shown in Figure 17.4, where a conducting bar of length L is sliding with velocity $\mathbf{v} = v\hat{\mathbf{i}}$ along fixed conducting rails in a region of uniform magnetic field $\mathbf{B}$, which is perpendicular to the plane of the rails and bar. The conducting path is completed by the resistor R, which represents the total resistance of the closed circuit. An external force $\mathbf{F}_{ext}$ pulls the bar along the rails at a constant speed v. Let us analyze this circuit, first without using Faraday's law, and then show that this analysis is equivalent to the use of the Faraday principle.

A positive charge q within the moving, conducting bar of Figure 17.4 experiences a magnetic force $\mathbf{F}_m$ in the $\hat{\mathbf{j}}$-direction, where

$$\mathbf{F}_m = q\mathbf{v} \times \mathbf{B}$$

Because the velocity $\mathbf{v}$ of the bar is perpendicular to $\mathbf{B}$, we may write

$$F_m = qvB \tag{17-6}$$

The external force $\mathbf{F}_{ext}$ that is pulling the sliding bar provides the work required to move the charges around the circuit. The work W done by $\mathbf{F}_m$ in moving a charge q through the length L of the bar is

$$W = F_m L = qvBL \tag{17-7}$$

Because the emf $\mathcal{E}_i$ induced in the conducting loop, that is, along the length of the bar, is equal to W/q, we may use Equation (17-7) to give

$$\mathcal{E}_i = \frac{W}{q} = BLv \tag{17-8}$$

and the current i_i induced in the circuit because of the motion of the bar is

$$i_i = \frac{\mathcal{E}_i}{R} = \frac{BLv}{R} \tag{17-9}$$

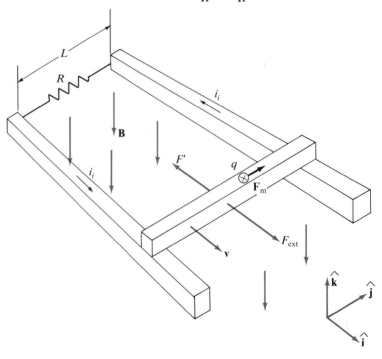

Figure 17.4
A conducting bar of length L sliding with velocity $\mathbf{v} = v\hat{\mathbf{i}}$ along fixed conducting rails in a region of uniform magnetic field $\mathbf{B}$, which is in a direction perpendicular to the plane of the bar and rails. The resistor R represents the total resistance of the closed circuit.

The rate P at which energy is dissipated as heat in the resistor is

$$P = i_i^2 R = \frac{B^2 L^2 v^2}{R} \tag{17-10}$$

Now we note that because a current is in the bar, which is moving through a region of magnetic field $\mathbf{B}$, the bar experiences a force

$$\mathbf{F'} = i_i \mathbf{L} \times \mathbf{B} \tag{17-11}$$

where $\mathbf{L}$ is a vector of magnitude L in the $\hat{\mathbf{j}}$-direction, the directional sense of the current in the bar. Because $\mathbf{L}$ and $\mathbf{B}$ are perpendicular, we have

$$F' = B i_i L \tag{17-12}$$

The force $\mathbf{F'}$ is directed oppositely to the external force $\mathbf{F}_{ext}$.

Because the bar moves at a constant velocity, the resultant force acting on the bar must be zero, i.e.,

$$\mathbf{F}_{ext} + \mathbf{F'} = 0$$

or

$$F_{ext} = F' = B i_i L$$

The rate P_{ext} at which the external force is doing work (on the bar) is given by

$$P_{ext} = \mathbf{F}_{ext} \cdot \mathbf{v} = F_{ext} v = B i_i L v$$

and because $i_i = BLv/R$, we get

$$P_{ext} = \frac{B^2 L^2 v^2}{R} \tag{17-13}$$

Thus, just as we expected, the external force provides energy to the system at precisely the rate at which energy is removed by the heat generation in the resistor.

E 17.7   If $R = 0.040\ \Omega$, $L = 25$ cm, $v = 1.5$ m/s, and $B = 0.080$ T in Figure 17.4, calculate the induced emf, the induced current, the magnitude of the external force, and the rate at which the external force is adding energy to the system.   Answers: 30 mV, 0.75 A, 0.015 N, 23 mW

As a result of the motion of the bar of Figure 17.4, the area $A = Lx$ of the surface bounded by the circuit is increasing at the rate

$$\frac{dA}{dt} = L \frac{dx}{dt} = Lv \tag{17-14}$$

Then the magnetic flux $\phi_m = BA$ through the circuit is changing at a rate

$$\frac{d\phi_m}{dt} = \frac{d}{dt}(BA) = B \frac{dA}{dt}$$

Substituting Equation (17-14) and using Faraday's law, $\mathcal{E}_i = -d\phi_m/dt$, gives

$$|\mathcal{E}_i| = \left| -\frac{d\phi_m}{dt} \right| = B \frac{dA}{dt} = BLv \tag{17-15}$$

the result obtained in Equation (17-8). Lenz's law requires that the induced current be in the directional sense indicated in Figure 17.4.

We may further illustrate the use of Faraday's law with a rigid loop moving through a nonuniform magnetic field.

Suppose a rigid rectangular conducting loop of length L and width w is oriented with its length parallel to a long wire (in the plane of the loop) carrying current i. If the loop starts at $t = 0$ with the side of length L nearest the wire positioned at a radial distance r_0 from the long wire and moves radially away from the wire at a constant speed v, write an expression for the emf induced in the loop as a function of time.

The problem situation may be visualized as shown in Figure 17.5 at an arbitrary time t, when the left side of the loop is at a radial distance r from the long wire carrying current i. The magnetic flux ϕ_m through the loop may be found by recognizing that $\mathbf{B}$ has constant magnitude along infinitesimal strips of length L and width dx, shown by the shaded area of Figure 17.5. Then we may write

$$\phi_m = \int \mathbf{B} \cdot d\mathbf{S} = \int B\,dS = \frac{\mu_0 iL}{2\pi} \int_r^{r+w} \frac{dx}{x} = \frac{\mu_0 iL}{2\pi} \ln\left(\frac{r + w}{r}\right) \qquad (17\text{-}16)$$

where we have used the facts that $\mathbf{B}$ is perpendicular to the surface bounded by the loop at every point on that surface and the magnitude of $\mathbf{B}$ is given by

$$B = \frac{\mu_0 i}{2\pi x} \qquad (17\text{-}17)$$

Faraday's law involves the time derivative of ϕ_m; this fact suggests that we should express ϕ_m as a function of time. Because the velocity of the loop has a constant magnitude v and is directed radially outward, we may write

$$r = r_0 + vt \qquad (17\text{-}18)$$

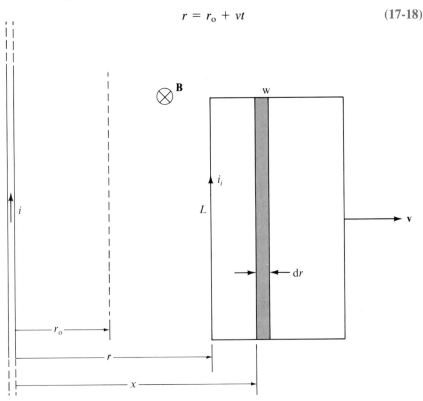

Figure 17.5
A rigid rectangular loop of wire moving with speed v. A long wire in the plane of the loop carries a constant current i.

Substituting Equation (17-18) into Equation (17-16) gives

$$\phi_m = \frac{\mu_o iL}{2\pi} \ln\left(\frac{r_o + vt + w}{r_o + vt}\right)$$

and the magnitude of $d\phi_m/dt$ is found by differentiation:

$$\left|\frac{d\phi_m}{dt}\right| = \frac{\mu_o iL}{2\pi} \left|\frac{-wv}{(r_o + vt)(r_o + vt + w)}\right|$$

Then the magnitude of the emf induced in the loop is given by Faraday's law, $\mathscr{E}_i = -d\phi_m/dt$, to be

$$|\mathscr{E}_i| = \frac{\mu_o iL}{2\pi}\left[\frac{wv}{(r_o + vt)(r_o + vt + w)}\right] \qquad (17\text{-}19)$$

The sense of the emf in the loop is quickly determined using Lenz's law: Because the magnetic flux through the loop is decreasing in magnitude as the loop moves, the current induced in the loop must be in the sense indicated in Figure 17.5, a directional sense that produces a magnetic flux through the loop that opposes the change.

Finally, we may check some aspects of our result. The units of the right side of Equation (17-19) are

$$\left(\frac{T \cdot m}{A}\right)(A)(m)\left[\frac{m^2}{s \cdot m^2}\right] = \frac{T \cdot m^2}{s}$$

and the unit of the induced emf is work per unit charge, or $J/C = V = (T \cdot m^2)/s$. Further, as $t \to \infty$, the emf of Equation (17-19) approaches zero, a value we expect because the magnetic field (and, therefore, the flux through the loop) approaches zero as r becomes very large, according to Equation (17-17). Finally, Equation (17-19) predicts that the induced emf is proportional to the speed of the loop. This prediction is consistent with the physical fact that flux through the loop changes more rapidly as the loop moves faster.

E 17.8   Calculate the magnitude of the induced emf in the loop at $t = 0$ if $i = 75$ A, $r_o = 1.0$ cm, $L = 1.0$ m, $w = 0.20$ m, and $v = 2.0$ m/s in Figure 17.5.
Answer: 2.9 mV

E 17.9   If the resistance of the loop of Exercise 17.8 is equal to 0.050 Ω, determine the magnitude of the induced current in the loop at $t = 0$.
Answer: 0.058 A

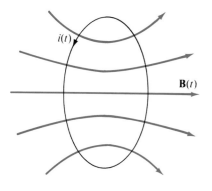

Figure 17.6
An isolated loop of wire carrying a time-dependent current $i(t)$, which produces a time-dependent magnetic field $\mathbf{B}(t)$ through the loop.

17.2 Inductance

If an isolated conducting loop, like that of Figure 17.6, carries a current that is changing with time, that is, a time-dependent current $i = i(t)$, the magnetic field $\mathbf{B} = \mathbf{B}(t)$ through the loop is changing. Then the magnetic flux through the loop is also changing, or $\phi_m = \phi_m(t)$. Because the magnetic field is proportional to the current in the loop, we may conclude that the magnetic flux ϕ_m is also proportional to the current, or

$$\phi_m = Li \qquad (17\text{-}20)$$

The quantity L, the constant of proportionality in Equation (17-20), is called the *self-inductance*, or simply the *inductance*, of the loop. A given loop with a particular size and shape has a fixed inductance L; in other words, L depends only on the geometry of the loop.

If a number N of turns of wire are closely spaced in a coil or wound in the form of a long solenoid, so that the magnetic flux ϕ_m is the same for each turn of wire, Equation (17-20) takes the form

$$N\phi_m = Li \tag{17-21}$$

In Equation (17-21), L is the inductance of the device and ϕ_m is the magnetic flux through a single turn of wire.

According to Faraday's law, a coil of wire will develop across its terminals an induced emf $\mathscr{E}_i$ given by

$$\mathscr{E}_i = -N\frac{\mathrm{d}\phi_m}{\mathrm{d}t} = -\frac{\mathrm{d}}{\mathrm{d}t}(Li)$$

and, because L is independent of time, the induced emf may be written

$$\mathscr{E}_i = -L\frac{\mathrm{d}i}{\mathrm{d}t} \tag{17-22}$$

The negative sign in Equation (17-22) is consistent with the sign conventions we have adopted; but, more practically, it is consistent with Lenz's law: The emf is induced in the directional sense that opposes any change in the current.

When a coil with inductance L is used as a component in a circuit, it is called an *inductor*. An inductor functions in a circuit to oppose any change of current in that circuit. Figure 17.7(a) shows a typical inductor, and Figure 17.7(b) shows the symbol used in circuit diagrams to represent an inductor having inductance L.

The unit of inductance is the henry (H), which is expressed in fundamental units as $(\mathrm{kg}\cdot\mathrm{m}^2)/(\mathrm{A}^2\cdot\mathrm{s}^2)$. It may be seen from Equations (17-20) and (17-22) that the unit of inductance may be expressed as

$$1\,\mathrm{H} = \frac{\mathrm{V}\cdot\mathrm{s}}{\mathrm{A}} = \Omega\cdot\mathrm{s} \tag{17-23}$$

E 17.10 At an instant when the current through a 2.5-mH inductor is equal to 0.80 A and is changing at a rate of 0.40 A/s, determine the magnitude of the induced emf across the terminals of the inductor. Answer: 1.0 mV

Figure 17.7
(a) An inductor. (b) The symbol used in circuit diagrams to represent an inductor having inductance L.

(a)

(b)

E 17.11   If the current i shown in the figure is increasing at a rate of 2.0 A/s, calculate the potential difference, $V_1 - V_2$. *Be sure you understand the sign of this result.*
Answer: 0.080 V

Exercise 17.11

E 17.12   At an instant when $i = 15$ mA and $di/dt = -40$ mA/s, calculate the potential difference, $V_A - V_B$, for the circuit segment shown in the figure.
Answer: -59 mV

Exercise 17.12

The calculation of the inductance of a coiled-wire device can usually be accomplished by the following procedure:

1. Assume the device carries a current i.

2. Determine the magnetic field **B** within the device.

3. Use **B** to find the magnetic flux $\phi_m = \int \mathbf{B} \cdot d\mathbf{S}$ within the device.

4. Use $L = N\phi_m/i$ to determine the inductance of the device.

The following problem illustrates this procedure.

Calculate the self-inductance of a toroid having a rectangular cross section. The toroid has N turns about a form having an inner radius a and an outer radius b. The toroid has thickness c.

Figure 17.8 shows a cutaway view of this toroid. The magnetic field within the interior of the toroid (see Chapter 16, "Magnetic Fields I," Section 16.4) has magnitude

$$B = \frac{\mu_o Ni}{2\pi r}$$

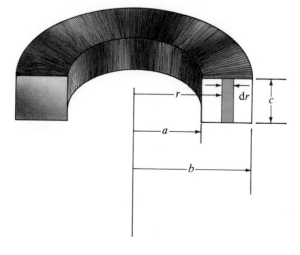

Figure 17.8
A cutaway view of a toroid.

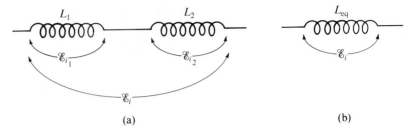

Figure 17.9
(a) Two inductors, L_1 and L_2, connected in series. **(b)** The equivalent inductance L_{eq} of inductors L_1 and L_2.

Then the magnetic flux ϕ_m within the interior of the toroid is found by noting that an infinitesimal area dS within the cross section of the toroid is equal to $c\,dr$. The magnetic flux for a cross-sectional area is, therefore,

$$\phi_m = \int \mathbf{B} \cdot d\mathbf{S} = \frac{\mu_o Nic}{2\pi} \int_a^b \frac{dr}{r} = \frac{\mu_o Nic}{2\pi} \ln\left(\frac{b}{a}\right) \qquad (17\text{-}24)$$

where we have used the fact that $\mathbf{B}$ is perpendicular to an element $d\mathbf{S}$ of area at every point of the interior of the toroid. Finally, the inductance L of the toroid is given by

$$L = \frac{N\phi_m}{i} = \frac{\mu_o N^2 c}{2\pi} \ln\left(\frac{b}{a}\right) \qquad (17\text{-}25)$$

E 17.13   Determine the inductance of the toroid of Figure 17.8 if $N = 1000$, $a = 1.0$ cm, $b = 2.0$ cm, and $c = 1.5$ cm.   Answer: 2.1 mH

Inductors as Circuit Components

Like series and parallel combinations of resistors or capacitors, combinations of inductors in a circuit may be replaced by an equivalent inductance L_{eq} without affecting the electrical function of the circuit. Figure 17.9(a) shows two inductors, L_1 and L_2, in series. The total emf $\mathcal{E}_i$ induced by the combination is equal to the sum of the emfs, $\mathcal{E}_{i1}$ and $\mathcal{E}_{i2}$, induced by L_1 and L_2, or

$$\mathcal{E}_i = \mathcal{E}_{i1} + \mathcal{E}_{i2} \qquad (17\text{-}26)$$

The emf $\mathcal{E}_i$ induced by the combination is equal to the emf induced by the equivalent inductor L_{eq} of Figure 17.9(b). Because the current i in each inductor of the combination is equal to the current in the equivalent inductor, we may write Equation (17-26) as

$$L_{eq}\frac{di}{dt} = L_1 \frac{di}{dt} + L_2 \frac{di}{dt}$$

$$L_{eq} = L_1 + L_2 \qquad (17\text{-}27)$$

The foregoing analysis may be extended to apply to N inductors in series:

$$L_{eq} = L_1 + L_2 + \ldots + L_N \quad \text{(inductors in series)} \qquad (17\text{-}28)$$

E 17.14   At an instant when the current i in the circuit segment shown in the figure has a value of 2.6 A and is increasing at a rate of 0.85 A/s, calculate (a) the potential difference, $V_A - V_B$, (b) the potential difference, $V_B - V_C$, (c) the potential difference, $V_A - V_C$, and (d) the equivalent inductance for the two inductors.   Answers: (a) 0.17 V; (b) 0.51 V; (c) 0.68 V; (d) 0.80 H

Exercise 17.14

$L_1 = 0.20$ H $L_2 = 0.60$ H

A B C

E 17.15   At an instant when the potential difference, $V_A - V_B$, is equal to $+15$ mV in the circuit segment shown, determine di/dt for the current i shown in the figure.   Answer: -0.18 A/s

$L_1 = 45$ mH $L_2 = 25$ mH $L_3 = 15$ mH

Exercise 17.15

A i B

Figure 17.10(a) shows two inductors, L_1 and L_2, connected in parallel. The current i in the combination is equal to the sum of the currents i_1 and i_2 in the individual inductors:

$$i = i_1 + i_2$$

so that

$$\frac{di}{dt} = \frac{di_1}{dt} + \frac{di_2}{dt} \tag{17-29}$$

Because the potential differences across components connected in parallel are equal, the induced emfs in each inductor must be equal to the emf $\mathscr{E}_i$ induced in the combination and, therefore, in the equivalent inductor L_{eq} of Figure 17.10(b). Then, using Equation (17-22), we may write

$$\frac{\mathscr{E}_i}{L_{eq}} = \frac{\mathscr{E}_i}{L_1} + \frac{\mathscr{E}_i}{L_2}$$

or

$$\frac{1}{L_{eq}} = \frac{1}{L_1} + \frac{1}{L_2} \tag{17-30}$$

Again, if this analysis is extended to apply to N inductors connected in parallel, we obtain

$$\frac{1}{L_{eq}} = \frac{1}{L_1} + \frac{1}{L_2} + \cdots + \frac{1}{L_N} \quad \text{(inductors in parallel)} \tag{17-31}$$

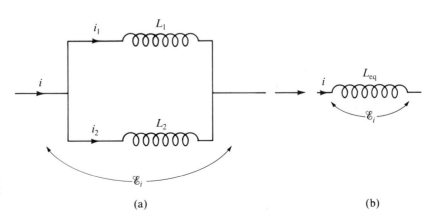

Figure 17.10
(a) Two inductors, L_1 and L_2, connected in parallel. **(b)** The equivalent inductance L_{eq} of inductors L_1 and L_2.

(a)   (b)

E 17.16   If $di/dt = +3.0$ A/s in the circuit segment shown in the figure, calculate the potential difference, $V_A - V_B$.   Answer: $+0.45$ V

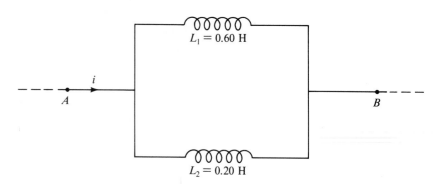

Exercise 17.16

E 17.17   If $V_A - V_B = +0.60$ V in the circuit segment shown in the figure, determine di/dt for the current i.   Answer: -0.60 A/s

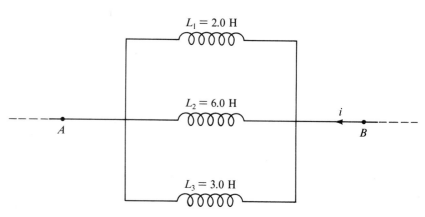

Exercise 17.17

Example 17.2   **PROBLEM**   When $I = 0.40$ A, $dI/dt = 30$ A/s, and $V_A - V_B = 6.0$ V in the circuit segment of Figure 17.11(a), calculate the currents I_1 and I_2 and the rate of change of each of these two currents.

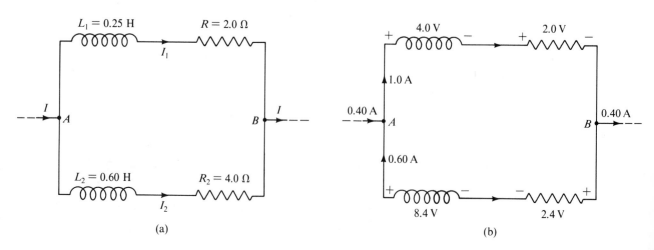

(a)   (b)

Figure 17.11   Example 17.2

SOLUTION   Applying the junction rule and its time derivative at point A gives

$$I_1 + I_2 = I$$

$$\frac{dI_1}{dt} + \frac{dI_2}{dt} = \frac{dI}{dt}$$

Summing potential changes along the I_1 path and along the I_2 path and equating each of the sums to $V_A - V_B$ gives

$$L_1\frac{dI_1}{dt} + R_1 I_1 = V_A - V_B$$

$$L_2\frac{dI_2}{dt} + R_2 I_2 = V_A - V_B$$

A solution of these four relationships for I_1, I_2, dI_1/dt, and dI_2/dt gives, after substituting the specified values for I, dI/dt, and $V_A - V_B$:

$$I_1 = 1.0 \text{ A}, \quad I_2 = -0.60 \text{ A}, \quad \frac{dI_1}{dt} = 16\frac{\text{A}}{\text{s}}, \quad \frac{dI_2}{dt} = 14\frac{\text{A}}{\text{s}}$$

Thus at the instant being considered, I_1 has a directional sense to the right, as shown in Figure 17.11(a), and a magnitude of 1.0 A that is increasing at a rate of 16 A/s. But I_2 has a directional sense to the left, which is opposite to that shown in Figure 17.11(a), and a magnitude of 0.60 A that is *decreasing* at the rate of 14 A/s. Figure 17.11(b) shows the current in each branch and the potential change across each component. ■

E 17.18   Verify the magnitude and polarity of each of the potential changes shown in Figure 17.11(b).

E 17.19   Verify that the junction rule is satisfied at points A and B of Figure 17.11(b) and that the loop rule is satisfied for the loop of Figure 17.11(b).

Energetics of Inductors

A circuit component having a potential difference ΔV across its terminals while carrying a current i is receiving or delivering energy at a rate given by $P = i\,\Delta V$, where P is the electrical power received or delivered by the component. In particular, an inductor L with an emf $\mathcal{E}_i = -L(di/dt)$ induced across its terminals at a given instant while carrying a current i is storing or discharging energy at a rate given by

$$P = \left|\mathcal{E}_i i\right| = \left|iL\frac{di}{dt}\right| = \left|\frac{d}{dt}\left(\frac{1}{2}Li^2\right)\right| \tag{17-32}$$

Because power is the time rate of change of energy, the energy U_L stored in an inductor at any instant is, according to Equation (17-32),

$$U_L = \frac{1}{2}Li^2 \tag{17-33}$$

Recall that the energy stored in a capacitor (see Chapter 15, *Direct Current Circuits*) is considered stored in the electric field between the conductors of the capacitor. In a similar way, the energy stored in an inductor is considered to reside in the magnetic field in and around the windings of the inductor. When the current in an inductor is increasing, energy is being delivered to the inductor, and its magnetic field is increasing in magnitude. When the current is decreasing, the inductor is returning its stored energy to the rest of the circuit, while the magnetic field associated with the inductor is decreasing in magnitude.

E 17.20   At an instant when the current i shown in the accompanying figure is equal to 0.70 A and is increasing at a rate of 0.50 A/s, calculate (a) the potential difference, $V_B - V_A$, (b) the energy stored in L, and (c) the rate at which the energy stored in L is changing.

Answers: (a) -40 mV; (b) 20 mJ; (c) $+28$ mW

Exercise 17.20

E 17.21   Repeat Exercise 17.20 for an instant when i is equal to 0.50 A and is decreasing at the rate of 0.60 A/s.

Answers: (a) $+48$ mV; (b) 10 mJ; (c) -24 mW

Example 17.3   **PROBLEM**   For the circuit segment shown in Figure 17.12(a), the potential differences $V_A - V_B$, $V_A - V_C$, and $V_A - V_D$ are, at a certain instant, equal to -10 V, $+15$ V, and $+12$ V, respectively. Determine the rate at which energy is being converted in each component.

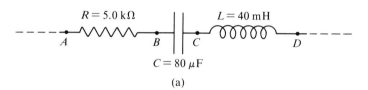

(a)

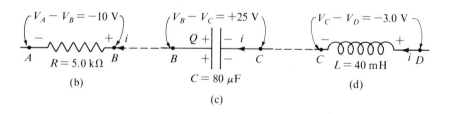

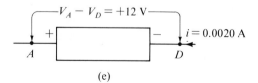

Figure 17.12   Example 17.3

SOLUTION   Figures 17.12(b), (c), and (d) show the resistor, capacitor, and inductor with appropriate potential differences (magnitudes and polarities), current i (sense), and charge Q (polarity). (Be sure you understand how each of these follows from the given information.) The magnitude of the current i is obtained using Ohm's law

$$i = \frac{\Delta V_R}{R} = \frac{10 \text{ V}}{5.0 \text{ k}\Omega} = 2.0 \text{ mA}$$

Because the rate at which energy is being converted in any circuit component is given by $|i\Delta V|$, we have

$$P_R = |i\Delta V_R| = (0.0020 \text{ A}) \cdot (10 \text{ V}) = 20 \text{ mW}$$

$$P_C = |i\Delta V_C| = (0.0020 \text{ A}) \cdot (25 \text{ V}) = 50 \text{ mW}$$

$$P_L = |i\Delta V_L| = (0.0020 \text{ A}) \cdot (3.0 \text{ V}) = 6.0 \text{ mW}$$

Thus energy is being *dissipated* as heat in the resistor at a rate of 20 mW, being *discharged* from the capacitor at a rate of 50 mW, and being *stored* in the inductor at a rate of 6.0 mW. A check on these results may be obtained by noting that because $V_A - V_D = +12$ V and $i = 2.0$ mA, as shown in Figure 17.12(e), the circuit segment is *providing* energy to the remainder of the circuit at a rate $P = |i\Delta V| = 24$ mW, a result in agreement with the 50 mW being added by C less the 26 mW being removed by R and L. ■

E 17.22   Determine the charge Q on the capacitor of Example 17.3.

Answer: 2.0 mC

E 17.23   Calculate the rate at which the current i is changing in Example 17.3.

Answer: +75 A/s

17.3 *LR* Circuits

The role of an inductor in a circuit may be illustrated by the analysis of a circuit like that of Figure 17.13(a). The inductor L is connected in series with a resistor R, a fixed emf $\mathcal{E}$, and an open switch S. Before the switch is closed, the current in the circuit is zero. Because of the presence of the inductor in the circuit, di/dt

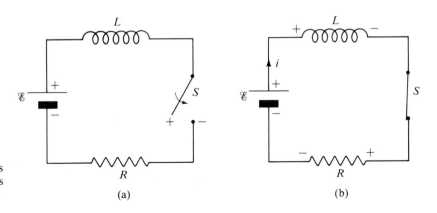

Figure 17.13
An LR circuit (a) before the switch S is closed, and (b) after the switch is closed.

(a)   (b)

cannot have an infinite value (or an infinite potential difference $\mathcal{E}_i$ would be induced across L). Therefore, immediately after the switch is closed, no current is in the circuit. As the current increases, after the switch is closed, from its zero value, the inductor responds to the increasing current by inducing an emf given by $\mathcal{E}_i = -L(di/dt)$ that appears as a potential difference across the terminals of the inductor. The induced emf opposes the change in current in the circuit; the sense of the emf induced in the inductor is indicated by the $+$ and $-$ signs on the inductor terminals in Figure 17.13(b).

The circuit may be analyzed by applying Kirchhoff's loop rule:

$$\mathcal{E} - L\frac{di}{dt} - iR = 0$$

or

$$L\frac{di}{dt} = \mathcal{E} - iR \tag{17-34}$$

Equation (17-34) may be rearranged as

$$\frac{di}{\mathcal{E} - iR} = \frac{dt}{L}$$

and integrated to give

$$\ln(\mathcal{E} - iR) = -\frac{R}{L}t + K \tag{17-35}$$

E 17.24 Perform the integration necessary to obtain Equation (17-35).

The integration constant K in Equation (17-35) may be evaluated using the initial condition for the current: $i = 0$ when $t = 0$. Then K has the value $K = \ln \mathcal{E}$, and Equation (17-35) becomes

$$i = \frac{\mathcal{E}}{R}(1 - e^{-(Rt/L)}) \tag{17-36}$$

E 17.25 Perform the algebraic operations necessary to obtain Equation (17-36) from Equation (17-35).

A graph of Equation (17-36) is shown in Figure 17.14. The current i in the circuit is equal to zero at $t = 0$ and asymptotically approaches the steady-state value $i = \mathcal{E}/R$ as t becomes large compared to L/R, the *time constant* of the *LR* circuit. The time constant of an *LR* circuit is a measure of the time required for the current, starting at a value equal to zero, to reach a value $(1 - 1/e) \cong 63\%$ of its steady-state value. It may be instructive to compare Equation (17-36) and the curve of Figure 17.14 to those corresponding to the *RC* circuit of Figure 15.18(b) in Chapter 15.

E 17.26 Verify that the quantity L/R has the SI unit of seconds.

Example 17.4   **PROBLEM** At $t = 0$ the switch S_1 of Figure 17.15(a) is closed. Later, at a time t_1, when the current in L is equal to 2.0 A, the switch S_2 is closed and then S_1 is immediately opened. Determine the current in L at a time 0.30 s after S_1 is closed.

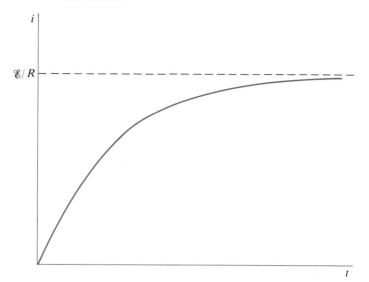

Figure 17.14
The current i as a function of time t in the circuit of Figure 17.13.

SOLUTION   After S_1 is closed, the circuit is as shown in Figure 17.15(b). This circuit is similar to the one just analyzed in the text, and, therefore, the current is given by

$$i = \frac{\mathcal{E}}{2R}(1 - e^{-(2Rt/L)}) = 2.5\,(1 - e^{-10t})\ \text{A}\quad (0 \leq t \leq t_1)$$

Because the current is equal to 2.0 A when $t = t_1$, we have

$$2.0\ \text{A} = 2.5\,(1 - e^{-10t_1})\text{A}$$

Solving for t_1 gives

$$t_1 = \frac{\ln 5}{10}\ \text{s} \approx 0.16\ \text{s}$$

For $t > t_1$, the circuit is as shown in Figure 17.15(c). The loop rule gives

$$-iR - L\frac{di}{dt} = 0$$

or

$$\frac{di}{i} = -\frac{R}{L}dt$$

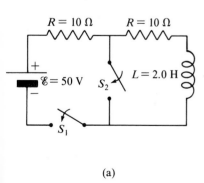

Figure 17.15   Example 17.4

Integrating and solving for i gives

$$i = Ke^{-Rt/L}$$

where K is an integration constant. Because $i = i_1 = 2.0$ A at $t = t_1$, we evaluate K to get

$$K = i_1 \, e^{Rt_1/L}$$

Substituting this value for K into the expression for i gives

$$i = i_1 \, e^{-(R/L) \cdot (t - t_1)} = 2.0 \, e^{-5(t - t_1)} \text{ A} \qquad (t_1 < t)$$

Finally, for $t = 0.30$ s, we get

$$i = 2.0 \, e^{-5(0.30 - \ln 5/10)} = 1.0 \text{ A} \quad \blacksquare$$

E 17.27   Sketch a graph of the current in the inductor of Example 17.4 for $0 \le t \le 1.0$ s.

Because the emf induced in an inductor is proportional to the rate of change of the current in that inductor, an inductor functions as a "short" when the current in the inductor reaches a steady value. The following problem illustrates how dc circuits containing inductors may be analyzed when either transient or steady-state currents are in the circuit.

In the circuit of Figure 17.16(a), the switch S is closed at $t = 0$. (a) Calculate the emf induced in L_1 immediately after the switch is closed. (b) What is the current in the resistor R when the currents reach their steady-state values?

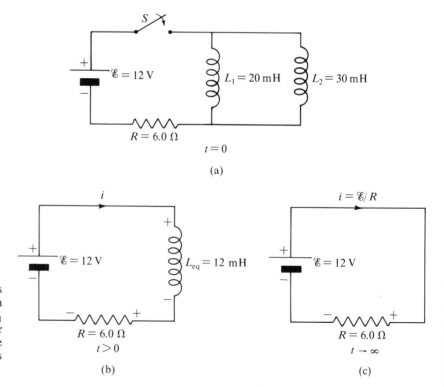

Figure 17.16
(a) An LR circuit before the switch S is closed. (b) The circuit after the switch has been closed and the inductors, L_1 and L_2, have been replaced by their equivalent inductance L_{eq}. (c) The effective circuit when the current has reached its steady-state value $i = \mathcal{E}/R$.

The inductors L_1 and L_2 are connected in parallel. Figure 17.16(b) shows the equivalent circuit in which L_1 and L_2 have been replaced by their equivalent inductance, calculated using Equation (17-30):

$$L_{eq} = \frac{L_1 L_2}{L_1 + L_2} = \frac{(20 \text{ mH})(30 \text{ mH})}{(50 \text{ mH})} = 12 \text{ mH} \qquad (17\text{-}37)$$

The loop rule for Figure 17.16 gives

$$\mathcal{E} - L_{eq}\frac{di}{dt} - iR = 0$$

Because $i = 0$ at $t = 0$, we get

$$\mathcal{E} - L_{eq}\frac{di}{dt}\Big|_{t=0} = 0$$

or

$$\frac{di}{dt}\Big|_{t=0} = \frac{\mathcal{E}}{L_{eq}} = \frac{12 \text{ V}}{12 \times 10^{-3} \text{ H}} = 1.0 \times 10^3 \frac{A}{s} \qquad (17\text{-}38)$$

Then the emf induced in L_{eq} is, using Equations (17-22) and (17-38),

$$\mathcal{E}_i = -L_{eq}\frac{di}{dt} = -(12 \times 10^{-3} \text{ H})\left(10^3 \frac{A}{s}\right) = -12 \text{ V} \qquad (17\text{-}39)$$

where the negative sign means that the induced emf in L_{eq} opposes the increasing current in the circuit. Because the potential difference across components connected in parallel is the same across each component, the emf across either L_1 or L_2 is equal to -12 V, the value we have determined in Equation (17-39) for the emf across the parallel combination. In particular, the emf induced in L_1 has a value of 12 V, opposing the flow of current in the circuit.

When the currents reach their steady-state values, the potential difference across each inductor is equal to zero, and both L_1 and L_2 function as shorts. The circuit effectively becomes that shown in Figure 17.16(c). Then, in the steady-state situation the current in R is given by Ohm's law:

$$i = \frac{\mathcal{E}}{R} = \frac{12 \text{ V}}{6.0 \text{ }\Omega} = 2.0 \text{ A} \qquad (17\text{-}40)$$

E 17.28   At $t = 0$ the switch S is closed in the circuit shown in the accompanying figure. Calculate (a) the current in the emf immediately after S is closed, that is, at $t = 0^+$, (b) the $t = 0^+$ potential difference across L, (c) the $t = 0^+$ rate of change of the current in L, and (d) the steady-state $(t \to \infty)$ current in the emf.   Answers: (a) 1.0 A; (b) 6.0 V; (c) 60 A/s; (d) 2.0 A

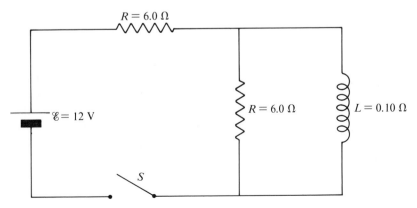

17.4 Magnetic Media

In this and the preceding chapter, we have considered numerous aspects of magnetic fields without having made any reference to our most common experience with magnetism—magnets. Most of us first encounter magnetic phenomena by seeing or playing with toy magnets that pick up nails; attach memos, notes, or cartoons to the refrigerator door; or exert forces on other magnets.

Permanent magnets exist because of special physical properties of a certain class of materials that includes iron, nickel, cobalt, and several rare earth elements. Here we consider briefly, mostly in a nonmathematical way, the magnetic nature of matter.

Magnetic Properties of Matter

Suppose a long, evacuated solenoid composed of n turns/m and carrying a current i has within its interior a magnetic field with magnitude $B_o = \mu_o n i$. Notice that B_o is proportional to μ_o, the permeability of free space. If the solenoid is now filled with a material substance, the magnetic field in that region changes in magnitude by a factor κ_m, the *relative permeability* of the particular substance. The magnitude B of the magnetic field within the material is

$$B = \kappa_m B_o \tag{17-41}$$

or

$$B = \kappa_m \mu_o n i \tag{17-42}$$

The quantity μ, defined by

$$\mu = \kappa_m \mu_o \tag{17-43}$$

is called the *permeability* of the material filling the solenoid. Then the magnitude of the magnetic field within the material becomes

$$B = \mu n i \tag{17-44}$$

The permeability μ, and, therefore, the relative permeability of a given substance, is *not* a constant value; it may depend upon temperature, the magnetic field present in the material, or even the history of the particular sample of the material.

Because κ_m is a pure number, μ has the same unit as μ_o, namely $(T \cdot m)/A$.

Every substance in a particular given state falls into one of a number of categories that are used to classify magnetic materials. Our concern is with three of those categories: paramagnetism, diamagnetism, and ferromagnetism. Substances for which κ_m is slightly greater than unity are called *paramagnetic* substances. Solid aluminum at room temperature, for example, has a relative permeability $\kappa_m = 1.000021$. Substances for which κ_m is slightly less than unity are called *diamagnetic* substances. Solid lead is a diamagnetic material at room temperature having $\kappa_m = 0.999984$. Because the relative permeability of paramagnetic substances and diamagnetic substances is not significantly different from that of free space, we may, for most practical purposes, ignore the magnetic effects of paramagnetic and diamagnetic materials. Finally, substances having values of κ_m that are much larger than unity are called *ferromagnetic* substances. Typically, ferromagnetic materials have values of κ_m in excess of 100 and sometimes in excess of 1000. We will consider ferromagnetism further in the next subsection.

The details of the microscopic origins of these magnetic effects—paramagnetism, diamagnetism, and ferromagnetism—are beyond the scope of this book. The gross nature of magnetism may, however, be apprehended in its essentials by assuming that tiny current loops, called *amperean currents*, are inherently associated with the subatomic structure of matter. Each of these current loops produces a local magnetic field, much like the field of a tiny bar magnet. In effect, the atoms of a substance have, in general, magnetic fields associated with them that can interact with the magnetic fields of other atoms.

Paramagnetism arises when the randomly directed magnetic fields of the amperean current loops are aligned slightly in the direction of an externally applied magnetic field within the substance. This partial alignment of magnetic fields enhances the applied field, thereby increasing κ_m for the material present to a value somewhat greater than $\kappa_m = 1$ for free space. The randomization of the directions of the amperean-loop fields is thermal in nature. Thus the paramagnetic effect is diminished with increasing temperature and augmented with decreasing temperature.

Diamagnetic effects in matter result from emfs induced in the amperean current loops when a magnetic field is applied to a material. In accordance with Faraday's law, the induced emf in each current loop changes the amperean current so as to oppose the applied magnetic field. Each new current then persists until a further change in the applied field occurs. Thus the induced magnetic field in the tiny current loops diminishes any applied field, and this effect is independent of the temperature of the material. Then diamagnetism, an effect that reduces an applied field in a material, competes with paramagnetism, which enhances the effect of the applied field. And because paramagnetic effects lessen with increasing temperature, while diamagnetic effects do not, we may conclude that all nonferromagnetic materials become diamagnetic at sufficiently high temperatures.

Ferromagnetism

Whereas the effects of paramagnetism and diamagnetism in a material are small and change κ_m only slightly from its value of unity for free space, the effect of ferromagnetism is very large. In materials like iron, nickel, and cobalt, called transition elements, there exist forces at a microscopic level that align the directions of the magnetic fields of the amperean current loops in these materials. These forces, called *exchange forces*, are of quantum-mechanical origin, and, therefore, cannot be explained in terms of classical physics; explanation of these forces is beyond the scope of our studies here. The alignment of the magnetic fields of amperean current loops by exchange forces is spontaneous; that is, it may occur without the application of any external magnetic field. This spontaneous alignment extends over volumes that are typically less than 1 mm³ but contain millions of atoms. Such a region of complete alignment of amperean magnetic fields is called a *ferromagnetic domain*.

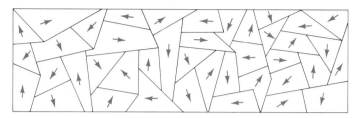

Figure 17.17
Randomly oriented ferromagnetic domains in a virgin sample of ferromagnetic material.

Consider now the domains within the macroscopic sample, illustrated in Figure 17.17, of a virgin ferromagnetic material, a material that has not been exposed to significantly large external magnetic fields. The magnetic-field directions of the domains are randomly oriented, so the net magnetic field within the sample is equal to zero. Figure 17.18(a) shows contiguous, or touching, domains of the virgin sample. Figure 17.18(b) shows the same region of this sample after a relatively small external magnetic field $\mathbf{B}_{ext}$ has been applied to the sample. The applied field has produced *domain wall motion*, which has caused the favorably oriented domains, those with fields aligned near the direction of $\mathbf{B}_{ext}$, to increase in volume at the expense of those less favorably aligned. The sample of Figure 17.18(b) is partially magnetized; the net magnetic field of the domains enhances the applied field. For small applied fields, domain wall motion is reversible; if the applied field were removed, the domain walls would resume their original (zero applied-field) positions, and the sample would be once again unmagnetized. But when a relatively strong magnetic field is applied to the sample, *domain rotation*, a situation indicated in Figure 17.18(c), occurs in addition to considerable domain wall motion. The external field has turned the direction of the amperean fields within the individual domains. Sufficiently large external fields cause complete alignment within the sample, and the sample is then said to be *saturated*.

When a ferromagnetic sample experiences domain rotation, or even considerable domain wall motion, the magnetizing effect on the sample becomes irreversible. That is, removal of the applied field does not restore the random orientation of the domain field directions, and the net magnetic field within the sample is not zero. A *remanent magnetic field* remains, and the sample is said to be magnetized. At this point, the ferromagnetic sample is a "permanent" magnet, like the toy magnets that are familiar to most of us.

Figure 17.18
Domains of a ferromagnetic sample when (a) no magnetic field has been applied, (b) a relatively small external magnetic field $\mathbf{B}_{ext}$ is applied, and (c) when a relatively strong external magnetic field $\mathbf{B}_{ext}$ is applied.

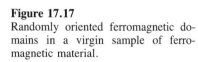

The word "permanent" is quite imprecise in this context. First, any permanent magnet can be completely demagnetized by raising its temperature above its *Curie point*, the temperature at which the randomizing effect of thermal motion in the sample is sufficient to overcome the magnetic alignment. Further, as we shall now see, permanent magnets can be altered; they may be demagnetized or have their magnetic fields changed or even reversed by changing the external field applied to the sample.

Suppose a sample of ferromagnetic material has been magnetized so that it has a remanent magnetic field in the x direction when no external field is applied to the sample. If now a magnetic field is applied to the sample in the negative x direction, opposite to the remanent field of the sample, the net field caused by domain alignment can be reduced to a zero value. Then if the magnitude of the external field is further increased in the negative x direction, the fields of the domains become aligned in a direction opposite to their original direction. And if the external applied field is removed, the sample is once again a "permanent" magnet, but the direction of its field has been reversed. It is now clear that the response of a ferromagnetic sample to an external applied field depends on the physical state of the sample, a state that depends, in turn, on the recent history of the sample.

A material with physical characteristics that depend on its magnetic history is said to exhibit *magnetic hysteresis*. A ferromagnetic sample "remembers" the direction (in its own reference frame) that it last experienced a significantly large magnetic field. Ferromagnetic materials are suitable memory elements in computers and calculators because they exhibit hysteresis.

The ferromagnetic toroid of Figure 17.19, for example, illustrates how a digital memory element may be constructed. In Figure 17.19(a), a current pulse through the coil labeled 1 in the directional sense indicated produces a remanent magnetic field **B** within the interior of the toroid in a clockwise sense, as shown. Similarly, as shown in Figure 17.19(b), a pulse of current through the coil labeled 0 produces a remanent **B** in the opposite sense. Either bit of information can be "permanently" stored in the domain-field orientation of the toroid. The coil labeled with a question mark in the figure may be used, in conjunction with appropriate circuitry, to interrogate, or sample, the field sense of the toroid without using sufficient current to change the sense of the remanent field of the toroid. Vast arrays of such memory elements can be assembled to form the memory banks of digital computers.

Figure 17.19
A toroid, made of a ferromagnetic material, used as a digital memory element. **(a)** The current in loop 1 produces a CW magnetic field. **(b)** The current in loop 0 produces a CCW magnetic field.

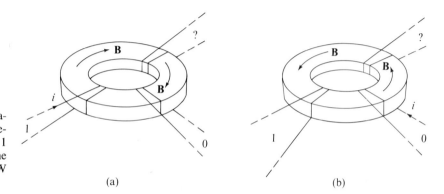

(a)   (b)

17.5 Maxwell's Equations

The principles of electromagnetism, as we have considered them to this point, may be summarized by four integral equations:

$$\oint_S \mathbf{E} \cdot d\mathbf{S} = \frac{q_{net}}{\epsilon_o} \tag{17-45}$$

$$\oint_S \mathbf{B} \cdot d\mathbf{S} = 0 \tag{17-46}$$

$$\oint_C \mathbf{E} \cdot d\mathbf{s} = -\frac{d\phi_m}{dt} \tag{17-47}$$

$$\oint_C \mathbf{B} \cdot d\mathbf{s} = \mu_o i_{net} \tag{17-48}$$

These are, in order, Gauss's law for electric fields, Gauss's law for magnetic fields, Faraday's law, and Ampère's law. Certain asymmetries among these relationships are apparent. In Equation (17-45), electric flux is related to the net charge within a closed surface, which establishes that electrostatic fields originate or terminate on electric charges. But in Equation (17-46), there is no magnetic equivalent to an isolated charge; no magnetic monopole exists. Equations (17-47) and (17-48) display a similar asymmetry. Faraday's law, Equation (17-47), provides that a changing magnetic flux produces an electric field, manifested as an induced emf. But Ampère's law, Equation (17-48), makes no provision for the production of a magnetic field as a result of a changing electric flux.

The first of these asymmetries must be accepted until the existence of magnetic monopoles is verified experimentally. But the lack of symmetry between the laws of Faraday and Ampère was unacceptable to James Clerk Maxwell (1831–1879). In a brilliant leap of imagination, he introduced an additional term, called a "displacement current," into Ampère's law, which then became

$$\oint_C \mathbf{B} \cdot d\mathbf{s} = \mu_o \left(i_{net} + \epsilon_o \frac{d\phi_e}{dt} \right) \tag{17-49}$$

where $d\phi_e/dt$ is the time rate of change of electric flux.

This alteration of Ampère's law provides that a changing electric flux produces a magnetic field, just as Faraday's law provides that a changing magnetic flux produces an electric field. With this modification, Maxwell used the fundamental relationships of electricity and magnetism to formulate a comprehensive theory of electromagnetism. The amended relationships of electricity and magnetism are called *Maxwell's equations*:

$$
\begin{array}{|ll|}
\hline
\oint \mathbf{E} \cdot d\mathbf{S} = \dfrac{q_{net}}{\epsilon_o} & \oint \mathbf{E} \cdot d\mathbf{s} = -\dfrac{d\phi_m}{dt} \\
\oint \mathbf{B} \cdot d\mathbf{S} = 0 & \oint \mathbf{B} \cdot d\mathbf{s} = \mu_o\!\left(i_{net} + \epsilon_o \dfrac{d\phi_e}{dt} \right) \\
\hline
\end{array} \tag{17-50}
$$

Not only do Maxwell's equations incorporate the fundamental principles of electricity and magnetism, but also they both predict and describe classical electromagnetic radiation. The significance and usefulness of electromagnetic radiation will be considered further in subsequent chapters.

17.6 Problem-Solving Summary

Induced emf is the central issue in two types of problems: (a) those problems in which the magnetic flux through a loop, usually a conducting loop, changes; and (b) those problems that involve inductors or devices the inductance of which is to be calculated. In the first case, the applicable physical principle is Faraday's law, $\mathscr{E}_i = -d\phi_m/dt$; in the latter case, $\mathscr{E}_i = -L\,di/dt$ is the operational relationship. Let us consider these categories separately.

In using Faraday's law to find the induced emf in a loop, the straightforward procedure is first to calculate the magnetic flux ϕ_m through the loop, then to express ϕ_m as a function of time, and finally to determine the time derivative of ϕ_m. The magnetic flux ϕ_m for an open surface S bounded by a closed curve is found using $\phi_m = \int_S \mathbf{B} \cdot d\mathbf{S} = \int_S B\,dS\cos\theta$, where θ is the angle between the magnetic field $\mathbf{B}$ and the element of surface area $d\mathbf{S}$ at each point on the surface. The magnetic flux ϕ_m for a given surface can change by any one (or more) of three ways: (a) The magnitude of $\mathbf{B}$ through that surface may change; (b) the direction of $\mathbf{B}$ may change; and (c) the area of the surface may change. But however the change in ϕ_m comes about, we must express ϕ_m as a function of time so that we may determine $d\phi_m/dt$. In practice, it is usually easiest to determine first the magnitude of the induced emf using Faraday's law, that is, use $|\mathscr{E}_i| = |-d\phi_m/dt|$ and then use Lenz's law to determine the direction of the emf. According to Lenz's law, an induced emf is in the directional sense that opposes the *change* in flux that produced the emf (and not necessarily in a direction opposite the field through the loop being considered).

If an emf $\mathscr{E}_i$ is induced in a conducting loop having resistance R, a current $i_i = \mathscr{E}_i/R$ will be induced in that loop. The induced current is in the same directional sense around the loop as the induced emf. Therefore, the sense of the induced current can be determined using the right-hand rule for magnetic fields caused by current in a wire.

The inductance L of a device with N turns, each of which has the same magnetic flux ϕ_m through it, obeys $\mathscr{E}_i = -L\,di/dt$, where $\mathscr{E}_i$ is the emf induced across the device carrying a current i. Because Faraday's law specifies the induced emf for N turns, each with a magnetic flux ϕ_m by $\mathscr{E}_i = -N\,d\phi_m/dt$, the inductance L may be expressed as $L = N\phi_m/i$. Then the inductance of a device of N turns may be calculated using the following procedure:

1. Assume a current i in each turn of the device.

2. Find an expression for B within each turn.

3. Calculate ϕ_m for each turn.

4. Use $L = N\phi_m/i$ to determine the inductance of the device.

An inductor L in a circuit functions to oppose any change in current in that circuit. The inductor produces an emf $\mathscr{E}_i = -L\,di/dt$ across its terminals, where i is the current in the inductor at any instant. An inductor acts as a short in a circuit carrying a steady current. When a circuit is completed, a circuit that is composed of an inductor L, a resistor R, and a fixed emf $\mathcal{E}$, in series, the current i in the circuit increases from a value of zero at $t = 0$ according to $i = (\mathcal{E}/R)(1 - e^{-Rt/L})$. The quantity of L/R is the time constant of the RL combination; when $t = L/R$, the current in the circuit reaches 63 percent of its steady-state value of $\mathcal{E}/R$.

For purposes of solving problems at the level of this text, we need to know two facts about the magnetic properties of materials: First, the presence of a material substance changes the magnitude of a magnetic field, whose magnitude in free space is B_o, to a value $\kappa_m B_o$, where κ_m is a unitless quantity called the relative permeability of that particular substance. Second, the permeability μ of a substance is equal to $\kappa_m \mu_o$, where $\mu_o = 4\pi \times 10^{-7}$ N/A^2 is the permeability of free space.

Problems

GROUP A

17.1 A metal airplane is flying at a speed of 600 km/h in a direction perpendicular to the magnetic field of the earth. That field has a magnitude of 5.4×10^{-5} T. What is the difference in potential between the wing tips, which are separated by 18 m?

17.2 If you run along the ground at a speed of 4.0 m/s holding a 100-turn loop that is connected to a sensitive voltmeter and the loop is oriented so that the plane of the loop (area $= 1.0$ m^2) is horizontal, what will the voltmeter read? Assume that the vertical component of the magnetic field of the earth at your location is equal to 4.0×10^{-5} T.

17.3 A long solenoid ($n = 80$ turns/cm) carries a current of 70 mA. If the interior of the solenoid is filled with a ferromagnetic material having an effective value of κ_m of 650, determine the magnitude of the magnetic field in the solenoid (a) before and (b) after the ferromagnetic material is inserted.

17.4 If $i = 1.5$ A and $V_2 - V_1 = 4.0$ V for the inductor shown in Figure 17.20, determine

(a) di/dt, the rate of change of i.

(b) the energy stored by the inductor.

(c) the rate at which the energy stored by the inductor is changing.

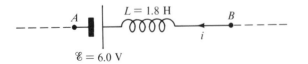

Figure 17.20   Problem 17.4

17.5 At an instant when a 50-mH inductor has stored 0.40 J of energy and is increasing this stored energy at a rate of 0.20 W, calculate

(a) the current i in the inductor.

(b) di/dt, the rate of change of i.

(c) the magnitude of the potential difference across the inductor.

17.6 A 1.0-mH inductor that has a resistance of 50 Ω is connected across an emf of 10 V at $t = 0$.

(a) What is the electric power loss in the circuit when the current reaches its steady-state value?

(b) How long after $t = 0$ does the current in the emf reach half its steady-state value?

(c) How much energy is stored in the inductor when the current is half its steady-state value?

17.7 If $i = 2.0$ A and $V_A - V_B = +4.0$ V in the circuit segment shown in Figure 17.21, determine di/dt.

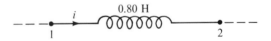

Figure 17.21   Problem 17.7

17.8 At an instant when the potential difference, $V_1 - V_3$, across the two inductors in Figure 17.22 is equal to $+36$ mV and the current i is equal to 6.0 A, calculate

(a) di/dt, the rate of change of i.

(b) the potential difference, $V_2 - V_3$.

(c) the total energy stored by the two inductors.

(d) the rate at which the energy stored by the 3.0-mH inductor is changing.

17.9 At an instant when $V_A - V_B = 20$ V and $V_A - V_C = 14$ V for the circuit segment (see Fig. 17.23), determine i and di/dt.

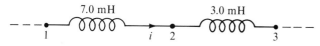

Figure 17.22   Problem 17.8

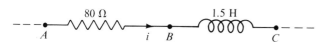

Figure 17.23   Problem 17.9

17.10   If the switch S in Figure 17.24 is closed at $t = 0$ in the circuit, determine

(a)   the magnitudes of the $t = 0^+$ potential differences across L and R.

(b)   the $t = 0^+$ current and rate of change of current.

(c)   the magnitudes of the $t \to \infty$ potential differences across L and R.

(d)   the $t \to \infty$ current and rate of change of current.

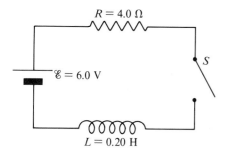

Figure 17.24   Problem 17.10

17.11   A plane loop of wire forms an area described by the vector $\mathbf{S} = (2\hat{\mathbf{i}} + 2\hat{\mathbf{j}})$ m² in a region where the magnetic field is changing according to $\mathbf{B} = (3t\hat{\mathbf{i}} - 2t^2\hat{\mathbf{j}}) \times 10^{-3}$ T, where t is in seconds. Calculate the magnitude of the emf induced in the loop when $t = 3.0$ s.

17.12   A conducting rod slides down parallel conducting rails that are 1.0 m apart and tilted at an angle of 30° from the direction of a uniform magnetic field of magnitude 5.0×10^{-3} T (see Fig. 17.25). If the resistor R joining the rails has a resistance of 5.0 Ω and the rod is moving at a speed of 10 m/s, calculate the current induced in the resistor.

17.13   For an instant when $i = 0.75$ A and $di/dt = -2.0$ A/s in the circuit segment depicted in Figure 17.26,

(a)   determine the potential difference, $V_A - V_B$.

(b)   is this segment adding or removing energy from the remainder of the circuit? At what rate?

17.14   A 60-cm rod is moved through a uniform magnetic field ($B = 0.070$ T) at a speed of 7.0 m/s (see Fig. 17.27). Determine the potential difference induced between the two ends of the rod.

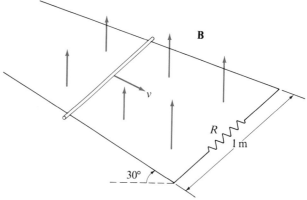

Figure 17.25   Problem 17.12

Figure 17.26   Problem 17.13

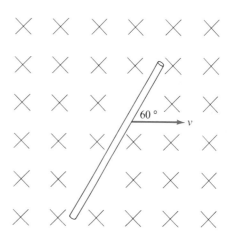

Figure 17.27   Problem 17.14

17.15   A coil (cross-sectional area = 12 cm², $N = 50$ turns) is oriented with its plane perpendicular to a 0.50-T magnetic field. If the coil is "flipped" so that in 0.20 s the direction of **B** relative to the coil is reversed, determine the average emf induced in the coil.

17.16   Determine di/dt for the circuit segment in Figure 17.28 if $V_A - V_B = -0.060$ V.

17.17   A conductor of length L moves with a velocity **v** near a long wire carrying a current i (see Fig. 17.29). The wire and conductor are coplanar. If $a = 0.50$ cm, $b = 8.0$ cm, $v = 20$ m/s, and $i = 50$ A, calculate the (induced) potential difference, $V_A - V_B$, between the two ends of the moving conductor.

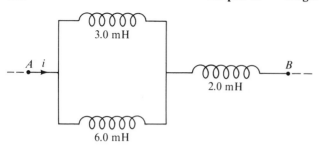

Figure 17.28   Problem 17.16

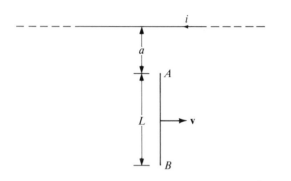

Figure 17.29   Problem 17.17

17.18   In the configuration shown in Figure 17.30 a long wire carries a constant current i. A sliding wire is moving at a speed v along parallel tracks that lie in a plane containing the long wire. If the sliding wire is at the left end of the track (a distance a from the long wire) at $t = 0$, write an expression for the emf induced in the closed loop as a function of time t.

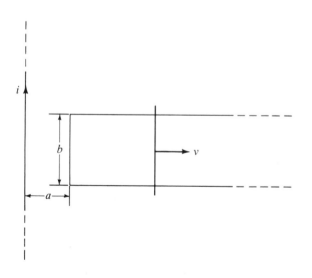

Figure 17.30   Problem 17.18

17.19   If $i = 0.75$ A and $V_A - V_B = 12$ V in the circuit segment shown in Figure 17.31,

(a)   determine di/dt.

(b)   at what rate is this segment adding or removing energy from the remainder of the circuit?

(c)   determine the rate at which each component of the circuit segment shown is adding or removing energy.

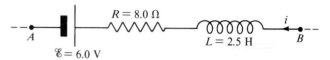

Figure 17.31   Problem 17.19

17.20   A 5.0-H inductor that has a resistance of 100 Ω is connected across an emf of 100 V at $t = 0$.

(a)   What is the value of the steady-state current in the emf?

(b)   How long after $t = 0$ is the current in the emf equal to 99 percent of its steady-state value?

(c)   What is the rate of increase of the current at $t = 0$?

(d)   At what rate is the current in the inductor increasing when the current is half its steady-state value?

17.21   A long solenoid with 20 turns per centimeter of length has a circular cross section with an area of 10 cm². Tightly wrapped around the outside of the solenoid is a 100-turn coil having a resistance of 6.0 Ω. If the current in the solenoid increases from zero to 8.0 A in a time interval of 0.40 s, what is the average power loss in the coil during that interval?

17.22   When an inductor having a resistance of 5.0 Ω is connected to a 15-V emf, the current reaches 50 percent of its final $(t \rightarrow \infty)$ value in 30 ms.

(a)   Determine the inductance of the device.

(b)   At what rate is the current changing when the current reaches 50 percent of its final value?

17.23   A coil of wire consisting of 100 turns is formed into a circle having an area of 6.0 m². The magnetic flux through the loop is described as a function of time t (in seconds) according to $\phi_m = (t^2 - 2t + 4)$ T·m².

(a)   What is the magnitude of the emf induced in the coil at $t = 3$ s?

(b)   What is the magnitude of the component of the uniform magnetic field that is perpendicular to the plane of the coil at $t = 1$ s?

17.24   For the circuit segment shown in Figure 17.32, calculate the potential difference, $V_1 - V_2$, at an instant when the current i shown and its rate of change di/dt are equal to

(a)   2.0 A, + 4.0 A/s

(b)   2.0 A, − 4.0 A/s

(c)   2.0 A, 0

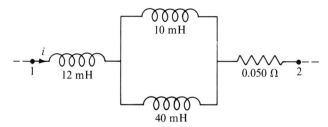

Figure 17.32   Problem 17.24

17.25   At an instant when $i = 0.60$ A in the circuit shown in Figure 17.33, determine

(a) di/dt.

(b)   the rate at which the emf is providing energy. Account for this energy.

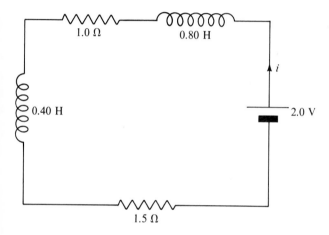

Figure 17.33   Problem 17.25

17.26   For an instant when $q = 8.0 \times 10^{-4}$ C, $i = 0.20$ mA, and $di/dt = -1.5$ A/s in the circuit segment shown in Figure 17.34, calculate

(a)   the potential difference, $V_B - V_A$.

(b)   the rate at which this segment is adding or removing energy from the remainder of the circuit.

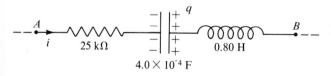

Figure 17.34   Problem 17.26

17.27   A solenoid with 1000 turns evenly wound around a thin cylindrical plastic tube that is 50 cm long and with a cross-sectional area of 10 cm² is bent into a toroid, which is filled with a ferromagnetic material. If the inductance of the toroid is 1.0 H, what is the relative permeability of the core material?

17.28   If the current in a long cylindrical solenoid having 1.2×10^3 turns/m and a diameter of 4.0 cm is changing at the rate of 5.0 A/s, determine the magnitude of the induced electric field as a function of r (distance from the axis of the solenoid) for $0 \le r < 0.020$ m.

17.29   When $i_1 = 2.0$ A, $i_2 = 3.0$ A, and $V_A - V_B = 40$ V for the circuit segment (*see* Fig. 17.35), determine

(a) di_1/dt.

(b) di_2/dt.

(c)   the rate at which energy is being added to or removed from the remainder of the circuit by the segment shown.

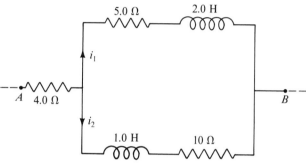

Figure 17.35   Problem 17.29

17.30   In the diagram shown in Figure 17.36, where $a = 0.5$ m, $b = 1.0$ m, and $c = 4.0$ m, the rectangular loop lies in a plane containing the long straight line. At an instant when the emf induced in the loop is 1.0 V, what is the time rate of change of the current in the long wire?

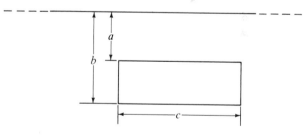

Figure 17.36   Problem 17.30

17.31   Switch S in Figure 17.37 is closed at $t = 0$. Determine

(a)   the $t = 0^+$ current in the 12-V emf.

(b)   the $t = 0^+$ rate of change of current in the 30-Ω resistor.

(c)   the $t \to \infty$ current in the 12-V emf.

17.32   The current in the long wire shown in Figure 17.38 varies as $i(t) = i_o \sin(2\pi\nu t)$ with $i_o = 8.0$ A and $\nu = 60$ Hz. If $a = 0.50$ cm, $b = 4.0$ cm, and $L = 16$ cm for the rectangular coil (N = 200 turns) shown, determine

(a)   the induced emf in the coil as a function of time.

(b)   the maximum value of this induced emf.

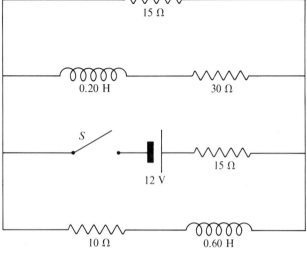

Figure 17.37   Problem 17.31

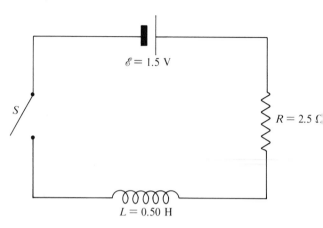

(a)   the initial ($t = 0^+$) rate of change of current in the emf.

(b)   the steady-state ($t \rightarrow \infty$) current in the emf.

Figure 17.39   Problem 17.33

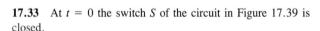

$i(t)$

Figure 17.38   Problem 17.32

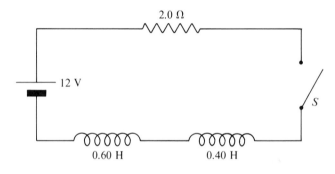

Figure 17.40   Problem 17.34

17.33   At $t = 0$ the switch S of the circuit in Figure 17.39 is closed.

(a)   Determine the current in the circuit at $t = 0.40$ s.

(b)   At what rate is the emf providing energy at $t = 0.40$ s? Account for this energy.

(c)   How much energy does the emf provide during the time interval $0 < t < 0.40$ s? Account for this energy.

17.34   If the switch S in Figure 17.40 is closed at $t = 0$,

(a)   write an expression for the current in the circuit for $t \geq 0$.

(b)   how much energy does the emf add to the circuit during the time interval $0 < t < 0.50$ s?

(c)   account for the energy calculated in part (b).

17.35   The switch S shown in Figure 17.41 is closed at $t = 0$. Determine

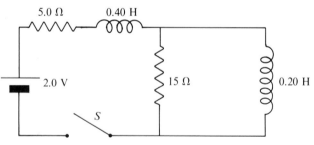

Figure 17.41   Problem 17.35

17.36   If the electric field in the x-y plane is given by

$$\mathbf{E} = K \left(-y\hat{\mathbf{i}} + x\hat{\mathbf{j}}\right)$$

where $K = 0.25$ V/m², calculate

(a) $\oint_C \mathbf{E} \cdot d\mathbf{s}$ for the closed path C shown in Figure 17.42.

(b) $|d\phi_m/dt|$ for the area in the x-y plane enclosed by C.

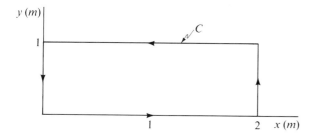

Figure 17.42   Problem 17.36

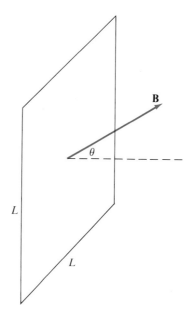

Figure 17.43   Problem 17.39

17.37 A circular (radius = a) loop of wire having a resistance R, initially oriented in a uniform magnetic field **B** with **B** in the plane of the loop, is rotated so that **B** is perpendicular to the plane of the loop. Show that the total charge ΔQ that flows through any cross section of the wire during the time interval required for this rotation is given by

$$\Delta Q = \frac{\pi B a^2}{R}$$

Verify that the units of this expression are correct.

17.38 An 80-cm length of wire (resistance = $2.0 \times 10^{-4}\ \Omega$), initially formed in the shape of a square oriented with its plane perpendicular to a uniform magnetic field having a magnitude of 60 G, is changed into a circular shape without altering the orientation of the surface of the loop relative to **B**. Calculate

   (a) the magnitude of the change in magnetic flux for the area bounded by the wire.

   (b) the magnitude of the total charge that flows through any cross section of the wire during the time required to reshape the wire.

17.39 A square ($L = 0.80$ m) loop of wire is fixed in a region of space where **B** is uniform in space but changing with time. Let θ be the angle between **B** and a normal to the plane of the loop, as shown in Figure 17.43.

   (a) At an instant when $\theta = \pi/6$ radians, $B = 0.50$ T, $d\theta/dt = 1.5$ radians/s, and $dB/dt = 0.80$ T/s, determine the magnitude of the emf induced in the loop.

   (b) Repeat part (a) for $\theta = \pi/6$ radians, $B = 0.50$ T, $d\theta/dt = -1.5$ radians/s, and $dB/dt = 0.80$ T/s.

   (c) Explain why the results for parts (a) and (b) differ.

17.40 If the switch S shown in Figure 17.44 is closed at $t = 0$,

   (a) determine the initial rate of change of current in the emf.

   (b) determine the current in the 0.60-H inductor at the time $t = 0.50$ s.

   (c) how much energy does the emf add to the circuit during the $0 < t < 0.050$ s time interval?

   (d) account for the energy calculated in part (c).

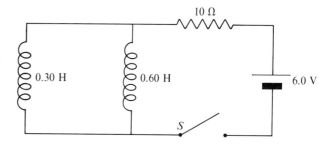

Figure 17.44   Problem 17.40

GROUP **B**

17.41 A solenoid of length l and radius a ($a \ll l$) has n turns per meter and carries a current i. If $l = 80$ cm, $a = 2.0$ cm, $n = 2.0 \times 10^3$ m^{-1}, and $i = 5.0$ A, estimate

   (a) the inductance of this device.

   (b) the magnetic energy stored by this device.

17.42 Two thin coaxial cylinders (radii a and b, $a < b$) carry uniformly distributed currents of equal magnitudes and opposite directional senses parallel to the axis of the cylinder. Determine the inductance of a length h of this *coaxial cable*.

17.43 The energy stored by an inductor is said to be stored in the magnetic field.

(a) Show that the energy stored in a length l of a long solenoid is given by

$$U_L = \frac{\pi B^2 a^2 l}{2\mu_0}$$

where B is the magnitude of the magnetic field within the solenoid and a is the radius of the solenoid.

(b) Show that the (magnetic) energy density u_m within the solenoid is given by

$$u_m = \frac{B^2}{2\mu_0}$$

(c) What is the energy density associated with a magnetic field having a magnitude of 1.0 G? 1.0 T?

17.44 A conducting disk (radius $= a$) is rotated with a constant frequency ν (about an axis coincident with that of the disk) in a region of uniform magnetic field **B** that is parallel to the axis of rotation, as shown in Figure 17.45. The leads of a voltmeter are contacted by brushes to the center and the outer edge of the disk. If $a = 8.0$ cm, $\nu = 60$ Hz, and $B = 85$ G,

(a) what reading is obtained from the voltmeter?

(b) which contact is at the lower potential if the rotation and magnetic field are as shown?

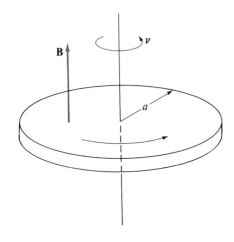

Figure 17.45　　Problem 17.44

17.45 At $t = 0$ the switch S in the circuit shown in Figure 17.46 is closed. Determine the values for i, the current in the emf $\mathcal{E}$; di/dt; and ΔV_C, the potential difference across the capacitor C

(a) initially, i.e., at $t = 0^+$.

(b) finally, i.e., as $t \to \infty$.

17.46 If the charge q on the capacitor C of the circuit segment shown in Figure 17.47 is given by

$$q(t) = C\mathcal{E}_0 \sin \omega t$$

where $\mathcal{E}_0$ and ω are constants, show that the potential difference, $\Delta V = V_1 - V_2$, is given by

$$\Delta V = V_0 \cos(\omega t - \phi)$$

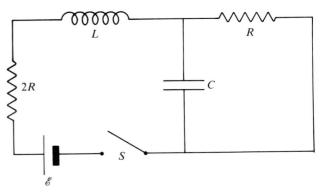

Figure 17.46　　Problem 17.45

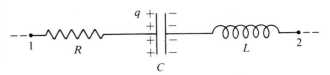

Figure 17.47　　Problem 17.46

where

$$V_0 = \sqrt{(\omega RC)^2 + (1 - \omega LC)^2}\, \mathcal{E}_0$$

and

$$\tan\phi = \frac{1}{\omega RC} - \frac{\omega L}{R}$$

17.47 If the switch S shown in Figure 17.48 is closed at $t = 0$, when will the current in the emf reach 50 percent of its steady-state value?

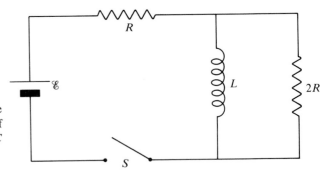

Figure 17.48　　Problem 17.47

17.48 At $t = 0$ the switch S shown in Figure 17.49 is closed. If i is the current through the emf, show that

$$\frac{di}{dt} + \frac{2R}{3L}i = \frac{2\mathcal{E}}{3L} \qquad (t > 0)$$

and that

$$i = \frac{\mathscr{E}}{R}\left(1 - \frac{2}{3}e^{-2Rt/3L}\right) \qquad (t > 0)$$

Sketch a graph of $i(t)$ for $t > 0$. When does the current reach one-half of its steady-state value?

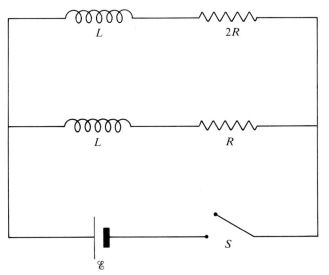

Figure 17.49   Problem 17.48

17.49 A capacitor C, charged to an initial potential difference V_0, is connected to an inductor L and a switch S, as shown in Figure 17.50. At $t = 0$ the switch is closed. If i is the ($t > 0$) current shown and q is the indicated ($t > 0$) charge, show that

(a) $q = LC\, di/dt$

(b) $d^2i/dt^2 = -i/LC$

(c)   the current varies sinusoidally with a frequency ν given by

$$\nu = \frac{1}{2\pi\sqrt{LC}}$$

and that

$$i = i_0 \sin(2\pi\nu t)$$

with $i_0 = V_0\sqrt{C/L}$

(d)   the charge q varies sinusoidally according to

$$q = q_0 \cos(2\pi\nu t)$$

with

$$q_0 = CV_0 = \frac{i_0}{2\pi\nu}$$

17.50 A capacitor C, charged to an initial potential difference V_0, is connected to an inductor L, a resistor R, and a switch S, as shown in Figure 17.51. At $t = 0$ the switch is closed. If i is the ($t > 0$) current shown, show that

(a) $LC\, d^2i/dt^2 + RC\, di/dt + i = 0$

(b)   if $R < 2\sqrt{L/C}$,

$$i = i_0 e^{-\alpha t} \sin(2\pi\nu t) \qquad \text{(underdamped oscillator)}$$

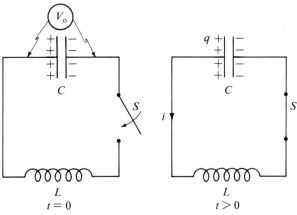

Figure 17.50   Problem 17.49

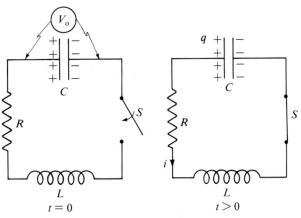

Figure 17.51   Problem 17.50

where

$$i_0 = \frac{V_0}{2\pi\nu L}, \; \alpha = \frac{R}{2L}, \; \text{and} \; \nu = \frac{1}{2\pi}\sqrt{\frac{1}{LC} - \left(\frac{R}{2L}\right)^2}$$

(c)   if $R > 2\sqrt{L/C}$,

$$i = i_0(e^{-\alpha_1 t} - e^{-\alpha_2 t}) \qquad \text{(overdamped oscillator)}$$

where

$$i_0 = \frac{V_0}{\sqrt{R^2 - 4L/C}}$$

$$\alpha_1 = \frac{R}{2L} - \sqrt{\left(\frac{R}{2L}\right)^2 - \frac{1}{LC}}$$

$$\alpha_2 = \frac{R}{2L} + \sqrt{\left(\frac{R}{2L}\right)^2 - \frac{1}{LC}}$$

(d)   if $R = 2\sqrt{L/C}$,

$$i = \frac{V_0 t}{L}e^{-(Rt/2L)} \qquad \text{(critically damped oscillator)}$$

18 Electromagnetic Oscillations

Back-and-forth motions of electric charges, and sometimes even the back-and-forth motions of electric and magnetic fields, are referred to as *electromagnetic oscillations*. Electric charges may be driven by sources of energy to oscillate in circuits consisting of resistors, capacitors, and inductors. Such alternating-current circuitry plays a prominent role in modern technology and, therefore, in our daily lives. We will consider some basic concepts associated with this circuitry. We will also see how oscillating electric and magnetic fields are produced by oscillating charges and how electromagnetic radiation may be described in terms of such oscillating fields.

18.1 Alternating-Current Circuits

Circuits carrying currents that change periodically in magnitude and directional sense are called *alternating-current circuits*, or, simply, *ac circuits*. Although other forms of ac exist, we will consider only alternating currents for which the current varies sinusoidally (as a sine or cosine function) with time. We are all familiar with sinusoidal sources of 60-cycle ac, like the electrical outlets in our homes and schools.

Components in AC Circuits

The most common form of alternating current is produced in circuits by an *alternating emf* that has a time-varying potential difference Δv across its terminals described by

$$\Delta v = V \sin \omega t = V \sin 2\pi \nu t \qquad (18\text{-}1)$$

In Equation (18-1), V is the amplitude of the alternating emf, that is, V is the maximum magnitude of Δv. The angular frequency ω is equal to $2\pi\nu$, where ν is the frequency of the periodic potential difference Δv. (In our discussions of ac circuits, lowercase letters represent sinusoidally varying quantities and capital letters represent their corresponding amplitudes.) Figure 18.1 shows the symbol

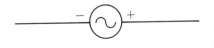

Figure 18.1
The symbol used in circuit diagrams to represent an alternating emf.

used in circuit diagrams to represent an alternating emf. Also shown in Figure 18.1 is the polarity convention that we use to indicate the polarity of the alternating emf: The polarity may be chosen arbitrarily, but once chosen that polarity corresponds to a positive value for Δv.

E 18.1   Write an expression similar to Equation (18-1) to describe an alternating emf with an amplitude of 150 V and a frequency of 60 Hz.

Answer: $\Delta v = 150 \sin(120\pi t)$ V

E 18.2   An alternating emf is described by

$$\Delta v = 60 \sin(50t) \text{ V}$$

What is the frequency of this emf?   Answer: 8.0 Hz

E 18.3
(a) What is the frequency of the alternating emf shown in the accompanying figure?   Answer: 0.50 Hz
(b) Write an expression similar to Equation (18-1) that describes the alternating emf shown.   Answer: $\Delta v = 15 \sin(\pi t)$ V

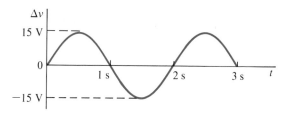

Exercise 18.3

Figure 18.2 is a graph of Δv as a function of time t. The figure also shows the reference circle, of radius V, that corresponds to Δv. The radius vector of length V in Figure 18.2 turns about the center of the reference circle at an angular speed ω. The projection of the head of the radius vector onto the Δv axis is equal to Δv at any instant of time t. The rotating radius vector of length V is called a *phasor*, and the angle ωt that designates the direction of the vector $\mathbf{V}$ relative to its direction at $t = 0$, is called the (instantaneous) *phase* of Δv.

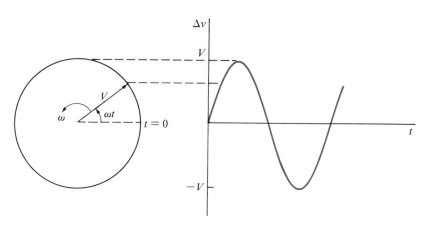

Figure 18.2
A graph of the sinusoidal potential difference Δv, produced by an alternating emf, as a function of time t. The reference circle for Δv has a radius V. The rotating vector $\mathbf{V}$, called a phasor, turns about the center of the reference circle at an angular speed ω.

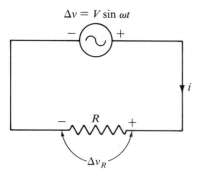

$$\Delta v = V \sin \omega t$$

Figure 18.3
A resistor R connected across an alternating emf Δv.

Figure 18.3 shows a resistor R connected across an alternating emf Δv. The positive sense of the current i and the potential difference, $\Delta v = iR$, across the resistor are as shown when the polarity of Δv is as shown in Figure 18.3. That is, if Δv has the polarity shown, the current i and the potential change across R should have the senses depicted. Now, using the loop rule, $\Delta v - iR = 0$, gives

$$i \left(= \frac{\Delta v_R}{R} \right) = \frac{\Delta v}{R} = \frac{V}{R} \sin \omega t = I \sin \omega t \qquad (18\text{-}2)$$

where $I = V/R$ is the amplitude of the sinusoidal current i. Because both Δv_R and i are proportional to $\sin \omega t$, Δv_R and i are said to be *in phase*, that is, the angle between their phasors is equal to zero. Figure 18.4 shows the graphs of Δv_R and i along with their corresponding phasors. Because they are in phase, Δv_R and i reach their maximum values at the same instants and become equal to zero at the same instants.

E 18.4 A 50-Ω resistor is connected across an alternating emf that has an amplitude of 120 V and a frequency of 60 Hz. Write an expression for the (alternating) current through the resistor.   Answer: $i = 2.4 \sin(120\pi t)$ A

Now let a capacitor C be connected across an alternating emf Δv, as shown in Figure 18.5. Again, the current i, the charge q, and the potential difference, $\Delta v_C = q/C$, across the capacitor are chosen to be consistent with the designated polarity of Δv. The loop rule, $\Delta v - q/C = 0$, gives

$$q \, (= C \Delta v_C) = C \Delta v = CV \sin \omega t \qquad (18\text{-}3)$$

The instantaneous current i through the capacitor is equal to the time derivative of q:

$$i = \frac{dq}{dt} = \omega CV \cos \omega t = \omega CV \sin(\omega t + 90°) = I \sin(\omega t + 90°) \qquad (18\text{-}4)$$

Figure 18.6 shows the graphs of i and Δv_C for the capacitor as functions of time t, along with their phasors. These graphs and phasors indicate that the current i (a cosine curve) is, at every instant, ahead in phase of the potential difference Δv_C by one-quarter cycle, or by 90°. We say that the current i in the capacitor *leads*

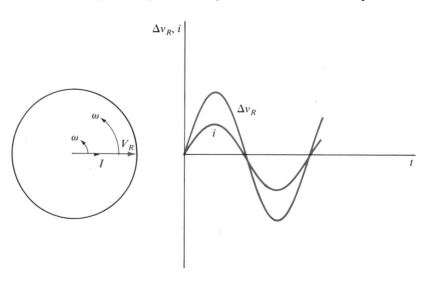

Figure 18.4
The potential difference Δv_R across the resistor of Figure 18.3 and the current i through the same resistor, each shown as a function of time t. Their phasors, V_R and I, are coincident at every instant, and Δv_R and i are in phase.

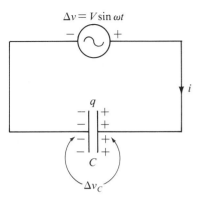

Figure 18.5
A capacitor C connected across an alternating emf Δv.

$\Delta v = V \sin \omega t$

the potential difference Δv_C by 90°. This lead-in phase is indicated in Equation (18-4), where $\cos \omega t$ has been replaced by its equivalent, $\sin(\omega t + 90°)$. But because it is convenient to refer the phase of potential differences across series components of ac circuits to the phase of their common current, we may equivalently say that in a capacitor the potential difference Δv_C across the capacitor *lags* the current i in the capacitor by 90°.

In Equation (18-4), the maximum value I of the current i in the capacitor C is equal to ωCV. And similarly to the way we relate the maximum current I and maximum potential difference V_R for a resistor by $I = V_R/R$, we may express the relationship between I and V_C, the maximum potential difference across the capacitor, by

$$I = \frac{V_C}{1/\omega C} = \frac{V_C}{X_C} \qquad (18\text{-}5)$$

where

$$X_C = \frac{1}{\omega C} = \frac{1}{2\pi\nu C} \qquad (18\text{-}6)$$

is called the *capacitive reactance* of the capacitor C. The *reactance* of a component in an ac circuit plays a similar role to that of resistance in a dc circuit. Thus, the reactance of a component is a measure of how effectively that component impedes the flow of charge in an ac circuit. According to Equation (18-6), the reactance X_C of a capacitor decreases with increasing frequency. In other words, a capacitor (which would completely halt the flow of charge in a dc circuit) impedes the flow of charge in an ac circuit less and less as the frequency ν of the ac increases.

E 18.5   If a 50-μF capacitor is connected across an alternating emf described by

$$\Delta v = 30 \sin(2000\pi t) \; V$$

(a) calculate the capacitive reactance for this circuit, (b) determine the maximum current through the capacitor, and (c) write an expression for the current through the capacitor.   Answers: (a) 3.2 Ω; (b) 9.4 A; (c) $i = 9.4 \cos(2000\pi t)$ A

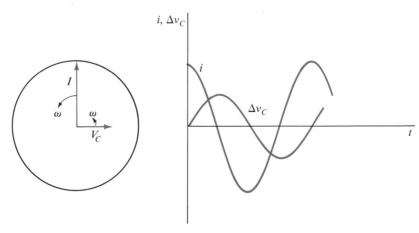

Figure 18.6
The potential difference Δv_C across the capacitor of Figure 18.5 and the current i through that capacitor, each shown as a function of time t. Their phasors, V_C and I, are separated by 90° at every instant, and Δv_C lags i by 90°.

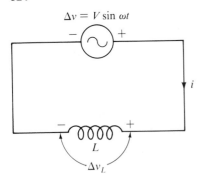

Figure 18.7
An inductor L connected across an alternating emf Δv.

Consider now the circuit of Figure 18.7, which shows an inductor L connected across an alternating emf Δv. The senses of the current i and the potential change across the inductor are shown. The loop rule gives

$$\Delta v - L \frac{di}{dt} = 0 \qquad (18\text{-}7)$$

or

$$\frac{di}{dt}\left(= \frac{\Delta v_L}{L}\right) = \frac{\Delta v}{L} = \frac{V}{L}\sin\omega t \qquad (18\text{-}8)$$

Integrating Equation (18-8) with respect to time t gives

$$i = -\frac{V}{\omega L}\cos\omega t \qquad (18\text{-}9)$$

where the integration constant has been set equal to zero so that the average value of i (over a period) is equal to zero.

Graphs of Δv_L and i for an inductor, along with their corresponding phasors, are shown in Figure 18.8. In this circuit, the potential difference phasor V_L is $90°$ ahead of the current phasor I at every instant. We say, therefore, that the potential difference Δv_L *leads* the current i in an inductor.

Again in a manner similar to the way we relate the current maximum I and the potential difference maximum V_R by $I = V_R/R$ in a resistive circuit, we may express the current maximum of Equation (18-9) as

$$I = \frac{V_L}{\omega L} = \frac{V_L}{X_L} \qquad (18\text{-}10)$$

where

$$X_L = \omega L = 2\pi\nu L \qquad (18\text{-}11)$$

is called the *inductive reactance* of an inductor L in an ac circuit. The reactance of an inductor, like the reactance of a capacitor, is a measure of its ability to impede the flow of charge in an ac circuit. From Equation (18-11), we may observe that inductive reactance becomes greater with increasing frequency ν.

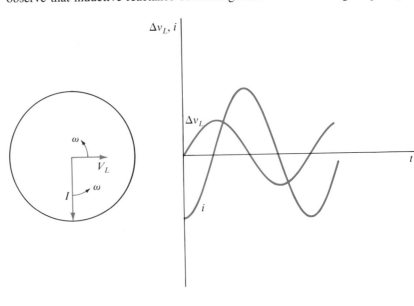

Figure 18.8
The potential difference Δv_L across the inductor of Figure 18.7 and the current i through that inductor, each shown as a function of time t. Their phasors, V_L and I, are separated by $90°$ at every instant, and Δv_L leads i by $90°$.

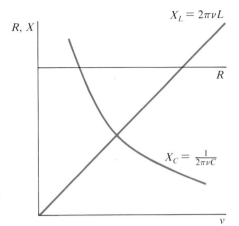

Figure 18.9
A graphical representation of the dependence of resistance R, capacitive reactance X_C, and inductive reactance X_L on frequency ν. While R is independent of ν, X_C is proportional to $1/\nu$; and X_L is directly proportional to ν.

E 18.6   A 15-mH inductance is connected across an alternating emf having a frequency of 60 Hz and an amplitude of 120 V.
(a) Calculate the inductive reactance for this circuit.   Answer: 5.7 Ω
(b) Determine the maximum current through the inductance.   Answer: 21 A

We may summarize the abilities of resistors, capacitors, and inductors to impede current in ac circuits by the graphs of Figure 18.9. These graphs show the resistance and reactances of the components as functions of the frequency ν of the alternating emf. Resistance R is independent of frequency, while the capacitive reactance X_C decreases with frequency in proportion to $1/\nu$ and inductive reactance X_L increases with frequency in direct proportion to ν.

Series *RLC* Circuits

The series *RLC* circuit of Figure 18.10 illustrates the use of phasor diagrams in simple ac circuits and permits the introduction of some further concepts that are useful in ac circuits.

In a series *RLC* circuit, the difference in potential across the alternating emf is, at any instant, equal to the sum of the potential differences across each of the other components:

$$\Delta v = \Delta v_R + \Delta v_L + \Delta v_C \qquad (18\text{-}12)$$

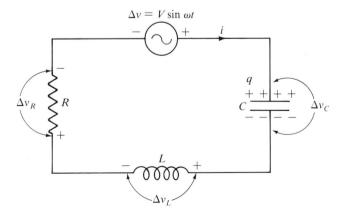

Figure 18.10
A series *RLC* circuit.

And because the components of the circuit are in series, the same current i is in each component of the circuit:

$$i = i_R = i_L = i_C \tag{18-13}$$

The complete phasor diagram for this series RLC circuit is shown in Figure 18.11 at a particular instant of time $t > 0$ when the current phasor I has turned through an angle ωt from its position at $t = 0$. Let us consider the phasor diagram, step by step.

Because the potential difference across R, Δv_R, is in phase with the current i, the phasor, $V_R = IR$, is in the same direction as the current phasor I. The projection of V_R onto the vertical axis at any instant is equal to the value of Δv_R at that instant. Because Δv_C lags i by 90°, the phasor, $V_C = IX_C$, is 90° behind I at every instant. The projection of V_C onto the vertical axis is equal to the instantaneous value of Δv_C. Because Δv_L leads i by 90°, the phasor, $V_L = IX_L$, is 90° ahead of I at every instant; and the projection of V_L onto the vertical axis is equal to the instantaneous value of Δv_L. According to Equation (18-12), Δv is the (algebraic) sum of Δv_R, Δv_L, and Δv_C; so Δv is the algebraic sum of the projections of V_R, V_L, and V_C onto the vertical axis. But the sum of these projections is just the projection of the *vector sum* of the phasors $\mathbf{V}_R$, $\mathbf{V}_L$, and $\mathbf{V}_C$. Then the vector sum $\mathbf{V} = \mathbf{V}_R + \mathbf{V}_L + \mathbf{V}_C$ is the phasor representing the potential difference across the alternating emf. And because $\mathbf{V}_L$ and $\mathbf{V}_C$ always lie along a straight line perpendicular to $\mathbf{V}_R$, we may form the vector $\mathbf{V}$ by first finding the vector sum, $\mathbf{V}_L + \mathbf{V}_C$, and adding that sum (vectorially) to $\mathbf{V}_R$ to give $\mathbf{V}$. Then the magnitude of $\mathbf{V}$ is

$$V = \sqrt{V_R^2 + (V_L - V_C)^2} = \sqrt{(IR)^2 + (IX_L - IX_C)^2}$$

$$V = I\sqrt{R^2 + (X_L - X_C)^2} = I\sqrt{R^2 + \left(\omega L - \frac{1}{\omega C}\right)^2} \tag{18-14}$$

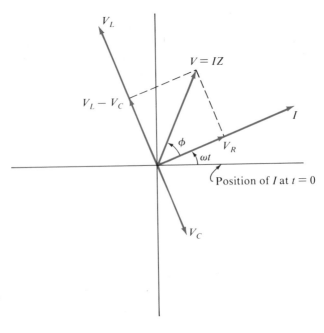

Figure 18.11
Phasor diagram for the *RLC* circuit of Figure 18.10.

In analogy to total resistance, which impedes the current in a dc circuit, we define the *impedance* Z of a series *RLC* circuit to be

$$Z = \sqrt{R^2 + (X_L - X_C)^2} = \sqrt{R^2 + \left(\omega L - \frac{1}{\omega C}\right)^2} \qquad (18\text{-}15)$$

With this definition we may write

$$V = IZ \qquad (18\text{-}16)$$

Although Equation (18-15) defines Z specifically for a series *RLC* circuit, the relationship, $Z = V/I$, from Equation (18-16) is generally true for any *RLC* network driven by an alternating emf.

The unit of impedance, according to Equation (18-16), is a volt per ampere (V/A), which is equal to an ohm (Ω).

Returning now to the *RLC* circuit of Figure 18.10, we may use the phasor diagram of Figure 18.11 to determine the *phase angle* ϕ between the current phasor *I* and the emf phasor *V*. Because the instantaneous current is $i = I \sin\omega t$, we may write the instantaneous emf as $\Delta v = V \sin(\omega t + \phi)$, where ϕ is the phase angle by which the emf leads the current. From the phasor diagram of Figure 18.11 we see that

$$\tan\phi = \frac{V_L - V_C}{V_R} = \frac{I(X_L - X_C)}{IR} = \frac{X}{R}$$

$$\phi = \tan^{-1}\left(\frac{X}{R}\right) \qquad (18\text{-}17)$$

where $X = X_L - X_C$ is the net reactance of the circuit. In Figure 18.11 we have assumed that the frequency is sufficiently great so that $X_L > X_C$. If, however, the frequency were such that $X_C > X_L$, the net reactance X would be negative; the phasor diagram would change accordingly so that V would lag I.

Example 18.1

PROBLEM   An alternating emf ($V = 50$ V, $\nu = 1.0$ kHz) drives a series *RLC* circuit ($L = 2.0$ mH, $R = 10\ \Omega$, $C = 80\ \mu$F). Determine the potential difference across each of the circuit components at an instant when the potential difference across the emf is equal to 50 V.

SOLUTION   Figure 18.12 shows the sense of the current and the potential differences across each of the components. The inductive reactance X_L, capacitive reactance X_C, total reactance X, impedance Z, current amplitude I, and phase angle ϕ for the circuit are

$$X_L = 2\pi\nu L = 2\pi(1000)(0.002) = 12.6\ \Omega$$

$$X_C = \frac{1}{2\pi\nu C} = \frac{1}{2\pi(1000)(8.0 \times 10^{-5})} = 2.0\ \Omega$$

$$X = X_L - X_C = 10.6\ \Omega$$

$$Z = (R^2 + X^2)^{1/2} = 14.6\ \Omega$$

$$I = \frac{V}{Z} = \frac{50}{14.6} = 3.42\ \text{A}$$

$$\phi = \tan^{-1}\left(\frac{X}{R}\right) = \tan^{-1}\left(\frac{10.6}{10}\right) = 46.7° = 0.815\ \text{rad}$$

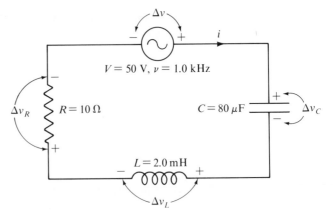

Figure 18.12   Example 18.1

Figure 18.13 shows the phasor diagram for this circuit at the instant when the emf has a maximum value. The phasor magnitudes V_L, V_C, and V_R are determined by

$$V_L = IX_L = (3.42)(12.6) = 43.1 \text{ V}$$

$$V_C = IX_C = (3.42)(2.0) = 6.8 \text{ V}$$

$$V_R = IR = (3.42)(10) = 34.2 \text{ V}$$

From the phasor diagram, the potential differences Δv_L, Δv_C, and Δv_R are given by

$$\Delta v_L = V_L \cos 43.3° = 31 \text{ V}$$

$$\Delta v_C = -V_C \cos 43.3° = -5.0 \text{ V}$$

$$\Delta v_R = V_R \cos 46.7° = 24 \text{ V}$$

The negative value for Δv_C indicates that the potential difference across C is opposite to the polarity shown in Figure 18.12. Finally we check to see that the sum

$$\Delta v_L + \Delta v_C + \Delta v_R = 31 - 5 + 24 = 50 \text{ V}$$

is indeed equal to Δv, as it should be. ■

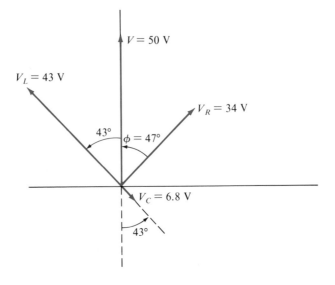

Figure 18.13   Example 18.1

E 18.7   At the instant under consideration in Example 18.1, determine (a) the current i in the circuit, (b) di/dt, the rate of change of i, and (c) the charge q on the capacitor.   Answers: (a) 2.4 A; (b) 1.6×10^4 A/s; (c) -4.0×10^4 C

Example 18.2   **PROBLEM**   For the circuit of Example 18.1 and Figure 18.12, determine the potential difference across each of the circuit components at an instant when the potential difference across the emf is equal to 30 V and decreasing.

SOLUTION   Figure 18.14(a) shows the two positions of the emf phasor that correspond to an emf potential difference of 30 V. Because phasors rotate CCW, position A corresponds to an increasing emf, and position B corresponds to a decreasing value for Δv. Figure 18.14(b) shows the inductive, capacitive, and resistive phasors in the orientation that corresponds to the specified condition. Using this diagram, we see that the desired potential differences are

$$\Delta v_L = -V_L \sin 6° = -43 \sin 6° = -4.5 \text{ V}$$

$$\Delta v_C = V_C \sin 6° = 6.8 \sin 6° = 0.70 \text{ V}$$

$$\Delta v_R = V_R \cos 6° = 34 \cos 6° = 33.8 \text{ V}$$

As a simple check on these results, we note that

$$\Delta v_L + \Delta v_C + \Delta v_R = -4.5 + 0.70 + 33.8 = 30 \text{ V}$$

a value equal to Δv, as expected.   ■

Resonance in AC Circuits

In a series RLC circuit, the impedance Z of the circuit depends on the frequency ν of the emf because $X_L = 2\pi\nu L$ and $X_C = 1/(2\pi\nu C)$ are both frequency dependent. The particular frequency, $\nu_0 = \omega_0/2\pi$, at which the impedance has its minimum value is called the *resonant frequency* of the circuit. According to

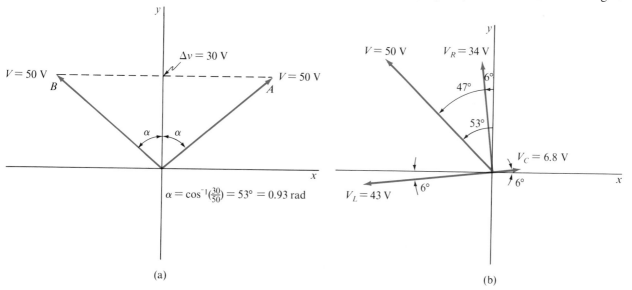

(a)

(b)

Figure 18.14   Example 18.2

Equation (18-15), the impedance has its minimum value when $X_L = X_C$, and Equation (18-15) becomes $Z = R$. Then Equation (18-16) becomes

$$I = \frac{V}{R} \quad \text{(at resonance)} \tag{18-18}$$

At the resonant frequency, the total reactance of the circuit is equal to zero because the capacitive reactance and the inductive reactance nullify each other. And because the impedance is minimal at ν_o, the current has its maximum amplitude at the resonant frequency. The sharpness of the current-vs.-frequency peak depends on the relative values of R and X_L (or X_C) at ν_o. For example, in Figure 18.15(a), where both X_L and X_C at ν_o are considerably larger than R, the current-vs.-frequency curve of Figure 18.15(b) peaks sharply at $\nu = \nu_o$. On the other hand, when X_L and X_C are small compared to R, as in Figure 18.16(a), the current peak is relatively broad, as shown in Figure 18.16(b).

The value of the resonant frequency of an ac circuit may be determined in terms of the values, L and C, of the reactive components of the circuit using the fact that, at resonance, $X_L = X_C$:

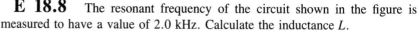

$$X_L - X_C = \omega_o L - \frac{1}{\omega_o C} = 2\pi\nu_o L - \frac{1}{2\pi\nu_o C} = 0$$

$$\nu_o = \frac{1}{2\pi\sqrt{LC}} \tag{18-19}$$

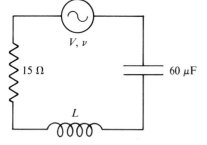

Exercise 18.8

E 18.8   The resonant frequency of the circuit shown in the figure is measured to have a value of 2.0 kHz. Calculate the inductance L.

Answer: 0.11 mH

E 18.9   If the current amplitude at resonance for the circuit of Exercise 18.8 is equal to 0.80 A, determine the amplitude V of the emf. Answer: 12 V

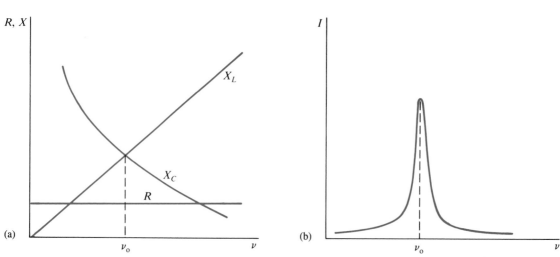

Figure 18.15
(a) Resistance R, inductive reactance X_L, and capacitive reactance X_C as functions of frequency ν when R is considerably less than both X_L and X_C at the resonant frequency ν_o.
(b) The current I peaks sharply at the resonant frequency ν_o.

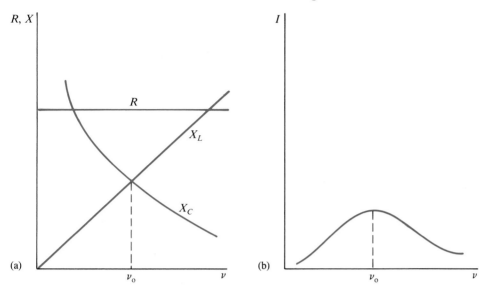

Figure 18.16
(a) Resistance R, inductive reactance X_L, and capacitive reactance X_C as functions of frequency ν when R is considerably greater than both X_L and X_C at the resonant frequency ν_0.
(b) The current I peaks broadly at $\nu = \nu_0$.

Example 18.3   **PROBLEM**   For a driven RLC circuit having $L = 8.0$ mH, $C = 20$ μF, and $V = 80$ V, draw a graph of I, the current amplitude, as a function of the frequency ν of the alternating emf for $R = 5.0$ Ω, 20 Ω, and 80 Ω.

SOLUTION   Figure 18.17(a) shows the circuit being considered. The impedance Z of this circuit is given by

$$Z = \sqrt{R^2 + \left(2\pi\nu L - \frac{1}{2\pi\nu C}\right)^2}$$

which may be written

$$Z = \sqrt{R^2 + \frac{L}{C}\left(\frac{\nu}{\nu_0} - \frac{\nu_0}{\nu}\right)^2}$$

where ν_0 is the resonant frequency of the circuit, i.e.,

$$\nu_0 = \frac{1}{2\pi\sqrt{LC}}$$

For this circuit, substituting $L = 8.0$ mH, $C = 20$ μF, and $V = 80$ V gives

$$\nu_0 = 400 \text{ Hz}$$

$$Z(\nu) = \sqrt{R^2 + 400\left(\frac{\nu}{400} - \frac{400}{\nu}\right)^2}$$

$$I(\nu) = \frac{V}{Z(\nu)} = \frac{80/R}{\sqrt{1 + \frac{400}{R^2}\left(\frac{\nu}{400} - \frac{400}{\nu}\right)^2}}$$

Figure 18.17(b) shows graphs of $I(\nu)$ vs. ν for $R = 5.0$ Ω, 20 Ω, and 80 Ω. In each case, both X_L and X_C are equal to 20 Ω for $\nu = \nu_0 = 400$ Hz. As expected,

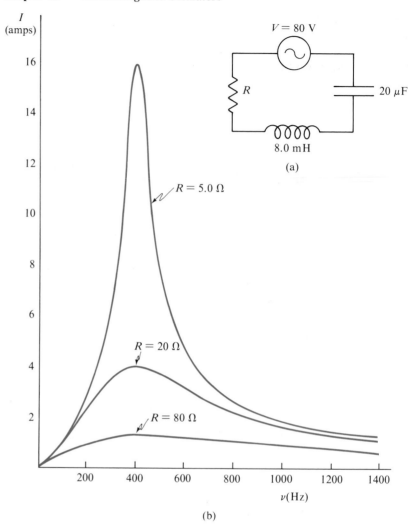

Figure 18.17   Example 18.3

(b)

the graph of $I(\nu)$ for $R = 5.0 \ \Omega$, a value one-fourth that of $X_L = 20 \ \Omega$ at $\nu = \nu_o$, is a sharply peaked, high-current resonance curve. Alternatively, for $R = 80 \ \Omega$, a value four times that for X_L at $\nu = \nu_o$, the graph of $I(\nu)$ is a very broad, low-current resonance curve. ■

Power and RMS Values in AC Circuits

The instantaneous electric power, $P = i^2R$, dissipated in the resistance R of an ac circuit varies with time, because the current i in R changes with time. We are usually interested in the average rate at which energy is dissipated in an ac circuit. Because $i = I \sin\omega t$ is the instantaneous current in the resistance R, we may express P as

$$P = i^2R = I^2R \sin^2\omega t \qquad (18\text{-}20)$$

Figure 18.18(a) shows a graph of the instantaneous current i in a resistance R, as a function of time t, and Figure 18.18(b) shows a graph of Equation (18-20), the instantaneous value of the electric power P dissipated in R as a function of t. To

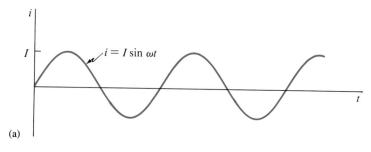

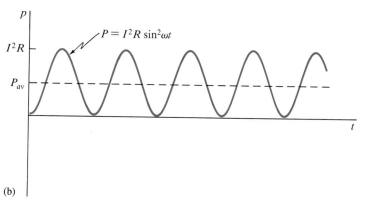

Figure 18.18
(a) A graph of the instantaneous current i in a resistance R as a function of time t. (b) A graph of the instantaneous electric power P dissipated in R as a function of time t. The average value P_{av} of the power dissipated in R is indicated.

find the average value P_{av} of P over any number of complete cycles, we determine the average value of $\sin^2\omega t$ by first recognizing that the squares of $\sin\omega t$ and $\cos\omega t$ are identical except for being displaced a quarter cycle along the t-axis. Then we may write

$$(\sin^2\omega t)_{av} = (\cos^2\omega t)_{av} \tag{18-21}$$

and because $\sin^2\omega t + \cos^2\omega t = 1$, Equation (18-21) becomes

$$(\sin^2\omega t)_{av} + (\cos^2\omega t)_{av} = (2\sin^2\omega t)_{av} = 1$$

$$(\sin^2\omega t)_{av} = \frac{1}{2} \tag{18-22}$$

Then the average power P_{av} dissipated in R is

$$P_{av} = I^2R(\sin^2\omega t)_{av} = \frac{I^2R}{2} \tag{18-23}$$

a value that is indicated in Figure 18.18(b).

It is convenient to define the *root-mean-square current* I_{rms} to be

$$I_{rms} = \sqrt{(i^2)_{av}} = \sqrt{I^2(\sin^2\omega t)_{av}} = \frac{I}{\sqrt{2}} \tag{18-24}$$

Because $P_{av} = I^2R/2$ from Equation (18-23) and, therefore, $P_{av} = I^2{}_{rms}R$, the *rms* value of the current is equal to the magnitude of a direct current that dissipates thermal energy in a resistor R at the same average rate that energy is dissipated in R by an alternating current with amplitude I.

Similarly, we define a *root-mean-square potential difference* V_{rms} for an alternating potential difference to be

$$V_{rms} = \sqrt{[(\Delta v)^2]_{av}} = \sqrt{V^2(\sin^2\omega t)_{av}} = \frac{V}{\sqrt{2}} \tag{18-25}$$

This value V_{rms} is equal to the magnitude of a dc difference in potential that, when connected across a resistance R, dissipates thermal energy in R at the same rate as would an alternating potential difference Δv with amplitude V.

E 18.10   Suppose the ac current through a 25-Ω resistor is given by $i = 0.20 \sin(500t)$ A.
(a) What is the average rate at which thermal energy is generated in the resistor?
Answer: 0.50 W
(b) Determine I_{rms} for the current through the resistor.   Answer: 0.14 A
(c) Determine V_{rms} for the potential difference across the resistor.
Answer: 3.5 V

Instruments used to measure currents and potential differences in ac circuits are usually calibrated to read root-mean-square values to facilitate the determination of average power values. Thus, most ac voltmeters and ammeters indicate V_{rms} or I_{rms}. Notice that Ohm's law for a resistance R is applicable whether we use instantaneous values, amplitudes, or rms values of current and potential difference:

$$R = \frac{\Delta v_R}{i} = \frac{v_R}{I} = \frac{V_{rms}}{I_{rms}} \qquad (18\text{-}26)$$

Because the instantaneous power P delivered by an emf Δv to any circuit carrying current i is given by $P = \Delta v\, i$, we may determine the average power P_{av} delivered to an ac circuit by finding the average value of P taken over a complete cycle, or over a period $T = 1/\nu = 2\pi/\omega$:

$$P_{av} = \frac{1}{T}\int_{t=0}^{t=T} \Delta v\, i\, dt = \frac{VI}{(2\pi/\omega)}\int_0^T \sin(\omega t + \phi)\,\sin\omega t\, dt \qquad (18\text{-}27)$$

Here we have used $i = I\sin\omega t$ and $\Delta v = V\sin(\omega t + \phi)$, where ϕ is the phase angle by which the emf leads the current. If, in Equation (18-27), we let $\alpha = \omega t$ and $d\alpha = \omega dt$, we have

$$P_{av} = \frac{VI}{2\pi}\int_0^{2\pi} \sin(\alpha + \phi)\,\sin\alpha\, d\alpha \qquad (18\text{-}28)$$

But since $\sin(\alpha + \phi) = \sin\phi\,\cos\alpha + \cos\phi\,\sin\alpha$, Equation (18-28) becomes

$$P_{av} = \frac{VI}{2\pi}\left\{\sin\phi\int_0^{2\pi}\sin\alpha\,\cos\alpha\, d\alpha + \cos\phi\int_0^{2\pi}\sin^2\alpha\, d\alpha\right\} \qquad (18\text{-}29)$$

in which the first integral is equal to zero and the second is equal to π. Then Equation (18-29) simplifies to

$$P_{av} = \frac{VI}{2}\cos\phi$$

which may be written, using Equations (18-24) and (18-25), as

$$P_{av} = V_{rms}I_{rms}\cos\phi \qquad (18\text{-}30)$$

In Equation (18-30), $\cos\phi$ is called the *power factor* of the ac circuit.

E 18.11   If the ac current through an alternating emf ($V = 8.0$ V) has an amplitude of 2.5 A and the emf delivers electrical energy at an average rate of 7.5 W, what is the power factor for this circuit?   Answer: 0.75

E 18.12   Using Equation (18-17), $\tan\phi = X/R$, show that $\cos\phi = R/Z$ for a series *RLC* ac circuit.

The practical importance of the power factor, $\cos\phi$, is illustrated by the following example.

Example 18.4   **PROBLEM**   A device operates at a power of 1.2 kW when connected to a 120-V (rms) 60-Hz ac-power source. The potential difference across the device leads the current by 60°, that is, $\phi = 60°$. What is the capacitance of a series capacitor that will change the power factor for the device to a value of one? What power will the device then absorb?

SOLUTION   Because $P_{av} = I_{rms}V_{rms}\cos\phi$ and $Z = V_{rms}/I_{rms}$, we have $P_{av} = (V^2_{rms}/Z)\cos\phi$. Thus the impedance for this device is

$$Z = \frac{V^2_{rms}\cos\phi}{P_{av}} = \frac{(120)^2\cos60°}{1200} = 6.0\ \Omega$$

Similarly, the resistance and (inductive) reactance of the device are

$$R = Z\cos\phi = (6.0\ \Omega)\cos60° = 3.0\ \Omega$$

$$X_L = R\tan\phi = (3.0\ \Omega)\tan60° = 5.2\ \Omega$$

If the power factor is to be changed to a value of one, then the reactance of the device may be changed to zero by adding a series capacitance C having a reactance equal to the (inductive) reactance, i.e.,

$$\frac{1}{2\pi\nu C} = X_C = X_L$$

or

$$C = \frac{1}{2\pi\nu X_L} = \frac{1}{2\pi(60)(5.1)} = 5.1 \times 10^{-4}\ \text{F}$$

Now, with $X_L - X_C = 0$, the impedance of the device is equal to the resistance, i.e., $Z = R = 3.0\ \Omega$; and the power absorbed by the device is now given by

$$P_{av} = \frac{(V_{rms})^2\cos\phi}{Z} = \frac{(120)^2(1.0)}{3.0} = 4.8\ \text{kW}\quad\blacksquare$$

18.2   Electromagnetic Radiation

An electromagnetic disturbance that is composed of time-varying electric and magnetic fields and can transport energy through space, even if no matter is present in that space, is called an *electromagnetic wave*. Once formed in free space, an electromagnetic wave propagates, or propels itself, at a characteristic speed called the speed of light. In fact, an electromagnetic wave transports momentum and energy at the speed of light without transporting either matter or charge. Visible light, radio and television transmissions, and X rays are all forms of *electromagnetic radiation*, which we shall abbreviate as EM radiation.

EM radiation results from electric charges that are accelerating. Oscillating charges, like those we have encountered in ac circuits, are repeatedly being accelerated and, therefore, are radiating. A circuit component that is designed to radiate EM waves into space is called an *antenna*. We will consider the nature of EM radiation in space far from an antenna, when the form of the wave is especially simple, but, first, let us characterize EM radiation in free space in a general way.

In the mid-nineteenth century, Maxwell predicted the existence of EM waves. Using what we now call Maxwell's equations (described in Chapter 17, *Magnetic Fields* II, Section 17.5), he cast these relationships into the form of differential equations from which he derived the wave equations that characterize EM radiation. Although we will not examine these formalisms here, we will present some of their more important results.

First, the wave equations developed by Maxwell predict the speed of propagation of EM radiation through space. In particular, Maxwell's equations predict that the speed of an EM wave through a medium depends on the permeability μ and the permittivity ϵ of that medium. In free space, for example, where there is no medium, the permeability is $\mu_o = 4\pi \times 10^{-7}$ T·m/A and the permittivity is $\epsilon_o = 8.85 \times 10^{-12}$ C^2/(N·m^2), the speed c of propagation of an EM wave in space is

$$c = \frac{1}{\sqrt{\mu_o \epsilon_o}} = 3.0 \times 10^8 \text{ m/s} \qquad (18\text{-}31)$$

which is commonly called the *speed of light*. This value of c coincides with the experimentally measured speed of visible light. In fact, a notable triumph of Maxwell's work was that his theory of EM radiation provided impressive evidence that light is an EM wave—an accepted fact now but a point of considerable contention in the nineteenth century.

Further, Maxwell's formulation of EM radiation describes the spatial and temporal relationships between the electric and magnetic fields that compose EM waves. But before we turn to these aspects of EM waves, let us consider how they may be generated.

A common source of EM radiation is a transmitting antenna (as opposed to a receiving antenna, like those attached to a radio or television receiver). Alternating current in the antenna is driven by a sinusoidal ac circuit, as indicated in Figure 18.19. As the oscillating charges accelerate during their back-and-forth motion in the antenna, electric and magnetic fields are created in a complicated geometry in the space around the antenna. The changing electric field produces a changing magnetic field, which in turn produces a changing electric field, and so on. In this way, an EM wave propagates away from the antenna. Near the antenna the electric and magnetic fields are quite complex and need not be considered here. But for the simple antenna (called a dipole antenna) of Figure 18.19, at a distance far from the antenna along a line perpendicular to the antenna, the fields form a relatively simple pattern called a *plane wave*. In a plane EM wave, the electric field **E** and the magnetic field **B** are, at any instant, constant in each plane perpendicular to the direction of propagation of the wave. Let us then consider some further characteristics of an EM wave using the relatively simple plane wave to illustrate its nature.

Figure 18.20 is one representation of the fields of a plane EM wave at a particular instant. At any point on the x axis, along which the wave is propagating through free space, the following facts are consistent with Maxwell's theory of EM radiation:

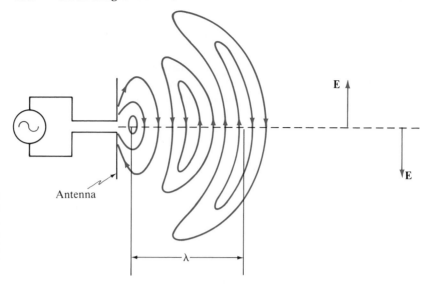

Figure 18.19
A representation of a portion of the electric field radiating from a simple (dipole) antenna, which is being driven by a sinusoidal ac circuit. At a far distance from the antenna, the radiated electromagnetic wave is approximately a plane wave.

1. The electric field **E** and the magnetic field **B** are mutually perpendicular. In Figure 18.20, the electric field is confined to the $\pm y$ direction; the magnetic field is confined to the $\pm z$ direction.

2. The magnitudes of **E** and **B** are related to the speed of propagation c by $|\mathbf{E}|/|\mathbf{B}| = c$.

3. Both the **E** and **B** fields are perpendicular to the direction of propagation of the wave.

    In Figure 18.20 the plane wave is proceeding in the positive x direction. The figure represents a "snapshot" of the plane wave at a particular instant. The sinusoidal curves are the envelopes of the electric and magnetic fields. At each point on the x axis, the fields change as time passes: They increase in magnitude, reach a maximum value, diminish in magnitude, become zero, increase in

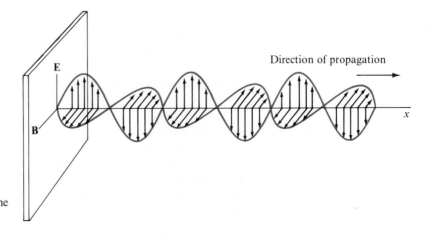

Figure 18.20
A representation of the fields of a plane electromagnetic wave.

magnitude in the opposite direction, and so on, as the wave moves in the positive x direction at a speed c.

An EM wave transports energy in the direction of its propagation. From Figure 18.20 it may be seen that the direction of energy flow is in the direction of $\mathbf{E} \times \mathbf{B}$ at every point along the wave. In fact, the magnitude and direction of the instantaneous flow of energy through a unit area in a unit time is given by the *Poynting vector* $\mathbf{S}$, defined by

$$\mathbf{S} = \frac{1}{\mu_o} \mathbf{E} \times \mathbf{B} \tag{18-32}$$

The Poynting vector $\mathbf{S}$ has the unit of energy per area per time interval, $J/(m^2 \cdot s)$, or, equivalently, power per unit area, W/m^2. For our purposes, the Poynting vector is a convenient device for determining the direction in which an EM wave is traveling and, therefore, transporting energy.

18.3 The Electromagnetic Spectrum

The parameters by which we characterize waves (of any kind) in any medium will be considered in detail in the next chapter. It is convenient, nevertheless, to recognize at this point that an EM wave may, at least in part, be characterized by its frequency ν and its speed of propagation (equal to c for EM waves in a vacuum). The *frequency* ν of an EM wave is the number of times per second that the changing electric (or magnetic) field at a point on the wave executes a complete cycle of its increase and decrease in magnitude. The unit of frequency is the hertz (Hz), or cycle per second. Further, the *wavelength* λ of an EM wave is the distance between successive maxima of the electric (or magnetic) field. The unit of wavelength is the meter (m). For any periodic EM wave in a vacuum, the frequency ν and wavelength λ are related to the speed of propagation c by

$$c = \nu\lambda \tag{18-33}$$

Thus, knowing the frequency of an EM wave in a vacuum is equivalent to knowing its wavelength and vice versa.

E 18.13
(a) Calculate the wavelength of an FM radio signal having a frequency of 100 MHz.   Answer: 3.0 m
(b) What is the frequency of light having a wavelength of 6.0×10^{-7} m?
                                                      Answer: 5.0×10^{14} Hz
(c) Determine the frequency of a 1.0-cm microwave. Answer: 3.0×10^{10} Hz

The *electromagnetic spectrum* is the range of frequencies (or wavelengths) that EM waves may have. Figure 18.21 is a graphical representation of the known electromagnetic spectrum.

The lowest frequencies (and, therefore, the longest wavelengths) of the EM spectrum we usually encounter are radio waves. Television frequencies are usually somewhat higher. Radio and television frequencies typically range from 10^3 to 10^8 Hz.

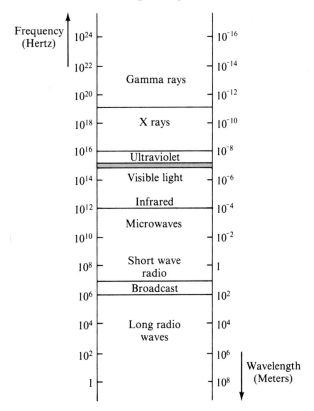

Figure 18.21
The electromagnetic spectrum.

The microwave region of the EM spectrum includes the range of frequencies between 10^8 and 10^{12} Hz. *Microwaves* are used in modern kitchen ovens and in radar systems. In weather radar systems, for example, EM waves with wavelengths of a few centimeters are beamed outward from an antenna, reflected back from dense clouds, and received by the antenna. The time delay between the transmitted and received pulse of EM energy serves as a measure of the distance to the target, because the speed of propagation (approximately equal to c in air) is a constant value.

The *infrared* portion of the EM spectrum is radiation that we sometimes sense as heat. Infrared (IR) radiation is produced by hot bodies, like radiant heaters or the sun. Infrared radiation is so called because its frequency range is immediately below (infra) that of red light in the visible portion of the EM spectrum. The *visible* spectrum is a very narrow range of wavelengths within the EM spectrum to which the human eye responds with the perception of vision. Red light is at the lower frequency end of the visible range; violet light is at the upper frequency end. *Ultraviolet* (UV) radiation is the frequency range beyond (ultra), or greater than, the frequency of violet light. Ultraviolet radiation is perhaps familiar from its use in sanitizing lamps, where it is used because of the capacity of UV radiation to kill many species of bacteria.

X rays are perhaps best known for their ability to penetrate soft tissue and expose film, thereby permitting X-ray pictures. *Gamma rays*, produced by nuclear rearrangements in atoms, are at the extreme upper frequency range of the EM spectrum.

Not only do the frequency regions named in Figure 18.21 overlap, but the radiation in the overlapping regions may also be produced by different means. For example, some X rays, which are produced by rearrangement of extra-nuclear electrons in atoms, have a higher frequency than some gamma rays, which are of nuclear origin. But whatever their origins, all electromagnetic waves of the same frequency are identical. Indeed, only the differences in frequency (or wavelength) distinguish radio waves, visible light, and X rays. Clearly, those frequency differences have a profound effect on the ways that EM waves of different frequencies interact with matter.

18.4 Problem-Solving Summary

Problems in ac circuitry are self-identifying. The most common ac circuits are driven by a sinusoidal alternating emf, $\Delta v = V \sin\omega t$, which may be conveniently represented by a rotating phasor of length V turning at an angular speed ω. In such ac circuits, the components R, L, and C have instantaneous values of potential difference Δv_R, Δv_L, and Δv_C, whose phasors V_R, V_L, and V_C have different phase relationships (phase angles) with their respective currents, i_R, i_L, and i_C. In particular, Δv_R and i_R are in phase, Δv_L leads i_L by 90°, and Δv_C lags i_C by 90°.

In an ac circuit, the reactance of a capacitance C or of an inductance L is a measure of the ability of that component to impede the flow of charge in that circuit. Reactance depends on the frequency ν of the alternating current. The reactance X_C of a capacitance C is equal to $1/\omega C = 1/2(\pi\nu C)$. The reactance X_L of an inductance L is equal to $\omega L = 2\pi\nu L$. The net reactance X of an ac circuit is given by $X = X_L - X_C$.

We begin the analysis of a series ac circuit by combining similar components when appropriate. If the equivalent components are then in a series RLC circuit, the same current is in each component at any instant. Therefore, it is convenient to relate the phasors of the circuit components to the current phasor I. In particular, the potential difference phasor, $V_R = IR$, for the resistance R is in phase with I, and the phasor V_R is in the same direction as the current phasor I. The potential difference phasor, $V_L = IX_L$, for the inductance L leads I by 90°, and the potential difference phasor, $V_C = IX_C$, for the capacitance lags I by 90°. We may then construct a phasor diagram with I drawn at an arbitrary time t so that I makes an angle ωt with the axis corresponding to the direction of I at $t = 0$. Then by drawing the phasors, V_R, V_L, and V_C, in their appropriate directions, we can construct the potential difference phasor V of the emf. The magnitude and direction of the phasor V is equal to the vector sum of the phasors, $\mathbf{V}_R$, $\mathbf{V}_L$, and $\mathbf{V}_C$. The phase angle ϕ between the emf and the current in the circuit is the angle between the phasors V and I, and the magnitude of ϕ is given by $\phi = \tan^{-1}(X/R)$.

The impedance $Z = V/I$ of an ac circuit plays the same role as the net resistance in a dc circuit. For a given series ac circuit, the impedance $Z = \sqrt{R^2 + (X_L - X_C)^2}$ depends on the resistance and reactance of the circuit. Because X_L and X_C depend on the frequency of the emf, impedance depends on frequency. The frequency at which the impedance is minimal is called the resonant frequency ν_0 of the circuit. Resonance occurs when the frequency is such that the net reactance, $X_L - X_C$, is equal to zero, that is, when $X_L = X_C$. The resonant frequency ν_0 may be expressed in terms of the inductance L and the capacitance C of the circuit by $\nu_o = 1/(2\pi\sqrt{LC})$. At the resonant frequency, the current in an ac circuit achieves its greatest amplitude.

Power measurements in ac circuits are conveniently described in terms of root-mean-square (rms) values of emf and current. Using $V_{rms} = V/\sqrt{2}$ and $I_{rms} = I/\sqrt{2}$, we may express the average power P_{av} delivered to an ac circuit by $P_{av} = V_{rms}I_{rms}\cos\phi$, where $\cos\phi$ is called the power factor of the circuit. Again ϕ is the phase angle between the emf phasor V and the current phasor I. The magnitude of ϕ is related to the net resistance R and the impedance Z of an ac circuit by $\cos\phi = R/Z$.

Electromagnetic (EM) waves are composed of mutually perpendicular electric and magnetic fields that propagate through free space at the speed of light ($c = 3.0 \times 10^8$ m/s). Both the electric field $\mathbf{E}$ and the magnetic field $\mathbf{B}$ of the EM wave are perpendicular to the direction of propagation of the wave. The direction of propagation is, at every point on the wave, in the direction specified by $\mathbf{E} \times \mathbf{B}$. The Poynting vector, $\mathbf{S} = (1/\mu_o)\, \mathbf{E} \times \mathbf{B}$, gives the magnitude and direction of the energy transported by an EM wave across a unit area in a unit time. Thus the magnitude of S represents power that an EM wave transmits through a unit area.

The wavelength λ of an EM wave is the distance between adjacent maxima of the electric field (or the magnetic field) measured along the direction of propagation of the wave. The wavelength λ and frequency ν of an EM wave in free space are related to the propagation speed c by $c = \nu\lambda$. Thus, specifying the wavelength of an EM wave, in effect, specifies its frequency; similarly, giving its frequency gives its wavelength.

The range of known wavelengths (or frequencies) of EM waves is called the electromagnetic spectrum. Radio and television waves, microwaves, infrared radiation, visible light, ultraviolet radiation, X rays, and gamma rays make up the EM spectrum. Although these radiations are produced in a variety of ways and interact very differently with matter, the essential difference between these EM radiations is their wavelength (or frequency).

Problems

GROUP A

18.1 Visible light is generally assumed to have a wavelength range from 0.40 μm to 0.70 μm. Determine the corresponding frequency range for visible light.

18.2 Determine the power factor and the average power dissipated in an RLC series circuit having $R = 25$ Ω, $L = 80$ mH, $C = 45$ μF, $V = 9.0$ V, and $\nu = 60$ Hz.

18.3 If a 70-μF capacitor is connected across an alternating emf having $V = 25$ V and $\nu = 80$ Hz, determine the maximum current through the emf.

18.4 Determine the impedance of the inductor-resistor combination shown in Figure 18.22 for a 1.5-kHz frequency.

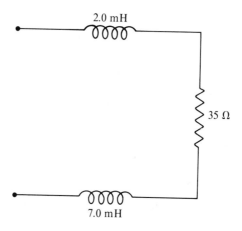

Figure 18.22 Problem 18.4

18.5 Determine the impedance of a series combination of a 55-Ω resistor, a 3.0-μF capacitor, and a 6.0-μF capacitor at a frequency of 2.5 kHz.

18.6 What is the impedance of a series circuit consisting of an alternating emf, a 20-Ω resistor, a 2.0-μF capacitor, and a 1.8-mH inductor (a) at 1.0 kHz and (b) at resonance.

18.7 If a 2.5-mH inductor is connected across an alternating emf having an amplitude equal to 35 V and a frequency equal to 2.5 kHz, calculate the maximum current in this circuit.

18.8 An *RLC* series circuit has $R = 15$ Ω, $L = 6.0$ mH, and $C = 15$ μF.

  (a) Determine the reactance and impedance for this circuit if the emf has a frequency equal to 0.65 kHz.

  (b) Calculate the resonant frequency for this circuit and the values for the reactance and impedance of this circuit at resonance.

18.9 A wire-wound inductor having resistance of 16 Ω is connected across an alternating emf having an amplitude of 30 V and a frequency of 1.5 kHz. If the maximum current in the circuit is equal to 1.2 A, determine the inductance of the device.

18.10 With $V = 120$ V and $\nu = 60$ Hz in a circuit (*see* Fig. 18.23), determine

  (a) the total reactance of the circuit.

  (b) the impedance of the circuit.

  (c) the maximum current.

  (d) the phase angle by which the emf Δv leads the current *i*.

  (e) an expression for the current as a function of time.

  (f) the maximum potential difference across the resistor.

  (g) the maximum potential difference across the capacitor.

  (h) an expression for the potential difference across the capacitor as a function of time.

  (i) the average power delivered by the emf.

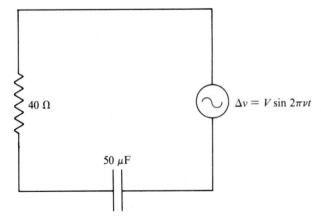

Figure 18.23   Problem 18.10

18.11 With $V = 50$ V and $\nu = 1.0$ kHz in the circuit shown in Figure 18.24, determine

  (a) the total reactance of the circuit.

  (b) the impedance of the circuit.

  (c) the maximum current.

  (d) the phase angle by which the emf Δv leads the current *i*.

  (e) an expression for the current as a function of time.

  (f) the maximum potential difference across the resistor.

  (g) the maximum potential difference across the inductor.

  (h) an expression for the potential difference across the inductor as a function of time.

  (i) the average power delivered by the emf.

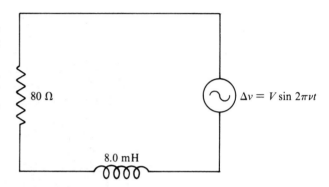

Figure 18.24   Problem 18.11

18.12 Determine the value of R in the circuit in Figure 18.25 if the rms-current through the inductor is equal to 0.28 A.

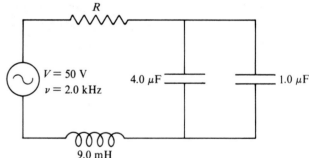

Figure 18.25   Problem 18.12

18.13 With $V = 15$ V and $\nu = 0.20$ kHz in the circuit (*see* Fig. 18.26), determine

  (a) the total reactance for the circuit.

  (b) the impedance for the circuit.

  (c) the maximum current.

  (d) the phase angle by which the emf Δv leads the current *i*.

  (e) an expression for the current as a function of time.

  (f) the maximum potential difference across the capacitor.

  (g) the maximum potential difference across the inductor.

(h) expressions for the potential differences across the capacitor and the inductor as functions of time.

(i) the average power delivered by the emf.

(j) the resonant frequency for this circuit.

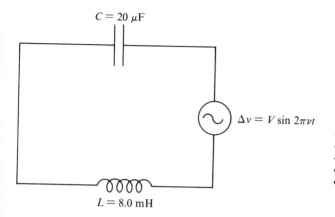

Figure 18.26   Problem 18.13

18.14   The graph in Figure 18.27 shows the potential difference $\Delta v_R(t)$ across the resistor ($R = 25\ \Omega$) and the alternating emf $\Delta v(t)$ for a series RLC circuit.

(a) What is the frequency of the emf?

(b) Determine the maximum current.

(c) Determine the impedance and the reactance of the circuit.

(d) Calculate the angle by which the emf leads the current.

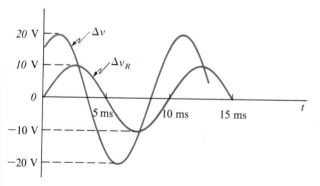

Figure 18.27   Problem 18.14

18.15   A series combination of a 20-Ω resistor and a 60-μF capacitor are placed across an alternating emf described by $\Delta v = V \sin \omega t$ with $V = 25$ V and $\omega = 600$ rad/s.

(a) Calculate the amplitude of the current.

(b) Determine the potential difference Δv_C across the capacitor as a function of the time t.

(c) At what (average) rate does the emf provide energy?

18.16   For the ac circuit shown in Figure 18.28,

(a) calculate the total impedance Z.

(b) determine the angle ϕ by which the alternating emf Δv leads the current i.

(c) calculate the amplitude of the current.

(d) write an expression for Δv if $i = I \sin(120\pi t)$.

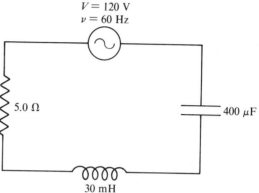

Figure 18.28   Problem 18.16

18.17   For the circuit of Problem 18.16 with $i = I \sin(2\pi\nu t)$,

(a) calculate the maximum potential difference across each of the components R, L, and C.

(b) write expressions for each of the alternating potential differences Δv_R, Δv_L, and Δv_C.

18.18   Calculate the power factor for, and the average power delivered by, the emf Δv of Problem 18.16.

18.19   For the circuit of Problem 18.16,

(a) for what frequency ν_o of the alternating emf is the net reactance of the circuit equal to zero?

(b) calculate the impedance Z and phase angle ϕ corresponding to the frequency ν_o of part (a).

(c) determine the amplitude of the current for the frequency ν_o (keeping $V = 120$ V).

(d) write an expression for i if $\Delta v = V \sin(2\pi\nu_o t)$.

18.20   If the maximum current through the emf in the circuit shown in Figure 18.29 is 85 mA, determine the amplitude of the alternating emf.

18.21   In a series RC circuit, the maximum potential differences across the resistor and the capacitor are $V_R = 75$ V and $V_C = 60$ V.

(a) What is the amplitude of the alternating emf?

(b) If $R = 30\ \Omega$ and $C = 25\ \mu$F, determine the *rms* current for the circuit, the frequency of the emf, and the power factor for the emf.

18.22   In a series RL circuit, the maximum potential differences across the resistor and inductor are $V_R = 45$ V and $V_L = 60$ V.

(a) What is the amplitude of the alternating emf?

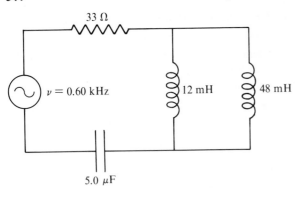

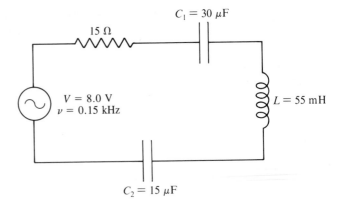

Figure 18.29   Problem 18.20

Figure 18.31   Problem 18.25

(b) If the rms current in the circuit is equal to 80 mA and the frequency of the emf is equal to 2.0 kHz, calculate R, L, and the average power delivered by the emf.

18.23 In a series *RLC* circuit, the maximum potential differences across the resistor, inductor, and capacitor are $V_R = 43$ V, $V_L = 62$ V, and $V_C = 98$ V.

(a) What is the amplitude of the alternating emf?

(b) If $R = 150$ Ω and $L = 4.5$ mH, determine the maximum current, the frequency of the emf, the impedance of the circuit, the total resistance of the circuit, and the power factor for the emf.

18.24

(a) Calculate the resonant frequency of the circuit shown in Figure 18.30.

(b) If the emf shown has an amplitude of 65 V and a frequency equal to the resonant frequency of the circuit, determine the maximum potential differences across R, L, and each of the capacitors.

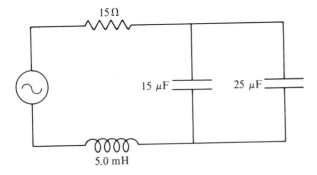

Figure 18.30   Problem 18.24

18.25 For the series *RLC* circuit shown in Figure 18.31,

(a) calculate the resonant frequency of the circuit.

(b) determine the amplitude of the current.

(c) what is the maximum potential difference across R? L? C_1? C_2?

18.26 A series *RL* circuit having $R = 35$ Ω and $L = 8.0$ mH is driven by a 0.50-kHz emf having an amplitude of 45 V. At an instant when the potential difference across the inductor is equal to 12 V and is increasing,

(a) what is the current in the circuit?

(b) what is the potential difference across R? the emf?

18.27 A series *RC* circuit having $R = 28$ Ω and $C = 80$ μF is driven by a 60-Hz emf having an amplitude of 25 V. At an instant when the current in the circuit is equal to 0.25 A and is decreasing,

(a) what is the potential difference across the emf? R? C?

(b) account for the (instantaneous) circuit energetics.

18.28 A series *RLC* circuit having $R = 9.0$ Ω, $L = 12$ mH, and $C = 30$ μF is driven by a 0.20-kHz emf having an amplitude of 50 V. At an instant when the potential difference across the emf is equal to 35 V and increasing,

(a) what is the current in the circuit?

(b) what is the potential difference across R? L? C?

(c) at what rate is the emf providing energy?

(d) account for the energy being provided by the emf.

18.29 An rms current of 2.0 A is observed in a series circuit consisting of a 20-Ω resistor and two 60-Hz emfs, each having an amplitude of 45 V. Determine the phase between the two emfs.

18.30 A device consisting of a series combination of resistors, inductors, and capacitors is connected in series with a variable capacitor C_1 and a 50-Hz emf that has an amplitude of 60 V. With $C_1 = 0$, the current amplitude is equal to 1.5 A. With C_1 properly adjusted, the current amplitude achieves a maximum value (at resonance) of 5.0 A.

(a) Determine the resistance, reactance, and impedance of the device.

(b) Calculate the value of C_1 for which the maximum current is achieved.

(c) What average power is the emf delivering with $C_1 = 0$ and with C_1 set up so as to maximize the current?

GROUP B

18.31 A capacitor C, initially charged to a potential difference V_o, is (at $t = 0$) connected across an inductor L, as shown in Figure 18.32.

(a) Use the loop rule to show that

$$i'' + \omega_o^2 i = 0$$

where

$$i'' = \frac{d^2i}{dt^2} \quad \text{and} \quad \omega_o = \frac{1}{\sqrt{LC}}$$

(b) The general solution to this (second order, linear, homogeneous) differential equation may be written

$$i(t) = A \cos\omega_o t + B \sin\omega_o t$$

If the initial ($t = 0$) values of i and i' are denoted by i_o and i_o', show that

$$i(t) = i_o \cos\omega_o t + \frac{i_o'}{\omega_o} \sin\omega_o t$$

(c) For the specific initial conditions specified here, i.e., $\Delta v_C = V_o$ and $i_o = 0$, show that

$$i_o' = -\frac{V_o}{L}$$

$$i(t) = -\frac{V_o}{\omega_o L} \sin\omega_o t$$

$$\Delta v_C = V_o \cos\omega_o t$$

(d) Determine the energy U_C stored by C and the energy U_L stored by L as functions of time, and show that the sum of these two energies remains constant. Can you explain why $U_C + U_L = $ constant?

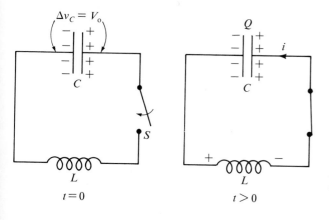

Figure 18.32   Problem 18.31

18.32 At $t = 0$ when the switch S shown in Figure 18.33 is closed, the potential difference Δv_C across C is equal to V_o.

(a) Use the loop rule to show that

$$i'' + 2\alpha i' + \omega_o^2 i = 0$$

with

$$i'' = \frac{d^2i}{dt^2}, \quad i' = \frac{di}{dt}, \quad \alpha = \frac{R}{2L}, \quad \text{and} \quad \omega_o = \frac{1}{\sqrt{LC}}$$

(b) The general solution to this (second order, linear, homogeneous) differential equation depends upon the relative values of α and ω_o. The three cases may be written:

(i) $0 < \alpha < \omega_o$ (underdamped oscillation)

$$i(t) = (A \cos\omega t + B \sin\omega t)e^{-\alpha t}$$

with

$$\omega = \sqrt{\omega_o^2 - \alpha^2}$$

(ii) $\alpha = \omega_o$ (critically damped oscillation)

$$i(t) = (A + Bt)e^{-\alpha t}$$

(iii) $0 < \omega_o < \alpha$ (overdamped oscillation)

$$i(t) = A e^{-\alpha_1 t} + B e^{-\alpha_2 t}$$

with

$$\alpha_1 = \alpha - \sqrt{\alpha^2 - \omega_o^2}$$

$$\alpha_2 = \alpha + \sqrt{\alpha^2 - \omega_o^2}$$

Verify that each of these results for $i(t)$ is indeed a solution to the differential equation obtained in part (a).

(c) For the specific initial conditions specified here, i.e., $\Delta v_C = V_o$ and $i_o = 0$, show that $i_o = -V_o/L$ and that for $0 <$

(i) $\alpha < \omega_o$, $i(t) = -(V_o/\omega_L) \sin\omega t \, e^{-\alpha t}$

(ii) $\alpha = \omega_o$, $i(t) = -(V_o t/L)e^{-\alpha t}$

(iii) $\omega_o < \alpha$, $i(t) = -[V_o/(\alpha_2 - \alpha_1)L](e^{-\alpha_1 t} - e^{-\alpha_2 t})$

(d) With $L = 0.10$ H, $C = 0.10$ F, and $V_o = 2.0$ V, determine $i(t)$ and sketch $i(t)$ for $0 < t < 1.0$ s if (i) $R = 0.20 \, \Omega$, (ii) $R = 2.0 \, \Omega$, (iii) $R = 2.2 \, \Omega$.

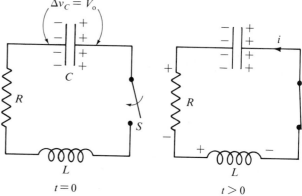

Figure 18.33   Problem 18.32

18.33 A series RC circuit is connected across an alternating emf, $\Delta v = V \sin\omega t$.

(a) Use the loop rule to show that

$$i' + \frac{1}{RC}i = \frac{\omega V}{R}\cos\omega t$$

(b) The general solution to this (inhomogeneous) differential equation is given by the sum of the (transient) solution i_T to the (homogeneous) differential equation

$$i' + \frac{1}{RC}i = 0$$

and the (steady-state) solution is determined in this chapter. Given this, show that the general solution to the differential equation of part (a) may be written

$$i(t) = A\,e^{-t/RC} + I\sin(\omega t - \phi)$$

where A is an arbitrary constant and

$$I = \frac{V}{\sqrt{R^2 + \left(\dfrac{1}{\omega C}\right)^2}}$$

$$\tan\phi = -\frac{1}{\omega RC}$$

(c) If, at $t = 0$, the capacitor C is charged to a potential difference V_o and the switch S shown in Figure 18.34 is closed, show that

$$i(t) = i_T(t) + i_S(t)$$

where the transient (i_T) and steady-state (i_S) currents are given by

$$i_T(t) = \left(I\sin\phi - \frac{V_o}{R}\right)e^{-t/RC}$$

$$i_S(t) = I\sin(\omega t - \phi)$$

(d) Determine a value for V_o of part (c) for which the transient current is zero. For this case, the general solution for the current is identical to the steady-state current.

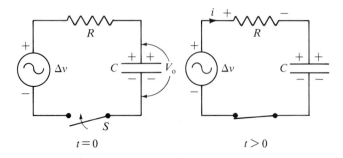

Figure 18.34   Problem 18.33

18.34 A series LR circuit is connected across an alternating emf, $\Delta v = V \sin\omega t$.

(a) Use the loop rule to show that

$$i' + \frac{R}{L}i = \frac{V}{R}\sin\omega t$$

(b) The general solution to this differential equation is equal to the sum of the transient solution i_T to the homogeneous differential equation

$$i' + \frac{R}{L}i = 0$$

and the steady-state solution i_S determined in this chapter. Show, then, that this general solution may be written

$$i(t) = Ae^{-Rt/L} + I\sin(\omega t - \phi)$$

where A is an arbitrary constant and

$$I = \frac{V}{\sqrt{R^2 + \omega^2 L^2}}$$

$$\tan\phi = \frac{\omega L}{R}$$

(c) If, at $t = 0$, the switch S shown in Figure 18.35 is closed, show that

$$i(t) = I\sin\phi\,e^{-Rt/L} + I\sin(\omega t - \phi)$$

where the transient and steady-state currents are

$$i_T(t) = I\sin\phi\,e^{-Rt/L}$$

$$i_S(t) = I\sin(\omega t - \phi)$$

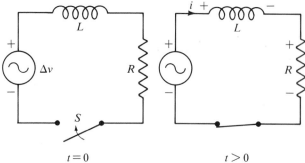

$t = 0$ $t > 0$

Figure 18.35   Problem 18.34

18.35 A series RLC circuit is driven by an alternating emf, $\Delta v = V\sin\omega t$.

(a) Use the loop rule to show that

$$i'' + 2\alpha i' + \omega_o^2\,i = \frac{\omega V}{L}\cos\omega t$$

where

$$\alpha = \frac{R}{2L}$$

$$\omega_o^2 = \frac{1}{LC}$$

(b) If $0 < \alpha < \omega_o$, write an expression for the general solution to the differential equation of part (a).

(c) Repeat part (b) for $\alpha = \omega_o$.

(d) Repeat part (b) for $0 < \omega_o < \alpha$.

(e) If $i = 0$ and $\Delta v_C = 0$ at $t = 0$ and if $\alpha = 0.6\,\omega_o$, determine the transient current for this circuit.

18.36 A series LC circuit is driven by an alternating emf, $\Delta v = V \sin\omega t$.

(a) Show that the loop rule for this circuit may be written

$$i'' + \omega_o^2\, i = \frac{\omega V}{L} \cos\omega t$$

where

$$\omega_o^2 = \frac{1}{LC}$$

(b) If $|\omega| \neq \omega_o$, show that the steady-state current is given by

$$i(t) = I(\omega) \sin(\omega t - \phi(\omega))$$

with

$$I(\omega) = \frac{\omega}{|\omega^2 - \omega_o^2|} \frac{V}{L}$$

and

$$\phi(\omega) = \begin{cases} -\pi/2 & \omega < \omega_o \\ \pi/2 & \omega_o < \omega \end{cases}$$

(c) Graph $I(\omega)$ and $\phi(\omega)$ for $0 < \omega < 5\omega_o$ ($\omega \neq \omega_o$). Compare these graphs with those for a series RLC circuit.

(d) If the frequency of the alternating emf is equal to the resonant frequency $\omega = \omega_o = (LC)^{-1/2}$, show that

$$i(t) = \frac{Vt}{2L} \sin\omega_o t$$

satisfies the "loop-rule" differential equation of part (a). Is this a steady-state current? What happened?

19 Wave Motion and Sound

Wave motion is a means of transferring energy from one place to another without the transfer of matter. A wave on the surface of a body of water, for example, can transfer energy over long distances while the particles within the medium, the water, move only by jiggling locally as the overall disturbance, the wave, passes. Thus we may define a *wave* as a disturbance that moves, or propagates, through a medium without the transport of matter. The media through which waves propagate may be liquids, solids, gases, or even a vacuum in the case of electromagnetic waves. In this chapter we concentrate on *mechanical waves* that propagate through *elastic media* composed of particles, which, after having been displaced from their undisturbed equilibrium positions with the passage of a wave, are restored to their equilibrium positions.

Physically, any mechanical wave passing through a solid, liquid, or gas may be considered an *acoustic wave*, or a *sound wave*. To physiologists and psychologists, sound waves are usually considered those waves to which the human ear responds. Our considerations of sound will center mainly on the physics of sound waves in air, usually within the range of human hearing.

19.1 Traveling Waves

Disturbances that move along or through a medium are called *traveling waves*. A wave moving along a taut rope is an example of a traveling wave. One form of a traveling wave, illustrated in Figure 19.1, is called an *impulsive wave*. In this case, a single upward thrust on the rope causes a local disturbance that then propagates along the rope. Another form of a traveling wave, illustrated in Figure 19.2, is a *periodic wave*. Here the source of the disturbance produces a continuing periodic displacement of particles at one end of the rope, and the periodic disturbance propagates along the rope.

As a wave passes, the resulting localized motion of a particle in the medium may at any instant have an arbitrary direction with respect to the direction of propagation of the wave itself. For example, when the motion of the particles in the medium is perpendicular (transverse) to the direction of propagation of the wave at every point along the wave, the disturbance is said to be a *transverse*

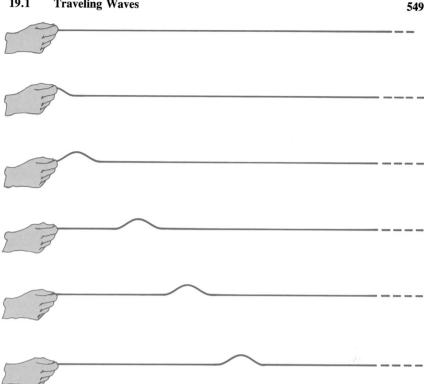

Figure 19.1
An impulsive transverse wave formed on a rope by a hand that moves up and down one time. The impulse then propagates along the rope.

wave. The wave on the rope in Figure 19.2 is a transverse wave. The particles in the rope vibrate vertically as the wave moves horizontally. Figure 19.3 shows a *longitudinal wave* moving along a spring. In this case, the particles of the medium vibrate along a direction parallel to the direction of propagation of the wave.

The speed of propagation of a wave (how fast a wave transfers energy through a medium) depends on the physical properties of the medium and on the nature of the disturbance. We will consider only simple media in which the speed of propagation has a constant value.

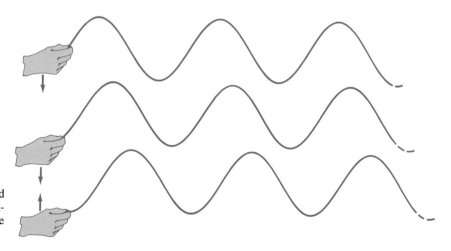

Figure 19.2
A periodic transverse wave produced on a rope by a hand moving periodically. The wave propagates along the rope to the right.

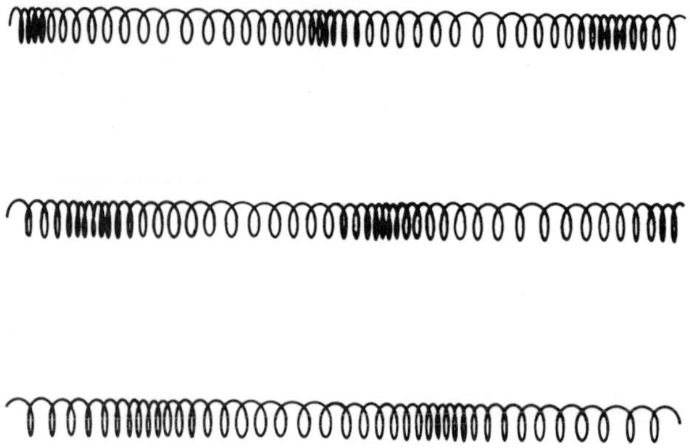

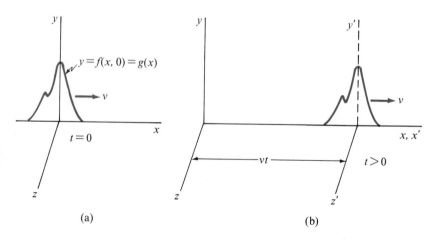

Figure 19.3
A longitudinal wave propagating along a spring. Here the particles of the medium, the spring, oscillate back and forth in the direction of propagation.

Let us now formulate a mathematical representation of traveling waves by considering a transverse wave moving in the positive x direction. Let the displacement of particles in the medium be in the y direction, and let t be the time after some arbitrarily chosen time $t = 0$. If x is the variable that locates a particular point along the medium in the direction of propagation, we may describe the wave by specifying the displacement y as a function of x and t:

$$y = f(x, t) \tag{19-1}$$

Suppose that the transverse disturbance, which is shown in Figure 19.4(a) at $t = 0$, is moving with a constant speed V in the positive x direction. At $t = 0$, the wave is at $x = 0$, and we may write

$$y = f(x, 0) = g(x) \quad \text{(at t = 0)} \tag{19-2}$$

Equation (19-2) specifies the shape g of the wave and locates the wave on the x axis. Figure 19.4(b) shows the disturbance at a time $t > 0$, when it has moved a distance Vt along the x axis. In a reference frame (x', y', z') that moves along with the disturbance, the disturbance appears to be fixed and, therefore, maintains the same form at all times, or

$$y' = g(x') \tag{19-3}$$

Now let us focus attention on a particular point P on the disturbance, as shown in Figure 19.4(b). The positions x and x' are related by

$$x' = x - Vt \tag{19-4}$$

Figure 19.4
A transverse disturbance, which moves at a constant speed v, shown at (a) $t = 0$, and (b) at a later time $t > 0$ when the disturbance has moved a distance vt. In the moving x', y', z' reference frame, the unchanging wave form appears to be fixed.

y

$y = f(x, 0) = g(x)$

v

$t = 0$

x

z

(a)

y

y'

v

vt

$t > 0$

x, x'

z

z'

(b)

The displacement of P is the same viewed from either frame, or $y = f(x, t)$ is the same as $y' = g(x')$. Then, using Equation (19-4), we obtain

$$y = f(x, t) = g(x - Vt) \qquad (19\text{-}5)$$

Thus, any wave having a fixed shape and moving in the positive x direction may be represented by expressing the displacement y as some function of $(x - Vt)$. By a similar analysis, a wave moving in the negative x direction may be represented by expressing the displacement y as a function of $(x + Vt)$.

E 19.1   The $t = 0$ shape of a transverse pulse moving 2.0 m/s in the positive x direction is shown in the accompanying figure. Using the definitions of $f(x, t)$ and $g(x)$ given, determine (a) $g(0)$, (b) $g(1)$, (c) $f(2, 1)$, (d) $f(3, 1)$, (e) $f(1, 1)$, and (f) $f(5, 1)$.

Answers: (a) $g(0) = f(0, 0) = 2$ cm; (b) $g(1) = f(1, 0) = 1$ cm;
(c) $f(2, 1) = g(0) = 2$ cm; (d) $f(3, 1) = g(1) = 1$ cm;
(e) $f(1, 1) = g(-1) = 1$ cm; (f) $f(5, 1) = g(3) = 0$

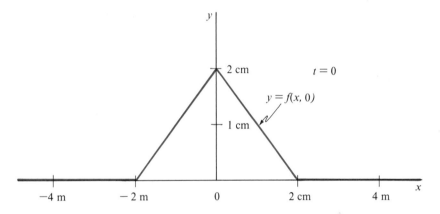

Exercise 19.1

E 19.2   Sketch the transverse disturbance of Exercise 19.1 for $t = 0.50$ s. Use this to assist you in determining the direction of motion at $t = 0$ of the particle located in the medium at (a) $x = +1.0$ m, and (b) $x = -1.0$ m.

Answers: (a) $+y$ direction; (b) $-y$ direction

The most common form of a traveling periodic wave is called a *harmonic wave*, or a *sinusoidal wave*. In the mathematical representation of a harmonic wave, y is a sinusoidal function of $(x \pm Vt)$, that is, $y = A \sin k(x \pm Vt)$ or $y = A \cos k(x \pm Vt)$, where A and k are constants (we will examine this subsequently). Each particle in a harmonic wave is executing simple harmonic motion (see Chapter 8, "Oscillations"), oscillating in the y direction.

Figure 19.5(a) is a graph of a harmonic wave of the form $y = A \cos k(x - Vt)$ at a particular instant chosen to be $t = 0$. Thus the wave form shown is a "snapshot" of the traveling wave at the instant $t = 0$. The wave is moving toward the right with speed V, and Figure 19.5(b) is another "snapshot" of the wave as it appears a short time later at an instant $t > 0$. The wave has moved a distance Vt in the positive x direction. Let us use this wave to illustrate how a harmonic wave may be characterized by the following quantities that were used to describe simple harmonic motion:

1. The *amplitude A*, indicated in Figure 19.5(b), is the maximum displacement of any particle from its equilibrium position.

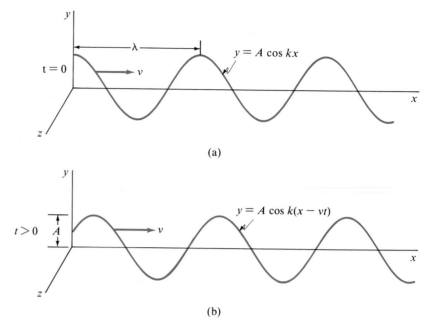

(a)

(b)

Figure 19.5
A harmonic wave $y = A \cos k(x - vt)$ shown **(a)** at $t = 0$ and **(b)** at a later instant $t > 0$. The wave has amplitude A and wavelength λ. It is propagating toward the right with a speed v.

2. The *wavelength* λ, indicated in Figure 19.5(a), is the distance between adjacent *crests*, points of maximum displacement on the wave.

3. The *period T* of the wave is the time required for each particle of the medium to execute one complete cycle of its periodic motion, and therefore T is also the time it takes for the wave to move a distance equal to one wavelength.

4. The *frequency* ν of the wave is equal to the number of cycles that any particle in the medium executes in one second. Thus the frequency ν is the reciprocal of the period, or $\nu = 1/T$. The unit of frequency is s^{-1}, or cycles/second = hertz (Hz). It is sometimes useful to use the *angular frequency* $\omega = 2\pi\nu$ in describing waves.

E 19.3   The $t = 0$ and $t = 0.25$ shapes of a sinusoidal wave are shown in the figure. The crest labeled P has moved from $x = 1$ m to $x = 3$ m. For this wave, determine (a) the amplitude A, (b) the wavelength λ, (c) the speed of propagation V, (d) the period T, (e) the frequency ν, and (f) the angular frequency ω.
Answers: (a) 3.0 cm; (b) 4.0 m; (c) 8.0 m/s; (d) 0.50 s; (e) 2.0 Hz; (f) 4π s^{-1}

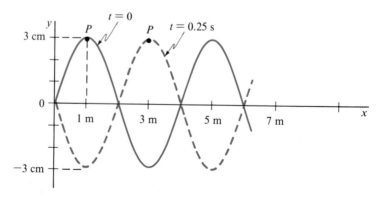

Exercise 19.3

Figure 19.5(a) shows a harmonic wave at a time $t = 0$. The displacement of any particle in the medium is given by

$$y = A \cos kx \qquad \text{(19-6)}$$

where A is the amplitude of the wave. At a later time $t > 0$, shown in Figure 19.5(b), the displacements of the particles are given by

$$y = A \cos k(x - Vt) \qquad \text{(19-7)}$$

where the wave is propagating in the positive x direction at a speed V. From the graph of Figure 19.5(a), we may see that the wave form begins repeating at $x = \lambda$, that is, when $kx = k\lambda = 2\pi$ radians. The constant

$$k = \frac{2\pi}{\lambda} \qquad \text{(19-8)}$$

is called the *wave number* of this wave. The unit of k is a reciprocal unit of length, like m^{-1}.

If we now consider the motion of the particle at $x = 0$ in Figure 19.5(a) as a function of time, Equation (19-7) becomes, for $x = 0$,

$$y = A \cos(-kVt) = A \cos kVt \qquad \text{(19-9)}$$

where we have set $x = 0$ and used the fact that $\cos(-\theta) = \cos\theta$. Figure 19.6 shows a graph of this displacement of the particle located at $x = 0$ as a function of time. After a time $t = T$, when that particle has executed one complete cycle of its motion, the argument of the cosine function of Equation (19-9) has increased by 2π radians, or

$$kVT = 2\pi \qquad \text{(19-10)}$$

Using $k = 2\pi/\lambda$ from Equation (19-8), we find

$$\frac{2\pi}{\lambda}VT = 2\pi$$

$$T = \frac{\lambda}{V} \qquad \text{(19-11)}$$

And because the period T is equal to $1/\nu$, the reciprocal of the frequency, we may write Equation (19-11) as

$$V = \nu\lambda \qquad \text{(19-12)}$$

The form of the traveling wave equation $y = A \cos k(x - Vt)$, in Equation (19-7), may be expressed in terms of other convenient quantities. In particular, using Equation (19-8), $k = 2\pi/\lambda$, and using $\omega = 2\pi\nu = 2\pi/T$, we may express the displacements of particles in a harmonic wave traveling in the positive x direction as

$$y = A \cos 2\pi\left[\frac{x}{\lambda} - \frac{t}{T}\right] \qquad \text{(19-13)}$$

or

$$y = A \cos(kx - \omega t) \qquad \text{(19-14)}$$

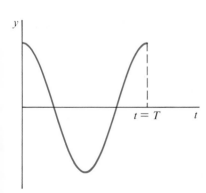

Figure 19.6
A graph of the displacement as a function of time of the particle located at $x = 0$ in Figure 19.5. The period T is the time required for one cycle of the motion.

E 19.4   A traveling wave is described by

$$y = 0.2 \cos\frac{\pi}{8}(x - 32t)$$

where x and y are measured in meters and t in seconds. For this wave, determine (a) the amplitude A, (b) the wave number k, (c) the wavelength λ, (d) the speed of propagation V, (e) the frequency ν, (f) the period T, and (g) the angular frequency ω.   Answers: (a) 0.2 m; (b) $\pi/8$ m^{-1}; (c) 16 m; (d) 32 m/s; (e) 2 Hz; (f) 0.5 s; (g) 4π s^{-1}

E 19.5   Write an equation of the form of Equation (19-13) that describes a traverse wave moving in the positive x direction with an amplitude of 0.15 m, a wavelength of 2.5 m, and a period of 0.50 s.
Answer: $y = 0.15 \cos 2\pi[(x/2.5) - (t/0.50)]$

E 19.6   What is the speed of propagation of the wave of Exercise 19.5?
Answer: 5.0 m/s

In our traveling wave equations, the argument of the cosine, $2\pi(x/\lambda - t/T)$ in Equation (19-13), for example, is called the *phase* ϕ of a point on the wave. The phase of a particular particle of the medium, that is, a point at a particular value of x, changes as time passes. On the other hand, a point on the wave, like a crest, maintains a constant phase as the wave moves. In Figure 19.5(a), for example, at $t = 0$ the crest located at $x = 0$ has a phase ϕ equal to zero, and the phase of this crest remains constant as the wave propagates through the medium. That a point on the moving wave maintains a constant phase may be seen by writing the displacement y as $y = A \cos \phi$. Then as the wave moves, a point on the wave (like the $\phi = 0$ crest of Figure 19.5) maintains a constant displacement y so that ϕ must remain constant.

E 19.7   At $t = 0$, the wave represented by Equation (19-13) has a crest at $x = 0$. What is the phase ϕ
(a) for the crest located at $x = 2\lambda$ at $t = 0$?
(b) for the zero-displacement point located at $x = \lambda/4$ at $t = 0$?
Answers: (a) 4π radians; (b) $\frac{1}{2}\pi$ radians

E 19.8
(a) For the wave represented by Equation (19-13), what is the phase difference, $\Delta\phi = \phi_B - \phi_A$, at $t = 0$ for the points $x_A = \lambda/4$ and $x_B = 3\lambda/2$?
Answer: $5\pi/2$ radians
(b) Determine the $t = 0$ value of $\phi_A - \phi_B$ for the points of part (a).
Answer: $-5\pi/2$ radians
(c) Repeat part (a) for $t = T/2$.   Answer: $5\pi/2$ radians
(d) Does the value for the phase difference, $\Delta\phi = \phi_B - \phi_A$, of part (a) depend upon the time t at which $\Delta\phi$ is calculated?
Answer: No, it has a constant value of $5\pi/2$ radians.

E 19.9   A harmonic wave is described by

$$y = 0.12 \cos(5x - 28t)$$

where x and y are measured in meters and t in seconds.
(a) What is the phase for the crest of this wave that is positioned at $x = (2\pi/5)$ m at $t = 0$?   Answer: 2π radians
(b) Use the fact that the phase for the crest of part (a) remains constant to determine the location of this crest at $t = 0.25$.   Answer: $x = 2.4$ m

Equations (19-13) and (19-14) describe a harmonic wave traveling in the positive x direction, a wave for which the particle at $x = 0$ is at its maximum displacement from equilibrium, that is, $y = A$, at $t = 0$. A more general form of the harmonic traveling wave equation, for a wave moving in either direction along the x axis, is

$$y = A \cos\left[2\pi\left(\frac{x}{\lambda} \pm \frac{t}{T}\right) + \delta\right] \qquad (19\text{-}15)$$

The sign between terms in the argument of the cosine is negative if the wave is traveling in the positive x direction; it is positive if the wave is traveling in the negative x direction. The *phase constant* δ may be chosen to position one point on the wave at any location x in space at any time t (usually at $t = 0$). Then for a given choice of δ, the subsequent motion of the wave is determined by Equation (19-15). If, for example, δ is chosen to be zero for the wave of Equation (19-15), the crest for which $\phi = 0$ is positioned at $x = 0$ when $t = 0$. A different choice of δ would place the crest at another location at $t = 0$.

E 19.10   Show that if $\delta = \pi/2$, Equation (19-15) is equivalent to

$$y = -A \sin 2\pi\left(\frac{x}{\lambda} \pm \frac{t}{T}\right)$$

E 19.11   Determine the speed and direction of propagation of each of the following sinusoidal waves, assuming that x is measured in meters and t in seconds.

(a) $y = 0.6 \cos(3x - 15t + 2)$ Answer: 5 m/s, $+x$
(b) $y = 0.4 \cos(3x + 15t - 2)$ Answer: 5 m/s, $-x$
(c) $y = 1.2 \sin(15t + 2x)$ Answer: 7.5 m/s, $-x$
(d) $y = 0.2 \sin(12t - x/2 + \pi)$ Answer: 24 m/s, $+x$

From physical considerations we may note that the frequency ν of a wave is determined by the source of the wave. The frequency of the oscillations on the rope in Figure 19.2, for example, is determined by the frequency at which the hand is vibrating. The propagation speed v of a traveling wave is determined by the properties of the medium. (The specific dependence of the propagation speed of a wave on the physical properties of particular media will be discussed later.) The frequency ν and the propagation speed V together determine the wavelength of any periodic traveling wave according to Equation (19-2), $V = \nu\lambda$.

Let us now examine the motion of particles in a medium as a transverse wave passes through that medium.

The equation, $y = 0.20 \sin \pi(x + 100t)$, where x and y are in meters and t is in seconds, represents a transverse traveling wave moving at a speed of 100 m/s in the negative x direction along a thin rope. Determine the maximum (transverse) speed experienced by any point on the rope.

Wave-equation analysis includes a rare class of problems that are not always facilitated by visualization: A drawing is not necessarily of assistance. Therefore, let us examine directly the given wave equation:

$$y = 0.20 \sin\pi(x + 100t) \qquad (19\text{-}16)$$

To investigate the velocity of particles in the medium, recall that the paticles move only in the y direction (the wave is transverse). Thus the velocity V_y of a point at any x and t is found by taking the time derivative of y:

$$V_y = \frac{dy}{dt} = 0.20 \,(100\pi) \cos \pi(x + 100t) \qquad (19\text{-}17)$$

The maximum velocity of any particle occurs when $\cos \pi(x + 100t)$ has its greatest positive value, namely $+1$. Then the greatest speed achieved by any particle in the medium is

$$|V_y|_{max} = 0.20 \,(100\pi)(1) = 20\pi = 63 \text{ m/s} \qquad (19\text{-}18)$$

We may check the units of Equation (19-18) by recognizing that each term in the argument of the cosine function in Equation (19-17) must be unitless. Then the factor 100π has the unit s^{-1}. And because the amplitude is 0.20 m, the unit for velocity is m/s.

E 19.12   A transverse wave described by $y = A \sin(kx - \omega t)$ moves along a string.

(a) Determine an expression for the (transverse) velocity V_y of any point on the string.   Answer: $V_y = -\omega A \cos(kx - \omega t)$

(b) Determine the maximum (transverse) speed for any point on the string.
     Answer: ωA

E 19.13

(a) For the wave described in Exercise 19.12, determine an expression for the acceleration, $a_y = dV_y/dt$, of any point on the string.
     Answer: $a_y = -\omega^2 A \sin(kx - \omega t)$

(b) Determine the maximum value for the acceleration of any point on the string.   Answer: $\omega^2 A$

(c) Verify that the answer for part (b) has the correct unit.

Example 19.1

PROBLEM   A sinusoidal wave on a string is represented by

$$y = 12 \sin \pi(x - 2t)$$

where y is measured in cm, x in meters, and t in seconds.

(a) Sketch the wave at $t = 0$ and at $t = 0.1$ s for $-1 \text{ m} < x < 4$ m. Label each position extremum (maximum or minimum) with its phase.

(b) Determine an expression for the velocity on each point on the string. Sketch a graph showing the wave at $t = 0$ and show velocity vectors for representative points on the string.

SOLUTION   The $t = 0$ and $t = 0.1$ s shapes of the wave are described by $y(x, t = 0) = 12 \sin \pi x$ and $y(x, t = 0.1) = 12 \sin \pi(x - 0.2)$. These two shapes are shown in Figure 19.7(a). The first extremum for $x > 0$ is the $\phi = \pi/2$ (radians) maximum, which at $t = 0$ is located at $x = 0.5$ m. Because the wave is moving with a velocity of $V_x = 2$ m/s, the $t = 0.1$ s position of this ($\phi = \pi/2$) maximum is $x = 0.7$ m, as shown. Each of the other position extrema ($\phi = -\pi/2, 3\pi/2, 5\pi/2, 7\pi/2$) shown at $t = 0$ and $t = 0.1$ s moves a distance of 0.2 m in the positive x direction during this 0.1-s time interval.

The transverse velocity V_y of a point on the string is obtained by differentiating the position y with respect to time, i.e.,

$$V_y = -24\pi \cos \pi(x - 2t)$$

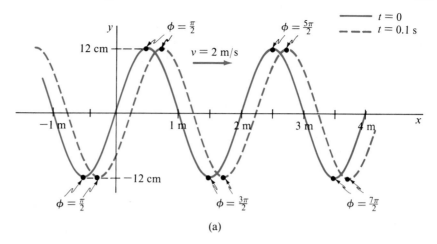

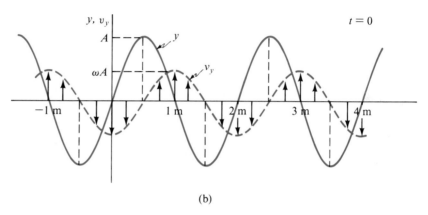

Figure 19.7   Example 19.1

where V_y is measured in cm/s. Thus at $t = 0$, the transverse velocity of any point on the string is given by $V_y = -24\pi \cos \pi x$. Figure 19.7(b) shows the $t = 0$ position y and velocity V_y for -1 m $< x < 4$ m. The velocity vectors shown emphasize that all points between any two adjacent position extrema move in the same (transverse) direction. For example, all points on the string between the $\phi = \pi/2$ maximum and the $\phi = 3\pi/2$ minimum are moving in the positive y direction. This should be obvious because the $\phi = \pi/2$ maximum is moving to the right, and slightly after $t = 0$, each point on the string just to the right of $x = 0.5$ m will have moved in the positive y direction. Can you offer a similar explanation as to why, at $t = 0$, each point on the string between $x = 1.5$ m and $x = 2.5$ m is moving in the negative y direction? ■

Example 19.2   **PROBLEM**   Write an equation describing a transverse sinusoidal wave ($\lambda = 2.0$ m, $\nu = 40$ Hz) traveling in the negative x direction if, at $t = 0$, the particle at $x = 0$ has a position of $y = 0.15$ m and a velocity of $V_y = -60$ m/s.

   SOLUTION   A sinusoidal wave moving in the negative x direction may be described by

$$y = A \cos(kx + \omega t + \delta)$$

where $k = 2\pi/\lambda = \pi\,\text{m}^{-1}$, $\omega = 2\pi\nu = 80\pi\,\text{s}^{-1}$, and A and δ are constants that must be determined from the $t = 0$ condition. The transverse velocity of any point in the medium is given by the derivative of y with respect to t:

$$V_y = -\omega A \sin(kx + \omega t + \delta)$$

Thus, the $t = 0$ position, $y_0 = 0.15$ m, and the $t = 0$ velocity, $V_{y0} = -60$ m/s of the particle at the point $x = 0$ are given by

$$y_0 = A \cos\delta$$

$$V_{y0} = -\omega A \sin\delta$$

These equations may be rewritten

$$A \cos\delta = y_0$$

$$A \sin\delta = -\frac{V_{y0}}{\omega}$$

The amplitude A may be determined by "squaring" each of these equations, adding, and using $\sin^2\delta + \cos^2\delta = 1$ to get

$$A^2 = y_0^2 + \left(\frac{V_{y0}}{\omega}\right)^2$$

Substituting the known values for y_0, V_{y0}, and ω gives

$$A = \sqrt{(.15)^2 + \left(\frac{-60}{80\pi}\right)^2} = 0.28 \text{ m}$$

The phase constant δ may be determined from

$$\cos\delta = \frac{y_0}{A} = \frac{0.15}{0.28}$$

which gives

$$\delta = 58° = 1.0 \text{ rad}$$

Thus, the desired wave equation is

$$y = 0.28 \cos(\pi x + 80\pi t + 1.0)$$

Figure 19.8 shows this wave at $t = 0$, i.e., $y = 0.28 \cos(\pi x + 1.0)$, and the $t = 0$ velocity, $V_y = -71 \sin(\pi x + 1.0)$. We may check the wave equation by verifying that it does reproduce the given $t = 0$ conditions, i.e.,

$$y_0 = 0.28 \cos(1.0) = 0.15 \text{ m}$$

$$V_{y0} = -71 \sin(1.0) = -60 \text{ m/s} \quad ■$$

To this point we have mostly been considering waves on thin ropes, essentially a one-dimensional solid medium. Waves may also travel on a two-dimensional surface, like that of a body of water, or through a three-dimensional medium, like air. Thus waves may propagate through a medium that is a solid, a liquid, or a gas. And, as was suggested earlier, the speeds with which waves propagate through different media depend on the physical properties of the particular medium through which a wave is moving.

The propagation speed of a disturbance in a medium depends, for example, on the *elastic properties* of that medium. The atoms in a medium interact much as if they were connected by springs, which may differ in stiffness in different media. In general, waves travel faster in media in which the interaction between atoms is stronger, that is, when the "springs" connecting the atoms are stiffer. The

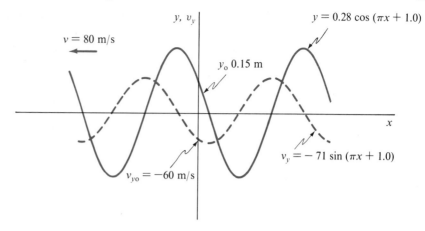

Figure 19.8   Example 19.2

propagation speed of a wave within a medium also depends on the *inertial properties* of that medium. We usually associate inertia with mass. In a one-dimensional medium, like a thin rope, the pertinent inertial property is expressed as *linear density* μ, the mass per unit length, measured in kg/m or slug/ft. For a three-dimensional medium, like air, the applicable inertial property is volume density ρ, measured in kg/m^3 or slug/ft^3.

Without derivation, we assert that theory predicts and experiment confirms that the speed V of a wave in a medium is related to the elastic and inertial properties of that medium by

$$V = \sqrt{\frac{\text{elastic property}}{\text{inertial property}}} \qquad (19\text{-}19)$$

Our considerations will concentrate on waves propagating along thin ropes or strings or propagating in air or water. The propagation speeds of waves in these media are given by

$$V = \sqrt{\frac{F}{\mu}} \qquad \text{(thin rope or string)} \qquad (19\text{-}20)$$

$$V = \sqrt{\frac{B}{\rho}} \qquad \text{(air or water)} \qquad (19\text{-}21)$$

In Equation (19-20), F is the tension, measured in N or lb, in a thin rope or string. (We use F to represent tension to avoid confusion with the symbol T used to represent the period of a wave.) The symbol B in Equation (19-21) represents the *bulk modulus* of a liquid or gas. Bulk modulus, which has the unit N/m^2 or lb/ft^2, is a quantity used in elastic studies.

E 19.14   Verify that the terms, $\sqrt{F/\mu}$ and $\sqrt{B/\rho}$, of Equations (19-20) and (19-21) have the appropriate unit, that is, m/s.

E 19.15   A 70-cm string has a mass of 1.4 g and is subjected to a tension of 18 N. Calculate
(a) μ, the mass per unit length, and
(b) V, the speed of propagation of a transverse disturbance.

Answers: (a) 2.0 g/m; (b) 95 m/s

In our considerations of waves on thin ropes and strings, Equation (19-20) will be useful. In our studies of sound waves in air (at 20°C and at atmospheric pressure) and water, however, we need only two results that are obtained from Equation (19-21), namely, the speed of propagation of sound waves through air and through water:

$$V_{air} = 340 \text{ m/s} = 1100 \text{ ft/s} \tag{19-22}$$

$$V_{water} = 1500 \text{ m/s} = 4900 \text{ ft/s} \tag{19-23}$$

19.2 Reflection, Superposition, and Standing Waves

Figure 19.9(a) shows a wave pulse, having an upward displacement, traveling to the right along a taut string fastened to a rigid wall. When the pulse reaches the wall, the string exerts an upward force on the wall, which, by Newton's third law, exerts a downward force on the string. This downward force causes a downward displacement of the string, thereby inverting the pulse. This inverted pulse then propagates toward the left. Thus, in encountering a fixed boundary, the wave is reflected and inverted. In analogy with a sinusoidal wave, we say the pulse undergoes a phase change of π radians (180°) upon reflection.

In Figure 19.9(b), a similar wave pulse is reflected from a "free end" of the string, which is attached to a vertical wire by a frictionless loop. In this case, no vertical force is exerted on the string when the pulse reaches the wire. And the

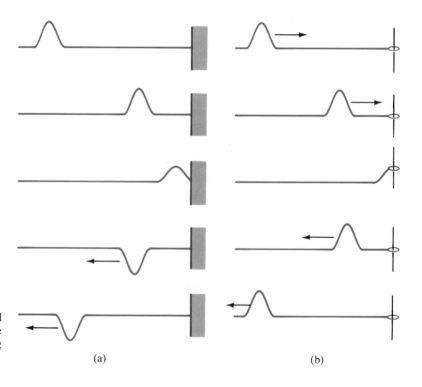

Figure 19.9
(a) A wave pulse reflected and inverted by a fixed boundary. (b) A wave pulse reflected by a "free end" without being inverted.

(a)   (b)

reflected pulse that propagates to the left remains upright. This reflected wave undergoes no phase change.

We may generalize these results to any traveling wave and summarize the *boundary conditions* on waves reflected at "fixed" and "free" boundaries: (a) Waves reflected from a fixed boundary experience a phase change of 180°, and (b) waves reflected from a free boundary experience no phase change.

Now suppose that two waves are traveling in opposite directions on a string simultaneously. If the amplitudes of the waves are not too great, two facts may be verified experimentally: (a) Each wave proceeds unimpeded by the presence of the other, and (b) the net displacement of any particle on the string is the algebraic sum of the displacements caused by each of the individual waves. The second of these facts is our statement of the *principle of superposition* for waves. This principle applies to any number of waves of any kind that are simultaneously present in a medium.

Figure 19.10(a) illustrates the superposition of two wave pulses, both with equal upward displacements, moving in opposite directions along a string. When the two pulses reach the same location on the string, the resultant pulse has twice the displacement of either individual pulse. When they coincide, these wave pulses experience *constructive interference*. After they coincide, each pulse proceeds without having been changed. In Figure 19.10(b), two wave pulses having opposite displacements of equal magnitudes move in opposite directions along a string. When these pulses coincide, the displacement of the particles of the string sum to zero. The wave pulses experience *destructive interference*. And again, as in constructive interference, after they coincide each wave proceeds as if it had not encountered the other.

Standing waves are produced when harmonic waves of equal frequencies and amplitudes travel in opposite directions in the same medium. The displacements of particles in the medium then form a *standing-wave pattern* (a pattern that does not propagate through the medium). Such standing waves are commonly seen in the vibrations of the strings of musical instruments, like guitars or violins. We may illustrate the analysis of standing waves by considering a similar situation: a taut string fixed at both ends, as shown in Figure 19.11.

Because the ends of the string are fixed, the end points of the string are *nodes*, points of the medium where the displacement is always equal to zero. Points on the string midway between nodes are called *antinodes*. The various forms of standing waves, three of which are shown in Figure 19.11, that are compatible with the boundary conditions (each end of the string is a node) are called the *natural modes* of vibration. The mode having the fewest number of nodes is called the *fundamental mode*. As we will demonstrate later, the distance between adjacent nodes is equal to $\lambda/2$, where λ is the wavelength of the two traveling waves that form the standing wave. Thus if L is the length of the string in Figure 19.11, the wavelength λ_1, associated with the fundamental mode, is equal to $2L$, as indicated in Figure 19.11(a). Generalizing from Figure 19.11, we may conclude that the wavelength λ_n associated with the nth natural mode is given by

$$\lambda_n = \frac{2L}{n} \quad ; \quad n = 1, 2, 3, \ldots \tag{19-24}$$

The *natural frequency* ν_n corresponding to the nth natural mode is given by

$$\nu_n = \frac{V}{\lambda_n} = \frac{n}{2L}\sqrt{\frac{F}{\mu}} \quad ; \quad n = 1, 2, 3, \ldots \tag{19-25}$$

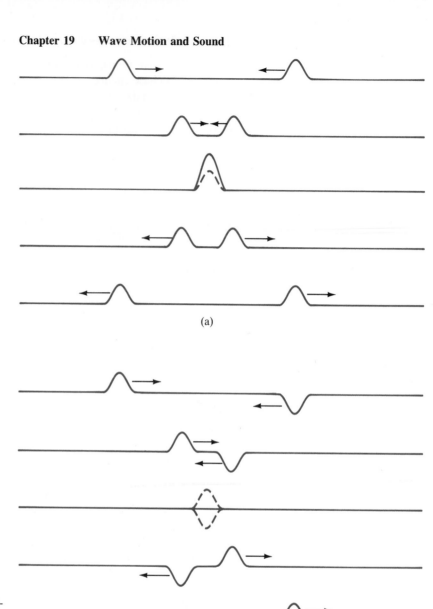

(a)

(b)

Figure 19.10
(a) Two wave pulses of equal amplitudes meeting and interfering constructively. (b) Two waves of equal amplitudes meeting and interfering destructively.

where $V = \sqrt{F/\mu}$ is the speed of propagation of a traveling wave on the string. The integers n number the natural modes in the order of increasing natural frequencies. Therefore, $n = 1$ corresponds to the fundamental mode, and $n = 2$ corresponds to the natural mode having the next higher frequency.

A sequence for which the nth member is n times the first member is called a *harmonic sequence*. The nth member of the sequence is called the nth harmonic of the first, or fundamental, member. Thus, for a string fixed at both ends, the nth natural frequency ν_n is the nth harmonic of the fundamental frequency ν_1, or $\nu_n = n\nu_1$.

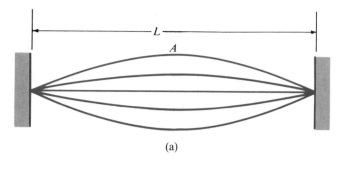

(a)

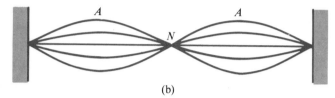

(b)

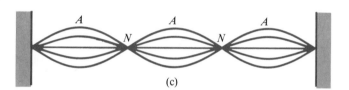

(c)

Figure 19.11
A taut string, fixed at each end, vibrating in its three longest wavelength modes.

E 19.16   The "middle-A" wire of a piano is to have a fundamental frequency of 440 Hz. If this wire has a length of 80 cm and a mass of 5.0 g, what is the required tension?   Answer: 3.1 kN ≈ 700 lb

Wave functions that describe standing waves may be derived by superposing two traveling waves of the same frequency ν and amplitude A moving in opposite directions on a string. Using a brief form of Equation (19-15), we may represent the displacements of the two traveling waves by

$$y_1 = A \cos(\omega t - kx + \delta_1) \tag{19-26}$$

$$y_2 = A \cos(\omega t + kx + \delta_2) \tag{19-27}$$

where y_1 is the displacement of the wave moving to the right, y_2 is the displacement of the wave moving to the left, $\omega = 2\pi\nu$, and $k = 2\pi/\lambda$. The displacement y of the resulting wave is

$$y = y_1 + y_2 = A[\cos(\omega t - kx + \delta_1) + \cos(\omega t + kx + \delta_2)] \tag{19-28}$$

Making use of the identity, $\cos\alpha + \cos\beta = 2\cos[\frac{1}{2}(\alpha + \beta)]\cos[\frac{1}{2}(\alpha - \beta)]$, and combining the phase constants as $\delta_x = \delta_1 + \delta_2$ and $\delta_t = \delta_1 - \delta_2$, we obtain

$$y = [2A\cos(kx + \delta_x)]\cos(\omega t + \delta_t) \tag{19-29}$$

The factor inside the brackets is the amplitude of the motion of a point located on the string at any position x, as that point executes simple harmonic motion. The values of the phase constants, δ_x and δ_t, may be chosen to allow any choice of origin ($x = 0$) and any initial ($t = 0$) configuration of the standing wave.

Example 19.3

PROBLEM   A transverse standing wave on a string is described by

$$y = 0.10 \sin(2\pi x) \cos(4\pi t)$$

where x and y are measured in meters and t in seconds.

(a) Determine the location of the nodes.

(b) Calculate the maximum transverse speed experienced by any point on the string.

(c) Sketch the standing wave for $t = 0$, $T/8$, and $T/4$, where T is the period of the motion. Show typical velocity vectors on each of these sketches.

(d) Determine the wavelength, frequency, and amplitude of the two traveling waves whose sum yields this standing wave.

SOLUTION   (a) The nodes of a standing wave are those points for which the transverse displacement is zero for all times, i.e.,

$$y = 0 = 0.10 \sin(2\pi x) \cos(4\pi t)$$

Because the value of $\cos(4\pi t)$ alternates continuously (with time) between -1 and $+1$, the nodal positions are determined by the values of x for which

$$\sin(2\pi x) = 0$$

$$2\pi x = n\pi \qquad (n = 0, \pm1, \pm2, \ldots)$$

$$x = \frac{n}{2} \qquad (n = 0, \pm1, \pm2, \ldots)$$

$$x = 0, \; \pm\frac{1}{2}\text{m}, \; \pm1 \text{ m}, \; \pm\frac{3}{2}\text{m}, \ldots$$

(b) The transverse velocity of a point on the string is given by the time derivative of y:

$$V_y = -0.4\pi \sin(2\pi x) \sin(4\pi t)$$

where V_y is measured in m/s. Because the value of $\sin(4\pi t)$ alternates continuously between -1 and $+1$, the maximum transverse speed of the point x on the string is given by $|0.4\pi \sin(2\pi x)|$. Thus, the greatest speed occurs for those points for which $|\sin(2\pi x)| = 1$. These points (antinodes) are halfway between adjacent nodes. The maximum transverse speed V_{max}, then, is given by

$$V_{max} = 0.4\pi \text{ m/s} \approx 1.3 \text{ m/s}$$

(c) The period T of the oscillatory motion of each point on the string is the period T of the function $\cos(4\pi t)$, or

$$4\pi T = 2\pi$$

$$T = \frac{1}{2}\text{s}$$

The transverse position y and transverse velocity V_y of points on the string are given for $t = 0$, $t = T/8$, and $t = T/4$ by:

$$t = 0 \begin{cases} y = 0.10 \sin(2\pi x) \cos(0) = \sin(2\pi x) \\ \text{or} \\ 4\pi t = 0 \quad V_y = -0.4\pi \sin(2\pi x) \sin(0) = 0 \end{cases}$$

$$t = \frac{T}{8} = \frac{1}{16}\,s \quad \left| \quad y = 0.10\ \sin(2\pi x)\ \cos(\pi/4) = 0.071\ \sin(2\pi x) \right.$$

or

$$4\pi t = \frac{\pi}{4} \quad \left| \quad V_y = -0.4\pi\ \sin(2\pi x)\ \sin(\pi/4) \approx -0.89\ \sin(2\pi x) \right.$$

$$t = \frac{T}{4} = \frac{1}{8}\,s \quad \left| \quad y = 0.10\ \sin(2\pi x)\ \cos(\pi/2) = 0 \right.$$

or

$$4\pi t = \frac{\pi}{2} \quad \left| \quad V_y = -0.4\pi\ \sin(2\pi x)\ \sin(\pi/2) \approx -1.3\ \sin(2\pi x) \right.$$

These positions and velocities are shown graphically in Figure 19.12.

(d) Comparing this standing wave equation, $y = 0.10\ \sin(2\pi x)\ \cos(4\pi t)$, to the general form of Equation (19-29), $y = 2A\ \cos(kx + \delta_x)\ \cos(\omega t + \delta_t)$, we see that $2A = 0.10$ m, $k = 2\pi\ \mathrm{m}^{-1}$, and $\omega = 4\pi\ \mathrm{s}^{-1}$. And so we have

$$A = 0.050\ \mathrm{m}$$

$$\lambda = \frac{2\pi}{k} = 1.0\ \mathrm{m}$$

$$\nu = \frac{\omega}{2\pi} = 2.0\ \mathrm{Hz}$$

$$T = \frac{1}{\nu} = 0.50\ \mathrm{s} \quad \blacksquare$$

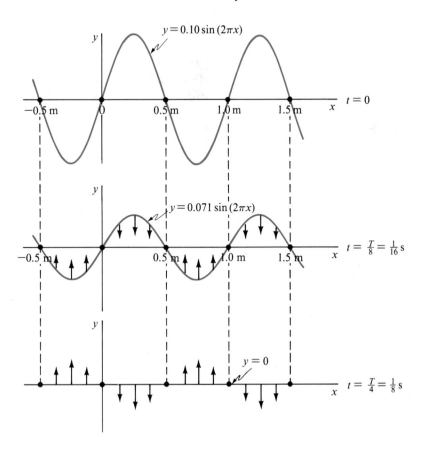

Figure 19.12   Example 19.3

Example 19.4

PROBLEM   A 0.60-m string is fixed at both ends and has a fundamental natural frequency ν_1 that is equal to 2.0 Hz. If the equilibrium position of the string lies along the positive x axis and the ($x = 0$) origin is coincident with one end of the string, write a standing wave equation that describes the third natural mode such that, at $t = 0$, the $x = 0.30$-m point on the string is moving through the equilibrium position ($y = 0$) with a velocity V_y equal to 6.0 m/s.

SOLUTION   Figure 19.13(a) shows the general form for the third natural mode of vibration for the string. Because the nodes are separated by 0.20 m, the associated wavelength λ_3 is equal to twice this distance, or $\lambda_3 = 0.40$ m. The frequency ν_3 for this ($n = 3$) mode is, according to Equation (19-25), given by

$$\nu_3 = 3\nu_1 = 6.0 \text{ Hz}$$

A general expression for a standing wave is given by Equation (19-29) to be

$$y = A \cos(kx + \delta_x) \cos(\omega t + \delta_t)$$

With $k = 2\pi/\lambda = 5\pi \text{ m}^{-1}$ and $\omega = 2\pi\nu = 12\pi \text{ s}^{-1}$, the standing wave equation (SWE) becomes

$$y = A \cos(5\pi x + \delta_x) \cos(12\pi t + \delta_t)$$

Because a node is located at $x = 0$, we have

$$0 = A \cos\delta_x \cos(12\pi t + \delta_t)$$

so that $A \cos\delta_x = 0$. Since $A \neq 0$ (why?), this condition requires that $\cos\delta_x = 0$, which is satisfied if $\delta_x = \pi/2$. The SWE is now

$$y = A \cos\left(5\pi x + \frac{\pi}{2}\right) \cos(12\pi t + \delta_t)$$

or

$$y = -A \sin(5\pi x) \cos(12\pi t + \delta_t)$$

since $\cos(5\pi x + \pi/2) = -\sin(5\pi x)$. The constants A and δ_t are determined by requiring that the transverse position y and transverse velocity V_y for $x = 0.30$ m at $t = 0$ satisfy the given conditions,

$$y(x = 0.30 \text{ m}, t = 0) = 0$$

$$V_y(x = 0.30 \text{ m}, t = 0) = 6.0 \text{ m/s}$$

Differentiating the SWE for y with respect to time gives for V_y:

$$V_y = 12\pi A \sin(5\pi x) \sin(12\pi t + \delta_t)$$

Thus, the given initial conditions require that

$$0 = A \sin(1.5\pi) \cos\delta_t = -A \cos\delta_t$$

$$6 = 12\pi A \sin(1.5\pi) \sin\delta_t = -12\pi A \sin\delta_t$$

The first of these two equations requires that $\cos\delta_t = 0$, which is satisfied by letting $\delta_t = \pm\pi/2$. Choosing $\delta_t = -\pi/2$, the second equation gives

$$12\pi A = 6$$

$$A = \frac{1}{2\pi} \approx 0.16 \text{ m}$$

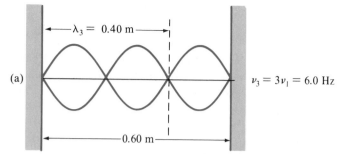

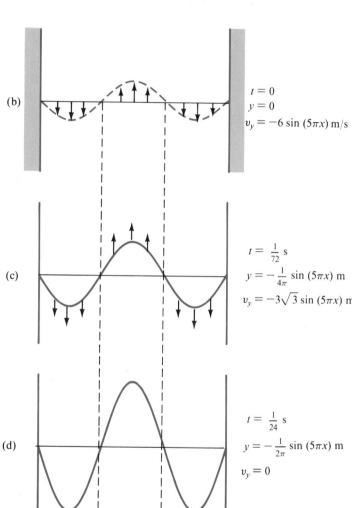

Figure 19.13   Example 19.4

(Would a choice of $\delta_t = \pi/2$ have been satisfactory? Why?) The desired standing wave equation is

$$y = -\frac{1}{2\pi} \sin(2\pi x) \cos\left(12\pi t - \frac{\pi}{2}\right)$$

and since $\cos(12\pi t - \pi/2) = +\sin(12\pi t)$, this may be written

$$y = -\frac{1}{2\pi} \sin(2\pi x) \sin(12\pi t)$$

The transverse velocity is given by

$$V_y = -6 \sin(2\pi x) \cos(12\pi t)$$

Figures 19.13(b), 19.13(c), and 19.13(d) show the $t = 0$, $t = (1/72)$ s, and $t = (1/24)$ s shapes of the string with a few representative velocity vectors also shown. ■

19.3 Sound Waves

An *acoustic wave* is any mechanical wave that propagates *through* an elastic medium, which may be solid, liquid, or gaseous. We use the term *sound wave* to refer to mechanical waves, usually in air, having frequencies within the range of human hearing, about 20 Hz to 20 kHz. Acoustic waves having frequencies less than 20 Hz are referred to as *infrasonic waves*; those having frequencies greater than 20 kHz are called *ultrasonic waves*. Our considerations will focus mainly on sound waves in air.

Sources of sound waves in air are usually vibrating solid materials that produce sound waves in the surrounding air. The vibrations of guitar or violin strings, clarinet reeds, or human vocal cords are familiar examples of sound sources.

Pressure Waves and Superposition

To illustrate the physical character of sound, consider first a single acoustic pulse, which may be produced by a piston fitted into a long column of air, as shown in Figure 19.14. Suppose the piston is suddenly thrust forward in the positive x direction and quickly returned to its original position. As the piston moves forward, the molecules of air near the piston are compressed, thereby reducing the volume they occupy. The compression takes place so fast that no significant heat can be exchanged with the compressed region (that is, the compression is said to be adiabatic), and the pressure is increased in that region. The air in the compressed region exerts a net force on the adjacent region. This second region becomes compressed, but not instantaneously, because of the inertia of the air molecules. Figure 19.14 shows the air column at equal time intervals. The pressure pulse (more precisely, it is a pulse of increase in pressure above the undisturbed, or equilibrium, air pressure) propagates along the air column. Behind the pressure pulse, the air is restored to its equilibrium pressure (the hallmark of an elastic medium). The graphs that accompany each drawing of the air column in Figure 19.14 indicate the pressure change ΔP in the air as a function of position x along the air column.

A harmonic sound wave, propagating in the positive x direction, may be produced in a similar air column by a piston that oscillates sinusoidally. This situation is shown in Figure 19.15(a), which represents a "snapshot" of the air column at one instant and indicates the density (proportional to the pressure) of air molecules as a function of position x along the column. Figure 19.15(b) is a graph of the x dependence of the corresponding pressure difference ΔP. Thus, the harmonic sound wave is a *pressure wave*, composed of alternate *regions of compressions* (regions of high density and high pressure) and *regions of rare-factions* (regions of low density and low pressure).

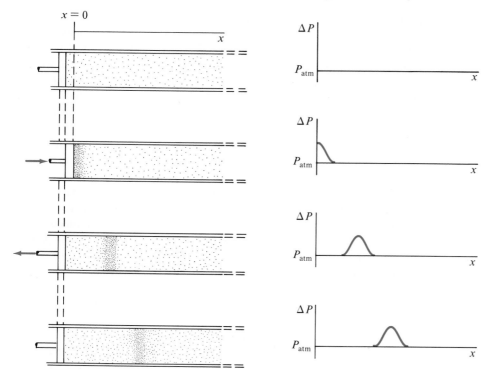

Figure 19.14
An acoustic pulse produced by the sudden thrust of a piston fitted in a tube. The accompanying graphs show the change ΔP in pressure from atmospheric pressure P_{atm} as a function of position x along the tube at approximately equal time intervals.

Sound waves in air are often characterized as longitudinal waves. But a longitudinal traveling wave is a propagating periodic disturbance in which the *particles of the medium* oscillate parallel to the direction of propagation. In air the particles of the medium are molecules, each of which is in chaotic motion and is colliding with other molecules about a billion times each second. Therefore, if we think of small, but macroscopic, volume elements of air as the "particles" of the medium, we may in this limited sense justify calling a sound wave a longitudinal wave. Nevertheless, it must be borne in mind that sound waves are, in a

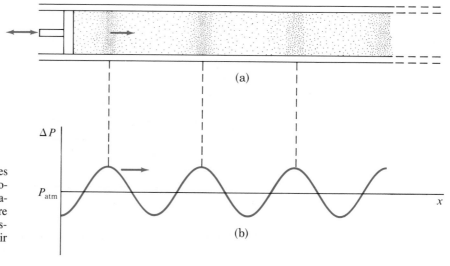

Figure 19.15
(a) Depicting the density of molecules in air as a harmonic sound wave, produced by an oscillating piston, propagates to the right. **(b)** The pressure difference ΔP from atmospheric pressure P_{atm} as a function of x in the air column at a particular instant.

fundamental physical sense, pressure waves. Because pressure is a scalar (non-directional) quantity while displacements have a directional character, a misconception of the basic nature of sound waves may lead to physical fallacies. Consider, for example, the principle of superposition as applied to sound waves.

In the system, called a *sound interferometer*, of Figure 19.16, a sound source S produces a harmonic sound wave in the air-filled glass tube, which divides the wave at point A. One path of length L_1 conducts one sound wave directly to point B, where there is a sound detector D, perhaps an ear. A second path, equipped with a trombonelike movable section of tube, conducts the other wave along a path of length L_2. The two waves that meet at point B will be in phase if the difference, $L_2 - L_1$, in path lengths is an integral number n of wavelengths. In that case the detector will receive a maximum sound signal. The pressure amplitudes of the two waves add constructively. If now the path length L_2 is adjusted so that it differs from L_1 by $(n + \frac{1}{2})$ wavelengths, the waves meeting at B will be 180° out of phase. Then the two waves destructively interfere, and point B is a node. In this case, the detector D receives a minimum (ideally zero) signal. The conditions for constructive and destructive interference in this device may be summarized by

$$L_2 - L_1 = n\lambda \quad ; \quad n = 0, 1, 2, 3, \ldots \qquad \text{(constructive interference)}$$
$$\text{(19-30)}$$

$$L_2 - L_1 = (n + \tfrac{1}{2})\lambda \quad ; \quad n = 0, 1, 2, 3, \ldots \qquad \text{(destructive interference)}$$
$$\text{(19-31)}$$

This apparatus may be used to measure the wavelength of a sound wave produced by the source S. Can you describe an experiment that will make such a measurement?

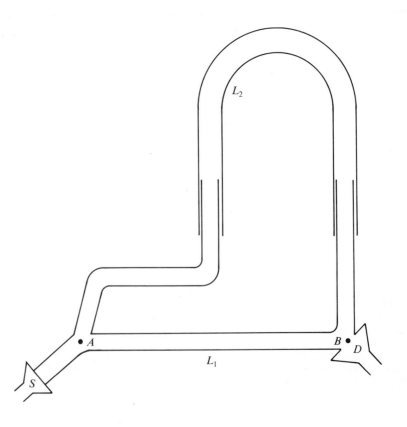

Figure 19.16
A sound interferometer.

E 19.17   The slide of the interferometer of Figure 19.16 moves 14 cm as the sound at point B changes from one minimum to the next minimum.
(a) Determine the corresponding change in the length of path L_2 shown in the figure.   Answer: 28 cm
(b) Calculate the wavelength of the sound being emitted by S. Answer: 28 cm
(c) What is the frequency of this sound?   Answer: 1.2 kHz

For the interferometer of Figure 19.16, notice that we cannot explain the observed complete destructive interference of two sound waves in terms of displacement of "particles" in the medium. Because at point B the displacements (vector quantities) associated with the two waves would be along perpendicular lines, the displacements could not result in a node. The displacements could not at all times sum to zero, no matter what phase relationship may exist between the two waves. Thus, a description of the sound waves in terms of "particle" displacements does not agree with the experimental facts that a node does exist at point B for an appropriate phase relationship between the two waves.

On the other hand, the observed complete destructive interference at point B is easily explained in terms of pressure waves. When the two waves at B are 180° out of phase, the pressure increase (a scalar quantity) of one wave at B is at all times equal in magnitude to the pressure decrease of the other wave at B. The pressure differences (from the equilibrium pressure of the air) of both waves sum to zero at all times at point B. In other words, a pressure node exists at B.

Therefore, because a pressure description of sound is more appropriate physically than a displacement description, we will henceforth consider sound waves pressure waves. In terms of pressure, then, the superposition principle for sound waves in a fluid (a liquid or gas) may be expressed as follows:

When two or more sound waves are at a point in a fluid simultaneously, the change in pressure from the equilibrium pressure of the fluid at that point is equal to the algebraic sum of pressure changes of the individual waves at that point.

Because harmonic sound waves are sinusoidal pressure waves, it is usually appropriate to formulate expressions describing sound waves in terms of pressure. If we let $p = \Delta P$ be the change in pressure from the undisturbed pressure P of the medium, we may describe a harmonic sound wave by

$$p = p_o \cos(kx \pm \omega t + \delta) \tag{19-32}$$

Standing Sound Waves

If sound waves are excited in a column of air within a length of pipe, the waves reflect from the ends of the pipe and may form standing sound waves. These standing waves may be described in much the same way as standing waves on taut strings, as we saw in Section 19.2, "Reflection, Superposition, and Standing Waves." They are two traveling waves with equal amplitudes and moving in opposite directions. Figure 19.17 shows an air-filled pipe, which has length L and is open at both ends. Because the air at or very near an open end of the pipe is always at atmospheric pressure, that is, the pressure does not change, an open end must be a pressure node in a standing sound wave. Then in the fundamental mode, shown in Figure 19.17(a), in which the standing wave has the longest wavelength consistent with having a pressure node at both ends of the pipe,

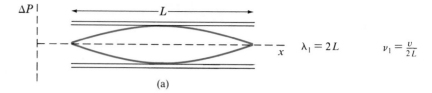

Figure 19.17
The difference in pressure ΔP as a function of position in a pipe, open at both ends, which is supporting a standing sound wave. (a) The fundamental mode has the longest wavelength consistent with a pressure node at each end of the pipe and (b) and (c) the next two higher frequency modes are consistent with the boundary conditions of the open ends of the pipe.

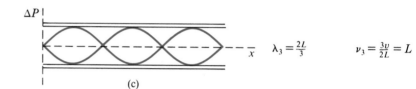

the wavelength λ_1 is equal to $2L$. The fundamental frequency ν_1 is given by $V/\lambda_1 = V/2L$, where $V = 340$ m/s is the speed of sound in air. The next two successively higher frequency modes of vibration that fit into the length L and have pressure nodes at either end are shown in Figures 19.17(b) and (c). The wavelengths and frequencies for the resonant modes in the air column are given by

$$\lambda_n = \frac{2L}{n} \quad ; \quad n = 1, 2, 3, \ldots \tag{19-33}$$

$$\nu_n = \frac{nV}{2L} = n\nu_1 , \quad ; \quad n = 1, 2, 3, \ldots \tag{19-34}$$

E 19.18   (a) Calculate the length of a pipe, open at both ends, which has a fundamental frequency of 100 Hz.   Answer: 1.7 m
(b) What are the next two highest natural frequencies of the air column in the pipe of part (a)?   Answer: 200 Hz, 300 Hz

Now consider a standing wave inside an organ pipe of length L shown in Figure 19.18(a). One end is effectively closed; the other end is open. Air forced through the small chamber in the organ pipe causes the sharp, thin lip to vibrate, thereby exciting standing waves in the air within the main body of the pipe. At the closed end of the long air column, the air molecules cannot move as the pressure waves reach the wall. The pressure at the wall, therefore, experiences maximum variation. In other words, a closed end of an air column is a pressure antinode. A pressure node is approximately at an open end of a pipe, although the exact location actually depends slightly on the diameter of the pipe. Figure 19.18(b) shows the pressure variation along the length of the pipe in three modes of vibration among the possible standing sound wave modes. The modes shown have the three longest wavelengths consistent with having a pressure node at the

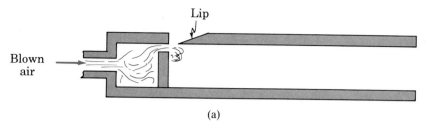

(a)

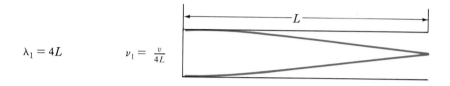

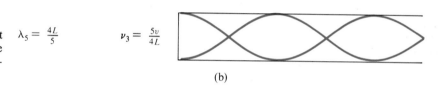

Figure 19.18
(a) An organ pipe, effectively closed at one end and open at the other. **(b)** The three longest wavelength modes of vibration in the organ pipe of length L.

(b)

open end and a pressure antinode at the closed end. The wavelengths and frequencies corresponding to the modes of vibration in this pipe are

$$\lambda_n = \frac{4L}{2n-1} \quad ; \quad n = 1, 2, 3, \ldots \qquad (19\text{-}35)$$

$$\nu_n = \frac{(2n-1)V}{4L} = (2n-1)\nu_1 \quad ; \quad n = 1, 2, 3, \ldots \qquad (19\text{-}36)$$

Notice that only *odd harmonics*, ν_1, $3\nu_1$, $5\nu_1$, and so on, constitute the natural frequencies of a pipe open at one end and closed at the other.

E 19.19   The three lowest frequencies of the air column in a 1.7-m pipe, open at both ends, were determined in Exercise 19.18 to be 100 Hz, 200 Hz, and 300 Hz. If the same length pipe is used as an organ pipe (closed at one end), determine the three lowest natural frequencies.

Answers: 50 Hz, 150 Hz, 250 Hz

Beats

When two sound waves of slightly different frequencies, ν_1 and ν_2 ($\nu_1 > \nu_2$) are superposed at a point, the waves interfere so that the amplitude of the combination pulsates (the sound alternately is louder and softer) at the difference frequency, $\nu_1 - \nu_2$. These amplitude pulsations are called *beats*. We can see the reason for this phenomenon at a point by considering two pressure waves having equal amplitudes p_o and slightly different frequencies ν_1 and ν_2:

$$p_1 = p_o \cos 2\pi \nu_1 t \qquad (19\text{-}37)$$

$$p_2 = p_o \cos 2\pi \nu_2 t \qquad (19\text{-}38)$$

According to the principle of superposition, the resultant wave at that point is given by

$$p = p_1 + p_2 = p_o(\cos 2\pi \nu_1 t + \cos 2\pi \nu_2 t) \qquad (19\text{-}39)$$

Using the identity, $\cos\alpha + \cos\beta = 2 \cos[\frac{1}{2}(\alpha + \beta)] \cos[\frac{1}{2}(\alpha - \beta)]$, we may write Equation (19-39) as

$$p = \left[2p_o \cos 2\pi\left(\frac{\nu_1 - \nu_2}{2}\right)t \right] \cos 2\pi\left(\frac{\nu_1 + \nu_2}{2}\right)t \qquad (19\text{-}40)$$

Figure 19.19 is a graph of Equation (19-40). It shows the resultant wave, which has a vibrational frequency $(\nu_1 + \nu_2)/2$. But the amplitude of the resultant wave is *modulated*, or controlled, by a sinusoidal envelope of frequency $(\nu_1 - \nu_2)/2$. In Equation (19-40), the factor in brackets represents this variation in the amplitude of the resultant wave. That amplitude reaches its maximum value when $\cos[2\pi(\nu_1 - \nu_2)/2]$ has values ± 1, that is, at a frequency of $2(\nu_1 - \nu_2)/2 = \nu_1 - \nu_2$, the *beat frequency*.

Orchestra conductors commonly use the phenomenon of beats when tuning instruments to a common frequency. For example, suppose an oboe plays a tone of frequency 440 Hz (middle A on the musical scale), and a cornet simultaneously plays a tone with frequency 442 Hz. The conductor hears a pulsating amplitude (an alternating loudness and softness) of the beat frequency $(442 - 440)$ Hz = 2 Hz. The cornetist may adjust his frequency by lengthening slightly the tubing of the cornet. The conductor can in this way, by listening to the frequency of the beats, tune the instruments precisely. When the beat disappears, that is, when the beat frequency becomes equal to zero, the two instruments are tuned to the same frequency.

When the frequency difference between two sounds that are heard simultaneously is greater than 5 or 6 Hz, the beats are no longer heard as distinct pulsations

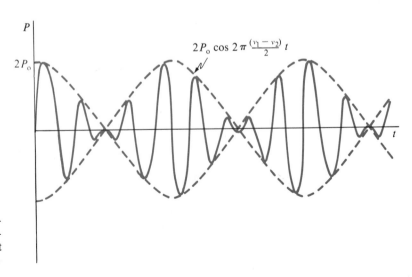

Figure 19.19
A graph of Equation (19-40). The envelope of the wave with vibrational frequency $(\nu_1 + \nu_2)/2$ modulates that wave at the beat frequency $(\nu_1 - \nu_2)/2$.

of amplitude. Beat frequencies between about 5 and 20 Hz are heard as a dissonance, or a "roughness" of the sound. Beat frequencies greater than about 20 Hz are heard as a third tone. In our considerations and in the problems for this chapter, we refer to any difference frequency between the frequencies of two simultaneously received sounds as a beat frequency.

E 19.20   Two nearby sound generators emit signals of equal amplitudes but different frequencies, $\nu_1 = 744$ Hz and $\nu_2 = 748$ Hz. What will you hear?

Answer: 746 Hz with a 4-Hz beat frequency

The Doppler Effect

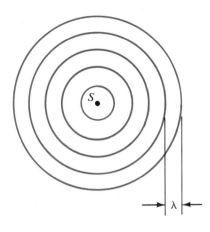

Figure 19.20
Wavefronts of a sound wave moving radially outward from a source S at rest. The radial distance between successive wavefronts is the wavelength λ of the sound waves.

If we stand beside a highway, the frequency we hear coming from the sounding horn of an approaching truck changes dramatically when the truck passes and recedes from us. This situation demonstrates the effect of a moving source of sound on the sound perceived by a listener. Careful measurements show that the frequency heard by a listener depends on the frequency emitted by a source and on the relative motions between the source, the listener, and the medium in which the sound moves. This change in frequency that a listener perceives because of relative motions of source, listener, and medium is called the *Doppler effect* for sound. To simplify our analysis of the Doppler effect, let us consider motions of a point source and listener that take place along a common straight line. Further, let us assume that the medium (air) is not moving relative to the earth, that is, no wind is blowing.

Sound waves travel radially outward from a point source of sound. A *wave front* is a surface composed of points lying along one compression maximum of a sound wave. Thus, wave fronts emanating from a point source S *at rest* are concentric spherical surfaces, as indicated in Figure 19.20. The wavelength λ of the sound wave is the distance between successive wave fronts.

Now consider a source S of sound moving to the right at a speed v, as shown in Figure 19.21. Here the successive spherical wave fronts emitted by the moving source are bunched in front of S and spread out behind S. The numbers on the wave fronts correspond to the numbered positions of the source where the corresponding wave fronts were emitted. The wavelength λ_f in front of the source is shorter than the wavelength λ of a source at rest. The wavelength λ_b behind the source is longer than λ.

To describe mathematically the Doppler effect, let us adopt the following sign convention: Velocities in the direction from the source S toward the listener L are positive. Then the velocity V of sound through the air is always a positive quantity ($V = 340$ m/s $= 1100$ ft/s). The velocities, V_S and V_L, of the source S and listener L are positive when directed from S toward L, negative when directed from L toward S.

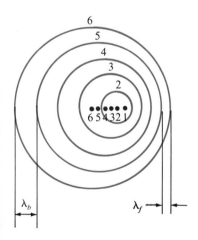

Figure 19.21
Wavefronts of a sound wave emitted from a source moving to the right. The numbers on the moving source indicate the position of the source when the correspondingly numbered wavefronts were emitted. In front of the source, the distance between wavefronts is λ_f; behind the source, the wavelength is λ_b.

Now consider the source, pictured in Figure 19.22 at point B at time t, moving with constant velocity V_S toward the right while emitting a sound of frequency ν_S. The source had been at point A at $t = 0$, so the distance $\overline{AB}$ is equal to $|V_S|t$. The point C lies on a wavefront that was emitted when the source was at point A, and during the time interval t, that wavefront has moved a distance Vt. Therefore, the distance $\overline{BC}$ is equal to $Vt - |V_S|t$. The number of wavefronts in the length $\overline{BC}$ is equal to $\nu_S t$, so the wavelength λ_f in front of the source is given by

$$\lambda_f = \frac{\overline{BC}}{\text{number of wavefronts in } \overline{BC}} = \frac{Vt - |V_S|t}{\nu_S t} = \frac{V - |V_S|}{\nu_S} \qquad (19\text{-}41)$$

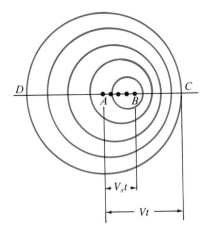

Figure 19.22
A source moving toward the right with speed V_S and emitting a sound wave that moves through the medium at speed V. The source, which is at point B at time t, was at point A at $t = 0$. In the time interval t, the source has moved a distance $V_S t = \overline{AB}$, and the wavefront emitted at point A has traveled a distance $Vt = \overline{AC}$.

Similarly, at point D behind the source, the distance $\overline{BD}$ is equal to $Vt + |V_S|t$, and the wavelength λ_b behind the source is

$$\lambda_b = \frac{\overline{BD}}{\text{number of wavefronts in } \overline{BD}} = \frac{Vt + |V_S|t}{\nu_S t} = \frac{V + |V_S|}{\nu_S} \quad (19\text{-}42)$$

Using our sign convention—the direction from source to listener is positive—we may write Equations (19-41) and (19-42) in a simple equation:

$$\lambda = \frac{V - V_S}{\nu_S} \quad (19\text{-}43)$$

E 19.21   An ambulance with a 2.6-kHz siren travels 72 km/hr (20 m/s). Calculate the wavelength of the sound from the siren (a) if the ambulance were at rest, (b) in front of the ambulance, and (c) behind the ambulance.

Answers: (a) 13 cm; (b) 12 cm; (c) 14 cm

The frequency ν_L that a listener perceives is equal to the number of wavefronts per second that the listener intercepts. Therefore, the frequency ν_L that the listener hears is equal to the velocity, $V - V_L$, of the wavefronts relative to the listener divided by the distance λ between adjacent wavefronts, or

$$\nu_L = \frac{V - V_L}{\lambda} \quad (19\text{-}44)$$

where we have again imposed our sign convention. Substituting Equation (19-43) for λ, we obtain the frequency ν_L heard by the listener in terms of the frequency ν_S emitted by the source:

$$\nu_L = \nu_S\left(\frac{V - V_L}{V - V_S}\right) \quad (19\text{-}45)$$

E 19.22   Determine the frequency heard by a person
(a) standing in front of the ambulance of Exercise 19.21,
(b) standing behind the ambulance, and
(c) riding in an auto following the ambulance at the same speed (20 m/s).

Answers: (a) 2.8 kHz; (b) 2.5 kHz; (c) 2.6 kHz

E 19.23   Determine the frequency heard by a person approaching the ambulance of Exercise 19.21 if this person is in a vehicle moving at a speed of 25 m/s.

Answer: 3.0 kHz

The following problem illustrates some of the uses of Equation (19-45).

A source of sound having a frequency of 1.0 kHz is moving eastward at 20 m/s. East of the source is a listener moving westward at 10 m/s. A vertical wall, located east of the listener, is moving westward at 20 m/s. If the listener hears one sound directly from the source and another reflected from the moving wall, what beat frequency is heard by the listener?

Figure 19.23(a) illustrates this physical situation. The sound that comes to the listener L directly from the source S is perceived by the listener to have a frequency ν_L that may be found by application of Equation (19-45). Notice that because the direction from S toward L is positive, $V_S = 20$ m/s is a positive

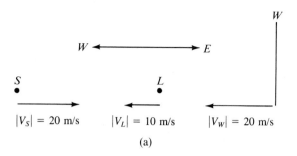

$|V_S| = 20$ m/s $|V_L| = 10$ m/s $|V_W| = 20$ m/s

(a)

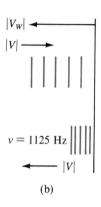

$v = 1125$ Hz

(b)

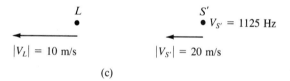

$|V_L| = 10$ m/s $|V_{S'}| = 20$ m/s

(c)

Figure 19.23
A visualization of the doppler-effect
problem discussed in the text.

quantity and $V_L = -10$ m/s is negative. Then the frequency ν_L heard by L because of the direct sound from S is given by

$$\nu_L = \nu_S\left(\frac{V - V_L}{V - V_S}\right) = (1000)\left(\frac{340 + 10}{340 - 20}\right) = 1094 \text{ Hz} \qquad \textbf{(19-46)}$$

To find the frequency, as heard by the listener, of the sound waves reflecting from the wall W, let us first determine the rate at which wavefronts are intercepted and, consequently, reflected by the wall. Because this rate, or frequency, is just the frequency ν_W that a listener on the wall would hear. We may treat the wall as a listener and use Equation (19-45) again:

$$\nu_W = \nu_S\left(\frac{V - V_W}{V - V_S}\right) = (1000)\left(\frac{340 + 20}{340 - 20}\right) = 1125 \text{ Hz} \qquad \textbf{(19-47)}$$

In Equation (19-47), the velocity V_W of the wall is toward the source and, therefore, a negative quantity. Figure 19.23(b) shows the wavefronts coming from the source being intercepted by the wall at a frequency of 1125 Hz. Because these waves are reflected from the wall at 1125 Hz, we may treat the sound moving westward from the wall toward the listener as if it were coming from a *moving* source S' of frequency 1125 Hz and located east of the listener L, as shown in Figure 19.23(c). Then the frequency ν_L, perceived by L because of the sound reflected from the wall is

$$\nu_L' = \nu_{S'}\left(\frac{V - V_{L'}}{V - V_{S'}}\right) = (1125)\left(\frac{340 - 10}{340 - 20}\right) = 1160 \text{ Hz} \qquad \textbf{(19-48)}$$

The beat frequency ν_B perceived by the listener is the difference of the two frequencies, ν_L and ν_L', that reach the listener simultaneously. Using Equations (19-46) and (19-48), we find the beat frequency to be

$$\nu_B = |\nu_L - \nu_L'| = 1160 - 1094 = 66 \text{ Hz} \qquad \textbf{(19-49)}$$

19.4 Sound and Human Hearing

In describing sounds as heard by humans, we often use terms like pitch, timbre, and loudness. These terms describe subjective qualities of sound, but each of these characteristics of perceived sound may be related to physical quantities we have encountered in this chapter.

Pitch is the characteristic of sound that is commonly referred to by terms like "a high note" or "low tones." Perceived pitch is directly related to the frequency of the sound wave that is heard. The higher the frequency of a sound wave, the higher is the pitch we perceive when we hear that sound. The frequency range of "normal" human hearing is approximately from 20 Hz to 20,000 Hz. Most people can distinguish a difference in pitch between two harmonic sound waves near a frequency of 1000 Hz that differ in frequency by about 3 Hz. Thus the typical human discrimination in pitch is about 0.3 percent at 1000 Hz.

All sound waves are not pure sinusoidal waves. Indeed, such pure tones rarely occur in nature. Most sounds have complex wave forms, like that of Figure 19.24(a). This complex wave, and, in fact, any periodic wave form, can be shown to be composed of a number of sinusoidal waves, each of which is a harmonic, or an integral multiple, of the fundamental frequency. Figure 19.24(b) shows the individual harmonic components of the complex wave of Figure 19.24(a). The pressure increase at any point along the complex wave is the algebraic sum of the pressure increases of its components at that position. That quality of sound called *timbre* is the characteristic by which we distinguish the sound of a violin from that of a trumpet when both are sounding the same fundamental frequency. Different musical instruments emit sound waves with different harmonic contents, that is, the various harmonics have different amplitudes and phases relative to those of the fundamental frequency. Indeed a single instrument, like a guitar, for example, can be made to sound "tinny" by plucking a string near the bridge (the device that fixes one end of the strings), thereby exciting more high-frequency components of the sound. The same string can be made to sound relatively "mellow" by plucking the string halfway between the bridge and the nut, which fixes the other end of the string; in this case, the fundamental frequency dominates the sound. Thus, the physical basis of timbre is the relative harmonic content in a complex sound wave.

The *loudness* of a given sound is related to the amplitude of that pressure wave that reaches our ears. But the response of the human ear to the amplitude of a pressure wave is nonlinear: Doubling the amplitude of a sound wave does not make it seem twice as loud. Of course, comparisons of loudness are subjective; different people make different judgments of what is "twice as loud." The

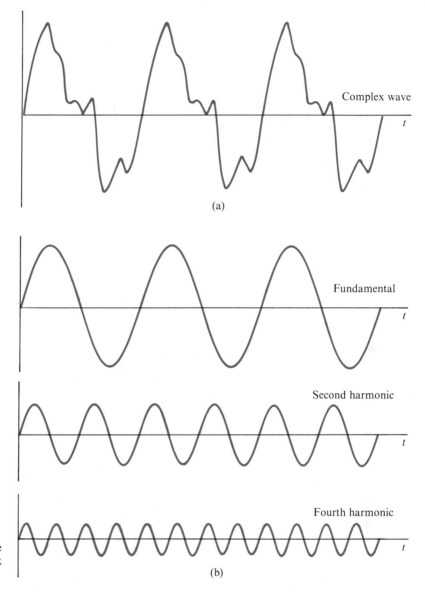

Figure 19.24
(a) A complex periodic wave. (b) Three harmonic components of the complex wave.

response of the human ear to changes in the amplitude of a pressure wave of a given frequency is roughly logarithmic. Consequently, the measure of loudness is usually expressed in terms of *sound pressure level L*, defined by

$$L = 20 \log\left(\frac{p_o}{p_{ot}}\right) \text{ decibels} \qquad (19\text{-}50)$$

where the *decibel* (db) is a unitless quantity that relates the amplitude p_o of a wave being measured to a reference amplitude p_{ot}. The amplitude p_{ot} is called the *threshold of hearing*, an arbitrary amplitude of a pressure wave chosen to have the value 2.0×10^{-5} N/m^2, which is the amplitude of the softest 1-kHz sound that the "average" ear can detect. According to Equation (19-50), a change by a factor of 10 in pressure amplitude of a sound wave corresponds to a change of 20 db in sound level. Table 19.1 gives the sound pressure levels and the corresponding pressure amplitudes of some typical situations.

TABLE 19.1   **A Scale of Relative Sound Pressure Levels**

Pressure (N/m²)	Pressure Level (db)	Example
0.00002	0	Threshold of hearing
	20	A whisper at about 4 feet
0.002	40	City noise at night
	60	Conversational speech
0.2	80	Loud radio; heavy traffic
	100	Riveter at 30 feet; rock band
20	120	Discomfort level; bad rock band
	140	Pain level
2000	160	Mechanical damage to ear

E 19.24   If the sound pressure level of a signal is increased by 5.0 db, by what (multiplicative) factor is the pressure amplitude of the signal increased?

Answer: 1.8

Figure 19.25 depicts the results of experiments in which human subjects judged sounds of different frequencies to be equally loud. Each curve shows the necessary variation in sound pressure level to make pure sinusoidal tones of different frequencies *seem* equally loud to the human ear. Each curve is labeled according to the value of its sound pressure level at 1 kHz. The unit of subjective loudness is called a *phon*. The loudness in phons of a sound at any frequency is equal to the sound pressure level, in db, above the threshold level (0 db) of a 1-kHz tone of "equal" loudness. For example, the curve labeled 40 phons passes

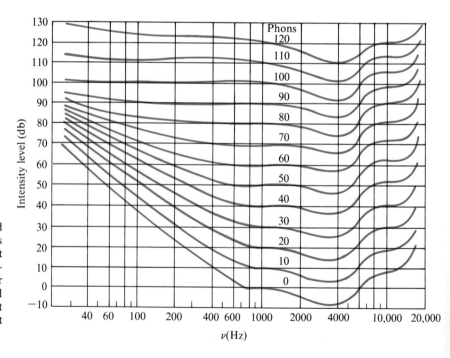

Figure 19.25
The response of the human ear to sound intensity level. Each curve represents the judgment of the average listener that the loudness seems constant as the frequency varies. The reference level for each curve is the actual intensity level of a 1000-Hz sound wave. See the text for the definition of the phon, the unit of subjective loudness.

through the sound pressure level of 40 db at a frequency of 1000 Hz. The same 40-phon curve passes through a level of 70 db at 60 Hz. Therefore, for a loudness level of 40 phons, a 60-Hz tone must have a sound pressure level that is 30 db greater than a 1-kHz tone in order to seem equally loud.

The curves of Figure 19.25 indicate that the human ear responds most sensitively to sound waves at about 3.4 kHz, regardless of the loudness of the sound. This fact may be explained physically by recognizing that the canal leading from the outside of the human ear to the eardrum has an average length of about 2.5 cm. Thus the external ear canal is a tube, open at one end and closed at the other. According to Equation (19-36), the fundamental frequency ν_1 to which such a tube is resonant is given by

$$\nu_1 = \frac{V}{4L} \tag{19-51}$$

Substituting $V = 340$ m/s for the speed of sound in air and $L = 2.5 \times 10^{-2}$ m for the length of the tube, Equation (19-51) gives

$$\nu_1 = \frac{340 \text{ m/s}}{4(2.5 \times 10^{-2} \text{ m})} = 3400 \text{ Hz} \tag{19-52}$$

which coincides with the frequency at which the human ear is most sensitive.

Further, the next higher mode at which the ear canal is resonant, according to Equation (19-36), is

$$\nu_2 = \frac{3V}{4L} = \frac{3(340 \text{ m/s})}{4(2.5 \times 10^{-2} \text{ m})} = 10,000 \text{ Hz} \tag{19-53}$$

The effect of this second resonance—the pressure wave in the ear canal is enhanced—is apparent in the curves of Figure 19.25. The rapidly decreasing sensitivity of the ear above 4000 Hz, indicated by the increasing slopes of the curves, is noticeably improved by the resonance at 10,000 Hz.

19.5 Problem-Solving Summary

A harmonic traveling wave may be characterized by its amplitude A, wavelength λ, and frequency ν. These quantities are related to other useful parameters of the wave, such as its period $T = 1/\nu$, its wave number $k = 2\pi/\lambda$, and its angular frequency $\omega = 2\pi\nu$. In problems that involve traveling wave equations and the parameters of those waves, it is usually helpful to cast the wave equation in a general form, for example, $y = A \cos[2\pi(x/\lambda \pm t/T) + \delta]$. In this wave equation, y is the displacement from equilibrium of a particle located at a position x in the medium. The $+$ sign indicates a wave moving in the negative x direction; the $-$ sign indicates a wave moving in the positive x direction. The phase constant δ may be chosen so that the wave equation applies for any origin $x = 0$ and initial condition $t = 0$.

In any traveling wave, the speed of propagation V is related to the wavelength λ and frequency ν of the wave by $V = \nu\lambda$. For transverse waves on strings or ropes, the speed of propagation V of a wave is related to the tension F in the string and the mass per unit length μ of the string by $V = \sqrt{F/\mu}$. The speed of sound has the values 340 m/s = 1100 ft/s in air and 1500 m/s = 4900 ft/s in water.

Traveling waves reflect from a fixed boundary of a medium with a phase change of 180°; they reflect from a free boundary without changing phase. The principle of superposition is generally applicable to waves that are simultaneously at the same location in a medium.

When two harmonic waves of equal frequency are simultaneously traveling in opposite directions in a medium, standing waves are formed. If two such traveling waves have identical amplitudes, the standing wave will have nodes, points in the medium that experience no displacement. The distance between adjacent nodes is equal to one-half wavelength of the traveling waves. When standing waves are on a string of length L, fixed at both ends, the natural (resonant) frequencies of the natural modes of vibration are given by $\nu_n = V/\lambda_n$, where n is a positive integer and $\lambda_n = 2L/n$. The displacement y of a particle located at a position x on a standing wave is described at any time t by

$$y = [2A \cos(kx + \delta_x)] \cos(\omega t + \delta_t)$$

where δ_x and δ_t are phase constants that may be chosen for any appropriate initial condition ($t = 0$) and any given origin ($x = 0$).

A harmonic sound wave in air is a pressure wave, composed of alternate regions of compression (high pressure) and rarefaction (low pressure). The wave may be represented by $p = p_o \cos(kx \pm \omega t + \delta)$, where p is the change in pressure from the equilibrium air pressure and p_o is the amplitude of the pressure wave. Standing sound waves, like those in an organ pipe, have resonant frequencies ν_n that depend on the length L of the pipe and the boundary conditions on the column of air in the pipe. For pipes open at both ends or closed at both ends, the wavelengths ν_n of the resonant modes of these pipes are given by $\lambda_n = 2L/n$, where $n = 1, 2, 3, \ldots$. The corresponding resonant frequencies are given by $\nu_n = nV/2L$, where $V = 340$ m/s is the speed of sound in air. For pipes open at one end and closed at the other, $\lambda_n = 4L/(2n - 1)$ and $\nu_n = (2n - 1)V/4L$, where only odd harmonics occur.

When two sound waves of different frequencies, ν_1 and ν_2, reach the human ear at the same time, the ear hears a beat frequency equal to $|\nu_2 - \nu_1|$.

The relative motions between a source of sound, a listener, and the medium (usually air) through which the sound travels affects the wavelength of the sound wave and the frequency heard by the listener. Changes in wavelength and perceived frequency because of these motions are called the Doppler effect. Doppler problems in one dimension (all motion along a straight line) can be subdued using only two relationships if we use a suitable sign convention. Let the direction from the source S toward the listener L be the positive direction. Then V, the velocity of sound, is always a positive value. The velocities V_S of the source and V_L of the listener are positive when they are directed from S toward L, negative when directed from L toward S. Then the wavelength of a sound wave in air at the position of a listener is given by $\lambda = (V - V_S)/\nu_S$, where ν_S is the frequency of the sound emitted by the source. The frequency ν_L heard by the listener is given by $\nu_L = \nu_S(V - V_L)/(V - V_S)$.

The subjective qualities of sound are pitch, loudness, and timbre. The underlying physical characteristics of sound waves that control these qualities are frequency, amplitude, and harmonic content, respectively. Loudness is measured in terms of sound pressure level L (in decibels) $= 20 \log(p_o/p_{ot})$, which relates the pressure amplitude p_o of a given sound to p_{ot}, the threshold of hearing, which has the value 2.0×10^{-5} N/m².

Problems

GROUP A

19.1 Show that the angular frequency ω, wave number k, and speed of propagation v of a sinusoidal wave are related by $\omega = kv$.

19.2

(a) What is the range of wavelengths in air of frequencies that are within the range of human hearing?

(b) Repeat part (a) for the range of wavelengths in water.

19.3 What frequency will you hear if you move with a speed of 65 m/s toward a stationary 3.0-kHz sound source?

19.4 A transverse traveling wave is described by

$$y = 0.12 \sin \pi(x + 12t)$$

Determine the velocity of propagation, amplitude, wavelength, and frequency of this wave.

19.5 If the pressure amplitude of a sound signal is increased by a (multiplicative) factor of 50, what is the corresponding increase in the sound pressure level of the signal?

19.6 Two signals of different frequencies combine to produce a 92-Hz sound with a beat frequency equal to 6.0 Hz. Determine the two frequencies.

19.7 If the air horn on a truck moving 80 km/h emits a 250-Hz signal, what frequency will you hear if the truck is (a) coming toward you? (b) going away from you?

19.8 Determine the constants, α and β, if

$$y = A \cos(\alpha x + \beta t)$$

is to represent a transverse wave traveling 35 m/s in the positive x direction with a frequency equal to 25 Hz.

19.9 If you are on a line between two stationary 200-Hz sound sources and are moving 3.4 m/s toward one of them, describe the sound you will hear.

19.10 A transverse wave on a string is described by

$$y = 0.12 \sin \pi(x/8 + 4t)$$

(a) Determine the transverse velocity and acceleration at $t = 0.20$ s for the point on the string at $x = 1.6$ m.

(b) What are the wavelength, period, and velocity of propagation of this wave?

19.11 What are the two lowest natural frequencies of oscillation for the air column in a glass tube that is 45 cm long and has one end closed?

19.12 What tension is required in a guitar string that has a length of 60 cm and a mass of 2.4 g if it is to have a fundamental frequency equal to 88 Hz?

19.13 A sound source moving 50 km/h emits a 5.0-kHz signal.

(a) What is the wavelength of the signal in front of the source? Behind the source?

(b) What frequency is heard by a stationary listener in front of the source? Behind the source?

19.14 The $t = 0$ shape of a transverse wave moving in the negative x direction with a speed of 100 m/s is shown in Figure 19.26. Write an expression that describes this wave.

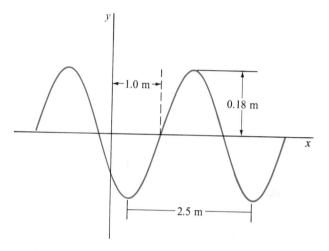

Figure 19.26 Problem 19.14

19.15 A 5.0-g string with ends fixed at $x = 0$ and $x = 2.0$ m oscillates in its fifth ($n = 5$) natural mode. The tension in the string is equal to 64 N. At $t = 0$, the $x = 0.60$ m point on the string is passing through the equilibrium position with a velocity v_y equal to $+65$ m/s. Write an equation describing the motion of the string.

19.16 The fundamental frequency for a 25-cm wire fixed at both ends is equal to 440 Hz.

(a) What is the speed of propagation of a transverse disturbance on this wire?

(b) If the tension in this wire is equal to 450 N, determine the mass of the wire.

19.17 A string (mass = 4.8 g, length = 2.0 m, and tension = 48 N), fixed at both ends, vibrates in its second ($n = 2$) natural mode. What is the wavelength in air of the sound emitted by this vibrating string?

19.18

(a) Show that

$$y = A(\sin\alpha x \cos\beta t + \cos\alpha x \sin\beta t)$$

describes a traveling wave.

(b) Determine the velocity of propagation, amplitude, wavelength, and period of this wave.

19.19 A 60-cm string fixed at both ends vibrates in its fundamental mode with a frequency of 25 Hz. If the maximum transverse displacement at the antinode is equal to 1.8 cm, calculate the maximum transverse speed of a point 12 cm from an end of the string.

19.20 A standing wave is described by

$$y = 0.080 \cos(\pi x + 1.2) \cos 20\pi t$$

(a) Determine the positions of the nodes.

(b) What are the amplitude, wavelength, and frequency of the two traveling waves that generate this standing wave?

19.21 A glass tube (open at both ends) of length L is positioned near an audio speaker of frequency $\nu = 0.68$ kHz. For what values of L will the tube resonate with the speaker?

19.22 Two identical strings, each fixed at both ends, are arranged near each other. If string A starts oscillating in its fundamental mode, it is observed that string B will begin vibrating in its third ($n = 3$) natural mode. Determine the ratio of the tension of string B to the tension of string A.

19.23 A listener moving 15 m/s hears a signal from a source moving 45 m/s in the same direction. If the source emits a 4.4-kHz signal, what frequency is heard if the listener is (a) in front of the source? (b) behind the source?

19.24 If two adjacent natural frequencies of an organ pipe are determined to be 0.55 kHz and 0.65 kHz, calculate the fundamental frequency and length of this pipe.

19.25 A 1.2-m string fixed at both ends vibrates in its fourth natural mode with a frequency equal to 0.68 kHz.

(a) Calculate the distance between adjacent nodes for this mode.

(b) What is the speed of propagation of a transverse disturbance on this string?

(c) If the amplitude of the oscillation at an antinode is equal to 1.5 cm, what is the maximum transverse speed of this (antinode) point on the string?

19.26 Ambulance A, having a 3.40-kHz siren, moves toward you from the east. Ambulance B, having a 3.30-kHz siren, moves toward you from the west. While at rest, you hear a frequency of 3.42 kHz with no beats. What is the velocity of each ambulance?

19.27 Two organ pipes that differ in length by 5.0 cm have a beat frequency of 4.0 Hz when excited in their fundamental modes. Determine the fundamental frequency of each pipe.

19.28 An observer positioned at a stationary 2.0-kHz sound source compares the signal reflected from an object moving away from the source to the source signal and observes a beat frequency of 86 Hz. How fast (in km/h) is his object moving?

19.29 A 5.0-kHz sound source S_o moving 20 m/s toward the east approaches an object moving with a speed V_o toward the west. If the reflected signal is determined by S_o to have a frequency of 6.5 kHz, determine V_o.

19.30 One end of a string (length = 40 cm, mass = 1.2 g) is tied to a ring of negligible mass. The ring moves transversely on a frictionless rod, as shown in Figure 19.27. If the other end of the string is fixed and the tension of the string is equal to 12 N, determine the natural frequencies for this system.

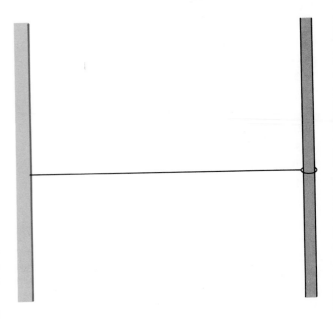

Figure 19.27   Problem 19.30

19.31 Two sound sources, S_1 and S_2, emit frequencies, ν_1 and ν_2. Source S_1 moves 17 m/s toward the east; S_2 moves 34 m/s toward the west. Both sources are moving toward you. When you are at rest, you hear a 551-Hz signal with a 5.0-Hz beat frequency. If you move toward the east, the beat frequency decreases. Determine ν_1 and ν_2 to three significant figures, assuming the speed of sound in air to be equal to 340 m/s.

19.32 A transverse sinusoidal wave travels in the positive x direction with a speed of propagation equal to 0.25 km/s. At $t = 0$, the $x = 0$ point is displaced $+8.0$ cm from its equilibrium position and is not moving. The maximum transverse speed for this point is equal to 24 m/s. Write an equation describing this wave.

19.33 A transverse disturbance on a string is described by

$$y = 0.040 \, e^{-(3x + 50t)^2}$$

(a) Sketch the shape of this disturbance at $t = 0$.

(b) What is the velocity of propagation of this disturbance?

(c) When does the point at $x = -50$ cm achieve a maximum displacement from equilibrium?

(d) What is the maximum transverse speed of any point on the string?

(e) When does the point at $x = -50$ cm achieve the maximum transverse speed determined in part (d)?

19.34

(a) Show that the transverse disturbance described by

$$y = A \sin(\alpha x + \beta t) + A \sin(\alpha x - \beta t)$$

is a standing wave.

(b) Determine the positions of the nodes.

(c) Determine the maximum transverse speed of any point in the supporting medium.

19.35 Two transverse disturbances described by

$$y_1 = 0.10 \sin(12x - 57t)$$

$$y_2 = 0.10 \cos(12x + 57t)$$

moves along a string.

(a) Show that the resulting disturbance is a standing wave.

(b) Determine the positions of the nodes.

19.36 Two transverse waves on a string are described by

$$y_1 = 0.050 \sin(6x - 100t)$$

$$y_2 = 0.080 \cos(6x - 100t)$$

(a) Show that the resulting disturbance is also a transverse wave that may be described by

$$y = A \cos(6x - 100t + \delta)$$

(b) Determine the amplitude A and phase constant δ of part (a).

19.37 Two transverse traveling waves in a medium are described by

$$y_1 = A_1 \sin(kx - \omega t + \delta_1)$$

$$y_2 = A_2 \sin(kx - \omega t + \delta_2)$$

(a) Show that the resulting disturbance is a traveling wave of the same wavelength, frequency, and direction of propagation, that is, $y = y_1 + y_2 = A \sin(kx - \omega t + \delta)$.

(b) Determine expressions for the amplitude A and phase constant δ of part (a).

19.38 Show that the amplitude A and phase constant δ determined in Problem 19.37 correspond to the magnitude and direction of a vector A that is the sum of the two vectors, A_1 and A_2, as shown in Figure 19.28.

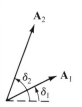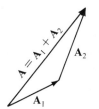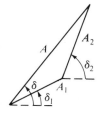

Figure 19.28   Problem 19.38

19.39 The two speakers, S_1 and S_2, shown in Figure 19.29 are in phase and emit 1.0-kHz signals. The distance d is equal to 1.2 m. Determine the values of x for which the signals from the speakers will be (a) in phase, (b) 180° out of phase.

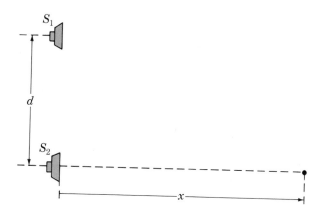

Figure 19.29   Problem 19.39

19.40 The two speakers, S_1 and S_2, shown in Figure 19.30 are in phase and emit 2.0-kHz signals. The distances d and D shown have the values $d = 0.60$ m and $D = 2.0$ m. Determine the values of y for which the signals from the speakers will be (a) in phase, (b) 180° out of phase.

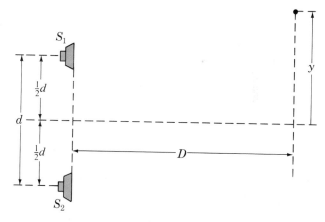

Figure 19.30   Problem 19.40

GROUP B

19.41 If N transverse traveling waves are described by

$$y_n = A_n \sin(kx - \omega t + \delta_n), \qquad n = 1, 2, \ldots, N$$

show that the resulting disturbance is described by

$$y = A \sin(kx - \omega t + \delta)$$

where

$$A = \sqrt{\left(\sum_{n=1}^{N} A_n \cos\delta_n\right)^2 + \left(\sum_{n=1}^{N} A_n \sin\delta_n\right)^2}$$

$$\tan\delta = \frac{\displaystyle\sum_{n=1}^{N} A_n \sin\delta_n}{\displaystyle\sum_{n=1}^{N} A_n \cos\delta_n}$$

19.42 Show that the amplitude A and phase constant δ of Problem 19.41 may be obtained from the sum of appropriate vectors. (*Hint:* See Problem 19.38.)

19.43 The *one-dimensional wave equation* may be derived by applying Newton's laws to a string on which a transverse disturbance is propagating in the x direction. To do so, consider the element of string shown in Figure 19.31, an element of length Δx and mass $\mu\Delta x$, where μ is the mass per unit length of the string. Assume that the tension F in the string remains constant and that each point on the string moves transversely (in the y direction) only.

(a) Use Newton's second law to show that

$$F[\sin\theta(x + \Delta x, t) - \sin\theta(x, t)] = \mu\Delta x\, a_y(x + \tfrac{\Delta x}{2}, t)$$

(b) Use the result of part (a) and show that

$$\cos\theta \frac{\partial\theta}{\partial x} = \frac{\mu}{F} a_y$$

(c) Because $(\partial y/\partial x) = \tan\theta$ (why?), show that

$$\frac{\partial\theta}{\partial x} = \cos^2\theta \frac{\partial^2 y}{\partial x^2}$$

(d) Show that

$$\cos^3\theta \frac{\partial^2 y}{\partial x^2} = \frac{\mu}{F} \frac{\partial^2 y}{\partial t^2}$$

(e) Finally, assume that for this disturbance $|\theta| \ll 1$ rad and show that

$$\frac{\partial^2 y}{\partial x^2} = \frac{\mu}{F} \frac{\partial^2 y}{\partial t^2} \qquad \text{(one-dimensional wave equation)}$$

19.44 Let $f(u)$ be any function for which $d^2 f/du^2$ exists and is continuous.

(a) Let $y(x, t) = f(x \pm vt)$. Show that (with $u = x \pm Vt$)

$$\frac{\partial y}{\partial x} = \frac{df}{du}$$

and

$$\frac{\partial y}{\partial t} = \pm v \frac{df}{du}$$

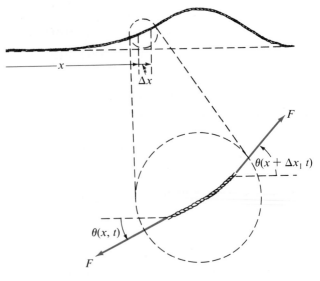

Figure 19.31   Problem 19.43

(b) Show that $y(x, t) = f(x \pm Vt)$ satisfies the one-dimensional wave equation of Problem 19.43 provided the speed of propagation v is given by

$$V = \sqrt{\frac{F}{\mu}}$$

19.45 If $y_1(x, t)$ and $y_2(x, t)$ are each solutions to the wave equation of Problem 19.43, show that

$$y(x, t) = y_1(x, t) + y_2(x, t)$$

is also a solution.

19.46 If $y = f(x, t)$ describes a transverse disturbance propagating on a string (mass per unit length $= \mu$, tension $= F$),

(a) show that the kinetic energy per unit length, that is, the *kinetic energy density* u_k at the point x and time t, is given by

$$u_k(x, t) = \frac{1}{2}\mu\left(\frac{\partial f}{\partial t}\right)^2$$

(b) show that the stretch of the string per unit length dl/dx at the point x and time t is given by

$$\frac{dl}{dx} = \sqrt{1 + \left(\frac{\partial f}{\partial x}\right)^2} - 1$$

(c) and if $\partial f/\partial x$ ($= \tan\theta$ where θ is the angle between a tangent to the string at (x, t) and the positive x direction) is $\ll 1$, then show that

$$\frac{dl}{dx} \approx \frac{1}{2}\left(\frac{\partial f}{\partial x}\right)^2$$

(d) and if the tension F is assumed to remain constant, show that the potential energy per unit length, that is, the *potential energy density* u_p, is given by

$$u_p = \frac{1}{2} F \left(\frac{\partial f}{\partial x} \right)^2$$

19.47 Consider a transverse disturbance $y = f(x, t)$ which propagates in either the positive or negative x direction, i.e., $f = f(x - Vt)$ or $f = f(x + Vt)$, on a string.

(a) Show that the kinetic energy density u_k and potential energy density u_p defined in Problem 19.46 are equal at each point x and time t, i.e.,

$$u_k(x, t) = u_p(x, t)$$

(b) If the *total energy density* u_E is defined as the sum of the kinetic and potential energy densities, $u_E = u_k + u_p$, show that, for a wave traveling in either the positive or negative x direction on a string,

$$u_E(x, t) = 2u_k(x, t) = 2u_p(x, t)$$

19.48

(a) If $y = A \sin k(x \pm Vt)$ describes a transverse wave on a string (mass per unit length $= \mu$, tension $= F$), show that

$$u_k = u_p = 2\pi^2 \mu \nu^2 A^2 \sin^2 k(x - Vt)$$

and

$$u_E = 4\pi^2 \mu \nu^2 A^2 \sin^2 k(x - Vt)$$

where ν is the frequency of the wave.

(b) Show that the total energy E that the sinusoidal traveling wave of part (a) propagates by the point x during one period is given by

$$E = 2\pi^2 \mu \lambda \nu^2 A^2 = \langle u_k \rangle \lambda = \langle u_p \rangle \lambda = \frac{1}{2} \langle u_E \rangle \lambda$$

where $\langle . . . \rangle$ indicates an average over one wavelength.

(c) Show that the average rate at which energy is transmitted, that is, the average power P_{av}, by the sinusoidal wave of part (a) is given by

$$P_{av} = 2\pi^2 \mu V \nu^2 A^2 = \langle u_k \rangle V = \langle u_p \rangle V = \frac{1}{2} \langle u_E \rangle V$$

19.49 Let

$$y = A \sin kx \sin \omega t$$

describe a (small amplitude) standing wave on a string (mass per unit length $= \mu$). Show that the total energy E "captured" between two adjacent nodes is given by

$$E = \frac{1}{4} \mu \lambda \omega^2 A^2$$

20 Light: Geometric Optics

Optics is the study of phenomena associated with light. In Chapter 18 we saw that light is an electromagnetic wave, and we know from everyday experience that waves can bend around the edges of obstructions, an occurrence called diffraction. For example, sound waves that are diffracted around the corners of buildings or hallways can be heard. Light waves may be diffracted, too; in Chapter 21 we will consider optical phenomena that depend expressly on the wave nature of light. In this chapter, however, we will use what is called the geometric approximation for the propagation of light: We will assume that light travels along straight-line paths and that light from a point source casts sharp shadows.

The geometric approximation in the analysis of optical systems permits us to take advantage of the concept of light rays. Figure 20.1 shows portions of the spherical wavefronts that move away radially from a point source S of light. *Light rays* are lines perpendicular to the wavefronts and are directed (as indicated by the arrowheads) at every point in the direction of propagation of the light.

Geometric optics is the study of how light rays are reflected and refracted, or bent, in passing from one medium to another. In geometric optics we use constructions of reflected and refracted light rays to analyze the formation of images (like the ones you see when you look at a mirror, for example). Our

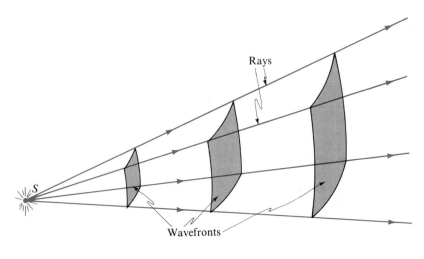

Figure 20.1
Portions of spherical wavefronts moving radially away from a point source S of light.

studies of optics will be mainly confined to the visible spectrum, that range of wavelengths of electromagnetic radiation that is visible to the human eye.

Visible light includes wavelengths between approximately 4×10^{-7}m and 7×10^{-7}m. The units of wavelength used in most optical measurements are the nanometer (1 nm $= 1 \times 10^{-9}$m) and the angstrom (1 Å $= 1 \times 10^{-10}$m). Therefore, we will consider the *visible spectrum* to include wavelengths between 400 nm and 700 nm or, equivalently, between 4000 Å and 7000 Å.

20.1 Fermat's Principle: The Law of Reflection

When light is incident upon a surface that is an interface between two media, some of the light is reflected and some is transmitted through the interface. Figure 20.2(a) shows (essentially parallel) rays of light from the sun incident on a *diffuse surface*, a surface with irregularities somewhat larger than the wavelength of the incident light. An eye looking at the reflected rays sees an image of the sun that is diffuse, or blurred. If the surface is very rough, the eye receives an unrecognizable image of the sun. In Figure 20.2(b), the light from the sun is incident on a *specular surface* having irregularities that are small compared to the wavelength of the light reflected from the surface. The image seen by the eye is sharp and clear. We usually call a specular surface a *mirror surface*, or simply a *mirror*.

Two principles form the theoretical basis of geometric optics. First, the principle of superposition for electromagnetic waves provides that the intersecting light rays do not alter the path of either ray. The second principle, called *Fermat's principle*, may be stated as follows:

In traveling from one point to another, light follows a path that requires the minimum time compared to the times that would be required along nearby paths.

Let us use Fermat's principle to establish the operational law governing reflections of light rays.

In Figure 20.3(a), points A and B are at vertical heights, a and b, above a horizontal plane mirror. Points A and B are separated by a horizontal distance d. A light ray traveling from A to B is reflected from the mirror at point P, which is in the vertical plane containing the points A and B and which is located at a horizontal distance x from point A. The angle θ_i is the *angle of incidence* of the ray; θ_r is the *angle of reflection*. Both angles are measured from the normal to the mirror at the point of reflection. The path length of the reflected ray from A to B

Figure 20.2
(a) Parallel rays of light reflected from a diffuse surface. (b) Parallel rays of light reflected from a specular, or mirror, surface.

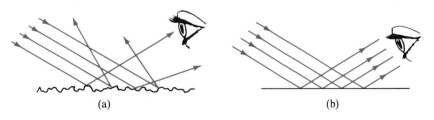

(a)   (b)

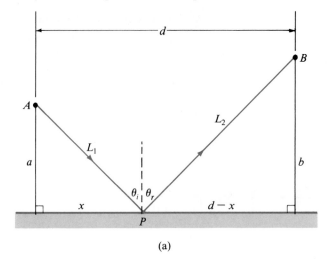

(a)

Figure 20.3
(a) Construction for the derivation of the law of reflection using Fermat's principle. (b) Construction demonstrating that a ray reflected at a plane interface lies in that plane containing the incident ray and the normal to the interface at the point P of reflection.

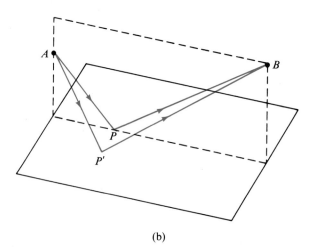

(b)

is equal to the sum, $L_1 + L_2$, and the time t of transit from A to B by way of the mirror is given by

$$t = \frac{L_1 + L_2}{c} \tag{20-1}$$

where c is the speed of light in the space (a vacuum, in this case) above the mirror. According to Fermat's principle, the ray follows the path having the minimum time of travel. The time of travel depends on where the ray strikes the mirror and, therefore, on x. Consequently, the time of travel is minimal when $dt/dx = 0$, or, using Equation (20-1),

$$\frac{dt}{dx} = \frac{1}{c} \left(\frac{dL_1}{dx} + \frac{dL_2}{dx} \right) = 0$$

or

$$\frac{dL_1}{dx} + \frac{dL_2}{dx} = 0 \tag{20-2}$$

But L_1 and L_2 are related to x by

$$L_1{}^2 = x^2 + a^2 \quad ; \quad L_2{}^2 = (d - x)^2 + b^2 \tag{20-3}$$

so dL_1/dx and dL_2/dx are found by taking derivatives of Equations (20-3):

$$2L_1 \frac{dL_1}{dx} = 2x \qquad\qquad 2L_2 \frac{dL_2}{dx} = -2(d - x)$$

$$\frac{dL_1}{dx} = \frac{x}{L_1} = \sin\theta_i \qquad\qquad \frac{dL_2}{dx} = -\frac{(d - x)}{L_2} = -\sin\theta_r \qquad (20\text{-}4)$$

Substituting Equations (20-4) into Equation (20-2) gives

$$\sin\theta_i = \sin\theta_r$$

$$\theta_i = \theta_r \qquad (20\text{-}5)$$

Here we have used Fermat's principle to obtain the *law of reflection*, Equation (20-5): *the angle of incidence θ_i is equal to the angle of reflection θ_r.*

Further, the law of reflection is extended to include the following experimental fact: When a light ray is reflected at an interface, the reflected ray lies in that plane containing the incident ray and the normal to the interface at the point of reflection. Figure 20.3(b) suggests that the path length $\overline{APB}$ is the shortest path only if P lies in the plane connecting A and B. Any other nearby path, like the path of length $\overline{AP'B}$, is greater in length than $\overline{APB}$.

E 20.1   Points A and B are positioned, as shown in the figure, relative to a flat reflecting surface.
(a) Calculate the propagation distance $\overline{APB}$ for the (reflected) light ray from A to B.
                                                            Answer: 60 cm
(b) Determine the time of propagation of a (reflected) light signal from A to B.
                                                            Answer: 2.0 ns
(c) What is the angle of reflection for the ray shown?   Answer: 37°

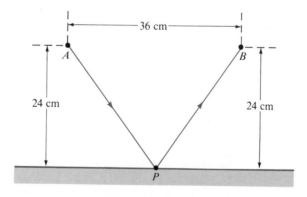

Exercise 20.1

Plane Mirrors

Using the law of reflection, Equation (20-5), we may discover the relationships between objects placed in front of a plane mirror and the corresponding images formed by light rays reflecting from the mirror. In Figure 20.4(a), point O is a *luminous object*, one from which light is either emitted or reflected. Point O is at a perpendicular distance p, the *object distance*, from the mirror. A small bundle of rays leaving the object O is shown reflecting from the mirror and entering the eye of an observer. Each ray of the bundle is constructed so that its angle of incidence at the mirror is equal to its angle of reflection. The rays diverging from

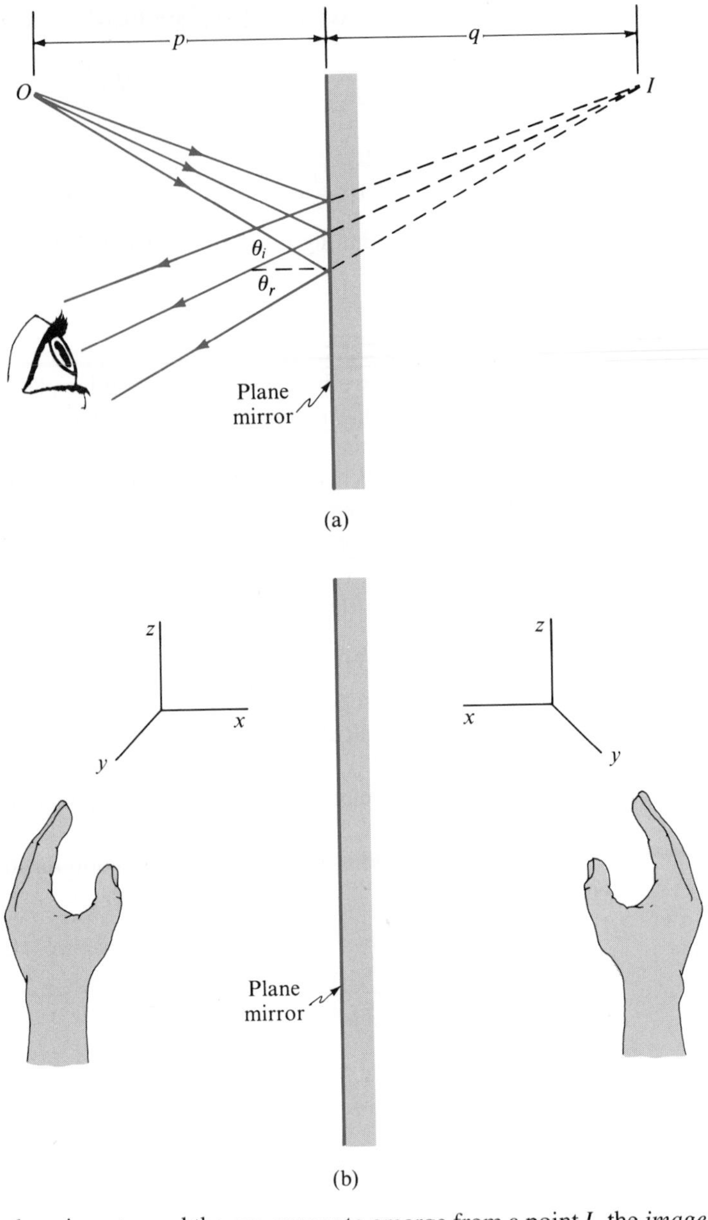

(a)

(b)

Figure 20.4
(a) A ray construction using the law of reflection, $\theta_i = \theta_r$, to show that the object distance p from an object O to a plane mirror is equal to the image distance q from the image I to the mirror. (b) The reflection of an extended object by a plane mirror. Left and right are interchanged by the reflection; a left-handed coordinate system is reflected as a right-handed system.

the mirror toward the eye appear to emerge from a point I, the *image* of the object O. From the geometry of the construction, we may see that the object distance p is equal to the *image distance q*. The image is the same distance behind the plane mirror as the object is in front of the mirror. The same construction may be carried out point by point on an extended object like the one shown in Figure 20.4(b). Notice that in a plane mirror the image is the same size as the object and that left and right are interchanged; a left-handed coordinate system becomes right-handed upon reflection in a plane mirror.

Optical images are characterized as either real or virtual. A *real image* is one for which the light rays that appear to come from the image actually pass through the image. Light rays diverge from, and appear to come from, a *virtual image* but do not actually emanate from that image. Thus images formed by diverging rays reflected from plane mirrors are always virtual images.

E 20.2   Each of the three figures shows a luminous arrow and a plane mirror. For each, sketch the image (of the arrow) formed by the mirror.

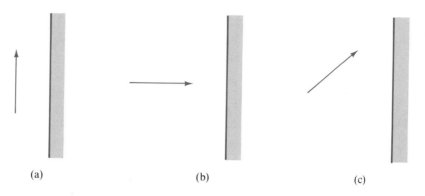

Exercise 20.2

(a)   (b)   (c)

Example 20.1

PROBLEM   A man, whose height is equal to 1.8 m, stands in front of a vertical, plane mirror. Determine the minimum vertical dimension of this mirror if he observes his whole image formed by the mirror.

SOLUTION   Figure 20.5(a) shows the man, the mirror, and the image formed by the mirror. The two limiting rays that determine the minimum vertical dimension for the mirror are shown in Figure 20.5(b). These two rays, one emanating from the man's highest point (A) and one emanating from his lowest point (B), must be reflected by the mirror into the person's eye (E) if he is to see his entire image. The law of reflection ensures that the reflecting point P_1 for the upper ray is (vertically) midway between points A and E. That is, if a is the vertical distance from A to E, then the vertical separation between points P_1 and E is equal to $\frac{1}{2}a$, as shown. Similarly, the reflecting point P_2 for the lower ray is vertically midway between E and B, as shown. Thus, the minimum vertical dimension of the mirror is equal to the distance $\overline{P_1P_2}$, which, according to Figure 20.5(b), is equal to $\frac{1}{2}(a + b)$. Because the man's height is equal to the length, $a + b$, the mirror must be at least 0.90 m in length. ■

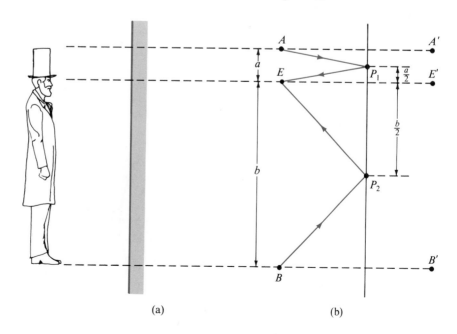

Figure 20.5   Example 20.1

(a)   (b)

Spherical Mirrors

Spherical reflecting surfaces are sometimes used as mirrors. Figure 20.6 shows (a) a *concave* spherical mirror and (b) a *convex* spherical mirror. These classifications are determined by the appearance from the point of view of an incident ray, so that a concave surface looks "caved in." In Figure 20.6, parallel rays are incident on each mirror. An *optical axis* of a mirror is a line perpendicular to the mirror through the center of curvature C of that mirror. The *vertex V* of a mirror is the point of intersection of the chosen optical axis and the surface of that mirror. When light rays parallel to the optical axis are incident on a concave mirror, like that of Figure 20.6(a), the reflected rays converge toward an image point called the *focal point F* of the mirror. Similarly, when rays parallel to the optical axis reflect from a convex mirror, as shown in Figure 20.6(b), the reflected rays diverge from the focal point F, which, in this case, lies behind the mirror.

Referring to Figure 20.7, we may use the law of reflection to relate the image distance q, the object distance p, and the radius R of a spherical mirror. A point object O is on the optical axis at a distance p from the vertex V of a spherical mirror. The mirror has a radius R and a center C. The ray traveling from O along the optical axis is reflected back along the axis, as shown. A second ray from O makes an angle α with the optical axis, strikes the mirror at point P, and reflects so that it passes through the point I on the axis. These two rays locate the image I (of the object O) at an image distance q from the vertex V of the mirror. Since point C is the center of the spherical surface of the mirror, the line CP is normal to the mirror. According to the law of reflection, we have $\theta_i = \theta_r = \theta$. Now consider the triangle CPI, where the exterior angle γ is equal to the sum of the opposite interior angles, β and θ:

$$\gamma = \beta + \theta \qquad (20\text{-}6)$$

Using the same geometric relationship for triangle OPC gives

$$\beta = \alpha + \theta \qquad (20\text{-}7)$$

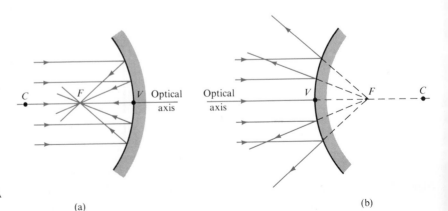

Figure 20.6
(a) A concave spherical mirror. **(b)** A convex spherical mirror.

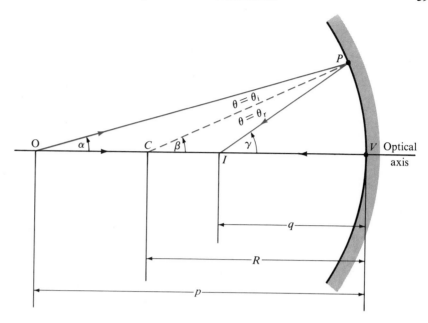

Figure 20.7
A ray construction using the law of reflection, $\theta_i = \theta_r$, to derive the relationship between object distance p, image distance q, and radius R of a concave spherical mirror.

Eliminating θ from Equations (20-6) and (20-7) leads to

$$\alpha + \gamma = 2\beta \qquad (20\text{-}8)$$

We will confine our considerations to *paraxial rays*, rays lying near the optical axis and making small angles with that axis. Then the angles α, β, and γ are small, that is, much less than one radian; and we may, therefore, use the following approximations:

$$\alpha \approx \tan\alpha \approx \frac{\overline{PV}}{p} \qquad (20\text{-}9)$$

$$\beta \approx \tan\beta \approx \frac{\overline{PV}}{R} \qquad (20\text{-}10)$$

$$\gamma \approx \tan\gamma \approx \frac{\overline{PV}}{q} \qquad (20\text{-}11)$$

E 20.4   Use your calculator to verify for $\alpha = 10°$ ($\pi/18$ radian), that

$$\frac{\tan\alpha}{\alpha} \approx 1.01$$

By choosing values of α even closer to zero, verify that, as α approaches zero, the quantity $(\tan\alpha/\alpha)$ approaches a limiting value of one.

Substituting Equations (20-9), (20-10), and (20-11) into Equation (20-8) and then dividing by $\overline{PV}$ yields

$$\frac{1}{p} + \frac{1}{q} = \frac{2}{R} \qquad (20\text{-}12)$$

Although Equation (20-12) was derived for an object distance greater than R, it applies to a concave mirror when the object is anywhere on the optical axis of the mirror. Equation (20-12) also applies to a convex mirror, and the derivation of Equation (20-12) may be reconstructed for that case using Figure 20.8. But if

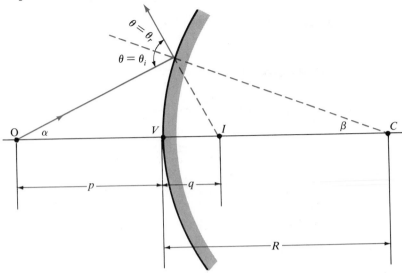

Figure 20.8
A ray construction using the law of reflection, $\theta_i = \theta_r$, for relating p, q, and R for a convex spherical mirror.

we are to use Equation (20-12) for all spherical mirrors, we must adopt suitable sign conventions for p, q, and R:

1. The object distance p is positive when (incident) rays from the object are diverging toward the mirror; p is negative when the incident rays are converging toward the mirror.

2. The image distance q is positive when the (reflected) rays from the mirror are converging toward the image; q is negative when the reflected rays are diverging from the image.

3. The radius R of the mirror is positive when the center of the mirror surface is on the same side of the mirror as are the reflected rays; R is negative when the center of the mirror surface is not on the same side as the reflected rays.

E 20.5   A point object O is positioned on the optical axis of a concave ($R = 5.0$ cm) mirror. The distance (p) from the object to the mirror is equal to 15 cm.
(a) What is the distance (q) from the mirror to the image I?   Answer: 3.0 cm
(b) Sketch this situation showing a few typical rays from O to I.
(c) Is the image I real or virtual?
                                                        Answer: Real

E 20.6   Repeat Exercise 20.5 for a convex ($R = -5.0$ cm) mirror.
                                    Answer: (a) -2.1 cm, (c) Virtual

We have seen that when light rays parallel to the optical axis are incident onto a spherical mirror, the image point is at the focal point F. We define the *focal length f* of a mirror to be the distance from the focal point to the vertex of that mirror. Because having incident light rays parallel to the optical axis corresponds to having an object at an infinite distance from the mirror, or $p \to \infty$, Equation (20-12) becomes, because here $f = q$,

$$\frac{1}{f} = \frac{2}{R}$$

$$f = \frac{R}{2} \qquad\qquad (20\text{-}13)$$

Then the spherical mirror equation may be written

$$\frac{1}{p} + \frac{1}{q} = \frac{1}{f} \tag{20-14}$$

where f has a positive value for a concave (convergent) mirror and a negative value for a convex (divergent) mirror.

E 20.7   What is the focal length of the mirror of
(a) Exercise 20.5?   Answer: $+2.5$ cm
(b) Exercise 20.6?   Answer: -2.5 cm

E 20.8   A spherical mirror forms a real image (8.0 cm from the mirror) of a real object (24 cm from the mirror). Determine the focal length and radius of this mirror.   Answer: $f = +6.0$ cm, $R = +12$ cm

E 20.9   Show that rays leaving from the focal point of a convergent mirror, i.e., $p = f$, are reflected parallel to the optical axis.

E 20.10   Show that rays converging toward the focal point of a divergent mirror are reflected parallel to the optical axis.

The image formed by a spherical mirror of an *extended object*, an object with finite extent, may be located graphically as well as by using Equation (20-14) or Equation (20-12). Figure 20.9 shows a luminous arrow (the traditional extended object in optics) in front of (a) a concave mirror and (b) a convex mirror. Two

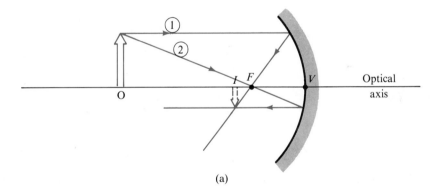

(a)

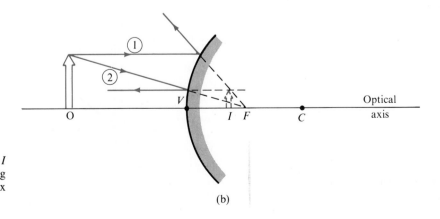

(b)

Figure 20.9
Construction rays locating the image I of an off-axis point object O using **(a)** a concave mirror and **(b)** a convex mirror.

construction rays are sufficient to locate the image of an off-axis point, like the arrow tip:

1. A ray (1) drawn from the arrow tip to the mirror parallel to the optical axis is reflected through (or in the case of a convex mirror, is extended from) the focal point.

2. A ray (2) drawn from the arrow tip through (or toward) the focal point is reflected parallel to the optical axis.

Example 20.2   **PROBLEM**   Show that graphical solutions obtained utilizing these two construction rays satisfy Equation (20-14), $1/p + 1/q = 1/f$, provided the (spherical) reflecting surface is drawn as a flat surface (oriented so that the optical axis is normal to the surface).

SOLUTION   Figure 20.10 shows the reflecting surface, its optical axis, and focal point F. Also shown is an off-axis point T on an extended object O. Ray 1 leaves T parallel to the optical axis and is reflected through F, as shown. Ray 2 leaves T, passes through F, and is reflected parallel to the axis. These two rays intersect at T', the image point of T. The object distance p, image distance q, and focal length f, are shown.

Since triangles TBA and FVA are similar (why?), we have

$$\frac{f}{p} = \frac{\overline{AV}}{\overline{AB}}$$

Similarly, using triangles $T'AB$ and FVB, we get

$$\frac{f}{q} = \frac{\overline{VB}}{\overline{AB}}$$

Combining these two equations by addition and using

$$\overline{AV} + \overline{VB} = \overline{AB}$$

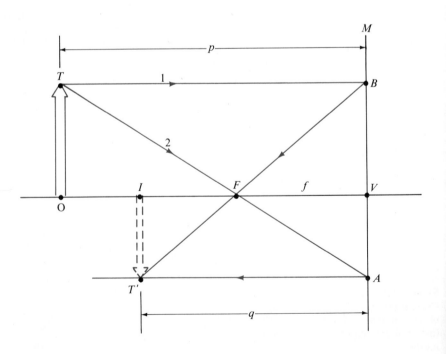

Figure 20.10   Example 20.2

gives

$$\frac{f}{p} + \frac{f}{q} = 1$$

or

$$\frac{1}{p} + \frac{1}{q} = \frac{1}{f}$$

What is this? How can a graphical solution that depicts a spherical surface as a flat surface yield an exact solution to the equation that describes the reflecting properties of that curved surface? Two comments may be helpful. First, Equation (20-14) was derived for paraxial rays, which necessarily strike the mirror near the vertex V in a region where the curved surface is "almost flat." (Remember that a calm ocean will appear almost flat to you unless a sufficiently large region is included in your view.) Secondly, the effect of the curvature of the surface on the reflection of paraxial rays is taken into account by requiring that the two construction rays utilize the properties of the focal point. ■

The relative sizes of corresponding images and objects may be determined from Figure 20.11, which shows an extended object and its corresponding image. A ray from T to the vertex V is reflected according to the law of reflection, so that $\theta_i = \theta_r = \theta$. The point T' is the image of point T. Then the similar triangles TVO and $T'VI$ relate the heights h of the object and h' of the image by

$$m = \frac{h'}{h} = -\frac{q}{p} \tag{20-15}$$

where m is the *lateral magnification* of the mirror. The *height* of a point on an object or an image is defined to be positive if that point lies above the optical axis and to be negative if that point lies below the optical axis. The negative sign in Equation (20-15) provides that a positive value of m indicates an *erect image*, that is, an image for which the image arrow points in the same direction as the object arrow, if p and q have opposite signs. A negative value of m indicates an *inverted image*, that is, an image for which the image arrow points oppositely to the direction of the object arrow, if p and q have the same sign.

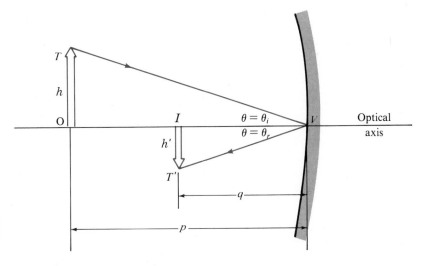

Figure 20.11

A construction using the law of reflection, $\theta_i = \theta_r$, to obtain a relationship between object distance p, image distance q, and lateral magnification $m = h'/h$ for a spherical mirror.

E 20.11   An extended object (lateral dimension = 1.5 cm) is 20 cm from a spherical mirror, which forms a real image of the object 8.0 cm from the mirror.

(a) What is the lateral magnification?   Answer: −0.40
(b) Determine the lateral size of the image.   Answer: 0.60 cm
(c) Is the image erect or inverted?   Answer: Inverted

Example 20.3

PROBLEM   Reflected rays from a spherical mirror ($f = -8.0$ cm) appear to emanate from an image that is 4.0 cm behind the mirror. Where is the object? Describe the image.

SOLUTION   Figure 20.12 depicts the problem. Because the image I is virtual, the image distance is negative, that is, $q = -4.0$ cm. The object distance p is obtained using

$$\frac{1}{p} + \frac{1}{q} = \frac{1}{f}$$

$$\frac{1}{p} = \frac{1}{f} - \frac{1}{q} = \frac{1}{-8} - \frac{1}{-4} = -\frac{1}{8} + \frac{1}{4} = \frac{1}{8}$$

$$p = 8.0 \text{ cm}$$

Thus the object is 8.0 cm from the mirror.

The lateral magnification by the mirror is given by

$$m = -\frac{q}{p} = -\frac{-4.0}{8.0} = +\frac{1}{2}$$

The image is, therefore, erect ($m > 0$) and has a height equal to one-half the height of the object.

A graphical solution to this problem is shown in Figure 20.13. Can you explain how rays 1 and 2 were constructed? ■

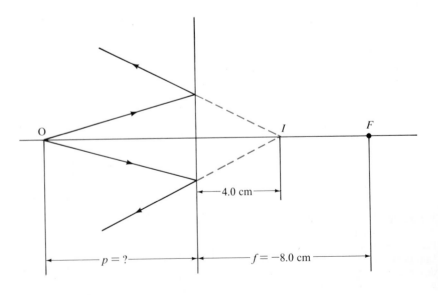

Figure 20.12   Example 20.3

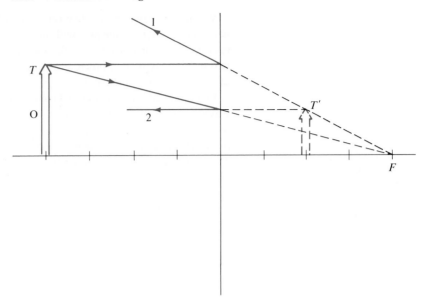

Figure 20.13   Example 20.3

20.2 Refraction of Light: The Law of Refraction

Light travels more slowly in a material medium than in free space. The ratio of the speed c of light in a vacuum to the speed v of light in any medium is called the *index of refraction n* of that medium, or

$$n = \frac{c}{v} \qquad\qquad (20\text{-}16)$$

Table 20.1 lists the indices of refraction of some common transparent materials for yellow light ($\lambda \simeq 600$ nm). Notice that for most purposes, we may use $n_{air} = 1$.

TABLE 20.1   **Typical Indices of Refraction of Some Transparent Materials**

Gases	
Air	1.0003
Carbon dioxide	1.0005
Liquids	
Carbon disulfide	1.64
Carbon tetrachloride	1.46
Ethyl alcohol	1.35
Water	1.33
Solids	
Diamond	2.42
Glass	1.50–1.66
Quartz	1.46

When light rays are incident on an interface between media having different indices of refraction, the rays will, in general, undergo a change in direction. The rays are bent, or *refracted*, at the interface. Refraction at an interface between two media may be interpreted physically in terms of the difference in the speeds of light in the two media.

In Figure 20.14 two rays, labeled 1 and 2, are perpendicular to successive wavefronts incident on an interface between two media. Suppose the upper medium has an index of refraction n_1, which is less than the index of refraction n_2 of the lower medium. Then the light travels more slowly in the lower medium. The effect of the wavefronts reaching the interface is much like lines of marchers reaching an interface between a concrete parade ground and a muddy field. The marchers are slowed as they pass into the mud because they take shorter steps while maintaining the same cadence, and the direction of march of the line is turned. After a time interval Δt, the wavefront AB has moved to become wavefront CD. During the interval Δt, ray 1 has traveled a distance $\overline{AC}$ at speed $v_1 = c/n_1$, and ray 2 has traveled a distance $\overline{BD}$ at speed $v_2 = c/n_2$. Then we may write for ray 1,

$$\Delta t = \frac{\overline{AC}}{v_1} = \frac{n_1}{c}\,\overline{AC} \tag{20-17}$$

and for ray 2,

$$\Delta t = \frac{\overline{BD}}{v_2} = \frac{n_2}{c}\,\overline{BD} \tag{20-18}$$

Since Δt is the same for each ray, we may equate Equations (20-17) and (20-18), which gives

$$n_1\overline{AC} = n_2\overline{BD} \tag{20-19}$$

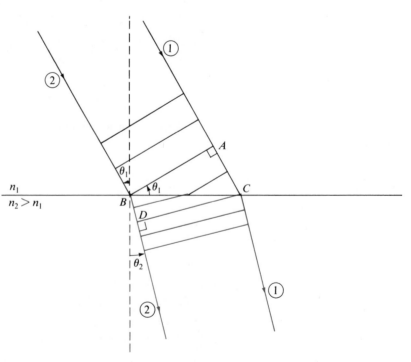

Figure 20.14
Wavefronts of light and their associated rays. These rays are incident at an angle θ_1 onto an interface between media with indices of refraction n_1 and $n_2 > n_1$. The refracted rays make an angle θ_2 with the normal to the interface. This construction permits the derivation of Snell's Law, $n_1 \sin\theta_1 = n_2 \sin\theta_2$.

And since $\overline{AC} = \overline{BC} \sin\theta_1$ and $\overline{BD} = \overline{BC} = \sin\theta_2$, Equation (20-19) becomes

$$n_1 \sin\theta_1 = n_2 \sin\theta_2 \qquad\qquad (20\text{-}20)$$

Equation (20-20) is called *Snell's law*. It relates the angle of incidence θ_1, the angle of refraction θ_2, and the indices of refraction of the two media.

The law of refraction includes both Snell's law, Equation (20-20), and the following experimental fact: When a light ray is refracted at the interface between two media, the refracted ray lies in that plane containing the incident ray and the normal to the interface at the point of refraction.

E 20.12   Light is refracted at an air-glass interface. If the angles of incidence and refraction are equal to 65° and 37°, respectively, determine the index of refraction of the glass.   Answer: 1.5

E 20.13   The angle of incidence (in air) for a light ray at an interface between air and a transparent plastic ($n = 1.80$) is equal to 45°. Calculate the angle of refraction.   Answer: 23°

Example 20.4 shows that the law of refraction, like the law of reflection, is a consequence of Fermat's principle.

Example 20.4   **PROBLEM**   Use Fermat's principle to obtain Snell's law.

SOLUTION   Consider a light ray passing from point A in medium 1 ($n = n_1$) to point B in medium 2 ($n = n_2$) through a horizontal interface between the media as shown in Figure 20.15. Points A and B are separated by a horizontal distance d. Let A be at a vertical distance a above the interface and B be at a vertical distance b below the interface. Let the point of refraction P be at a horizontal distance x from point A; then P is a horizontal distance $d - x$ from point B. Let L_1 be the path length of the ray in medium 1, where the speed of light is v_1, and L_2 be its path length in medium 2, where the speed of light is v_2. Then the total transit time t of the ray from A to B is given by

$$t = \frac{L_1}{v_1} + \frac{L_2}{v_2} = \frac{1}{c}(n_1 L_1 + n_2 L_2)$$

According to Fermat's principle, the light path will be that for which t is minimal, or $dt/dx = 0$:

$$\frac{dt}{dx} = n_1 \frac{dL_1}{dx} + n_2 \frac{dL_2}{dx} = 0$$

Now L_1 and L_2 are related to x by

$$L_1{}^2 = x^2 + a^2 \quad ; \quad L_2{}^2 = (d - x)^2 + b^2$$

so

$$2L_1 \frac{dL_1}{dx} = 2x \qquad\qquad 2L_2 \frac{dL_2}{dx} = -2(d - x)$$

$$\frac{dL_1}{dx} = \frac{x}{L_1} = \sin\theta_1 \qquad\qquad \frac{dL_2}{dx} = -\frac{(d - x)}{L_2} = -\sin\theta_2$$

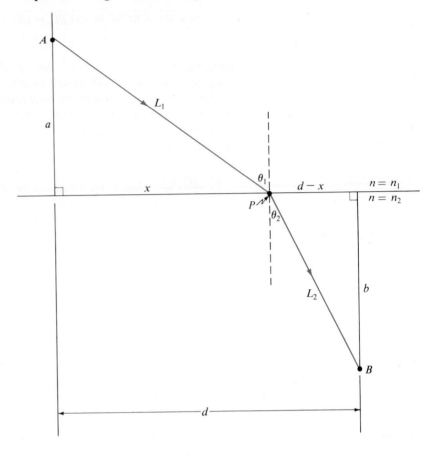

Figure 20.15   Example 20.4

Then because $dt/dx = 0$, it follows that

$$n_1 \sin\theta_1 = n_2 \sin\theta_2 \qquad \text{(Snell's law)}$$

where θ_1 and θ_2 are the angles of incidence and refraction, respectively.   ■

Total Internal Reflection

When light is incident on an interface between two media, part of the light energy is, in general, reflected back into the medium of the incident ray and part is refracted and transmitted into the other medium. Figure 20.16(a) shows a ray

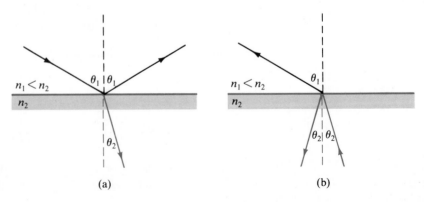

Figure 20.16
Illustrating that the path of a light ray refracted at an interface between media is reversible.

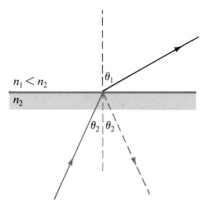

Figure 20.17
A light ray refracted at an interface between media having indices of refraction n_1 and $n_2 > n_1$.

being reflected and refracted at an interface between media. Experiment shows that the path of a refracted ray is *reversible*: If, as in Figure 20.16(b), the ray in the lower medium is reversed so that it is incident on the interface at the same angle θ_2 as in Figure 20.16(a), that ray is refracted at the angle θ_1. In both Figures 20.16(a) and 20.16(b), the reflected rays obey the law of reflection.

Let us now consider certain conditions under which Snell's law *cannot* apply. Suppose, as in Figure 20.17, a light ray is refracted at an interface separating media having indices of refraction n_1 and $n_2 > n_1$. If θ_2 is the angle of incidence and θ_1 is the angle of refraction, we may write Snell's law as

$$\sin\theta_1 = \frac{n_2}{n_1}\sin\theta_2 \qquad (20\text{-}21)$$

Because the sine of an angle must be equal to or less than 1, Equation (20-21) cannot be satisfied if its right-hand side has a value greater than 1. When θ_2 has the value θ_c, called the *critical angle*, the angle of refraction θ_1 is equal to 90° so that $\sin\theta_1 = 1$. Then the angle of incidence θ_2 is critical when

$$1 = \frac{n_2}{n_1}\sin\theta_c$$

$$\theta_c = \sin^{-1}\left(\frac{n_1}{n_2}\right) \quad ; \quad n_1 < n_2 \qquad (20\text{-}22)$$

Figure 20.18 illustrates that if the angle of incidence is greater than the critical angle θ_c, Snell's law is not applicable. Ray 1 is incident on the interface with $\theta_2 < \theta_c$. This ray is refracted at an angle θ_1 and is partially reflected. Ray 2 strikes the interface at the critical angle θ_c and is refracted at 90°. Ray 3 meets the interface at an angle $\theta_2 > \theta_c$, and this ray does not refract into the upper medium; it is completely reflected at the interface. When a ray is incident on an interface between media at an angle greater than the critical angle, the ray is said to undergo *total internal reflection*. Of course, total internal reflection occurs only when the incident ray is in the medium of higher index of refraction.

E 20.14   Calculate the critical angle for light refracted at an interface separating water ($n = 1.33$) and glass ($n = 1.60$). If total internal reflection is to occur, which medium must be the medium of incidence?   Answer: 56°, glass

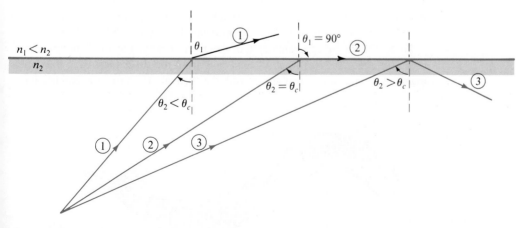

Figure 20.18
Illustrating the condition for total internal reflection.

An important practical application of total internal reflection is the *light pipe* used in the modern technological area of fiber optics. A transparent material with a high index of refraction may be used to conduct light through flexible pathways by forming the material into a cylindrical filament like that shown in Figure 20.19. Light enters the filament nearly parallel to the cylinder axis and subsequently strikes the interface between the filament and air at angles of incidence greater than the critical angle if the filament is not bent too sharply. In the figure, the angles of incidence, θ_1, θ_2, and θ_3, are all greater than the critical angle.

Refraction at Plane Surfaces

Let us illustrate some typical uses of the law of refraction at plane surfaces by considering two problems.

A scuba diver 20 m beneath the smooth surface of a clear lake looks upward and judges the sun to be 40° from directly overhead. At the same time, a fisherman is in a bass boat directly above the diver. (a) At what angle from the vertical would the fisherman measure the sun? (b) If the fisherman looks downward, at what depth below the surface would he judge the diver to be?

The physical situation of this problem is suggested by Figure 20.20. A light ray from the sun is incident on the air-water interface at an angle of incidence θ_1. That ray, after being refracted at the surface of the water, passes through the water to the diver at an angle $\theta_2 = 40°$ from the vertical, which is normal to the air-water interface. Using $n_1 = 1$ for air and $n_2 = 4/3$ for water, we may apply Snell's law, Equation (20-20), to this ray:

$$n_1 \sin\theta_1 = n_2 \sin\theta_2$$

$$(1)\ \sin\theta_1 = \left(\frac{4}{3}\right) \sin 40°$$

$$\sin\theta_1 = \frac{4}{3}(0.643) = 0.857$$

$$\theta_1 = 59° \tag{20-23}$$

Thus the ray from the sun is passing through the air at 59° from the vertical. Because the rays from the sun are virtually parallel anywhere immediately above the lake surface, the fisherman would measure the sun to be 59° from the vertical.

To determine the apparent depth of the diver as judged by the fisherman let us note (*see* Fig. 20.21(a)) that all light rays from the diver that reach the fisherman's eye make very small angles with the vertical. Then for convenience let us

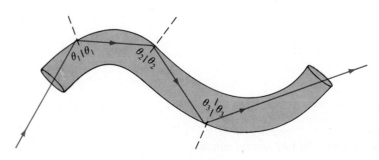

Figure 20.19
Total internal reflection in a light pipe.

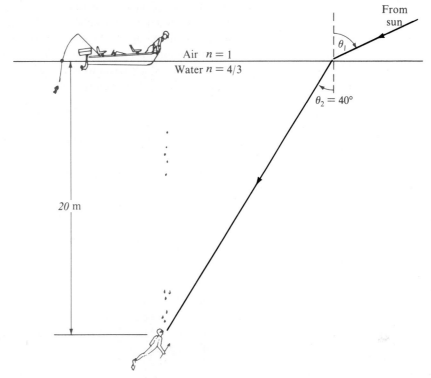

Figure 20.20
A light ray from the sun refracted at the surface of a lake toward a scuba diver beneath the surface. The figure illustrates a problem in the text.

reconstruct the situation in Figure 20.21(b) with the angles from the vertical greatly exaggerated (but let us remember that they are actually extremely small angles). The diver D is 20 m below the surface of the lake. A ray labeled 1 leaves the diver at an angle θ_2 and reaches the surface at point P, where the angle of incidence is θ_2. Ray 1, after being refracted, makes an angle θ_1 with the vertical. Ray 2 is vertical and strikes the lake surface normally. Snell's law requires that a

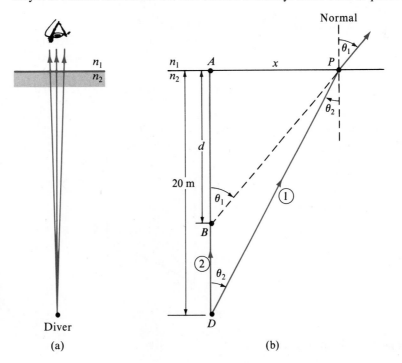

Figure 20.21
Illustrating a problem in the text. **(a)** Light rays from a diver beneath the lake surface making very small angles with the vertical are being intercepted by the eye of the fisherman directly above the diver. **(b)** Two rays from the diver are shown with exaggerated angles to facilitate analysis of the problem.

ray striking an interface normally is transmitted undeviated, regardless of the indices of refraction on either side of the interface; and ray 2 enters the air vertically. Together rays 1 and 2 in air form an image of the diver at point B, which is a distance d below the surface. The angle that the extension of ray 1 makes with ray 2 is equal to θ_1 (because BP is a transversal intersecting parallel lines). Applying Snell's law to ray 1, we obtain

$$n_1 \sin\theta_1 = n_2 \sin\theta_2$$

$$(1)\ \sin\theta_1 = \left(\frac{4}{3}\right)\sin\theta_2 \tag{20-24}$$

Now let x represent the distance $\overline{AP}$. Using triangle DAP, we find

$$\tan\theta_2 = \frac{x}{20} \simeq \sin\theta_2 \tag{20-25}$$

and using triangle BAP, we have

$$\tan\theta_1 = \frac{x}{d} \simeq \sin\theta_1 \tag{20-26}$$

In Equations (20-25) and (20-26) we have used the fact that for small angles (remember?), $\sin\theta \simeq \tan\theta$, an excellent approximation in this situation. Substituting the values of $\sin\theta_2$ and $\sin\theta_1$ from Equations (20-25) and (20-26) into Equation (20-24) gives

$$\frac{x}{d} = \frac{4}{3}\left(\frac{x}{20}\right)$$

$$d = 15\ \text{m} \tag{20-27}$$

Because of refraction of light at the surface of the lake, then, the diver appears to the fisherman to be only 15 m below the surface, three-fourths of his actual depth.

Example 20.5   **PROBLEM**   Light, initially propagating in air, is incident upon a slab (thickness $= t$) of refracting material ($n > 1$), as shown in Figure 20.22(a). After being refracted at two interfaces, the direction of propagation is the same as the initial direction, but the ray is displaced a distance d, as shown. Determine an expression for d in terms of t, n, and θ (the initial angle of incidence).

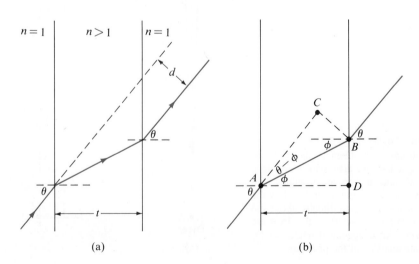

Figure 20.22   Example 20.5

(a)   (b)

SOLUTION   Figure 20.22(b) shows the details of the first refraction at point A and the second refraction at point B. We see, using triangle ABC, that the displacement d is given by

$$d = \overline{AB}\ \sin(\theta - \phi)$$

and from triangle ABD, we get

$$\overline{AB} = \frac{t}{\cos\phi}$$

Thus, the displacement of the ray may be expressed

$$d = t\,\frac{\sin(\theta - \phi)}{\cos\phi}$$

This relationship may be simplified using the trigonometric identity

$$\sin(\theta - \phi) = \sin\theta\,\cos\phi - \cos\theta\,\sin\phi$$

to get

$$d = t(\sin\theta - \cos\theta\,\tan\phi)$$

Now, using Snell's law to relate θ and ϕ, gives

$$\sin\phi = \frac{\sin\theta}{n}$$

$$\cos\phi = \sqrt{1 - \sin^2\phi} = \sqrt{1 - \frac{\sin^2\theta}{n^2}}$$

$$\tan\phi = \frac{\sin\phi}{\cos\phi} = \frac{\sin\theta}{\sqrt{n^2 - \sin^2\theta}}$$

Substituting this result into the expression for the displacement gives

$$d = t\,\sin\theta \left(1 - \frac{\cos\theta}{\sqrt{n^2 - \sin^2\theta}}\right)$$

which is the desired result.   ■

E 20.15   For a given angle of incidence in Example 20.5, it is reasonable to expect that the displacement d of the light ray would approach zero as the thickness t of the refracting slab approaches zero. Show that the result obtained in Example 20.5 supports this premise.

Now consider another problem involving refraction at plane surfaces.

An optical prism, surrounded by air and having an index of refraction of 1.5, has an apex angle of 40°. An incident light ray strikes one face of the prism at an angle of incidence of 45°. Calculate the total deviation of the ray in passing through the prism.

First, let us examine the terminology used in the problem. An *optical prism*, or simply a *prism*, shown in Figure 20.23, is a transparent five-sided solid. It has a rectangular *base* and two sides that are isosceles triangles oriented perpendicular to the base. The *optical faces* are rectangles inclined to each other at the *apex angle A*. In general, a light ray is deviated from its original path each

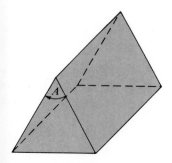

Figure 20.23
An optical prism.

time it passes through (or is reflected from) a prism face. Figure 20.24(a) shows the deviation δ of a ray caused by refraction. The deviation is equal to the magnitude of the difference between θ_1 and θ_2, the angles of incidence and refraction. Figure 20.24(b) shows that the deviation δ caused by a reflection is equal to $180° - (\theta_1 + \theta_2)$, where θ_1 and θ_2 are the angles of incidence and reflection.

Figure 20.25 shows the tentative problem situation. We say tentative because, although we are certain that the ray incident on the first prism face will pass from the $n_1 = 1$ medium to the $n_2 = 1.5$ medium, we must test for total internal reflection at the second face. In other words, unless we are sure that ϕ_2 is less than the critical angle, we cannot be sure the ray exits the prism as indicated in the figure. Applying Snell's law to the first prism face gives

$$(1)\ \sin 45° = 1.5\ \sin\theta_2$$

$$\sin\theta_2 = \frac{0.707}{1.5} = 0.470$$

$$\theta_2 = 28° \tag{20-28}$$

Angle α in Figure 20.25 is equal to the prism angle (40°), because α is formed by normals to the prism faces, which are inclined to each other at the prism angle. Then the angle β is equal to the supplement of α, or $\beta = 140°$. The sum of the angles in triangle ABC is equal to 180°, so $\phi_2 = 180° - (140° + 28°) = 12°$. Now we compare ϕ_2 to the critical angle ϕ_c at the second prism face:

$$\phi_c = \sin^{-1}\left(\frac{n_1}{n_2}\right) = \sin^{-1}(0.667) = 42° \tag{20-29}$$

Because $\phi_2 = 12°$ is less than $\phi_c = 42°$, we may use Snell's law at the second prism face:

$$1.5\ \sin 12° = (1)\ \sin\phi_1$$

$$\sin\phi_1 = (1.5)(0.208) = 0.312$$

$$\phi_1 = 18° \tag{20-30}$$

The deviation δ_1 of the ray at the first face is equal to $\theta_1 - \theta_2 = 45° - 28° = 17°$. The deviation δ_2 at the second face is equal to $\phi_1 - \phi_2 = 18° - 12° = 6°$. Then

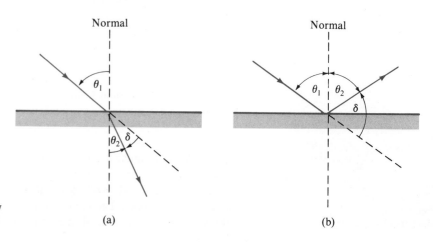

Figure 20.24
The deviation δ of **(a)** a refracted ray and **(b)** a reflected ray.

(a)   (b)

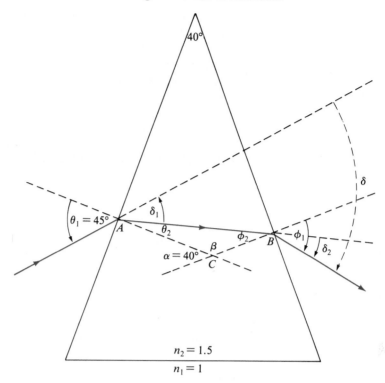

Figure 20.25
Diagram of a light ray passing through an optical prism. The particular values of the quantities shown refer to a problem in the text.

the total deviation δ, shown in Figure 20.25, is the sum of deviations at each face, or

$$\delta = \delta_1 + \delta_2 = 17° + 6° = 23°$$ (20-31)

Optical prisms are often used to demonstrate dependence of the index of refraction for a given medium on the wavelength of light being refracted by that medium, a phenomenon called *dispersion*. A typical optical glass, for example, has an index of refraction n that varies continuously from about 1.69 for violet light ($\lambda = 400$ nm) to about 1.66 for red light ($\lambda = 700$ nm). When white light, which contains all wavelengths of light within the visible spectrum, is incident on a prism, each wavelength (color) is deviated by a different amount. Thus, a thin beam of white light incident on a prism, as shown in Figure 20.26, produces a continuous spectrum of colors ranging from red to violet.

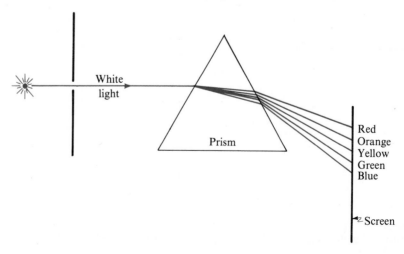

Figure 20.26
An optical prism dispersing white light into its component colors.

Dispersion of light manifests itself other than in prisms. The formation of rainbows, for example, results from the dispersion of white sunlight by droplets of water suspended in air. And dispersion of sunlight by the atmosphere itself contributes to the reddish colors of sunsets. At dusk, the white sunlight passes through a long atmospheric path, and the shorter wavelengths are deviated more than the longer ones. The bluer colors are bent into the earth, leaving the redder colors visible.

Refraction at Spherical Surfaces

Suppose in Figure 20.27 that an object O is placed at a distance p from the vertex V of a convex spherical interface of radius R. The medium containing the object has an index of refraction n_1; the other side of the interface is a medium having an index of refraction $n_2 > n_1$. Ray 1 (from O) that passes through the center of curvature C of the spherical surface forms an optical axis. Ray 1 crosses the interface at V without being deviated. A paraxial ray, ray 2, makes a small angle α with the optical axis. Ray 2 is refracted at the interface and crosses the axis at I, which is the image of O, formed by rays 1 and 2. Let q be the distance from the vertex V to the image I. For paraxial rays, the angles, θ_1, θ_2, α, β, and γ, are small and may be approximated by their sines or tangents. Then ray 2 is refracted at the interface according to Snell's law, $n_1 \sin\theta_1 = n_2 \sin\theta_2$, which may be approximated by

$$n_1\theta_1 \simeq n_2\theta_2 \qquad (20\text{-}32)$$

The geometry of Figure 20.27 shows that

$$\theta_1 = \alpha + \beta \qquad (20\text{-}33)$$

$$\theta_2 = \beta - \gamma \qquad (20\text{-}34)$$

Elimination of θ_2 from Equations (20-32) and (20-34) gives

$$\frac{n_1}{n_2}\theta_1 = \beta - \gamma$$

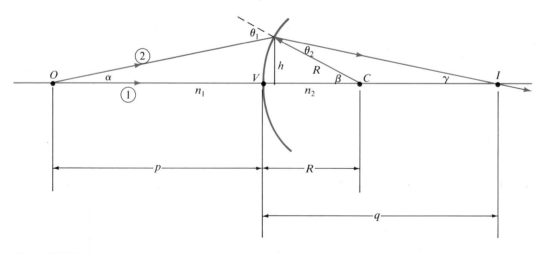

Figure 20.27
Diagram used in the analysis of refraction at a spherical surface between media having indices of refraction n_1 and n_2.

which, when combined with Equation (20-33) leads to

$$n_1\alpha_1 + n_2\gamma = (n_2 - n_1)\beta \qquad (20\text{-}35)$$

Referring to Figure 20.27 and again using the small angle approximations, we may express α, β, and γ in radian measure as

$$\alpha \simeq \frac{h}{p} \quad ; \quad \beta \simeq \frac{h}{R} \quad ; \quad \gamma \simeq \frac{h}{q} \qquad (20\text{-}36)$$

Substitution of Equations (20-36) into Equation (20-35) gives

$$\frac{n_1}{p} + \frac{n_2}{q} = \frac{n_2 - n_1}{R} \qquad (20\text{-}37)$$

Equation (20-37) applies to refraction at any spherical surface. But in order that this equation apply to real and virtual objects, to real and virtual images, and to convex and concave spherical surfaces, we must adopt a set of conventions:

1. The medium in which the rays from the object are impinging on the interface is taken to be the medium with index of refraction n_1.

2. The object distance p is positive if rays that impinge upon the interface are diverging.

3. The image distance q is positive if rays that emerge from the interface are converging.

4. The radius R is positive if the center of the spherical interface is on the same side as the rays that have been refracted at the interface.

The sign conventions (2, 3, and 4 above) are consistent with those we used for spherical mirrors.

Example 20.6

PROBLEM   A point source of light is placed 12 cm from the surface of a glass ($n = 1.5$) sphere (radius = 3.0 cm). Where are paraxial rays from the source focused by the sphere?

SOLUTION   Figure 20.28(a) depicts the given situation. Rays from the source O are initially refracted by the convex surface. For this first refraction, we have $p_1 = +12$ cm and $R_1 = +3.0$ cm. Using Equation (20-37), we get

$$\frac{1.0}{12} + \frac{1.5}{q_1} = \frac{1.5 - 1.0}{3}$$

where q_1 is the image distance for the first refraction. A solution for q_1 gives

$$q_1 = +18 \text{ cm}$$

After the first refraction, paraxial rays would form a real image I_1 at a distance of 18 cm beyond the first refracting surface, as shown in Figure 20.28(b). Before this image is formed, however, a second refraction occurs at the other (concave) surface of the sphere. For this refraction, the incident rays are converging, that is, $p_2 = -12$ cm, onto a concave surface so that $R_2 = -3.0$ cm. Using Equation (20-37) with $n_1 = 1.5$ and $n_2 = 1.0$, we get

$$\frac{1.5}{-12} + \frac{1.0}{q_2} = \frac{1.0 - 1.5}{-3}$$

Solving for the image distance q_2 gives

$$q_2 = 3.4 \text{ cm}$$

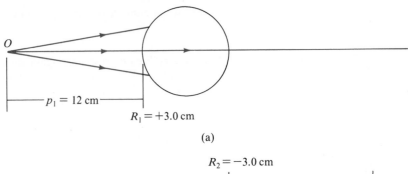

$p_1 = 12$ cm

$R_1 = +3.0$ cm

(a)

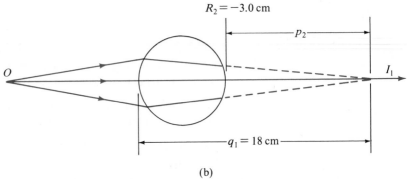

$R_2 = -3.0$ cm

p_2

O

I_1

$q_1 = 18$ cm

(b)

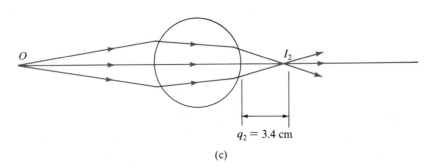

O

I_2

$q_2 = 3.4$ cm

(c)

Figure 20.28   Example 20.6

Consequently, paraxial rays emerge from the sphere and form a real image I_2 at a distance of 3.4 cm beyond the second surface, as shown in Figure 20.28(c). ■

E 20.16   One end of a long glass ($n = 1.5$) rod is polished to a convex spherical ($R = 8.0$ cm) shape, as shown in the figure. Light rays parallel to the optical axis are incident (in air) onto the surface, as shown. Where are these rays focused?   Answer: On the optical axis, 24 cm to the right of the vertex

E 20.17   For the glass rod of Exercise 20.16, where must a point object be placed so that the refracted rays (in the glass) are parallel to the optical axis?
                    Answer: On the optical axis, 16 cm to the left of the vertex

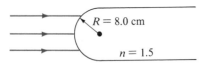

$R = 8.0$ cm

$n = 1.5$

Exercise 20.16

20.3 Thin Lenses

An *optical lens* is a device formed from a transparent material, usually glass, that is designed to converge or diverge light rays. A lens has one or more curved surfaces. We will consider simple lenses having two surfaces, each of which is a portion of a sphere, as shown in Figure 20.29. A line passing through the centers

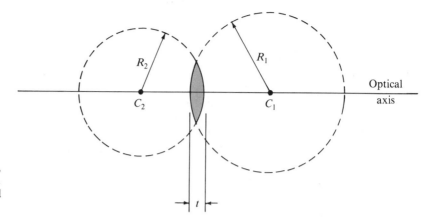

Figure 20.29
An optical lens defined by two surfaces, each of which is a portion of a sphere. The spheres have radii R_1 and R_2, and the lens has thickness t.

of the two spheres is called the *optical axis* of the lens. We will concentrate on thin lenses designed for use in air. A *thin lens* is one with a thickness t that is negligibly small compared to the radius, R_1 or R_2, of either of the spherical surfaces.

We may relate the optical properties of a thin lens in air ($n_{air} = 1$) to the geometry and index of refraction n of the lens by using Equation (20-37) to describe refractions at each surface of the lens. Figure 20.30 shows an arbitrary ray from an object O on the optical axis of a thin lens impinging on the first surface, which has radius R_1. The object distance to the first surface is p. That ray is refracted at the first surface toward an image I_1, located at a distance q_1 from the first surface. Applying Equation (20-37) to this refraction gives

$$\frac{1}{p} + \frac{n}{q_1} = \frac{n-1}{R_1} \tag{20-38}$$

We now use the image I_1, formed by the first surface, as the object of the refraction at the second surface with radius R_2. The object distance p_2 for the refraction at the second surface is, therefore, equal to $-(q_1 - t)$, a negative quantity because the rays (the arbitrary ray and the one along the optical axis) impinging on the second surface are converging toward I_1. Because the lens is thin, we may neglect the lens thickness t. Then applying Equation (20-37) to refraction at the second surface yields

$$\frac{n}{-q_1} + \frac{1}{q} = \frac{1-n}{R_2} \tag{20-39}$$

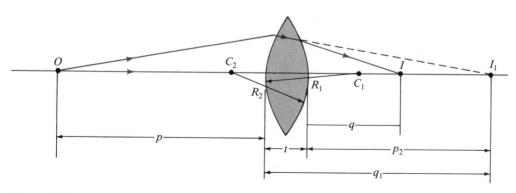

Figure 20.30
Diagram used in deriving the lens-maker's equation.

where q is the distance from the second surface to the final image I formed by the lens. Elimination of q_1 from Equations (20-38) and (20-39) results in

$$\frac{1}{p} + \frac{1}{q} = (n - 1)\left(\frac{1}{R_1} - \frac{1}{R_2}\right) \tag{20-40}$$

We have derived Equation (20-40) for a particular lens for which R_1 is positive and R_2 is negative. This equation, however, relates object distance p, image distance q, index of refraction n of the lens material, and the spherical radii R_1 and R_2 for any thin lens in air—provided we apply our sign conventions to p, q, R_1, and R_2.

Let us now define the *focal length f* of a thin lens to be equal to the image distance q that results when the incident rays are parallel to the optical axis, that is, when $p \rightarrow \infty$. Thus, letting $p \rightarrow \infty$ and $q = f$ in Equation (20-40), gives

$$\frac{1}{f} = (n - 1)\left(\frac{1}{R_1} - \frac{1}{R_2}\right) \tag{20-41}$$

Equation (20-41) is called the *lens-maker's equation* because it specifies how a lens may be formed from a material having an index of refraction n by grinding the surfaces to radii R_1 and R_2 to produce a lens with a given focal length f. Lenses are characterized as either *converging lenses*, for which the focal lengths f have positive values, or *diverging lenses*, which have negative focal lengths. Figure 20.31(a) shows rays parallel to the optical axis incident on a converging lens. The lens converges the rays to the *focal point F* of the lens. Figure 20.31(b) shows a diverging lens that causes incident rays parallel to the optical axis to diverge away from the focal point F of that lens.

Combining Equations (20-40) and (20-41) gives

$$\frac{1}{p} + \frac{1}{q} = \frac{1}{f} \tag{20-42}$$

Equation (20-42) is called the *thin-lens equation* and is identical to Equation (20-14), the spherical mirror equation.

The following example demonstrates that thin lenses have a second focal point that will be useful in obtaining graphical solutions.

Example 20.7   **PROBLEM**   Show that if the rays incident onto a thin lens form an object at a distance p that is equal to the focal length f of the lens, then the emerging rays will be parallel.

SOLUTION   Letting $p = f$ in the thin-lens equation, Equation (20-42), gives

$$\frac{1}{f} + \frac{1}{q} = \frac{1}{f}$$

or

$$\frac{1}{q} = 0$$

Thus, the image is at infinity, that is, the emerging rays are parallel. Figure 20.32(a) illustrates this situation for a converging lens. For such a lens, the focal length f is positive; the object distance, $p = f$, is positive; and the incident rays, therefore, diverge from a real object at the point F', as shown. For a diverging ($f < 0$) lens, the object distance, $p = f$, is negative; and the incident rays must be converging onto the lens toward the point F', as shown in Figure 20.32(b). In either case, the point F' is called the *secondary focal point* of the lens.

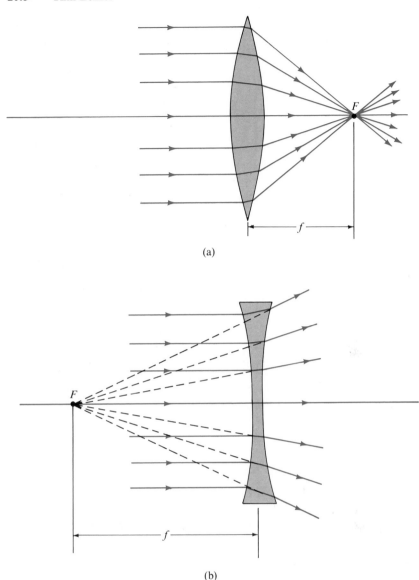

Figure 20.31
Light rays parallel to the optical axis of a lens **(a)** converging toward the focal point F of a converging lens and **(b)** diverging away from the focal point F of a diverging lens. The focal length f of each lens is the distance from the lens to the focal point.

In summary the (primary) focal point F of a thin lens is the image point that results when the incident rays are parallel to the optical axis of the lens. The secondary focal point F' of a thin lens is that object point for incident rays that results in emergent rays that are parallel to the optical axis of the lens. ∎

Although the thin-lens equation, Equation (20-42), permits our locating images analytically for a given lens or lens system, graphical construction of images are equally important. Figure 20.33(a) shows an extended object of height h located a distance p from a converging lens. Two construction rays from the tip of the object arrow are sufficient to locate the image of height h' at a distance q from the lens:

1. A ray (1) parallel to the optical axis is refracted by the lens so that is passes through the (primary) focal point F of that lens.

2. A ray (2) through the secondary focal point F' is refracted by the lens and emerges parallel to the optical axis.

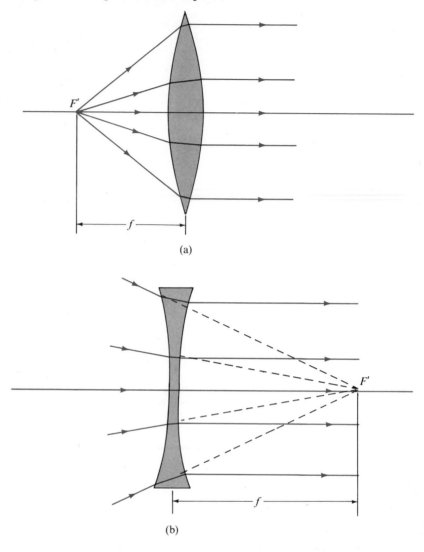

(a)

(b)

Figure 20.32   Example 20.7

A third ray (3), which can serve to check the solution obtained by the interaction of rays 1 and 2, through the vertex of the lens passes undeviated. Here we use the results of Example 20.5 and Exercise 20.15. At the vertex of a thin lens, the two refracting surfaces are parallel (the emergent ray is parallel to the incident ray) and the thickness of the lens is assumed to be negligible (the displacement of the emergent ray relative to the incident ray is zero).

Similarly, an image formed by a diverging lens may be located and checked by constructing the three rays shown in Figure 20.33(b):

1. A ray (1) parallel to the optical axis is refracted by the lens so that it diverges from the (primary) focal point F of the lens.

2. A ray (2) directed toward the secondary focal point F' will emerge parallel to the optical axis.

3. A ray (3) through the vertex passes through the lens undeviated.

Just as for spherical mirrors, the lateral magnification, $m = h'/h$, produced by

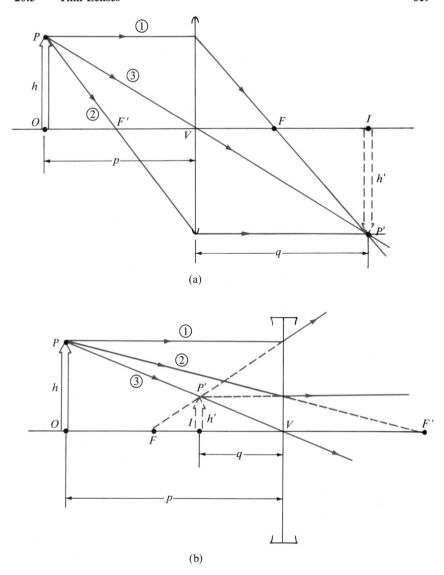

Figure 20.33
Using construction rays to locate the image I of an extended object O for **(a)** a converging lens and **(b)** a diverging lens. In each case, the height of the object is h, and the height of the image is h'.

a lens may be found from either Figure 20.33(a) or (b), where the similar triangles PVO and $P'VI$ show that

$$m = \frac{h'}{h} = -\frac{q}{p} \qquad (20\text{-}43)$$

In Figure 20.33(a), both p and q are positive, so m is negative, indicating that the image is inverted (pointing oppositely to the direction of the object arrow). In Figure 20.33(b) where p is positive but q is negative, m has a positive value, indicating an erect image (oriented in the same direction as the object arrow).

Graphical constructions for lenses and lens systems clearly show whether images are real or virtual. The rays actually pass through the real image of Figure 20.33(a), but in Figure 20.33(b) the rays that diverge from the virtual image do not actually originate at the image. The following problem illustrates how both analytical and graphical techniques are useful in characterizing the image formed by a system of lenses.

An object 3.0 cm high is placed 30 cm to the left of a diverging lens ($f = -20$ cm). A converging lens ($f = +10$ cm) is placed coaxially with and 10 cm to the right of the diverging lens. (a) Locate the image produced by this lens system. (b) What is the height of that image? (c) What is the nature of that image?

A graphical representation of this optical system is shown in Figure 20.34. We begin by drawing the construction rays labeled 1 and 2 from the tip of the object arrow to form the image I_1 produced by the first (diverging) lens. Ray 1, parallel to the optical axis, is refracted at the first lens to diverge as ray 3 from the focal point F_1 of the first lens. Ray 2, passing through the vertex of the first lens, is undeviated. Rays 1 and 3 are discontinued at the second lens because in the region between the lenses they are neither parallel to the optical axis nor directed toward the vertex of the second lens. Therefore, rays 1 and 3 are not appropriate construction rays for forming an image produced by the second lens. But we can construct a ray incident on the second lens that is parallel to the axis by recognizing that every ray between the lenses must be in a direction that is diverging from the image I_1. Ray 4 is constructed between the lenses so that it is diverging from I_1 and is parallel to the axis. Then ray 4 is refracted by the second lens so that it passes through the focal point F_2 of the second lens as ray 6. Similarly, ray 5 is constructed so that it diverges from I_1 and passes through the vertex of the second lens undeviated. The intersection of rays 5 and 6 locate the final image I.

The graphical solution provides the required qualitative answer to part (c) of the problem. Figure 20.34 indicates that the image is inverted (with respect to the original object) and real (because each of the construction rays actually passes through the image). Further, assuming the construction sketch of Figure 20.34 to be approximately to scale, we may estimate the quantitative answers to parts (a) and (b). In particular, the final image appears to be located somewhat less than 20 cm from the converging lens and looks to be about one-third the height of the object. A careful construction permits an accurate determination of the answers by measurement with a scale. But let us determine the location and height of the image analytically.

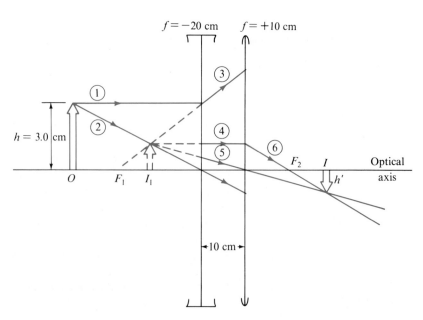

Figure 20.34
Graphical solution to a problem in the text.

In the first lens, which has a focal length $f_1 = -20$ cm, the object distance is $p_1 = +30$ cm. Then the image distance q_1 from the first lens to the image I_1 may be found using the thin-lens equation:

$$\frac{1}{p_1} + \frac{1}{q_1} = \frac{1}{f_1}$$

$$\frac{1}{+30 \text{ cm}} + \frac{1}{q_1} = \frac{1}{-20 \text{ cm}}$$

$$q_1 = -12 \text{ cm} \tag{20-44}$$

This result indicates that the image I_1 formed by the first lens is a virtual image located 12 cm to the left of the first lens. In a lens system, the image of the first lens serves as the object for the next lens. Here the rays from the image I_1 are diverging onto the second lens from a distance of 22 cm (the distance from I_1 to the first lens plus the distance separating the lenses). Then the object distance p_2 to the second lens is positive ($+22$ cm). Applying the thin-lens equation to the second lens ($f_2 = +10$ cm), we find

$$\frac{1}{+22 \text{ cm}} + \frac{1}{q_2} = \frac{1}{+10 \text{ cm}}$$

$$q_2 = \frac{55}{3} \text{ cm} = 18 \text{ cm} \tag{20-45}$$

This positive value for q_2 indicates that the final image is real and is located 18 cm to the right of the converging lens. This solution is consistent with our estimate obtained from the graphical solution.

To determine the height of the final image, let us calculate the lateral magnification provided by each lens separately:

$$m_1 = -\frac{q_1}{p_1} = -\frac{(-12 \text{ cm})}{(+30 \text{ cm})} = +\frac{2}{5} \tag{20-46}$$

$$m_2 = -\frac{q_2}{p_2} = -\frac{\left(\frac{55}{3} \text{ cm}\right)}{(22 \text{ cm})} = -\frac{5}{6} \tag{20-47}$$

The total lateral magnification m of the lens system is equal to the product $m_1 m_2$ of the magnification caused by each lens, or

$$m = m_1 m_2 = \left(+\frac{2}{5}\right)\left(-\frac{5}{6}\right) = -\frac{1}{3} \tag{20-48}$$

From Equation (20-43), we may find the height h' of the final image in terms of m and the given value of the object height ($h = 3.0$ cm):

$$h' = mh = \left(-\frac{1}{3}\right)(3.0 \text{ cm}) = -1.0 \text{ cm} \tag{20-49}$$

The negative sign confirms that the final image is inverted and one-third the object height, results we have anticipated from the graphical solution.

Example 20.8   PROBLEM   If the lenses of the problem just completed are interchanged, determine the final image using graphical techniques.

SOLUTION   Figure 20.35(a) depicts the problem with a real object placed 30 cm to the left of a lens system consisting of a converging lens ($f_1 = +10$ cm)

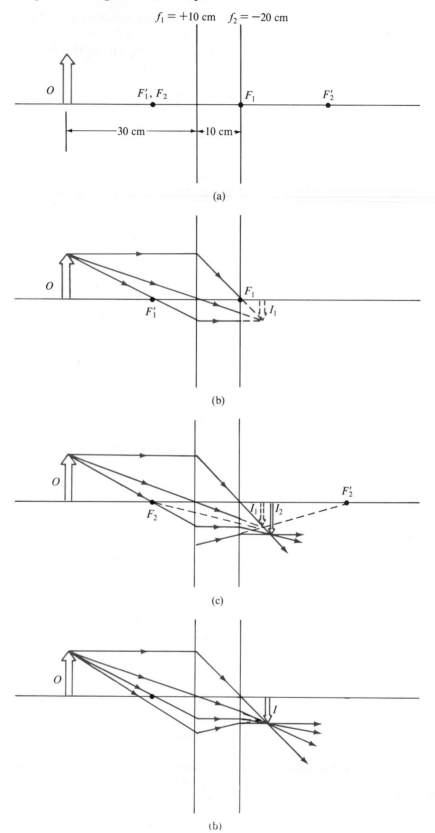

$f_1 = +10$ cm $f_2 = -20$ cm

(a)

(b)

(c)

(b)

Figure 20.35   Example 20.8

and a diverging lens ($f_2 = -20$ cm) placed 10 cm to the right of the converging lens.

Figure 20.35(b) shows the ray construction that determines the image I_1 formed by the first lens. (Can you explain how each ray is constructed?) If the second lens were not present, this first image would be real, inverted, and approximately 15 cm beyond the first lens.

Figure 20.35(c) shows the ray construction that determines the image I_2 formed by the second lens. This final image is real, inverted, approximately two-thirds the size of the original object, and approximately 7 cm to the right of the second lens.

Each ray used in this solution is shown in Figure 20.35(d) from its origin at the tip of the object O until after it passes through the final image. A calculation, which you might like to verify, indicates that the final image is 6.7 cm beyond the second lens and that this image is indeed two-thirds the size of the object. ■

20.4 Optical Instruments

The optical instruments we consider here are those intended to assist or correct the vision of the human eye. In a sense, the eye is itself an optical instrument. The eye may also be considered part of the systems we will describe: magnifiers, microscopes, and telescopes. From either point of view (no pun intended!), the capabilities and limitations of the human eye are pertinent to our considerations.

Light and Human Vision

The human visual apparatus includes the eyeball, the optic nerve, and the visual centers of the brain. The eyeball forms an image on special receptors, much like a camera forms an image on film. Electrical impulses created by light on the receptor cells in the eyeball are transmitted along the optic nerve to the brain, where those impulses are interpreted as the perception of a visual image. The optical instrumentation of interest here is the eyeball.

The human eyeball is approximately spherical with a diameter of about 25 mm. A horizontal section of the eyeball is shown in Figure 20.36. The refracting media of the eye are the cornea, the aqueous humor (a watery fluid), the crystalline lens, and the vitreous humor (a jellylike medium). Light entering the eyeball is refracted by these media, much as if they functioned as a thin lens (they are not, of course); even the crystalline lens is not a thin lens. The effect of refraction through the eyeball is to form a sharp image on the retina, where special cells, called rods and cones, respond chemically to the light.

The crystalline lens can change shape by the action of tiny muscles attached to the lens. In this manner, the lens may change its focal length, thereby allowing the eye to focus on objects at various distances from the eye. This ability of the eye is called *accommodation*. The smallest object distance for which the eye can form a sharply focused image on the retina is called the *near point*, which varies from about 10 cm for normal young children to about 100 cm for the elderly. The *iris* is a muscular tissue shaped like a flattened doughnut. It is arranged around an opening called the *pupil*. The diameter of the pupil may be changed by involuntary muscular movements of the iris. In this way the eye regulates the amount of light passing into the interior of the eyeball.

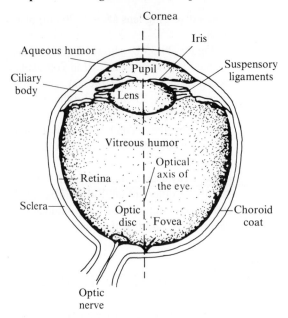

Figure 20.36
A horizontal section of the human eye-ball.

The human brain is adapted to interpret light signals from sharp images formed on the retina of the normal eye. Abnormalities in the eye, therefore, produce deficiencies in vision. Some deficiencies can be corrected by relatively simple optical aids in the form of eyeglasses. For example, *farsightedness* is a common visual defect whereby a person cannot focus sharply on nearby objects because the eyeball cannot refract light sufficiently to form their images sharply on the retina. This situation is illustrated in Figure 20.37(a). An appropriate converging lens, as shown in Figure 20.37(b), provides additional refraction, focusing nearby objects on the retina. *Nearsightedness*, illustrated in Figure 20.37(c), is a defect in which a person cannot see distant objects clearly because the eyeball focuses the incoming rays in front of the retina. An appropriate diverging lens, shown in Figure 20.37(d), focuses light from distant objects on the retina.

Lenses, especially those used as eyeglasses, may be characterized by their

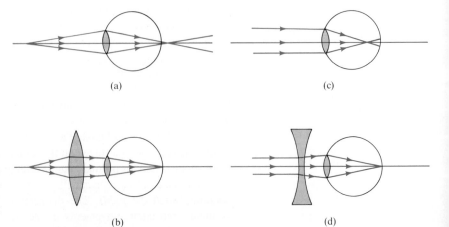

Figure 20.37
(a) A farsighted eye, incapable of focusing light rays from a nearby object onto the retina of the eyeball. (b) Farsightedness corrected by a converging lens. (c) A nearsighted eye, incapable of bringing the light rays from a distant object to focus on the retina. (d) Nearsightedness corrected by a diverging lens.

optical strength, or *optical power*. The strength S of a lens, measured in *diopters*, is equal to the reciprocal of its focal length f, measured in meters:

$$S \text{ (in diopters)} = \frac{1}{f \text{ (in meters)}} \tag{20-50}$$

Prescriptions for eyeglasses usually specify lens strengths in diopters, positive values indicating converging lenses and negative values indicating diverging lenses.

E 20.18   Estimate the strength of the lens system of a human eye that focuses incident parallel rays on the retina. Assume the lens-to-retina distance to be equal to 25 mm.   Answer: +40 diopters

E 20.19   Estimate the strength of the lens system of a human eye that forms a sharp image (on the retina) of an object that is 25 cm from the eye. What, then, is the approximate accommodation range (in diopters) of a human eye that has a near point of 25 cm and a far point at infinity?

Answer: 44 diopters, 40–44 diopters

Some optical imperfections in spherical lenses—both the lenses of the human eye and those of other optical systems—are inherent. These defects do not result from poorly formed lens surfaces or faulty workmanship but are consequences of the law of refraction. Images formed by actual lenses, however perfectly formed into spherical surfaces, are imperfect because real lenses are called upon to use nonparaxial rays. Our theoretical treatment of lenses was based on the assumption that all rays passing through a lens make small angles with the optical axis. Departures of actual images formed by a lens from those predicted by our simplified theory, therefore, are defects, called *aberrations*, of the lens.

Perhaps the most common lens defect is *spherical aberration*. This imperfection occurs because nonparaxial rays from an object point on the optical axis fail, in general, to form a point image after being refracted by a lens. Figure 20.38(a) illustrates spherical aberration for a converging lens. A spherical mirror, shown in Figure 20.38(b), similarly suffers from spherical aberration. A similar aberration, called *coma*, occurs when rays from a point off the axis do not form a point image after refraction by a lens. Coma is so called because an off-axis image of a point is shaped like a comet.

The effects of spherical aberration may be reduced by limiting the usable portion of a lens to a circular region near the optical axis. A mask for this purpose covering the outer portion of a lens is called a *stop*. In the eye the iris functions as a stop. We see objects more clearly when they are brightly illuminated because the iris reflexively reduces the diameter of the pupil, thereby restricting the incoming light to paraxial rays and effectively eliminating spherical aberration.

Let us consider one further common defect of lenses. *Chromatic aberration*, illustrated in Figure 20.39, occurs because of the variation in the index of refraction of a lens material with wavelength. Thus rays of different colors from an object form images at different points. Chromatic aberrations may be minimized by using combinations of lenses made from materials that have different dispersion characteristics.

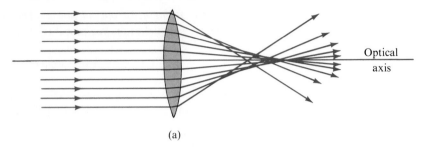

(a)

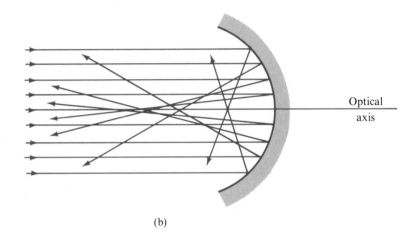

Figure 20.38
Spherical aberration in **(a)** a converging
lens and **(b)** a spherical mirror.

(b)

Magnifying Instruments

Some optical instruments are intended to assist the eye in viewing objects. We
will briefly consider three such devices, all of which are designed to increase the
apparent size of an object.

A single converging lens can be used as a *simple magnifier*, as shown in Figure
20.40(a). The lens is positioned at a distance slightly less than its focal length f

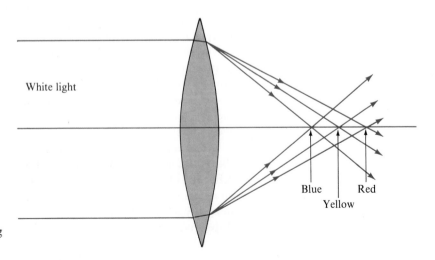

Figure 20.39
Chromatic aberration in a converging
lens.

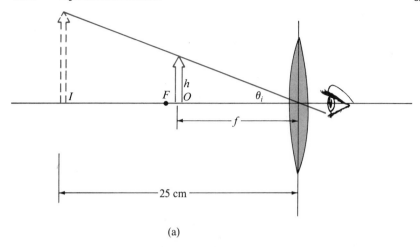

(a)

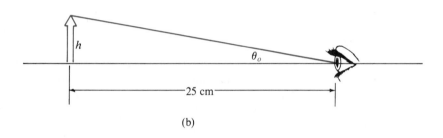

(b)

Figure 20.40
A simple magnifier.

from an object of height h so that a virtual image is formed at the near point of the eye, which for optical instruments is *defined* to be 25 cm. The angle θ_i is subtended by the image at the lens (and, therefore, approximately at the eye). From Figure 20.40(b), we may see that θ_i is greater than θ_o, the angle subtended by the object when the object is viewed directly at a distance of 25 cm from the eye.

The ratio θ_i / θ_o is defined to be the *angular magnification M* of the lens used as a magnifier:

$$M \equiv \frac{\theta_i}{\theta_o} \qquad (20\text{-}51)$$

Because the eye is close to the lens, we may, for maximum magnification, take the image distance q for the virtual image to be equal to -25 cm. Then the lens equation, Equation (20-42), may be used to find the object distance p:

$$\frac{1}{p} + \frac{1}{-25 \text{ cm}} = \frac{1}{f}$$

$$p = \frac{25 f}{f + 25} \qquad (p \text{ and } f \text{ in cm}) \qquad (20\text{-}52)$$

Then if the angles θ_i and θ_o are small, they may be approximated as

$$\theta_i \cong \tan\theta_i = \frac{h}{p} = \frac{h(f + 25)}{25f} \qquad (h \text{ and } f \text{ in cm}) \qquad (20\text{-}53)$$

$$\theta_o \cong \tan\theta_o = \frac{h}{25} \qquad (h \text{ in cm}) \qquad (20\text{-}54)$$

Substituting Equations (20-53) and (20-54) into Equation (20-51) gives the maximum magnification

$$M_{max} = \frac{\theta_i}{\theta_o} = \frac{f + 25}{f} = 1 + \frac{25}{f} \qquad (f \text{ in cm}) \qquad (20\text{-}55)$$

If now the lens is positioned so that the image is at infinity, that is, $1/q = 0$, and the eye can relax while viewing the image, the object distance p becomes equal to f. Then for most comfortable viewing, the most comfortable magnification M_c is given by

$$M_c = \frac{\theta_i}{\theta_o} = \frac{h/f}{h/25 \text{ cm}} = \frac{25}{f} \qquad (f \text{ in cm}) \qquad (20\text{-}56)$$

E 20.20   A person whose near point is 25 cm requires a +8.0-diopter lens used as a simple magnifier to view comfortably a map with fine detail. What is the magnification M_c in this case?   Answer: 2.0

Two converging lenses can be arranged to form a *compound microscope*, as shown in Figure 20.41. The lens nearer the object O is called the *objective lens*, or simply the *objective*; the lens nearer the eye is called the *ocular lens*, or the *eyepiece*. The object is placed just outside the focal point of the objective so that a real image I_1 is formed between the lenses. The relative position of the two lenses is adjusted so that I_1 is approximately coincident with the focal point of the eyepiece. The eyepiece then functions as a simple magnifier, forming an enlarged virtual image by using I_1 as an object of the eyepiece.

The lateral magnification, $m = -q_o/p_o$, provided by the objective is equal to h'/h, where h' is the height of the image and h is the height of the object. Let us assume that the object distance p_o is approximately equal to f_o, the focal length of the objective. The image distance q_o from the objective is made to be as large as the length L of the microscope housing permits. Then the lateral magnification m_o for the objective becomes

$$m_o \simeq -\frac{L}{f_o} \qquad (20\text{-}57)$$

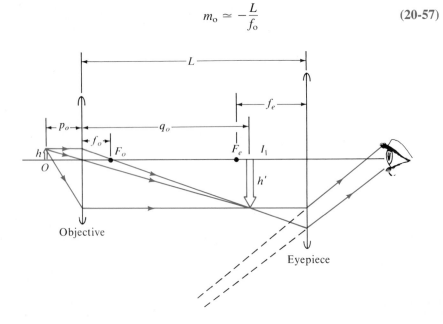

Figure 20.41
A compound microscope.

The eyepiece, with focal length f_e, is used as a simple magnifier to view the final image, using the first image I_1 as its object. For most comfortable viewing, the angular magnification M_e provided by the eyepiece is, according to Equation (20-56),

$$M_e = \frac{25}{f_e} \tag{20-58}$$

The total magnification M of the compound microscope is equal to the product of m_o and M_e, or, using Equations (20-57) and (20-58),

$$M = m_o M_e = -\frac{25L}{f_o f_e} \quad \text{(all distances in cm)} \tag{20-59}$$

The negative sign indicates that the final image is inverted with respect to the object.

E 20.21   If you wish to construct a compound microscope having a length of 24 cm, a magnification of 100, and a 100-diopter objective lens, what is the power required for the eyepiece?   Answer: 25 diopters

Two converging lenses may also be used as a *telescope*, an optical device used to view objects that are very far from the eye. Figure 20.42 is a schematic representation of a telescope. An image I_1 is formed by the objective lens at the focal point of the eyepiece, which functions as a simple magnifier. And because the object being viewed is far from the telescope, the image I_1 is formed approximately at the focal point of the objective. Then the distance between the lenses is equal to $f_o + f_e$, where f_o and f_e are the focal lengths of the objective and the eyepiece, respectively.

To determine the angular magnification M of the telescope, let θ_1 be the angle subtended by the object at the objective, as shown in Figure 20.42. The angle θ_1 is also the angle subtended by the object at the unaided eye, because the object is very far away. Let θ_2 be the angle subtended at the eye by the image I_1. Letting h be the height of the image I_1, we may use the approximation for small angles to write

$$\theta_1 \simeq \tan\theta_1 = -\frac{h}{f_o} \tag{20-60}$$

$$\theta_2 \simeq \tan\theta_2 = \frac{h}{f_e} \tag{20-61}$$

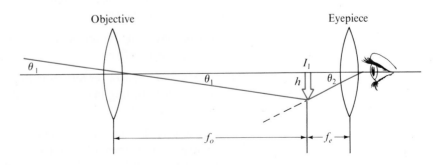

Figure 20.42
A telescope.

where the negative sign in Equation (20-60) provides that h is negative when θ_1 is positive. The angular magnification M of the telescope may be written as

$$M = \frac{\theta_2}{\theta_1} \simeq \frac{h/f_e}{-h/f_o}$$

$$M \simeq -\frac{f_o}{f_e} \tag{20-62}$$

According to Equation (20-62), significant magnification may be achieved with a telescope when the objective has a long focal length much greater than the focal length of the eyepiece.

E 20.22   A telescope having a 1.0-diopter objective lens and a 40-diopter eyepiece is used to view a 2.0-cm-high object at a distance of 100 m. What is the angular size of this object when viewed without the telescope? With the telescope?   Answer: 2.0×10^{-4} radian, 8.0×10^{-3} radian

20.5 Problem-Solving Summary

Problems that involve mirrors or lenses, or both, usually can be analyzed using geometric optics. The techniques of geometric optics treat light rays, directed lines perpendicular to wavefronts of visible electromagnetic radiation ($\lambda = 400$–700 nm). The theoretical basis of geometric optics is Fermat's principle, which, practically speaking, assumes that light rays proceed between two points along the path (compared to nearby paths) of least time. The practical consequences of Fermat's principle are (a) the law of reflection, and (b) the law of refraction.

Reflections of light at an interface between media obey the law of reflection: The angle θ_i of incidence is equal to (and in the same plane as) the angle θ_r of reflection, where both angles are measured from the normal to the interface. All mirror problems rely on this law, which is sufficient for complete analysis of plane mirror problems. It is sometimes helpful, nevertheless, to recall that the virtual images (the rays that form a virtual image do not pass through the image) formed by plane mirrors are as far behind the mirror surface as the object is in front.

If an object is placed at a distance p in front of the vertex (center) of a spherical mirror of radius R, an image is formed at a distance q from the vertex, where p, q, and R are related by $1/p + 1/q = 2/R$, provided we use the following sign conventions, which also apply to lenses:

1.   If rays from an object are diverging onto an interface, p is a positive value. Otherwise, p is negative.

2.   If rays are converging toward an image from an interface, q is a positive value. Otherwise, q is negative.

3.   The radius R of a mirror (or refracting interface) is positive where the center of the mirror (or refracting interface) is on the same side as the rays leaving the surface (or interface).

Paraxial rays that are incident onto a spherical mirror and are parallel to the optical axis (a line perpendicular to the mirror at its vertex) form a point image at the focal point of the mirror. The focal length, $f = R/2$, of a spherical mirror is the distance from its vertex to its focal point. The lateral magnification m of a spherical mirror is equal to $-q/p$. A positive value of m indicates an erect image; a negative value of m indicates an inverted image.

The index of refraction n of a medium is equal to the ratio of the speed c of light in a vacuum to the speed v of light in that medium. Light rays refract at an interface between media indices of refraction n_1 and n_2; they obey the law of refraction, $n_1 \sin\theta_1 = n_2 \sin\theta_2$ (Snell's law); and the incident and refracted rays lie in the same plane. The angles θ_1 and θ_2 are measured from the normal to the interface between media 1 and 2, which have indices of refraction n_1 and n_2.

When a light ray in a medium having an index of refraction n_2 encounters a medium having an index $n_1 < n_2$, the ray will be totally internally reflected (no light passes to the n_1 medium) if the angle of incidence θ_2 is greater than the critical angle $\theta_c = \sin^{-1}(n_1/n_2)$. In tracing rays through interfaces between media, as we do in prism problems, for example, it is important that we test for internal reflection at every interface where a ray is (or may be) passing from a medium of higher index to one of lower index.

Rays from an object within a medium ($n = n_1$) and at a distance p from a spherical (radius = R) interface with a second medium ($n = n_2$), form an image at a distance q from that interface. The quantities p, q, n_1, n_2, and R are all related by $n_1/p + n_2/q = (n_2 - n_1)/R$. Our usual sign conventions apply to p, q, and R.

Most thin lenses have two spherical surfaces of radii, R_1 and R_2, and are constructed of a transparent material having an index of refraction n for use in air ($n = 1$). Then the lens-maker's equation, $1/f = (n - 1)(1/R_1 - 1/R_2)$, relates the radii of the lens surfaces to the focal length f, which is equal to the image distance q that results when incident rays are parallel to the optical axis, i.e., when $p \to \infty$. The lens equation, $1/p + 1/q = 1/f$, relates the object distance p, the image distance q, and the focal length f, where f is positive for a converging lens and negative for a diverging lens. Again, our sign convention applies to p and q. And, as is the case for spherical mirrors, the lateral magnification m of a thin lens is equal to $-q/p$. A positive magnification indicates an erect image; a negative magnification indicates an inverted image.

Two ray constructions are sufficient to form ray diagrams for a thin-lens system: (a) Rays parallel to the optical axis are refracted at a lens to converge toward or diverge away from the focal point of that lens, and (b) rays incident on the vertex of a lens pass through that lens undeviated. The lateral magnification of a lens system is equal to the product of the individual magnifications produced by the lenses of the system.

When the human eye is used as part of a direct-viewing optical instrument, the near-point of the eye is taken to be equal to 25 cm. The strengths of lenses used to correct nearsightedness and farsightedness in humans are measured in diopters. The strength of a lens (in diopters) is equal to the reciprocal of the focal length (in meters) of that lens.

A single converging lens used as a magnifier has a maximum angular magnification (image at 25 cm) given by $M = 1 + 25/f$, where f is the focal length (in cm) of the lens. The magnification of a single lens, adjusted for most comfortable viewing (image at infinity), is given by $M = 25/f$. A compound microscope uses two lenses, an objective with focal length f_o and an eyepiece with focal length f_e. The net magnification of a microscope having approximately a length L between its lenses is given by $M = m_o M_e \simeq -25L/(f_o f_e)$. A telescope with an objective of focal length f_o and an eyepiece of focal length f_e has a net magnification $M \simeq -f_o/f_e$.

Problems

GROUP A

20.1 While swimming at night, an underwater diver determines the moon to be 35° from the vertical. What value will a person riding on the surface get for this angle?

20.2 A point object is placed midway between two parallel plane mirrors separated by 20 cm.

  (a) Determine the distances from the object to the four nearest images.

  (b) Sketch a ray diagram showing these images.

20.3 A thin lens made of glass having $n = 1.60$ is to be flat on one side and have a strength of +4.0 diopters.

  (a) Sketch a cross section of the lens.

  (b) What is the magnitude of the radius of curvature of the curved side of the lens?

20.4 Determine the index of refraction for the optical prism shown in Figure 20.43.

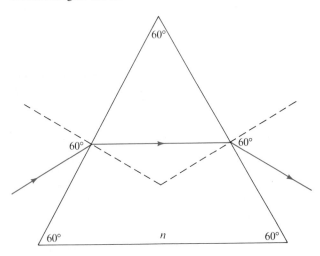

Figure 20.43   Problem 20.4

20.5 A diver, whose eyes are 10 m above the surface of a pool, uses an optical range finder to determine that the bottom of the pool is apparently 13 m from his eyes. Determine the depth of the water ($n = 1.33$) in the pool.

20.6 A layer of water is placed on a glass plate (*see* Fig. 20.44). Determine the path for each of the four rays shown.

20.7 What optical strength is required for an eyepiece if it is to be used with an objective lens ($f = 150$ cm) to fabricate a telescope with an angular magnification equal to 80?

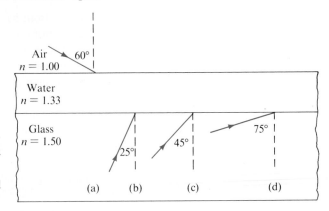

Figure 20.44   Problem 20.6

20.8 A 5.5-diopter lens is to be used as a simple magnifier. What range of angular magnification is available to the user?

20.9 Estimate the length of a 400-power microscope that has a 200-diopter objective lens and a 25-diopter eyepiece.

20.10 What is the accommodation range (in diopters) of a nearsighted human eye that cannot focus objects that are farther than 20 cm or closer than 15 cm from the surface of the eye. (Assume that the eye has a diameter of 25 mm and that the entire optical system is approximated by a thin lens at the surface of the eye.)

20.11 A 1.5-cm-high object is placed 15 cm from a lens ($S = 10$ diopters). Locate and describe the resulting image. Be sure to check your result with a graphical solution.

20.12 If a thin lens forms an image midway between the lens and an object placed 12 cm from the lens, what is the optical strength of this lens?

20.13 An object is placed on the optical axis of a thin lens at a distance of 25 cm from this lens. The real image formed by the lens is 40 cm from the object.

  (a) Calculate the optical strength of this lens.

  (b) Determine the magnification for this configuration.

20.14 An object is positioned in front of a concave mirror in such a way that a virtual image three times the size of the object is formed 9.0 cm from the vertex of the mirror.

  (a) What is the focal length of this mirror?

  (b) Is the image erect or inverted?

  (c) Construct a ray diagram depicting this problem.

20.15 If a person's eye can resolve (distinguish) objects having an angular separation greater than 1.0×10^{-4} radian,

  (a) what is the minimum separation between two objects at a distance of 100 m if the person is to distinguish them?

  (b) what is this minimum separation for two objects on the moon (earth-moon distance = 3.8×10^8 m)?

(c) determine the minimum magnification for a telescope if this person is to resolve two lunar objects separated by 400 m.

(d) calculate the minimum focal length for an objective lens if it is to be used with a 50-diopter eyepiece to fabricate the telescope of part (c).

20.16 A point object is placed between two parallel plane mirrors separated by 20 cm. The object is 7.0 cm from one of the mirrors.

(a) Determine the distances from the object to the five nearest images.

(b) Sketch a ray diagram showing these images.

20.17 A camera has a lens system of variable focal length. The lens-to-film distance is fixed at 35 mm, and the device will focus all objects no closer than 25 cm. Using a thin-lens model, estimate the focal length range for the lens system of this camera.

20.18 If $R_2 = 8.0$ cm and $n = 1.55$ for the thin lens shown in Figure 20.45, determine R_1 if the lens is to have a strength of -3.0 diopters.

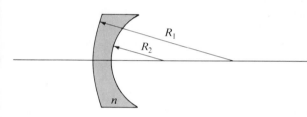

Figure 20.45 Problem 20.18

20.19 An object (height $h = 2.0$ cm) is placed with its center on the optical axis of a thin lens ($f = +15$ cm) that forms a virtual image ($h' = 4.0$ cm) of this object. Determine the object and image distances for this system.

20.20 A luminous object (length $= 4.0$ cm) lies along (is coincident with) the optical axis of a thin lens ($f = 8.0$ cm). The center of the object is 14 cm from the lens. Describe the resulting image.

20.21 Two plane mirrors are joined with the angle between the two equal to $90°$. A point object is positioned so that it is 5.0 cm from each reflecting surface.

(a) How many images are formed?

(b) Calculate the distances from the object to the images determined in part (a).

(c) Sketch a ray diagram showing these images.

20.22 Two plane mirrors are joined with the angle between the two equal to $90°$. A point object is placed so that it is 6.0 cm from one reflecting surface and 8.0 cm from the other.

(a) How many images are formed?

(b) Calculate the distances from the object to the images determined in part (a).

(c) Sketch a ray diagram showing these images.

20.23 A certain farsighted person requires corrective lenses ($f = +20$ cm) to focus an object held 15 cm from her eyes. What is this person's near point? (Assume that the corrective lenses are placed 1.0 cm from the surfaces of the eyes.)

20.24 A certain nearsighted person cannot focus objects that are farther than 12 cm or closer then 9.0 cm from the surface of one eye.

(a) Determine the optical strength of a corrective lens (placed 1.0 cm from the surface of this eye) that will permit the person to see distant objects.

(b) What is the near point for this person when he is wearing the lens of part (a)?

20.25 Two thin lenses, each having a focal length $f > 0$, are mounted with their optical axes coincident and with a distance $d < f$ separating them. Where will incident rays, parallel to the optical axis, be focused?

20.26 Two thin lenses L_1 ($f_1 = 15$ cm) and L_2 ($f_2 = -15$ cm) are positioned coaxially and separated by 5.0 cm. An object O is placed 30 cm from L_1, as shown in Figure 20.46.

(a) Locate and describe the final image.

(b) What is the total magnification?

(c) Solve this problem graphically.

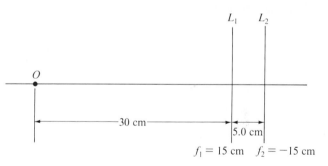

Figure 20.46 Problem 20.26

20.27 Thin lenses L_1 ($f_1 = 12$ cm) and L_2 ($f_2 = 18$ cm) are positioned coaxially and separated by 24 cm. An object O is placed 36 cm from L_1 (see Fig. 20.47).

(a) Locate and describe the final image.

(b) What is the total magnification?

(c) Solve this problem graphically.

20.28 A plane mirror is 15 cm behind a lens ($f = 12$ cm). The mirror is oriented so that the optical axis of the lens is normal to the surface of the mirror. If an object is placed 30 cm in front of the lens, determine the location and nature of the final image.

20.29 A glass ($n = 1.5$) rod of length L is ground so that each end of the rod is convex with a radius of curvature equal to R. Rays that enter parallel to the optical axis of the rod emerge to

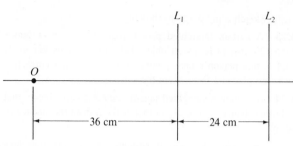

Figure 20.47   Problem 20.27

form a real image at a distance of $2R$ from the rod. Determine the length L.

20.30   Two thin lenses are placed in contact to form a composite lens. Show that

(a) the optical strength S of the composite lens is given by

$$S = S_1 + S_2$$

where S_1 and S_2 are the optical strengths of the two component lenses.

(b) the focal length f of the composite lens is given by

$$f = \frac{f_1 f_2}{f_1 + f_2}$$

where f_1 and f_2 are the focal lengths of the two component lenses.

20.31   An object is 15 cm from a lens ($f_1 = +10$ cm). A mirror ($f_2 = -15$ cm) is 5.0 cm behind the lens, and the two devices have coincident optical axes.

(a) Locate and describe the final image.

(b) Solve this problem graphically.

20.32   If $n = 1.5$ for the optical prism shown in Figure 20.48, determine the total angular deviation of the ray shown.

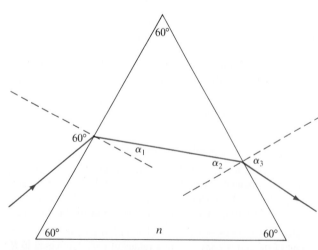

Figure 20.48   Problem 20.32

20.33   Repeat Problem 20.32 with $n = 2.0$.

20.34   A mirror ($f = +10$ cm) is placed with its optical axis perpendicular to a plane reflecting surface. The distance separating these two mirrors (measured along the optical axis) is equal to 10 cm. A point object is placed on the optical axis midway between these two mirrors.

(a) How many images are formed by this system?

(b) Where are the images located?

(c) Sketch a ray diagram showing these images.

GROUP **B**

20.35   If the index of refraction of the glass used in a thin lens changes from 1.630 to 1.625 as the wavelength of light changes from 400 nm to 700 nm, calculate the corresponding percentage change in the focal length of the lens.

20.36   Three thin lenses are positioned coaxially. The first and third lenses have equal focal lengths of +20 cm and are separated by 20 cm. Midway between these two lenses is a diverging lens ($f = -15$ cm). A real object is positioned 60 cm from the first lens.

(a) Describe the final image.

(b) Solve this problem graphically.

20.37   For the three-lens system of Problem 20.36, determine

(a) the image point for incident rays that are parallel to the axis.

(b) the object point for which rays leave the system parallel to the axis.

20.38   A thin lens, made of a material having an index of refraction n_1, has spherical surfaces with radii R_1 and R_2. The lens is immersed in a liquid (index of refraction $= n_2$).

(a) Show that image and object distances still obey the thin-lens equation

$$\frac{1}{p} + \frac{1}{q} = \frac{1}{f}$$

where the focal length f is given by

$$\frac{1}{f} = (n_1 - n_2)\left(\frac{1}{R_1} - \frac{1}{R_2}\right)$$

(b) If $R_1 > 0$ and $R_2 < 0$, is the lens a converging lens or a diverging lens?

20.39   The angular deviation δ of a ray by a prism is shown in Figure 20.49. Show that

(a) the deviation has a minimum value δ_m for $\phi = A/2$, where A is the prism angle.

(b) the index of refraction of the prism is given by

$$n = \frac{\sin\left(\dfrac{A + \delta_m}{2}\right)}{\sin\left(\dfrac{A}{2}\right)}$$

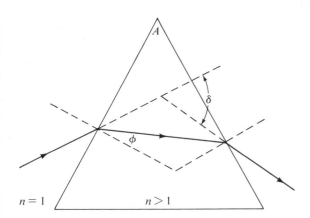

Figure 20.49   Problem 20.39

20.40 A converging mirror ($f = +20$ cm) is placed with its optical axis perpendicular to a plane mirror. The distance separating these two mirrors (measured along the optical axis) is equal to 10 cm. A point object is placed on the optical axis midway between these two mirrors.

(a) How many images are formed by these two surfaces?

(b) Where are the images located?

(c) Sketch a ray diagram showing these images.

20.41 Two rays are leaving an object O and striking a convex spherical refracting (radius = R) surface (*see* Fig. 20.50). Ray 1 is not deviated because it is directed toward the center of curvature C. Ray 2 strikes the surface at point P, is refracted, and intersects ray 1 at point I. If point I were the same for all values of the angle α (for which rays actually strike the refracting surface), then I would be a point image of O. But this is not the case, and this failure of the refracted rays to form a point image

is called spherical aberration. The purpose of this problem is to demonstrate a procedure for determining q if the quantities n_1, n_2, R, p, and α are specified.

(a) Use triangle OPC to show that

$$\sin(\alpha + \gamma) = \left(1 + \frac{p}{R}\right)\sin\alpha$$

(b) Show that $n_2 \sin(\gamma - \beta) = n_1 \sin(\alpha + \gamma)$.

(c) Use triangle IPC to show that

$$q = R\left(1 + \frac{\sin(\gamma + \beta)}{\sin\beta}\right)$$

(d) If the angles α, β, and γ are each small compared to one radian, that is, if ray 2 is a paraxial ray, show that parts (a), (b), and (c) combine to give Equation (20-37):

$$\frac{n_1}{p} + \frac{n_2}{q} = \frac{n_2 - n_1}{R}$$

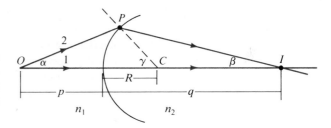

Figure 20.50   Problem 20.41

20.42 Let $n_1 = 1.000$, $n_2 = 1.500$, $R = 5.00$ cm, and $p = 15.00$ cm in the previous problem.

(a) Determine q for paraxial rays.

(b) Determine q for $\alpha = 1.0°, 3.0°, 5.0°, 7.5°, 10.0°$, and $12.5°$.

(c) Determine the limiting value of α for which rays are refracted. What is q for this value of α?

(d) Graph q as a function of α.

21 Light: Physical Optics

Physical optics treats those phenomena associated specifically with the wave nature of light. In contrast, geometric optics, discussed in Chapter 20, uses the ray approximation, in which light is assumed to travel in straight lines and cast sharp shadows. No such restrictions apply to physical optics. Some optical interference phenomena are most apparent when light passes through slits or circular openings having dimensions of a few wavelengths of light. Such phenomena can be explained only in terms of the wave-like characteristics of light. The description of another category of optical phenomena, polarization effects, depends on the wave nature of light. Polarization is related specifically to the transverse character of light waves. This chapter, then, is concerned with topics of physical optics: interference of light and polarization of light.

21.1 Optical Interference

Optical interference may occur when two light waves are simultaneously present in the same region of space. The effects of *interference*, when they exist, may usually be observed as alternating regions of bright and dim light on a screen. At those locations on the screen that appear relatively bright, we say *constructive interference* occurs; at locations that appear relatively dim, or perhaps completely dark, we say *destructive interference* takes place.

Optical interference phenomena are observable only under special conditions. Before we describe those conditions and investigate interference, let us introduce some concepts of physical optics that will facilitate our considerations.

When waves of any kind pass through an opening or pass the edge of an obstacle, the waves spread into the region that is not exposed to the direct path of the incident wave. This spreading, or bending, of a wave is called *diffraction*. The diffraction of sound waves, for example, at the edges of walls or doorways permits us to hear sounds around corners in a building. Similarly, on the surface of water, waves that reach an opening in a dam spread radially outward from the opening, as shown in the photograph of Figure 21.1. Light, as we will see, may be diffracted by obstacles into regions lying within the "geometric shadow" of those obstacles, that is, within regions that would be completely dark if light traveled only in straight lines, as we assume in geometric optics.

Figure 21.1
Diffraction of water waves by an opening in a dam.

Because light is an electromagnetic wave, as we saw in Chapter 18, we may correctly assume that the diffraction of light is predicted and described by electromagnetic theory. A rigorous electromagnetic description of most diffraction phenomena would, however, require very complicated solutions of Maxwell's equations. Fortunately, far simpler treatments of many optical diffraction situations are possible because of a description of wave propagation that was suggested in the seventeenth century by Christian Huygens. Called *Huygens' principle*, this description asserts that *every point on a wavefront of light may be considered a source of secondary wavefronts, called wavelets, each of which propagates in all directions at the speed of light in the medium of the wavelet.* At a later time, the position of a new wavefront is a surface tangent to these wavelets.

The general character of light propagation is explained by Huygens' principle. For example, in Figure 21.2(a) a wavefront, shown at time t_1, moves radially away from a source S. Points on that wavefront are treated as sources of wavelets that propagate in all directions at a speed v. Then after a time interval Δt, the wavelets are at a maximum distance $v\Delta t$ from the original wavefront. The envelope of these wavelets is a spherical surface that represents the wavefront at a time $t_1 + \Delta t$. (In his description of wave propagation, Huygens ignored the fact that if points on a wavefront were actual point sources, the wavelets would radiate backward as well as forward, the actual direction of propagation. This fact does not, however, lessen the usefulness of Huygens' principle.)

A more specific application of Huygens' principle that will be useful for our purposes is illustrated in Figure 21.2(b). Wavefronts of light from a point source S impinge upon a circular opening in an opaque mask. If the opening is small compared to the wavelength of light, we may treat the opening as a point source S of Huygens wavelets. Then spherical wavelets proceed past the mask, as indicated in the figure. If the opening in the mask is a long narrow slit, the Huygens wavelets form cylindrical wavefronts that proceed outward from an axis along the slit.

(a)

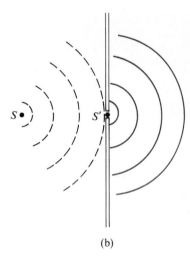

(b)

Figure 21.2
(a) A spherical wavefront, moving radially outward from a source S, shown at time t_1. Points on the wavefront act as sources of Huygens wavelets propagating at a speed v; and after a time interval Δt, the envelope of the wavelets form the wavefront at time $t_1 + \Delta t$. **(b)** Wavelets radiate outward from S', a slit in an opaque mask. The point S' is on a wavefront that has reached the slit from the point source S.

Three further physical concepts—monochromaticity, superposition, and coherence—are important in the understanding of optical interference phenomena. Let us describe these ideas in the context of a specific interference situation.

Double-Slit Interference

The experimental arrangement represented schematically in Figure 21.3(a) was first used in 1800 by Thomas Young to demonstrate optical interference. Young's experiment was the first compelling evidence that light is a wave.

In Figure 21.3(a), S is a source of *monochromatic* light, which literally means one-color light. Thus monochromatic light is an electromagnetic wave having a single wavelength (or frequency). For many years reasonably bright light that is approximately monochromatic has been available from sodium-vapor, mercury-vapor, or other lamps that produce light containing only a few wavelengths. Most unwanted wavelengths may be absorbed, or removed, by passing the light from such lamps through appropriate colored-glass filters. Today relatively inexpensive *lasers* (see Chapter 24, "Topics in Quantum Physics") are sources of very bright light that is virtually monochromatic. The common helium-neon laser, for example, produces red light having a wavelength of 6328 Å, and the range over which the wavelength varies is only about 10^{-5} Å.

In Figure 21.3(a), the monochromatic light from S reaches the first opaque mask M_1, which has a narrow slit S_o. Huygens wavelets then emerge from S_o toward the mask M_2, which has two slits, S_1 and S_2, spaced at equal distances from the system axis that passes through S and S_o. The light reaching S_1 and S_2 has traveled equal distances from S_o, and as a consequence the light waves impinging on S_1 and S_2 are in phase. The definite phase difference (zero, in this case) between the light waves at S_1 and S_2 is a special relationship that is crucial for the existence of observable interference phenomena. Let us consider such relationships.

Two sources of light are said to be *coherent* if they produce waves of the same frequency and maintain a constant phase difference between themselves. Two independent sources of light cannot, in general, be coherent, because light from one source is emitted from atoms in a random fashion that is unrelated to the emissions from another source. An especially useful device that may be used to provide coherent light sources is the laser, in which the emissions of light from atoms within the laser are synchronized. Different points on the cross section of a laser beam have a fixed phase relationship. Thus different parts of the beam from a laser may be used as two or more coherent light sources.

Returning now to Figure 21.3(a), we may observe that the slits, S_1 and S_2, act as coherent sources because they are located at fixed distances from S_o. And because S_1 and S_2 are located at equal distances from S_o, they are not only coherent but are also in phase. The wavefronts from S_1 and S_2 travel equal path lengths, $r_1 = r_2$, to reach the viewing screen at point C, which lies on the system axis. Because both waves are in phase when they leave S_1 and S_2 and both travel equal distances to reach point C, the two waves are in phase at point C. Let us pause now to consider the physical significance of having two in-phase waves at the same point.

Because light is an electromagnetic wave, the principle of superposition for electric and magnetic fields is valid for light waves. Thus, if two light waves are

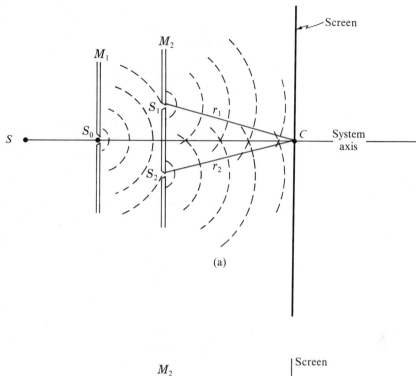

Figure 21.3
(a) Young's double-slit interference system. **(b)** Diagram illustrating the geometry of a double-slit system.

simultaneously at the same point, their electric fields add vectorially. In particular, if two light waves from coherent sources are in phase at a point, the amplitude of the electromagnetic wave at that point is greater than the amplitude of either individual wave. In terms of human vision, a greater amplitude of a light wave corresponds to increased brightness when the light is viewed. If two light waves from coherent sources are superimposed at a point and if they are *in phase*, that point will appear brighter than it would if it were illuminated by either of the individual waves (assuming we place a screen at that point to observe the light). The two waves are said to *interfere constructively* at that point. And if two equal-amplitude light waves from coherent sources are superposed at a point, but are 180° out of phase, that point appears dark. In this case, the waves *interfere destructively* at that point.

We may conclude now that observable optical interference effects require the simultaneous presence of two light waves from coherent sources. Two incoherent light waves incident on a region of a screen do not maintain a constant phase difference at any point on the screen. Consequently, we could not observe constructive or destructive interference, and at every point we would see light that is brighter than it would be if only one of the waves were present.

Returning again to Young's experiment, we see in Figure 21.3(a) that C is a point of constructive interference: Point C is a bright spot. Let us now examine the physical situation at other points on the screen.

In Figure 21.3(b), the slits, S_1 and S_2, are separated by a distance d; and the screen is at a distance $L(\gg d)$ from the mask M_2. The point P on the screen is located a distance y from the system axis. Let θ be the angle between the axis and the line passing through P and B, the midpoint between S_1 and S_2. If, using P as a center, we construct an arc (radius r_1) from S_1 to A, that arc is approximately a straight line because $L \gg d$. Then the angle between this line S_1A and the line S_1S_2 is also equal to θ, and the path difference, $\Delta r = r_2 - r_1$, is equal to $d\sin\theta$. Now realizing that the phase difference δ between two coherent waves in the same medium increases by 2π radians for each wavelength λ of path-length difference Δr, we may write

$$\delta = 2\pi\left(\frac{\Delta r}{\lambda}\right) \tag{21-1}$$

The waves with path lengths, r_1 and r_2, are in phase at S_1 and S_2; and thus they will be in phase at P, producing a bright spot, or an *interference maximum*, provided

$$\Delta r = d\sin\theta = n\lambda \qquad n = 0, 1, 2, \ldots \tag{21-2}$$

On the other hand, the waves will be 180° out of phase at P, producing a dark spot, or an *interference minimum*, when

$$\Delta r = d\sin\theta = \left(n + \frac{1}{2}\right)\lambda \qquad n = 0, 1, 2, \ldots \tag{21-3}$$

Equations (21-2) and (21-3) indicate that an *interference pattern*, consisting of alternating regions of brightness and darkness, results from a double-slit system such as we have described. Because S_1 and S_2 are parallel slits, the screen will have alternating bright and dark regions, called *fringes*, that are parallel to the slits S_1 and S_2. Figure 21.4 is a photograph showing the fringes in an interference pattern produced by a double-slit system.

E 21.1   Determine the angular separation between the $n = 0$ and $n = 1$ maxima for a double-slit system ($d = 0.20$ mm) utilizing light having a wavelength of 580 nm.   Answer: 0.17° = 0.0029 rad

In Figure 21.3(b) we see that

$$y = L\tan\theta \tag{21-4}$$

where y is the distance from the axis to the point P. Because $L \gg d$, the angle θ is quite small, and we may use the approximation, $\tan\theta \simeq \sin\theta$. Then Equation (21-4) becomes

$$y \cong L\sin\theta \tag{21-5}$$

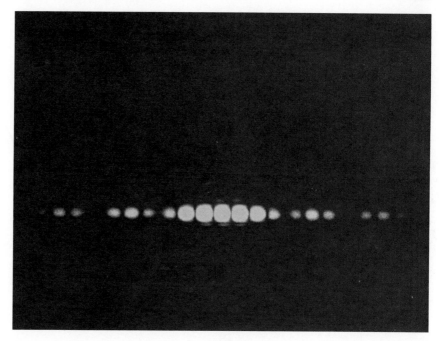

Figure 21.4
Photograph of the fringes of a double-slit interference pattern.

If we now suppose that point P is the center of the nth bright fringe, we obtain, using Equation (21-2),

$$y_n \simeq \frac{nL\lambda}{d}$$

or

$$\lambda = \frac{y_n d}{nL} \qquad (21\text{-}6)$$

Notice that the distances, y_n, d, and L, are quantities that are easy to measure. Therefore, Equation (21-6) provides a simple, inexpensive means for determining the wavelength of a monochromatic source of light.

E 21.2   If $d = 0.20$ mm and $L = 1.5$ m in the double-slit arrangement of Figure 21.3(b), the distance y between the $n = 0$ and $n = 1$ maxima is 4.6 mm. Determine the wavelength of the light used.   Answer: 610 nm

Thin-Film Interference

The vivid display of colors we see when sunlight reflects from a soap bubble or from a thin layer of oil floating on the surface of water is the result of optical interference. Figure 21.5 shows a section of a thin soap film with air on each side of the film. Incident light reaching each surface of the film is partly reflected and partly transmitted through that surface. Consequently, the reflected light reaching the eye is composed of light reflected directly from surface 1 and light that

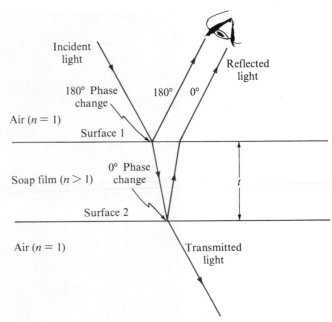

Figure 21.5
A section of a thin soap film with air on either side of the film. Light reaching each surface of the film is partly reflected and partly transmitted. The phase changes that occur upon reflection are indicated at each surface in the diagram.

has reflected from surface 2 and passed twice through the film. Those wavelengths of light that interfere constructively appear bright at the eye; those that interfere destructively appear dim or are completely missing. To analyze interference associated with thin films, we first need to be aware of the phase changes that take place when light is reflected at interfaces between media of different indices of refraction.

Electromagnetic theory predicts and experiment confirms that the phase of a light wave, traveling in a medium with index of refraction n_1 and reflecting from an interface with a medium having an index of refraction n_2, is affected as follows:

1. If $n_1 < n_2$, the phase of the light wave is shifted by 180° upon reflection.

2. If $n_1 > n_2$, the phase of the light wave is unchanged upon reflection.

That portion of a light wave that is transmitted through an interface between media undergoes no phase shift in crossing the interface.

In Figure 21.5 light of wavelength λ is incident nearly perpendicularly to the surface of a soap film of thickness t. According to the rules governing phase shifts at media interfaces, that part of the incident light wave that is reflected at surface 1 undergoes a phase shift of 180°. And that part, transmitted through the film, reflected from surface 2, and transmitted through surface 1 toward the eye, undergoes no phase change at either surface.

The phase shifts associated with these reflected waves are indicated in Figure 21.5 on the rays proceeding toward the eye. But a further phase difference exists between these rays because one has traveled an additional path length $2t$ (if we assume the light is incident normal to the film). When the path length $2t$ is equal to an integral number m of wavelengths λ_{med} in the medium of the soap film, the extra path length does not affect the phase difference between the viewed waves. These waves differ in phase by 180° because of reflections and, therefore, interfere destructively when

$$2t = m\lambda_{\text{med}} \quad ; \quad m = 0, 1, 2, 3, \ldots \tag{21-7}$$

and interfere constructively when

$$2t = \left(m + \frac{1}{2}\right)\lambda_{\text{med}} \quad ; \quad m = 0, 1, 2, 3, \ldots \qquad \textbf{(21-8)}$$

The wavelength λ_{med} in the film is equal to V/ν, where V is the speed of light in the medium of the film and ν is the frequency of the light wave. Then because the index of refraction n of the medium is equal to c/V, we may express the wavelength λ_{med} in terms of the wavelength λ in a vacuum (or in air) as

$$\lambda_{\text{med}} = \frac{V}{\nu} = \frac{1}{\nu}\left(\frac{c}{n}\right) = \frac{1}{n}\left(\frac{c}{\nu}\right) = \frac{\lambda}{n} \qquad \textbf{(21-9)}$$

E 21.3   Calculate the wavelength in water ($n = 1.33$) of yellow light ($\lambda = 580$ nm).   Answer: 440 nm

Thus if light of wavelength λ is incident normally onto a thin film of thickness t and index of refraction n, the conditions for constructive and destructive interference of the reflected light waves are:

The sum of the phase changes caused by both reflections is equal to $0°$ or $360°$.

$$2t = m\frac{\lambda}{n} \qquad \text{(Constructive)} \qquad \textbf{(21-10)}$$

$$2t = \left(m + \frac{1}{2}\right)\frac{\lambda}{n} \qquad \text{(Destructive)} \qquad \textbf{(21-11)}$$

The sum of the phase changes caused by both reflections is equal to $180°$.

$$2t = m\frac{\lambda}{n} \qquad \text{(Destructive)} \qquad \textbf{(21-12)}$$

$$2t = \left(m + \frac{1}{2}\right)\frac{\lambda}{n} \qquad \text{(Constructive)} \qquad \textbf{(21-13)}$$

Consider now the following problem:

White sunlight is incident normally onto a thin (550 nm) film of oil ($n = 1.2$), which is floating on water. Which wavelengths within the visible spectrum are minimally transmitted through the film?

The physical situation of this problem is visualized in Figure 21.6. Because the index of refraction of the film is intermediate between those of air and water, the reflection at each surface of the film introduces a phase change of $180°$. Then the sum of the phase changes caused by both reflections is equal to $360°$, which is equivalent to having no net phase change for the reflected light. In the present situation, light that is minimally transmitted through the film is maximally reflected so that we seek the condition for constructive interference in the reflected light. Because the wavelength in the medium of the film is equal to λ/n, this condition for constructive interference in the reflected waves occurs when the path length $2t$ through the film is an integral number of wavelengths $\lambda_{\text{med}} = \lambda/n$, or

$$2t = m\frac{\lambda}{n} \quad ; \quad m = 1, 2, 3, \ldots$$

$$2(550 \text{ nm}) = \frac{m\lambda}{1.2}$$

$$\lambda = \frac{1320 \text{ nm}}{m} \quad ; \quad m = 1, 2, 3, \ldots \qquad \textbf{(21-14)}$$

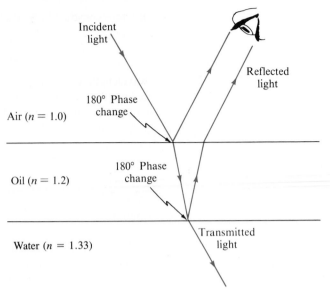

Figure 21.6
A thin film of oil floating on water. The light, which is incident nearly normally onto the film from air, undergoes a phase change of 180° at each interface.

Substitution of integral values of m into Equation (21-14) gives those wavelengths that interfere constructively when reflected from the film:

$$m = 1: \qquad \lambda = 1320 \text{ nm}$$

$$m = 2: \qquad \lambda = 660 \text{ nm}$$

$$m = 3: \qquad \lambda = 440 \text{ nm}$$

$$m = 4: \qquad \lambda = 330 \text{ nm} \qquad \qquad \textbf{(21-15)}$$

Because the visible spectrum includes wavelengths between 400 nm and 700 nm, we may conclude that the required wavelengths are 660 nm and 440 nm.

Example 21.1

PROBLEM   What is the minimum thickness of a soap bubble ($n = 1.3$) that will maximally reflect normally incident light having a wavelength of 520 nm?

SOLUTION   Figure 21.7 depicts the physical situation. Because light reflected from surface 1 has a phase change of 180° and light reflected from surface

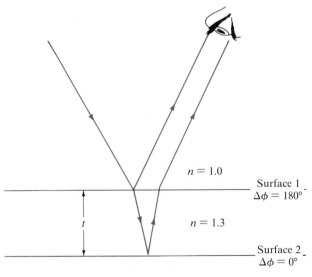

Figure 21.7   Example 21.1

2 has no phase change, the reflected light will be brightest if the extra path length $2t$ introduces an additional phase difference equal to 180° (or, more generally, 180° + n(360°), where $n = 0, 1, 2, . . .$). Because this phase difference is proportional to $2t$, the minimum value of t corresponds to a phase difference equal to 180°, or

$$2t = \frac{1}{2}\,\lambda_{\text{med}} = \frac{1}{2}\frac{\lambda}{n} = \frac{520 \text{ nm}}{2(1.3)} = 200 \text{ nm}$$

$$t = 100 \text{ nm} \quad \blacksquare$$

E 21.4   For Example 21.1, calculate the next smaller bubble-thickness for which maximum reflection occurs.   Answer: 300 nm

21.2 Optical Diffraction

In section 21.1 we introduced the concept of diffraction, the spreading or bending of light when it passes through an opening with dimensions comparable to the wavelength of the light. We have also seen how Huygens' principle permits us to use a simplified analysis of diffraction phenomena. In this section we will use similar techniques to examine the basic characteristics of diffraction patterns that result from narrow slits.

In the preceding section we treated the slits as if their widths were much less than the wavelength of the incident light. Actual slits have widths that are equal to many wavelengths of visible light. We will now consider the numerous Huygens sources across the width of a slit in order to analyze the diffraction and the resulting interference pattern from a single slit and then from multiple-slit systems.

Diffraction phenomena involving small openings are conveniently divided into two categories. When either the source of light or the viewing screen is close to the opening that causes diffraction, the phenomena is called *Fresnel diffraction*. When both the source and screen are sufficiently far from the opening that the light rays incident on and exiting from the opening may be considered parallel, the phenomenon is called *Fraunhofer diffraction*. Analysis of Fresnel diffraction is complex; we will consider only the Fraunhofer case.

Single-Slit Diffraction

Fraunhofer diffraction from a single slit involves the superposition of parallel rays from different points along the width of the diffraction slit. Figure 21.8(a) shows parallel rays incident onto a slit and diffracted parallel rays that are superposed at a point far from the slit. In practice we may observe the effects of Fraunhofer diffraction using the apparatus shown in Figure 21.8(b). A convergent lens L is positioned between the point source S and the slit. This lens is at a distance from S equal to the focal length f_1 of the lens. Then plane wavefronts are incident onto the slit. A second convergent lens L_2 is placed at a distance equal to its focal length f_2 from the screen. The parallel rays leaving the slit at an angle θ are then focused at a point P on the screen.

Let us now analyze the diffraction pattern produced by a single slit using Figure 21.8(c). According to Fermat's principle, the transit time is the same for

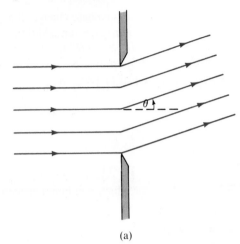

(a)

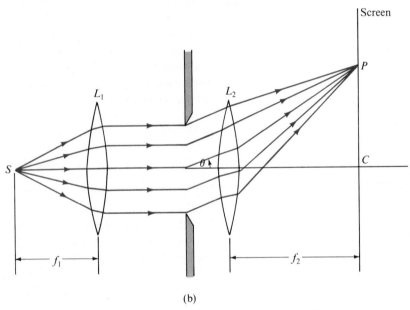

(b)

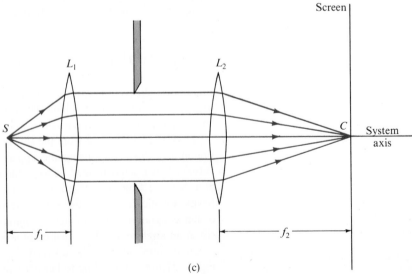

(c)

Figure 21.8
(a) Parallel rays that emerge from a diffracting slit at an angle θ. These rays may be considered to converge to a point at a distance very far from the slit. (b) System for studying Fraunhofer diffraction. (c) Formation of the central maximum in a Fraunhofer diffraction system.

each ray that emerges from an object point, passes through a converging lens, and then converges on an image point. In Figure 21.8(c) the parallel rays incident on the lens L_2 are equivalent to rays from a point source at an infinite distance from L_2. Therefore, the time of transit is the same for each ray reaching point C on the system axis. The frequency of each ray is identical, because the source is monochromatic, so the phase difference between any pair of rays that reaches point C is equal to zero. Then point C is a point of constructive interference, a bright spot in the *central maximum* of the diffraction pattern on the screen.

Similarly, the phase differences of the light rays diffracted by the slit and converged toward another point, like point P in Figure 21.8(a), are not affected by the lens L_2. Then the minima, or dark lines, in the single-slit diffraction pattern may be determined by first examining two rays like the ones shown exiting the slit in Figure 21.9. The Huygens sources of these rays are spaced half the slit width apart, or $w/2$ apart. When the path difference of these two rays is equal to $\lambda/2$, where λ is the wavelength of the light, they arrive at the screen with a phase difference of π radians and, therefore, interfere destructively. Every other pair of Huygens sources that is spaced $w/2$ apart produces rays that similarly interfere destructively, and, as a result, a dark line appears on the screen where these rays coincide. If these rays make an angle θ_1 with the system axis, the construction of Figure 21.9 shows that the path difference between each pair of rays with sources separated by $w/2$ is equal to $(w/2)\sin\theta_1$. Then the condition for the first minimum (nearest the central maximum) of the diffraction pattern is

$$\frac{w}{2}\sin\theta_1 = \frac{\lambda}{2}$$

$$\sin\theta_1 = \frac{\lambda}{w} \qquad\qquad (21\text{-}16)$$

The angle θ_m that locates the mth minimum may be determined by considering pairs of sources that lie along the slit width and are separated by $w/(2m)$, where $m = 1, 2, 3, \ldots$. Here the effect is as though the width w of the slit were divided into m regions, each of width $w/(2m)$. In each of these regions, light from the upper half interferes destructively with light from the lower half of that region, just as we have demonstrated in locating the first minimum. Figure 21.10 illustrates the conditions required for the production of minima in the $m = 1$ and $m = 2$ cases. Each pair of these rays interferes destructively at the screen when

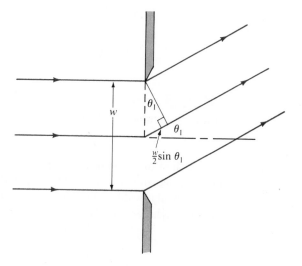

Figure 21.9
Diagram for analysis of the path difference between pairs of rays from Huygens sources that are separated by half the width w of a single slit.

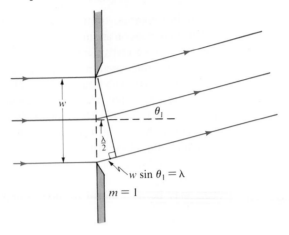

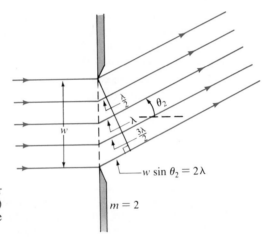

Figure 21.10
Illustrating the conditions required for producing minima in the first ($m = 1$) and second ($m = 2$) orders for a single slit.

θ_m is such that the path lengths of these rays differ by $\lambda/2$. Since that path difference is equal to $[w/(2m)]\sin\theta_m$, the condition for the mth minimum of the diffraction pattern is

$$\frac{w}{2m} \sin\theta_m = \frac{\lambda}{2}$$

$$\sin\theta_m = \frac{m\lambda}{w} \quad ; \quad m = \pm1, \pm2, \pm3, \ldots \qquad (21\text{-}17)$$

E 21.5   Calculate the angular separation between the $m = 1$ and $m = 2$ diffraction minima of a 0.20-mm slit for light having $\lambda = 500$ nm.
Answer: $0.14° = 0.0025$ rad

A graph of the brightness of the single-slit diffraction pattern as a function of angle θ is shown in Figure 21.11(a) when the slit width w is equal to 6λ. Notice that the angular width of the central maximum is approximately twice that of the adjacent bright regions. This characteristic of a single-slit pattern may be under-

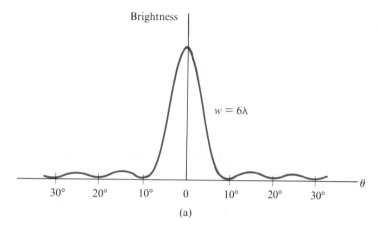

(a)

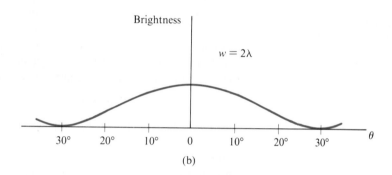

(b)

Figure 21.11
Graphs of brightness in a single-slit diffraction pattern as a function of θ for **(a)** a slit width $w = 6\lambda$ and **(b)** a slit width $w = 2\lambda$. **(c)** A photograph of a single-slit diffraction pattern when the slit width $w \gg \lambda$.

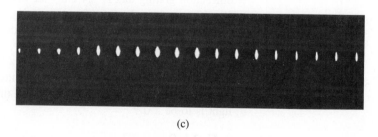

(c)

stood from Equation (21-17), in which the angular widths between minima are obtained by using successive integers (except $m = 0$, which yields the central maximum). Thus the width of the central maximum corresponds to a two-integer difference in m, from $m = -1$ to $m = +1$.

E 21.6   Determine the angular width of the central maximum of the single-slit diffraction pattern of Exercise 21.5, in which $w = 0.20$ mm and $\lambda = 500$ nm.
Answer: $0.29° = 0.0050$ rad

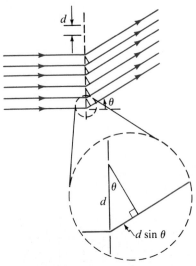

Figure 21.12
Fraunhofer diffraction at an angle θ from a plane diffraction grating having slits separated by a distance d.

Figure 21.11(b) shows a graph of the brightness of a single-slit diffraction pattern as a function of θ for $w = 2\lambda$. Here the central maximum extends over a greater angle than for the wider slit of Figure 21.11(a). In fact, Equation (21-16) indicates that when $w = \lambda$, θ_1 becomes equal to 90°, and the central maximum extends over the entire length of the plane of the screen. Figure 21.11(c) is a photograph of a single slit diffraction pattern when the slit width w is considerably greater than λ.

Diffraction Gratings

An opaque mask having a large number of equally spaced, parallel slits through which light can pass is called a *diffraction grating*, a device that can be used to measure the wavelengths of light. Typical gratings contain several thousands of slits (usually called lines) per cm. The spacing d between slits is equal to the reciprocal of the number of lines per unit length. Thus a grating with 2000 lines/cm has a spacing of $d = (1/2000)$ cm $= 5.0 \times 10^{-4}$ cm.

A plane diffraction grating is schematically illustrated in Figure 21.12. Let us consider the Fraunhofer diffraction at each slit, if the incident light of wavelength λ is normal to the plane of the grating. Each diffracted ray interferes with parallel rays from the other slits. If θ is the angle that rays make with the system axis, we may see in Figure 21.12 that constructive interference occurs when

$$d \sin\theta = m\lambda \qquad m = 0, \pm 1, \pm 2, \ldots \qquad \text{(21-18)}$$

where d is the spacing between slits and m is called the *order* of the maximum produced by the grating. The brightness of light in the diffraction pattern of a grating is plotted as a function of $\sin\theta$ in Figure 21.13(a). Figure 21.13(b) is a photograph of the pattern of a diffraction grating.

E 21.7   Light having a wavelength of 600 nm is incident normally onto a diffraction grating that is 2.4 cm wide and has 2000 uniformly spaced lines.
(a) What is the value of d for this grating?   Answer: 1.2×10^{-3} cm
(b) Determine the angular separation of the $m = 1$ and $m = 3$ maxima.
Answer: $5.8° = 0.10$ rad

E 21.8   How many lines per cm must be ruled on a diffraction grating if the angular separation of the central and first maxima for 500-nm wavelength light is to be 2.0°?   Answer: 700 lines/cm

E 21.9   A grating having 1800 lines/cm is used to study light having a wavelength of 630 nm. What is the highest order maximum that can be observed?   Answer: $m = 8$

The interference pattern produced by a grating with a large number N of slits, as shown in Figure 21.13, is noticeably different from the pattern of a two-slit system like that of Figure 21.4 in two significant ways. First, the brightness of each maximum of the N-slit pattern is greater than that of the two-slit pattern. Also, each maximum of the grating pattern is much sharper than a two-slit maximum; each narrow bright region is separated from an adjacent maximum by a wide dark region. We may explain these differences physically by the following considerations.

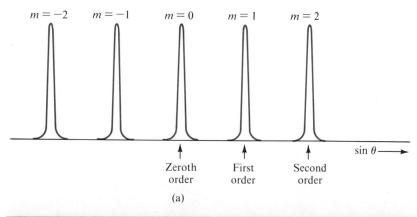

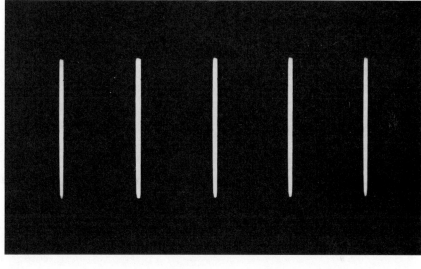

Figure 21.13
(a) Brightness of light in the diffraction pattern of a plane grating plotted as a function of $\sin\theta$. **(b)** A photograph of the pattern produced by a plane diffraction grating.

(b)

The brightness of each maximum in a multislit pattern is greater than one in the two-slit case because N waves are superposed constructively at each maximum instead of only two. The sharper maxima in the pattern of the multislit system are perhaps best explained using a specific example, say a 50-slit grating, as shown in Figure 21.14. If a maximum appears on a screen at the point P_{max}, a minimum occurs at P_{min} when the waves from slits 1 and 26 are out of phase by 180° or when the path difference of these waves is equal to $\lambda/2$. These waves destructively interfere at P_{min}. Similarly, the waves from slits 2 and 27, slits 3 and 28, and so on to slits 25 and 50 destructively interfere at P_{min}, which is, therefore, dark. Then if the distance d between adjacent slits is the same in both cases, the half-width of a maximum (the distance from P_{max} to P_{min}) in the 50-slit system is smaller by a factor of $2/50 = 1/25$ than that for the two-slit system. And, in general, for an N-slit grating, the maxima of its interference pattern are more narrow than that of a two-slit system by a factor of $N/2$.

Although diffraction gratings are used primarily to measure the wavelength of a light source, two characteristics of a given grating are related to how the grating affects light having wavelengths that differ by a small amount. First, the *angular dispersion D* of a grating is a measure of the angular separation $\Delta\theta$ that a given

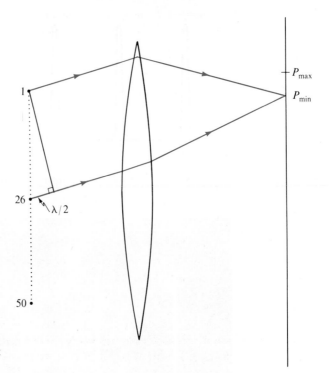

Figure 21.14
Diagram for the analysis of a 50-slit diffraction grating.

grating produces between two normally incident monochromatic waves having wavelengths that differ by a small interval $\Delta\lambda$. We may obtain an expression for dispersion by differentiating Equation (21-18) with respect to λ, treating θ as a function of λ:

$$\cos\theta \frac{d\theta}{d\lambda} = \frac{m}{d} \qquad (21\text{-}19)$$

Then the dispersion D of a grating is given by

$$D = \frac{d\theta}{d\lambda} = \frac{m}{d\cos\theta} \simeq \frac{\Delta\theta}{\Delta\lambda} \qquad (21\text{-}20)$$

where we have approximated $d\theta/d\lambda$ by the ratio of the small angular separation $\Delta\theta$ of two nearly equal wavelengths separated by $\Delta\lambda$.

E 21.10   A 2000-line/cm grating is illuminated with monochromatic light having $\lambda = 550$ nm.
(a) Determine θ for the first order ($m = 1$) maximum.

Answer: 0.11 rad = 6.3°

(b) Calculate the (first-order) dispersion for this situation.

Answer: 2.0×10^5 rad/m = 1.2×10^7 °/m

E 21.11   For the situation of Exercise 21.10, determine the angular separation of the first-order maxima for two wavelengths that are near 550 nm but differ by 1.0 nm.   Answer: 0.011° = 2.0×10^{-4} rad

A second characteristic of a grating depends on its ability to *resolve*, or to permit our distinguishing between the patterns produced by, two waves separated by a small wavelength interval $\Delta\lambda$. The *resolving power R* of a diffraction grating is defined to be the wavelength λ divided by the smallest wavelength difference $\Delta\lambda = \lambda - \lambda'$, where $\lambda \simeq \lambda'$, that can be resolved by the grating, i.e.,

$$R = \frac{\lambda}{\Delta\lambda} \qquad (21\text{-}21)$$

And we assert without proof that the maximum resolving power of a diffraction grating is directly proportional to the order m of diffraction and to the total number N of lines in the grating, or that

$$R = mN \qquad (21\text{-}22)$$

E 21.12   A grating with a first-order resolving power of 2000 is to be used to study visible radiation having wavelengths near 600 nm. Determine the smallest wavelength difference that can be resolved for $m = 1$.

Answer: 0.30 nm

E 21.13   What is the maximum second-order resolving power of a 2.0-cm-wide grating having 1500 lines/cm?   Answer: 6000

Example 21.2   **PROBLEM**   A collimated light beam, which has a circular cross section that is 1.0 mm in diameter, is known to consist of two monochromatic signals having wavelengths of approximately 580 nm that differ by 0.10 nm. Will a 2.0-cm-wide diffraction grating having 5000 lines/cm resolve these two wavelengths?

SOLUTION   The resolving power R necessary to distinguish between the two wavelengths must satisfy

$$R > \frac{\lambda}{\Delta\lambda} = \frac{580 \text{ nm}}{0.10 \text{ nm}} = 5800$$

Since the light beam is only 1.0 mm wide, the effective number of lines on the grating is given by

$$N = (0.10 \text{ cm})(5000 \text{ lines/cm}) = 500$$

And because the resolving power for a grating is given by $R = mN$, we have

$$mN > \frac{\lambda}{\Delta\lambda}$$

or

$$m > \frac{5800}{500} = 11.6$$

That is, if the two wavelengths are to be resolved, the $m = 12$ maxima must be observed. But for a grating, we have

$$m\lambda = d \sin\theta \leq d$$

so that

$$m \leq \frac{d}{\lambda} = \frac{(1/500000)\text{m}}{580 \times 10^{-9}\text{m}} = 3.4$$

Since the $m = 3$ maxima are the highest order that may be observed, we conclude that the grating will not resolve the two wavelengths.   ■

E 21.14   If the light beam of Example 21.2 were as wide as the grating, could the two lines be resolved?   Answer: Yes, for both $m = 2$ and $m = 3$

21.3 Polarization of Light

Throughout our considerations of optics, we have repeatedly appealed to our earlier assertion that, according to electromagnetic theory, light is a transverse electromagnetic wave. The interference phenomena that we have described in this chapter are considered conclusive experimental evidence that light is indeed a wave. The polarization phenomena that are the subject of this section constitute evidence that light waves are transverse, that is, both the electric and magnetic fields of a light wave are perpendicular to the direction of propagation of that wave.

Most light sources produce waves in which the electric fields, although confined to planes perpendicular to the direction of propagation, are directed randomly (and are, therefore, uniformly distributed) throughout those transverse planes. Illustrating this situation, Figure 21.15 shows the symmetrical distribution of electric-field directions in the plane of the page, which is normal to the direction of propagation (out of the page). Such light is said to be *unpolarized*. Then light that displays any asymmetry in its electric-field distribution in the plane perpendicular to the propagation direction is said to be *polarized*.

We may distinguish between several types of optical polarization. If, for example, the electric fields of a light wave are confined to one plane, the wave is said to be *plane-polarized* (or *linearly polarized*, because at any point on the wave, oscillation of the electric field is along a line). Light is said to be *partially polarized* if it is a mixture of polarized and unpolarized light. If the electric-field vector of a light wave rotates about the direction of propagation and its tip traces out a circle, the wave is said to be *circularly polarized*. If the tip traces out an ellipse, the wave is *elliptically polarized*. We are concerned here primarily with linearly polarized light. We will represent linearly polarized rays by diagrams like those of Figure 21.16. Figure 21.16(a) is a perspective view of a schematic version of a light ray that is linearly polarized along the *y* axis and propagating in the *z* direction. Figures 21.16(b) and 21.16(c) show the same linearly polarized ray viewed looking in the *y* direction and in the *x* direction, respectively.

Unpolarized light can be polarized (at least partially and, in some cases, completely) by a number of processes. Let us illustrate by considering one of these processes: selective absorption. Certain natural crystals, like tourmaline and a manufactured product called Polaroid, which contains microscopic crystals imbedded and aligned in nitrocellulose, exhibit *selective absorption*. They strongly absorb light waves with electric fields perpendicular to a particular crystalline direction but pass light relatively undiminished when its electric fields are parallel to that direction. Figure 21.17 shows unpolarized light incident on a sheet of polarizing material, which selectively absorbs light with its electric field perpendicular to the polarizing direction of the sheet. The light transmitted through the *polarizer*, as the first sheet is called, is linearly polarized. This fact may be verified by using a second sheet of the same material oriented in the same direction as the first, as shown in Figure 21.18(a). The polarized light passes through the second sheet, called an *analyzer*, and experiences negligible absorption. If, however, the analyzer is rotated 90°, as indicated in Figure 21.18(b), virtually no light passes through the analyzer.

Some of the other processes that can produce polarized light include reflection and double refraction. Let us consider each of these processes briefly.

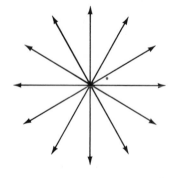

Figure 21.15
Symmetrical distribution of electric field directions in unpolarized light that is propagating in a direction normal to the plane of the page.

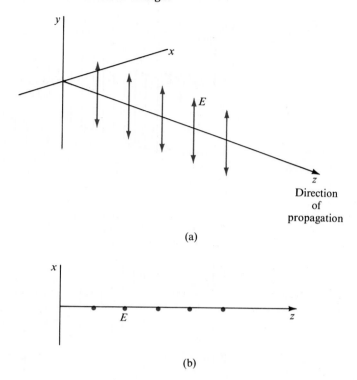

(a)

(b)

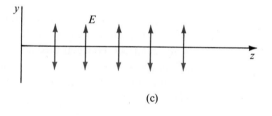

Figure 21.16
Schematic representations of light that
is linearly polarized in the *y* direction
and propagating in the *z* direction.

(c)

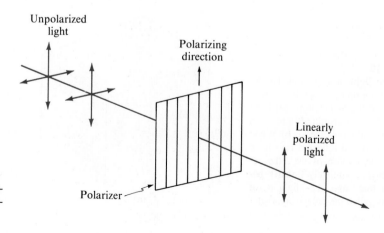

Figure 21.17
Polarization of light by selective absorption as it passes through a polarizer.

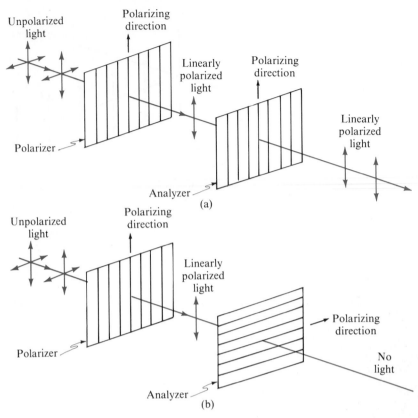

Figure 21.18
Using an analyzer to determine that a polarizer has produced linearly polarized light.

When light is reflected from a dielectric surface like glass the reflected light is, in general, partially polarized. In Figure 21.19(a) the unpolarized light incident on the interface between media with indices of refraction n_1 and n_2 at the incident angle θ_1 has a reflected component at angle θ_1 and a refracted component at an angle θ_2. The reflected ray is unpolarized when θ_1 is equal to 0° or 90°, corresponding to light reaching the surface at normal incidence or grazing inci-

Figure 21.19
Unpolarized light in a medium having index of refraction $n_1 < n_2$ reflects from the surface of the n_2 medium. **(a)** Both the reflected and transmitted light is, in general, partially polarized. **(b)** When the transmitted and reflected rays have an angular separation of 90°, the reflected light is completely linearly polarized. In that case, the angle θ_p of incidence is called the polarizing angle.

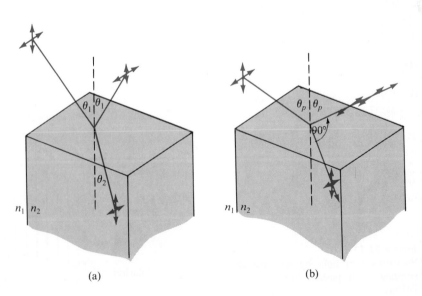

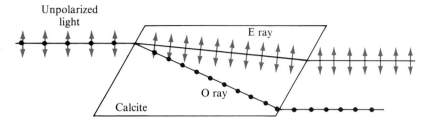

Figure 21.20
A ray of unpolarized light, incident onto a birefringent crystal of calcite, splits into two linearly polarized rays. The directions of polarization of the rays in the crystal are mutually perpendicular.

dence. At all other angles of incidence the reflected light is polarized predominantly in a direction parallel to the interface. Sir David Brewster discovered that when the reflected and refracted rays are 90° apart, the reflected light is completely polarized, as indicated in Figure 21.19(b). The incident angle θ_o at which complete polarization of the reflected light takes place is called the *polarizing angle* θ_p. From the geometry of Figure 21.19(b), we may relate θ_p and θ_2 by

$$\sin\theta_2 = \cos\theta_p \qquad (21\text{-}23)$$

Using Equation (21-23) in conjunction with Snell's law, $n_1 \sin\theta_1 = n_2 \sin\theta_2$, and letting $\theta_1 = \theta_p$, we obtain

$$n_1 \sin\theta_p = n_2 \cos\theta_p$$

$$\tan\theta_p = \frac{n_2}{n_1}$$

or

$$\theta_p = \tan^{-1}\left(\frac{n_2}{n_1}\right) \qquad (21\text{-}24)$$

Equation (21-24) is called *Brewster's law*, which relates the polarizing angle of reflection to the indices of refraction of media on either side of an interface.

E 21.15   For what angle of reflection will light reflected from the surface of a smooth lake ($n = 1.33$) be completely polarized?   Answer: 53°

Polarization by double refraction depends on special optical characteristics of certain transparent crystals like calcite and quartz. Unlike glass these materials are *anisotropic*, that is, they have different optical characteristics along different crystalline directions. In particular, the indices of refraction of light have different values in different crystalline directions. Thus when a ray of unpolarized light is incident on a face of a calcite crystal, as shown in Figure 21.20, two rays proceed through the crystal. One ray is called the *ordinary ray*, labeled O, which is refracted in accordance with Snell's law for an index of refraction n_O for the crystal. The other ray, called the *extraordinary ray*, is refracted at an angle consistent with an index of refraction n_E. The two rays are linearly polarized in mutually perpendicular directions, as indicated in Figure 21.20. As they exit the crystal, two linearly polarized beams are spatially separated. Because such anisotropic crystals are characterized by two indices of refraction, they are sometimes called *doubly refracting*, or *birefringent*, crystals. Figure 21.21 is a photograph that demonstrates the birefringent character of calcite, which produces a double image when an object is viewed through the crystal.

Figure 21.21
Photograph showing the double image produced when an object is viewed through a birefringent crystal.

21.4 Problem-Solving Summary

Problems in physical optics involve the interference, diffraction, or polarization of light waves. Such problems depend specifically on the wave nature of light. The analysis of problems in physical optics is simplified by Huygens' principle, which permits the treatment of every point on a wavefront as if it were a source of secondary wavelets.

Optical interference occurs when two or more light waves are simultaneously at the same point. An observable interference pattern occurs only if the light sources are monochromatic sources having equal wavelengths and are coherent (that is, have a constant phase relationship).

A double-slit interference pattern is produced by two slits separated by a distance d and illuminated by monochromatic light of wavelength λ. The pattern on a screen at a distance L from the slits consists of alternating bright and dark bands, or fringes. Because the phase difference δ between two coherent waves increases by 2π radians for each wavelength λ of path-length difference Δr, or $\delta = (2\pi\Delta r)/\lambda$, the condition for an interference maximum (a bright line) is $d\sin\theta = n\lambda$ ($n = 0, 1, 2, \ldots$). A dark line occurs when $d\sin\theta = (n + \frac{1}{2})\lambda$. If y_n is the distance from the center of the pattern to the nth bright fringe, the wavelength λ of the light may be related to the easily measured quantities d and L by $\lambda = y_n d/(nL)$, provided $y_n \ll L$.

Analysis of the interference of light reflected from thin films is based on the following facts:

1. A light wave traveling in a medium of lower index of refraction, when it encounters and reflects from an interface with a medium of higher index of refraction, undergoes a phase change of 180°.

2. No phase change occurs when light waves traveling in a medium of a higher index of refraction encounter and reflect from an interface with a medium of lower index of refraction.

3. Light waves passing through an interface between media with different indices of refraction experience no phase change.

4. The wavelength λ_{med} of light in a medium with index of refraction n is equal to λ/n, where λ is the wavelength of light in a vacuum (or in air).

If the net phase change caused by reflections from both surfaces of a thin film is equal to 0° or 360°, *constructive* interference occurs for light reflected normally to the film of thickness t when the length $2t$ is equal to an integral multiple of λ_{med}. If the net phase change is equal to 180°, *destructive* interference occurs for the reflected light when $2t$ is equal to an integral multiple of λ_{med}.

Diffraction is the bending of light passing near an edge or through an opening with dimensions comparable to the dimensions of the wavelength of the light. Diffraction patterns caused by light passing through narrow slits result from the interference of large numbers of light waves from coherent Huygens sources within the slits. Fraunhofer diffraction occurs when the interfering rays of light are approximately parallel.

Fraunhofer diffraction from a single slit of width w produces minima (dark lines) when $w\sin\theta = m\lambda$ ($m = \pm1, \pm2, \ldots$), where m is called the order of diffraction.

A diffraction grating is an opaque mask with a large number of identical slits spaced an equal distance d apart. When light of wavelength λ is incident normally on the grating, interference maxima occur when $d \sin\theta = m\lambda$, where m is the order of diffraction. The angular dispersion D of a grating measures its ability to provide angular separation $\Delta\theta$ per unit wavelength interval $\Delta\lambda$ according to $D = \Delta\theta/\Delta\lambda = m/(d \cos\theta)$. The resolving power R of a grating measures its ability to resolve, or permit our distinguishing between, patterns produced by a wavelength λ and another wavelength that differs from λ by a small wavelength interval $\Delta\lambda$. The resolving power, $R = \lambda/\Delta\lambda$, is given by $R = mN$, where m is the order of the pattern and N is the total number of slits in the grating. Notice that the resolving power is greater for higher orders of the pattern.

The polarization of light refers to any asymmetry in the distribution of the electric fields in a light wave in planes perpendicular to the direction of propagation of the wave. Unpolarized light may be partially, and sometimes completely, linearly polarized by selective absorption in certain polarizing materials (like tourmaline or Polaroid sheets), by reflection, and by double refraction from birefringent crystals (like calcite or quartz). In the case of polarization by reflection, when unpolarized light in a medium with index of refraction n_1 is reflected from the surface of a dielectric material with index n_2, the wave is, in general, partially polarized in a plane parallel to the reflecting surface. Complete linear polarization occurs when the unpolarized light is incident on (and reflected from) the surface at the polarizing angle θ_p, given by Brewster's law: $\theta_p = \tan^{-1}(n_2/n_1)$.

Problems

GROUP A

21.1 In a double-slit interference system, what happens to the locations of maxima in the interference pattern if (a) the separation between the slits is increased? (b) the wavelength of incident light is increased?

21.2 For single-slit diffraction, what happens to the locations of maxima in the interference pattern if (a) the width of the slit is decreased? (b) the wavelength of the incident light is decreased?

21.3 Light beams of wavelengths λ_1 and λ_2 are simultaneously incident onto a single slit. The first interference minimum of λ_1 coincides with the second interference minimum of λ_2. How are the wavelengths λ_1 and λ_2 related?

21.4 A 450-nm wavelength light signal is incident in air onto a transparent material ($n = 1.5$).

(a) Determine the frequency of this light wave in the transparent material.

(b) What is the wavelength of this light wave as it propagates in the material?

21.5 A fourth-order maximum in a double-slit interference pattern is 1.0 cm from the central maximum of the pattern. The separation between the two slits is equal to 600 wavelengths of the monochromatic light incident onto the slits. What is the distance between the plane of the slits and the viewing screen?

21.6 Unpolarized light is reflected from a certain smooth surface and then viewed through a Polaroid sheet. The reflected light can be "turned off" with the polarizer when the angle of reflection is equal to 56°. What is the index of refraction of the reflecting material?

21.7 What slit width is required so that a single-slit interference pattern will produce a central maximum that subtends an angle of 2.0° at the slit when light of 540 nm is incident normally onto the slit?

21.8 A grating having 2000 lines/cm is illuminated with a light signal containing only two wavelengths, $\lambda_1 = 480$ nm and $\lambda_2 = 540$ nm. The signal is incident perpendicularly onto the grating. What is the angular separation for the third order maxima of these two wavelengths?

21.9 Red light of wavelength 600 nm is incident on a double-slit system that is 10 m from the viewing screen. If the second-order bright fringe is 5.0 cm from the system axis, what is the separation between the two slits?

21.10 A transparent thin film (thickness = t, index of refraction = 1.4) reflects normally incident light (λ = 560 nm).

(a) What is the wavelength of the light as it propagates in the film?

(b) Determine the two smallest values of t for which the reflected light is a maximum.

21.11 What minimum thickness of an oil film (n = 1.2) floating on water in a parking lot will minimally reflect light of wavelength 500 nm incident normally onto the film from the air?

21.12 What is the minimum thickness of a transparent coating (n = 1.4) that must be applied to the outer surface of a camera lens (n = 1.6) in order that the lens transmit the maximum amount of light at a wavelength of 500 nm?

21.13 The third-order maximum of a diffraction pattern occurs at an angle of 30° when light of wavelength 400 nm is incident normally onto a diffraction grating. This grating is composed of how many lines per cm?

21.14 A uniform film of oil (n = 1.31) is floating on water. When sunlight in air is incident normally onto the oil film, an observer (in air) finds that the reflected light has a brightness maximum at λ = 450 nm and a brightness minimum at 600 nm. What is the thickness of the oil film?

21.15 Estimate the minimum width for a diffraction grating (2000 lines/cm) that is to be used to resolve (in the first order) two light waves having wavelengths of approximately 550 nm that differ in wavelength by 0.25 nm.

21.16

(a) What is the wavelength of light that, when incident normally onto a transmission grating (5000 lines/cm), is deviated by 30° in the second order?

(b) What is the first-order deviation for light of this wavelength?

(c) What is the total number of maxima that can be seen in this diffraction pattern?

21.17 White light is incident normally onto a soap film (n = 1.4), which is 200 nm thick and has air on both sides. What wavelengths within the visible spectrum have minimum brightness to an observer who views the light transmitted through the film?

21.18 When white light is incident normally onto a diffraction grating that has 2000 lines/cm, for what value of m is the mth order maximum for blue light (λ = 400 nm) closer to the central maximum than the (m − 1)th-order maximum of red light (λ = 650 nm)?

21.19

(a) Show that the angular dispersion (D = $d\theta/d\lambda$) for a grating that is illuminated normally with monochromatic light of wavelength λ is given by

$$D = \frac{\tan\theta}{\lambda}$$

where θ is the angular deviation (of a maximum) from the original direction of propagation.

(b) Estimate the angular separation of the maxima (of the same order) at θ = 45° for two monochromatic light signals (wavelengths $\approx$ 550 nm) that differ in wavelength by 4.0 nm.

21.20 Show that the resolving power R for a grating of width W obeys

$$R < \frac{W}{\lambda}$$

where λ is the wavelength of the light to be studied.

21.21 A slice (thickness = t) of transparent material (n = 1.6) is placed over one of the two slits of a double-slit experiment. If the wavelength of the light used is equal to 480 nm, determine the two smallest values of the thickness t for which the central point on the screen will be (a) a maximum; (b) a minimum.

21.22 A piece of tissue paper is inserted between two horizontal microscope slides lying one on top of the other, thereby forming a wedge of air between them. When light of 600-nm wavelength shines vertically onto the plates, an observer sees six dark interference fringes per cm of length along the slides. At what angle is one slide inclined to the other?

21.23 A diffraction grating is constructed with 5000 lines/cm over a total length of 3.0 cm. Two monochromatic beams of light with a mean wavelength of 600 nm are barely resolved in the second order by this grating. What is the wavelength difference for these two waves?

21.24 What is the minimum thickness of a thin film (n = 1.4) that will maximally transmit a 630-nm light signal while maximally reflecting a 560-nm signal? The film has air on both sides.

21.25 Using a technique similar to that by which the minima of a single-slit interference pattern were obtained, derive an expression giving the angular position of the maxima of a single-slit interference pattern in terms of the slit width w and the wavelength λ of light that is normally incident onto the slit. Find the angular separation between the central maximum and the first two successive maxima when light of 650 nm is normally incident onto a slit having a width of 0.10 mm.

21.26 Two lines in the visible spectrum of hydrogen have wavelengths of 410 nm and 434 nm. How many lines/cm must a transmission grating have in order that these hydrogen lines will be separated by 5.0° in the second order?

21.27 Suppose light of wavelength λ in air is incident onto a thin transparent film with index of refraction n at an angle θ (not necessarily near θ = 0) measured from the normal to the surface of the film. Air is on both sides of the film. Derive an expression for the thickness t of the film, in terms of λ, n, and θ, so that the light reflected from the film will constructively interfere.

GROUP B

21.28 Maxwell's equations may be used in conjunction with the Poynting vector to show that the intensity I of a (sinusoidal) electromagnetic wave is proportional to the square of the amplitude of the electric field. Suppose two electromagnetic waves of the same frequency ω produce, at a point P in space, electric fields given by

$$E_1(t) = E_o \sin \omega t$$

$$E_2(t) = E_o \sin(\omega t + \delta)$$

where E_o is the amplitude of each field and δ is the phase difference between the two disturbances.

(a) Show that the magnitude of the resultant electric field at point P may be written

$$E(t) = A \sin(\omega t + \delta/2)$$

where the amplitude A is given by

$$A = 2E_o \cos(\delta/2)$$

(b) Show that the intensity of the electromagnetic signal at point P may be written

$$I(\delta) = I_o \cos^2(\delta/2)$$

where I_o is the intensity corresponding to a phase difference of zero, that is, $\delta = 0$.

(c) Show that the amplitude A obtained in part (a) is equal to the magnitude of the sum of two vectors (phasors) of amplitude E_o whose directions (phases) differ by the angle δ, as shown in Figure 21.22.

(d) Sketch the vector triangle of part (c) for $\delta = 0, 2\pi/3$, and π. For what values of δ is A equal to $2E_o$? E_o? zero? a maximum? a minimum?

(e) Sketch a graph of $I(\delta)$ for $0 \le \delta \le 2\pi$.

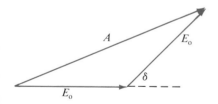

Figure 21.22   Problem 21.28

21.29 The double-slit interference pattern that is shown in Figure 21.4 may be predicted by combining the results of Problem 21.28 and the double-slit geometry shown in Figure 21.3.

(a) Show that the intensity at point P in Figure 21.3 is given by

$$I = I_o \cos^2\left(\frac{\pi d \sin\theta}{\lambda}\right)$$

where I_o is the intensity at the central ($\theta = 0$) maximum.

(b) Show that the locations of the maxima and minima of the intensity relation obtained in part (a) agree with those predicted by Equations (21-2) and (21-3).

(c) Sketch a graph of the double-slit intensity I versus the unitless quantity $\delta = 2\pi d \sin\theta/\lambda$ for $-4\pi < \delta < 4\pi$.

21.30 Suppose three electromagnetic waves of the same frequency ω produce, at a point P in space, electric fields described by

$$E_1(t) = E_o \sin(\omega t + \delta)$$

$$E_2(t) = E_o \sin \omega t$$

$$E_3(t) = E_o \sin(\omega t - \delta)$$

where E_o is the amplitude of each field.

(a) Show that the magnitude of the resultant electric field at point P may be written

$$E(t) = A \sin \omega t$$

where the amplitude A is given by

$$A = (1 + 2\cos\delta)E_o$$

(b) Show that the intensity of the electromagnetic signal at point P may be written

$$I(\delta) = I_o\left(\frac{1 + 2\cos\delta}{3}\right)^2$$

where I_o is the intensity for $\delta = 0$.

(c) Show that the amplitude A obtained in part (a) is equal to the magnitude of the sum of three vectors (phasors) of amplitude E_o that have directions (phases) that differ by the angle δ, as shown in Figure 21.23.

(d) Sketch the vector triangle of part (c) for $\delta = 0, 2\pi/3; \pi, 4\pi/3$, and 2π.

(e) Sketch a graph of $I(\delta)$ for $0 \le \delta \le 2\pi$.

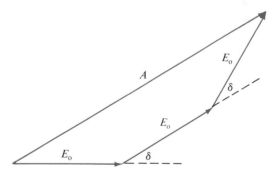

Figure 21.23   Problem 21.30

21.31 Consider the three-slit experiment suggested by Figure 21.24.

(a) Use the results of Problem 21.30 to show that the intensity of light at point P in the figure is given by

$$I = \frac{1}{9}\left[1 + 2\cos\left(\frac{2\pi d \sin\theta}{\lambda}\right)\right]^2 I_o$$

where I_o is the intensity of the central ($\theta = 0$) maximum.

(b) Show that principal maxima ($I = I_o$) occur for the same locations as do the maxima of the corresponding double-slit experiment with slit separation d and wavelength λ, that is, principal maxima occur for values of δ that satisfy $\cos\delta = 1$, or

$$d \sin\theta = n\lambda \quad ; \quad n = 0, \pm1, \pm2, \ldots$$

(c) Show that secondary maxima ($I = I_o/9$) occur for values of δ that satisfy $\cos\delta = -1$, or

$$d \sin\theta = \left(n + \frac{1}{2}\right)\lambda \quad ; \quad n = 0, \pm1, \pm2, \ldots$$

(d) For what values of δ do minima ($I = 0$) occur?

(e) Let α be the angular width of the (central) principal maximum, i.e.,

$$\alpha = 2\theta_o$$

where θ_o is the value of θ for the first minimum beyond the central maximum. Show that for the double-slit experiment

$$\sin\left(\frac{\alpha}{2}\right) = \frac{\lambda}{2d} \quad ; \quad \text{double-slit}$$

while for the three-slit experiment

$$\sin\left(\frac{\alpha}{2}\right) = \frac{\lambda}{3d} \quad ; \quad \text{three-slits}$$

Thus, the primary effects of adding one slit to the double-slit experiment are (i) to introduce a secondary maximum between each of the principal maxima and (ii) to sharpen and intensify each of the principal maxima.

21.32 Consider the four-slit experiment shown in Figure 21.25.

(a) Show that the intensity of light at point P in the figure is given by

$$I = I_o \cos^2(\delta/2)\cos^2\delta$$

where

$$\delta = \frac{\Delta r}{\lambda} = \frac{2\pi d \sin\theta}{\lambda}$$

and I_o is the intensity for the central ($\theta = 0$) maximum.

(b) Show that principal maxima ($I = I_o$) occur for the same locations as do the maxima of the corresponding double-slit experiment, that is, principal maxima occur for values of δ that satisfy $\cos\delta = 1$, or

$$d \sin\theta = n\lambda \quad ; \quad n = 0, \pm1, \pm2, \ldots$$

(c) Show that two secondary maxima ($I = 2I_o/27$) occur between each pair of adjacent principal maxima and that these secondary maxima occur for values of δ that satisfy $\cos\delta = -2/3$.

(d) For what values of δ do minima ($I = 0$) occur?

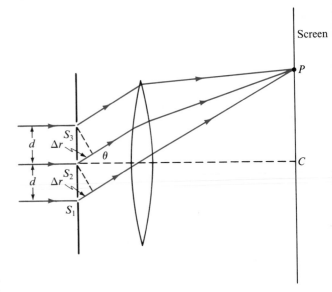

Figure 21.24   Problem 21.31

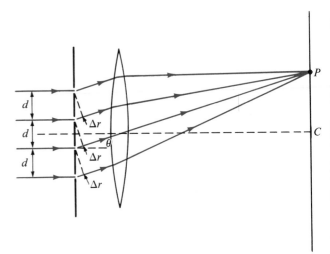

Figure 21.25   Problem 21.32

(e) Show that the angular width α of the (central) principal maximum is determined by

$$\sin\left(\frac{\alpha}{2}\right) = \frac{\lambda}{4d} \quad ; \quad \text{four slits}$$

(f) Compare these results to those summarized in part (e) of Problem 21.31. Figure 21.26 shows the intensity as a function of δ for the two-, three-, and four-slit interference patterns.

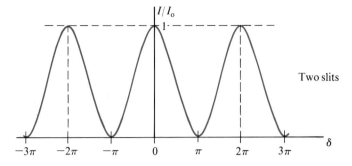

Two slits

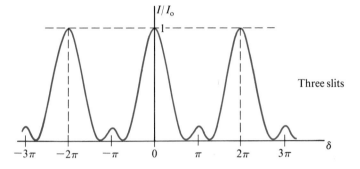

Three slits

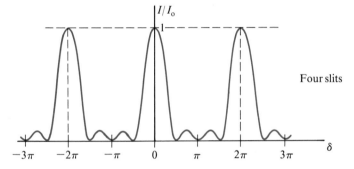

Four slits

Figure 21.26   Problem 21.32

22 Special Relativity

In the early twentieth century Albert Einstein (1879–1955) proposed the *special theory of relativity*, a revolutionary thesis that changed the conception of space and time. In this chapter, we examine some of the bases of Einstein's proposal and consider a few of its consequences that are strikingly contrary to our everyday experience.

For many years before 1900 it had been understood that Newton's laws of mechanics were applicable to an inertial reference frame. It was also clear that any reference frame moving at a constant velocity relative to an inertial frame was also an inertial frame. In other words, the Newtonian laws of mechanics were assumed to be valid in any inertial frame. But the laws of electrodynamics were not as clear with respect to reference frames. In particular, Maxwell's equations, the basic relationships of electromagnetic theory, predict that the speed of light ($c = 3.0 \times 10^8$ m/s) is a constant value—but constant with respect to what reference frame? Our experience with other forms of waves, like sound waves or mechanical waves on ropes, suggests that the speed of propagation of a wave should be measured relative to the medium in which that wave is moving. But what is the medium of a light wave traveling through a vacuum? One conjecture, which had been popular since Maxwell's time, was the presence of a medium that is "fixed" in space, that is, a medium fixed within an absolute reference frame. This proposed medium, through which light was presumed to travel at a speed c, was called the *ether*.

The notion of a fixed ether was considerably discredited by an ingenious experiment conducted by A. A. Michelson and E. W. Morley in 1887. They used a sensitive device, called a Michelson interferometer, which was demonstrably capable of detecting the motion of the earth (orbital speed $\simeq$ 30 km/s) relative to the fixed ether. Their experiment detected no relative motion between the earth and the ether. Then the *Michelson-Morley experiment*, perhaps the most famous of all "negative-result" experiments, implies that the motion of an observer does not affect the speed of light relative to that observer.

In 1905 Einstein proposed a solution to the dilemma that has been suggested here. But before we consider Einstein's assumptions and their spectacular consequences, let us reexamine (as Einstein did) some of the most fundamental of all physical concepts: space and time.

22.1 Space, Time, and the Galilean Transformation

We inherently recognize much of the physical significance of space and time. Nevertheless, it will be convenient to formalize and summarize some of our perceptions of space and time, concepts with definitions we have more or less assumed up to now.

We may perceive *space* as a three-dimensional extent, a construct, in which we can locate objects. We may, in an abstract sense, imagine space to be what would remain of our universe if everything were removed. Then space is *homogeneous*, which is to say that, in the absence of anything, every point in space has the same properties as every other point. Similarly, space is *isotropic*; that is, no direction in space is inherently different from any other direction. And because no point or direction in space is unique, we may select any point to serve as the origin of a reference frame, a coordinate system with three perpendicular axes; and we may orient one of those axes in an arbitrary direction. Such a reference frame may then be used to locate points in space relative to the origin of the frame.

Time is an abstraction conceived to order the occurrence of events. One may, therefore, observe that A occurs before B in a one-dimensional continuum called time. Since no instant of time inherently has different properties from any other, time is homogeneous. No unique time origin exists, and we may designate any arbitrary instant to be $t = 0$, a time origin. Although time is one-dimensional, the passage of time has a unique "direction," or sense, that one may recognize as proceeding from one instant to a later instant. (In a similar way, we have seen in Chapter 10 that "thermodynamic time" proceeds in that sense for which the entropy of the universe increases.)

The measurement of spatial distances may be accomplished in terms of any standard unit of length. And time is measured by a *clock*, a device calibrated in terms of any periodic motion that is sustained over a large number of cycles. Physical observations are made in terms of *events*, discernible occurrences (like the flash of a light or the firing of a gun) that happen at a particular location in space at a particular time. Thus one may identify an event uniquely by a quartet (x, y, z, t), which includes the rectangular coordinates, x, y, and z and a time t.

The study of relativity is concerned with comparisons of measurements of space and time in reference frames that are moving relative to one another. To make such comparisons, it is convenient to establish two reference frames, or coordinate systems, which we shall refer to as S and S', as illustrated in Figure 22.1. Because it is the effects of the relative motion between inertial frames that we shall investigate, we may assume, without loss of generality, that

1. the S frame, with perpendicular coordinate axes, x, y, and z, is "fixed,"

2. the S' frame, with perpendicular axes, x', y', and z', oriented parallel to the x, y, and z axes such that the x' axis remains colinear with the x axis, moves with a constant velocity v_x in the x direction, and

3. the time origins, $t = 0$ and $t' = 0$, of both reference frames are taken to be the instant at which their spatial origins, O and O', coincide.

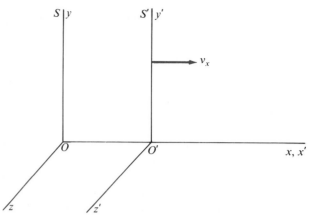

Figure 22.1
Two reference frames, S and S', moving relative to each other. Frame S' moves with a constant velocity v_x relative to S. By convention, the x and x' axes of the two frames are coincident, and the origins O and O' coincide at $t = t' = 0$.

Let us use these inertial frames, S and S', to formulate the basic relationships of *classical relativity*, the familiar relationship of kinematics (Chapter 2) that result from relative motion between inertial frames. Suppose an event E, indicated in Figure 22.2, occurs at coordinates (x, y, z, t) in the frame S. We wish to establish a *transformation* between the coordinates of E in S and S'. Such a coordinate transformation is a set of equations that permits the calculation of the coordinates (of an event E) in one reference frame, given the coordinates (of the same event) in another reference frame. In the frame S', which is moving at a constant velocity relative to S, the coordinates of that same event E are specified by (x', y', z', t'). And since the origins, O and O', of S and S' were coincident when $t = t' = 0$, the x coordinate of O', as measured in S, at the time of the event is equal to $v_x t$. Then from Figure 22.2 we may relate the coordinates of the event as measured in S and S' by

$$x' = x - v_x t \tag{22-1}$$

$$y' = y \tag{22-2}$$

$$z' = z \tag{22-3}$$

$$t' = t \tag{22-4}$$

These equations constitute a *galilean transformation*, the basic relationships of relative motion in *classical mechanics*.

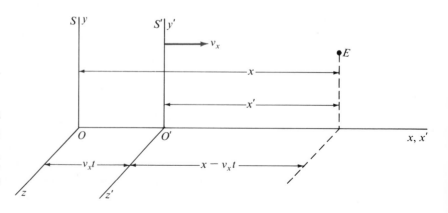

Figure 22.2
The graphical representation of an event E, which has coordinates (x, y, z, t) in S and (x', y', z', t') in S'. Since S' is moving at a velocity v_x relative to S, the origins are separated at a time t by a distance equal to $v_x t$, according to classical physics.

E 22.1   Observers in S record the coordinates of an event to be (10 m, 4.0 m, 3.0 m, 2.0 s). If $v_x = 4.0$ m/s, determine the coordinates of this event as measured in S'.   Answers: (2.0 m, 4.0 m, 3.0 m, 2.0 s)

E 22.2   Two events are determined to have S coordinates given by (13 m, 0, 0, 2.0 s) and (21 m, 0, 0, 4.0 s). If $v_x = 4.0$ m/s, what are the S'-coordinates of these two events?   Answers: (5.0 m, 0, 0, 2.0 s); (5.0 m, 0, 0, 4.0 s)

We may note that in Equation (22-4) we have *assumed* that the times at which E occurs in both reference frames are the same. In other words, if two events are simultaneous in one reference frame, they are simultaneous in the other frame. Thus the time interval between any two events has the same value in all inertial frames. Once clocks in S and S' are synchronized to an initial event (the co-incidence of O and O', for example), the time of any event is the same whether determined by clocks in S or S'. Because clocks in S and S' always agree, an S' clock moving at a constant velocity relative to S clocks "ticks" at the same rate as S clocks. This fact is consistent with our daily experience: We do not have to "correct" the reading of a wristwatch that has been in motion most of the day.

A second important aspect inherent in galilean relativity is that the distance between two simultaneous events is the same for all observers. To see this, suppose that events A and B occur simultaneously at points x_A and x_B on the x axis at a time $t_o = t_o'$. The distance d' between these events, as measured in S', is equal to $\sqrt{(x_A - x_B)^2}$. If now we use the galilean transformation, Equations (22-1, 2, 3, 4), we find that

$$d' = \sqrt{(x_A - v_x t_o - x_B + v_x t_o)^2}$$
$$d' = \sqrt{(x_A - x_B)^2}$$
$$d' = d \tag{22-5}$$

where d is the distance between A and B as measured in S. Thus both S' and S observers agree on the distance separating any two simultaneous events. We conclude, then, that in galilean relativity, the distance between two simultaneous events is a universal quantity.

This universal distance between simultaneous events, agreed upon by observers in all inertial frames, leads us to conclude further that the length of a moving object is the same in all inertial frames. Measurement of the length of a moving stick, for example, is accomplished in a given reference frame, say S, by having S observers who are adjacent to the ends, A and B, of the stick at a prescribed time place marks at each end of the stick. The distance between these two simultaneous events (the marking of the ends) is taken to be the length of the moving object. And, as we have seen, the distance (in this case, the length of the moving stick) between simultaneous events is the same for all observers.

E 22.3   The simultaneous markings of ends A and B of a moving stick are observed in frame S to have coordinates (12 m, 0, 0, 2.0 s) and (15 m, 0, 0, 2.0 s).
(a) What is the length of the stick as measured in S?   Answer: 3.0 m
(b) If $v_x = 3.0$ m/s, what are the coordinates of the two marking events as determined in S'?   Answer: (6.0 m, 0, 0, 2.0 s), (9.0 m, 0, 0, 2.0 s)
(c) What length do S' observers measure?   Answer: 3.0 m

E 22.4   A carpenter hammers a nail (located in S at $x = 4.0$ m, $y = z = 0$) at $t = 0$ and $t = 2.0$ s. If $v_x = 5.0$ m/s, what distance will S' observers measure between these two events that occurs at the same point in S?

Answer: 10 m

As we shall see in the next section, the assumptions of Einstein's theory of special relativity led to a transformation that is very different from the galilean transformation. But how different can another valid transformation be if our conclusions drawn from the galilean transformation are borne out by our experience (at speeds much less than c)? In fact, as we will see, it can be startlingly different! Yet any valid transformation must reduce to the classical form, that is, it must, in effect, become the galilean transformation when the relative speed v between reference frames is much less than the speed c of light. In other words, any valid transformation, however different it may be from the galilean transformation, must, when $v \ll c$, differ negligibly from the well-confirmed classical results.

22.2 The Einstein Postulates, Synchronization, and Simultaneity

Einstein formulated his theory of special relativity to reconcile Maxwell's electromagnetic theory with experimental observation. Einstein's theory, based on two postulates, requires a careful examination of the appropriate means by which time intervals (and, as a consequence, lengths) measured in different inertial frames may be compared. The postulates of the special theory of relativity may be stated as follows:

1. The laws of physics have the same mathematical form in every inertial frame.

2. The speed of light in a vacuum has the same value c ($= 3.0 \times 10^8$ m/s) when measured by any observer. Therefore, that speed is independent of any motion of the observer or of the source of the light.

The implications and consequences of these postulates will be appreciated only if we recognize some essential aspects of the measurement of time. In particular, the assumption of the constancy of the speed of light as measured by any observer suggests that we consider carefully what is meant by measurements of a time interval using clocks. For example, what does it mean to say that two clocks are synchronized? And what does it mean to say that two events occur simultaneously? Let us examine these concepts.

In "thought experiments" like those that Einstein introduced to illustrate and clarify the concepts of relativity, it is often convenient to imagine that several clocks are positioned at different points within a single reference frame. Then they must be synchronized if all of them are to indicate the correct time associated with that reference frame.

Two clocks within the same reference frame may be synchronized, in principle at least, by the following procedure, which utilizes the constancy of the speed of light: Each clock is set to a predetermined time when a light pulse from a point midway between the clocks reaches that clock. Any number of clocks within the same reference frame may be synchronized in this manner.

But even when we use synchronized clocks, what does it mean to say that two events occur simultaneously when observers of the events are in reference frames moving relative to each other? Two events that are simultaneous in one reference frame may *not* be simultaneous in a second frame moving relative to the first. Let us illustrate this disagreement about simultaneity by using a traditional thought experiment that was devised by Einstein himself.

Imagine a train, illustrated in Figure 22.3, moving toward the right with a constant nonzero speed v. Suppose that two lightning bolts strike opposite ends of the train. These two events leave marks on the ground at points A and B and on the train at points A' and B'. An observer O is on the ground at a point midway between A and B. Another observer O', midway between A' and B', is moving along with the train. Each observer uses light signals from the lightning to observe the two events (lightning striking each end of the train). Now suppose the two events are simultaneous to the observer O on the ground; the light signals from A and B travel equal distances in O's reference frame to reach O at the same time. But the observer O' is moving away from B and toward A, so the light signal from the event at A' reaches O' before it reaches O. Similarly, the light signal from the event at B' reaches O' after it reaches O. The moving observer O' therefore receives the light signal from A' before receiving the signal from B'. And since A' and B' are equidistant from O', the moving observer concludes that the two events did *not* occur simultaneously.

The two observers moving relative to one another clearly disagree as to whether or not the two events were simultaneous. Because we have no basis for assuming that one reference frame is preferable to, or more "correct than," the other, we must conclude that each observer has drawn the correct conclusion in that observer's own reference frame. We may further conclude that whether or not events that occur at different locations are simultaneous depends on the motion of the observer. It then follows that *two observers moving relative to each other will, in general, measure different time intervals between events that occur at different locations.*

As a consequence of these considerations of time, we may now formulate prescriptions for the appropriate measurement of both times and lengths in frames of reference that are moving with constant velocity relative to one another. First, the time of an event in a given reference frame must be measured by a "local" clock, a clock that is at the location of the event, at rest in the reference frame of the observer, and synchronized with all clocks used in the reference frame of the observer.

The length of an object is measured in a reference frame, say S, by the following procedure. Observers in S make marks in their frame that coincide with the ends of the object. These marks must be made in the S frame at the same instant as determined by clocks synchronized in S. If we let the x coordinates of

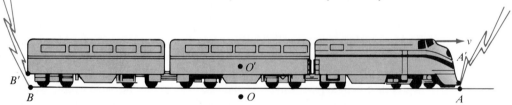

Figure 22.3
Illustrating the concept of simultaneity.

the two ends of the object (like a long, thin rod aligned parallel to the x axis of the S frame) be x_1 and x_2, then the length L of the object, as measured in the S frame, is

$$L = x_2 - x_1 \qquad (x_2 > x_1) \tag{22-6}$$

22.3 The Lorentz Transformation: Relativistic Kinematics

We now seek a transformation of the coordinates of an event, a transformation consistent with the Einstein postulates of special relativity. Let us recall that a transformation is a relationship of the coordinates (x', y', z', t') of an event as measured by an observer in a reference frame S' and the coordinates (x, y, z, t) of the same event as measured by an observer in S. As usual, we assume that S' and S are rectangular coordinate systems oriented with their x' and x axes coincident and that S' is moving in the x direction with a constant velocity v_x relative to S. We further assume that observers in both reference frames are supplied with ample identical clocks synchronized in the appropriate frame and ample identical measuring rods. The appropriate transformation may be obtained by considering the following situation.

Suppose the origins of S and S' coincide at $t = t' = 0$. Further suppose that when the origins are coincident, a pulse of light is emitted in all directions from the momentarily coincident origins. Now suppose the light pulse, which travels at a speed c, strikes a mirror. The light pulse striking the mirror has S coordinates (x, y, z, t). Because the distance $(x^2 + y^2 + z^2)^{1/2}$ traveled by the light from the origin to the mirror is equal to ct, we may write

$$x^2 + y^2 + z^2 = c^2t^2 \tag{22-7}$$

If S' observers determine the coordinates of this event to be (x', y', z', t'), the light pulse has traveled an S' distance $(x'^2 + y'^2 + z'^2)^{1/2}$, which is equal to ct', or

$$x'^2 + y'^2 + z'^2 = c^2t'^2 \tag{22-8}$$

Notice that in both relationships, Equations (22-7) and (22-8), we have invoked the Einstein postulate that the speed c of light has the same value in any reference frame.

Let us now assume that we can develop a transformation of the form

$$x' = a_1x + a_2t$$
$$y' = y$$
$$z' = z$$
$$t' = b_1x + b_2t \tag{22-9}$$

Such a transformation is called a linear transformation, characterized by the fact that the quantities a_1, a_2, b_1, and b_2 are constants, that is, quantities that are independent of x, y, z, and t.

Because both Equations (22-7) and (22-8) relate to the same event, we may substitute Equations (22-9) into Equation (22-8). Then, after considerable algebraic manipulations and using the two further conditions that (a) the origin of S' is located relative to the origin of S by $x = v_xt$, and (b) the transformation we seek must reduce to the galilean transformation as $v_x \to 0$, one may obtain the desired transformation:

$$x' = \gamma(x - v_x t)$$

$$y' = y$$

$$z' = z$$

$$t' = \gamma\left(t - \frac{v_x x}{c^2}\right) \qquad \text{(22-10)}$$

where

$$\gamma = \frac{1}{\sqrt{1 - v_x^2/c^2}}$$

These equations, called the *Lorentz transformation*, are the operational kinematic relationships of special relativity. They permit us to make calculations that relate positions and times as measured in reference frames that are moving relative to each other.

E 22.5   Show that the Lorentz transformation reduces to the galilean transformation for $v \ll c$.

E 22.6   Event A is observed in frame S to occur at $x = 2700$ km on the x axis and at $t = 0$, that is, S observers determine that event A is simultaneous with the $t = t' = 0$ coincidence of the origins of S and S'. Let $v_x = 0.80\ c$.
(a) Where does event A occur S'?   Answer: $x_A' = 4500$ km
(b) When does event A occur in S'.   Answer: $t_A' = -12$ ms

E 22.7   Consider again the two events of Exercise 22.6. Recall that since both events occur at $t = 0$, S observers conclude that the events occur simultaneously.
(a) Are the two events also simultaneous in S'?   Answer: No.
(b) Which event occurs first in S'?
                        Answer: Event A occurs 12 ms before the two origins coincide.

Example 22.1   **PROBLEM**   What conditions, if any, ensure that if two events are simultaneous in one inertial frame, these events will be simultaneous in other inertial frames?

SOLUTION   Suppose events A and B are specified in frame S by (x_A, y_A, z_A, t_A) and (x_B, y_B, z_B, t_B). The times for these events in any other frame S' are given by

$$t_A' = \gamma(t_A - v_x x_A/c^2)$$

$$t_B' = \gamma(t_B - v_x x_B/c^2)$$

Thus the time interval, $\Delta t' = t_B' - t_A'$, in S' is related to the time interval, $\Delta t = t_B - t_A$, in S by

$$\Delta t' = t_B' - t_A'$$

$$\Delta t' = \gamma(t_B - v_x x_B/c^2) - \gamma(t_A - v_x x_A/c^2)$$

$$\Delta t' = \gamma(t_B - t_A) - \gamma v_x(x_B - x_A)/c^2$$

$$\Delta t' = \gamma \Delta t - \gamma v_x \Delta x/c^2$$

where $\Delta x = x_B - x_A$ is the x separation in S of events A and B. If now A and B are simultaneous events in S, that is, if $\Delta t = 0$, then

$$\Delta t' = \gamma v_x \Delta x / c^2$$

And the two events will be simultaneous in S', that is, $\Delta t' = 0$, if and only if $\Delta x = 0$. We conclude, therefore, that if two events occur (i) simultaneously in S and (ii) in the same y-z plane in S, then these two events will be simultaneous in any frame S' moving with a constant velocity $\mathbf{v} = v_x \hat{\mathbf{i}}$ relative to S. More generally, any two events that occur simultaneously in one frame, say S, will occur simultaneously in all inertial frames that move relative to S with a velocity oriented perpendicularly to the S line connecting the two events. ∎

Time Dilation

Suppose a clock C_0, fixed at the origin of frame S', moves with a constant velocity v_x relative to two synchronized S clocks, C_1 and C_2. Figure 22.4(a) depicts event A, the coincidence of clocks C_0 and C_1. Event B, the coincidence of clocks C_0 and C_2, is shown in Figure 22.4(b). Clock C_0 indicates t'_A and t'_B for events A and B, clock C_1 reads t_A for event A, and clock C_2 reads t_B for event B. Both events occur in S' at the same point $x'_0 = 0$, the x' coordinate of C_0. The S-time interval, $\Delta t = t_B - t_A$, and the S'-time interval, $\Delta t' = t'_B - t'_A$, are related by

$$\Delta t = t_B - t_A = \gamma(t'_B + v_x x'_0/c^2) - \gamma(t'_A + v_x x'_0/c^2)$$

$$\Delta t = \gamma(t'_B - t'_A)$$

$$\Delta t = \gamma \Delta t' = \frac{\Delta t'}{\sqrt{1 - v_x^2/c^2}} > \Delta t' \tag{22-11}$$

So the "moving" clock C_0 measures a time interval $\Delta t'$ that is less than the time interval Δt measured by the "fixed" clocks, C_1 and C_2. A clock moving through S runs at a slower rate than the S clocks it passes! Similarly, if an S clock is observed as it passes S' clocks, this "moving" S clock will run at a slower rate than the "fixed" S' clocks it passes. This phenomenon, a moving clock running at a slower rate than the fixed clocks to which it is compared, is called *time dilation*. In the classical limit, $v \to 0$, this effect becomes negligible, that is, clocks moving at nonrelativistic speeds "tick" at the same rate as fixed clocks.

Figure 22.4
(a) Event A, the coincidence of a clock C_0, which is at the origin of S' and moving relative to S with a constant velocity $v_x = v$, with a clock C_1 fixed at the origin of S. **(b)** Event B, the coincidence of C_0 with a clock C_2 fixed on the x axis of S.

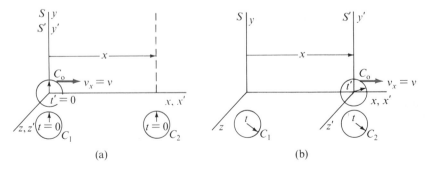

(a)   (b)

E 22.8   A clock moves past fixed clocks at a speed of 0.60 c. What time interval will the fixed clocks measure for a one-second interval indicated on the moving clock?

Answer: 1.3 s

E 22.9   Two events occur at the same point in S'. An S' clock at this point indicates the time interval between these two events to be equal to 2.0 μs. If the velocity of S' relative to S is $v_x = 0.60\ c$, what will S clocks indicate for this time interval?

Answer: 2.5 μs

E 22.10   The two events of Exercise 22.9 occur at the same point in S'. What will S observers measure for the distance between these events?

Answer: 450 m

Length Contraction

The Lorentz transformation predicts another effect that is strikingly different from our everyday experience and, therefore, different from the predictions of classical mechanics. To illustrate this effect, consider Figure 22.5(a), which shows a stick that has a length L_o when measured by observers in the S' reference frame in which the stick is at rest. The stick is oriented parallel to the x' axis in S', which is moving at a constant velocity $v_x = v$ relative to the fixed reference frame S. Thus the stick is moving at a velocity $v_x = v$ relative to S.

Because the stick is at rest in the S' frame, observers in S' can measure the coordinates x'_A and x'_B of the ends A and B of the stick at any times. In so doing, they measure the *rest-length* L_o of the stick to be

$$L_o = x'_A - x'_B \qquad (22\text{-}12)$$

But in S, observers, who see the stick moving, must locate the ends of the stick simultaneously. As suggested in Figure 22.5(b), S observers measure the coordinates of two events: (1) the x coordinate x_A of A at time t and (2) the x coordinate x_B of B at the same time t. Then the length L of the stick, as measured in the frame S, is

$$L = x_A - x_B \qquad (22\text{-}13)$$

But what are the coordinates in the moving frame S' that correspond to these two events? According to the Lorentz transformation, Equations (22-10), the x' coordinate corresponding to event (1) is given by $\gamma(x_A - vt)$. And this value

Figure 22.5
(a) A stick of rest length L_o, as measured by observers in the S' frame in which the stick is at rest. (b) The same stick, measured to be a length L by observers in the S frame, in which the stick has a velocity v_x.

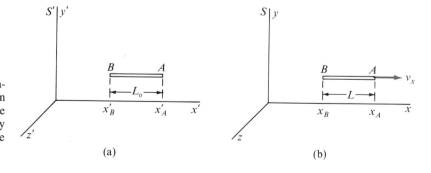

(a)   (b)

must be identical to x'_A, because the coordinate of A in S' is x'_A at all times. Hence we may write

$$x'_A = \gamma(x_A - vt) \tag{22-14}$$

and similarly

$$x'_B = \gamma(x_B - vt) \tag{22-15}$$

Then the rest-length, $L_o = x'_A - x'_B$, may be found by subtracting Equation (22-15) from Equation (22-14) and using Equation (22-13):

$$L_o = x'_A - x'_B = \gamma(x_A - x_B) = \frac{L}{\sqrt{1 - v^2/c^2}}$$

or

$$L = L_o\sqrt{1 - v^2/c^2} \tag{22-16}$$

Thus the length L of the moving stick as measured in S is less than the rest-length L_o of the stick measured in S'. This effect, the foreshortening of the length of an object along the direction of relative motion as measured by observers not in the rest-frame of the object, is called *length contraction*. Again, the effect is greater at relative speeds v approaching the speed c of light, and the effect becomes negligible as $v \to 0$.

E 22.11   What length will observers measure for a spaceship moving relative to them at a speed of $0.99\,c$ if the occupants of the ship determine the length of the ship to be 50 m?   Answer: 7.1 m

E 22.12   The length of a moving object is determined to be one-half of its rest-length. How fast is the object moving?   Answer: $0.87\,c$

Relativistic Velocity Transformation

Suppose we know the velocity **u** of an object as measured in a given reference frame, say S. The relationships that permit our determining the velocity **u'** of that object as measured in S', a reference frame moving with a constant velocity v_x relative to S, are called a *velocity transformation*. In relativistic kinematics, we may obtain the appropriate velocity relationships by using the definitions of velocity components and the Lorentz transformation for position coordinates and time.

If an object has a velocity **u**, as measured in S, the components of that velocity are

$$u_x = \frac{dx}{dt} \quad ; \quad u_y = \frac{dy}{dt} \quad ; \quad u_z = \frac{dz}{dt} \tag{22-17}$$

Similarly, the velocity **u'** of that object, as measured in S', has components

$$u'_x = \frac{dx'}{dt'} \quad ; \quad u'_y = \frac{dy'}{dt'} \quad ; \quad u'_z = \frac{dz'}{dt'} \tag{22-18}$$

We may find the desired velocity transformation as follows:

$$u'_x = \frac{dx'}{dt'} = \frac{d}{dt'}\left[\frac{x - v_x t}{\sqrt{1 - v_x^2/c^2}}\right]$$

Since both v_x and c are constants, u'_x may be written as

$$u'_x = \gamma \left[\frac{dx}{dt'} - v_x \frac{dt}{dt'} \right]$$

$$u'_x = \gamma \left[\frac{dx}{dt} \frac{dt}{dt'} - v_x \frac{dt}{dt'} \right]$$

$$u'_x = \gamma \left[u_x - v_x \right] \frac{dt}{dt'} \tag{22-19}$$

But for dt/dt' we may write, using the Lorentz transformation, Equation (22-10),

$$\frac{dt}{dt'} = \frac{1}{dt'/dt} = \frac{1}{\dfrac{d}{dt}[\gamma(t - v_x x/c^2)]} = \frac{\sqrt{1 - v_x^2/c^2}}{1 - v_x u_x/c^2} = \frac{1}{\gamma(1 - v_x u_x/c^2)} \tag{22-20}$$

Substituting Equation (22-20) into Equation (22-19) gives

$$u'_x = \frac{u_x - v_x}{1 - v_x u_x/c^2} \tag{22-21}$$

Similar substitutions into the definitions of u'_y and u'_z yield

$$u'_y = \frac{u_y}{\gamma(1 - v_x u_x/c^2)} \tag{22-22}$$

$$u'_z = \frac{u_z}{\gamma(1 - v_x u_x/c^2)} \tag{22-23}$$

The following problem illustrates the use of the relativistic velocity transformation and suggests an important inference that we may draw concerning velocity measurements.

A rocket is fired from the earth toward a spaceship that is traveling toward the earth. Observers on earth measure the speed of the rocket to be equal to 0.90 c and measure the speed of the spaceship to be 0.90 c. What speed for the approaching rocket is measured by observers in the spaceship?

Figure 22.6 visualizes the problem situation. The spaceship is at the origin of the S' frame, which is moving toward the earth with a velocity $v_x = -0.90\ c$. The rocket is proceeding along the x axis of S with a velocity, relative to the earth, of $u_x = +0.90\ c$. The velocity u'_x with which the rocket moves relative to the spaceship may be found by substitution of the given values for v_x and u_x into

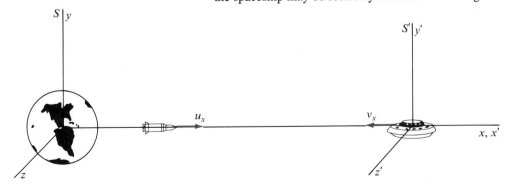

Figure 22.6
A rocket moving at a velocity u_x relative to the earth toward a spaceship that is approaching the earth with a velocity v_x relative to the earth.

the relativistic velocity transformation, Equation (22-21), giving appropriate attention to the signs of the quantities involved:

$$u'_x = \frac{u_x - v_x}{1 - v_x u_x/c^2} = \frac{+0.90\ c - (-0.90\ c)}{1 - \dfrac{(-0.90\ c)(+0.90\ c)}{c^2}} = \frac{1.8\ c}{1.81} = +0.99\ c \qquad (22\text{-}24)$$

The positive value we obtain for u'_x indicates that the rocket is moving in the positive x' direction relative to an observer in the spaceship. The most striking feature of this result, however, is its obvious difference from what we might have expected had we used classical kinematics. Classically, we would expect the rocket to approach the spaceship at a speed equal to $1.8\ c$. Because the relativistic treatment indicates that observers in the spaceship measure the speed of the rocket to be less than c, we may suspect that observers in any reference frame never measure the speed of any object to be greater than the speed of light. Later we will see that is precisely the case: The speed of light is the ultimate speed limit for objects in our universe, regardless of the reference frame in which those speeds are measured.

E 22.13   A light signal is observed in S to propagate in the positive x direction with a speed c. Use the velocity transformation to determine the velocity of the signal in S if $v_x = 4c/5$.   Answer: $u'_x = c,\ u'_y = u'_z = 0$

E 22.14   The velocity of a particle in S is $\mathbf{u} = (2.4\hat{\mathbf{i}} + 1.0\hat{\mathbf{j}}) \times 10^8$ m/s. If $v_x = 3c/5$, determine
(a) the x component in S' of this velocity,
(b) the y component in S' of this velocity.
        Answers: (a) 1.2×10^8 m/s; (b) 1.5×10^8 m/s

E 22.15   Determine the speed in S and in S' for the particle of Exercise 22.14.   Answer: $u = 2.6 \times 10^8$ m/s, $u' = 1.9 \times 10^8$ m/s

E 22.16   If the direction of a vector in the x-y plane is specified by the angle between that vector and the positive x direction, determine the direction in S and in S' of the velocity of Exercise 22.14.   Answer: $\theta = 23°;\ \theta' = 53°$

Example 22.2   **PROBLEM**   A light signal is observed in S to have a velocity of propagation given by

$$\mathbf{u} = c(\cos\theta\ \hat{\mathbf{i}} + \sin\theta\ \hat{\mathbf{j}})$$

where θ is the angle between $\mathbf{u}$ and $\hat{\mathbf{i}}$. If $v_x = 3c/5$, show that observers in S' will, as expected, determine the speed of propagation of this light signal to be c.

SOLUTION   Using $\gamma = (1 - 0.36)^{-1/2} = 5/4$ in the velocity transformations, we get for the S'-velocity components:

$$u'_x = \frac{u_x - v_x}{1 - u_x v_x/c^2} = \left(\frac{\cos\theta - \dfrac{3}{5}}{1 - \dfrac{3}{5}\cos\theta}\right) c = \left(\frac{5\cos\theta - 3}{5 - 3\cos\theta}\right) c$$

$$u'_y = \frac{u_y}{\gamma(1 - u_x v_x/c^2)} = \frac{4\ c\ \sin\theta}{5(1 - \dfrac{3}{5}\cos\theta)} = \left(\frac{4\sin\theta}{5 - 3\cos\theta}\right) c$$

Squaring each of these components and adding gives

$$u_x'^2 + u_y'^2 = \frac{(5\cos\theta - 3)^2 + (4\sin\theta)^2}{(5 - 3\cos\theta)^2}\,c^2$$

$$u_x'^2 + u_y'^2 = \frac{25\cos^2\theta - 30\cos\theta + 9 + 16\sin^2\theta}{25 - 30\cos\theta + 9\cos^2\theta}\,c^2$$

Setting $\sin^2\theta = 1 - \cos^2\theta$ in the numerator of this expression, we obtain

$$u_x'^2 + u_y'^2 = \left(\frac{25\cos^2\theta - 30\cos\theta + 9 + 16 - 16\cos^2\theta}{25 - 30\cos\theta + 9\cos^2\theta}\right)c^2$$

$$u_x'^2 + u_y'^2 = \left(\frac{25 - 30\cos\theta + 9\cos^2\theta}{25 - 30\cos\theta + 9\cos^2\theta}\right)c^2 = c^2$$

$$u' = \sqrt{u_x'^2 + u_y'^2} = c$$

Thus, S' observers do measure the speed of this light signal to be c. ■

Example 22.3   **PROBLEM**   If θ' is the angle between $\mathbf{u}'$, the S'-velocity of propagation of the light signal of Example 22.2, and the positive x direction, express θ' in terms of θ, the S direction of propagation of this signal.

SOLUTION   Since $\tan\theta' = u_y'/u_x'$, we get, using the expressions for u_x' and u_y' obtained in Example 22.2,

$$\tan\theta' = \frac{u_y'}{u_x'} = \frac{4\sin\theta}{5 - 3\cos\theta} \cdot \frac{5 - 3\cos\theta}{5\cos\theta - 3}$$

$$\tan\theta' = \frac{4\sin\theta}{5\cos\theta - 3}$$

or

$$\theta' = \tan^{-1}\left(\frac{4\sin\theta}{5\cos\theta - 3}\right)$$

which is the desired expression. ■

E 22.17   Show that the result obtained in Example 22.3 gives the expected answer for a light signal propagating in the positive x direction in S.
Answer: $\theta = 0$ implies $\theta' = 0$

E 22.18   For the light signal of Examples 22.2 and 22.3, find the direction θ of propagation in S if S' observers determine the light signal to be propagating in the positive y direction.   Answer: $\theta = \cos^{-1}(3/5) = 53°$

22.4 Relativistic Momentum, Mass, and Energy

The first postulate of special relativity requires that the laws of physics be the same in all inertial frames. In particular, the principle of conservation of momentum must apply in both of any two inertial frames, which are necessarily moving relative to each other at constant velocity. More specifically, if two particles collide, the total momentum of that two-body system is unchanged by the collision. In other words, the momentum of that isolated two-body system must be conserved, as observed and measured from any inertial frame.

Let us suppose that such a two-body collision is described from the point of view of observers in an inertial frame S, in which the total linear momentum of the two-particle system is conserved. If now we suppose that same collision is analyzed from the point of view of observers in a second reference frame S' moving with constant velocity relative to S (see Problem 22.24), using the Lorentz transformations appropriately, one finds that linear momentum (as we have previously defined it) is *not* conserved in S'. That is, momentum is not conserved if we use the classical definition of linear momentum, $\mathbf{p} = m\mathbf{u}$. Here we use $\mathbf{u}$ to represent the velocity of a moving particle in order to distinguish that quantity from $\mathbf{v}$, the velocity of one reference frame relative to another.

But again, the first postulate of special relativity requires that linear momentum conservation, if it is to remain a fundamental law, must apply in all inertial frames. Then if we assume the validity of the Lorentz transformation, we must, as Einstein suggested, redefine linear momentum appropriately so that linear momentum conservation applies in all inertial frames. Such a relativistic momentum must, of course, approach the classical value $m\mathbf{u}$ as $u \to 0$. Einstein showed, using his elegant thought experiments, that the appropriate expression for the *relativistic linear momentum* $\mathbf{p}$ of a particle in a reference frame where it has a velocity $\mathbf{u}$ is

$$\mathbf{p} = \frac{m\mathbf{u}}{\sqrt{1 - u^2/c^2}} = \gamma_u\, m\mathbf{u} \tag{22-25}$$

where m is the mass of the particle and γ_u is defined by

$$\gamma_u = \frac{1}{\sqrt{1 - u^2/c^2}} \tag{22-26}$$

E 22.19   Calculate the magnitudes of the classical and relativistic linear momenta of an electron (mass $= 9.1 \times 10^{-31}$ kg) moving at a speed equal to 1.8×10^8 m/s.   Answer: 1.6×10^{-22} kg $\cdot$ m/s, 2.0×10^{-22} kg $\cdot$ m/s

The relativistic definition of linear momentum $\mathbf{p}$ in Equation (22-25) results in a new expression for the relativistic force $\mathbf{F}$, defined so that Newton's second law takes the form

$$\mathbf{F} = \frac{d\mathbf{p}}{dt} = \frac{d}{dt}(\gamma_u m\mathbf{u}) = m\frac{d}{dt}\left[\frac{\mathbf{u}}{\sqrt{1 - u^2/c^2}}\right] \tag{22-27}$$

We now use this relativistic force to find an expression for relativistic kinetic energy.

Let us consider a particle of mass m that has a motion restricted to the x axis. The work done on this particle by a net force $\Sigma\mathbf{F} = \Sigma F_x \hat{\mathbf{i}}$, as the particle proceeds from rest to a final speed u_f, is given by

$$W = \int_{u=0}^{u_f} \Sigma F_x\, dx \tag{22-28}$$

We may use Equation (22-27), $\Sigma F_x = dp/dt$, and the chain rule for differentiation, $dp/dt = (dp/du)(du/dx)(dx/dt)$, to write

$$W = \int_{u=0}^{u_f} \Sigma F_x \, dx = \int_{u=0}^{u_f} \frac{dp}{du} \frac{du}{dx} \frac{dx}{dt} \, dx = \int_{u=0}^{u_f} \frac{dp}{du} \, u \, du$$

We may express dp/du by differentiating Equation (22-25). If we recognize that u_f represents an arbitrary final speed u, then we may write

$$W = \int_{u=0}^{u_f} \frac{mu \, du}{(1 - u^2/c^2)^{3/2}}$$

$$W = \frac{mc^2}{\sqrt{1 - u^2/c^2}} - mc^2 = (\gamma_u - 1) \, mc^2 \qquad (22\text{-}29)$$

This work W (required to bring the particle from rest to a speed u) on the particle is defined to be the relativistic kinetic energy K of a particle having a mass m, or

$$K = \frac{mc^2}{\sqrt{1 - u^2/c^2}} - mc^2 = (\gamma_u - 1) \, mc^2 \qquad (22\text{-}30)$$

E 22.20   Calculate the classical and relativistic kinetic energies of a proton (mass $= 1.7 \times 10^{-27}$ kg) moving with a speed of 1.8×10^8 m/s.

Answer: 2.8×10^{-11} J, 3.8×10^{-11} J

E 22.21   Use the binomial expansion

$$(1 - \epsilon)^r = 1 - r\epsilon + \frac{r(r - 1)}{2} \epsilon^2 + \dots$$

to show that the relativistic kinetic energy of a particle differs negligibly from the classical kinetic energy of that particle provided the speed u of the particle is much less than c.

Dependence of kinetic energy of a particle on its speed, as given in Equation 22-30, has been repeatedly confirmed experimentally using particle accelerators, devices that accelerate atomic particles, like electrons and protons, to speeds comparable to the speed of light. Figure 22.7 shows a graph of Equation (22-30), which indicates that the kinetic energy K of a particle increases dramatically as the speed u of that particle approaches c, the speed of light.

Figure 22.7
A graph of the relativistic kinetic energy $K = (\gamma_u - 1) \, mc^2$ of a particle as a function of the square of the particle speed u. The dashed curve shows the classical kinetic energy $K_{classical} = \frac{1}{2}mu^2$ as a function of u^2.

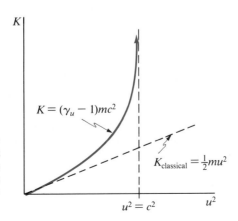

Because the kinetic energy of a particle increases so that $K \rightarrow \infty$ as the particle speed $u \rightarrow c$, no amount of force applied to the particle and no expenditure of energy can cause that particle to achieve a speed equal to the speed of light. Therefore, as we anticipated, the speed of light represents the ultimate speed for an object. Indeed, since electromagnetic waves in a vacuum travel at a speed c, regardless of the inertial frame in which c is measured, the speed of light in a vacuum represents the absolute speed limit for the transmission of objects, energy, or information.

This fact, that c is the ultimate speed, constitutes a profound limitation imposed by nature. Even though our science fiction literature has traditionally encouraged anticipation of a future technology that will permit us to travel at speeds that are multiples of c ("proceed to Star-fleet base at warp 8, Mr. Spock"), our present understanding of nature emphatically denies that possibility.

Because K is an energy, we may recognize that each of the other two terms in Equation (22-30) are energies. By rewriting Equation (22-30) as

$$\gamma_u m c^2 = K + m c^2 \tag{22-31}$$

we may interpret each of the terms as follows: $\gamma_u m c^2$ is the *total relativistic energy* of a particle with mass m; K is the relativistic kinetic energy of the particle, the energy associated with the motion of the particle; and mc^2 is the *rest energy* of the particle, an energy inherently associated with the mass. Using E to represent total relativistic energy and E_o to represent the rest energy, we may write

$$E = \gamma_u m c^2 \tag{22-32}$$

and

$$E_o = m c^2 \tag{22-33}$$

which are the famous Einstein equations that establish the fundamental relationship between mass and energy. (Some treatments of relativity define a relativistic mass m_r such that $m_r = \gamma_u m$, from which it follows that $E = m_r c^2$.)

E 22.22   A neutron (mass $= 1.7 \times 10^{-27}$ kg) moves with a speed of 1.8×10^8 m/s. Determine (a) the rest energy of the neutron, (b) the kinetic energy of the neutron, and (c) the total relativistic energy of the neutron.
Answers: (a) 1.5×10^{-10} J; (b) 3.8×10^{-11} J; (c) 1.9×10^{-10} J

E 22.23   For what speed will the kinetic energy of a particle be equal to twice the rest energy of the particle?   Answer: $2\sqrt{2}\, c/3 \approx 2.8 \times 10^8$ m/s

Example 22.4   **PROBLEM**   A particle of mass m is moving at a speed $v = 4c/5$ when it collides with a particle of mass $2m$ initially at rest. The two particles "stick together" and form a composite particle. Assuming that total relativistic momentum and total relativistic energy are conserved, determine the mass and speed of the composite particle.
SOLUTION   The initial and final states of this problem are shown in Figure 22.8. Conservation of linear momentum ensures that

$$\gamma_u\, M u = \gamma_v\, m v$$

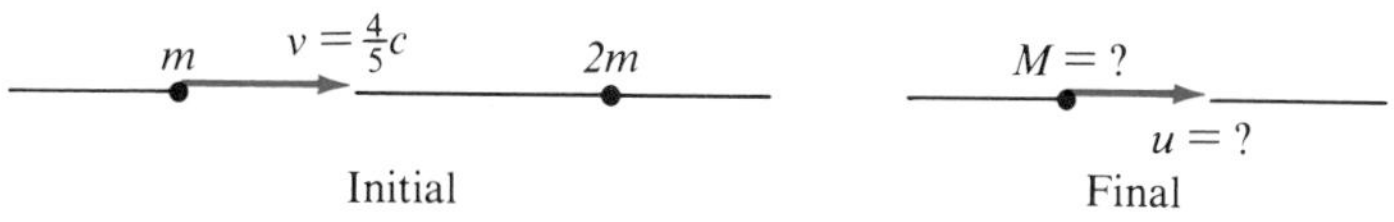

Figure 22.8   Example 22.4

with $v = 4c/5$ and $\gamma_v = 5/3$. Conservation of total (relativistic) energy gives

$$\gamma_u\, Mc^2 = \gamma_v\, mc^2 + 2mc^2 = \frac{11}{3}\, mc^2$$

Solving the momentum equation for $\gamma_u\, M$ and substituting this result into the energy equation gives

$$u = \frac{4}{11}\, c$$

Using this value for u in the energy equation yields

$$M = \frac{\sqrt{105}}{3}\, m$$

Thus the total mass of the system after the collision exceeds the total mass before the collision by

$$\Delta m = m_{\text{final}} - m_{\text{initial}} = \left(\frac{\sqrt{105}}{3} - 3\right) m \approx 0.42m$$

Can you account (using energy considerations) for this increase in mass? ■

E 22.24   Determine the change in kinetic energy that occurs as a result of the collision of Example 22.4. Can you account for this lost kinetic energy?
$$\text{Answer: } \Delta K = K_f - K_i = -\left(\frac{\sqrt{105}}{3} - 3\right) mc^2$$

Example 22.5   **PROBLEM**   A particle of mass m moving with a speed of $4c/5$ in the positive x direction collides with and "sticks to" a particle of mass $2m$ moving with a speed of $3c/5$ in the positive y direction. Using conservation of energy and momentum, determine the mass and velocity of the composite particle.
    SOLUTION   Figure 22.9 depicts the physical situation just before and after the collision. Because momentum is a vector quantity, momentum conservation ensures that both x and y components are conserved during the collision, i.e.,

$$\gamma_U\, MU \cos\theta = \gamma_v\, mv = \frac{4}{3}\, mc$$

$$\gamma_U\, MU \sin\theta = \gamma_u\, (2m)u = \frac{3}{2}\, mc$$

And energy conservation gives

$$\gamma_U\, Mc^2 = \gamma_v\, mc^2 + \gamma_u\, mc^2 = \frac{25}{6}\, mc^2$$

Dividing the second momentum equation by the first gives

$$\tan\theta = \frac{9}{8}$$

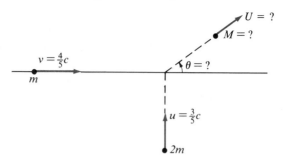

Figure 22.9   Example 22.5

or

$$\theta = \tan^{-1}\left(\frac{9}{8}\right) \approx 48°$$

By squaring each side of the two momentum equations and then adding, we get

$$\gamma_U{}^2 \, M^2 U^2 = \frac{16}{9} \, m^2 c^2 + \frac{9}{4} \, m^2 c^2 = \frac{145}{36} \, m^2 c^2$$

or

$$\gamma_U \, MU = \frac{\sqrt{145}}{6} \, mc$$

Solving this result for $\gamma_U \, M$ and substituting into the energy equation yields

$$U = \frac{\sqrt{145}}{25} \, c \approx 0.48 \; c$$

and

$$\gamma_U = \frac{25}{\sqrt{480}} = \frac{5\sqrt{30}}{24}$$

Substituting this result into the energy equation, we obtain

$$M = \frac{25}{6 \, \gamma_U} \, m = \frac{\sqrt{480}}{6} \, m = \frac{2}{3} \, \sqrt{30} \, m$$

and again the collision results in a mass increase of

$$\Delta m = M - 3m = \left(\frac{2\sqrt{30}}{3} - 3\right) m \approx 0.65 \, m \quad \blacksquare$$

E 22.25   Calculate the change in kinetic energy that occurs during the collision of Example 22.5. Account for this lost kinetic energy.
                    Answer: $\Delta K = -[(2\sqrt{30}/3)-3] \, mc^2$

22.5 Experimental Confirmation of Relativity

In this chapter we have described numerous physical phenomena predicted by the special theory of relativity, phenomena that cannot be explained by the classical mechanics of Newton. We have seen that relativistic mechanics does not contradict the validity of classical mechanics in the realm of speeds that are small compared to the speed of light, in those situations for which classical mechanics has been tested successfully for centuries. In fact, we have seen that the mechanics of relativity becomes identical to classical mechanics for speeds that are vanishingly small. Thus the theory of relativity generalizes classical mechanics: Relativity extends the range of validity of mechanics to include speeds that are comparable to the speed of light. The results of this extension, as we have seen, are often contrary to our experience and our physical intuition. Just as we do for any proposed physical theory, we expect experimental confirmation of the theory of relativity before we accept it as an appropriate description of nature.

Experimental Confirmation of Time Dilation

Perhaps the most direct experimental confirmation of the time dilation predicted by the theory of relativity has been accomplished using extremely accurate chronometers, which are sometimes called "atomic clocks." A cesium clock is a device that has a timekeeping precision of better than one part in 1 billion, a precision corresponding to an error of 1 second in about 30 years. Such devices were placed in airplanes that circled the earth. The time intervals indicated by the airborne clocks were then compared to those of another cesium clock fixed on the earth. Appropriate corrections were made for the effects of gravity and for the fact that neither of the clocks was in an inertial frame. The corrected difference in the time intervals measured by the "moving" and "stationary" clocks was predicted, according to relativistic time dilation, to be about 3×10^{-7} s. The measured time difference agreed with the theoretical prediction within an experimental error of about 0.2×10^{-7} s.

Another experimental test of relativistic time dilation is based on the decay, or the breaking up, of elementary particles called muons. These unstable particles have an "average expected lifetime" of about 2.2×10^{-6} s when measured in a reference frame in which the muons are at rest. Muons are formed in the upper atmosphere, about 10,000 m above the surface of the earth, by cosmic radiation from outer space. They have been measured to move toward the earth at a speed of about $0.999\ c$, a speed that is obviously relativistically significant. If, however, the effects of relativity were ignored, we would expect particles that exist for an average time interval of 2.2×10^{-6} s and that are traveling at approximately $c = 3.0 \times 10^{8}$ m/s to move a distance $d = 3.0 \times 10^{8}$ m/s $\times 2.2 \times 10^{-6}$ s $= 660$ m during their average lifetime. Because 660 m is considerably less than 10,000 m, we should expect that virtually none of the muons reaches the surface of the earth. Experimentally, however, it is found that a significant fraction of cosmic muons reach the surface of the earth before they decay. Then let us examine what is expected when relativistic effects are taken into account.

Let us call $t = 2.2 \times 10^{-6}$ s the average lifetime of cosmic muons as measured in the rest-frame of those muons. Then to observers on the earth, who measure the speed of the muons to be $v = 0.999\ c$, the dilated time interval t' during which these particles exist (on the average) is given by $t' = t/\sqrt{1 - v^2/c^2}$. Using $v/c = 0.999$, we find the average lifetime t' of muons, as measured by observers on the earth, to be

$$t' = \frac{2.2 \times 10^{-6} \text{ s}}{\sqrt{1 - 0.998}} = \frac{2.2 \times 10^{-6} \text{ s}}{0.045} = 49 \times 10^{-6} \text{ s} \qquad (22\text{-}34)$$

a value that is more than 22 times greater than the average lifetime of muons measured in their own rest-frame. The travel time t_t' required for a particle to negotiate a distance of 10,000 m at a speed of 3.0×10^8 m/s is

$$t_t' = \frac{1.0 \times 10^4 \text{ m}}{3.0 \times 10^8 \text{ m/s}} = 33 \times 10^{-6} \text{ s} \qquad (22\text{-}35)$$

a value less than the lifetime $t' = 49 \times 10^{-6}$ s. Therefore, we expect that most of the "average muons" reach the earth before decaying. This result is consistent with what is observed. And, in fact, numerous experiments, analyzed using appropriate decay statistics, have shown that the predictions of relativistic time dilation agree quantitively with observations.

Nuclear Fission and Fusion

The relativistic relationship between mass and rest energy, $E_o = mc^2$, has been confirmed in uniquely dramatic ways. The conversion of mass into energy is the physical basis for the controlled release of energy in nuclear reactors and the uncontrolled explosions of atomic and hydrogen bombs. The processes by which reactors and atomic bombs convert mass into energy are called *nuclear fission*, the breaking up of large atoms into smaller ones. The processes by which the sun generates energy and by which hydrogen bombs release energy are called *nuclear fusion*, the merger of small atoms to form larger ones. Let us consider briefly how the relativistic equivalence of mass and energy predicts and accounts for the enormous energies produced in fission and fusion processes.

A typical fissionable *isotope*, or species of atom characterized by the total number of protons and neutrons in its nucleus, is $^{235}_{92}\text{U}$. This isotope of uranium contains, in addition to the 92 protons that define that element to be uranium, 143 ($= 235 - 92$) neutrons within its nucleus. The $^{235}_{92}\text{U}$ isotope of uranium may fission spontaneously, breaking into two smaller atoms and releasing free neutrons. One such reaction, in which uranium fissions to form barium, krypton, and two free neutrons is represented by

$$^{235}_{92}\text{U} \rightarrow {}^{141}_{56}\text{Ba} + {}^{92}_{36}\text{Kr} + 2{}^{1}_{0}\text{n} + Q \qquad (22\text{-}36)$$

where Q represents the energy released by this reaction. The energy Q, which appears predominantly as kinetic energy of the reaction products, is provided by the difference in mass between the uranium atom and the final products. In this case, the total mass of the fission products is smaller than the mass of ^{235}U by 0.22 unified mass units (u). A *unified mass unit* is defined to be one-twelfth of the mass of the carbon isotope ^{12}C, so that 1 u is equal to 1.66×10^{-27} kg. Then the energy Q released by the fission of one uranium atom is

$$Q = \Delta E_o = \Delta mc^2 = (0.22 \text{ u})\left(1.66 \times 10^{-27} \frac{\text{kg}}{\text{u}}\right)(3.0 \times 10^8 \text{ m/s})^2$$

$$Q = 3.3 \times 10^{-11} \text{ J} = 200 \text{ MeV} \qquad (22\text{-}37)$$

Lest this quantity of energy seem small, consider that if one gram of uranium-235 were permitted to fission, over 80 billion joules of energy would be released. That much energy purchased from power companies in the form of electrical energy costs about $2500 at 1984 rates.

A substantial succession of fission processes may occur in a sizable mass of

^{235}U when one of the neutrons released is absorbed by the nucleus of another uranium atom. That absorbing nucleus is stimulated by the presence of the additional neutron to fission immediately (within about 10^{-15} s), thereby producing more free neutrons that cause other uranium atoms to fission. In this way, a *chain reaction* occurs, which, if uncontrolled, produces awesome quantities of energy in extremely short intervals of time. In less technical terms, an explosion takes place: An atomic bomb detonates. In nuclear reactors, the chain reaction is controlled by the presence of nonfissionable material that absorbs an appropriate fraction of the emitted neutrons. Thus energy is produced in reactors at carefully controlled rates.

Fusion processes depend similarly upon the conversion of mass into energy. Of the many possible fusion processes, one takes place when two atoms of *deuterium*, a " heavy hydrogen" atom that contains one proton and one neutron in its nucleus, combine to form helium:

$$\text{$_1^2$H} + \text{$_1^2$H} \rightarrow \text{$_2^4$He} + Q \qquad \text{(22-38)}$$

In this reaction, the mass of ^{4}He is less than the mass of the two ^{2}H atoms by 0.026 u, or by 4.3×10^{-29} kg. This mass difference Δm is converted into the reaction energy Q. In accordance with the relativistic equivalence between mass and energy, the energy produced by each such fusion is

$$Q = \Delta E_\text{o} = \Delta m c^2 = (4.3 \times 10^{-29} \text{ kg})\left(3.0 \times 10^8 \frac{\text{m}}{\text{s}}\right)^2$$

$$Q = 3.9 \times 10^{-12} \text{ J} = 24 \text{ MeV} \qquad \text{(22-39)}$$

Again, since each gram of hydrogen contains more than 10^{23} atoms, the conversion of relatively small masses of hydrogen into helium produces enormous quantities of energy. Indeed, our basic energy source, the sun, is fueled by this and similar fusion processes. Nuclear fusion as an energy source has been exploited in uncontrolled fusion devices—hydrogen bombs. Considerable current research is directed toward the development of controlled fusion reactors, which are perhaps the utilitarian energy sources of the future.

22.6 Problem-Solving Summary

We should be alerted that a problem requires a relativistic treatment if it describes an object moving at a speed that is an appreciable fraction of the speed of light. Problems of this kind that compare length or time intervals suggest that the Lorentz transformation is the appropriate problem-solving tool. Velocity comparisons in such problems suggest the use of the relativistic velocity transformation. Finally, relativistic problems that involve dynamics or energetics require that we use the relativistic expressions and relationships for linear momentum and kinetic energy.

Most problems involving special relativity can be formulated in terms of a "fixed" system S, where an event is described by coordinates x, y, z, and t, and a second system S', in which that event is specified by x', y', z', and t'. We usually assume that the corresponding axes of the two systems are parallel, that the x' axis is colinear with the x axis, and that the S' frame moves in the x direction with a constant velocity v_x relative to S.

Comparisons of lengths and time intervals as measured in two frames, S and S', are accomplished by recording the locations and times of specific events as they are measured by observers in each of these frames. These events, so determined within a given reference frame, must be measured using "local" clocks, each synchronized within that frame and positioned at the location of each event being measured. Remember that two observers moving relative to each other will, in general, measure different time intervals between events that occur at different locations. Thus observers in S and S' will, in general, disagree as to what constitutes two simultaneous events. Having appropriately measured the coordinates of each pertinent event in one frame, say S, we may relate these coordinates to those in S' using the Lorentz transformations: $x' = \gamma(x - v_x t)$, $y' = y$, $z' = z$, and $t' = \gamma(t - v_x x/c^2)$.

Comparisons of lengths and time intervals in frames of reference that are moving relative to each other produce two particularly significant results. First, a moving clock runs at a rate less than that of fixed clocks to which it is compared. This phenomenon is called time dilation. The second result is called length contraction. The length of a moving object appears to fixed observers to be contracted, or shortened, in the direction of its motion.

An object moving with a velocity $\mathbf{u}$, as measured by observers in S, has velocity components u_x, u_y, and u_z. The components u'_x, u'_y, and u'_z, of the velocity of that object, when measured by observers in S', which is moving in the x direction at a constant velocity v_x relative to S, are given by the relativistic velocity transformation:

$$u'_x = \frac{u_x - v_x}{1 - v_x u_x/c^2} \quad ; \quad u'_y = \frac{u_y}{\gamma(1 - v_x u_x/c^2)} \quad ; \quad u'_z = \frac{u_z}{\gamma(1 - v_x u_x/c^2)}$$

These relationships ensure that the speed of an object moving at a speed $u < c$ in a given reference frame will always be measured in any other reference frame to be moving at a speed u', which is less than c. This is true regardless of the relative speed $v < c$ between the two reference frames.

In order that linear momentum be conserved in any two inertial frames that are moving relative to each another, we define a relativistic momentum $\mathbf{p} = \gamma_u m \mathbf{u}$, where u is the speed of a particle having a mass m.

Using the relativistic momentum $\mathbf{p}$ for a particle, we define a relativistic force, $\mathbf{F} = d\mathbf{p}/dt$, which in turn may be used to determine for that particle a relativistic kinetic energy $K = (\gamma_u - 1) mc^2$. The quantity $E = \gamma_u mc^2$ is the total relativistic energy of the particle, and $E_o = mc^2$ is the rest energy of the particle. These relationships indicate the fundamental equivalence of mass and energy.

Nuclear fission, the splitting of large atoms into smaller ones, and nuclear fusion, the merging of small atoms to form larger ones, both result in reaction products having a total mass less than the mass of the original reactants. That mass difference Δm is converted into an energy Q of the reaction according to $Q = \Delta mc^2$. Most calculations of mass difference in nuclear fission and fusion processes are made in terms of the unified mass unit (u), which has the value 1 u $= 1.66 \times 10^{-27}$ kg.

Problems

GROUP A

22.1

(a) How much energy would be required to accelerate a 1000-kg automobile to a speed equal to $4c/5$?

(b) If energy costs $0.10/kWh, determine the energy cost to accomplish the acceleration of part (a).

22.2 The S coordinates of two events are (6.0 m, 0, 0, 20 ns) and (12 m, 0, 0, 30 ns). If S' moves with a velocity $v_x = c/2$ relative to S,

(a) what are the S' coordinates of these events?

(b) which event occurs first in S'?

22.3 S coordinates of events A and B are (12 m, 0, 0, 15 ns) and (24 m, 0, 0, 20 ns), respectively. If S' moves relative to S at a constant velocity, $v_x = 4c/5$,

(a) what are the S' coordinates of these events?

(b) which event occurs first in S? in S'?

22.4 Show that the energy equivalent to one unified mass unit is 931 MeV.

22.5 A typical fission process that is induced when uranium-235 absorbs a neutron is given by

$$^{235}_{92}U + ^{1}_{0}n \rightarrow ^{95}_{42}Mo + ^{139}_{57}La + 2\,^{1}_{0}n + Q$$

The masses (in unified mass units) of the reactants are

U: 235.12 u

Mo: 94.94 u

La: 138.95 u

n: 1.01 u

What is Q for this reaction?

22.6 A neutron decays into a proton and an electron according to

$$n \rightarrow p + e$$

If the masses of these particles are $m_n = 1.67495 \times 10^{-27}$ kg, $m_p = 1.67265 \times 10^{-27}$ kg, and $m_e = 9.1 \times 10^{-31}$ kg, how much energy (in MeV) is released by this reaction?

22.7 The position and velocity of a particle are observed in reference frame S at $t = 60$ ns to be $\mathbf{r} = (8.0\hat{\mathbf{i}} + 6.0\hat{\mathbf{j}})$m and $\mathbf{u} = (1.5\hat{\mathbf{i}} - 2.0\hat{\mathbf{j}}) \times 10^8$ m/s. Determine the corresponding S' quantities if $v_x = 3c/5$.

22.8 With $\gamma = (1 - v^2/c^2)^{-1/2}$, $E = \gamma mc^2$, $E_o = mc^2$, and $p = \gamma mv$, show that

(a) $\gamma^2 c^2 = c^2 + \gamma^2 v^2$

(b) $E^2 = E_o^2 + p^2 c^2$

22.9 A 100-kW (130-hp) engine, which converts captured "interstellar-radiation" energy to propulsion energy, is used to accelerate a 10,000-kg spaceship. How much time would be required to give this ship a speed of 0.90 c?

22.10 If a rocket ship were powered by a "mass-to-energy" conversion engine that could convert 10 percent of the mass of the ship directly to energy, determine the limiting speed of this ship.

22.11 Two particles, each traveling at a speed v, move toward each other. Determine v if the speed of either of the particles relative to the other is equal to $c/2$.

22.12 An astronaut circles the earth with an orbital speed of 8.0 km/s. How much time must this astronaut spend in orbit if his watch is to lose 1.0 s relative to earth clocks?

22.13 Determine the "decrease in length" of a space shuttle (rest-length = 50 m) traveling 17,000 mi/h relative to earth observers.

22.14 The velocity of frame S' relative to frame S is given by $v_x = 2.7 \times 10^8$ m/s. A light signal is observed in S to be propagating in the positive y direction. Determine the S' direction of propagation of this light signal.

22.15 A pulse of light is emitted (event 1) at $t = t' = 0$ from the momentarily coincident origins of frames S and S'. The pulse subsequently reflects (event 2) from an object located on the y axis of S 240 m from the origin. Determine the coordinates of event 2 in S and S'. Assume S' moves with a constant velocity $v_x = 4c/5$ relative to S.

22.16 A particle is confined to move along the x axis. If the S position of the particle at time t_1 is x_1 and at time t_2 is x_2, the average speed s_{av} in S during this time interval is defined to be

$$s_{av} = \frac{x_2 - x_1}{t_2 - t_1}$$

Show that the corresponding S' quantity is given by

$$s'_{av} = \frac{s_{av} - v_x}{1 - \frac{v_x}{c^2}s_{av}}$$

22.17 Two events are specified in frame S by (5.0 m, 3.0 m, 0, 10 ns) and (8.0 m, 7.0 m, 0, 15 ns). Observers in frame S', which moves in the x direction at a constant speed relative to S, determine these two events to be simultaneous.

(a) What is v_x, the velocity of S' relative to S?

(b) When do these two events occur in S'?

(c) What is the distance between these two events in S? in S'?

(d) Is there a reference frame S'' moving in the x direction relative to S in which these two events have the same x'' coordinate? Explain.

22.18 Two events are observed in frame S to have coordinates (5.0 m, 3.0 m, 0, 10 ns) and (8.0 m, 7.0 m, 0, 30 ns). Observers in frame S', which moves at a constant speed in the x direction relative to S, determine these two events to have the same x' coordinate.

(a) What is v_x, the velocity of S' with respect to S?

(b) Where and when do these two events occur in S'?

(c) What time interval lapses between these events in S? in S'?

(d) Is there a reference frame S'' moving in the x direction relative to S in which these two events are simultaneous?

22.19 The S coordinates of two events are (30 m, 0, 0, 50 ns) and (18 m, 16 m, 0, 75 ns). If the velocity of frame S' relative to frame S is given by $v_x = 2.4 \times 10^8$ m/s,

(a) what is the distance between these two events in S? in S'?

(b) what is the time interval between these two events in S? in S'?

22.20 The S positions of simultaneous (in S) events A and B are $(-10, 0, 0)$ m and $(30, 30, 0)$ m, respectively. If the velocity of S' relative to S is given by $v_x = -1.8 \times 10^8$ m/s,

(a) determine the time interval, $t'_B - t'_A$, measured in S'.

(b) which event occurs first in S'?

22.21 A particle (mass $= 1.2 \times 10^{-26}$ kg), initially at rest, decays into two identical particles. Each of the particles moves at a speed of 1.8×10^8 m/s just after this "explosion."

(a) Determine the mass of each particle.

(b) Account for the gain in kinetic energy that occurs during this explosion.

22.22 Supertrain (rest-length $= 100$ m) travels at a speed of $0.95\ c$ as it passes through a tunnel (rest-length $= 50$ m).

(a) What length of supertrain is measured by observers at rest in the tunnel?

(b) What tunnel length is measured by train observers?

(c) Are there (tunnel) times when the train is entirely within the tunnel?

(d) Are there (train) times when the train is entirely within the tunnel?

(e) If the answer is yes to either (or both) part (c) and part (d), determine the length of time for which this condition exists.

22.23 A particle (mass $= M$), initially at rest, decays into two particles. One of these particles is observed to have a mass equal to $M/4$ and a speed of $4c/5$ just after the decay. Determine the mass and speed of the other particle.

22.24 A particle moving with speed $u_1 = 4c/5$ collides with and sticks to an identical particle initially at rest, as shown in Figure 22.10.

(a) Use classical considerations to determine the mass M and speed u_3 of the composite particle.

(b) Use the galilean velocity transformation to determine the S' speeds (u'_1, u'_2, u'_3) corresponding to the S speeds (u_1, u_2, u_3) if $v_x = -4c/5$. Show that with this velocity transformation, classical momentum is also conserved in S'.

(c) Use the relativistic velocity transformation to determine (u'_1, u'_2, u'_3) if $v_x = -4c/5$. Show that classical momentum is not conserved in S' when the relativistic velocity transformation is used.

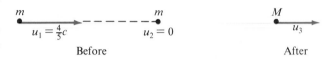

Before   After

Figure 22.10   Problem 22.24

22.25 Analyze the collision of Problem 22.24 relativistically.

(a) First determine M and u_3 using conservation of energy and momentum in S.

(b) If $v_x = -4c/5$, determine S' speeds, u'_1 and u'_2, which correspond to the S speeds, u_1 and u_2.

(c) Use energy and momentum conservation in S' to determine M and u'_3. Do these correspond to the values expected from the results of part (a)?

(d) What is the loss of kinetic energy in S? in S'?

22.26 A particle (mass $= M$), initially moving with a speed of $3c/5$, decays into two particles. One of the particles, which has a mass of $M/4$, is observed to move with a speed of $4c/5$ in a direction that is perpendicular to the original direction of motion. Determine the mass and speed of the other particle.

22.27 Determine the magnitude of the acceleration of a 2.0-kg object that is acted upon by a 6.0-N force that is parallel to the velocity of the object if the speed of that object is equal to

(a) 2.0 m/s

(b) 2.0 km/s

(c) 2.0×10^6 m/s

(d) 2.0×10^8 m/s

22.28 Determine the magnitude of the acceleration of a 2.0-kg object that is acted upon by a 6.0-N force that is perpendicular to the velocity of the object if the speed of that object is equal to

(a) 2.0 m/s

(b) 2.0 km/s

(c) 2.0×10^6 m/s

(d) 2.0×10^8 m/s

22.29 When observed in frame S, two events are separated in space by 4.0 m and in time by 5.0 ns. Is there a reference frame in which both events occur (a) simultaneously? (b) at the same point, say the origin? (c) If the answer to either or both of the above is yes, describe the reference frame.

22.30 Two events occur simultaneously in frame S when the distance separating the two events is equal to d_o.

(a) If d is the distance between these two events in any other frame moving at a constant velocity with respect to S, show that $d \geq d_o$.

(b) What conditions, if any, ensure that $d = d_o$?

22.31 Two events occur at the same point in frame S where the time interval between the two events is equal to T_o. If T is the time interval in any other frame moving at a constant velocity with respect to S, show that $T > T_o$.

GROUP B

22.32 A particle (mass $= 2M$) moving with a speed of $4c/5$ collides with another particle of mass M initially at rest. After the collision, the particles move as shown in Figure 22.11. Determine the speeds, V and v, assuming that relativistic momentum and energy are conserved.

Before   After

Figure 22.11   Problem 22.32

22.33 The collision shown in Figure 22.12 is elastic. Determine the speeds V and v.

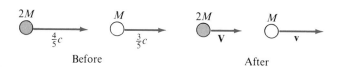

Before   After

Figure 22.12   Problem 22.33

22.34 At the moment when S' clock C_1 and S clock C_2 are adjacent, they indicate $t = t' = 0$. Later C_1 passes an S'' clock C_3 and these two indicate $t' = t'' = 30$ s at the moment of their coincidence. If S' moves relative to S with a velocity $v'_x = 4c/5$ and S'' moves relative to S with a velocity $v''_x = -3c/5$,

(a) what time is assigned by S observers to the moment of coincidence of C_1 and C_3?

(b) what times do C_2 and C_3 indicate at the moment of their coincidence?

(c) what time is assigned by S' observers to the moment of coincidence of C_2 and C_3?

22.35 Observers in frame S use a 1.0-m measuring rod that is at rest in the x-y plane and makes an angle of $60°$ with the x axis. If frame S' moves relative to S with a constant velocity $v_x = 4c/5$,

(a) what angle between the measuring rod and the x' axis is measured by S' observers?

(b) what is the S' length of the measuring rod?

22.36 Consider the S line, $y = mx + b$, with slope m and y-intercept b.

(a) Show that S' observers agree that these points are a straight line.

(b) Determine the S'-slope m' and S'-intercept b'. Which, if either of these, changes with time?

22.37 The *space-time interval S^2* between two events separated in S-time by $\Delta t (>0)$ and S-distance by $\Delta r (>0)$ is defined to be

$$S^2 = c^2(\Delta t)^2 - (\Delta r)^2$$

This time-space interval is said to be *timelike* if $S^2 > 0$, *null* if $S^2 = 0$, or *spacelike* if $S^2 < 0$.

(a) Show that the value of S^2 is the same in all inertial frames. A quantity, such as S^2, which has the same value in all inertial frames is said to be *invariant* under a Lorentz transformation. This invariance of S^2 ensures that, even though the two events being considered will, in general, have different space-time coordinates in different inertial frames, all inertial observers agree on the (numerical) value and (timelike, null, or spacelike) nature of the space-time interval S^2 separating the two events.

(b) If the space-time interval separating two events is timelike, show that:

(i) The distance (in any inertial frame) between these events is less than the distance light will travel during the time interval (in the same inertial frame) between the events;

(ii) there is an inertial frame, say S_o, in which both events occur at the origin;

(iii) the S_o-time-interval Δt_o (called the *proper-time interval*) between the two events is less than the time interval in any other inertial frame moving relative to S_o;

(iv) $$\Delta t_o = \frac{\sqrt{S^2}}{c}.$$

(c) If the space-time interval separating two events is spacelike, show that:

(i) the distance (in any inertial frame) between these events is greater than the distance light will travel during the time interval (in the same inertial frame) between the events;

(ii) there is an inertial frame, say S_o, in which the two events are simultaneous;

(iii) the S_o-distance Δr_o (called the *proper length*) between the two events is less than or equal to the separation distance in any other inertial frame moving relative to S_o;

(iv) $$\Delta r_o = \sqrt{-S^2}$$

22.38 For each of the pairs of events with S coordinates that are given below, determine the space-time interval S^2 (*see* Problem 22.37), the nature (timelike, null, or spacelike) of this interval, the frame S_o, if appropriate, and the proper time Δt_o or proper length Δr_o, as appropriate.

(a) (4.0 m, 0, 0, 10 ns), (6.0 m, 0, 0, 20 ns)

(b) (4.0 m, 0, 0, 10 ns), (8.0 m, 0, 0, 20 ns)

(c) (4.0 m, 0, 0, 10 ns), (7.0 m, 0, 0, 20 ns)

(d) (4.0 m, 0, 0, 10 ns), (7.0 m, 4.0 m, 0, 30 ns)

(e) (4.0 m, 0, 0, 10 ns), (10.0 m, −8.0 m, 0, 30 ns)

23 Early Quantum Physics

After two centuries the *classical physics* of newtonian mechanics, maxwellian electromagnetics, and thermodynamics had, by 1900, been developed into a coherent system that accurately described the natural laws of the macroscopic world. The microscopic aspects of nature, however, remained a mystery. The atomic and subatomic nature of matter had not yet yielded to the scientific method.

By the beginning of the twentieth century experimental techniques had become quite sophisticated and remarkably precise. Careful experimentation began to reveal phenomena that could not be explained by classical physics, and considerable experimental data began to accumulate. These data suggested that they held clues to the structure of atomic matter and the rules that govern subatomic interactions.

The scientists of the early twentieth century were particularly challenged by this natural puzzle. Doubtless they were at the same time tantalized and excited by the prospects of new and revolutionary concepts that would be required to decipher this microscopic mystery. During the first three decades of the twentieth century, they not only resolved many of the major experimental-theoretical dilemmas, but they also revolutionized physics in the process. This chapter and the next outline some of the crucial steps in the creation of the new physics, which has come to be called *modern physics* for obvious reasons or sometimes *quantum physics* for reasons we will soon discover.

23.1 The Blackbody Dilemma: Planck's Hypothesis

One physical system that was the subject of careful experimental measurements before 1900 was the cavity radiator, more commonly called a blackbody. An ideal *blackbody* is an object that absorbs all wavelengths of electromagnetic radiation incident upon it. Reflecting no radiation, the object appears to be black. At thermal equilibrium, a blackbody is characterized by radiating a distribution of wavelengths that depends only on the temperature of the blackbody. Such an ideal radiator, or absorber, can be approximated experimentally by a cavity surrounded by *any material*. We may observe this radiation through a small opening through which electromagnetic radiations are emitted. The blackbody radiator commanded considerable scientific attention, because physicists usually suspect that if a physical phenomenon is independent of the material involved, that phenomenon probably involves fundamental interactions and, therefore, deserves careful study. Thus blackbody radiation was carefully studied—both experimentally and theoretically.

Extensive, careful measurements were made of the intensity of the various wavelengths of radiated energy from blackbodies at a number of fixed temperatures. Figure 23.1 shows experimental curves of the *spectral energy density* $f(\lambda, T)$ in each wavelength interval $d\lambda$ of the radiation from a blackbody at several absolute temperatures T. The spectral energy density is a function of wavelength λ and absolute temperature T. Notice that the peak of the radiated energy moves toward shorter wavelengths with increasing temperature. These curves agree with our observations that solid objects being heated first glow cherry red, then orange, and finally a brilliant white, which includes all the visible wavelengths of light.

Using the tools of classical physics, eminent physicists of the period made theoretical calculations of $f(\lambda, T)$. Their results were then compared to the accurately measured experimental values. Boltzmann and Wien, for example, used a thermodynamic approach and obtained theoretical results that fit the experimental measurements at short wavelengths but failed to give correct values at the longer wavelengths. Figure 23.2 shows a comparison of Wien's predictions with a typical experimental curve. Later, Rayleigh treated the blackbody radiation problem by assuming that the moving charges in and on the walls of a cavity radiator should be treated as harmonic oscillators, each capable of radiat-

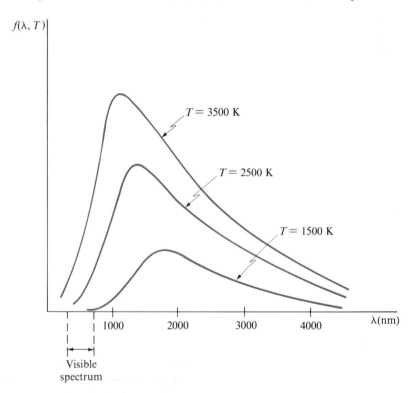

Figure 23.1
The spectral density $f(\lambda, T)$ emitted from a blackbody at three different temperatures.

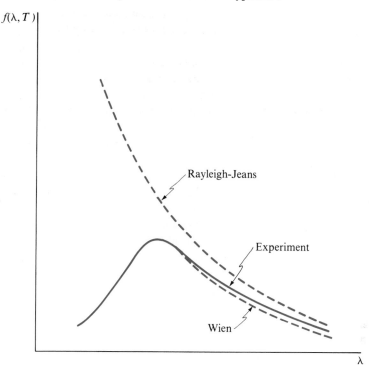

Figure 23.2
Comparisons of the theoretical results of the calculations of Rayleigh and Jeans and of Wien with the experimental measurements of blackbody radiation.

ing electromagnetic waves. Rayleigh's results, corrected by Sir James Jeans, fit the experimental data at long wavelengths. But, as seen in Figure 23.2, the Rayleigh-Jeans treatment failed miserably in the short-wavelength region of the spectrum. In fact, this discrepancy between theory and experiment was so drastic at short wavelengths that it became known as the *ultraviolet catastrophe*. This embarrassing failure of classical theory was indeed a catastrophe to physicists. A new approach was indicated, and a radically new concept was soon forthcoming.

In 1900 Max Planck made some mathematical changes to the Rayleigh-Jeans treatment that, surprisingly, produced an expression that perfectly fit the experimental data! Instead of integrating over a continuum of energy states for the assumed harmonic oscillators, as Rayleigh had done, Planck discovered the exact solution by summing over *discrete* ranges of energy.

Planck himself was reluctant to accept the radical implications of his discovery. His results clearly specified that the allowed energies of a harmonic oscillator with a natural frequency ν are not continuous, as permitted by classical physics, but are restricted to discrete quantities $0, h\nu, 2h\nu, 3h\nu, \ldots$, where h is a constant. In other words, Planck's results required that the energies of a harmonic oscillator are *quantized*, that is, the energy can assume *only* specific discrete values. Thus if ν is the natural frequency of a harmonic oscillator, the quantized energies E_n of the oscillator are given by

$$E_n = nh\nu \quad ; \quad n = 0, 1, 2, \ldots \tag{23-1}$$

Planck's determination of the constant h was made by comparing his expressions for $f(\lambda, T)$ to the experimental data. He found a value for h, which is now called *Planck's constant*, that was very close to the value now accepted,

$$h = 6.63 \times 10^{-34} \text{ J} \cdot \text{s} \tag{23-2}$$

Planck further hypothesized that the minimum quantity of energy ΔE, called a *quantum*, which can be absorbed or radiated by a harmonic oscillator is given by

$$\Delta E = h\nu \qquad (23\text{-}3)$$

We may reasonably ask, if the energy of a harmonic oscillator is quantized, why do we not observe evidence of discrete energies in swinging pendula or in oscillating masses connected to springs? The following example indicates why the quantum nature of macroscopic systems is not observable.

Example 23.1

PROBLEM   A simple pendulum consists of a 10-g mass suspended from a string 20 cm long. The pendulum makes an angle of 0.20 radians with the vertical when released from rest. As friction decreases the amplitude of the oscillations, with what precision must we be able to measure the energy of the pendulum in order to detect the quantum jumps in the energy of the pendulum?

SOLUTION   The frequency of a simple pendulum of mass m and length L is given by

$$\nu = \frac{1}{2\pi}\sqrt{\frac{g}{L}}$$

so that

$$\nu = \frac{1}{2\pi}\sqrt{\frac{9.8 \text{ m/s}^2}{0.20 \text{ m}}} = 1.1 \text{ s}^{-1}$$

Then, according to the Planck relationship, the energy change (quantum jump) between adjacent energy states of the oscillator is

$$\Delta E = h\nu = (6.63 \times 10^{-34} \text{ J} \cdot \text{s})(1.1 \text{ s}^{-1}) = 7.4 \times 10^{-34} \text{ J}$$

The initial energy of the oscillating system is equal to the gravitational potential energy of the pendulum at its greatest angular displacement θ from the vertical, or

$$E = mgL(1 - \cos\theta)$$

$$E = (0.01 \text{ kg})(9.8 \text{ m/s}^2)(0.20 \text{ m})(1 - \cos 0.20)$$

$$E = 3.9 \times 10^{-4} \text{ J}$$

We must then be able to detect the ratio $\Delta E/E$ in order to observe the quantization of the pendulum energy, or we must detect

$$\frac{\Delta E}{E} = \frac{7.4 \times 10^{-34} \text{ J}}{3.9 \times 10^{-4} \text{ J}} = 1.9 \times 10^{-30}$$

In other words, detection of quantum jumps in the pendulum energy would require energy measurement techniques with a precision of better than 2 parts in 10^{30}, a requirement that far exceeds any available experimental capabilities.

Thus, quantum phenomena are not observable in macroscopic systems because Planck's constant is so small. Although quantum effects exist in numerous physical systems, they are noticeable only in systems at the microscopic, atomic level. ∎

The departure of Planck's hypothesis from classical physics was, of course, immediately obvious. But the fundamental impact of the notion of energy quantization was not appreciated until Einstein seized upon the concept to elucidate another dilemma that had arisen between classical theory and experiment, the photoelectric effect.

23.2   The Photoelectric Effect and Photons

Before 1900 Heinrich Hertz had demonstrated that electrons are emitted from a clean metal surface when light of sufficiently short wavelength shines on that surface. This phenomenon is called the *photoelectric effect*, and the emitted electrons are called *photoelectrons*.

Other investigators used circuits similar to the one in Figure 23.3 to study the photoelectric effect in more detail. When light of sufficiently short wavelength from the light source L is incident on the metal M within the evacuated photoelectric cell, electrons are emitted from M. With the switch S in the position shown in the figure, the emitted photoelectrons are attracted to the anode An, the positive terminal of the cell. The potential difference ΔV across the terminals of the cell is measured by the voltmeter V. The ammeter A measures any photocurrent present in the circuit because of the flow of photoelectrons across the cell.

Figure 23.4 shows experimental curves of the photocurrent I_p as a function of ΔV for two different brightnesses of the light source. When ΔV has a sufficiently large positive value, the flat curves indicate that all the photoelectrons emitted from M are collected at the positive anode. When ΔV is negative (that is, when the switch S in Figure 23.3 has been thrown), only those photoelectrons emitted from M with kinetic energies greater than $e|\Delta V|$ can reach the anode. The photocurrent reaches a value equal to zero when $-\Delta V$ is equal to V_o, the *stopping potential*. When $-\Delta V = V_o$, even the most energetic photoelectrons emitted from M do not reach the anode; and so the maximum kinetic energy K_{max} of the most energetic photoelectron is given by

$$K_{max} = eV_o \tag{23-4}$$

These and similar experiments using photoelectric cells produced a number of

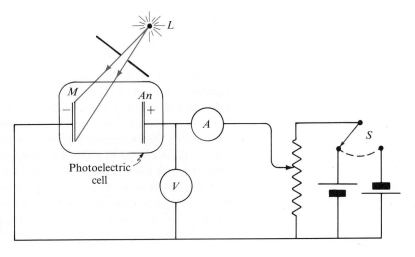

Figure 23.3
A photoelectric cell and its associated circuitry.

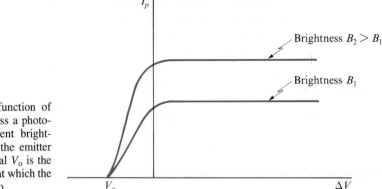

Figure 23.4
The photocurrent I_p as a function of potential difference ΔV across a photo-electric cell for two different bright-nesses of light incident on the emitter metal. The stopping potential V_o is the negative of the value of ΔV at which the photocurrent is equal to zero.

results that are not consistent with classical theory. Among the most important of these results are the following:

1. The maximum kinetic energy of photoelectrons emitted from a metal is inde-pendent of the brightness of the light shining on the metal. Classically, we expect that brighter light, having electric fields of greater magnitude, will produce more energetic electrons than a less bright light.

2. No photoelectrons are emitted from a metal surface until the frequency of the incident light, regardless of its intensity, exceeds a critical frequency ν_c. Classi-cal theory suggests that any frequency of light should produce photoelectrons from a metal surface, provided that light is sufficiently bright.

Einstein provided a fitting explanation of the results of the photoelectric experiments. Extending Planck's quantization concept for harmonic oscillators to include all electromagnetic radiations, Einstein assumed that light energy is composed of discrete packets of energy, now called *photons*. He assumed that each photon has an energy $E = h\nu$, where ν is the frequency of the light and h is Planck's constant. Then, when a photon with energy $h\nu$ is absorbed by an electron within a metal, all the energy $h\nu$ of that photon is converted to additional kinetic energy of that electron. A quantity of energy ϕ, called the *work function* of the metal, is required to free the most energetic of the bound electrons from the surface of the metal. Therefore, according to Einstein, the emitted photo-electrons should have maximum kinetic energy $K_{max} = eV_o$ given by

$$K_{max} = eV_o = h\nu - \phi \qquad (23\text{-}5)$$

Equation (23-5) is called the *Einstein photoelectric equation*. Each term in the equation is a quantity of energy. According to Equation (23-5), a plot of K_{max} as a function of light frequency ν should be a straight line having a slope equal to Planck's constant h. Subsequent experiments confirmed this prediction. Figure 23.5 shows the experimental results for a few typical metals, each of which, when K_{max} is plotted as a function of ν, produces a straight line of slope h but each of which has a different work function ϕ.

A convenient and commonly used unit of energy that is appropriate to most quantum physics applications is the *electron volt* (eV), defined to be the kinetic energy acquired by an electron that has been accelerated through a difference in

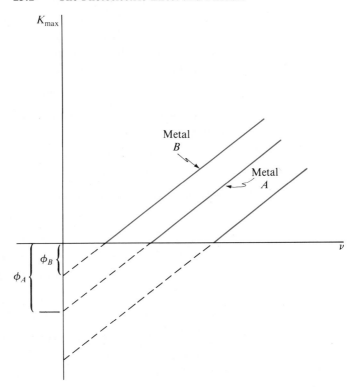

Figure 23.5
The maximum kinetic energy K_{max} of photoelectrons plotted as a function of the frequency ν of incident light on several different metals. The work function ϕ of each metal is equal to the negative of the K_{max}-intercept of the line corresponding to that metal. The slope of each line is equal to Planck's constant h.

potential having a magnitude of 1 V. Thus, 1 eV is equivalent to 1.6×10^{-19} J. In a photoelectric experiment, then, if a stopping potential V_o is measured to be 3.0 V, the maximum kinetic energy K_{max} of any photoelectron is equal to $eV_o = 3.0$ eV. Although light is typically characterized by its wavelength λ, it is often convenient to determine the energy $h\nu$ of a photon, whose wavelength is λ, in electron volts. Because the photon energy $h\nu$ is equal to hc/λ, we may express the product hc in appropriate units so that the photon energy $h\nu$, measured in eV, is related to the wavelength λ of the light, where λ is measured in nm, by

$$h\nu \text{ (in eV)} = \frac{1.24 \times 10^3}{\lambda \text{ (in nm)}} \tag{23-6}$$

E 23.1   Find the energy of a photon having a wavelength of (a) 470 nm and (b) 1840 Å.   Answers: (a) 2.6 eV; (b) 6.7 eV

E 23.2   What is the wavelength of a photon with an energy of 4.0 eV?
Answer: 310 nm

Example 23.2   PROBLEM   A metal used as the emitter in a photoelectric cell has a work function of 2.4 eV. (a) What is the *threshold wavelength* (the maximum wavelength) of light that will produce photoelectrons from this metal? (b) If monochromatic light of wavelength 400 nm is incident on the emitter, what is the stopping potential across this cell?

SOLUTION (a) The threshold wavelength corresponds to the minimum-energy photons that will cause an electron (with zero kinetic energy) to escape the emitter metal. In other words, photons at the threshold wavelength have energy equal to the work function ϕ of the metal, or

$$h\nu = 2.4 \text{ eV}$$

The threshold wavelength λ_t corresponding to this photon energy is given in nm by Equation (23-6):

$$\lambda_t = \frac{1.24 \times 10^3 \text{ eV} \cdot \text{nm}}{h\nu} = \frac{1.24 \times 10^3 \text{ eV} \cdot \text{nm}}{2.4 \text{ eV}} = 520 \text{ nm}$$

(b) Light of wavelength 400 nm has an energy (in eV) given by Equation (23-6):

$$h\nu = \frac{1.24 \times 10^3 \text{ eV} \cdot \text{nm}}{400 \text{ nm}} = 3.1 \text{ eV}$$

The stopping potential is determined by first obtaining the maximum kinetic energy, $K_{max} = eV_o$, of any photoelectron by direct substitution into the photoelectric equation, Equation (23-5):

$$K_{max} = eV_o = h\nu - \phi = 3.1 \text{ eV} - 2.4 \text{ eV} = 0.70 \text{ eV}$$

Then the stopping potential V_o, measured in volts, is numerically equal to K_{max}, measured in electron volts, or

$$V_o = 0.70 \text{ V} \quad \blacksquare$$

The quantum nature of electromagnetic radiation was clearly established by Einstein's explanation of the photoelectric effect. Other experimental results provided further evidence confirming that electromagnetic radiation is composed of photons, each having an energy $E = h\nu$. For example, A. H. Compton demonstrated that x rays, high-frequency electromagnetic radiation, when scattered from electrons, experience a change in wavelength that can be explained by assuming the particlelike nature of electromagnetic radiation. This phenomenon, called *Compton scattering*, was quantitatively explained using the conservation principles of energy and momentum. To do so, Compton recognized that a massless photon possesses not only an energy E given by

$$E = h\nu = \frac{hc}{\lambda} \tag{23-7}$$

but also a quantum of linear momentum having magnitude p given by

$$p = \frac{E}{c} = \frac{\left(\frac{hc}{\lambda}\right)}{c} = \frac{h}{\lambda} \tag{23-8}$$

Thus both the energy and momentum of photons may be expressed in terms of their wavelengths λ and Planck's constant h, as given in Equations (23-7) and (23-8). These properties of photons will be useful in our subsequent descriptions of quantum phenomena.

E 23.3 How much momentum does a photon of wavelength 300 nm possess?

Answer: $2.2 \times 10^{-27} \text{ kg} \cdot \text{m} \cdot \text{s}^{-1}$

23.3 Atomic Models, Spectra, and Atomic Structure

By the beginning of the twentieth century the laws of chemistry had securely confirmed the atomic nature of matter. It was clearly established that very small atoms composed the building blocks of matter. For example, about 6×10^{23} atoms of carbon make up a 12-g sample of carbon. Even so, the basic physical structure of atoms was still a mystery in 1900.

Soon after J. J. Thompson discovered the existence of the electron in 1897, he proposed a structural model of the atom. He suggested that the negatively charged electrons were particles distributed like "plums" throughout a continuous, positively charged "pudding," as suggested by Figure 23.6. This *plum-pudding model* of an atom explained how a neutral atom may be composed of equal amounts of (continuous) positive charge and negative charge. Thus, if one or more electrons were removed from the atom, an ionized atom would be produced.

In 1911 Ernest Rutherford and his coworkers performed an experiment, the results of which compelled a change in the accepted notions of atomic structure. They directed *alpha particles*, positively charged helium atoms with their orbital electrons removed, at a very thin sheet of gold foil. By studying the spatial distribution of the scattered alpha particles—some of them were bounced back in directions near the incident beam—the experimenters were able to show that most of the mass and all of the positive charge of an atom are concentrated in a small central region of the atom called its nucleus. Such a *nuclear atom*, illustrated in Figure 23.7, has a massive, positive nucleus and negatively charged electrons distributed throughout a considerable volume about the nucleus. This nuclear model of the atom, developed from the Rutherford experiments, is consistent with what we currently understand to be the general structure of atoms. However, the Rutherford model provides no details of atomic structure that permit our understanding of how various species of atoms differ physically and chemically. The details of atomic structure were obtained from the study and interpretation of atomic spectra, which we shall now consider.

Atomic Line Spectra

For many years before Rutherford developed the nuclear model of the atom, physicists and chemists had observed and carefully measured the various wavelengths of electromagnetic radiation emitted or absorbed by different species of

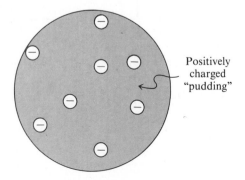

Positively charged "pudding"

Figure 23.6
The plum-pudding model of the atom.

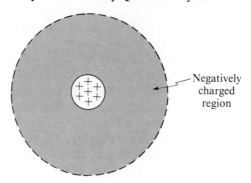

Figure 23.7
The nuclear atom.

atoms. Records of these radiations, called *spectra*, may be obtained using an apparatus called a *spectrometer*, like that of Figure 23.8.

In a spectrometer, the light emitted by atoms that have been excited electrically, as shown in Figure 23.8(a), is passed through a thin slit and then through a prism, which disperses the different wavelengths of the light in different directions. A film strip in the spectrometer records those wavelengths present as *lines*, which are separate images of the slit. The position of each spectral line corresponds to a wavelength, and the position, or wavelength, can be measured

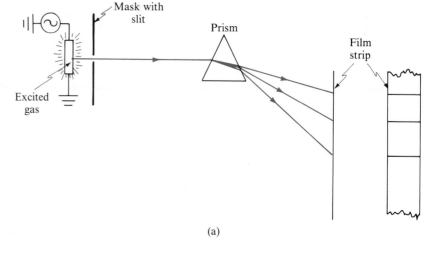

(a)

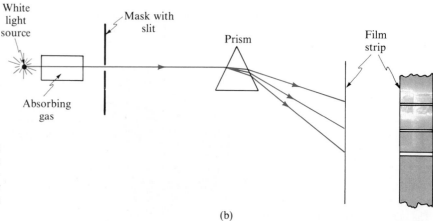

Figure 23.8
A prism spectrometer. **(a)** An emission spectrum is produced by the light from an excited gas. **(b)** An absorption spectrum is formed by having particular wavelengths absorbed in a gas through which white light is passed.

(b)

with considerable accuracy. Such a spectral record on the film is called an *emission spectrum*. In a similar way, *absorption spectra* of gases may be obtained by passing white light through a sample of gas before the light enters the prism, as illustrated in Figure 23.8(b). In this case, the spectral lines absorbed by the gas are produced at the same positions as those wavelengths of light produced by emission when the same gas sample is excited. Figure 23.9 shows the line spectra of typical gases.

Because the line spectra of most elements are extremely complicated, that is, they are composed of complex patterns of a very large number of lines, the search for a theory to explain the origins of spectral lines was concentrated on hydrogen, the simplest element. A large number of wavelengths of the lines of the hydrogen spectrum had been precisely measured considerably earlier than 1900, but no theory was available that could explain the pattern of wavelengths present in the hydrogen spectrum.

The first breakthrough in the puzzle of the hydrogen spectrum came in 1885, when a Swiss high school teacher, Johann Balmer, discovered an equation that precisely predicts the wavelengths of a group of lines (called a *series* of lines) in the hydrogen spectrum. The relationship that Balmer discovered, which gives the wavelengths of lines in the *Balmer series* of hydrogen, is

$$\frac{1}{\lambda} = R\left(\frac{1}{2^2} - \frac{1}{n^2}\right) \quad ; \quad n = 3, 4, 5, \ldots \qquad \text{(23-9)}$$

where $R = 1.1 \times 10^7 \text{ m}^{-1}$ is now called the *Rydberg constant*. Similar relationships, named for their discoverers, were later found for other series in the hydrogen spectrum:

$$\text{Lyman series: } \frac{1}{\lambda} = R\left(\frac{1}{1^2} - \frac{1}{n^2}\right) \quad ; \quad n = 2, 3, 4, \ldots$$

$$\text{Paschen series: } \frac{1}{\lambda} = R\left(\frac{1}{3^2} - \frac{1}{n^2}\right) \quad ; \quad n = 4, 5, 6, \ldots$$

$$\text{Brackett series: } \frac{1}{\lambda} = R\left(\frac{1}{4^2} - \frac{1}{n^2}\right) \quad ; \quad n = 5, 6, 7, \ldots$$

$$\text{Pfund series: } \frac{1}{\lambda} = R\left(\frac{1}{5^2} - \frac{1}{n^2}\right) \quad ; \quad n = 6, 7, 8, \ldots \qquad \text{(23-10)}$$

It finally became clear that the wavelengths of any and every line of the hydrogen spectrum could be found from a relationship that combined all of the relationships of Equations (23-9) and (23-10):

$$\frac{1}{\lambda} = R\left(\frac{1}{\ell^2} - \frac{1}{n^2}\right) \quad ; \quad \ell, n, \text{ positive integers such that } n > \ell \qquad \text{(23-11)}$$

E 23.4   What is the longest wavelength line of the Balmer series of the hydrogen spectrum?   Answer: 650 nm

E 23.5   What is the shortest wavelength line of the Lyman series of the hydrogen spectrum?   Answer: 910 nm

The empirical relationship of Equation (23-11), which so accurately predicts all the wavelengths of the hydrogen spectrum, became the goal that theorists

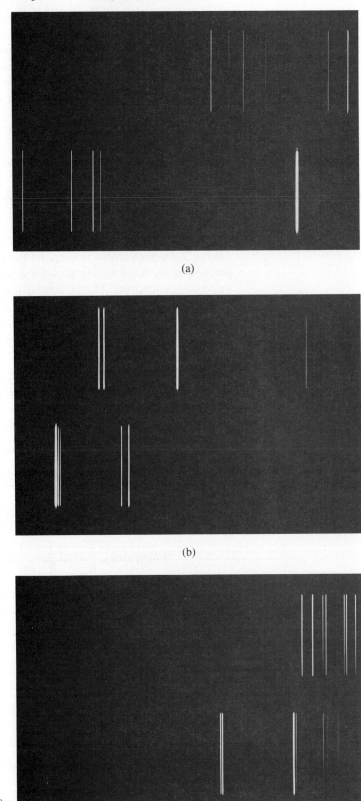

(a)

(b)

(c)

Figure 23.9
Photographs of the emission spectra of
some typical gases: **(a)** helium, **(b)** mercury, and **(c)** krypton.

sought to explain from fundamental principles. The next significant advance in the understanding of atomic structure came when Niels Bohr proposed a detailed model of a nuclear atom, a model from which he was able to obtain Equation (23-11) using the basic physical relationships of mechanics and electricity.

The Bohr Model of the Hydrogen Atom

Extending Rutherford's concept of the nuclear atom, Bohr proposed a model of the hydrogen atom that successfully predicts the wavelengths of all the lines of the hydrogen spectrum. In 1913 Bohr proposed a planetary model of the hydrogen atom, depicted in Figure 23.10, in which an electron with a negative charge of magnitude e revolves in a circular orbit, somewhat like a planet revolving around the sun, about a *proton* that has a positive charge of magnitude e. The electron is held in its orbit by the attractive coulombic force between the electron and proton. The proton, because it is nearly 2000 times as massive as the electron, is assumed to be fixed as the electron revolves about it. As we shall see, the Bohr model incorporates elements of classical physics, the notion of quantization suggested by Planck, and the photon concept introduced by Einstein.

In order to establish his planetary model of hydrogen, Bohr found it necessary to assert the following postulates:

1. The revolving electron exists in stable, nonradiating circular orbits, called *stationary orbits*, about the central proton. According to classical theory, an accelerating electron, like one in a circular orbit, would radiate electromagnetic energy and would, therefore, spiral into the nucleus of the atom. This postulate contends, in effect, that at least some classical electromagnetic theory is not applicable at the atomic level.

2. The electron can exist only in orbits for which the angular momentum of the electron about the proton is equal to an integral multiple of $h/2\pi$, where h is Planck's constant. Mathematically, the magnitude mvr of the angular momentum of the electron is restricted to the quantized values

$$mvr = \frac{nh}{2\pi} \quad ; \quad n = 1, 2, 3, \ldots \qquad (23\text{-}12)$$

where m is the mass of the electron, v is its speed, and r is the radius of its stationary orbit.

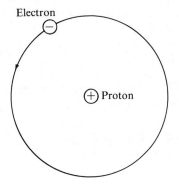

Figure 23.10
A planetary model of the hydrogen atom.

3. The atom radiates or absorbs a photon with energy $h\nu$ when the electron undergoes a transition, or "jumps," between any two allowed stationary orbits. If E_i and E_f represent the total energies of an electron in its initial and final states, the photon energy is given by

$$h\nu = |E_f - E_i| \tag{23-13}$$

Using these postulates, we may reconstruct Bohr's development of the planetary model of the hydrogen atom and arrive at a relationship that specifies the wavelengths of the hydrogen spectrum. Again, we assume that the electron of a hydrogen atom has mass m, moves with speed v in an orbit of radius r, and has charge e.

The coulombic force $\mathbf{F}$ on the electron has magnitude

$$F = \frac{e^2}{4\pi\epsilon_0 r^2} \tag{23-14}$$

and because this force is the centripetal force on the electron in its circular orbit, we may write

$$\frac{mv^2}{r} = \frac{e^2}{4\pi\epsilon_0 r^2} \tag{23-15}$$

If we now solve Equations (23-15) and (23-12) simultaneously for r and v, we obtain the radius r_n of the nth orbit and the speed v_n of the electron in the nth orbit:

$$r_n = \frac{\epsilon_0 n^2 h^2}{\pi m e^2} \quad ; \quad n = 1, 2, 3, \ldots \tag{23-16}$$

and

$$v_n = \frac{e^2}{2\epsilon_0 n h} \quad ; \quad n = 1, 2, 3, \ldots \tag{23-17}$$

Here we may note that the stationary orbit having the least radius, corresponding to $n = 1$ in Equation (23-16), gives

$$r_1 = \frac{\epsilon_0 h^2}{\pi m e^2} \tag{23-18}$$

If values of the constants, $\epsilon_0 = 8.85 \times 10^{-12}$ C^2/(N·m^2), $h = 6.6 \times 10^{-34}$ J·s, $m = 9.1 \times 10^{-31}$ kg, and $e = 1.6 \times 10^{-19}$ C, are inserted into Equation (23-18), we find the radius r_1 of the first Bohr orbit to be

$$r_1 = 5.3 \times 10^{-11} \text{ m} \tag{23-19}$$

a value that agrees with other experimental estimates of atomic radii. Notice also that the radii r_n of all the other stationary orbits are related to the radius r_1 of the first Bohr orbit by

$$r_n = n^2 r_1 \tag{23-20}$$

E 23.6   What is the ratio of the radius of the second Bohr orbit to the radius of the fourth Bohr orbit?   Answer: 1/4

E 23.7   What is the ratio of speeds of electrons in the second and fourth Bohr orbits?   Answer: 2

Consider now the energy of the orbital electron in the planetary model of a hydrogen atom. Using the speed v_n of an electron in the nth orbit, as expressed in Equation (23-17), the kinetic energy K_n of an electron in the nth orbit is

$$K_n = \frac{1}{2}mv_n^2 = \frac{me^4}{8\epsilon_o^2 n^2 h^2} \tag{23-21}$$

Similarly, the electric potential energy U_n of an electron in the nth orbit may be written, using the radius r_n given by Equation (23-16), as

$$U_n = -\frac{e^2}{4\pi\epsilon_o r} = -\frac{me^4}{4\epsilon_o^2 n^2 h^2} \tag{23-22}$$

Here the potential energy of the electron is negative because the potential energy is chosen to be zero when the electron is an infinite distance from the proton. Then the total energy E_n of an electron in the nth orbit is

$$E_n = K_n + U_n = -\frac{me^4}{8\epsilon_o^2 n^2 h^2} \tag{23-23}$$

Suppose now that an electron initially in the nth orbit has energy E_i given by

$$E_i = -\frac{me^4}{8\epsilon_o^2 n^2 h^2}$$

and this electron undergoes a transition to the ℓth orbit, where its energy E_f is

$$E_f = -\frac{me^4}{8\epsilon_o^2 \ell^2 h^2} \quad ; \quad \ell < n$$

Then, according to the third Bohr postulate, when the electron jumps from the nth orbit to the ℓth orbit, the atom emits a photon with energy equal to $E_i - E_f$. The wavelength of the emitted photon is related to the *quantum numbers*, ℓ and n, of the Bohr orbits by

$$E_i - E_f = h\nu = \frac{hc}{\lambda} = \frac{me^4}{8\epsilon_o^2 h^2}\left(\frac{1}{\ell^2} - \frac{1}{n^2}\right)$$

or

$$\frac{1}{\lambda} = \frac{me^4}{8\epsilon_o^2 h^3 c}\left(\frac{1}{\ell^2} - \frac{1}{n^2}\right) \tag{23-24}$$

If we substitute the numerical values of the constants into Equation (23-24), we find

$$\frac{1}{\lambda} = (1.1 \times 10^7 \text{ m}^{-1})\left(\frac{1}{\ell^2} - \frac{1}{n^2}\right)$$

or

$$\frac{1}{\lambda} = R\left(\frac{1}{\ell^2} - \frac{1}{n^2}\right) \tag{23-25}$$

which is identical to the empirical expression of Equation (23-11) that so precisely predicts the wavelengths of the lines of the hydrogen spectrum.

We may calculate any wavelength of the hydrogen spectrum directly from Equation (23-25) using the appropriate values of the quantum numbers ℓ and n.

The various energy levels of the hydrogen atom, that is, the possible energies of an electron in any of the allowed stationary orbits, may be computed and expressed in eV by using Equation (23-23) with the constants in that equation expressed in appropriate units:

$$E_n = -\frac{13.6}{n^2} \, eV \tag{23-26}$$

Figure 23.11(a) shows an *energy-level diagram* for hydrogen. The lowest energy level, called the *ground state*, corresponds to $n = 1$. According to Equation (23-26), the energy E_1 of an electron in the ground state is equal to -13.6 eV. Similarly, the $n = 2$ state has an energy E_2 equal to -3.40 eV, the $n = 3$ state has energy $E_3 = -1.51$ eV, and so on. As indicated in Figure 23.11(a), the various series of the hydrogen spectrum, the Lyman series, the Balmer series, and the others are caused by transitions between the $n = 1$ level, the $n = 2$ level, and so on, and the higher energy levels. Figure 23.11(b) is a representation of the possible Bohr orbits about the proton in a hydrogen atom and some of the transitions between orbits.

Example 23.3   **PROBLEM**   Calculate the frequency of the least energetic photon that must be absorbed by the electron of a hydrogen atom in its ground state in order to ionize that atom.

SOLUTION   A hydrogen atom in its ground state, or lowest energy state, has its electron in the first ($n = 1$) Bohr orbit. To ionize such an atom requires that the electron receive enough energy to remove it from the proton: The electron must acquire at least the energy to elevate it to the energy level corresponding to $n = \infty$. Then the minimum energy required is obtained using Equation (23-26):

$$\Delta E = E_\infty - E_1 = 0 - (-13.6 \text{ eV}) = 13.6 \text{ eV}$$

This energy difference is equal to the energy of a photon of wavelength

$$\lambda = \frac{1.24 \times 10^3 \text{ eV} \cdot \text{nm}}{13.6 \text{ eV}} = 91.2 \text{ nm} = 9.1 \times 10^{-8} \text{ m}$$

and a photon of this wavelength has a frequency

$$\nu = \frac{c}{\lambda} = \frac{3.0 \times 10^8 \text{ m/s}}{9.1 \times 10^{-8} \text{ m}} = 3.3 \times 10^{15} \text{ Hz} \quad \blacksquare$$

The success of the Bohr theory was a significant advance for quantum physics and for the understanding of atomic structure. Still, the success of this merger of classical and quantum concepts was limited. The Bohr theory and its extensions proved to be inadequate to explain the spectra of atoms more complex than hydrogen. Furthermore, the postulates of the theory had no known physical basis other than the imposing fact that they led to the correct results. Finally, the Bohr theory offered no physical insight into the electronic transition process: What causes the transition or what takes place during the transition? About a decade after Bohr introduced this theory, further advances in the understanding of atomic structure were forthcoming.

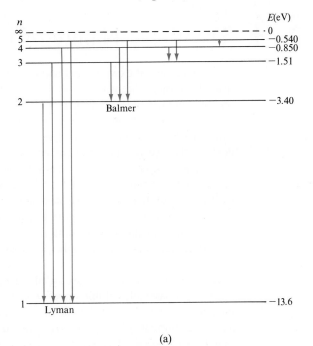

(a)

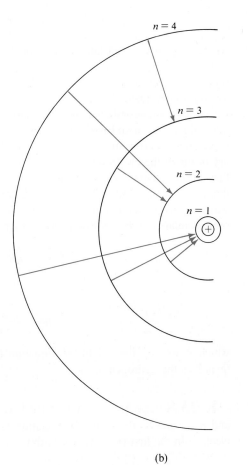

Figure 23.11
(a) An energy-level diagram for hydrogen. (b) Some transitions between orbits in the hydrogen atom.

(b)

23.4 The Wave Nature of Particles

The progression of experimental and theoretical evidence convincingly suggests that electromagnetic waves, or simply light, occur as discrete, particlelike photons. In view of this evidence, we seem constrained to accept the particlelike nature of light. But our earlier studies of interference and diffraction, each of which has been accurately explained in terms of the wave nature of light, seem to demand that we accept that light is a wave. Then what is the real nature of electromagnetic radiation? Is light a wave or is it composed of discrete packets of energy? Consideration of all the evidence now requires that we recognize the *dual nature* of light. We must now acknowledge that the physical nature of electromagnetic radiation is more subtle than may be indicated by a single experimental situation. Depending on the particular conditions of a given experiment, light may exhibit either its wave nature or its particlelike, photon character.

Is there a similar duality associated with particles? Does an electron, or, for example, a baseball, exhibit wavelike characteristics under appropriate experimental conditions? In 1923 Louis de Broglie proposed that these questions be answered affirmatively. Specifically, he suggested that an electron of mass m moving with speed v has associated with it a wavelength λ given by

$$\lambda = \frac{h}{p} \qquad (23\text{-}27)$$

where h is Planck's constant, and $p = mv$ is the linear momentum of the electron.

This unusual assertion—it was not taken particularly seriously when first proposed—had the immediate theoretical advantage that it explained the Bohr assumption of quantized angular momentum for electrons in the hydrogen atom. De Broglie suggested that the stationary Bohr orbits were those that had an integral number of de Broglie wavelengths. Just as the wavelengths of a standing wave on a vibrating string are restricted to discrete quantized values, the stationary orbits of the electron in hydrogen are restricted to lengths (circumferences) that are an integral number of de Broglie wavelengths. Figure 23.12 illustrates the $n = 3$ Bohr orbit in hydrogen; the orbit contains exactly three de Broglie wavelengths in its circumference. Then the circumference $2\pi r_n$ of the nth Bohr orbit is related to the de Broglie wavelength λ by

$$2\pi r_n = n\lambda = \frac{nh}{mv}$$

or

$$mvr = \frac{nh}{2\pi}$$

which is exactly the condition, Equation (23-12), of the Bohr postulate that quantizes the hydrogen orbits.

E 23.8 (a) Use $e = 1.6 \times 10^{-19}$ C, $\epsilon_o = 8.85 \times 10^{-12}$ C$^2 \cdot$N$^{-1} \cdot$m^{-2}, and $h = 6.6 \times 10^{-34}$ J$\cdot$s in Equation (23-17) to calculate the speed of an electron in the first ($n = 1$) Bohr orbit.   Answer: 3.3×10^6 m/s

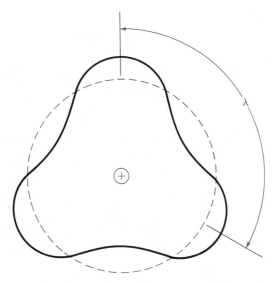

Figure 23.12
The $n = 3$ Bohr orbit of hydrogen. This orbit contains exactly three de Broglie wavelengths λ of the planetary electron.

(b) Given that the mass of an electron is $m = 9.1 \times 10^{-31}$ kg, calculate the de Broglie wavelength of the ground-state electron of a hydrogen atom.

Answer: 3.4×10^{-10} m

(c) Assuming that this de Broglie wavelength is equal to the circumference of the first Bohr orbit, calculate the radius of the first Bohr orbit in a hydrogen atom. Compare your result with Equation (23-19).

Answer: 5.3×10^{-11} m

The singular notion proposed by de Broglie, that moving particles inherently possess characteristics of waves, was soon supported by experiment. In 1927 Davisson and Germer found that electrons, which are reflected from a crystal of nickel, emerged in preferred directions relative to the incident electron beam. The intensity maxima of the reflected electrons occurred at specific angles corresponding to those angles that would be expected if the regular spacing between atoms of the crystal are treated as a diffraction grating encountered by the de Broglie waves associated with the electrons. The Davisson-Germer experiment brought immediate acceptance of the de Broglie hypothesis. Later, diffraction experiments with other particles, protons, neutrons, and small neutral atoms, confirmed the wave nature of particles in accordance with the de Broglie relationship, $\lambda = h/p$.

Before 1940 the physical perception of both electromagnetic radiation and matter was profoundly changed. Light possesses a particulate aspect along with its familiar wave nature. Similarly, particles sometimes exhibit the characteristics of waves. Furthermore, both the wavelike and particlelike aspects of a particle cannot be demonstrated simultaneously in a single measurement. Bohr, in his *principle of complementarity* emphasized that neither a particle nor an electromagnetic wave is adequately described in all experimental situations without recognizing that each is an entity having the dual characteristics of both waves and particles, characteristics that complement each other.

23.5 Uncertainty and Probability

In classical physics, before evidence of photons and matter waves imposed upon science the concept of wave-particle duality, we assumed (usually without making an issue of it) that physical quantities could, in principle, be measured with arbitrary accuracy. We took it for granted that, by using appropriate care and sufficiently accurate instrumentation, quantities like position, speed, and mass could be measured to any desired degree of accuracy. The advent of wave-particle duality caused a reassessment of the measurement process itself.

Physicists came to recognize that the very act of making a measurement affects the result of that measurement. This fact is perhaps most easily seen by considering the use of light to observe the position of a very small particle. Each photon of the light possesses momentum given by Equation (23-8). Then each photon that strikes the particle whose position is being measured transfers momentum to that particle and thereby changes the position of the particle. Thus the measurement process itself alters any quantity being measured.

Werner Heisenberg was the first to quantify this fundamental limitation inherent in the measurement process. In 1927 he proposed the *uncertainty principle*, which states that *the exact position and the exact momentum of a particle cannot be measured simultaneously.* Heisenberg showed that if Δx and Δp_x represent the uncertainties at a particular instant in the measured values of the x position and the x component of the linear momentum of a particle, the product $\Delta x \cdot \Delta p_x$ cannot be less than a value that is approximately the order of Planck's constant. The uncertainty principle, stated mathematically, is

$$\Delta x \cdot \Delta p_x \gtrsim h \qquad (23\text{-}28)$$

The uncertainty principle means, physically, that the more accurately we know the position of a particle at a given instant, the less accurately we can simultaneously know its momentum, and vice versa.

The uncertainty principle may also be expressed in terms of the other two rectangular coordinates:

$$\Delta y \cdot \Delta p_y \gtrsim h \qquad (23\text{-}29)$$

$$\Delta z \cdot \Delta p_z \gtrsim h \qquad (23\text{-}30)$$

Further, the fundamental uncertainty of the measurement process may be expressed in terms of other sets of variables. If, for example, E represents the energy of a physical system at a time t, it may be shown that

$$\Delta E \cdot \Delta t \gtrsim h \qquad (23\text{-}31)$$

where ΔE is the uncertainty in our knowledge of the value of E and Δt is the uncertainty in our knowledge of t. The following physical situation will serve both to illustrate the uncertainty principle and to introduce a probabilistic interpretation of the measurement of physical variables.

Consider the experimental arrangement of Figure 23.13. A beam of constant-velocity electrons passes through a narrow slit, and the electrons proceed to a screen located some distance from the slit. A photographic plate on the screen records the position of each electron that arrives at the screen. Just as when light passes through a narrow slit, a diffraction pattern is formed on the screen. In Figure 23.13 a graph of the intensity of electrons (number of electrons per unit area per unit time) at the screen as a function of y is shown to the right of the

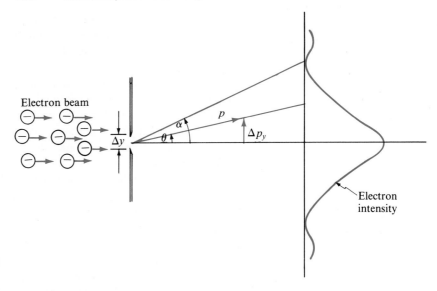

Figure 23.13
Diffraction of an electron beam that
passes through a narrow slit.

screen. If we designate the width of the slit to be Δy, then the y coordinate of
each electron passing through the slit is indeterminate, or uncertain, by an
amount Δy. The appearance of electrons outside the direct path of electrons
through the slit indicates that those electrons somehow acquired linear momentum in a direction parallel to the slit, that is, in the y direction.

Let p be the magnitude of the linear momentum of an electron between the slit
and the screen. Then the magnitude of the component p_y of its momentum in the
y direction is equal to $p \sin\theta$, where θ is the angle at which the electron is
deviated in passing through the slit. An arbitrary electron that has passed through
the slit may appear anywhere within the diffraction pattern. If we consider very
narrow slit widths, the central maximum will include virtually all the pattern,
because the half-width α of the central maximum for a single slit is related to the
wavelength λ of the incident beam and the slit width Δy by

$$\sin\alpha = \frac{\lambda}{\Delta y}$$

or

$$\Delta y = \frac{\lambda}{\sin\alpha} \tag{23-32}$$

Then we may express the uncertainty Δp_y in our knowledge of the y component
of the electronic momentum as

$$\Delta p_y \simeq p \sin\alpha \tag{23-33}$$

According to the de Broglie hypothesis, the magnitude p of the momentum of an
electron is related to the wavelength λ associated with that electron by

$$p = \frac{h}{\lambda} \tag{23-34}$$

Then the product $\Delta y \cdot \Delta p_y$ is obtained from Equations (23-32), (23-33), and
(23-34) to be

$$\Delta y \cdot \Delta p_y \simeq h \tag{23-35}$$

which is consistent with the uncertainty principle.

E 23.9   If the uncertainty in the position of a particle is equal to its de Broglie wavelength, show that the uncertainty in the speed of the particle is approximately equal to its speed.

The occurrence of single-slit diffraction patterns for electrons suggests that, in addition to the fact that a wavelength may be associated with electrons, these particles exhibit interference phenomena associated with waves. Recall that analysis of the single-slit diffraction of light uses the concept that wavelets of light from different regions of the slit interfere with each other. But in electron diffraction experiments a single electron passing through a slit may appear at a random position in the diffraction pattern. With what, then, does a single electron interfere? In some way the wave associated with the electron must "interfere with itself." Obviously an electron sometimes behaves very differently from the classical concept of a well-defined particle moving in a well-defined trajectory.

Let us consider further the single-slit diffraction of electrons. Even if one electron at a time is sent through the slit, the diffraction pattern accumulates when a large number of electrons are used. Although we cannot predict where any one electron will strike the screen, the diffraction pattern that results from a large number of electrons may be interpreted as a probability distribution of the electron position along the screen. In other words, in a region where the intensity of electrons is large, there is a large probability that a given electron will proceed from the slit in the direction of that region. A region of low electron intensity on the screen corresponds to a small probability that an electron will arrive at that location.

Such a probabilistic interpretation of particle location, which is based on the wave nature of particles, led to the idea of describing particles in terms of a function that can be associated with probability. In particular, Max Born in 1926 suggested that a *wave function* $\psi(x, y, z, t)$ be used to represent the inherent wavelike character of a particle. This wave function ψ itself is called the *probability amplitude* of a particle, but the physically measurable property associated with the wave function is the square of the absolute magnitude of ψ. Further, the quantity $|\psi|^2$ is proportional to the probability that a measurement will locate the particle to be at or near the point (x, y, z).

In the same year, 1926, Erwin Schrödinger developed a differential equation having solutions that provide the analytical form of ψ. The Schrödinger equation became the basis of *wave mechanics*. Now accepted almost universally as the most accurate and complete description of the microscopic realm of atomic and molecular physics, wave mechanics (at least wave mechanics in its most elementary aspects) is the topic of Chapter 25.

23.6 Problem-Solving Summary

The presence of Planck's constant, $h = 6.6 \times 10^{-34}$ J·s, in a problem identifies that problem as being in the category of quantum physics. Planck's constant appears in association with photons, the particlelike aspect of electromagnetic waves; it also appears in association with the de Broglie wavelength, the wavelike characteristic attributed to particles in quantum physics.

A photon is a particlelike packet of electromagnetic energy. The energy E of a photon is equal to $h\nu = hc/\lambda$, where ν is the frequency of the electromagnetic radiation and λ is its wavelength. Energies in quantum physics are commonly expressed in electron volts (1 eV $= 1.6 \times 10^{-9}$ J) and wavelengths are typically given in nanometers (1 nm $= 10^{-9}$ m). A handy relationship between photon energy E in eV and radiation wavelength λ in nm frequently simplifies calculations in quantum problems: $E(\text{eV}) = 1.24 \times 10^3/\lambda(\text{nm})$. Although a photon with wavelength λ has no mass, it possesses linear momentum of magnitude $p = h/\lambda$. This momentum, or part of it, may be transferred to particles with which photons collide.

When light of sufficiently high frequency is incident upon a metal surface, photoelectrons are emitted from that surface. This phenomenon is called the photoelectric effect. The most energetic electrons inside a given metal are bound to that metal by an amount of energy called the work function ϕ of that particular metal. When a photon of energy $h\nu > \phi$ is incident upon a given metal surface, photoelectrons that have a maximum kinetic energy $K_{\max} = h\nu - \phi$ are emitted. This photoelectric equation is sometimes written with $K_{\max}$ expressed as eV_o, where e is the electronic charge and V_o is the stopping potential, the minimum potential difference (across the electrodes of a photoelectric cell) that will stop the most energetic photoelectrons. Notice that the value of the stopping potential V_o, measured in volts (V), is numerically equal to the maximum kinetic energy $K_{\max}$, measured in electron volts (eV).

The spectral lines of hydrogen are records of the wavelengths of radiations emitted or absorbed by hydrogen atoms. All wavelengths of the hydrogen spectrum are accurately predicted by the Bohr theory, which assumes that the single electron in hydrogen is held in stable, nonradiating circular orbits by the coulombic attraction of the stationary central proton of the hydrogen atom. The Bohr theory postulates that the angular momentum mvr of the orbital electron is quantized: The angular momentum is restricted to values $mvr = nh/2\pi$, where n is the quantum number associated with the nth stationary orbit. Using the coulombic force $e^2/4\pi\epsilon_o r^2$ exerted on the electron by the photon as the centripetal force holding the electron in its circular orbit, it is found that the radius r_n of the nth Bohr orbit is proportional to n^2, or $r_n = n^2 r_1$, where r_1 is the radius of the first Bohr orbit. Similarly, the speed v_n of the electron in the nth Bohr orbit is proportional to $1/n$. The total energy of an electron in the nth orbit is negative and proportional to $1/n^2$; the energy E_n of an electron in the nth orbit is expressed in electron volts by $E_n = (-13.6/n^2)$ eV. Bohr assumed that photons of energy $|E_f - E_i| = h\nu = hc/\lambda$ are emitted or absorbed when electrons "jump" from an orbit with energy E_i to one of energy E_f. The wavelength λ of a line in the hydrogen spectrum is related to the quantum numbers n and ℓ of the transitional orbits according to

$$\frac{1}{\lambda} = R\left(\frac{1}{\ell^2} - \frac{1}{n^2}\right) \quad ; \quad n > \ell$$

where $R = 1.1 \times 10^7$ m^{-1} is called the Rydberg constant.

Any particle having mass m and speed v has associated with it a de Broglie wavelength λ given by $\lambda = h/p$, where $p = mv$ is the magnitude of the linear momentum of that particle. This wavelike characteristic of particles provides a basis for understanding and analyzing situations in which particles undergo interference or diffraction.

The uncertainty principle expresses a fundamental limitation on the measurement process. If Δx represents the uncertainty in position of a particle and Δp_x represents the uncertainty in the x component of its linear momentum, the product $\Delta x \cdot \Delta p_x$ cannot be less than a value approximately equal to Planck's constant h. The uncertainty principle also applies to other sets of variables. In particular, the simultaneous measurement of energy E and time t is limited by the relationship $\Delta E \cdot \Delta t \gtrsim h$.

Problems

GROUP A

23.1 What is the energy of a photon radiated by an FM radio station operating at 90 MHz?

23.2 A microwave oven heats with electromagnetic radiation having a wavelength of 3.0 cm. What is the energy of each photon?

23.3 (a) What is the frequency of the most energetic photon in the hydrogen spectrum? (b) What is the energy in each photon of that frequency?

23.4 (a) What is the ratio of the radii of the fourth and sixth Bohr orbits? (b) What is the ratio of the energies of electrons in the fourth and sixth Bohr orbits? (c) What is the ratio of the speeds of electrons in the fourth and sixth Bohr orbits?

23.5 Compare the de Broglie wavelengths of an electron and a proton, each of which is moving with a speed of 10^4 m/s.

23.6 What is the energy (in eV) of a photon that has a wavelength equal to the de Broglie wavelength of an electron moving at a speed of 100 m/s?

23.7 A line in the Balmer series of the hydrogen spectrum has a wavelength of 434 nm. This wavelength corresponds to an electronic transition between energy states of hydrogen having what two quantum numbers?

23.8 What is the de Broglie wavelength of an electron in the ground state of the hydrogen atom?

23.9 According to the Bohr theory, what is the ratio of the centripetal force on an electron in hydrogen in the $n = 2$ state to that of an electron in the $n = 3$ state?

23.10 A stopping potential of 4.6 V causes the photocurrent to be reduced to zero in a cell in which light shines on a metal having a work function of 2.1 eV. What is the minimum wavelength of the light incident on the metal? What is the threshold wavelength for photoemission from this metal?

23.11 (a) What is the frequency of a photon having an energy of 1.0×10^{-18} J? (b) What is the wavelength of a photon with energy equal to 300 eV? (c) What is the energy (in eV) of a photon with a wavelength of 10 m? (d) What is the energy (in J)

of a photon of frequency 5.0×10^{14} Hz? (e) What is the energy range (in eV) of photons that are within the visible spectrum?

23.12 Lithium, iron, aluminum, and tungsten have work functions equal to 2.3, 3.9, 4.2, and 4.5 eV, respectively. From which of these metals will light of wavelength 440 nm produce photoelectrons?

23.13 Aluminum has a work function of 4.2 eV. (a) What is the lowest frequency of light that will produce photoelectrons from aluminum? (b) If ultraviolet light of wavelength 200 nm is incident on aluminum, what is the maximum speed of any photoelectrons emitted?

23.14 What is the de Broglie wavelength of a proton that has an energy of 0.50 MeV?

23.15 For an electron in the second Bohr orbit of a hydrogen atom, calculate (a) the radius of the circular orbit, (b) the centripetal force on the electron, (c) the speed of the electron, and (d) the total energy of the electron.

23.16 In a photoelectric cell light with a minimum wavelength of 340 nm produces photoelectrons that are completely stopped by a difference in potential of 4.8 V across the cell. (a) What is the work function of the emitter metal in the cell? (b) What is the maximum speed of the photoelectrons?

23.17 (a) Electrons in the $n = 6$ and $n = 3$ orbits of hydrogen have what energy difference? (b) What is the frequency of a photon emitted when an electron makes a transition from the $n = 6$ orbit to the $n = 3$ orbit?

23.18 The distance between adjacent atoms in crystals is of the order of 1.0 Å. The use of electrons in diffraction studies of crystals requires that the de Broglie wavelength of the electrons is of the order of the distance between atoms of the crystals. What must be the minimum energy (in eV) of electrons to be used for this purpose?

23.19 (a) Find the energies of electrons in each of the first three Bohr orbits of hydrogen and use these energies to calculate the wavelengths of the first two lines of the Lyman series. (b) Use Equation (23-11) with the appropriate quantum numbers to determine the same wavelengths.

23.20 A metal that has a photoelectric threshold wavelength of 250 nm has monochromatic light of wavelength 200 nm incident on it in a photoelectric cell. (a) What is the work function of the metal? (b) What stopping potential is required across the cell to reduce the photocurrent to zero? (c) What is the energy of the most energetic electrons emitted from the metal?

23.21 (a) An electron is moving at 200 m/s, and its speed is measured with an accuracy of 0.10%. What is the minimum uncertainty with which its position may be determined? (b) Suppose a small pellet of mass 0.50 g is moving at the same speed, measured with the same accuracy. What is the minimum uncertainty in the position of the pellet?

23.22 At a particular instant, the position of a proton is measured to lie between $x = -0.50 \times 10^{-8}$ m and $x = 0.50 \times 10^{-8}$ m. What is the uncertainty in the x component of the velocity of the proton at that instant?

23.23 The electron in a hydrogen atom in the $n = 2$ state absorbs a photon of wavelength 327 nm. (a) Does the hydrogen atom become ionized? (b) If the atom is ionized, what is the kinetic energy of the freed electron assuming the proton is fixed?

23.24 An electron is confined to a region of diameter 10^{-8} m. Calculate the minimum uncertainty in the kinetic energy (in eV) of that electron.

23.25 Suppose the Bohr theory is applied to the He^{+1} ion, that is, to a single planetary electron assumed to be in a circular orbit about a nucleus containing two protons. (a) Show that the radius of the nth Bohr orbit in He^{+1} is half that of the corresponding orbit in hydrogen, while the speed of the electron in the nth orbit of helium is twice that in hydrogen. (b) Show that the total energy of an orbital electron in He^{+1} is four times that of the corresponding electron in H. (c) What is the ratio of the wavelengths of lines in the He^{+1} spectrum to the wavelengths of corresponding lines in the hydrogen spectrum?

GROUP **B**

23.26 The change in wavelength of a photon scattered from an electron is called the *Compton shift*. Consider a photon of wavelength λ_p initially moving in the x direction, as shown in Figure 23.14. After it collides with an electron at rest, the photon has wavelength λ_s and is scattered at an angle θ with the x axis. The recoiling electron with rest mass m_o is then moving with a velocity $\mathbf{v}$ that makes an angle ϕ with the x axis. The energy of the photon before the collision is $E_p = h\nu = hc/\lambda_p$; and because the recoiling electron may acquire a speed comparable to the speed of light c, (a) show that the conservation of energy requires that

$$\frac{hc}{\lambda_p} = \frac{hc}{\lambda_s} + (m - m_o)c^2$$

where $m = m_o/\sqrt{1 - (v/c)^2}$. (b) Because the rest mass of a photon is equal to zero and the magnitude p of the linear momen-

tum of a photon of wavelength λ is given by $p = E/c = h/\lambda$, use the principle of conservation of linear momentum in component form to show that

$$\frac{h}{\lambda_p} = \frac{h}{\lambda_s}\cos\theta + \frac{m_o v}{\sqrt{1 - (v/c)^2}}\cos\phi$$

and

$$\frac{h}{\lambda_s}\sin\theta = \frac{m_o v}{\sqrt{1 - (v/c)^2}}\sin\phi$$

(c) Eliminate v and ϕ from the equations expressing the conservation of energy and momentum to obtain

$$\Delta\lambda = \lambda_s - \lambda_p = \frac{h}{m_o c}(1 - \cos\theta)$$

which is the *Compton shift equation*. (d) Assume that photons of wavelength 0.15 nm are scattered at 60° from the incident beam by electrons at rest. Find the Compton wavelength shift $\Delta\lambda$ of the photons and the energy of the recoiling electrons.

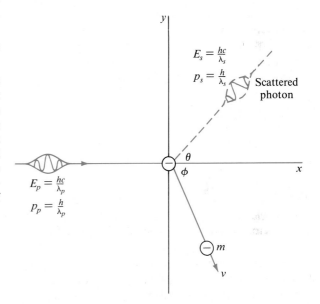

Figure 23.14   Problem 23.26

23.27 X rays are electromagnetic radiations that are produced when moving charged particles, usually electrons, strike a *target material* and are suddenly accelerated to rest. In general, part of the kinetic energy of a moving electron is converted into heat and part into X radiation. The most energetic photons are produced when all the kinetic energy of an electron is converted into X radiation. (a) What is the shortest wavelength of the photons produced by an X ray tube in which electrons are accelerated from rest by a potential difference of 50 kV before striking a target? Other intense X radiation is produced when the incident electrons knock orbital electrons from atoms of the target producing a vacancy in the orbit. X radiation is produced when electrons in higher energy orbits drop to a lower orbit to fill this vacancy. The lowest energy orbit (which has the least radius) of an atom is called the *K shell*, the next lower energy orbit is called the *L shell*, and so on alphabetically. X rays produced when electrons jump to the *K* shell from any higher energy orbit are called *K X rays*, and those that jump to the L shell from a higher energy shell are called *L X rays*. X rays in the K series of an element are called K_α, K_β, K_γ, and so on, when they are produced by electronic transitions to the K shell from the higher energy L, M, N, and so on, shells; and L X rays are similarly labeled L_α, L_β, etc. Figure 23.15 is an energy level diagram depicting the *binding energies* E_B of electrons in the atoms of tungsten, a common target material in X ray tubes. (b) Calculate the wavelengths of the K_α, K_β, and L_α X rays of tungsten.

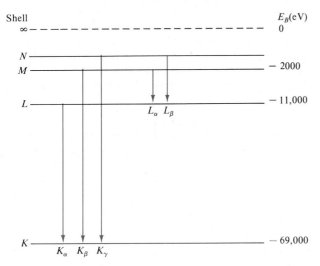

Figure 23.15   Problem 23.27

23.28 If you have considered the previous problem, consider the energy level diagram (Fig. 23.16) shown for a target material of an X ray tube. The wavelength of the L_α X ray is 0.113 nm. (a) Calculate the binding energy of an L-shell electron in this material. (b) Calculate the wavelength of an L_β X ray for this material. (c) What is the minimum potential difference across an X ray tube (using this material as a target) that will produce K X rays?

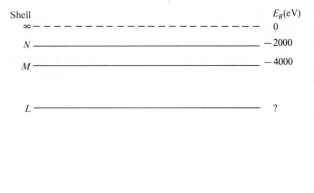

Figure 23.16   Problem 23.28

23.29 The *intensity I* of a beam of X rays is a measure of the radiation energy passing normally through a unit area per unit time. X rays are absorbed as they pass through matter in accordance with the relationship

$$I = I_o \, e^{-\mu x}$$

where I_o is the initial intensity of a beam of X rays, I is the intensity of the beam after having passed through a thickness x of a material having an *absorption coefficient* μ, a quantity that characterizes the ability of a material to absorb X rays. Notice that the unit of μ is the reciprocal of the unit in which the thickness x is given. Since μ is a function of wavelength of the X rays involved, it has a specific value only for a specific X ray wavelength. (a) If the absorption coefficient of a given X ray absorber at a particular wavelength is 0.40 cm^{-1}, what thickness of this absorber is required to reduce the intensity of an X ray beam to 1.0% of its incident intensity? (b) If the incident beam of X rays having a certain wavelength is reduced to half its incident intensity by a 2.0-cm thickness of a given substance, what is the absorption coefficient of that substance at this particular wavelength? (c) By what factor is the intensity of an X ray beam reduced if it passes through 1.0 cm of a material with an absorption coefficient for these X rays of 2.5 cm^{-1}?

24 Topics in Quantum Physics

The quantum theory brought to physics many fundamental conceptual changes. In Chapter 23 we discussed the wave-particle duality, the uncertainty principle, and the probabilistic interpretation of microscopic particles, like electrons. These innovations require that we now conceive of such particles in terms that are foreign to our macroscopic experience. Quantum theory was brought to fruition in 1925 by Schrödinger's formalism, wave mechanics (see Chapter 25, "Introduction to Wave Mechanics"), which remains our best and most complete representation of phenomena involving microscopic particles. In Chapter 25 the basic rules of wave mechanics and the interpretation of the results obtained by using those rules will be presented in a brief, introductory form. But here we will state and use some of the results of wave mechanics that are applicable to the understanding of atomic and nuclear structure.

24.1 Atomic Structure

The Bohr theory of the hydrogen atom introduced and suggested the significance of the quantum number n in that description of the simplest atom. The appearance of n in Bohr's model was an ad hoc assumption: The quantization of the angular momentum was introduced with appropriately chosen values because these choices produced results consistent with spectroscopic observations. When the hydrogen atom is treated by wave mechanics, however, the analogous quantum number, still called n, arises automatically from the solution of the basic equation of wave mechanics. In the following exposition, we will describe the quantum numbers that arise from the wave mechanical treatment of hydrogen, along with the rules that govern them, without concerning ourselves with the mathematical procedures by which they are obtained.

Quantum Numbers of Atomic Electrons

A wave-mechanical treatment of an electron moving in the electric field of the much more massive proton requires not one but three quantum numbers to specify a permitted state of the hydrogen atom. As in the Bohr theory, the *principal quantum number n* may assume any of the positive integral values, that is, $n = 1, 2, 3, \ldots$; and the energy E_n of any particular state of a hydrogen atom is determined by the value of n to which that state corresponds according to

$$E_n = -\frac{E_o}{n^2} \quad ; \quad n = 1, 2, 3, \ldots \tag{24-1}$$

where the energy E_o is approximately 13.6 eV. And whereas the angular momentum **L** of the electron about the proton is determined in the Bohr theory solely by the value of n, wave mechanics relates this angular momentum to the *orbital angular momentum quantum number l*. This second quantum number, which may be equal to zero or any positive integer less then n, specifies that the magnitude L of this (orbital) angular momentum is given by

$$L = \sqrt{l(l+1)}\, \hbar \quad ; \quad l = 0, 1, \ldots, n-1 \tag{24-2}$$

The quantity $\hbar = h/(2\pi)$ appears frequently in quantum calculations and is called "h-bar." The third quantum number m_l, the *orbital magnetic quantum number*, is related to the component of **L** that lies along a particular direction in space. Although there is no special direction in space associated with an isolated hydrogen atom, we may create a distinctive direction, say the z direction, by orienting a magnetic field along the z axis. Then the component L_z of the angular momentum of the electron is related to the magnetic quantum number m_l by

$$L_z = m_l \hbar \tag{24-3}$$

where m_l is restricted to positive or negative integral values, including zero, from $-l$ to $+l$.

The quantum numbers, n, l, and m_l, result naturally from the wave mechanical treatment of the hydrogen atom; they are related to dynamic variables of the system, like energy and angular momentum. One further quantum number was proposed by Uhlenbeck and Goudsmidt to complete the physical basis necessary to explain observed spectral phenomena. A *spin quantum number s* = 1/2 was associated with the electron. It is assumed that the electron has an intrinsic (inherently built-in) angular momentum called its *spin*, a property that is independent of any orbital angular momentum the electron may have associated with it. The magnitude of the spin angular momentum **S** of an electron is related to the spin quantum number s by

$$S = \sqrt{s(s+1)}\, \hbar = \frac{\sqrt{3}}{2}\, \hbar \tag{24-4}$$

Like **L**, **S** is quantized in space, and a *spin magnetic quantum number* m_s is used to specify this spatial quantization. The quantum number m_s can have only one of the two values, $\pm 1/2$. Thus if a magnetic field is used to define the z direction for the system containing an electron, the component S_z of the spin angular momentum **S** is restricted to one of the values

$$S_z = m_s \hbar = \pm\frac{1}{2}\, \hbar \tag{24-5}$$

We may summarize the quantization rules for the four quantum numbers associated with an orbital electron as follows:

1. $n = 1, 2, 3, 4, \ldots$.

2. $l = 0, 1, 2, \ldots, (n-1)$ for a given n.

3. $m_l = 0, \pm 1, \pm 2, \pm 3, \ldots, \pm l$ for given n and l.

4. $m_s = \pm 1/2$ for a given set of n, l, and m_l.

Note that the value of m_s, $\pm 1/2$, is independent of the values of the other quantum numbers, n, l, and m_l.

E 24.1

(a) If the principal quantum number n for a certain electronic state is equal to 3, what are the possible values of the orbital (angular momentum) quantum number l?   Answer: 0, 1, 2

(b) What are the corresponding values for L, the magnitude of the orbital angular momentum?   Answer: $0, \sqrt{2}\hbar, \sqrt{6}\hbar$

E 24.2

(a) If the orbital quantum number l for a certain electronic state is equal to 2, what are the possible values for the orbital magnetic quantum number m_l?
                                                    Answers: $-2, -1, 0, +1, +2$

(b) What are the corresponding values for L_z, the z component of the angular momentum L?   Answers: $-2\hbar, -\hbar, 0, +\hbar, +2\hbar$

E 24.3

(a) How many distinct electronic states are there with $n = 1$?   Answer: 2

(b) List these states using (n, l, m_l, m_s) notation.
                Answer: $(1, 0, 0, +1/2)$, $(1, 0, 0, -1/2)$, or $(1, 0, 0, \pm 1/2)$

E 24.4   Repeat Exercise 24.3 (a) and (b) for $n = 2$.
        Answers: (a) 8; (b) $(2, 0, 0, \pm 1/2)$, $(2, 1, 0, \pm 1/2)$, $(2, 1, \pm 1, \pm 1/2)$

Electronic Configurations in Atoms

How do the quantization rules apply to the arrangements of the numerous electrons in the orbits of more massive atoms? If every electron in an arbitrary atom were in the lowest possible Bohr energy state ($n = 1$), we would expect to observe only gradual changes in the chemical properties of the elements as the number of electrons (and thus the atomic number) in an atom increases. Such a gradual change does not occur. Hydrogen, for example, is chemically active, while helium, the element with two orbital electrons, is chemically inert.

An empirical principle, one based on observed facts rather than derived from more fundamental laws, was proposed by Wolfgang Pauli. This rule, which provides a logical explanation of the electronic configurations of the chemical elements, is called the *Pauli exclusion principle:*

In a given atom, no more than one electron can be in the same quantum state.

A *quantum state* of an electron is the state of that electron characterized by the four quantum numbers, n, l, m_l, and m_s.

Early in the history of atomic spectroscopy, electronic transitions to levels with different values of the orbital angular momentum quantum number l became associated with various descriptions of the observed spectral lines. Lines associated with $l = 0$ were termed *s*harp, and $l = 0$ came to be designated the s-state.

Similarly, other lines came to be described as *principal* ($l = 1$ designates the p-state), *diffuse* ($l = 2$ designates the d-state), and *fine* ($l = 3$ designates the f-state). These and other of the traditional designations of states with different values of l may be summarized as:

Value of l	0	1	2	3	4	5	6
Spectral Designation	s	p	d	f	g	h	j

In a given atom, a group of electrons in quantum states that have the same values of n forms a *shell* of that atom. Similarly, a group of electrons whose quantum states have the same values of both n and l forms a *subshell* of that atom. When all the available quantum states within a shell are occupied, that shell is said to be filled. Similarly, when all the quantum states of a subshell are occupied, that subshell is said to be filled.

E 24.5 How many electronic states are there in the (a) $n = 1$ shell? (b) $n = 2$ shell?   Answers: (a) 2; (b) 8

E 24.6 How many electronic states are there in an (a) $l = 0$ subshell? (b) $l = 1$ subshell? (c) $l = 2$ subshell?   Answers: 2; 6; 10

When all the electrons in a given atom are in the lowest-energy quantum states available, that atom is said to be in its *ground state*. In general, the energies of electronic states of an atom depend predominantly on the values of n and l; for both n and l, the lower the quantum number, the lower is the energy of the corresponding atomic state. Thus we may expect that as the atomic numbers of elements increase, the electrons in the atoms of those elements will occupy the states with the lowest possible values of n and l. But the Pauli exclusion principle restricts the number of electrons in each subshell. Because m_l can assume integral values from $-l$ to $+l$, there are $2l + 1$ values of m_l; and because each of the values of m_l may have values of $m_s = \pm 1/2$, there are $2(2l + 1)$ electrons in each subshell. We may now see how the configuration of electrons is patterned in the various elements.

Table 24.1 tabulates the quantum states available to electrons in the first three shells of atoms. Recall now that the electrons of elements with increasing atomic number (the number of protons in the nucleus and, therefore, the number of orbital electrons in a normal, un-ionized, atom) occupy the states with lowest energy. The lower energy states are, in general, the quantum states with the lowest values of n and l; Table 24.2 is a guide to the electronic configurations of the elements.

Table 24.2 lists the first 13 elements in ascending order of atomic number (and, therefore, in the number of orbital electrons contained in their atoms). The table shows the distribution of the orbital electrons in the various shells and subshells.

Among the heavier elements, the lowest available energy state does not always occur when the electron spins are paired ($\uparrow\downarrow$). Consequently, the ground-state electronic configuration of the more massive atoms is somewhat more complicated than we have indicated. Nevertheless, we may now see how the quantum theory, along with the Pauli exclusion principle, provides a physical basis for understanding the systematic properties of groups of elements that were

TABLE 24.1   **Available Electronic Quantum States in Atoms**

Shell n	Subshell l	m_l	m_s	Spectroscopic Designation	Total Number of Available States	
					Subshell	Shell
1	0	0	+1/2			
1	0	0	−1/2	1s	2	2
2	0	0	+1/2			
2	0	0	−1/2	2s	2	
2	1	−1	+1/2			
2	1	−1	−1/2			
2	1	0	+1/2			
2	1	0	−1/2			
2	1	1	+1/2			
2	1	1	−1/2	2p	6	8
3	0	0	+1/2			
3	0	0	−1/2	3s	2	
3	1	−1	+1/2			
3	1	−1	−1/2			
3	1	0	+1/2			
3	1	0	−1/2			
3	1	1	+1/2			
3	1	1	−1/2	3p	6	
3	2	−2	+1/2			
3	2	−2	−1/2			
3	2	−1	+1/2			
3	2	−1	−1/2			
3	2	0	+1/2			
3	2	0	−1/2			
3	2	1	+1/2			
3	2	1	−1/2			
3	2	2	+1/2			
3	2	2	−1/2	3d	10	18

TABLE 24.2   **Ground State Configurations of Electrons in the First 13 Elements**

Element	Number of Electrons (Atomic Number)	Electronic Configuration
Hydrogen	1	$1s^1$
Helium	2	$1s^2$
Lithium	3	$1s^2 2s^1$
Beryllium	4	$1s^2 2s^2$
Boron	5	$1s^2 2s^2 2p^1$
Carbon	6	$1s^2 2s^2 2p^2$
Nitrogen	7	$1s^2 2s^2 2p^3$
Oxygen	8	$1s^2 2s^2 2p^4$
Fluorine	9	$1s^2 2s^2 2p^5$
Neon	10	$1s^2 2s^2 2p^6$
Sodium	11	$1s^2 2s^2 2p^6 3s^1$
Magnesium	12	$1s^2 2s^2 2p^6 3s^2$
Aluminum	13	$1s^2 2s^2 2p^6 3s^2 3p^1$

recognized and incorporated by Mendeleev into the *periodic table of elements* more than a half century before the advent of quantum physics.

The configuration of electrons in atoms explains a great deal of the chemical properties of the elements. For example, those electrons that are in lower energy states are more tightly bound to the atomic nucleus, and they are, therefore, less likely to take part in chemical reactions between atoms. Electrons in the outermost shells of atoms are less tightly bound than those that have lower values of n. Therefore, the most loosely bound electrons of an atom are those lying outside the outermost filled shell. These electrons are called *valence electrons*. Atoms with equal numbers of valence electrons have similar chemical properties. The alkali metals (Li, Na, K, Rb, and Cs), for example, all have a single valence electron and all have similar chemical properties. (See, for example, the electronic configurations of lithium and sodium in Table 24.2.)

Example 24.1

PROBLEM   The inert elements, also called the noble gases, helium (He), neon (Ne), argon (Ar), krypton (Kr), xenon (Xe), and radon (Rn), have the atomic numbers 2, 10, 18, 36, 54, and 86, respectively. Using these facts, propose a sequence of filled subshells for the periodic table of the elements.

SOLUTION   The general rule that we utilize here is that an electronic energy level is usually increased when either the principal quantum number n or the orbital quantum number l is increased. The 2 orbital electrons of helium ($Z = 2$), then, fill the subshell with $n = 1$ and $l = 0$. The additional 8 electrons of neon ($Z = 10$) reside in the $n = 2$ shell filling the $l = 0$ and $l = 1$ subshells. Surprisingly, the next inert element, argon, does not occur with the filling of the $n = 3$ shell, which has 18 electronic states (2 in the $l = 0$ subshell, 6 in the $l = 1$ subshell, and 10 in the $l = 2$ subshell). Because argon ($Z = 18$) has only 8 more orbital electrons than neon, it appears that this inert element occurs not with the filling of the $n = 3$ shell but rather with the full occupancy of the ($n = 3$) $l = 0$ and $l = 1$ subshells. The 18 additional electrons of krypton ($Z = 36$) occupy the 10 states of the $l = 2$ subshell, which corresponds to $n = 3$, and the 8 states of the $l = 0$ and $l = 1$ subshells having $n = 4$. Similarly, the inert element xenon ($Z = 54$) has its additional 18 electrons in the $l = 2$ subshell for $n = 4$ and the $l = 0$ and $l = 1$ subshells with $n = 5$. The next 32 electrons required to constitute radon ($Z = 84$) fill the 14 states of the $n = 4$ and $l = 3$ subshell, the 10 states of the $n = 5$ and $l = 2$ subshell, and the 8 states of the ($n = 6$) $l = 0$ and $l = 1$ subshells.

Generally, then, the filling of an $l = 1$ subshell marks that an element demonstrates extreme chemical inactivity. This result disagrees with the Bohr theory (electronic energies depend upon the principal quantum number only), which suggests that an inert element will result each time a principal shell is fully occupied. This apparent conflict is resolved and may be understood when one realizes that the Bohr theory includes only the attractive electron-nucleus interactions but ignores the repulsive electron-electron interactions. When this electron-electron repulsion is correctly included, the energy levels show a strong (non-Bohr) dependence on the orbital quantum number l. For relatively small values of l, say $l = 0$ or $l = 1$, the corresponding electronic orbitals are highly elliptical. As a result, an outershell electron at times penetrates the repulsive electron "cloud" shielding it from the strongly attractive positive nucleus, and this causes a significant lowering of the potential energy for this low-angular-momentum orbit. As l increases toward its maximum value of $n - 1$, the orbital becomes more nearly spherical and a significantly higher energy results. ■

Lasers

We have seen in Chapter 23 how an electron in the orbit of an atom may absorb a photon of electromagnetic radiation and jump to a higher energy orbit of that atom. An atom with such energized electrons is said to be in an *excited state*. We have also seen how an atom in an excited state may emit a photon when the electron of the excited atom falls to a lower energy level of that atom.

Suppose we could devise an atomic system in which we "pump" electromagnetic energy into one species of atom and thereby continually raise orbital electrons of those atoms to an energy level where they remain for a relatively long time—until we supplied them with a nudge, a signal that causes them to jump to a lower energy level. The result would be an intense source of radiation at a very precise wavelength: an extremely bright, monochromatic source of electromagnetic radiation. What if we could arrange it so that the radiation from all of those atomic transitions were in a definite phase relationship with one another? Then the radiation would be coherent (see Chapter 21, "Physical Optics"). We would then have a very special device: a source of intense, monochromatic, coherent radiation.

Such a device would be a technological wonder. With such characteristics, the radiation could be collimated (formed into parallel rays) so that we could bounce its signals from small reflectors placed on the moon by astronauts. The considerable energy in the beam of such intense radiation could be focused with a simple lens into a minute volume, thereby enabling the beam to slice through steel. These physical characteristics could permit ophthalmologists to "weld" a detached retina into place in a human eyeball by using the lens of the affected eye to focus the beam onto the retinal surface. If the single wavelength of the radiation from such a device happened to lie within the visible spectrum, we could produce dazzling displays of light that would find a wide range of uses from experimental optics laboratories to the enhancement of visual effects at rock concerts.

The device we have been describing, of course, is the *laser*. Its name is an acronym formed from the initial letters of light amplification by stimulated emission of radiation. Today laser beams are used as frictionless styluses to read the information from video discs; they are used to read the labels of merchandise in supermarkets; and they are used as the single carrier beam for thousands of simultaneous telephone conversations. Our technology continues to find new ways to exploit the characteristics of lasers.

Lasers may be constructed of atomic systems, which are solids, liquids, or gases. They can be made to provide radiation in the visible, ultraviolet, infrared, or microwave region of the electromagnetic spectrum (in the microwave region, however, they are called *masers*).

What, then, are the requirements of a system of atoms in order that it may function as a laser? First, a suitable system must contain atoms that have *metastable* energy levels into which orbital electrons may be excited. In metastable levels, the electrons take a relatively long time before they spontaneously jump to lower energy levels. Generally, pumping electromagnetic energy into a laser system causes electrons in the ground states of atoms to be raised to a high-energy state from which they fall rapidly to a metastable level. Those electrons remain in the metastable energy level until more electrons are in the higher (metastable) state than are in the lower (and normally more populous) ground state. Such a configuration is called a *population inversion*. Figure 24.1(a) illustrates the population inversion of two energy levels, E_1 and E_2, in an atom, where E_2 is a metastable state.

Next, it is necessary that the excited electrons with energy E_2 be stimulated to jump back to a lower energy level, say E_1. In doing so they emit photons with frequency ν and energy $h\nu = E_2 - E_1$, as shown in Figure 24.1 (b). Fortunately, it is a fact that when atoms containing electrons in the state with energy E_3 are bathed in radiation of the frequency $\nu = (E_2 - E_1)/h$, the probability that the electron will jump from E_2 to E_1 is significantly increased. Even more fortunately, the photons that are emitted because of the stimulation are in phase with the stimulating radiation. These propitious circumstances were treated quantitatively in 1916 by Einstein.

E 24.7   If a laser is to emit radiation with a wavelength of 550 nm, what is the energy separation, $E_2 - E_1$, of the initial (metastable) and the final states of the laser?   Answer: 2.3 eV

E 24.8   Repeat Exercise 24.7 for microwave radiation with a 2.0-cm wavelength.   Answer: 62 μeV

The remaining requirement for producing laser action in an atomic system is that the emitted radiation must be confined for a sufficiently long time so that it can stimulate further transitions. In most laser systems this is accomplished by

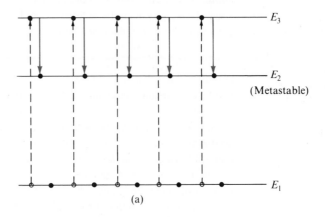

(a)

Figure 24.1
(a) The population inversion of atomic energy levels. External excitations raise electrons from the ground state E_1 to E_3, from which they rapidly fall to the metastable state E_2. When population inversion takes place, there are a greater number of electrons in E_2 than in E_1. (b) When the electrons in E_2 are stimulated by appropriate radiation, they jump to a lower energy level E_1, thereby emitting photons of frequency $\nu = (E_2 - E_1)/h$.

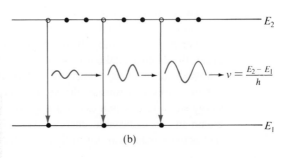

(b)

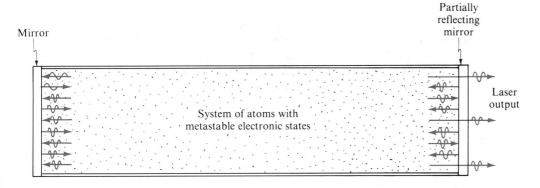

Figure 24.2
Mirror arrangement in a typical gaseous laser. The mirror on the left completely reflects the light, while the mirror on the right is partially reflecting. This arrangement maintains stimulating radiation within the system, yet permits an output laser beam.

placing the system of atoms between two mirrors. Figure 24.2 illustrates how the mirrors of a laser may be arranged. The mirror on the left reflects essentially all photons reaching it back through the laser system. The partially reflecting mirror on the right reflects enough of the photons back into the system to maintain stimulation of emission yet permits part of the photons to escape the system, thereby providing an output beam of laser radiation.

24.2   Molecular Structure and Solids

Chemical compounds are aggregates of individual atoms joined by chemical bonds. The distribution of the orbital electrons of the elements, which we considered in the previous section, suggests how the elements may be categorized according to their abilities to bond chemically. For example, the noble gases like He, Ne, A, Kr, and Xe with completely filled subshells, are inert: They form no compounds. An element like the alkali metals, Li, Na, K, Rb, and Cs, that has a single electron outside of completely filled subshells, combines most easily with an element like Cl that has one electron missing from its outermost, otherwise filled subshell. Similar systematic combinations of elements throughout the periodic table are consistent with the distribution of orbital electrons as specified by the Pauli exclusion principle.

Molecular Bonds

Individual atoms form compounds because a net attractive force forms a molecular bond between those atoms. The potential energy of a system of bound atoms is less than the potential energy of the individual unbound atoms. The physical mechanism of molecular bonding may be classified in descending order of bond strength as: (a) the ionic bond, (b) the covalent bond, and (c) the hydrogen bond. A fourth type of bond, the metallic bond, will be considered later.

The *ionic bond* is typified by the bond between the Na^+ ion and the Cl^- ion in

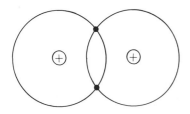

Figure 24.3
Schematic representation of the configuration of orbital electrons in the H_2 molecule.

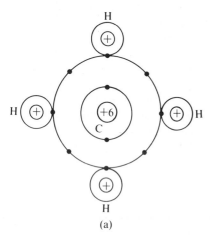

(a)

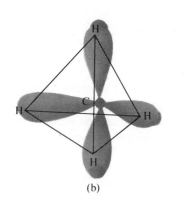

(b)

Figure 24.4
(a) Schematic representation of the covalent bonding of hydrogen to carbon in CH_4. (b) The electron cloud configuration in CH_4. The hydrogen atoms are at the four corners of a regular tetrahedron, and the carbon atom is at the center of that figure.

NaCl, common table salt. A sodium atom becomes more stable when it gives up its 3s valence electron and becomes an Na^+ ion. Similarly, a chlorine atom becomes more stable by filling its outermost shell with an electron, thereby becoming a Cl^- ion. The coulombic attraction between these two ions, each of which has a net charge of magnitude $|e|$, provides a net negative potential energy that binds the two-atom system together. This negative potential energy is dominated by a positive repulsive core potential when the atoms become too closely spaced. The core potential provides a short-range repulsion. Thus there results a spacing, about 2.4 Å, between the Na and Cl nuclei at which the potential energy of the molecular system is minimal. At that position there is a net attractive potential energy of about -5 eV that binds the Na^+ and Cl^- ions into a stable NaCl molecule.

Some ionic bonds involve the exchange of two electrons between atoms. In MgO, for example, a magnesium Mg^{++} ion is formed when two valence electrons are removed from the neutral atom and two electrons are accepted into the outermost shell of oxygen, forming an O^{--} ion. Again the dominant force that binds the Mg^{++} and O^{--} ions is the electrostatic coulombic force. Atoms with more than two valence electrons rarely combine through the mechanism of the ionic bond. In those cases, other types of bonds become dominant.

The *covalent bond* involves the sharing of electrons between atoms that form a molecule. When atoms of the same element combine to form a diatomic molecule, a molecule composed of two atoms, like H_2, F_2, or N_2, the valence electrons are shared by both atoms of a molecule to effect a closed shell configuration.

The electronic configuration of the H_2 molecule is suggested by Figure 24.3, where two atoms of hydrogen share their single 1s electrons to form a "closed-shell" configuration. Covalent bonds are probably the most prevalent in all natural compounds. They are the principal form of bonding in organic compounds. Figure 24.4(a) shows schematic diagrams of a typical organic molecule, CH_4, or methane, in which a covalent bond is formed between the carbon atom and each of four hydrogen atoms. The spatial distributions, sometimes called electron clouds, of the shared electrons associated with each covalent bond are indicated in Figure 24.4(b), which shows the four hydrogen nuclei at the four corners of a regular tetrahedron and the carbon nucleus at the center.

A hydrogen atom has only one orbital electron, and, therefore, can participate in only one covalent bond. But hydrogen atoms are sometimes involved in bonds between two atoms or ions. Hydrogen forms a distinctively polar covalent bond; the bound orbital electron is drawn toward such a bond; and away from that bond, the positive proton is exposed. The proton end of such a polar bond attracts the negative regions of other nearby molecules with sufficient force to serve as a weak chemical bond, called a *hydrogen bond*. The hydrogen bond is especially important in biological molecules like DNA, in which hydrogen bonds between O and N atoms bind the two strands of the double helix of DNA. Because the hydrogen bonds are weak, only a small quantity of energy is required to unzip the DNA double helix when DNA is in a process of either replicating itself or coding RNA for the production of proteins. Hydrogen bonds are also important binding agents in liquids. For example, water and alcohol mix so completely because of the polar character of a water molecule, which permits the positive hydrogen sites to form hydrogen bonds with the relatively negative regions of the alcohol molecules.

Crystalline Solids

A *crystalline solid* is a regular array of large numbers of atoms bound into a relatively rigid structure. The atoms of some solids, like NaCl, are bound by ionic bonds. Other solids, like silicon, germanium, and the diamond form of C, are bound by covalent bonds. Yet another form of chemical bond, which has no counterpart among individual molecules, occurs in solids. This type of bond, called a *metallic bond*, is not as strong as the ionic or covalent bonds but is considerably stronger than the hydrogen bond. The metallic bond plays an important role in the electrical conductivity of solids, so let us consider it further.

As the name suggests, the metallic bond is responsible for the binding together of the atoms of metals. As a metallic solid crystallizes into a regular three-dimensional lattice, the relatively fixed atoms release one or more valence electrons that become part of a sea of highly mobile electrons, which move throughout the solid. This fluid of delocalized electrons serves as the glue that binds the atomic sites together. The mobility of these electrons is responsible for the large electrical conductivity associated with metals. And it follows that crystalline solids with ionic or covalent bonds, which strongly bind electrons to their atomic sites, are not good electrical conductors in general but are insulators.

An important and useful concept related to the electrical conductivity of solids is a result of quantum theory. According to the wave-mechanical treatment of regular arrays of large numbers of atoms, the mobile electrons of a metallic solid belong not to an individual atom but to a large number of atoms collectively. As a result, instead of the electrons having sharply defined quantized energy levels usually associated with individual atoms, the electrons of a solid may occupy any of a continuous range of energies that lie within one of a number of permitted *energy bands*. Figure 24.5(a) shows the sharply defined energy levels of an individual atom, and Figure 24.5(b) illustrates how those quantum states are broadened into energy bands when the atoms are in a crystalline solid. The energy gaps that lie between the bands of permitted energies are said to be forbidden regions of energy. Electrons of a solid may not have energies that lie within the forbidden gaps. (An elementary treatment of the quantum basis of band theory is developed in Chapter 25.)

According to the *band theory of solids*, the electrical conductivity of a solid is

Figure 24.5
(a) The sharply defined energy levels of unbound atoms. **(b)** The energy bands that develop for such atoms when they are incorporated into a solid.

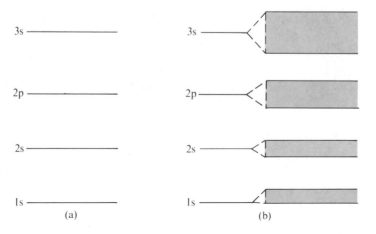

3s ————————   3s ———<

2p ————————   2p ———<

2s ————————   2s ———<

1s ————————   1s ———<

        (a)   (b)

determined by the relative occupancy of the adjacent energy bands of that solid and by the spacing between energy bands. Figure 24.6(a) shows an energy level diagram for a solid conductor. The uppermost occupied energy band, called the *conduction band*, is partially filled, and all the energy bands of lesser energy are completely filled. Thus when an electric field is applied to a conductor, the electrons in that conduction band have nearby unoccupied energy states within the conduction band into which these electrons can move and thereby acquire additional kinetic energy and, therefore, velocity. Figure 24.6(b) depicts the energy distribution of electrons in an insulator, a material with a small electrical conductivity. Here the uppermost occupied band, called the *valence band*, is completely filled and is separated from the next higher band by a considerable energy gap, typically about 5 eV wide for good electrical insulators. Figure 24.6(c) represents the energy level configuration for electrons in a *semiconductor* at low temperature. No electrons occupy the conduction band, and the valence band is completely filled, just as it is for an insulator. But in a semiconductor, the energy gap between the conduction and valence bands is small, perhaps a few tenths of an eV. As the temperature of a semiconductor increases, electrons near the top of the valence band may acquire sufficient thermal energy to make transitions into the conduction band. The conductivity of the semiconductor therefore increases with increasing temperature as more and more electrons are raised into the conduction band.

Conduction in a semiconductor takes place by two processes. First, the electrons elevated into the conduction band move oppositely to the direction of an electric field applied to the semiconductor. Because electrons have vacated some of the energy states of the valence band, the empty states of the valence band, called *holes*, respond to an applied electric field as if they were positive charges. The motion of a hole may be likened to a vacant space in an otherwise filled parking lot: As the cars (electrons) move in one direction to fill the vacancy repeatedly, the empty space (a hole) moves in the opposite direction. Thus conduction in a semiconductor is composed of both the motion of electrons in the conduction band and the motion of holes in the valence band.

Semiconductor Devices

The semiconductors we have been describing are typified by pure samples of crystalline silicon (Si) or germanium (Ge), which are called *intrinsic semiconductors*. The technological revolution in electronics over the past few dec-

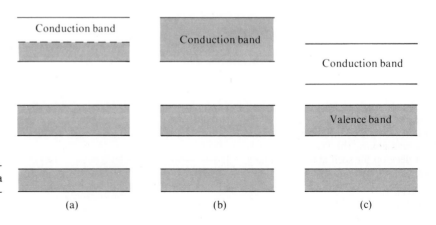

Figure 24.6
Typical energy bands of **(a)** a solid conductor, **(b)** a solid insulator, and **(c)** a solid semiconductor at low temperature.

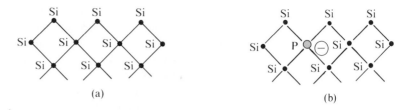

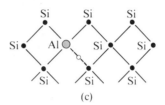

Figure 24.7
A representation of the lattice of solid silicon **(a)** in its pure state, **(b)** when it is doped with a donor atom (phosphorus, in this case), and **(c)** when it is doped with an acceptor atom (aluminum, in this case).

ades has been made possible by our capability to introduce controlled quantities of impurities into the lattices of intrinsic semiconductors in order to obtain specific conduction characteristics. For example, an intrinsic semiconductor like Si has a crystalline lattice, depicted schematically in Figure 24.7(a), in which each atom of Si is bound by four bonds to other Si atoms. If an impurity atom, like phosphorus (P) or arsenic (As) with five valence electrons, is introduced into the lattice, as shown in Figure 24.7(b), the fifth electron of P is loosely bound (about 0.01 eV) to the impurity atom and is easily elevated into the conduction band. The silicon sample is then said to be *doped* with *donors*, impurity atoms that donate conduction electrons. Semiconductors that contain donor impurities conduct by the flow of negative charges and are, therefore, called *n-type semiconductors*. If an impurity atom with three valence electrons, like aluminum (Al) or gallium (Ga), is introduced into the silicon lattice, as shown in Figure 24.7(c), the impurity atom can then accept an electron from the nearby (about 0.01 eV) valence band. In this case, the silicon is said to be doped with *acceptors;* and because the current in these materials consists of the flow of holes (the equivalent of positive charges) in the valence band, a semiconductor doped with acceptors is called a *p-type semiconductor.*

A *p-n junction diode*, illustrated in Figure 24.8, is a device composed of a semiconductor, half of which is *p*-type while the other half is *n*-type. In the region where the *n*- and *p*-types meet, the electrons and holes of each region diffuse throughout a narrow region called a *p-n junction* until the electric field

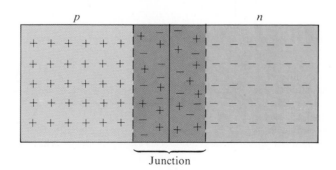

Figure 24.8
A *p-n* junction.

caused by these layers of opposite charges prevents further diffusion. If now an emf E and a load resistor R are connected to the device as shown in Figure 24.9(a), the electrons flow across the junction in the sense indicated. The electrons that move away from the junction are replaced by the vast number of electrons available in the n-type portion of the device. At the same time, holes flow across the junction in the opposite sense. The result of the electron flow and the hole flow is a large net current in the circuit as indicated by the ammeter A in the circuit. With the polarity of the emf in the sense indicated, the diode is said to be *forward-biased*. Under these conditions the diode permits a large current in the circuit; the current is usually limited by the value of the load resistance R.

In Figure 24.9(b) the polarity of the emf has been reversed. Both the electrons and the holes are forced away from the junction, which now contains virtually no mobile electrons or holes. The junction region is, therefore, essentially nonconducting; and practically no current is in the circuit for this *reverse-bias*

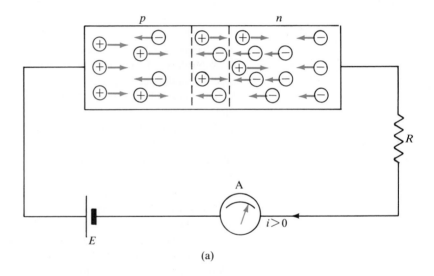

(a)

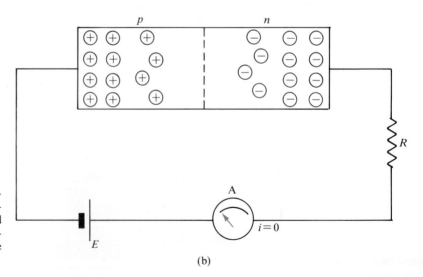

Figure 24.9
A circuit containing a *p-n* junction diode. **(a)** The forward-biased diode permits current in the circuit, as indicated by the ammeter A. **(b)** The reverse-biased diode permits no current in the circuit.

(b)

arrangement. If an alternating emf is imposed on a *p-n* junction diode, current is permitted through the diode only during the forward-bias portion of the cycle. In such an arrangement, the passive diode functions as a *rectifier*.

A *transistor* is a device constructed with a sandwichlike arrangement of *p-n-p* or *n-p-n* type semiconductors. A circuit using a *p-n-p* transistor is shown in Figure 24.10(a), and Figure 24.10(b) shows this circuit using the standard symbol for a transistor. In this arrangement, the thin *n*-type region is called the *base*, the *p*-type region on the left is called the *emitter*, and the *p*-type region on the right is called the *collector*. When ΔV_{in} is such that there is no current through the emf E_e in the emitter branch of the circuit, there is almost no current in the collector branch, which includes an emf E_c and a load resistor R, because the base-collector junction of the transistor is reverse biased. But when the potential difference ΔV_{in} is changed, holes (positively charged) pass into and through the thin base to the base-collector junction. At that junction the holes are forced through the collector branch of the circuit by E_c, which is usually much greater than E_e. And this initially small current in the collector branch causes a potential drop across R, which in turn decreases the reverse bias across the base-collector junction, thereby increasing further the collector current. In this way, small changes in ΔV_{in} produce large changes in the collector current i_c and, therefore, large changes in ΔV_{out} for appropriate values of R. Thus transistor circuits may be used to amplify potential difference changes. For judiciously chosen values of R, the amplification $\Delta V_{out}/\Delta V_{in}$ may be equal to several thousand.

Transistors are small, inexpensive, long lasting, and efficient compared to the vacuum tubes they have almost completely replaced. Microscopically small transistors are now incorporated into *integrated circuits* (IC's), which are formed by successive deposition and etching of carefully doped semiconductors and oxides. A single IC may have virtually an arbitrary number of transistors, diodes, capacitors, resistors, and conducting pathways ingeniously formed from the semiconductor and oxide materials. Thus a single *chip*, hardly larger than a pinhead, may combine a large number of electronic functions into one miniature device, like the one shown in Figure 24.11. Such IC chips have made possible the economical manufacture of compact calculators, microprocessors, and microcomputers.

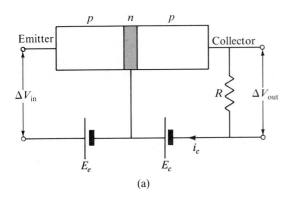

(a)

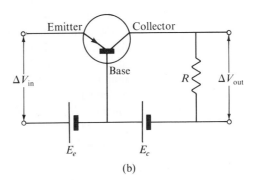

(b)

Figure 24.10
A *p-n-p* transistor circuit.

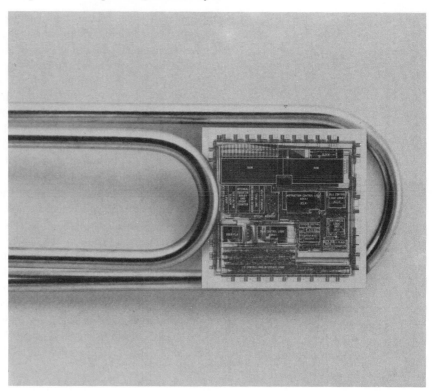

Figure 24.11
Photograph of an integrated circuit (IC) chip.

24.3 Nuclear and Particle Physics

Nuclear Structure and Stability

All atoms that are more massive than hydrogen have a central *nucleus* composed of *protons* and *neutrons*, which are particles of nearly equal mass. Protons (p) have a positive charge equal to the electronic charge $e = 1.6 \times 10^{-19}$ C. Neutrons (n) have no net charge. The *atomic number* Z of a given atom is equal to the number of protons in its nucleus. Atoms with different atomic numbers, that is, with different numbers of protons, are different chemical elements. The *atomic mass number* A of an atom is equal to the sum of the total number Z of protons and the total number N of neutrons in the nucleus. Each nuclear species, which has a specific value of Z and a specific value of $A = (Z + N)$, is called a *nuclide*. We symbolize a given nuclide by using the chemical symbol for the element (determined by Z) with Z specifically expressed in a left subscript and A expressed in a left superscript. For example, ${}^{18}_{8}O$ represents a particular nuclide of oxygen that has 8 protons and $18 - 8 = 10$ neutrons.

E 24.9   What are the values of A, N, and Z for the nuclide ${}^{17}_{8}O$?

Answer: 17, 9, 8

Nuclides that have the same value for Z but a different A are called *isotopes* of the element specified by Z. Then $^{16}_8O$, $^{17}_8O$, and $^{18}_8O$ are three isotopes of oxygen, each nucleus of which contains 8 protons and 8, 9, and 10 neutrons, respectively. Nuclides that have the same value for A but different values of Z are called *isobars*, like $^{40}_{18}A$ and $^{40}_{20}Ca$.

Certain small nuclei have special names. The proton (p) itself is the nucleus of the most common isotope of hydrogen, ^1_1H. The *deuteron* (d) is the nucleus of an isotope (^2_1H) of hydrogen that is called *deuterium*. A deuteron is composed of one proton and one neutron. The *triton* (t) is the nucleus of the isotope (^3_1H) called *tritium*. The *alpha particle* (α) is the nucleus of the helium isotope ^4_2He and is composed of two protons and two neutrons.

Nucleons (a generic name for a nuclear particle, either a proton or a neutron) are either positively charged (p) or neutral (n). Therefore, the coulombic repulsive force between protons tends to disrupt a given nucleus. The cohesive forces that act between all nucleons and bind them together despite the coulombic repulsion between protons are called *nuclear forces*. These strong forces, which are found only in interactions between nucleons, are short-range forces that are exerted only over very short distances (of the order of 10^{-15} m). The net negative potential energy associated with the attractive force that holds the nucleus together is called the *binding energy* E_B of that nucleus. The relativistic mass equivalent ($E_B = \Delta mc^2$) of this binding energy is called the *mass decrement* Δm of the nucleus. The mass decrement is so-called because the total mass of a nucleus is less than the sum of the masses of the particles that comprise that nucleus. We may express the binding energy E_B in terms of the mass m of an atomic nucleus and the masses m_p of a proton and m_n of a neutron:

$$E_B = (Zm_p + Nm_n - m)c^2 \qquad (24\text{-}6)$$

The mass defect Δm of a given nucleus is equal to the quantity in parentheses in Equation (24-6), the difference between the sum of the masses of its constituent nucleons and the total mass of the nucleus, or

$$\Delta m = Zm_p + Nm_n - m \qquad (24\text{-}7)$$

Because there are A nucleons in a given nucleus, the ratio E_B/A is the *average binding energy per nucleon* of a given nuclide. The average binding energy per nucleon is a measure of the stability of a nuclide: The greater the value of E_B/A for a given nuclide, the more stable is that nuclide.

Nuclear masses and their associated equivalent energies are conveniently expressed in terms of the *unified mass unit u*, defined to be one-twelfth of the mass of $^{12}_6C$, or

$$1 \text{ u} = 1.660559 \times 10^{-27} \text{ kg} \leftrightarrow 931.50 \text{ MeV}$$

Table 24.3 lists the masses and energy equivalents of some particles used in nuclear calculations. The following problem illustrates how some of the quantities affecting the stability of nuclides are calculated.

The atomic mass of $^{16}_8O$ is 15.99491 u. Calculate the average binding energy per nucleon in this nuclide.

This nuclide of oxygen contains eight protons ($Z = 8$) and eight neutrons

TABLE 24.3   Particle Masses and Energy Equivalents

Particle	Symbol	Mass (kg)	Mass (u)	Energy Equivalent (MeV)
Electron	e	9.110×10^{-31}	0.000549	0.511
Proton	p	1.673×10^{-27}	1.007276	938.28
Neutron	n	1.675×10^{-27}	1.008665	939.57
Deuteron	d	3.344×10^{-27}	2.013553	1875.62
Alpha particle	α	6.647×10^{-27}	4.002603	3728.42

($N = A - Z = 6 - 8 = 8$). Using Equation (24-7) and the data of Table 24.3, we determine the mass defect Δm to be

$$\Delta m = Zm_p + Nm_n - m$$

$$\Delta m = 8(1.007276 \text{ u}) + 8(1.008665 \text{ u}) - 15.99491 \text{ u}$$

$$\Delta m = 1.132618 \text{ u} \tag{24-8}$$

The total binding energy of $^{16}_8\text{O}$ is equal to the energy equivalent of its mass defect, or

$$E_B = \Delta mc^2 = (0.132618 \text{ u})(931.50 \text{ MeV/u}) = 123.53 \text{ MeV} \tag{24-9}$$

Finally, because this nuclide contains 16 nucleons, the average binding energy per nucleon is

$$\frac{E_B}{A} = \frac{123.53 \text{ MeV}}{16 \text{ nucleons}} = 7.7 \text{ MeV/nucleon} \tag{24-10}$$

E 24.10   If the binding energy of a certain nucleus with $A = 40$ is 320 MeV, determine (a) the average binding energy per nucleon, and (b) the mass decrement for this nucleus.

Answers: (a) 8.0 MeV/nucleon; (b) 0.34 u = 5.7×10^{-28} kg

Figure 24.12 shows how the average binding energy per nucleon, E_B/A, varies with mass number A. The quantity E_B/A increases irregularly and rapidly for low values of A, reaches a maximum value of about 8.7 MeV/nucleon at about $A = 60$, and gradually decreases to about 7.6 MeV/nucleon for uranium ($A = 235$). The nearly constant value of E_B/A over a wide range of A reflects the short-range nature of the nuclear force that acts between nucleons. A nucleon is bound by the nuclear force of only its nearest neighbors within a nucleus. Therefore, in a roughly spherical nucleus composed of more than about 25 nucleons, the short-range interactions between the nucleons in the interior of the nucleus does not depend on the total number of nucleons in that nucleus. This effect is called the saturation of the nuclear force.

As the mass number A increases, the effects of the saturation of the nuclear force between nucleons and the increasing coulombic force between protons favors the addition of neutrons instead of protons in order to maintain stability of the nuclides. As A becomes sufficiently large (about $A = 210$), however, the coulombic repulsion between protons becomes so strong that a nucleus is unstable. This instability may result in radioactive decay.

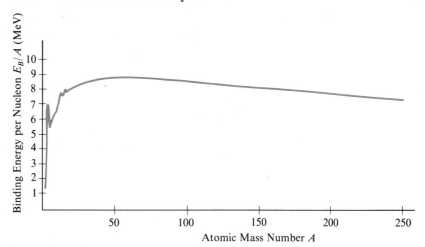

Figure 24.12
A graph of the average binding energy per nucleon, E_B/A, as a function of atomic mass number A.

Radioactive Decay

When an atomic nucleus reorganizes itself into a more favorable (lower) energy state by emitting particles or photons, that process is called *radioactive decay*. The radioactive nuclei may emit (a) an alpha particle (α), a stable helium nucleus composed of two protons and two neutrons; (b) a beta particle (β), an electron of nuclear origin; or (c) a gamma ray (γ), a high-energy photon of nuclear origin.

In each radioactive decay process, the following conservation principles apply:

1. Energy is conserved.

2. Linear momentum is conserved.

3. Angular momentum is conserved.

4. Electric charge is conserved.

5. The total number of nucleons, A, is conserved.

Radioactive decay is a statistical process. We cannot predict when a particular nucleus in a given sample of a radioactive nuclide will decay. But because even a small sample of a radioactive substance contains enormous numbers of individual atoms, we can accurately predict when a given fraction of the nuclei in that sample will have undergone radioactive decay. Suppose, for example, that atoms of a *parent nuclide*, like $^{235}_{92}U$, each decay by the emission of an alpha particle. This process reduces the number of protons in each uranium nucleus by two and the number of neutrons by two so that the remaining *daughter nuclide* becomes $^{231}_{90}Th$. Thus in this decay process, atoms of uranium are *transmuted*, changed into another element, to become atoms of thorium:

$$^{235}_{92}U \rightarrow {}^{231}_{90}Th + {}^{4}_{2}He$$

which may be written

$$^{235}_{92}U \rightarrow {}^{231}_{90}Th + \alpha$$

E 24.11   If a radioactive nucleus ($Z = 92$, $A = 238$) decays by emitting an alpha particle, what are the values of Z and A for the daughter nucleus?

Answer: $Z = 90$, $A = 234$

E 24.12   If a radioactive nucleus ($Z = 90$, $A = 233$) decays by emitting a beta particle, what are the values of Z and A for the daughter nucleus?

Answer: $Z = 91$, $A = 233$

The following example illustrates the applicability of the conservation laws to radioactive processes.

Example 24.2

PROBLEM   An isotope of plutonium decays by alpha-particle emission to uranium according to

$$^{239}_{94}\text{Pu} \rightarrow \,^{235}_{92}\text{U} + \,^{4}_{2}\text{He}$$

If the plutonium atom is assumed to be at rest just before the decay, determine the speed of the uranium atom just after the decay. Assume the isotopic masses of the plutonium, uranium, and helium atoms to be 239.05124 u, 235.04301 u, and 4.00261 u, respectively.

SOLUTION   Figure 24.13 depicts the decay. We will assume that all of the mass that is lost during the decay is converted to the kinetic energy of the two particles that remain after the decay. This energy E_0 is given by

$$E_0 = (M^* - M - m)\, c^2$$

where M^*, M, and m are the masses of the plutonium, uranium, and alpha particles, respectively. Substitution of the given mass values yields

$$E_0 = (0.00562 \text{ u}) \times (931.50 \text{ MeV/u}) = 5.24 \text{ MeV} = 8.38 \times 10^{-13} \text{ J}$$

Conservation of momentum requires that

$$MV - mv = 0$$

while conservation of energy gives

$$\frac{1}{2}MV^2 + \frac{1}{2}mv^2 = E_0$$

where V is the speed of the uranium atom and v is the speed of the alpha particle after the decay. Solving for these two speeds gives

$$V = 2.7 \times 10^5 \text{ m/s}$$

$$v = 1.6 \times 10^7 \text{ m/s}$$

During this calculation, we have assumed that nonrelativistic kinematics is appropriate. Is this assumption valid?   ∎

Although the number of parent nuclei in a given sample of radioactive material

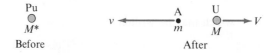

Figure 24.13   Example 24.2

decreases as some of the nuclei disintegrate, or decay, the rate at which the decrease occurs varies considerably for different parent nuclides. If N is the number of parent nuclei in a sample at time t, the rate at which N is changing is equal to dN/dt, a negative quantity because N is decreasing. The greater the number N of nuclei in the sample, the greater will be the number of disintegrations. Then dN/dt is proportional to N, or

$$\frac{dN}{dt} = -\lambda N \tag{24-11}$$

in which λ is a constant of proportionality called the *disintegration constant*. Equation (24-11) may be rearranged and integrated to give

$$\frac{dN}{N} = -\lambda \, dt$$

$$\ln N = -\lambda t + C \tag{24-12}$$

where C is a constant of integration. If we let N_o be the number of undecayed nuclei when $t = 0$ and evaluate Equation (24-12) at $t = 0$, we find that $\ln N_o = C$, or

$$\ln(N/N_o) = -\lambda t$$

$$N = N_o e^{-\lambda t} \tag{24-13}$$

which is the *decay law* for the transformation of a sample containing a single radioactive isotope.

A useful characterization of a radioactive isotope is its *half-life* $t_{1/2}$, the time interval required for half the nuclei in a sample to disintegrate. Thus when $t = t_{1/2}$, N/N_o is equal to 1/2, and substitution of these values into Equation (24-13) gives

$$\ln(1/2) = -\lambda t_{1/2}$$

$$\ln 2 = \lambda t_{1/2}$$

$$t_{1/2} = \frac{\ln 2}{\lambda} = \frac{0.693}{\lambda} \tag{24-14}$$

The half-life of a radioactive nuclide may be determined experimentally by measuring its *activity* R, the number of disintegrations per unit time that occurs in a sample of that nuclide. The SI unit of activity, named for the discoverer of radioactivity, is the *becquerel* (Bq), which is defined to be equal to 1 disintegration per second (dps). A more commonly used unit of activity is the *curie* (Ci), which is approximately the activity of 1 g of radium, or

$$1 \text{ Ci} = 3.70 \times 10^{10} \text{ Bq} = 3.70 \times 10^{10} \text{ dps} \tag{24-15}$$

The activity $R = -dN/dt$ at which a given radioactive sample is decaying at any time t after a chosen $t = 0$, may be determined by taking the time derivative of Equation (24-13) to give

$$\frac{dN}{dt} = -\lambda N_o \, e^{-\lambda t}$$

$$R = R_o e^{-\lambda t} \tag{24-16}$$

where $R_o = \lambda N_o$ is the rate of decay at $t = 0$.

The graph of Equation (24-16) in Figure 24.14 shows how the activity of a sample of radioactive material reduces exponentially as a function of time t. The activity is reduced by a factor of two in a time interval equal to one half-life, by a factor of four in two half-lives, by a factor of eight in three half-lives, and so on. Because activity is a quantity that is easily measured, let us illustrate how we may use the decay law and its associated relationships to predict activity.

What is the initial activity of a 10-μg sample of pure $^{222}_{86}Rn$, which has a half-life of 3.8 days? What is the decay rate of this sample after a time interval of 10 days? How much time (after the pure sample is prepared) must pass before the activity of this sample is reduced to 10% of its initial value?

The number N_o of radioactive atoms in the pure sample is found by recognizing that the atomic mass number ($A = 222$) is equal to the molecular weight (in g/mol) of radon. Then using the fact that there are Avogadro's number, 6.02×10^{23}, of atoms in a mole of a substance, we find

$$N_o = \frac{(10^{-5} \text{ g})(6.02 \times 10^{23} \text{ atoms/mol})}{222 \text{ g/mol}} = 2.71 \times 10^{16} \qquad (24\text{-}17)$$

The disintegration constant λ for radon is obtained from Equation (24-14):

$$\lambda = \frac{0.693}{t_{1/2}} = \frac{0.693}{(3.82 \text{ d})(8.64 \times 10^4 \text{ s/d})} = 2.10 \times 10^{-6} \text{s}^{-1}$$

Then the initial rate of decay is

$$R_o = \lambda N_o = (2.10 \times 10^{-6}\text{s}^{-1})(2.71 \times 10^{16})$$

$$R_o = (5.69 \times 10^{10} \text{ Bq})\left(\frac{1 \text{ Ci}}{3.70 \times 10^{10} \text{ Bq}}\right) = 1.5 \text{ Ci} \qquad (24\text{-}18)$$

After 10 days ($= 8.64 \times 10^5$ s), the activity R is obtained using Equation (24-16):

$$R = R_o e^{-\lambda t} = (1.5 \text{ Ci}) \, e^{-(2.1 \times 10^{-6}\text{s}^{-1})(8.64 \times 10^5\text{s})}$$

$$R = (1.5 \text{ Ci})(0.164) = 0.25 \text{ Ci} \qquad (24\text{-}19)$$

Finally, when the decay rate has become 10% of its initial value, Equation (24-16) gives

$$\frac{R}{R_o} = 0.1 = e^{-\lambda t}$$

$$10 = e^{\lambda t}$$

$$\ln 10 = \lambda t$$

$$t = \frac{\ln 10}{\lambda} = \frac{2.30}{2.1 \times 10^{-6} \text{ s}^{-1}} = 1.1 \times 10^6 \text{ s}$$

or about 13 days.

E 24.13   If the activity of a sample containing 4.0×10^{20} radioactive nuclei is equal to 6.0×10^{11} disintegrations per second, determine (a) the disintegration constant for this sample, and (b) the half-life for this nucleus.

Answers: (a) 1.5×10^{-9} s^{-1}; (b) 15 years

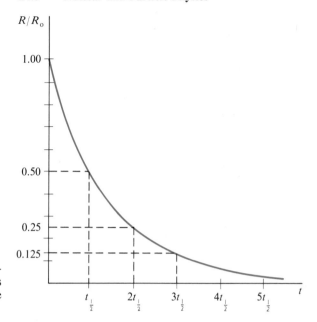

Figure 24.14
Illustrating how the activity R of a sample of radioactive material is reduced as a function of time t from its initial value R_o.

E 24.14   If the activity of a certain radioactive sample decreases by 15% in 2.0 days, determine (a) the disintegration constant for this sample, and (b) the half-life for this sample.   Answers: (a) 9.4×10^{-7} s^{-1}; (b) 8.5 days

Nuclear radiations may be detected using a variety of devices. One of the oldest and simplest radiation detectors is a gas-filled container, called a *Geiger-Mueller tube*, like the one illustrated in Figure 24.15. The central electrode, usually a thin wire, is maintained at a positive potential relative to the conducting inner wall of the container. When a nuclear radiation passes through the gas in the container, it ionizes the gas, producing electrons and positive ions of the gas. The electrons formed in this way are attracted to the positively charged central wire, and the positive ions are attracted to the negative wall of the container. Thus a pulse of current passes through the resistor R, and the amplifier-counter amplifies and records the event. If the potential difference across the container is large, the electrons produced by the ionizing radiation are accelerated as they approach the central wire and produce more ion pairs as they collide with the atoms of the gas. The accelerating ions thus tend to cascade, producing a large number of ion pairs, which result in a large pulse to the amplifier-counter. When used in this way, the detection system is called a *Geiger counter*. And although more sophisticated detectors are available, the Geiger counter is useful because it is reliable and inexpensive.

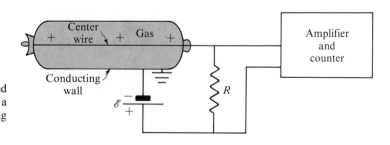

Figure 24.15
A Geiger-Mueller tube incorporated into a circuit that may be used as a Geiger counter, a detector of ionizing radiation.

Radiocarbon Dating

One of the more interesting applications of the radioactive decay law is *radiocarbon dating*. This process is a technique that depends on the careful assay, or the quantitative measurement of activity, in the radioactive isotope $^{14}_6C$. Radiocarbon dating is the determination of the time of death of any once-living artifact. Because it permits the dating of mummies, wood, linen, petrified or fossilized animals and plants, paper, and bones, this technique is a valuable tool of archaeologists, historians, and scholars in many disciplines.

Every living thing exchanges carbon with the atmosphere. Plants ingest carbon in the form of CO_2 during photosynthesis, the process by which glucose is manufactured from sunlight; and animals either ingest plants or other animals that have eaten plants. Then during the lifetime of an organism, the carbon in its body has the same relative abundance of the various isotopes of carbon as that of the atmosphere. The CO_2 in the atmosphere has a virtually constant proportion of stable $^{12}_6C$ and radioactive $^{14}_6C$, which is a β-emitter with a half-life of 5568 years. Although the radioactive isotope decays, it is replenished by nuclear reactions that take place when high-energy gamma rays, called cosmic rays, enter the upper reaches of the atmosphere. In life, then, an organism has within it the same proportion of $^{12}_6C$ and $^{14}_6C$ as the atmosphere, but upon the death of the organism, the exchange of carbon between the atmosphere and the organism ceases. From that moment, the $^{14}_6C$ concentration in the organism decays, becoming halved every 5568 years.

Samples of once-living materials can be dated by comparing their $^{14}_6C$ activity to that of the atmosphere. Specimens that died more than 25,000 years ago have been dated using this technique—but only by exceedingly careful measurements. One gram of pure $^{14}_6C$ has an activity of only about 15 disintegrations per minute, so counting rates of less than 1 per minute must be accurately measured to date samples (which originally contained 1 g of pure $^{14}_6C$) that are 25,000 years old.

Techniques for counting the extremely low activities required in radiocarbon dating have been highly developed. Specimens are often prepared by burning, after which the carbon residue in the form of soot is used to line the inside wall of a detector tube (similar to the Geiger-Mueller tube described earlier). Because of the low counting rates, it is sometimes necessary to count for months to accumulate a significant number of counts compared to the background radiation count, which is caused by cosmic rays of incidental radiation in the vicinity of the detector. Of course, the detectors are carefully shielded, but even that does not screen out the very penetrating cosmic radiation. Clever *anticoincidence* arrangements have been contrived to eliminate most of the cosmic ray background. Figure 24.16 illustrates one way in which pairs of detectors may be arranged around a counter so that a cosmic ray passing through the counter has a great probability of firing a pair of the detectors. The electronic circuitry associated with this system eliminates any counts resulting from radiations that trigger a pair of detectors "simultaneously," that is, within a predetermined very short span of time.

The effectiveness of radiocarbon dating has been confirmed repeatedly by comparing $^{14}_6C$ data with the known ages of specimens, like sequoia trees, whose rings may be counted, and like the mummies of particular pharaohs whose death dates are accurately known.

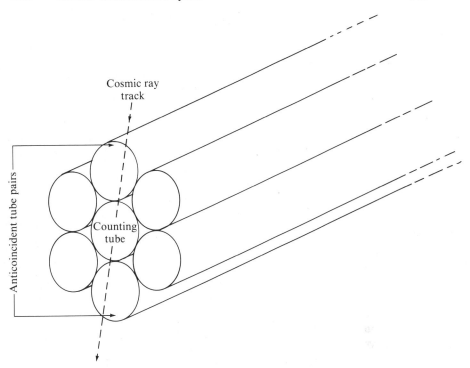

Cosmic ray track

Anticoincident tube pairs

Counting tube

Figure 24.16
An anticoincidence arrangement of detectors.

Elementary Particles

Early in the twentieth century electrons and protons were established as the building blocks of matter by the model of J. J. Thompson, the scattering experiments of Rutherford, and the success of the Bohr model of the hydrogen atom. In 1930 Chadwick demonstrated the existence of the neutron. With a positive, a negative, and a neutral particle to constitute matter, it seemed that the fundamental particles of nature had been established. In a few years, however, C. D. Anderson demonstrated the existence—a brief but definite existence—of the *positron* (e^+), a particle with the same mass as an electron but with a charge of $+e$. A positron is produced when a high-energy photon interacts with matter, disappears in a process called *pair production*, and converts into an electron-positron pair. The energy of the photon that is converted into the e^+-e^- pair must have more than enough energy (about 2×0.51 MeV) to account for the energy equivalent of the rest mass $(2mc^2)$ of the particle pair. Any additional energy that the photon possesses appears as kinetic energy of the particles produced. The positron is the *anti-particle* of the electron, a "mirror image" of the electron with the opposite charge. The discovery of the positron stirred conjecture about the possible existence of other anti-particles.

Beta-decay studies brought a further surprise. When an electron is emitted in a nuclear process like the decay of $^{14}_6\text{C}$ that we encountered in radiocarbon dating,

$$^{14}_6\text{C} \rightarrow {}^{14}_7\text{N} + e^-$$

the emitted electron is expected to have a fixed energy, namely 0.156 MeV. The measured values of the β-particle energies in such decay processes are, in fact, found to span a range of energies from very nearly zero to a maximum value of 0.156 MeV. Is energy not conserved in the β-decay process? Rather than accept that the conservation of energy is violated, Wolfgang Pauli suggested that another particle is created in the β-decay process. Such a particle would have to have no charge and no appreciable mass (because some of the emitted electrons have the expected kinetic energy of 0.156 MeV). This "particle" was named the *neutrino* (little neutral one). Its existence was more or less taken for granted, although several decades passed before neutrinos were actually detected.

Are these, then, the elementary particles of nature: electrons, protons, neutrons, neutrinos, and perhaps their anti-particles? Theorists were concerned about this issue. Further, they found themselves hard pressed to explain and understand the strong nuclear force between nucleons. In 1935 Hideki Yukawa proposed that the strong nuclear force is mediated by a particle with a rest mass about 300 times as massive as the electron. The particle, referred to as a *meson*, is assumed to be associated with the force between nucleons somewhat as two people tossing a medicine ball back and forth, or pulling it away from each other, are interacting by means of the ball. Two years later a particle with mass intermediate betweeen that of an electron and a proton was discovered. The mass that was discovered was not the Yukawa meson; that was not observed until 1946. The first of these particles that was discovered is called a *muon;* the Yukawa meson is called a π *meson* or *pion*.

Since the discovery of mesons, particle accelerators have become highly developed and are capable of producing very high-energy protons, deuterons, electrons, and other particles that can be directed to collide with other matter. Such devices have been the tools of the discovery of a large number (actually hundreds!) of "elementary" particles. Many of the particles that have been observed are far more massive than the proton, and others have masses intermediate between those of the electron and the proton. But are these particles really fundamental? Are they the ultimate building blocks of matter, or is there further substructure? And how may we understand the forces through which this vast variety of particles interacts? These were (and are) the questions that confronted the physicists in that branch of modern physics called high-energy physics or particle physics.

The myriad particles that had been detected by the decade of the 1960s were first classified according to mass. Electrons, muons, and neutrinos were called *leptons* (from leptos meaning small); particles with mass intermediate between those of muons and the proton were labeled mesons (mesos means mid); and protons and particles more massive than the proton were named *baryons* (barytes means weight). Particles were sometimes classified by various quantum numbers that came to be associated with interactions between particles. These quantum quantities were called *baryon number*, *isospin*, and *strangeness*.

The classification of elementary particles shifted to reflect an emphasis on the interactions between them, on the kinds of forces that act between particles. It became more or less accepted that four basic interactions may occur between particles. These interactions, in descending order of their strengths are:

1. Strong

2. Electromagnetic

3. Weak

4. Gravitational

Those particles that interact through the short-range *strong nuclear force*, like the nucleons, for example, and in general the more massive particles, are called *hadrons* (from hadros meaning bulky). The electromagnetic force acts between electric charges; this force is weaker than the strong nuclear force and acts over a much greater distance. The electromagnetic force is a long-range force. The *weak interactions* function in β-decay processes and in the decay of many of the unstable particles, like the decay of pions into muons. The gravitational interaction is many orders of magnitude weaker than the other three forces. Although responsible for important macroscopic forces in nature, gravitational forces apparently do not play a significant role in the interactions of elementary particles.

In the 1960s and 1970s theorists speculated that the vast array of "elementary" particles is composed of a substructure of particles that is consequently more fundamental. In the most recent formulations of this approach, leptons are considered fundamental: Leptons are not made up of more elementary components. On the other hand, hadrons are considered to be composed of *quarks*, particles with fractional electronic charge ($\pm e/3$; $\pm 2e/3$) and of undetermined mass. Originally, combinations of three quarks and three *anti-quarks* were assumed to be sufficient to accommodate descriptions of known hadrons. They were called **u** (*up*), **d** (*down*), and **s** (*strange*); and their associated anti-quarks were identified as $\bar{\mathbf{u}}$, $\bar{\mathbf{d}}$, and $\bar{\mathbf{s}}$. High-energy experiments in the 1970s suggested three additional quarks, which were given the fanciful names *charm* (**c**), *truth* (**t**), and *beauty* (**b**). These newer quarks, along with their anti-quarks, $\bar{\mathbf{c}}$, $\bar{\mathbf{t}}$, and $\bar{\mathbf{b}}$, are said to exist in three forms, called, with equal whimsy, *colors*.

Photons are the "particles" that mediate the electromagnetic interaction. The particles that are presumed to mediate the strong nuclear interaction are called *gluons* because they serve to "glue" together the quarks that constitute hadrons. Similarly, *gravitons* are the particles assumed to mediate the gravitational interaction. A particle called a *weak boson*, or simply a *W-particle*, was predicted to mediate the weak interaction. Although gluons and gravitons have yet to be detected, the W-particle was observed in 1983.

As is always the case in research, new discoveries raise further questions. Is there a single, unified theory in the offing that will incorporate the four interactions as special manifestations of a single universal interaction? Are the fundamental particles of nature limited in number, and, if so, what are they? Current research in particle physics continues the search for new particles. Investigations, both theoretical and experimental, are simultaneously pointed toward the understanding of the interactions between particles and toward a definitive conception of the fundamental particles in nature.

24.4 Problem-Solving Summary

The rules of quantization from the wave-mechanical treatment of atomic electrons provide that three quantum numbers are associated with each electron in an atom. Another quantum number associated with the intrinsic spin of an electron provides a fourth quantum number, and the quantum state of an electron is specified by a set of those four quantum numbers. Quantum numbers of atomic electrons are subject to the following constraints: The principle quantum number n may have any positive integral value, or $n = 1, 2, 3, \ldots$; the orbital angular momentum quantum number l may have any nonnegative integral value up to $(n - 1)$; the magnetic quantum number m_l may have any positive or negative integral value, including zero, from $-l$ to $+l$; and the spin magnetic quantum number m_s may have one of the values $\pm 1/2$.

The Pauli exclusion principle states that no two electrons in a given atom may be in the same quantum state. Therefore, using the permitted quantum numbers for an atomic electron along with the Pauli principle allows us to understand the foundations of the periodic table of elements. Groups of electrons in the same atom that have the same value of n are in the same atomic shell; those with the same values of both n and l are in the same subshell. The occupancy of atomic subshells are expressed using a symbol like $2p^3$, in which the leftmost number is the value of n, the letter specifies the value of l (s : $l = 0$, p : $l = 1$, d : $l = 2$, f : $l = 3$, and so on), and the superscript specifies the number of electrons in the subshell.

Lasers are intense sources of coherent light from the stimulated emission of radiation from atoms that have metastable states and in which the populations of atomic energy levels are inverted.

Molecular bonds between the atoms of solids may be categorized in order of decreasing strengths as ionic, covalent, and hydrogen bonds. Further, a metallic bond is formed between the atoms of metallic conductors. The sharply defined atomic energy levels are broadened in solids to form energy bands. The uppermost occupied energy band of a solid is called the conduction band. When the conduction band is partially filled, the electrons in the conduction band are free to move in response to an electric field, and that solid is a conductor. If the conduction band is empty and the gap between the conduction band and the next lower band (the valence band) is relatively large, the solid is an electrical insulator. On the other hand, if the conduction band is empty and the gap between the conduction and valence bands is relatively small, the solid is an intrinsic semiconductor.

When relatively pure intrinsic semiconductors are doped with donor or acceptor atoms, they become n-type or p-type semiconductors, respectively. Junction diodes are formed from the fusion of n- and p-type semiconductors, which readily conduct current in one direction but not in the other. Transistors are n-p-n or p-n-p sandwiches of semiconductors that can be used in circuits that function as amplifiers of current or potential difference.

The sum of the number Z of protons and the number N of neutrons in the nucleus of an atomic species determines the atomic number A of that nuclide. Isotopes of a nucleus have the same value of Z but different values of N (and, therefore, different values of A). Nuclear isobars have the same value of A but different values of Z.

Nucleons, either protons or neutrons, are bound together in a nucleus by strong nuclear forces that compete with the coulombic forces that tend to repel the protons of that nucleus. The binding energy E_B of a nucleus is the net attractive potential energy that holds that nucleus together. If m is the mass of a nucleus, then the relativistic mass equivalent, $E_B = \Delta mc^2$, is the mass decrement Δm of that nucleus. If m_p and m_n are the masses of the proton and neutron, the mass decrement Δm of a nucleus with a given Z and N is $(Zm_p + Nm_n - m)$ and the binding energy E_B of that nucleus is equal to Δmc^2. The average binding energy per nucleon is expressed as E_B/A.

In problems concerning nuclear energetics it is often useful to relate the masses of the particles involved using the energy equivalent of the unified mass unit u $= 1.066 \times 10^{-27}$ kg, which has an energy equivalent of 931.5 MeV.

Radioactive decay occurs when a parent nucleus emits a particle or a photon and becomes a daughter nucleus. In a decay process, energy, linear momentum, angular momentum, electric charge, and the total number of nucleons are conserved. The radioactive decay law, $N = N_o e^{-\lambda t}$, relates the number N of undecayed parent nuclei present at a time t to the number N_o of parent nuclei present at an arbitrarily chosen initial time, $t = 0$. The constant λ is the disintegration constant of the particular nuclide that is decaying. The half-life $t_{1/2}$ of a radioactive nuclide is the time required for half of the nuclei present in a sample to disintegrate. The half-life $t_{1/2}$ of a nuclide is related to the disintegration constant λ of that nuclide by $t_{1/2} = 0.693/\lambda$. The activity R, the number of disintegrations that occur per unit time, of a radioactive sample at any time is related to the initial activity R_o of that sample at $t = 0$ by $R = R_o e^{-\lambda t}$. The unit of activity is the becquerel, Bq, which is equal to 1 disintegration per second, or the curie, Ci, which is equal to 3.70×10^{10} Bq.

Problems

GROUP A

24.1 What are the values for A, Z, and N for the nuclide $^{239}_{94}$Pu?

24.2 If the half-life of the $^{131}_{53}$I nuclide is 8.0 days, determine the disintegration constant.

24.3 What fraction of a collection of radioactive nuclei will remain after a 30-minute time interval if the half-life of the material is 5.0 minutes?

24.4 If the half-life of an isotope of radon is 3.8 days, what fraction of a collection of these nuclei will decay during a 6-day time interval?

24.5 How many electronic states are there in an $l = 3$ subshell?

24.6 Determine the energy, the magnitude of the angular momentum, and the z component of the angular momentum for a hydrogen atom in the quantum state specified by $n = 2$, $l = 1$, and $m_l = -1$.

24.7 How many electronic states correspond to $n = 4$?

24.8 What percentage of a radioactive sample decays during an elapsed time of 2.5 half-lives?

24.9 If $E_1 = 0$, $E_2 = 2.0$ eV, and $E_3 = 3.0$ eV in Figure 24.1 (*see* text), determine

(a) the wavelength of the radiation from the laser.

(b) the minimum input power to the laser if the laser is to provide 1.5 mW radiation.

24.10 What percentage of a radioactive sample remains after an elapsed time of 4.0 half-lives?

24.11 Calculate the binding energy of the $^{235}_{92}$U nucleus if its atomic mass is equal to 235.043925 u. Recall that the atomic mass of a hydrogen atom and a neutron are 1.007825 u and 1.008665 u, respectively.

24.12 Determine the binding energy per nucleon of the $^{56}_{26}$Fe nucleus if the atomic mass of this nuclide is 55.934939 u.

24.13 The nucleus $^{235}_{92}$U decays (half-life $= 7.0 \times 10^8$ yr) by emitting an α particle. Assume that the α particle is emitted from the surface of the nucleus with a relatively small kinetic energy and obtain an estimate for the kinetic energy of this particle when it is far from the nucleus. The radius R of any nucleus may be estimated using

$$A = (R/a)^3$$

where A is the atomic mass number for the nucleus and the constant a is given by

$$a = 1.2 \times 10^{-15} \text{ m}$$

How does your result compare to the measured value of 4.2 MeV for the α particle emitted from this nuclide? Can you offer an explanation for this discrepancy?

24.14 The fusion of hydrogen to make helium is proposed as a source of energy. For simplicity suppose that the process of interest is

$$2\,^1_1\text{H} + 2\,^1_0\text{n} \rightarrow \,^4_2\text{He}$$

(a) How much energy is released each time this process occurs? (The atomic mass of the helium isotope is 4.003860 u.)

(b) Estimate the "energy available" per kilogram of water if this fusion process could be controlled. Make a similar estimate for fissioning a kilogram of uranium and for burning a kilogram of gasoline.

GROUP **B**

24.15 A radioactive nuclide with disintegration constant λ_1 decays into a stable nuclide. At $t = 0$, the number of radioactive atoms is N and the number of stable atoms is zero.

(a) If $x_1(t)$ is the fraction of radioactive atoms remaining at $t > 0$, and $x_2(t)$ is the number of stable atoms (at $t > 0$) divided by N, show that

$$\frac{dx_2}{dt} = \lambda_1 x_1$$

(b) Show that

$$x_2(t) = 1 - e^{-\lambda_1 t}$$

(c) Sketch graphs of x_1 and x_2 as functions of time.

(d) Show that $x_1 + x_2 = 1$. Why?

24.16 Consider the conditions of the previous problem, but let the daughter nuclide be radioactive with decay constant λ_2.

(a) Show that

$$\frac{dx_2}{dt} = \lambda_1 x_1 - \lambda_2 x_2$$

(b) Show that

$$x_2 = \frac{\lambda}{\lambda_1 - \lambda_2}(e^{-\lambda_2 t} - e^{-\lambda_1 t})$$

(c) Show that the result of part (b) reduces to the expression for x_2 obtained in the previous problem where $\lambda_2 = 0$.

(d) Define the function $f(t)$ by $f = x_1 + x_2$ for $t > 0$. Show that

$$f(0) = 1;$$
$$f'(t) < 0 \qquad \text{for } t > 0$$
$$f(t) \to 0 \qquad \text{for } t \gg 0$$
$$0 < f(t) < 1 \qquad \text{for } t > 0$$

(e) What is the physical significance of each of the results of part (d)?

24.17 Consider the conditions of the previous problem.

(a) Show that the daughter nuclide achieves a maximum population at the time

$$t_2 = \frac{\ln(\lambda_2/\lambda_1)}{\lambda_2 - \lambda_1}$$

(b) Show that this maximum value for x_2 is given by

$$x_{2,m} = \left(\frac{\lambda_2}{\lambda_1}\right)^{\lambda_2/(\lambda_1 - \lambda_2)}$$

(c) If $\lambda_1 = 1.00$ and $\lambda_2 = 0.50$, sketch graphs of $x_1(t)$ and $x_2(t)$ for $0 < t < 8$.

24.18 Consider a radioactive chain that starts with a nuclide (decay constant $= \lambda_1$) that disintegrates into a second nuclide (decay constant $= \lambda_2$). This second nuclide then decays into a third nuclide (decay constant $= \lambda_3$), and so on. Suppose that the nth member of this chain is stable ($\lambda_n = 0$).

(a) If at $t = 0$ the number of radioactive atoms of type 1 is N and if $x_k(t)$ represents the number of atoms (of type k) divided by N, show that

$$\frac{dx_1}{dt} = -\lambda_1 x_1$$

$$\frac{dx_k}{dt} = \lambda_{k-1} n_{k-1} - \lambda_k x_k \qquad 1 < k < n$$

$$\frac{dx_n}{dt} = \lambda_{n-1} x_{n-1}$$

(b) Suppose that $n = 3$. If $x_2(0) = x_3(0) = 0$, show that

$$x_3(t) = 1 - \frac{\lambda_2 e^{-\lambda_1 t} - \lambda_1 e^{-\lambda_2 t}}{\lambda_2 - \lambda_1}$$

(c) Show that $x_1 + x_2 + x_3 = 1$. Why?

(d) Show that $0 < x_3 < 1$ for $0 < t$. Why?

(e) If $\lambda_1 = 1.00$ and $\lambda_2 = 0.50$, sketch $x_3(t)$ for $0 < t < 8$.

24.19 Suppose $\lambda_3 > 0$ in the previous problem.

(a) Show that

$$x_3(t) = \frac{\lambda_1 \lambda_2}{(\lambda_2 - \lambda_1)(\lambda_3 - \lambda_1)(\lambda_3 - \lambda_2)}[(\lambda_3 - \lambda_2)e^{-\lambda_1 t}$$
$$+ (\lambda_1 - \lambda_3)e^{-\lambda_2 t} + (\lambda_2 - \lambda_1)e^{-\lambda_3 t}]$$

(b) Show that x_3 achieves a maximum at the time t_3 which satisfies

$$\lambda_1(\lambda_3 - \lambda_2)e^{-\lambda_1 t_3} + \lambda_2(\lambda_1 - \lambda_3)e^{-\lambda_2 t_3} + \lambda_3(\lambda_2 - \lambda_1)e^{-\lambda_3 t_3} = 0$$

(c) If $x_1(0) = 1$, $x_2(0) = x_3(0) = 0$, $\lambda_1 = 1.00$, $\lambda_2 = 0.50$, and $\lambda_3 = 0.25$, sketch x_1, x_2, and x_3 for $0 < t < 8$. Show that $t_3 \cong 4.020$ and $x_3 \cong 0.464$.

25 Introduction to Wave Mechanics

The new physics that evolved after 1900 imposes a new set of ground rules for the description of physical particles. The new description differs from that of classical physics in three essential ways. First, the de Broglie hypothesis, confirmed by the experiments of Davisson and Germer, requires that we recognize the wavelike nature of particles and suggests that we may describe particles using wavelike functions. Next, the uncertainty principle imposes limitations on simultaneous measurements of certain physical quantities. Finally, because particle-interference experiments demonstrate that particle waves extend throughout appreciable space and yet, when those particles are detected, they are quite localized, the description of particles must yield to a probabilistic interpretation.

Conforming to these requirements, Erwin Schrödinger devised a formalism—a set of rules—that replaces the classical newtonian dynamics. This formalism, called *wave mechanics*, may be applied to a physical particle to achieve the following:

1. Wave mechanics specifies a wave function Ψ which permits our predicting from Ψ the results of any physical measurements on a particle described by Ψ.

2. Wave mechanics predicts how the wave function Ψ, and, therefore, how a particle it describes, is affected by interactions with the remainder of the universe.

In this chapter we will assert some of the more basic rules of wave mechanics and stress the interpretation of the results obtained when those rules are applied to physical situations. Finally, we will apply those rules to an especially simplified set of problem situations. Although these problem situations are purposely designed to be simplistic, the analyses of these problems suggest many of the essential wave-mechanical characteristics of important actual physical systems, like atoms, molecules, and solids.

25.1 Wave Functions and the Schrödinger Equation

In wave mechanics all measurable information about a particle moving along the x axis may be predicted from a wave function $\Psi(x, t)$ associated with this particle. This wave function, which is a solution of a wave equation developed by Schrödinger, is in fact a probability amplitude defined at each space-time point (x, t). The interpretation of Ψ as a probability-amplitude function is made because $|\Psi|^2 dx$ is interpreted to be proportional to the probability that a measurement at time t will locate the particle within the space interval between x and $x + dx$.

During the 1910–1930 era, when wave mechanics was being formulated, Schrödinger and others found it necessary to allow the wave function Ψ to be a complex function of the real quantities x and t. In other words, $\Psi(x, t)$ may, in general, have both a real part $R(x, t)$ and an imaginary part $I(x, t)$ such that $\Psi(x, t) = R(x, t) + iI(x, t)$, where i is the imaginary number defined by $i = \sqrt{-1}$. But how can a function like Ψ, which is to describe a physical particle, have an imaginary component? The answer lies in recognizing that $|\Psi|^2$ has physical significance while Ψ itself does not. The quantity $|\Psi|^2$, the square of the absolute value of Ψ, is obtained from Ψ according to $|\Psi|^2 = \Psi^* \Psi$, where Ψ^* (called the complex conjugate of Ψ) is determined from Ψ by changing the algebraic sign of the imaginary component of Ψ. Thus, if $\Psi = R + iI$, we get $\Psi^* = R - iI$; and it follows that $|\Psi|^2 = R^2 + I^2$.

The wave function $\Psi(x, t)$ associated with a particle of mass m may be obtained from the solutions of the one-dimensional *Schrödinger equation*,

$$-\frac{\hbar^2}{2m}\frac{\partial^2 \Psi}{\partial x^2} + U\Psi = i\hbar\frac{\partial \Psi}{\partial t} \tag{25-1}$$

where U is the potential energy of the particle of mass m; h is Planck's constant divided by 2π, or $\hbar = h/2\pi$; and i is equal to $\sqrt{-1}$. Fortunately, we will be considering a class of problems in wave mechanics that will permit our using a simplified version of Schrödinger's equation, which has relatively simple solutions. In particular, we will consider only those physical situations in which the energy of a particle has a precise value, say E. In such cases the Schrödinger equation has solutions of the form

$$\Psi(x, t) = \psi(x)\, e^{-iEt/\hbar} \tag{25-2}$$

where $\psi(x)$ is a function of position only. The function $\psi(x)$ is independent of time and is a solution of the *time-independent Schrödinger equation*,

$$-\frac{\hbar^2}{2m}\frac{d^2\psi}{dx^2} + U\psi = E\psi \tag{25-3}$$

where E is the precise energy of the particle described by ψ. Then because $\psi(x)$, the solution of the time-independent Schrödinger equation, may be inserted into Equation (25-2) to give the time-dependent wave function $\Psi(x, t)$, solving Equation (25-3) is sufficient to determine $\Psi(x, t)$.

E 25.1   Show that the time-independent wave equation, Equation (25-3), results if Equation (25-2) is substituted into the time-dependent wave equation, Equation (25-1). Recall that the partial derivative $\partial \Psi/\partial x$ is the derivative of Ψ with respect to x treating t as though it is a constant. Similarly, $\partial \Psi/\partial t$ is the derivative of Ψ with respect to t treating x as a constant.

Let us now illustrate how $\psi(x)$ may be found for what is perhaps the simplest physical situation of interest: a free particle that has mass m and a given specific kinetic energy K. Classically, a free particle is a particle on which no forces act.

Equivalently, a free particle is one that has a constant potential energy U, and no generality is lost if we let the constant value of U be equal to zero. Then the (positive) kinetic energy K of the particle is equal to the total energy E of the particle, and Equation (25-3) becomes

$$\frac{\hbar^2}{2m}\frac{d^2\psi}{dx^2} + E\psi = 0 \qquad (25\text{-}4)$$

One form of the general solution of Equation (25-4) is

$$\psi = C_1 \sin\!\left(\frac{\sqrt{2mE}\,x}{\hbar}\right) + C_2 \cos\!\left(\frac{\sqrt{2mE}\,x}{\hbar}\right) \qquad (25\text{-}5)$$

where C_1 and C_2 are arbitrary constants. That this equation is a solution of Equation (25-4) may be verified by taking the second derivative of the solution, Equation (25-5), and substituting $d^2\psi/dx^2$ and ψ directly into Equation (25-4).

E 25.2 Show, using the procedure suggested, that Equation (25-5) does satisfy Equation (25-4).

Example 25.1 **PROBLEM** Show that the wave function ψ of Equation (25-5) has an associated wavelength λ that is the de Broglie wavelength for the free particle described by ψ.

SOLUTION The wavelength λ associated with the sinusoidal functions of Equation (25-5) corresponds to the change in x that results in a change of 2π in the argument of the sine and cosine functions of Equation (25-5), i.e.,

$$\frac{\sqrt{2mE}\,(x+\lambda)}{\hbar} = \frac{\sqrt{2mE}\,x}{\hbar} + 2\pi$$

$$\frac{\sqrt{2mE}\,\lambda}{\hbar} = 2\pi$$

$$\lambda = \frac{2\pi\hbar}{\sqrt{2mE}}$$

Because the energy E of the free particle is equal to the kinetic energy of the particle (recall that $U = 0$), the energy E and the linear momentum of magnitude p of the particle are related by

$$\frac{p^2}{2m} = E$$

$$p = \sqrt{2mE}$$

Using this result and the identity, $h = 2\pi\hbar$, in the equation for λ gives

$$\lambda = \frac{h}{p}$$

which is the de Broglie relationship between the wavelength and momentum of a particle. ■

E 25.3 Rewrite Equation (25-5) in terms of the linear momentum $p = \sqrt{2mE}$ of the free particle.   Answer: $\psi = A\,\sin(px/\hbar) + B\,\cos(px/\hbar)$

Using the relationships, $e^{i\theta} = \cos\theta + i\sin\theta$ and $e^{-i\theta} = \cos\theta - i\sin\theta$, we may write Equation (25-5) in the convenient form,

$$\psi = A\,e^{ipx/\hbar} + B\,e^{-ipx/\hbar} \tag{25-6}$$

where A and B are constants and p, the magnitude of the linear momentum of the particle, is equal to $\sqrt{2mE}$. Simplifying further, we may use $k = 2\pi/\lambda = 2\pi p/h = p/\hbar$ and write Equation (25-6) as

$$\psi = A\,e^{ikx} + B\,e^{-ikx} \tag{25-7}$$

E 25.4   Show that

$$\sin\theta = \frac{e^{i\theta} - e^{-i\theta}}{2i} \quad ; \quad \cos\theta = \frac{e^{i\theta} + e^{-i\theta}}{2}$$

E 25.5   Show that the constants, A and B, of Equations (25-6) and (25-7) are related to the constants, C_1 and C_2, of Equation (25-5) by

$$A = \frac{C_1 - iC_2}{2} \quad ; \quad B = \frac{C_1 + iC_2}{2}$$

Thus the solution of the wave equation is composed of one term, $A\,e^{ikx}$, which corresponds (see Example 25.2) to a wave having an amplitude A and traveling in the positive x direction, and another term, $B\,e^{-ikx}$, which corresponds to a wave (amplitude $= B$) traveling in the negative x direction. Each term represents one of two possible momentum states of a free particle with total energy E: That particle may be moving in the $+x$ direction ($B = 0$) with momentum p or in the $-x$ direction ($A = 0$) with momentum $-p$, and in either case, the particle has energy $E = p^2/2m$. Indeed, the particle may be in a mixed momentum state, in which the wave function is composed of two parts: One part is a wave moving in the $+x$ direction ($A \neq 0$), and the other part is a wave moving in the $-x$ direction ($B \neq 0$). Thus if a particle is in a state defined by Equation (25-6), the probability P_+ that a measurement will determine the linear momentum to be $+p$ is given by

$$P_+ = \frac{|A|^2}{|A|^2 + |B|^2} \tag{25-8}$$

And the probability P_- that a momentum measurement will yield $-p$ is given by

$$P_- = \frac{|B|^2}{|A|^2 + |B|^2} \tag{25-9}$$

E 25.6   A particle of mass m is in a state described by

$$\psi(x) = e^{ix/a}$$

where a is a positive constant. Determine
(a) the de Broglie wavelength for this particle,
(b) the magnitude of the linear momentum for this particle,
(c) the kinetic energy of the particle,
(d) the probability that a measurement of the linear momentum will give a positive value.

   Answers: (a) $\lambda = 2\pi a$; (b) $p = \hbar/a$; (c) $K = \hbar^2/(2ma^2)$; (d) $P_+ = 1$

E 25.7   An electron (mass = 9.1×10^{-31} kg) is in a state described by

$$\psi(x) = 3\, e^{ix/L} - 4\, e^{-ix/L}$$

where $L = 0.10$ nm. Calculate
(a) the de Broglie wavelength for this electron,
(b) the kinetic energy of this electron,
(c) the probability that a measurement will determine the electron to be moving in the negative x direction.

Answers: (a) 0.63 nm; (b) 6.1×10^{-19} J = 3.8 eV; (c) 0.64

Example 25.2   **PROBLEM**   Show that the wave function $\Psi(x, t)$ for a free particle of energy E corresponds to two sinusoidal waves propagating in opposite directions.

SOLUTION   Substituting Equation (25-6) into Equation (25-2) gives for $\Psi(x, t)$

$$\Psi(x, t) = A\, e^{i(px/\hbar - Et/\hbar)} + B\, e^{-i(px/\hbar + Et/\hbar)}$$

$$\Psi(x, t) = A\, e^{2\pi i(px/h - Et/h)} + B\, e^{-2\pi i(px/h + Et/h)}$$

Using the de Broglie relation $\lambda = h/p$ and defining a frequency ν by $\nu = E/h$, the equation for Ψ may be written

$$\Psi(x, t) = A\, e^{2\pi i(x/\lambda - \nu t)} + B\, e^{-2\pi i(x/\lambda + \nu t)}$$

The first term (coefficient = A) is a (complex) sinusoidal wave traveling in the positive x direction, and the second is a sinusoidal wave traveling in the negative x direction. Each wave has a wavelength equal to the de Broglie wavelength associated with the particle, and each has a frequency ν that satisfies $E = h\nu$.

■

25.2 A Special Potential Function: Barrier Penetration

We have seen how a free particle, which is confined to motion along the x axis and has a constant potential energy U, may be represented by a wave function $\psi(x)$ that is a simple combination of two sinusoidal functions. If, however, a particle is not free but interacts with its surroundings, we represent that interaction by a nonconstant potential-energy function. Thus the wave functions of a particle in a physically interesting situation is the solution of Schrödinger's equation containing the potential-energy function U that is characteristic of that physical situation. Unfortunately, the mathematics involved in the solutions of Schrödinger's equation in which the potential-energy function characterizes some of the most significant problems of physics is beyond the scope of this book. We may, nevertheless, understand the basic wave-mechanical aspects of a wide variety of physical situations by considering combinations of one simple type of potential-energy function called a delta function (δ-function).

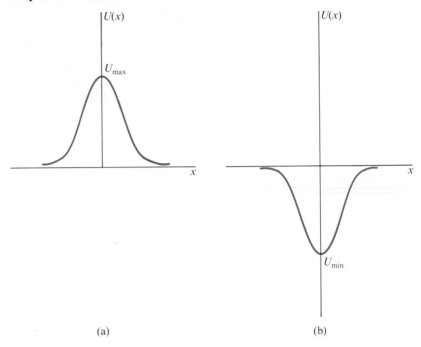

(a)   (b)

Figure 25.1
(a) A repulsive potential-energy bump.
(b) An attractive potential-energy bump.

The Delta-Function Potential Energy

A localized bump in a potential-energy function may be either repulsive or attractive. Figure 25.1(a) shows a positive bump that is repulsive. In a classical situation, where $U(x)$ represents a gravitational potential energy, for example, a ball having a total energy less that U_{max} cannot pass the potential hill; the ball is repelled by the potential bump. A ball with energy greater than U_{max} is slowed (repelled) as it approaches the hill; and once it passes the top of the hill, it is pushed away, or sped up (again repelled), by the hill. Thus a positive bump in the potential-energy function is always repulsive. On the other hand, the negative potential-energy bump of Figure 25.1(b) is attractive. If the bump also represents a gravitational potential energy, a ball approaching or receding from the bump is accelerated toward the position ($x = 0$) of the potential-energy minimum U_{min}. In other words, the ball is always attracted toward the center of the negative bump in the potential energy.

Let us now define a special kind of potential-energy bump, which, though it may be either positive (repulsive) or negative (attractive), we will assume to be positive. Suppose a potential-energy function $U(x)$ is equal to zero at all values of x except at one position, $x = x_1$, where the potential energy spikes to an arbitrarily large value. Figure 25.2 shows such a potential-energy function, which is called a *delta function*, or δ-function. We will be concerned throughout the remainder of this chapter with how the wave functions of particles are affected by such δ-functions. Therefore, we will now determine the rules, sometimes called *boundary conditions*, that tell us how wave functions behave at the position of and in the neighborhood of a δ-function.

Consider first the potential-energy function $U(x)$ in Figure 25.3. Here $U(x)$ has the value zero except in the very narrow range of positions between $x = x_1 - \varepsilon$

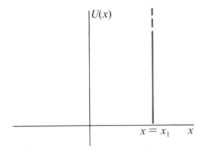

Figure 25.2
A potential-energy delta function.

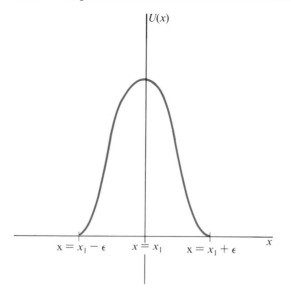

Figure 25.3
A potential-energy function that has a value equal to zero except in the narrow region between $x_1 - \epsilon$ and $x_1 + \epsilon$.

and $x = x_1 + \varepsilon$, where ε is a small distance. Let us define the area under the potential curve to be the strength S of the potential bump, so that

$$S = \int_{x_1 - \varepsilon}^{x_1 + \varepsilon} U(x)\, dx \qquad (25\text{-}10)$$

If now a particle of mass m and total energy E has the potential energy $U(x)$ shown in Figure 25.3, the Schrödinger equation for that particle is

$$-\frac{\hbar^2}{2m}\frac{d^2\psi(x)}{dx^2} + U(x)\psi(x) = E\psi(x)$$

or

$$\frac{d^2\psi(x)}{dx^2} = \frac{2m}{\hbar^2}U(x)\psi(x) - \frac{2mE}{\hbar^2}\psi(x) \qquad (25\text{-}11)$$

Integrating each term of Equation (25-11) over the region that includes the potential bump, we obtain

$$\int_{x_1 - \varepsilon}^{x_1 + \varepsilon}\frac{d^2\psi}{dx^2}\, dx = \frac{2m}{\hbar^2}\int_{x_1 - \varepsilon}^{x_1 + \varepsilon} U(x)\psi(x)dx - \frac{2mE}{\hbar^2}\int_{x_1 - \varepsilon}^{x_1 + \varepsilon}\psi(x)dx \qquad (25\text{-}12)$$

Suppose now that the potential bump is sufficiently narrow that the wave function does not change appreciably between $x_1 - \varepsilon$ and $x_1 + \varepsilon$. Then for sufficiently small ε, $\psi(x)$ is very nearly equal to $\psi(x_1)$ throughout that range of x, and Equation (25-12) becomes

$$\psi'(x_1 + \varepsilon) - \psi'(x_1 - \varepsilon) = \frac{2m}{\hbar^2}\,\psi(x_1)\int_{x_1 - \varepsilon}^{x_1 + \varepsilon} U(x)dx - \frac{2mE}{\hbar^2}\psi(x_1)\int_{x_1 - \varepsilon}^{x_1 + \varepsilon} dx$$

$$(25\text{-}13)$$

where $\psi' = d\psi/dx$ is the slope of the wave function and $\psi'(x_1 + \varepsilon) - \psi'(x_1 - \varepsilon)$ represents the change in the slope of the wave function across the potential bump.

We now let the potential bump become arbitrarily narrow ($\varepsilon \to 0$) while its height becomes greater in such a way that the strength S of the bump, the area under the bump, remains constant. In other words, the bump becomes a δ-function of strength S. Then the integral in the last term of Equation (25-13) becomes equal to zero. The other integral is, according to Equation (25-10), equal to the strength S of the bump. Then Equation (25-13) may be written

$$\psi'(x_1^+) - \psi'(x_1^-) = \frac{2mS}{\hbar^2}\,\psi(x_1) \qquad (25\text{-}14)$$

in which $\psi'(x_1^+) - \psi'(x_1^-)$ represents the change in the slope of the wave function $\psi(x)$ between points immediately to the right of x_1, (x_1^+), and immediately to the left of x_1, (x_1^-).

We now have the two rules that are the boundary conditions on a wave function $\psi(x)$ and on the slope $\psi'(x)$ of a wave function at the location x_1 of a δ-function of strength S:

1. A wave function $\psi(x)$ is continuous at a δ-function potential:

$$\psi(x_1^+) = \psi(x_1^-) \qquad (25\text{-}15)$$

2. The change in slope of a wave function $\psi(x)$ is proportional to the strength S of the δ-function potential and is given by

$$\psi'(x_1^+) - \psi'(x_1^-) = \frac{2mS}{\hbar^2}\,\psi(x_1) \qquad (25\text{-}16)$$

Figure 25.4(a) shows how a wave function $\psi(x)$ is affected by a repulsive (positive) δ-function. The wave function is continuous at $x = x_1$, the location of the repulsive δ-function, and the slope $\psi'(x)$ of the wave function abruptly *increases* as x increases through the value x_1, the position of the δ-function. In Figure 25.4(b), the wave function $\psi(x)$ is again continuous at $x = x_1$; but because the attractive δ-function is negative, and therefore the strength S of the δ-function is negative, the slope $\psi'(x)$ of the wave function abruptly *decreases* as x increases through x_1.

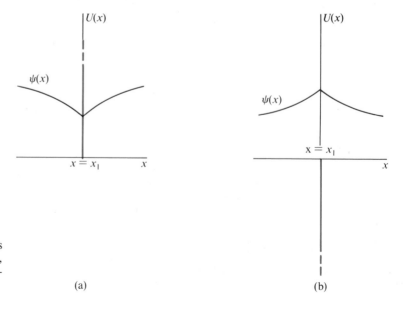

Figure 25.4
Demonstrating how a wave function is affected by (a) a positive, or repulsive, δ-function and (b) a negative, or attractive, δ-function.

(a)   (b)

The boundary conditions, Equations (25-15) and (25-16), permit us to analyze the wave mechanics of many representative physical situations. Let us begin our considerations of a few such situations with a particle that encounters a single repulsive potential barrier.

Transmission, Reflection, and Tunneling

Classically, when a particle having total energy E encounters a potential barrier with a maximum potential energy $U_{max} > E$, the particle can never surmount the barrier. Recall that a ball with total energy E at the bottom of a hill with a height that corresponds to a potential energy $U_{max} > E$ cannot reach the top of the hill: The ball is confined to one side of the hill. But now let us consider the predictions of wave mechanics when a particle with total energy E encounters a potential-energy barrier in the form of a δ-function. Classically, this particle could never penetrate the δ-function barrier, which is infinitely high.

The wave-mechanical analysis of a particle (mass $= m$, total energy $= E$) encountering a δ-function barrier is accomplished by solving the Schrödinger equation in which $U(x) = 0$ except at $x = 0$, where the potential energy is a δ-function characterized by a strength S. Figure 25.5(a) graphically depicts the situation. At all locations except at $x = 0$, the Schrödinger equation is

$$-\frac{\hbar^2}{2m}\frac{d^2\psi(x)}{dx^2} = E\psi(x)$$

$$\frac{d^2\psi(x)}{dx^2} - \frac{2mE}{\hbar^2}\psi(x) = 0$$

$$\frac{d^2\psi(x)}{dx^2} - k^2\psi(x) = 0 \qquad (25\text{-}17)$$

where $k = \sqrt{2mE}/\hbar = p/\hbar = 2\pi/\lambda$ is the wave number of the particle. The general solution of Equation (25-17) is

$$\psi(x) = A\,e^{ikx} + B\,e^{-ikx} \qquad \text{for } x < 0 \qquad (25\text{-}18)$$

$$\psi(x) = C\,e^{ikx} + D\,e^{-ikx} \qquad \text{for } x > 0 \qquad (25\text{-}19)$$

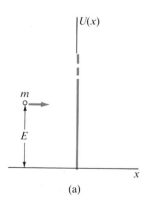

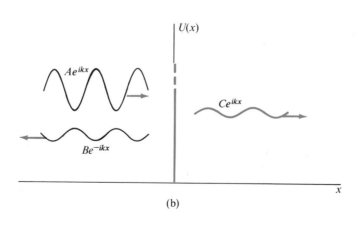

(a)   (b)

Figure 25.5
(a) A repulsive potential-energy δ-function for a particle of mass m and energy E. (b) The wave $A\,e^{ikx}$ associated with a particle approaching the potential-energy barrier from the left, the wave $B\,e^{-ikx}$ associated with the particle reflected by the barrier, and the wave $C\,e^{ikx}$ associated with the particle after it has been transmitted through the barrier.

which may be verified by direct substitution of the solution and its second derivative into Equation (25-17). As indicated in Figure 25.5(b), $A\ e^{ikx}$ represents a wave (associated with the particle approaching the barrier from the left) of amplitude A; $B\ e^{-ikx}$ represents a wave (associated with the particle after it has been reflected from the barrier and is moving toward the left) of amplitude B; and $C\ e^{ikx}$ represents a wave (associated with the particle after it has been transmitted through the barrier and is moving toward the right) of amplitude C. Because we are considering a particle approaching the barrier from the left, we assume that there is no wave traveling toward the left in the interval to the right of the barrier. Therefore, we set the coefficient D in Equation (25-19) equal to zero.

We now apply the boundary conditions, Equations (25-15) and (25-16), to ψ and ψ' at $x = 0$ and obtain the following relationships between the remaining amplitudes A, B, and C:

$$\psi(0^+) = \psi(0^-) \qquad \Rightarrow \qquad C = A + B \tag{25-20}$$

$$\psi'(0^+) = \psi'(0^-) + \frac{2mS}{\hbar^2}\psi(0) \Rightarrow ikC = ik(A - B) + \frac{2mS}{\hbar^2}C \tag{25-21}$$

E 25.8   Perform the operations that yield the results of Equations (25-20) and (25-21).

From Equations (25-20) and (25-21) we may deduce, after a bit of algebra, the ratios C/A and B/A of the wave amplitudes:

$$\frac{C}{A} = \frac{1}{1 + imS/\hbar^2 k} \quad ; \quad \frac{B}{A} = \frac{-imS/\hbar^2 k}{1 + imS/\hbar^2 k} \tag{25-22}$$

E 25.9   Solve Equations (25-20) and (25-21) for the results given in Equation (25-22).

Because A is the amplitude of the wave function associated with the particle incident on the barrier and C is the amplitude of the wave function associated with the particle transmitted through the barrier, we may express the probability that the particle is transmitted through the barrier by $|C/A|^2$. Then the *transmission coefficient*, $T(E) = |C/A|^2$, specifies the probability that a particle of mass m and energy E will be transmitted through a barrier of strength S according to

$$T(E) = \left|\frac{C}{A}\right|^2 = \left(\frac{C}{A}\right)^*\left(\frac{C}{A}\right) = \frac{E}{E + E_o} \tag{25-23}$$

where $E_o = mS^2/(2\hbar^2)$.

Similarly, the *reflection coefficient*, $R(E) = |B/A|^2$, specifies the probability that a particle of mass m and energy E will be reflected by a barrier of strength S, or

$$R(E) = \left|\frac{B}{A}\right|^2 = \left(\frac{B}{A}\right)^*\left(\frac{B}{A}\right) = \frac{E_o}{E + E_o} \tag{25-24}$$

E 25.10   Verify the results given in Equations (25-23) and (25-24).

Inspection of Equations (25-23) and (25-24) shows that the sum of $T(E)$ and $R(E)$ is equal to 1 for a given value of E, or

$$T(E) + R(E) = 1 \qquad (25\text{-}25)$$

The expressions for $T(E)$ and $R(E)$ in Equations (25-23) and (25-24) are shown plotted as functions of the energy E in Figure 25.6, where we may see from the graphs that Equation (25-25) is valid at each value of E.

The relationship of Equation (25-25) arises from the fact that a probability equal to 1 represents a certainty and a probability equal to zero represents an impossibility. Then Equation (25-25) is a statement that for any energy E the portion of the wave not transmitted through the barrier must be reflected from the barrier. Further, the probabilistic interpretation of wave mechanics means, in this case, that when a large number of experiments are performed to detect whether or not a particle of mass m and energy E has penetrated a barrier of strength S, the fraction of experiments that indicate transmission through the barrier is given by $T(E)$. Similarly, the fraction of experiments that indicates reflection from the barrier is given by $R(E)$.

The nonclassical result of our analysis, that a particle has a nonzero probability of penetrating a potential-energy barrier whose height exceeds the energy of the particle, is a wave-mechanical phenomenon called *tunneling*. This name is suggested by the fact that, because the particle cannot surmount such a potential-energy barrier in the classical sense, the resulting penetration is as though the particle had "tunneled" through the barrier.

In actual physical situations in which particles tunnel through potential-energy barriers, the barriers are not, of course, δ-functions. Figure 25.7 shows two kinds of potential-energy barriers that are sometimes used to approximate actual physical situations. The detailed wave-mechanical analysis of barriers like these are beyond the scope of this book, but we may examine some of the interesting results of such analyses.

The "square," or "top-hat," barrier of Figure 25.7(a) is a rough approximation of the potential-energy barrier encountered by electrons in certain solid-state circuit components, like a *tunnel diode*. As its name suggests, the tunnel diode is useful because electrons in this device penetrate, or tunnel through, a potential-

Figure 25.6
The transmission coefficient $T(E)$ and the reflection coefficient $R(E)$ plotted as a function of E for a particle encountering a repulsive potential-energy δ-function.

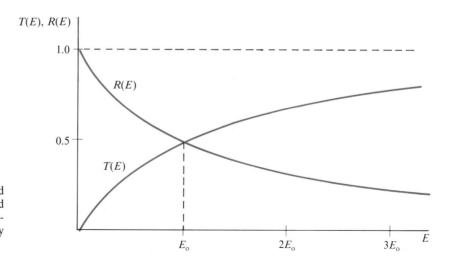

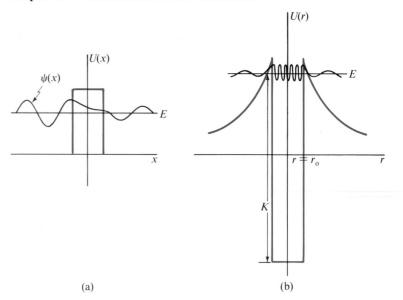

Figure 25.7
(a) The wave function $\psi(x)$ of a particle that encounters a "top-hat" potential-energy barrier. (b) A representation of the wave function of an α-particle in a deep nuclear potential-energy well.

(a)   (b)

energy barrier. These electrons pass through a barrier that, classically, would prevent their passage. That they actually pass through the barrier is demonstrated by the nonzero amplitude of the wave function on the right-hand side of the barrier, as seen in Figure 25.7(a). In a tunnel diode, a difference in electric potential imposed across the device has an effect similar to raising or lowering the height of the potential-energy barrier, thereby decreasing or increasing the probability that electrons penetrate the barrier. In Figure 25.7(a), the electronic wave functions, the solutions of the Schrödinger equation, are sinusoidal on either side of the barrier. Inside the barrier, however, the solution is an exponentially decaying function. The rules of wave mechanics require that both the wave function $\psi(x)$ and its slope $d\psi(x)/dx$ be continuous at a *finite* discontinuity in the potential energy. The wave function depicted in Figure 25.7(a) satisfies these requirements, and the probabilistic interpretation of the wave function explains the presence of electrons in the region past the barrier, electrons that have tunneled through the barrier.

The wave-mechanical tunneling of particles through a potential-energy barrier explains another physical phenomenon that has no classical basis: the spontaneous emission of alpha particles from radioactive (unstable) nuclei. An *alpha particle* (α-particle) is a helium nucleus composed of two protons and two neutrons bound together by nuclear forces. Near and inside the nucleus of a high-atomic-weight atom, the α-particle has a potential energy that may be approximated by a function like $U(r)$, shown in Figure 25.7(b), where r is the radial distance from the center of the nucleus. Because the nuclear potential-energy well is very deep, the α-particle, whose energy E is indicated in the figure, has considerable kinetic energy K, as suggested by the short de Broglie wavelength (large momentum) of the wave function indicated within the nuclear radius r_0. Because the wave function of the α-particle has a nonzero value outside the nuclear potential-energy barrier, there is a finite probability that the particle will escape the nucleus. Thus we may interpret α-particle emission in the following way: The α-particle, with a large kinetic energy, bounces rapidly back and forth between the confining "walls" of the nuclear potential-energy barrier before finally, after what may be a relatively long time, tunneling through the barrier to become a free particle.

25.3   An Attractive Potential: The Bound State and Atoms

Recall that when a potential-energy bump is negative relative to the potential energy in the surrounding region, that bump is said to be attractive. We have already discussed the classical situation in which a ball is attracted by a localized depression in an otherwise level surface. Now consider a ball released from rest on the slope of a local depression. The ball can never escape the depression; the ball is *bound* by the attractive potential energy of the depression. The ball is said to be in a *bound state*.

Now consider the wave-mechanical analysis of a particle that has mass m and a precise negative energy $E < 0$ and encounters a single negative (attractive) potential-energy δ-function, as shown in Figure 25.8. If the δ-function is at $x = 0$, then $U(x)$ is equal to zero at all values of x except $x = 0$, and the Schrödinger equation for the particle at all $x \neq 0$ is

$$\frac{-\hbar^2}{2m} \frac{d^2\psi(x)}{dx^2} = E\psi(x)$$

$$\frac{d^2\psi(x)}{dx^2} = -\frac{2mE}{\hbar^2}\psi(x) = \alpha^2\psi(x) \tag{25-26}$$

where

$$E = -\frac{\hbar^2\alpha^2}{2m} \tag{25-27}$$

Then for $x \neq 0$, Equation (25-26) has a solution

$$\psi(x) = A\,e^{-\alpha|x|} = \begin{cases} A\,e^{\alpha x} & ; \quad x < 0 \\ A\,e^{-\alpha x} & ; \quad x > 0 \end{cases} \tag{25-28}$$

where $\alpha > 0$.

E 25.11   Verify that $\psi(x) = A\,e^{-\alpha|x|}$ with $\alpha > 0$ is a solution of

$$\frac{d^2\psi}{dx^2} = -\alpha^2\psi$$

E 25.12   Verify that $\psi(x) = B\,e^{\alpha|x|}$ with $\alpha > 0$ is also a solution of

$$\frac{d^2\psi}{dx^2} = -\alpha^2\psi$$

Can you suggest why we have not included this solution as part of Equation (25-28)? (*Hint:* We expect that the particle will most likely be found near the attractive potential-energy well to which it is bound.)

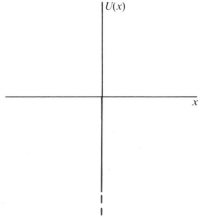

Figure 25.8
A single attractive (negative) potential-energy δ-function.

The boundary condition for $\psi(x)$, Equation (25-15), is satisfied at $x = 0$:

$$\psi(0^+) = \psi(0^-) = A \qquad (25\text{-}29)$$

To impose the boundary condition on $\psi'(x)$, we first find the derivative of $\psi(x)$ using Equation (25-28):

$$\psi'(x) = \frac{d\psi(x)}{dx} = \begin{cases} \alpha A\, e^{\alpha x} = \alpha\psi(x) & ; \quad x < 0 \\ -\alpha A\, e^{-\alpha x} = -\alpha\psi(x) & ; \quad x > 0 \end{cases} \qquad (25\text{-}30)$$

Then the rule of Equation (25-16) gives, if we use $-S$ to be the strength of the δ-function,

$$\psi'(0^+) = \psi'(0^-) + \frac{2m(-S)}{\hbar^2}\psi(0)$$

$$-\alpha A = \alpha A - \frac{2mS}{\hbar^2}A$$

$$\alpha = \frac{mS}{\hbar^2} \qquad (25\text{-}31)$$

Substituting Equation (25-31) into Equation (25-27), we find the value of the energy E of the single bound state of the particle to be

$$E = -\frac{mS^2}{2\hbar^2} = -E_o \qquad (25\text{-}32)$$

The classical and the wave-mechanical results for a particle bound by a single negative (attractive) δ-function potential energy differ in two important aspects. First, a single attractive δ-function with a given strength permits only one bound energy state for a particle. That precise energy, expressed by Equation (25-32), is determined by the strength S of the δ-function and the mass m of the particle. Classically, a continuous range of energy states, $-U_o < E < 0$, is permitted for a particle bound by a potential-energy well with a depth equal to U_o. The second difference may be seen if we plot the wave function $\psi(x)$, Equation (25-28), of the particle, as shown in Figure 25.9(a), and $|\psi(x)|^2$, as shown in Figure 25.9(b). Because $|\psi(x)|^2 dx$ is proportional to the probability that the particle lies within the interval between x and $x + dx$, Figure 25.9(b) indicates that the particle will

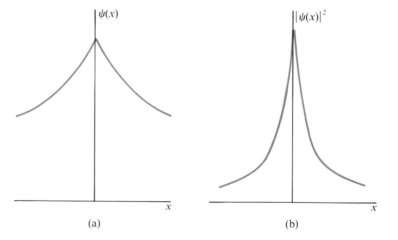

Figure 25.9
(a) The wave function $\psi(x)$ of a particle bound by a single attractive potential-energy δ-function. (b) The value $|\psi(x)|^2$, plotted as a function of x, for the same particle.

(a)   (b)

be found outside the confines of the attractive potential that is "binding" it to a localized region of space. In accordance with the probabilistic interpretation of wave mechanics, Figure 25.9(b) shows that if a large number of measurements are made to locate the particle, most such measurements would find the particle relatively near the δ-function. Nevertheless, some of the measurements would locate the particle far from $x = 0$—a very nonclassical result.

Our treatment of a single attractive δ-function potential energy is in many ways similar to the more complicated treatment of the "square-well" potential energy. Shown along with a δ-function in Figure 25.10 are several forms of a square-well potential energy. Each square well has the same strength, $-S$, as the δ-function. The figure illustrates roughly how the number of bound energy states associated with a particle depends on the relative shape of the potential well. In Figure 25.10(a), the single bound energy state associated with a δ-function is essentially the same as for a thin, deep square well, which is shown in Figure 25.10(b). A somewhat wider and shallower well, like that of Figure 25.10(c), permits an additional energy state. A very wide, shallow well, like the one in Figure 25.10(d) is relatively dense with allowed energy levels.

The square-well potential energy function may be considered an approximation of the potential energy of an electron bound to the nucleus of an atom. Figure 25.11 shows, for example, the coulombic potential energy function that binds the orbital electron to the proton in a hydrogen atom. The most obvious wave-mechanical feature of this physical situation is the quantization of the energy of the electron. Further, the spacing of these electronic energy levels of hydrogen is consistent with the characteristic spacing in square wells of different widths: The lower energy states are far apart where the hydrogen well is narrow and close together where the well is wider. A more extensive wave-mechanical treatment of complex atoms provides our most accurate means for the analysis of atomic structure.

Wave mechanics provides even more tools for the analysis of atomic phenomena. Just as the emission or absorption of photons occurs when electrons make transitions between allowed energy levels in hydrogen (Chapter 23, "Early Quantum Physics"), the atomic spectra of complex atoms result from the tran-

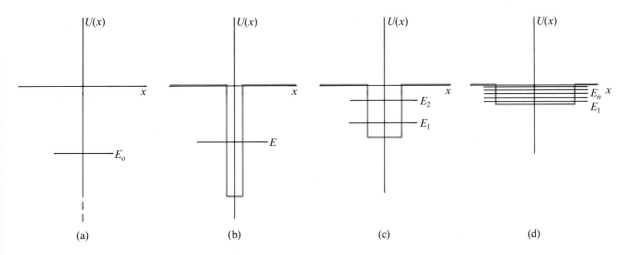

Figure 25.10
Illustrating how the number of bound energy states associated with a particle depends on the shape of the potential-energy well.

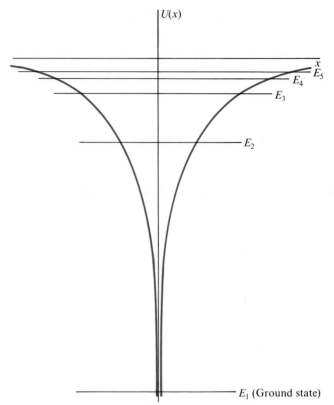

Figure 25.11
The coulombic potential-energy function for an electron in a hydrogen atom. The energy levels are spaced far apart where the potential-energy well is narrow and close together where the potential-energy well is wide.

sitions of orbital electrons between allowed energy states. And although we will not consider here the wave-mechanical treatment of transitions between energy states, the accurate calculation of atomic spectra has been one of the major successes of wave mechanics.

25.4 A Double Attractive Potential: Multiple Bound States and Molecules

We may introduce additional physically significant characteristics of wave mechanics by considering two attractive δ-functions separated by a distance a, as shown in Figure 25.12. A particle with this potential energy, a mass m, and a (negative) energy $E < 0$ is described by the solution of the Schrödinger equation,

$$\frac{d^2\psi(x)}{dx^2} = -\frac{2mE}{\hbar^2}\psi(x) = \alpha^2\psi(x) \quad ; \quad x \neq \pm\frac{a}{2} \tag{25-33}$$

where

$$E = -\frac{\hbar^2\alpha^2}{2m} \tag{25-34}$$

A solution of Equation (25-33) may be expressed as

$$\psi(x) = A\,e^{-\alpha|x + a/2|} + B\,e^{-\alpha|x - a/2|} \quad ; \quad \alpha > 0 \tag{25-35}$$

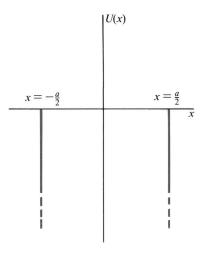

Figure 25.12
Two attractive potential-energy δ-functions separated by a distance a.

which may be written explicitly in the three regions of interest as

$$\psi(x) = \begin{cases} A\, e^{\alpha(x + a/2)} + B\, e^{\alpha(x - a/2)} & ; \quad x < -\dfrac{a}{2} \\[2mm] A\, e^{-\alpha(x + a/2)} + B\, e^{\alpha(x - a/2)} & ; \quad -\dfrac{a}{2} < x < \dfrac{a}{2} \\[2mm] A\, e^{-\alpha(x + a/2)} + B\, e^{-\alpha(x - a/2)} & ; \quad x > \dfrac{a}{2} \end{cases} \tag{25-36}$$

E 25.13   Show that the function ψ of Equations (25-35) and (25-36) satisfies

$$\psi\!\left(-\frac{a^{+}}{2}\right) = \psi\!\left(-\frac{a^{-}}{2}\right) = A + B\, e^{-\alpha a}$$

$$\psi\!\left(\frac{a^{+}}{2}\right) = \psi\!\left(\frac{a^{-}}{2}\right) = A\, e^{-\alpha a} + B$$

Exercise 25.13 verifies that the boundary condition, Equation (25-15), for the continuity of $\psi(x)$ at $x = \pm(a/2)$ is satisfied by Equation (25-36). So that we may apply the rule, Equation (25-16), on $\psi'(x)$, we find the expressions for $\psi'(x)$ in each interval of x:

$$\psi'(x) = \begin{cases} \alpha A\, e^{\alpha(x + a/2)} + \alpha B\, e^{\alpha(x - a/2)} & ; \quad x < -\dfrac{a}{2} \\[2mm] -\alpha A\, e^{-\alpha(x + a/2)} + \alpha B\, e^{\alpha(x - a/2)} & ; \quad -\dfrac{a}{2} < x < \dfrac{a}{2} \\[2mm] -\alpha A\, e^{-\alpha(x + a/2)} - \alpha B\, e^{-\alpha(x - a/2)} & ; \quad x > \dfrac{a}{2} \end{cases} \tag{25-37}$$

Applying the boundary condition, Equation (25-16), to $\psi'(x)$ at each δ-function, we obtain

$$\psi'\!\left(\pm\frac{a^{+}}{2}\right) = \psi'\!\left(\pm\frac{a^{-}}{2}\right) + \frac{2m(-S)}{\hbar^{2}}\psi\!\left(\pm\frac{a}{2}\right)$$

or

$$x = -\frac{a}{2} \quad : \quad -\alpha A + \alpha B\, e^{-\alpha a} = \alpha A + \alpha B\, e^{-\alpha a} - \frac{2mS}{\hbar^{2}}(A + B\, e^{-\alpha a})$$

$$\left(\alpha - \frac{mS}{\hbar^{2}}\right)A - \frac{mS}{\hbar^{2}} e^{-\alpha a}\, B = 0 \tag{25-38}$$

$$x = +\frac{a}{2} \quad : \quad -\alpha A\, e^{-\alpha a} - \alpha B = -\alpha A\, e^{-\alpha a} + \alpha B - \frac{2mS}{\hbar^{2}}(A\, e^{-\alpha a} + B)$$

$$-\frac{mS}{\hbar^{2}} e^{-\alpha a}\, A + \left(\alpha - \frac{mS}{\hbar^{2}}\right)B = 0 \tag{25-39}$$

The simultaneous solution of Equations (25-38) and (25-39) for α may be obtained by the simple algebraic procedure of solving one equation for A and substituting that expression for A into the other equation. A more general procedure, one that is useful when more than two equations (corresponding to more than two δ-functions in the potential-energy function) must be solved simultaneously, is to equate the determinant of the coefficients of A and B in Equations

(25-38) and (25-39) to zero. In the present case, either procedure gives two equations whose solutions are α_1 and α_2:

$$\alpha_1 = \frac{mS}{\hbar^2}(1 + e^{-\alpha_1 a}) \qquad \text{(25-40)}$$

$$\alpha_2 = \frac{mS}{\hbar^2}(1 - e^{-\alpha_2 a}) \qquad \text{(25-41)}$$

The direct substitution of either of these values of α into Equations (25-38) and (25-39) will confirm that α_1 corresponds to the case for which $A = B$ and α_2 corresponds to the case for which $A = -B$.

E 25.14
(a) Using Equation (25-40), substitute for α in Equation (25-38) and show that $A = B$ for $\alpha = \alpha_1$.
(b) What happens if Equation (25-40) is substituted into Equation (25-39)?

Answer: $A = B$

(c) Verify that if α_2 of Equation (25-41) is used in either Equation (25-38) or Equation (25-39), then $A = -B$.

Example 25.3

PROBLEM   Verify that equating the determinant of the coefficients of A and B in Equations (25-38) and (25-39) to zero does yield the results given by Equations (25-40) and (25-41).

SOLUTION   Equating this determinant to zero, we get

$$\begin{vmatrix} \alpha - \dfrac{mS}{\hbar^2} & -\dfrac{mS}{\hbar^2}e^{-\alpha a} \\[2ex] -\dfrac{mS}{\hbar^2}e^{-\alpha a} & \alpha - \dfrac{mS}{\hbar^2} \end{vmatrix} = 0$$

Expanding the determinant gives

$$\left(\alpha - \frac{mS}{\hbar^2}\right)^2 - \left(\frac{mS}{\hbar^2}e^{-\alpha a}\right)^2 = 0$$

and by factoring this difference in squares, we get

$$\left(\alpha - \frac{mS}{\hbar^2} - \frac{mS}{\hbar^2}e^{-\alpha a}\right)\left(\alpha - \frac{mS}{\hbar^2} + \frac{mS}{\hbar^2}e^{-\alpha a}\right) = 0$$

Equating each of the two factors to zero and labeling the solutions as α_1 and α_2 gives

$$\alpha_1 = \frac{mS}{\hbar^2}\left(1 + e^{-\alpha_1 a}\right)$$

$$\alpha_2 = \frac{mS}{\hbar^2}\left(1 - e^{-\alpha_2 a}\right)$$

which is the desired result. ∎

The solutions of Equations (25-40) and (25-41) for α_1 and α_2 will, because α is related to E by Equation (25-34), provide the desired values of the energies, E_1 and E_2, that are permitted for the particle we are considering. Because the equations in α_1 and α_2 are transcendental equations, their solutions are perhaps most easily obtained graphically. To do this, let us write Equations (25-40) and (25-41) as

$$\frac{1}{\gamma}y = 1 \pm e^{-y} \qquad (25\text{-}42)$$

where we have made the substitutions

$$y = \alpha a \qquad (25\text{-}43)$$

$$\gamma = \frac{mSa}{\hbar^2} \qquad (25\text{-}44)$$

Now we set some variable, say w, equal to each side of Equation (25-42) and plot the curve $w = (1/\gamma)y$ and the curves $w = 1 \pm e^{-y}$ on the same set of axes, w versus y, as shown in Figure 25.13(a). The points of intersection of the curve $w = (1/\gamma)y$ with the curves $w = 1 \pm e^{-y}$ occur at those values of $y = \alpha a$ that are the solutions of Equation (25-42). Because a is the (known) separation between the two δ-functions in the potential-energy function of the particle being considered, these solutions for y ($= \alpha a$) specify the permitted values of α and, therefore, the permitted energies.

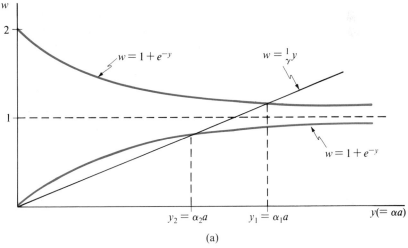

(a)

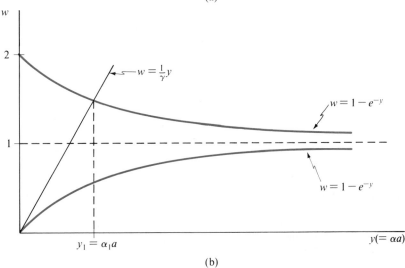

(b)

Figure 25.13
Graphical solutions for y ($= \alpha a$), which specify the permitted energies of a particle bound by two attractive potential-energy δ-functions. **(a)** When $w = (1/\gamma)y$ has sufficiently small slope, two solutions, y_1 and y_2, exist. **(b)** When $w = (1/\gamma)y$ has sufficiently large slope, only one solution, y_1, exists.

In Figure 25.13(a), the slope of the straight line, $w = (1/\gamma)y$, is sufficiently small—this means, according to Equation (25-44), $\gamma = mSa/\hbar^2$, that the separation a between δ-functions is sufficiently large—that there are two intersections and, therefore, two possible energy levels, E_1 and E_2. If a is small enough, however, the straight line $w = (1/\gamma)y$ intersects only the curve $w = 1 + e^{-y}$ as shown in Figure 25.13(b); and only one energy is permitted. The energies, E_1 and E_2 are, in terms of α_1 and α_2,

$$E_1 = -\frac{\hbar^2\alpha_1^2}{2m} \quad ; \quad E_2 = -\frac{\hbar^2\alpha_2^2}{2m} \tag{25-45}$$

The state having the lower energy E_1 is called the *ground state* of the particle; the higher-energy state with energy E_2 is called the *excited state*.

The relationships between the energy levels, E_1 and E_2, and the separation a between attractive centers (the δ-functions) are important, both now and in our subsequent considerations. Let us, then, examine Figure 25.14, which shows how E_1 and E_2 depend on the separation a (for a given δ-function strength $-S$ and particle mass m). At small values of a, only one energy level, namely E_1, exists, as we have previously noted. When a is sufficiently large, two energy levels, E_1 and E_2, are present; and the energy difference between E_1 and E_2 becomes smaller as the separation a increases. For very large separations, E_1 and E_2 converge to the value $-E_o = -mS^2/(2\hbar^2)$, which is the only permitted energy of a particle when it is bound by a single δ-function of strength $-S$.

Corresponding to the two energy levels, E_1 and E_2, of a particle bound by two attractive potential-energy δ-functions are two wave functions, $\psi_1(x)$ and $\psi_2(x)$, which are the solutions of Schrödinger's equation for this system. These wave functions are shown in Figure 25.15. In Figure 25.15(a), where $\psi_1(x) = \psi_1(-x)$, the ground-state wave function is *symmetric*. In Figure 25.15(b), where $\psi_2(x) = -\psi_2(-x)$, the excited-state wave function is *antisymmetric*.

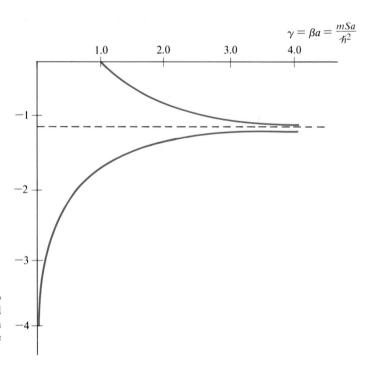

Figure 25.14
Illustrating the dependence of the two energies, E_1 and E_2, of a particle bound by two attractive δ-functions on $\gamma = mSa/\hbar^2$, which is a measure of the separation a between the δ-functions.

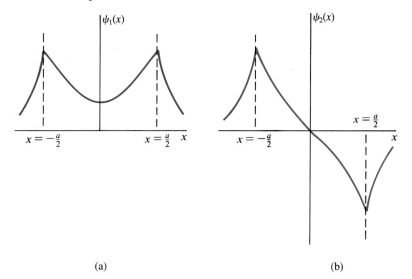

Figure 25.15
Wave functions, **(a)** $\psi_1(x)$ and **(b)** $\psi_2(x)$, corresponding to the two energies, E_1 and E_2, of a particle bound by two attractive δ-functions.

(a)   (b)

The system that we have treated here, a particle bound by two δ-functions, illustrates many of the characteristics of actual physical systems that have two centers of attraction. A diatomic molecule, for example, which has two attractive nuclei, is similar in many ways to the system that we have been considering. Figure 25.16 compares the potential-energy functions of a particle bound by two δ-functions and of the electron bound by the two nuclei of a singly ionized hydrogen molecule. Figure 25.16(b) shows the coulombic potential energy of the electron in an ionized hydrogen molecule as a function of position x along the line through the two nuclei. In our treatment of δ-functions in one dimension, we have seen that the introduction of more than one attractive center causes more than one energy level to be permitted. Further, the permitted energies are quantized. These characteristics are also true of actual two-centered attractive systems, like diatomic molecules. Notice also in Figure 25.16(b) that the large number of energy levels in the hydrogen molecule are spaced far apart at the lower energies, where the potential wells are narrow, and close together at the higher energies, where the wells are wide. In the analysis of molecular structure and spectra, just as was the case with other systems we have considered, wave mechanics is the modern tool of physical analysis by which molecular systems may be most completely understood.

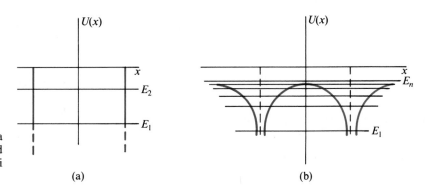

Figure 25.16
The potential-energy functions of **(a)** a particle bound by two δ-functions, and **(b)** an electron bound by the two nuclei of a singly ionized hydrogen atom.

(a)   (b)

Example 25.4

PROBLEM   Determine the bound-state energy levels, E_1 and E_2, for an electron ($m = 9.1 \times 10^{-31}$ kg) bound by two δ-functions of strength $-S = -1.6 \times 10^{-28}$ J·m separated by $a = 0.15$ nm.

SOLUTION   For the given values, the parameter γ has the value

$$\gamma = \frac{9.1 \times 10^{-31} \times 1.6 \times 10^{-28} \times 0.15 \times 10^{-9}}{(1.05 \times 10^{-34})^2} = 2.0$$

Figure 25.17 shows a graph of the curves

$$w_1 = 1 + e^{-y}$$

$$w_2 = 1 - e^{-y}$$

$$w_3 = \frac{1}{\gamma} y = \frac{1}{2} y$$

The intersections of the w_3 curve with the w_1 and w_2 curves indicate values of y_1 and y_2 given by

$$y_1 = 2.2$$

$$y_2 = 1.6$$

Thus, because $y = \alpha a$, we get for α_1 and α_2:

$$\alpha_1 = \frac{y_1}{a} = \frac{2.2}{0.15 \times 10^{-9}} = 1.5 \times 10^{10} \text{ m}^{-1}$$

$$\alpha_2 = \frac{y_2}{a} = \frac{1.6}{0.15 \times 10^{-9}} = 1.1 \times 10^{10} \text{ m}^{-1}$$

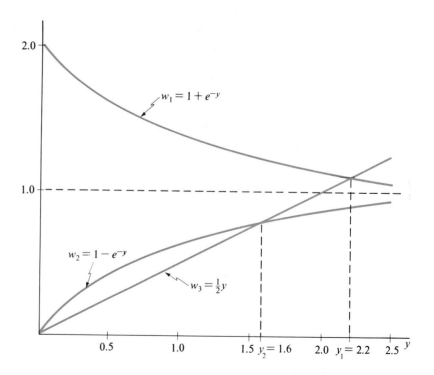

Figure 25.17   Example 25.4

And the energies, E_1 and E_2, are given by

$$E_1 = -\frac{\hbar^2 \alpha_1^2}{2m} = -\frac{(1.05 \times 10^{-34} \times 1.5 \times 10^{10})^2}{2 \times 9.1 \times 10^{-31}} = -1.4 \times 10^{-18}\,\text{J} = -8.5\,\text{eV}$$

$$E_2 = -\frac{\hbar^2 \alpha_2^2}{2m} = -\frac{(1.05 \times 10^{-34} \times 1.1 \times 10^{10})^2}{2 \times 9.1 \times 10^{-31}} = -7.3 \times 10^{-19}\,\text{J} = -4.6\,\text{eV}$$

■

E 25.15   Write out the wave functions, ψ_1 and ψ_2, for the two bound states of Example 25.4.

Answer: $\psi_1 = A(e^{-\alpha_1 x + 1.1} + e^{-\alpha_1 x - 1.1})$

$\psi_2 = A(e^{-\alpha_2 x + 0.8} - e^{-\alpha_2 x - 0.8})$

where $\alpha_1 = 1.5 \times 10^{10}\,\text{m}^{-1}$ and $\alpha_2 = 1.1 \times 10^{10}\,\text{m}^{-1}$.

25.5 Multiple Attractive Potentials: Band Theory and Solids

We have seen how the simplistic model of one attractive δ-function is suggestive of the potential energy of an atom and how that model predicts some of the wave-mechanical characteristics of actual atoms. Further, we have seen how the δ-function model, when extended to include two attractive δ-functions, imitates the important features of a diatomic molecule. Then would it not be interesting and instructive to extend the model to 4, 10, or, finally, a very large number of equally spaced attractive δ-functions with the expectation that such a model will mimic a crystalline solid? In pursuing this question, we will omit most of the mathematical details, which become more cumbersome (though quite straightforward) as the potential-energy function for a particle includes additional attractive δ-functions.

Let us begin by outlining how we may proceed in analyzing a particle that has mass m and is bound by a potential-energy function that contains four attractive δ-functions, each with strength $-S$, spaced apart by a distance a. Figure 25.18 shows the potential energy of the particle in this situation. The solution of Schrödinger's equation and its analysis for this situation are procedurally identical to the treatment of the two δ-function case:

1. The solution of the Schrödinger equation at all positions x other than at the δ-functions is

$$\psi(x) = A\,e^{-\alpha|x + 3a/2|} + B\,e^{-\alpha|x + a/2|} + C\,e^{-\alpha|x - a/2|} + D\,e^{-\alpha|x - 3a/2|} \quad \text{(25-46)}$$

where A, B, C, and D are constants. The quantity α is related to the total energy E of the particle by

$$E = -\frac{\hbar^2 \alpha^2}{2m} \qquad \text{(25-47)}$$

2. The boundary condition on the wave function ($\psi(x)$ is continuous at each

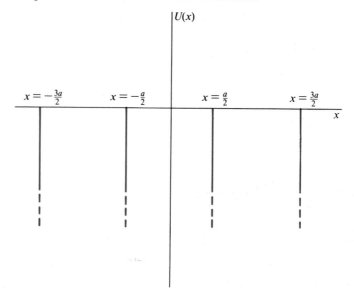

Figure 25.18
The potential-energy function $U(x)$ of a particle bound by four attractive δ-functions, each of strength $-S$, separated by a distance a.

δ-function) is satisfied. The boundary condition on $\psi'(x)$, Equation (25-16), when applied at each δ-function, provides that

$$\psi'(na^+) - \psi'(na^-) = -\frac{2mS}{\hbar^2}\psi(na) \quad ; \quad n = \pm\frac{1}{2}; \ \pm\frac{3}{2} \quad (25\text{-}48)$$

Application of this boundary condition at the four δ-function locations gives four equations:

$$\left(\frac{\alpha}{\beta} - 1\right)A - e^{-\alpha a}B - e^{-2\alpha a}C - e^{-3\alpha a}D = 0$$

$$-e^{-\alpha a}A + \left(\frac{\alpha}{\beta} - 1\right)B - e^{-\alpha a}C - e^{-2\alpha a}D = 0$$

$$-e^{-2\alpha a}A - e^{-\alpha a}B + \left(\frac{\alpha}{\beta} - 1\right)C - e^{-\alpha a}D = 0$$

$$-e^{-3\alpha a}A - e^{-2\alpha a}B - e^{-\alpha a}C + \left(\frac{\alpha}{\beta} - 1\right)D = 0 \quad (25\text{-}49)$$

In Equation (25-49) we have used $\beta = mS/\hbar^2$.

3. The simultaneous solution of Equations (25-49) may be accomplished by setting the determinant of the coefficients of A, B, C, and D equal to zero. In terms of the variables used in the two-δ-function case, namely $y = \alpha a$ and $\gamma = mSa/\hbar^2$, we obtain at least one and not more than four values of y such that $y_1 > y_2 > y_3 > y_4$. Then the permitted energies E_n may be expressed as

$$E_n = -\frac{\alpha_n^2\hbar^2}{2m} = \frac{-\hbar^2 y_n^2 \beta^2}{2m\gamma^2} = -\left(\frac{y_n}{\gamma}\right)^2 E_o \quad ; \quad n = 1, 2, 3, 4 \quad (25\text{-}50)$$

where

$$E_o = \frac{\hbar^2\beta^2}{2m} = \frac{mS^2}{2\hbar^2} \quad (25\text{-}51)$$

and $-E_o$ is the energy of a particle bound by a single attractive potential.

The energy solutions for the four-δ-function case are plotted in Figure 25.19, which shows the permitted energies (expressed in terms of E/E_o) as a function of $\gamma = mSa/\hbar^2$, a variable that is proportional to the separation a between the δ-functions. Notice that for sufficiently small values of γ, when the δ-functions are closely spaced, only one energy level exists. As the spacing increases, additional energy levels appear until all four energies are permitted. As the separation between δ-functions becomes very large, the energy levels coalesce toward a common value $-E_o$.

The four wave functions associated with the four allowed energies of this system are plotted in Figure 25.20 for $\gamma = 2$, which corresponds to a spacing between δ-functions for which all four energy levels are present. The wave functions, $\psi_1(x)$ and $\psi_3(x)$, corresponding to the energies, E_1 and E_3, are symmetric; $\psi_2(x)$ and $\psi_4(x)$ are antisymmetric. And it is generally true in systems with multiple evenly spaced δ-functions that the wave functions associated with successive energy levels are alternately symmetric and antisymmetric.

If this same analytical procedure is carried out for a particle which has a potential-energy function composed of N equally spaced δ-functions of equal strength, we obtain as many as N allowed energies and an equal number of wave functions that are solutions of the Schrödinger equation. Figure 25.21 shows, for $N = 10$, how the allowed energies, $E_1, E_2, \ldots, E_{10}$, vary as γ, which is proportional to the separation a between δ-functions, increases. Comparing Figures 25.21 and 25.19, we can see that the curves of the allowed energies are filling the region between E_1 and E_N as N increases.

Finally, as $N \to \infty$, the infinite number of allowed energies form a dense *band* of energies, as indicated in Figure 25.22. The energy levels within this band once again converge to the value $-E_o$ as the separation between the δ-functions

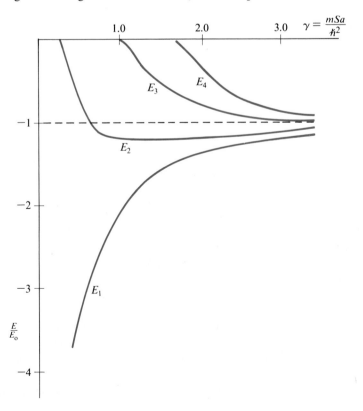

Figure 25.19
The four permitted energies of a particle bound by four attractive δ-functions plotted as a function of γ, which is proportional to the separation a between the δ-functions.

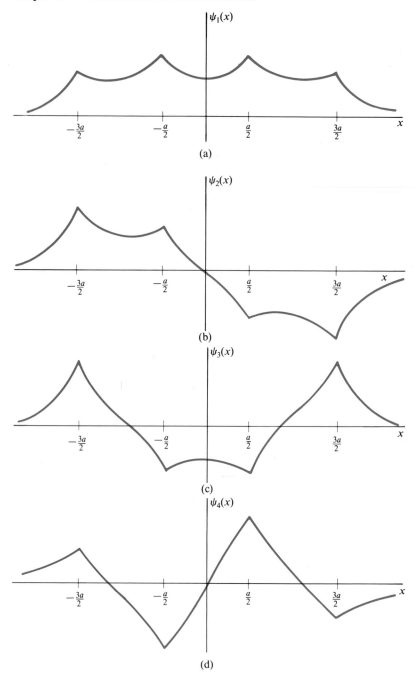

Figure 25.20
The four wave functions associated with the four permitted energies of a particle bound by four attractive δ-functions.

becomes large. A particle bound by an infinite set of equally spaced δ-functions may have an energy that lies anywhere within the allowed energy band.

Our treatment in one dimension of an infinite number of evenly spaced δ-functions exhibits some of the important characteristics of real three-dimensional solids. A *crystalline solid* is a regular three-dimensional array of atoms with nuclei that are attractive centers for the orbital electrons of these atoms. The outermost orbital electrons of each atom in a solid are loosely bound to the nucleus of that atom; and when the spacing between the nuclei is small, the electronic wave functions overlap significantly between nuclear sites. The outermost electrons can then move about from atom to atom; they are bound by the

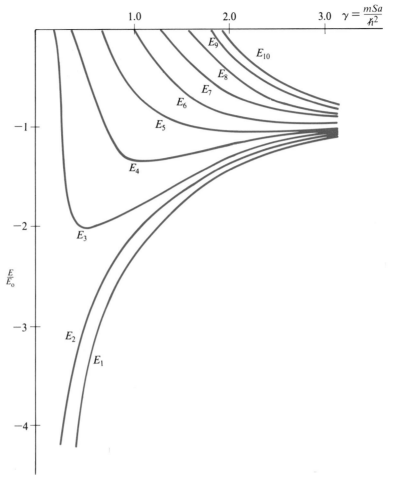

Figure 25.21
The 10 permitted energies of a particle bound by 10 equally spaced attractive δ-functions plotted as a function of γ, which is proportional to the separation a between δ-functions.

potential energy of the solid as a whole. But because each atom of a solid produces many electronic energy levels (whereas a δ-function produces only one), a solid lattice exhibits numerous energy bands similar to the single band we have seen using our simple model. Figure 25.23 shows two such bands that are typical of real solids.

Energy bands play an important role in our present understanding of the physical properties of solids. For example, the electrical conductivity of several classes of solids may be summarized in terms of energy bands as follows:

1. A *metallic conductor*, like copper or silver, is characterized by having its highest occupied energy band partially filled, while its lower energy bands are completely filled, as shown in Figure 25.24(a). A current in a solid is caused by an external agent, like a battery, which creates an electric field in the solid. And even a weak electric field can impart momentum to a large number of electrons with a relatively low expenditure of energy in a conductor in which the occupied and unoccupied energy levels are verly closely spaced.

2. A solid *insulator*, like sulfur or diamond, is a crystal in which the uppermost occupied energy band is completely filled and in which a considerable energy gap separates the uppermost filled band from the next higher band. A very large electric field, and therefore considerable energy, would be required to excite an electron from the filled band across the gap to the next band. This situation is illustrated in Figure 25.24(b).

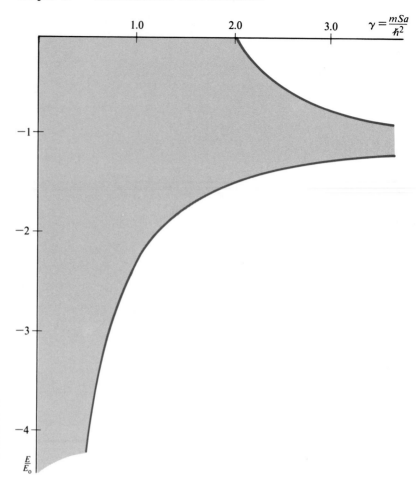

Figure 25.22
The dense band of allowed energies for a particle bound by an infinite number of equally spaced δ-functions plotted as a function of γ, which is proportional to the separation a between the δ-functions.

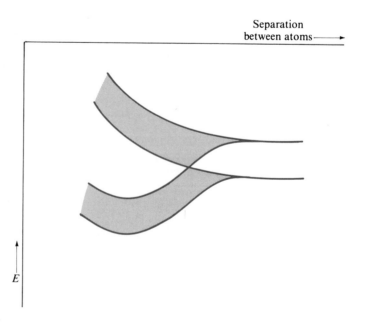

Figure 25.23
Typical energy bands in an actual solid.

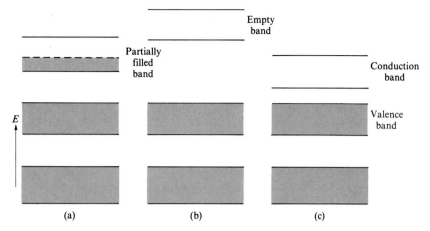

Figure 25.24
Energy band configurations in a solid **(a)** metallic conductor, **(b)** insulator, and **(c)** intrinsic semiconductor.

3. An *intrinsic semiconductor*, like silicon or germanium, has a filled band, called a *valence band*, and a nearby higher but empty band, called a *conduction band*, as shown in Figure 25.24(c). At low temperatures there are no electrons in the conduction band, and such a semiconductor is an insulator. As the temperature of a semiconductor increases, more and more electrons are thermally excited into the conduction band, and the semiconductor is a better and better conductor.

25.6 Two Special Examples

This section presents two examples from wave mechanics that substantiate many of the quantum features previously presented in this chapter. Both of these problems illustrate the energy quantization that is characteristic of wave-mechanical descriptions of bound states. The first example determines the energy levels and corresponding wave functions for a particle that is free except when it bounces off either of two infinitely hard walls. The two lowest energy states of the simple harmonic oscillator are considered in the second example.

Example 25.5   PROBLEM   A particle of mass m moves along the x axis and is confined to the interval $0 < x < a$ by the potential energy

$$U(x) = \begin{cases} \infty & x < 0 \\ 0 & 0 < x < a \\ \infty & a < x \end{cases}$$

Determine the energy levels and corresponding wave functions for this particle.
SOLUTION   Here we seek energy values E and corresponding functions $\psi(x)$ that satisfy the Schrödinger equation

$$-\frac{\hbar^2}{2m}\frac{d^2\psi(x)}{dx^2} + U(x)\psi(x) = E\psi(x)$$

with the potential energy $U(x)$ given above. First, we assert that $\psi(x) = 0$ in the regions where the potential energy is infinite, that is, $\psi(x) = 0$ for $x < 0$ or $x > a$. This assertion may be justified by considering solutions to the Schrödinger equation for $x > a$ with $U(x) = U_\text{o} > E$. In this case, the Schrödinger equation may be written

$$\frac{\mathrm{d}^2\psi}{\mathrm{d}x^2} = -\frac{2m}{\hbar^2}(U_\text{o} - E)\psi = -\beta^2\psi$$

with

$$\beta = \sqrt{2m(U_\text{o} - E)}/\hbar$$

A general solution to this equation is

$$\psi(x) = K_1 e^{-\beta x} + K_2 e^{\beta x}$$

where K_1 and K_2 are arbitrary constants. The second term, $K_2 e^{\beta x}$, in this solution increases without bound as $x \to \infty$. To avoid this physically unacceptable result, that is, ψ, and therefore $|\psi|^2$, increasing without bound in the interval where the particle would never be found classically, we set $K_2 = 0$. Thus in the interval $x > a$, we have

$$\psi(x) = K_1 e^{-\beta x}$$

But as $U_\text{o} \to \infty$, the constant β becomes arbitrarily large; and consequently ψ approaches zero at each point in this interval. We see then that, as we have asserted, $\psi(x) = 0$ for $x < 0$ and $a < x$.

For $0 < x < a$ where $U = 0$, the Schrödinger equation

$$-\frac{\hbar^2}{2m}\frac{\mathrm{d}^2\psi}{\mathrm{d}x^2} = E\psi$$

$$\frac{\mathrm{d}^2\psi}{\mathrm{d}x^2} = -\frac{2mE}{\hbar^2}\psi = -\alpha^2\psi$$

with $\alpha = \sqrt{2mE}/\hbar$. A general solution for ψ in this interval is

$$\psi(x) = A \sin \alpha x + B \cos \alpha x$$

Thus we have for the full range of x,

$$\psi(x) = \begin{cases} 0 & ; \quad x < 0 \\ A \sin \alpha x + B \cos \alpha x & ; \quad 0 < x < a \\ 0 & ; \quad a < x \end{cases}$$

If we require ψ to be continuous at $x = 0$ and at $x = a$, we obtain

$$B = 0 \qquad \text{for } x = 0$$

$$A \sin \alpha a + B \cos \alpha a = 0 \qquad \text{for } x = a$$

Using $B = 0$ in the $x = a$ equation gives

$$A \sin \alpha a = 0$$

Consequently, either $A = 0$, in which case $\psi = 0$ for all x, or $\sin \alpha a = 0$. Clearly the $A = 0$ case is unacceptable, and so we have solutions to the Schrödinger equation for those positive values of α that satisfy

$$\sin \alpha a = 0$$

$$\alpha_n a = n\pi \quad ; \quad n = 1, 2, 3, \ldots$$

$$\alpha_n = n\pi/a \quad ; \quad n = 1, 2, 3, \ldots$$

Because $E = \hbar^2 \alpha^2/2m$, we find the energy levels to be

$$E_n = \frac{\hbar^2 \alpha_n^2}{2m} = n^2 \frac{h^2}{8ma^2} = n^2 E_1 \quad ; \quad n = 1, 2, 3, \ldots$$

where the ground-state energy E_1 is given by

$$E_1 = \frac{h^2}{8ma^2}$$

The wave functions $\psi_n(x)$ that correspond to these energy values are

$$\psi_n(x) = \begin{cases} 0 & ; \quad x < 0 \\ A \sin \dfrac{n\pi x}{a} & ; \quad 0 < x < a \\ 0 & ; \quad a < x \end{cases}$$

where A is an arbitrary constant. Figure 25.25 shows the lower energy levels and the corresponding wave functions. Notice that once again the wave functions are alternately symmetric and antisymmetric about the (symmetry) center, $x = a/2$, of the potential energy $U(x)$. ■

If we think of the wave functions of Example 25.5 as standing waves with zero displacement at $x = 0$ and $x = a$ and if the corresponding wavelengths are interpreted as de Broglie wavelengths, the energy levels and wave functions may be obtained using Figure 25.26 as follows:

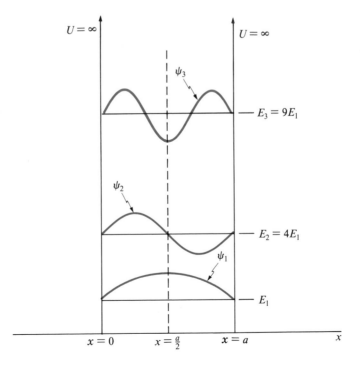

Figure 25.25   Example 25.5

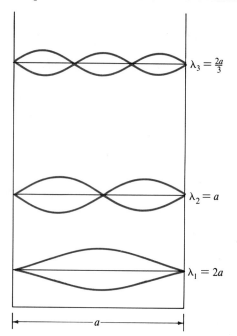

Figure 25.26
Interpreting the wave functions of a particle in an "infinite square well" as standing waves.

1. The wavelengths for the first three standing waves are

    $$\lambda_1 = 2a$$

    $$\lambda_2 = a = \frac{2a}{2}$$

    $$\lambda_3 = \frac{2}{3}a = \frac{2a}{3}$$

    and in general

    $$\lambda_n = \frac{2a}{n} \quad ; \quad n = 1, 2, 3, \ldots \tag{25-52}$$

2. Using the de Broglie relation, the magnitude of the linear momentum of the particle is

    $$p_n = \frac{h}{\lambda_n} = n\left(\frac{h}{2a}\right) \tag{25-53}$$

    and the corresponding (kinetic) energy is

    $$E_n = \frac{p_n^2}{2m} = n^2\frac{h^2}{8ma^2} = n^2E_1 \tag{25-54}$$

    which is identical to the result of Example 25.4.

3. Using the wavelengths of step 1, the standing-wave shapes shown are described by

    $$\psi_n(x) = A \sin\frac{2\pi x}{\lambda_n} = A \sin\frac{n\pi x}{a} \tag{25-55}$$

    which is identical to the wave functions of Example 25.4.
    The ground state energy E_1 of Example 25.5 represents the minimum (kinetic)

energy of a particle of mass m that is localized to an interval of length a. Consequently, this energy is called the *energy of localization*.

E 25.16   Calculate the energy of localization for

(a) an electron localized to an atomic dimension of 0.10 nm,
(b) an electron localized to a nuclear dimension of 1.0×10^{-14} m,
(c) a neutron localized to a nuclear dimension of 1.0×10^{-14} m,
(d) a 1.0-g particle localized to a macroscopic dimension of 1.0 cm.

Answers: (a) 6.0×10^{-18} J $= 38$ eV; (b) 6.0×10^{-10} J $= 3800$ MeV;
(c) 3.3×10^{-13} J $= 2.1$ MeV; (d) 5.5×10^{-61} J

Example 25.6

PROBLEM   A particle of mass m has the one-dimensional simple-harmonic-oscillator potential energy given by

$$U(x) = \frac{1}{2}kx^2$$

where k is a positive constant. Show that

$$\psi_1(x) = e^{-\alpha x^2}$$

and

$$\psi_2(x) = x\, e^{-\alpha x^2}$$

are solutions to the associated Schrödinger equation provided α is chosen correctly. Determine the energy value that corresponds to each of these wave functions.

SOLUTION   The Schrödinger equation for this particle is

$$-\frac{\hbar^2}{2m}\psi''(x) + \frac{1}{2}kx^2\psi(x) = E\psi(x)$$

With ψ_1 as given, we get

$$\psi_1'(x) = -2\alpha x\, e^{-\alpha x^2}$$

$$\psi_1''(x) = (-2\alpha + 4\alpha^2 x^2)\, e^{-\alpha x^2}$$

If we substitute for ψ_1 and ψ_1'' into the Schrödinger equation and let $E = E_1$, then we get

$$\left(\frac{\hbar^2\alpha}{m} - \frac{2\hbar^2\alpha^2 x^2}{m} + \frac{1}{2}kx^2\right)e^{-\alpha x^2} = E_1\, e^{-\alpha x^2}$$

which may be rewritten as

$$E_1 - \frac{\hbar^2\alpha}{m} + \left(\frac{2\hbar^2\alpha^2}{m} - \frac{1}{2}k\right)x^2 = 0$$

Because this equation is to be true for all x, the coefficient of x^2 must be zero, i.e.,

$$\frac{2\hbar^2\alpha^2}{m} - \frac{1}{2}k = 0$$

$$\alpha^2 = \frac{km}{4\hbar^2}$$

$$\alpha = \frac{\sqrt{km}}{2\hbar}$$

And with the x^2 term no longer present, we get

$$E_1 = \frac{\hbar^2 \alpha}{m} = \frac{\hbar}{2}\sqrt{\frac{k}{m}}$$

Recalling that the frequency ν for the classical harmonic oscillator is

$$\nu = \frac{1}{2\pi}\sqrt{\frac{k}{m}}$$

we get

$$E_1 = \frac{1}{2}h\nu$$

Repeating this procedure using ψ_2, we get

$$\psi_2' = (1 - 2\alpha x^2)e^{-\alpha x^2}$$

$$\psi_2'' = (-6\alpha x + 4\alpha^2 x^3)e^{-\alpha x^2}$$

$$E_2 - \frac{3\alpha\hbar^2}{m} + \left(\frac{2\hbar^2\alpha^2}{m} - \frac{1}{2}k\right)x^2 = 0$$

$$\alpha = \frac{\sqrt{km}}{2\hbar}$$

$$E_2 = \frac{3}{2}h\nu = 3E_1$$

Summarizing, the two energy levels and the associated wave functions are

$$E_1 = \frac{1}{2}h\nu \quad ; \quad \psi_1(x) = e^{-\alpha x^2}$$

$$E_2 = \frac{3}{2}h\nu \quad ; \quad \psi_2(x) = x\,e^{-\alpha x^2}$$

$$\alpha = \frac{\sqrt{km}}{2\hbar}$$

$$\nu = \frac{1}{2\pi}\sqrt{\frac{k}{m}}$$

Figure 25.27 shows these energy levels and wave functions that are, in fact, the ground state and first excited state of the simple harmonic oscillator. We assert that all energy levels and associated wave functions are given by

$$\left.\begin{array}{c} E_n = \left(n - \dfrac{1}{2}\right)h\nu \\ \psi_n(x) = P_n(x)e^{-\alpha x^2} \end{array}\right\} \quad n = 1, 2, 3, \ldots$$

where $P_n(x)$ is (i) a polynomial of degree n and (ii) symmetric, if n is odd, or antisymmetric, if n is even, about $x = 0$. ∎

These two examples reiterate a general wave-mechanical property we have seen before: As the width of a confining potential-energy well increases, the relative spacing between adjacent energy levels decreases. For the infinite potential-energy well (Example 25.5), which has a fixed width a, the energy levels are given by $E_n = n^2 E_1$; and the spacing between adjacent levels is

$$\Delta E = E_{n+1} - E_n = (n + 1)^2 E_1 - n^2 E_1$$

$$\Delta E = (2n + 1)E_1 \qquad \text{(infinite well)} \qquad \text{(25-56)}$$

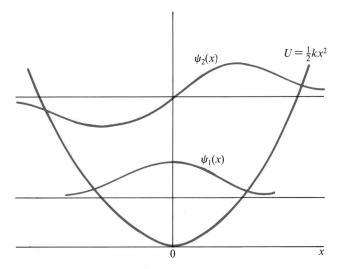

Figure 25.27   Example 25.6

Thus the spacing between levels increases as the energy increases. But for the potential energy of the harmonic oscillator (Example 25.6), which "opens up" as the energy increases, the energy levels are given by $E_n = (n - 1/2)h\nu$; and the spacing between adjacent levels is

$$\Delta E = E_{n+1} - E_n = \left(n + \frac{1}{2}\right)h\nu - \left(n - \frac{1}{2}\right)h\nu$$

$$\Delta E = h\nu \quad \text{(harmonic oscillator)} \quad (25\text{-}57)$$

The potential energy of the harmonic oscillator opens up in such a way that the spacing between adjacent levels remains constant. Atomic and molecular potential energies, which are essentially electrostatic, open up much more rapidly; and the spacing between atomic and molecular energy decreases as the energy increases. Figure 25.28 summarizes these results.

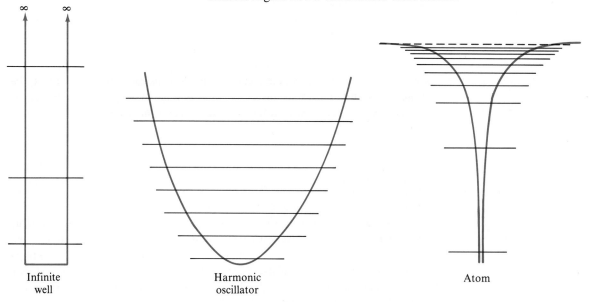

Figure 25.28
Comparing the spacing of energy levels in (a) an infinite square well, (b) a harmonic oscillator, and (c) an atom.

25.7 Problem-Solving Summary

The title of this section more nearly reflects adherence to pattern than the actual content of this section. This chapter, unlike all the others, is not primarily oriented toward teaching the student to solve problems. Rather it is intended to introduce the basic concepts of the wave mechanics and to illustrate how the solutions of some simple systems of δ-function potential energies can mimic the wave-mechanical characteristics of numerous important physical systems. Therefore, this section is actually a descriptive synopsis of the present chapter.

Wave mechanics is a description of particles that reflects the wave nature of particles and that is consistent with the de Broglie hypothesis, the uncertainty principle, and a probabilistic interpretation of measurement processes. In the wave mechanics, a particle is represented by (or associated with) a wave function Ψ, which is a solution of the Schrödinger equation, an equation that includes the interaction of the particle with its surroundings. From Ψ, which may be a complex quantity, we may predict the results of physical measurements on the particle Ψ describes. The wave function Ψ has no direct physical significance, but $|\Psi|^2 dx$ is proportional to the probability of finding the particle associated with Ψ in the interval between x and $x + dx$.

For those one-dimensional physical situations in which a particle has a precise energy E, the wave function $\Psi(x, t)$ may be expressed by a product of $\psi(x)$, a complex function of x only, and $e^{-iEt/\hbar}$. Then $\psi(x)$ is a solution of the time-independent Schrödinger equation,

$$-\frac{\hbar^2}{2m}\frac{d^2\psi(x)}{dx^2} + U\psi(x) = E\psi(x)$$

where m is the mass of the particle and U is the potential energy of the particle.

The interpretation of $\psi(x)$ as a wave follows from the general solution of the Schrödinger equation for a free particle with energy $E = p^2/(2m)$, a particle for which U is a constant. Then $\psi(x) = A\,e^{ikx} + B\,e^{-ikx}$, where $k = 2\pi/\lambda = p/\hbar$, represents a wave of amplitude A moving in the positive x direction and a wave of amplitude B moving in the negative x direction. Thus $\psi(x)$ may represent a particle of energy E with momentum p in the positive x direction ($B = 0$), a particle of energy E with momentum $-p$ in the negative x direction ($A = 0$), or a particle of energy E in a mixed momentum state ($A \neq 0$ and $B \neq 0$).

The wave-mechanical treatment of a particle having a single repulsive (positive) δ-function potential energy illustrates that:

1. A particle encountering a potential-energy barrier that has a height exceeding the energy of the particle has a finite probability of tunneling through that barrier.

2. The probability that a particle will pass through a barrier of a given strength may be expressed in terms of $T(E)$, the transmission coefficient of the particle. And $T(E)$ is related to the reflection coefficient $R(E)$ by $T(E) + R(E) = 1$.

The system of a single repulsive δ-function mimics the actual physical system in which a particle tunnels through a barrier of a given strength. Such nonclassical tunneling is exemplified by the flow of electrons in semiconductor devices, like the tunnel diode, and in the emission of α-particles from radioactive nuclei.

The treatment of a particle having a single attractive (negative) δ-function potential energy illustrates that particles bound by negative potential-energy bumps, or potential-energy wells, may be in bound energy states. Particles in actual potential-energy wells of a given strength have one or more bound energy levels, depending on the shape of the well: The energy levels are more numerous and more closely spaced as the width of the well becomes greater. These characteristics are exemplified by the energy levels of a single atom, whose nucleus serves as a single center of attraction for an orbital electron.

A particle that has more than one attractive δ-function in its potential energy illustrates how the addition of attractive centers in a potential-energy function increases the number of permitted energy levels for that particle. Further, these multiple-δ-function systems demonstrate that the permitted energies are quantized and lie within energy ranges that are determined by the separations between attractive centers. A two-δ-function system, for example, has many of the wave-mechanical characteristics of a diatomic molecule, in which the two nuclei are the attractive centers for electrons.

Multiple-δ-function systems illustrate how large numbers of attractive centers produce energy bands, which become more nearly dense with energy levels as the number of attractive centers increases. A system with an infinite number of δ-functions mimics that of a crystalline solid, which, because each atom in the solid has more than one energy level, exhibits many energy bands. The energy bands associated with solids constitute the basis of our present understanding of the electrical properties of solid conductors, insulators, and semiconductors.

Problems

GROUP A

25.1 A free particle is described by the wave function

$$\psi(x) = 2\, e^{ikx} + 3\, e^{-ikx}$$

What is the probability that a measurement will determine the particle to be moving in the negative x direction, that is, to have a negative x momentum?

25.2 A 6.0-eV electron is scattered by a repulsive δ-function potential energy ($S = 2.0 \times 10^{-28}$ J·m). What is the probability that the electron will be reflected by this barrier?

25.3 An alpha particle ($m = 6.7 \times 10^{-27}$ kg) is bound by a single attractive δ-function potential energy that has a strength $-S = -2.0 \times 10^{-27}$ J·m. Determine the bound-state energy of the α-particle and the corresponding wave function.

25.4 An electron is localized to a one-dimensional region of width $a = 0.20$ nm by an infinite well. Determine the three lowest energy values and the corresponding wave functions.

25.5 A free particle of mass m is described by the wave function

$$\psi(x) = A\, e^{ikx}$$

(a) Determine the (kinetic) energy of this particle.

(b) Write the corresponding time-dependent wave function $\Psi(x, t)$ for this particle.

25.6 A free neutron is described by the wave function

$$\psi(x) = A\, e^{i(2 \times 10^{12}\, m^{-1})x}$$

Calculate for this neutron

(a) the de Broglie wavelength.

(b) the magnitude of the linear momentum.

(c) the (kinetic) energy.

(d) the probability that a measurement will determine the linear momentum to be in the negative x direction.

25.7 A free electron is described by the wave function

$$\psi(x) = 3\, e^{ikx} + 4\, e^{-ikx}$$

with $k = 2.0 \times 10^9$ m^{-1}. Determine

(a) the de Broglie wavelength for this electron.

(b) the magnitude of the linear momentum of this electron.

(c) the (kinetic) energy of this electron.

(d) the probability that a measurement will determine that this electron is moving in the positive x direction.

25.8 A 2.0-keV neutron interacts with a repulsive δ-function located at $x = 0$. The appropriate wave function is

$$\psi(x) = \begin{cases} 25\, e^{ikx} - (9 + 12i)e^{-ikx} & ;\quad x < 0 \\ (16 - 12i)e^{ikx} & ;\quad x > 0 \end{cases}$$

(a) Determine k.

(b) Calculate the strength S of the δ-function.

25.9 A neutron is described by the wave function

$$\psi(x) = \sin(\alpha x)$$

where $\alpha = 4.5 \times 10^{10}$ m^{-1}. Calculate

(a) the de Broglie wavelength for this particle.

(b) the (kinetic) energy of this neutron.

(c) the probability that a measurement will determine the linear momentum to be in the negative x direction.

25.10 Write a time-dependent wave function $\Psi(x, t)$ that describes a free 2.0-MeV proton that has a momentum in the negative x direction.

25.11 Determine the strength S of the δ-function potential energy that approximates each of the potential-energy functions shown in Figure 25.29.

25.12 An electron has the potential energy shown in Figure 25.30. With $U_0 = 50$ eV and $a = 0.050$ nm, use a δ-function approximation to U and estimate the ground-state energy of the electron in this well.

25.13 A proton has the potential energy shown in Figure 25.31. With $U_0 = 5.0$ MeV and $a = 5.0 \times 10^{-15}$ m, use a δ-function approximation to U and estimate the ground-state energy of the proton in this well.

25.14 A particle of mass m is bound by an attractive δ-function (strength $= -S$) at $x = 0$. Determine the probability that a measurement will find this particle in the region $|x| < \hbar^2/(mS)$.

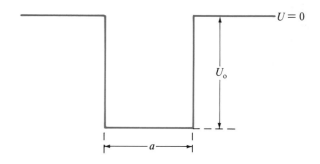

Figure 25.30   Problem 25.12

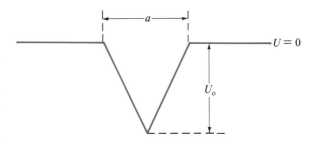

Figure 25.31   Problem 25.13

25.15 A free particle of mass m is scattered from an attractive δ-function of strength $-S$. Show that the reflection and transmission coefficients are given by

$$R(E) = \frac{E_0}{E + E_0}$$

$$T(E) = \frac{E}{E + E_0}$$

where

$$E_0 = mS^2/(2\hbar^2).$$

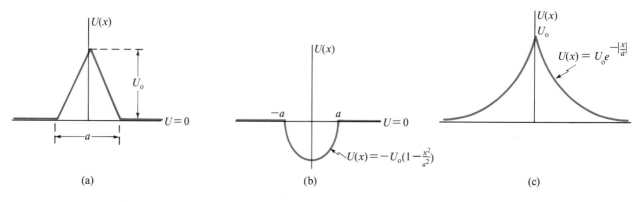

(a)   (b)   (c)

Figure 25.29   Problem 25.11

25.16 A beam of 0.50-nm-wavelength electrons is incident upon the potential-energy barrier shown in Figure 25.32. If $U_o = 20$ eV and $a = 15$ pm,

(a) calculate the (kinetic) energy of each incident electron.

(b) from a classical viewpoint, would the electron in this beam be transmitted beyond or reflected by this potential-energy barrier?

(c) use a δ-function potential-energy approximation to predict the probability that an electron in the beam will be transmitted through the barrier.

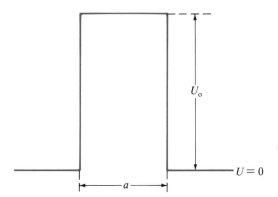

Figure 25.32   Problem 25.16

25.17 A free proton is incident upon the potential-energy barrier shown in Figure 25.33. Use a δ-function approximation to determine the de Broglie wavelength of this proton if there is to be a probability of 0.60 that the proton will be reflected by this barrier.

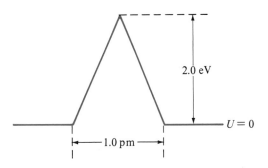

Figure 25.33   Problem 25.17

25.18 The $n = 3$ wave function for the harmonic oscillator is of the form

$$\psi_3(x) = (1 + ax^2)e^{-\alpha x^2}$$

Determine a, α, and the corresponding energy E_3.

25.19 Determine the bound-state energy levels, E_1 and E_2, for a proton bound by two δ-functions, each with a strength of $-S = -5.0 \times 10^{-28}$ J·m, separated by $a = 3.0 \times 10^{-14}$ m.

25.20 Determine the ground-state energy E_1 and corresponding wave function ψ_1 for an electron bound by two attractive δ-functions of strength -4.0×10^{-29} J·m separated by 0.20 nm. Is there another bound state?

GROUP **B**

25.21 A particle of mass m is bound by two attractive δ-functions, each having a strength $-S$, separated by a distance a. Show that there is only one bound state for $a < \hbar^2/(mS)$, or equivalently, $\gamma < 1$.

25.22 Let $\psi_1(x)$, $\psi_2(x)$, ... be the bound-state wave functions for a particle that has the corresponding energies, $E_1 < E_2 < \ldots$

(a) These wave functions can generally be *normalized* so that

$$\int_{-\infty}^{\infty} |\psi_n(x)|^2 \, dx = 1$$

Show that the normalized wave functions for a particle of mass m confined by an infinite well of width a are

$$\psi_n(x) = \sqrt{\frac{2}{a}} \sin \frac{n\pi x}{a} \quad ; \quad n = 1, 2, 3, \ldots$$

(b) These wave functions are always mutually *orthogonal*, that is, for $n \neq m$

$$\int_{-\infty}^{\infty} \psi_n^*(x) \, \psi_m(x) \, dx = 0$$

Show that the bound-state wave functions for a particle of mass m confined by an infinite well are orthogonal.

25.23 The three lowest energy values and corresponding wave functions for a particle of mass m having the simple harmonic potential energy, $U(x) = \frac{1}{2}kx^2$, have been shown to be

$$E_1 = \frac{1}{2}h\nu \quad ; \quad \psi_1 = A_1 e^{-\alpha x^2}$$

$$E_2 = \frac{3}{2}h\nu \quad ; \quad \psi_2 = A_2 x e^{-\alpha x^2}$$

$$E_3 = \frac{5}{2}h\nu \quad ; \quad \psi_3 = A_3(1 - 4x)e^{-\alpha x^2}$$

with

$$\alpha = \frac{\sqrt{km}}{2\hbar} \quad \text{and} \quad \nu = \frac{1}{2\pi}\sqrt{\frac{k}{m}}$$

Show that, as claimed in Problem 25.22(b), these wave functions are mutually orthogonal, that is, for $n \neq m$

$$\int_{-\infty}^{\infty} \psi_n(x) \, \psi_m(x) \, dx = 0$$

25.24 A particle of mass m is bound by two attractive δ-functions separated by a distance a. The strengths of the two δ-functions are $-S$ and $-\lambda S$ with $\lambda \geq 1$.

(a) Show that the ground-state energy E_1 and the excited-state energy E_2 are given by

$$E_1 = -\frac{\hbar^2}{2ma^2}y_+^2 \quad ; \quad E_2 = -\frac{\hbar^2}{2ma^2}y_-^2$$

where y_+ and y_- are the solutions of

$$y = \frac{\gamma}{2}[(1 + \lambda) \pm \sqrt{(1 + \lambda)^2 - 4\lambda(1 - e^{-2y})}]$$

with $\gamma = mSa/\hbar^2$.

(b) Show that there is only one bound state for

$$\gamma < \frac{1 + \lambda}{2\lambda}$$

25.25 A particle of mass m interacts with three equal-strength $(-S)$ δ-functions at $x = -a$, 0, and $+a$.

(a) Use the general solution

$$\psi(x) = A\,e^{-\alpha|x + a|} + B\,e^{-\alpha|x|} + C\,e^{-\alpha|x - a|}$$

of the Schrödinger equation ($E = -\hbar^2\alpha^2/(2m)$) and the three boundary conditions

$$\psi'(-a^+) = \psi'(-a^-) - \frac{2mS}{\hbar^2}\psi(-a)$$

$$\psi'(0^+) = \psi'(0^-) - \frac{2mS}{\hbar^2}\psi(0)$$

$$\psi'(a^+) = \psi'(a^-) - \frac{2mS}{\hbar^2}\psi(a)$$

to show that bound-state solutions occur for

$$\begin{vmatrix} y - \gamma & -\gamma e^{-y} & -\gamma e^{-2y} \\ -\gamma e^{-y} & y - \gamma & -\gamma e^{-y} \\ -\gamma e^{-2y} & -\gamma e^{-y} & y - \gamma \end{vmatrix} = 0$$

with $y = \alpha a$ and $\gamma = mSa/\hbar^2$.

(b) Expand the determinant of part (a) and factor the resulting expression to show that three bound states correspond to y_1, y_2, and y_3, which satisfy

$$y_1 = \gamma[1 + \frac{1}{2}e^{-2y_1}(1 + \sqrt{1 + 8e^{2y_1}})]$$

$$y_2 = \gamma(1 - e^{-2y_2})$$

$$y_3 = \gamma[1 + \frac{1}{2}e^{-2y_3}(1 - \sqrt{1 + 8e^{2y_3}})]$$

with corresponding energies

$$E_n = -\frac{\hbar^2}{2m}\left(\frac{y_n}{a}\right)^2$$

(c) Show that the number of bound states is determined by:

$$0 \leq \gamma \leq 1/2 \leftrightarrow \text{One bound state}$$

$$1/2 < \gamma \leq 3/2 \leftrightarrow \text{Two bound states}$$

$$3/2 < \gamma \qquad\quad \leftrightarrow \text{Three bound states}$$

25.26 Consider a particle that interacts with three equally spaced attractive δ-functions of equal strengths, as in Problem 25.25. Show that for $\gamma = 2$, the values of y and E are

$$y_1 = 2.30 \quad ; \quad E_1 = -1.32\,E_o$$

$$y_2 = 1.96 \quad ; \quad E_2 = -0.96\,E_o$$

$$y_3 = 1.30 \quad ; \quad E_3 = -0.42\,E_o$$

where $E_o = mS^2/(2\hbar^2)$.

Appendix

Trigonometry

The most commonly used trigonometric functions are the sine, cosine, and tangent, which are abbreviated sin, cos, and tan. These functions are simply related to the sides and angles of a right triangle. In Figure A-1 the angle C is the right angle, and the sides a, b, and c, are opposite the angles A, B, and C, respectively. The basic trigonometric functions may be expressed in terms of the components of that right triangle by

$$\sin A = \frac{a}{c} = \frac{\text{side opposite } A}{\text{hypotenuse}}$$

$$\cos A = \frac{b}{c} = \frac{\text{side adjacent to } A}{\text{hypotenuse}}$$

$$\tan A = \frac{a}{b} = \frac{\text{side opposite } A}{\text{side adjacent to } A} \tag{A-1}$$

The functions secant, cosecant, and cotangent, abbreviated sec, csc, and cot (or ctn), respectively, are defined by

$$\sec A = \frac{1}{\cos A} = \frac{c}{b}$$

$$\csc A = \frac{1}{\sin A} = \frac{c}{a}$$

$$\cot A = \frac{1}{\tan A} = \frac{b}{a} \tag{A-2}$$

Corresponding functions of the angle B in Figure A-1 are given by a similar set of relationships:

$$\sin B = \frac{b}{c} = \frac{\text{side opposite } B}{\text{hypotenuse}}$$

$$\cos B = \frac{a}{c} = \frac{\text{side adjacent to } B}{\text{hypotenuse}}$$

$$\tan B = \frac{b}{a} = \frac{\text{side opposite } B}{\text{side adjacent to } B} \tag{A-3}$$

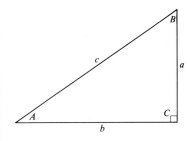

Figure A-1
A right-angle triangle.

787

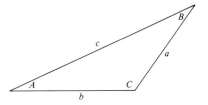

Figure A-2

We may use Equations (A-1) and (A-2) in conjunction with the Pythagorean theorem, $c^2 = a^2 + b^2$, to obtain the following trigonometric identities:

$$\sin^2 A + \cos^2 A = 1 \tag{A-4}$$

$$\tan^2 A + 1 = \sec^2 A \tag{A-5}$$

$$1 + \cot^2 A = \csc^2 A \tag{A-6}$$

Equations (A-4), (A-5), and (A-6) are also correct if B is substituted for A.

Relationships between the lengths of the sides and the angles of arbitrary triangles, including triangles that do not contain a right angle, are summarized by two trigonometric laws: the law of cosines and the law of sines. In Figure A-2 consider the triangle with sides of lengths a, b, and c opposite the angles A, B, and C, respectively.

The law of cosines, useful when two sides and an included angle are known, specifies the length of the remaining side according to

$$c^2 = a^2 + b^2 - 2ab \cos C \tag{A-7}$$

or, permuting the sides and angles,

$$b^2 = a^2 + c^2 - 2ac \cos B \tag{A-8}$$

$$a^2 = b^2 + c^2 - 2bc \cos A \tag{A-9}$$

Further, the law of sines relates the sides and angles of the triangle according to

$$\frac{a}{\sin A} = \frac{b}{\sin B} = \frac{c}{\sin C} \tag{A-10}$$

Other useful trigonometric identities are listed here:

$$\sin 2A = 2 \sin A \cos A$$

$$\cos 2A = \cos^2 A - \sin^2 A$$

$$\tan 2A = \frac{2 \tan A}{1 - \tan^2 A}$$

$$\sin^2\left(\frac{A}{2}\right) = \frac{1}{2}(1 - \cos A)$$

$$\cos^2\left(\frac{A}{2}\right) = \frac{1}{2}(1 + \cos A)$$

$$\tan^2\left(\frac{A}{2}\right) = \frac{1 - \cos A}{1 + \cos A}$$

$$\sin(A \pm B) = \sin A \cos B \pm \cos A \sin B$$

$$\cos(A \pm B) = \cos A \cos B \mp \sin A \sin B$$

$$\tan(A \pm B) = \frac{\tan A \pm \tan B}{1 \mp \tan A \tan B}$$

Answers to Odd-Numbered Problems

Chapter 1

1. (a) 1.4×10^4 kg/m^3; (b) No way! It weighs 1300 lb.
3. (a) $12\ \underline{/-48°}$; (b) $6.0\hat{\mathbf{i}} - 10\hat{\mathbf{j}}$;
   (c) $14\hat{\mathbf{i}} - 19\hat{\mathbf{j}} = 24\ \underline{/-54°}$; (d) 140
5. (a) 18 m; (b) 34° S of E;
   (c) $(15\hat{\mathbf{i}} - 10\hat{\mathbf{j}})$ m = 18 m $\underline{/-34°}$
7. (a) 7.9, 7.1; (b) 11; (c) 7.0; (d) 0.13; (e) 83°
9. (a) 0; (b) $280\hat{\mathbf{i}} + 130\hat{\mathbf{j}}$; (c) $-370\ \underline{/295°}$
11. (a) 18; (b) 39°; (c) $18\ \underline{/171°}$
13. $9.8\ \underline{/153°}$
15. 1.1 km $\underline{/15°}$
17. (a) 6.2 ft; (b) 36°
19. (a) $(2\hat{\mathbf{i}} - \hat{\mathbf{j}} + 2\hat{\mathbf{k}})/3$; (b) $(9\hat{\mathbf{i}} + 20\hat{\mathbf{j}} + 12\hat{\mathbf{k}})/25$
    (c) $-0.73\hat{\mathbf{i}} - 0.084\hat{\mathbf{j}} + 0.68\hat{\mathbf{k}}$; (d) Not quite. If $\hat{\mathbf{c}}$ is perpendicular to both **a** and **b**, so is $-\hat{\mathbf{c}}$.
21. $\alpha = 5, \beta = 7$

Chapter 2

1. (a) 0.78 s; (b) 7.7 m/s
3. (a) 5.0 s; (b) 28 m/s; (c) 120 m
5. (a) 3.6 ft/s^2, down; (b) 3.6 ft/s^2, up
7. (a) 120 s; (b) 30 m/s; (c) 1800 m
9. 92 mi/h, 22° E of N
11. (a) 90 ft/s = 61 mi/h; (b) 1.8 s
13. (a) 91 ms; (b) 9.7×10^2 ft/s^2 west = 30g west.
15. (a) 100 mi/h; (b) $+\hat{\mathbf{i}}$
17. (a) 220 s, 4.0×10^3 ft = 0.92 mi;
    (b) 440 s, 1.9×10^4 ft = 3.7 mi, 88 ft/s = 60 mi/h
19. 270 km/h, 14° N of W
21. (a) 2.0 ft/s; (b) -1.3 ft/s
25. (a) 160 ft; (b) 1.8 s;
    (c) $v_{yA} = -59$ ft/s, $v_{yB} = 59$ ft/s
27. (a) 0.86 s; (b) 12 ft; (c) 1.9 s; (d) 67 ft;
    (e) 48 ft/s
29. (a) $(10 + 20t - t^2)$ ft/s, $(25 + 10t + 10t^2 - t^3/3)$ ft;
    (b) 10 s; (c) 110 ft/s; (d) 930 ft
31. (a) 280 ft/s $\underline{/44°}$; (b) 570 ft; (c) 240 ft/s $\underline{/-33°}$
33. 44 m
35. (a) 90°; (b) 0.13 h;

(c) $(5.0\hat{\mathbf{i}} + 12\hat{\mathbf{j}})$ km/h = 13 km/h $\underline{/67°}$;
(d) 1.6 km $\underline{/67°}$ = $(0.63\hat{\mathbf{i}} + 1.5\hat{\mathbf{j}})$ km;
(e) 115°, 0.14 h
37. (a) $-6.0\hat{\mathbf{i}}$ ft/s; (b) 9.4 ft/s; (c) 0, 7.9 ft/s
41. (a) 8.0 m; (b) -2.0 m; (c) -12 m; (d) 14 m;
    (e) 12 m; (f) 0
43. (a) $+40$ ft/s; (b) -40 ft/s; (c) 60 ft/s; (d) 20 ft/s
49. 0.23 h
51. (d) 300 ft/s, 55°; (e) 21°, 75°

Chapter 3

1. $110\hat{\mathbf{j}}$ N
3. (a) $-52\hat{\mathbf{i}}$ N; (b) $-150\hat{\mathbf{i}}$ N
5. (a) $P = 160$ N, $T = 95$ N; (b) $P = 38$ N, $T = 23$ N
7. (a) $29\hat{\mathbf{i}}$ N; (b) 12 N; (c) $12\hat{\mathbf{i}}$ N; (d) $-12\hat{\mathbf{i}}$ N;
   (e) $-49\hat{\mathbf{j}}$ N
9. (a) 3.9 lb; (b) 5.0 lb; (c) 3.9 lb
11. (a) 39 lb; (b) 22 lb, down.
13. (a) 1.3 s; (b) 3.0 m/s
15. (a) 33 N; (b) 4.6 kg
17. $(-5.1\hat{\mathbf{i}} + 4.8\hat{\mathbf{j}})$ N
19. (a) $3.4\hat{\mathbf{i}}$ ft/s^2; (b) $-3.0\hat{\mathbf{i}}$ lb; (c) $17\hat{\mathbf{j}}$ lb; (d) $28\hat{\mathbf{j}}$ lb
21. (a) 5.4 m/s^2, downward; (b) 9.7 N;
    (c) 9.7 N to the right; (d) 12 N, downward.
23. (a) 24 lb; (b) 20 lb
25. (a) 5.2 m/s^2; (b) 18 N; (c) 8.7 N
27. (a) 33 lb; (b) 21 lb
29. (a) 14 kg; (b) 2.1 m/s^2 to the left; (c) 110 N
31. (a) 0.84 m/s^2 up the plane; (b) 23 N
33. (a) 11 ft/s^2; (b) 6.0 lb
35. 120 lb upward
37. (a) 0.86 kg; (b) 4.7 N
39. (a) 1.7$\hat{\mathbf{i}}$ m/s^2; (b) $(3.4\hat{\mathbf{i}} + 20\hat{\mathbf{j}})$ N;
    (c) $(-3.4\hat{\mathbf{i}} - 20\hat{\mathbf{j}})$ N
41. (a) 21 ft/s^2; (b) 3.4 lb
43. (a) 6.4 ft/s^2 to the right; (b) 13 ft/s^2 to the right;
    (c) 16 lb; (d) 8.0 lb
45. $P = (\frac{2}{3})(W_1 + W_2), T = (\frac{2}{3})W_1$

47. (a) $a_1 = 7.1$ m/s^2, right, $a_2 = 6.2$ m/s^2, down,
$a_3 = 8.0$ m/s^2, down; (b) $T_1 = 36$ N, $T_2 = 18$ N
49. (d) 7.6 s; (e) 740 ft; (f) $v_o(1 + 2e^{-gt/v_o})$

Chapter 4

1. (b) 63 lb; (c) 37 lb
3. 1.1 N $\diagup 56°$
5. (a) 1.2 lb east; (b) 1.2 lb east;
(c) 1.2 lb west, 100 lb downward.
7. (a) 150 lb upward; (b) 96 lb upward.
9. 0.55
11. (a) 64 lb radially inward; (b) 0.40;
(c) 70 ft/s = 47 mi/h
13. (a) 0.57 kN; (b) 0.37 kN $\diagup 221°$
15. (a) 1.5 m/s^2 radially inward; (b) 1.6 km/s;
(c) 7100 s = 2.0 h
17. 87 lb
19. (a) 5.6 lb $\leq P \leq$ 15 lb; (b) 3.8 ft/s^2 up the plane;
(c) 5.0 ft/s^2 down the plane.
21. (a) 2.1 m/s; (b) 2.2 m/s^2 radially inward;
(c) 150 N, radially inward; (d) 0.22; (e) 3.6 m
23. (a) 74 lb; (b) 40 lb
25. (a) 140 lb; (b) 190 lb $\diagup 114°$
27. (a) $0 \leq P \leq 49$ N; (b) $(-27\hat{i} + 130\hat{j})$ N;
(c) 1.7$\hat{i}$ m/s^2
29. (a) 9.2 lb; (b) 29 lb radially inward; (c) 15 ft/s
31. (a) 20 N $\leq P \leq$ 140 N;
(b) 1.8 m/s^2, 93 N; (c) 1.7 m/s^2, 65 N
33. (a) 230 N $\diagup 180°$; (b) 460 N $\diagup 59°$
35. (a) $(-12\hat{i} + 68\hat{j})$ N; (b) $(-24\hat{i} + 65\hat{j})$ N;
(c) $(-18\hat{i} + 59\hat{j})$ N; (d) 0, 0, 17$\hat{i}$ N
37. (a) 14 N down the inclined surface; (b) 0.16;
(c) 14 N up the incline, 3.5 m/s^2 down the incline.
39. (a) 1.1×10^{26} kg; (b) 4.5×10^{12} m;
(c) 16 h; (d) 5.4 km/s
41. (a) 280 lb; (b) 540 lb $\diagup 64°$
43. (a) 3.7 ft; (b) yes; (c) 7.6 ft/s
45. (a) 1.6×10^{24} kg; (b) 6.8 m/s^2
49. (a) $0 \leq P \leq 26$ lb;
(b) $a_{15} = a_{50} = 9.9\hat{i}$ ft/s^2, $(4.6\hat{i} + 15\hat{j})$ lb;
(c) $a_{15} = 6.4\hat{i}$ ft/s^2, $a_{50} = 21\hat{i}$ ft/s^2, $(3.0\hat{i} + 15\hat{j})$ lb
51. (a) $0 \leq P \leq 15$ lb; (b) 8.6 ft/s^2, 4.7 lb
53. (a) 50 ft/s^2 $\diagup 90°$, 3.1 lb; (b) 31 ft/s^2 $\diagup 257°$, 0.27 lb
57. $\mu_s > \sqrt{3}/3$
59. (a) $(4\pi/3)d_oR^3 \ln 2$; (b) $d_o \ln 2$;
(c) $(16\pi^2/81)Gd_o^2R^4(\ln 2)^2$

Chapter 5

1. 74 J
3. 330 J
5. 28 ft·lb

7. (a) 16 J; (b) 16 J; (c) −8.0 J
9. (a) 140 W = 0.18 hp; (b) −140 W = −0.18 hp
11. −8.0 J
13. 5.9 J
15. (a) 470 J; (b) 58 W
17. (a) $U_g = 2.0y$ J; (b) $E = 40$ J;
(c) $K = (40 − 2.0y)$ J, $0 \leq y \leq 20$ m; (d) 20 m
19. 8.9×10^8 J
21. (a) 49 ft·lb; (b) 240 ft·lb/s = 0.44 hp
23. (a) 0.12 m; (b) 9.0 J; (c) −18 J
27. (a) 90 ft/s; (b) −340 ft·lb/s; (c) zero;
(d) 340 ft·lb/s
29. (a) 63 ft·lb/s = 0.12 hp;
(b) −63 ft·lb/s = −0.12 hp
31. (a) 0.61 ft; (b) 5.8 ft/s; (c) $K = 4.7 − 13x^2$
33. 11 km/s = 7.0 mi/s = 2.5×10^4 mi/h
35. (a) 8.2 m/s; (b) 7.1 m/s; (c) −350 W
37. (a) 5.0 J; (b) 6.1 m/s
39. (a) 1.2×10^8 J; (b) 6.7×10^6 m
41. (a) 400 J; (b) 16 m
43. (a) 1.5 ft; (b) 3.3 ft·lb;
(c) −66 ft·lb/s = −0.12 hp
45. 6.0 kg
47. 2.7 J
49. (a) 2.5 m/s^2, left; (b) 0.71 m/s
51. (a) 1.2 m; (b) 0.91 m; (c) 2.7 J
53. (a) 0.30 m; (b) 1.5 m/s; (c) $K = (30 − 100y)y$
57. (a) infinite; (b) 8.8×10^8 J; (c) 10 km/s
59. 130 m/s, 1.9 h
61. (a) 60 m/s^2; (b) 1.7 m/s

Chapter 6

1. 0.91 m
3. $-28\hat{j}$ m/s
5. 3400 lb
7. (a) $14\hat{i}$ N·s; (b) 100 m/s; (c) $34\hat{i}$ kg·m/s
9. (a) $(1.4\hat{i} + 3.2\hat{j})$ m/s; (b) $(-11\hat{i} + 9.6\hat{j})$ N·s;
(c) $(11\hat{i} − 9.6\hat{j})$ N·s
11. 7.2 ft
13. 0.42 m, 55 m
15. 16 N
17. 2.4 ft·lb
19. (a) 2.6 slug·ft/s; (b) 0.026 s
21. (a) 7.5 kg·m/s, upward; (b) 19 ms; (c) 7.8 J
23. 6.0 ft/s, west.
25. $M = 0.67$ kg, $v = (6.0\hat{i} − 3.0\hat{j})$ m/s
27. (a) 1.6 m; (b) 99.75% lost in collision, 0.25% dissipated by friction.
29. (a) 11 m; (b) 49 J
31. 56°
33. (a) $(v_x)_m = 2v$, $(v_x)_{3m} = 0$;
(b) $(\Delta p_x)_m = 3mv$, $(\Delta p_x)_{3m} = −3mv$; (c) $-2v\hat{i}$, $+2v\hat{i}$
35. Two solutions: 58 m/s, 63°, 3400 J and 19 m/s, 17°, 410 J

37. (a) 6.0 m/s; (b) 6.3 m/s
39. $v_1 = 10$ m/s, $v_2 = 16$ m/s, $v_3 = 96$ m/s
41. 4.9 s
43. (a) $\lambda_o L$, $L/2$; (b) $\lambda_o L/2$, $2L/3$
49. (c) $v_x(t) = v_o |\ln(1 - t/2\tau)|$, $v_{\max} = v_o \ln 2$

Chapter 7

1. (a) 0.050 rad/s^2 CW; (b) 0.99 rad/s^2 CW
3. (a) 180 ft/s; (b) 44 ft/s
5. (a) 18; (b) 12
7. (a) 15 cm/s, 25 m/s^2; (b) 51 cm/s^2
9. (a) 9.0 rad/s; (b) 75 rad; (c) 14 rad/s
11. (a) 11 rad/s^2; (b) 26 rad
13. (a) 15 rad/s^2; (b) 15 m/s^2 = 1.5g;
(c) 5.4 m/s, 29 m/s^2 upward.
15. (a) 6700 ft·lb; (b) 0.39
17. (a) 9.8 rad/s; (b) 0.38; (c) 0.38
19. (a) 19 slug·ft^2; (b) 340 hp; (c) 550 hp
21. (a) 260 J; (b) 27 kg·m^2/s
23. (a) 53 J; (b) 3.6 m; (c) 3.5 m/s^2 down the surface;
(d) 1.4 s
25. (a) 50 rad/s^2; (b) 2.4 J; (c) 3.1 m/s^2, 44 m/s^2 up
27. (a) 1.4 m/s; (b) $K_m = 4.8$ J, $K_I = 190$ J
29. (a) 0.60 m/s^2; (b) 2.6 s; (c) 1.5 m/s; (d) 0.88
31. (a) 9.6 m/s^2 down; (b) 19 m/s^2 up
33. (a) 0.044 slug·ft^2; (b) 15 rad/s^2; (c) 2.4 ft;
(d) 0.96
35. (a) 27 rad/s^2; (b) 8.0 m/s^2 downward; (c) 7.3 rad/s;
(d) 16 m/s^2 upward;
(e) 12 N downward, 24 N upward;
(f) 2.7 N upward, 39 N up
37. (a) 3.3 m/s^2; (b) 6.5 N; (c) 26 J, 13 J
41. (a) 11 ft/s^2; (b) 5.8 ft/s^2; (c) 7.2 rad/s^2;
(d) 6.4 lb; (e) 2.1 lb to the right;
(f) 7.1 ft·lb, 8.6 ft·lb, 4.3 ft·lb
43. (a) $[64/(128 - 24t + 3t^2)]$ rev/s;
(b) 1.6×10^4 J, 2.5×10^4 J
45. (a) 0.75 kg; (b) 2.5 N up the inclined surface.
47. (a) $2\pi\alpha LR^3/3$; (b) $3MR^2/5$

Chapter 8

1. (a) −2.2 m/s, −16 m/s^2; (b) 10 cycles.
3. $x(t) = 0.1 \cos(4\pi t)$ m
5. (a) 1.9 Hz; (b) $v_x(t) = 180 \sin(12t)$ cm/s
7. $v_x(t) = 3 \cos(9t)$ m/s
9. (a) 9.8 ft/s; (b) 0.57 ft = 6.9 in
11. (a) 0.94 m/s; (b) 0.50 s; (c) 0.59 N
13. (a) 38 cm/s; (b) 95 cm/s^2
15. (a) $x(t) = -0.12 \sin(5\pi t)$ m; (b) 0.10 s;
(c) −5.9 N
17. (a) 1.1 m; (b) 25 J; (c) 0.71 Hz

23. 1.7 s
25. (a) 0.56 m; (b) 0.84 rad/s; (c) 3.5 rad/s^2;
(d) $\theta = 0.20 \cos(4.2t)$ rad
27. (a) 0.80 Hz; (b) 0.83 m; (c) 4.1 m/s; (d) 0.34 J;
(e) $x(t) = 0.83 \cos(5t - 1.3)$ m =
$[0.2 \cos(5t) + 0.8 \sin(5t)]$ m
29. (a) 1.0 s; (b) 0.27 m below initial position;
(c) 0.43 m; (d) 0.16 m, 0.70 m; (e) 44 J; (f) 17 J
33. (a) 0.12 s; (b) −0.34 ft; (c) 2.9 ft/s; (d) 9.6 ft
43. (b) ω_o, $\sqrt{5}\omega_o$

Chapter 9

1. 5.9 m/s
3. 8.6 N down.
5. (a) 2.7×10^4 Pa; (b) 130 N; (c) 130 N down.
7. 40 cm^3
9. 160 N
11. 0.75 atm
13. 4000 ft^3
15. 1.5 kW
17. 1600 m^3
19. 72 N
21. 1.6×10^{-2} m^3/s; (b) 16 kg/s; (c) 500 J/s
23. (a) 1.9×10^5 N; (b) 8.5×10^5 N down;
(c) $P_A = 2.7 \times 10^5$ Pa, $P_B = 2.1 \times 10^5$ Pa;
(d) 6.4×10^5 Pa
25. 0.73 kg
27. 27 cm
29. (a) 17 m/s, 0.99 m/s; (b) 3.5×10^5 Pa
31. 1.4×10^3 kg/m^3
33. 2.3 m
35. (a) 2.5 cm; (b) 1.5×10^5 Pa; (c) 7.1 m/s;
(d) 250 W
37. 1.9 m/s^2 upward.
39. $T_1 = 5.8 \times 10^3$ N, $T_2 = 2.6 \times 10^3$ N
41. 1.2×10^3 kg/m^3
43. (a) 3.8×10^5 Pa; (b) 0.63 m^3
45. (b) 2.1 m; (c) 210 N
47. (a) 4.6 m; (b) 170 N; (c) 480 N
49. 2.0 s
51. $\rho b(b - 3h)/3\rho_o$
53. (a) 5.2×10^6 N; (b) 1.1×10^7 N
55. 20 min

Chapter 10

1. 22 L
3. 0.21 cal/(g·K)
5. 0.22 ft
7. 290 cal/K
11. (a) 1.5 cal/K; (b) 1.3 cal/(g·K); (c) 0.75 cal/(g·K)
13. (a) 32 cal/K; (b) −18 cal/K; (c) +14 cal/K;
(d) No.

15. $+0.11$ cm^2

17. (a) 80 cal;   (b) 67 cal

19. (a) 0.55;   (b) 460 J;   (c) 0.82

21. 2.8×10^8 cal

23. (a) 30 cm;   (b) 1.6×10^4 cm

25. (a) 1.1 kJ;   (b) 0.73 kcal;   (c) 0.47 kcal;
    (d) 650 K $= 380°$C

27. 0.66 kcal/K

29. (a) 0.31 kcal;   (b) 0.62 kcal;   (c) -1.0 kcal

31. (a) 1.0 kJ;   (b) 0

33. -0.99 kcal

35. 9.8 cm^3

37. (a) 1.5;   (b) 2.8 kJ

39. 7.3 kcal

41. 3.1 m

43. 0.029 cal/(g·K)

45. -20%

47. (a) 400 cal;   (b) -240 cal;   (c) 640 cal

49. (a) 2.9 kcal;   (b) 8.2 cal/K

51. 3.5 kW

53. 26 cal

55. 56 g of ice and 114 g of water at 0°C.

57. 4.0 cal/K, No.

59. (a) 0.30;   (b) 0.50

61. (a) $(k_1 d_2 T_H + k_2 d_1 T_L)/(k_1 d_2 + k_2 d_1)$;
    (b) $[k_1 k_2 (T_H - T_L)]/(k_1 d_2 + k_2 d_1)$

63. 0.28

65. (a) 0.18;   (b) 0.94;   (c) 4.5

67. (b) 0.58

73. (b) 0.92 J/(g·K) $= 0.22$ cal/(g·K)

Chapter 11

1. 6.3×10^5 m/s^2 toward proton, 340 m/s^2 toward electron.

3. $-1.1 \times 10^4 \hat{\mathbf{k}}$ N/C

5. 5.8×10^{11} N/C

7. 1.7 mm

9. $mv^2/(2qd)$

11. (a) 93 m/s;   (b) 0.017 s

13. 720 N $/84°$

15. 122°

17. 14 J

19. 6.8 m

21. $q_3 = -\frac{8}{9} \mu$C, $x_3 = 2.0$ m

23. $q = 1.3$ μC, $T = 0.035$ N

25. 4.3 J

27. 14 N

29. (a) $(qE_o/m)\hat{\mathbf{i}}$;   (b) 0.79 d;   (c) $\sqrt{3qE_o d/4m}$

31. 0.22 J

33. $0.16\hat{\mathbf{j}}$ N

35. $Q_1 = 12$ μC, $Q_2 = -9.7$ μC

37. (a) 1.4 N;   (b) 0.40 μC

39. $kqQ \sin(\beta/2)/(R^2\beta/2)$

41. $(9.5\hat{\mathbf{i}} - 1.9\hat{\mathbf{j}})$ N $= 9.7$ N $/-11°$

43. 56°

45. (a) $Q = \alpha L^2$;   (b) $(2kQ/L^2)(1 - R/\sqrt{R^2 + L^2})$

47. $2k\lambda_1\lambda_2 \ln(b/a)$

49. (a) $\lambda a \sqrt{1 + \alpha^2}(e^{\alpha\pi} - 1)/\alpha$;
    (b) $F_x = -kq\lambda\alpha(1 + e^{-\alpha\pi})/(a\sqrt{1 + \alpha^2})$,
       $F_y = -kq\lambda(1 + e^{-\alpha\pi})/(a\sqrt{1 + \alpha^2})$

Chapter 12

1. 6.6×10^{-10} C

3. (a) $[2kQx/(x^2 + d^2)^{3/2}]\hat{\mathbf{i}}$

5. 0.070 N·m^2/C

7. (a) $-1.3 \times 10^4 \hat{\mathbf{j}}$ N/C;   (b) $2.6 \times 10^4 \hat{\mathbf{j}}$ N/C;
    (c) $1.1 \times 10^4 \hat{\mathbf{j}}$ N/C;   (d) 1.8×10^3 N/C $/250°$

9. 25 N/C $/16°$

11. -11 μC

13. (a) $[2kQd^2(3r^2 - d^2)/r^2(r^2 - d^2)^2]\hat{\mathbf{i}}$;   (b) $(6kQd^2/r^4)\hat{\mathbf{i}}$

17. (a) 5.3×10^{-11} C/m^2;   (b) 1.1×10^{-12} C

19. (a) 0.90 N/C;   (b) 1.8×10^4 N/C

21. -1.3×10^{-9} C

23. 0.58×10^{-6} C

25. (a) -3.0×10^{-8} C;   (b) -3.0×10^{-8} C

29. (a) zero;   (b) 6.6×10^2 N/C;   (c) 1.0×10^3 N/C

31. $\rho a/2\epsilon_o$, radially outward;   (b) $-\rho a$;   (c) $2\rho a/3$

33. 5.2 cm

35. $2k\sigma \ln[(d + b)/(d - b)]\hat{\mathbf{i}}$

37. (a) $(2\lambda/\pi\epsilon_o R)\hat{\mathbf{i}}$;   (b) $(\lambda/4\pi\epsilon_o R)(\hat{\mathbf{i}} + \hat{\mathbf{k}})$

39. (a) $2kQL/x(x^2 - L^2)$

41. $\alpha r^4/6\epsilon_o$ for $0 \leq r \leq R$, $\alpha R^6/6\epsilon_o r^2$ for $R < r$

43. $(\lambda/2\epsilon_o R)\hat{\mathbf{j}}$

45. $40\hat{\mathbf{i}}$ N/C

47. A volume charge density $= 4\epsilon_o \alpha r^4$.

49. Point charge $= 4\pi\alpha\epsilon_o$ at origin and a uniform surface charge of density $-2\alpha\epsilon_o/R^2$ on the $r = R$ surface.

51. (a) $1.8\hat{\mathbf{i}}$ N/C;   (b) $2.2\hat{\mathbf{i}}$ N/C;   (c) $2.3\hat{\mathbf{i}}$ N/C;
    (d) $2.4\hat{\mathbf{i}}$ N/C

53. (a) $0.25 \, kQ/R^2$;   (b) $0.18 \, kQ/R^2$;   (c) $0.26 \, kQ/R^2$

55. (a) 54 N/C;   (b) 52 N/C

Chapter 13

1. 3200 V

3. -220 V

5. (a) $-kQ^2/\sqrt{2}R$;   (b) $-kQ^2/2R$

7. (a) $2kQa^2/r(r^2 - a^2)$;   (b) $2kQa^2/r^3$

9. -600 V

11. (a) $2kQ/\sqrt{x^2 + a^2}$;   (b) $2kQx/(x^2 + a^2)^{3/2}$

13. (a) 68 V;   (b) 68 V;   (c) 4.5 V

15. 7.1×10^{-8} C/m^2

17. 1100 V

19. 6.1 J

21. 7.4×10^4 m/s

23. -8.5×10^{-9} C

25. 3.3 J

27. (a) $2(\sqrt{3} - 1)kqQ/L$; (b) $2\sqrt{3}kqQ/L$

29. -38 V

31. 710 V

33. 3500 V

35. (a) 69 m/s; (b) 0.28 m; (c) 72 m/s

37. (a) $\alpha L^2/2$; (b) $2kQ(\sqrt{y^2 + L^2} - y)/L^2$;
    (d) $2kQ[1 - (y/\sqrt{y^2 + L^2})]/L^2$

39. 680 m/s

41. $\sigma_1 = \sigma_4 = 8.0 \times 10^{-10}$ C/m^2,
    $\sigma_2 = -\sigma_3 = 4.0 \times 10^{-10}$ C/m^2, $|\Delta V| = 0.68$ V

43. $47kQ/(64R)$.

47. (a) $\lambda a \sqrt{\alpha^2 + 1} \, (e^{2\pi\alpha} - 1)/\alpha$; (b) $2\pi k\lambda \sqrt{\alpha^2 + 1}$

Chapter 14

1. 8.0 Ω

3. (a) 8.0 Ω in series; (b) 47 Ω in parallel.

5. (a) 0.75 μF; (b) 64 μC

7. 4

9. -15 V

11. 5.0 μF, 10 μF, 23 μF, 45μF

13. 6.0 Ω, 11 Ω, 15 Ω, 25 Ω, 38 Ω, 60 Ω

15. (a) 50 μC; (b) 42 μJ

17. 19 mJ

19. (a) 3.2×10^7 A/m^2; (b) 1.1 V; (c) 0.25 A

21. $Q_1 = Q_2 = Q_3 = 4.0 \times 10^{-4}$ C, $U = 0.080$ J

23. $Q_1 = Q_2 = 1.4$ mC, 55 mJ

25. (a) 0.90 A; (b) 18 V; (c) 1.8 W

27. (a) -1.8 V; (b) 0.18 μJ; (c) -75 mJ/s

29. (a) 4.8×10^{-2} J; (b) 1.2×10^2 V

31. (a) 35 μF; (b) 20 V; (c) 6.0×10^{-4} C;
    (d) 2.0 mJ

33. $q_1 = 0.25$ μC, $q_2 = 0.15$ μC

35. $q_1 = 3.3 \times 10^{-4}$ C, $q_2 = 1.7 \times 10^{-4}$ C

37. 49 mJ and -45 J/s, or 11 mJ and 21 J/s

39. (a) 41 Ω; (b) 5.0×10^2 V

41. $\alpha^2 \rho_o L/2\pi[\alpha b - \ln(1 + \alpha b)]$

43. $2\kappa C_o/(\kappa + 1)$

45. $(\kappa + 1)C_o/2$

47. $(\kappa + 1)C_o/2$

49. (e) 4.4×10^{-6} J/cm^3

Chapter 15

1. 2.0 V

3. 2.8 A, B to A

5. (a) 1.4 A; (b) 71 V

7. (a) 10 W; (b) 6.0 V; (c) 4.5 W

9. -40 V

11. 1.9×10^{-4} C

13. 3.6×10^{-3} C, right plate is positive.

15. (a) 2.0 mA; (b) 2.7 s

17. 20 V

19. (a) 10^6 Ω; (b) 4.0 μA

21. 0

23. (a) 0.33 A; (b) 4.0 V; (c) 0.53 W

25. (a) 0.44 kW; (b) 24 V; (c) 96 W

27. 0

29. (a) 2.0 A; (b) 750 W

31. -1.0×10^{-4} C/s

33. (a) 81 mW; (b) 5.4×10^{-4} C; (c) 49 mW

35. (a) 0.83 mA; (b) 30 V; (c) 2.0 mA

37. (a) 1.5 A; (b) -10 V; (c) $\mathscr{E}_1$ and $\mathscr{E}_2$ supply 110 W,
    $\mathscr{E}_1$ absorbs 5.0 W, resistors generate 110 W of heat, and
    outside agent provides 5.0 W.

39. (a) $5\mathscr{E}/(3R)$; (b) $\mathscr{E}/(2R)$; (c) $5C\mathscr{E}^2/8$

41. $i_1 = (8/11)$ A, $i_2 = (-1/11)$ A, $i_3 = (-7/11)$ A;
    5$-$V emf adds (40/11) W, 4$-$V emf absorbs (4/11) W,
    3$-$V emf absorbs (21/11) W, and resistors dissipate
    (15/11) W as heat.

43. (a) $7\mathscr{E}/(37R)$; (b) $37R/23$; (c) $R/2$

Chapter 16

1. 2.6×10^{-11} N

3. (a) 1.1×10^{-3} T; (b) 3.1×10^{-8} s

5. $-(2.6\hat{\mathbf{i}} + 29\hat{\mathbf{j}}) \times 10^{-18}$ N

7. $B(r) = (10r)$ T, $0 < r < 0.0010$ m;
   $B(r) = (1.0 \times 10^{-5}/r)$ T, 0.0010 m $< r$

9. (a) 8.9 m; (b) 1.2 m; (c) 0

11. (a) $2.6 \times 10^{-4}\hat{\mathbf{i}}$ N; (b) $-2.8 \times 10^{-4}\hat{\mathbf{i}}$ N

15. 2.0 N

19. $6.0\hat{\mathbf{i}}$ G

21. (a) 1.7 cm; (b) 1.1×10^3; (c) 0.20 T

23. 1.6×10^{-15} N parallel to wire with same sense as
    current.

25. $7.9 \times 10^{-5}\hat{\mathbf{k}}$ T

27. $2\sqrt{2}\mu_o i/\pi d$

29. (a) $B = \mu_o ir/(2\pi a^2)$;
    (b) $B = (\mu_o i/2\pi r)[(b^2 - r^2)/(b^2 - a^2)]$, $a < r < b$;
    (c) $B = 0$

31. (a) $-2.8 \times 10^{-11}\hat{\mathbf{j}}$ N; (b) 0

33. (a) 2; (b) 27 W

35. $-4.0 \times 10^{-5}\hat{\mathbf{j}}$ T

37. 2.5 ms, $(0.36\hat{\mathbf{i}} + 0.28\hat{\mathbf{j}})$ m, $(0.72\hat{\mathbf{i}} + 1.9\hat{\mathbf{j}}) \times 10^2$ m/s

43. $9\mu_o i/(2\pi a)$

45. $\mu_o v \sigma \pi(b - a)$

47. 7.7 G

51. (a) $\mathbf{r} = [\pi^2 mE_o/(2qB_o^2)]\hat{\mathbf{i}} - [2mv_o/(qB_o)]\hat{\mathbf{k}}$,
    $\mathbf{v} = (\pi E_o/B_o)\hat{\mathbf{i}} - v_o\hat{\mathbf{j}}$;
    (b) A helical path with axis parallel to $\hat{\mathbf{i}}$, a "pitch" that
    increases with time, and a radius given by
    $r = mv_o/(qB_o)$;
    (c) $K = mv_o^2/2 + q^2E_o^2t^2/(2m)$

Chapter 17

1. 0.16 V
3. (a) 7.0×10^{-4} T;　(b) 0.46 T
5. (a) 4.0 A;　(b) $+1.0$ A/s;　(c) 50 mV
7. -5.6 A/s
9. 0.25 A, -4.0 A/s
11. 18 mV
13. (a) $+2.8$ V;　(b) Removing, 2.1 W
15. 30 mV
17. -0.55 mV
19. (a) -9.6 A/s;　(b) Adding, 9.0 W;
    (c) L adding 18 W, R and $\mathscr{E}$ each removing 4.5 W.
21. 4.2 μW
23. (a) 400 V;　(b) 0
25. (a) 0.42 A/s;
    (b) 1.2 W, 0.90 W in resistors, 0.30 W in inductors.
27. 400
29. (a) $+5.0$ A/s;　(b) -10 A/s;　(c) Removed, 200 W
31. (a) 0.40 A;　(b) $+30$ A/s;　(c) 0.60 A
33. (a) 0.52 A;　(b) 0.78 W, 0.67 W in R, 0.11 W in L;
    (c) 0.20 J, 0.14 J in R, 0.067 J in L
35. (a) 5.0 A/s;　(b) 0.40 A
39. (a) 0.20 V;　(b) 0.68 V
41. (a) 13 mH;　(b) 16 mJ
43. (c) 4.0×10^{-3} J/m^3, 4.0×10^5 J/m^3
45. (a) 0, $\mathscr{E}/L$, 0;　(b) $\mathscr{E}/(3R)$, 0, $\mathscr{E}/3$
47. $0.14L/R$

Chapter 18

1. 4.3×10^{14} Hz $< \nu < 7.5 \times 10^{14}$ Hz
3. 0.88 A
5. 64 Ω
7. 0.89 A
9. 2.0 mH
11. (a) 50 Ω;　(b) 94 Ω;　(c) 0.53 A;
    (d) $32°$;　(e) $i(t) = 0.53 \sin(2000\pi t - 32°)$ A;
    (f) 42 V;　(g) 27 V;
    (h) $\Delta v_L = 27 \cos(2000\ \pi t - 32°)$ V;　(i) 11 W
13. (a) -30 Ω;　(b) 30 Ω;　(c) 0.50 A;　(d) $-90°$;
    (e) $i(t) = 0.50 \cos(400\pi t)$ A;　(f) 20 V;　(g) 5.1 V;
    (h) $\Delta v_C = +20 \sin(400\pi t)$ V,
    　　$\Delta v_L = -5.1 \sin(400\pi t)$ V;
    (i) 0;　(j) 0.40 Hz
15. (a) 0.73 A;　(b) $\Delta v_C = 20 \sin(600t - 0.62$ rad) V;
    (c) 5.3 W
17. (a) $V_R = 88$ V, $V_L = 198$ V, $V_C = 116$ V;
    (b) $\Delta v_R = 88 \sin(120\pi t)$ V, $\Delta v_L = 198 \cos(120\pi t)$ V,
    　　$\Delta v_C = -116 \cos(120\pi t)$ V
19. (a) 46 Hz;　(b) 5.0 Ω, 0;　(c) 24 A;
    (d) $i = 24 \sin(92\pi t)$ A
21. (a) 96 V;　(b) 1.7 A, 0.27 kHz, 0.78

23. (a) 56 V;
    (b) 0.29 A, 7.6 kHz, 0.20 kΩ, -0.13 kΩ, 0.77
25. (a) 0.21 kHz;　(b) 0.14 A;　(c) 2.1 V, 7.4 V, 15 V
27. (a) $\Delta v = 24$ V, $\Delta v_R = 7.0$ V, $\Delta v_C = 17$ V;
    (b) $P_{\mathscr{E}} = 6.1$ W, $P_R = -1.8$ W, $P_C = -4.3$ W
29. $102°$
35. (b) $i(t) = (A \cos\omega't + B \sin\omega't)e^{-\alpha t} + I \sin(\omega t - \phi)$,
    　　$\omega' = \sqrt{\omega_o^2 - \alpha^2}$;
    (c) $i(t) = (A + Bt)e^{-\alpha t} + I \sin(\omega t - \phi)$;
    (d) $i(t) = Ae^{-\alpha_1 t} + Be^{-\alpha_2 t}$, $\alpha_1 = \alpha - \sqrt{\alpha^2 - \omega_o^2}$,
    　　$\alpha_2 = \alpha + \sqrt{\alpha^2 - \omega_o^2}$;
    (e) $i(t) = I\{\sin\phi \cos(0.8\omega_o t) + [5\omega/(4\omega_o)\cos\phi + (\frac{3}{4})\sin\phi)\sin(0.8\omega_o t)]\}e^{-0.6\omega_o t}$

Chapter 19

3. 3.6 kHz
5. 34 db
7. (a) 0.27 kHz;　(b) 0.23 kHz
9. 200-Hz sound with a 4.0-Hz beat frequency.
11. 0.19 kHz, 0.57 kHz
13. (a) 6.5 cm, 7.1 cm;　(b) 5.2 kHz, 4.8 kHz
15. $y = 0.052 \sin(2.5\pi x)\sin(400\pi t + \pi)$
17. 4.8 m
19. 1.7 m/s
21. $L_n = n(0.25$ m), $n = 1, 2, 3, \ldots$
23. (a) 4.8 kHz;　(b) 4.1 kHz
25. (a) 0.30 m;　(b) 0.41 km/s;　(c) 64 m/s
27. 80 Hz, 84 Hz
29. 25 m/s
31. $\nu_1 = 526$ Hz, $\nu_2 = 494$ Hz
33. (b) $-(50/3)\hat{\mathbf{i}}$ m/s;　(c) 30 ms;　(d) 1.7 m/s;
    (e) 16 ms, 44 ms
35. (a) $y = 0.20 \sin(12x + \pi/4)\cos(57t + \pi/4)$;
    (b) $x_n = (4n - 1)\pi/48$, $n = 0, \pm1, \pm2, \ldots$
37. (b) $A = \sqrt{A_1^2 + A_2^2 + 2A_1 A_2\cos(\delta_1 - \delta_2)}$,
    $$\tan \delta = \frac{A_1 \sin\delta_1 + A_2 \sin\delta_2}{A_1 \cos\delta_1 + A_2 \cos\delta_2}$$
39. (a) 0.20 m, 0.72 m, 1.8 m;　(b) 0.42 m, 1.2 m, 4.2 m

Chapter 20

1. $50°$
3. (b) 15 cm
5. 4.0 m
7. 53 diopters.
9. 32 cm
11. $q = 30$ cm, real, inverted, 3.0 cm high.
13. (a) 11 diopters;　(b) -0.60
15. (a) 1.0 cm;　(b) 38 km;　(c) 95;　(d) 1.9 m
17. 31–35 mm

19. $p = 7.5$ cm, $q = -15$ cm
21. (a) 3; (b) 10 cm, 10 cm, 14 cm
23. 48 cm
25. At a distance equal to $f(f - d)/(2f - d)$ beyond the second lens.
27. (a) Between lenses, 9.0 cm from L_2, virtual, inverted; (b) -0.75
29. $1.5R$
31. (a) Real, inverted, $0.92 \times$ the object size, 13 cm away from the lens, 18 cm from the mirror.
33. $180°$
35. $\Delta f/f = +0.80\%$
37. (a) 10 cm beyond the third lens; (b) 10 cm from the first or third lens.

Chapter 21

1. (a) Pattern is compressed; (b) Pattern is expanded.
3. $\lambda_1/\lambda_2 = 2$
5. 1.5 m
7. 0.031 mm
9. 0.24 mm
11. 180 nm
13. 4200 lines/cm
15. 1.1 cm
17. 370 nm
19. (b) 7.3 mrad = $0.42°$
21. (a) 800 nm, 1600 nm; (b) 400 nm, 1200 nm
23. 0.020 nm
25. $\sin \theta_m = 2m\lambda/w$, $0.74°$, $1.5°$
27. $t = [(m + \frac{1}{2})\lambda\sqrt{n^2 - \sin^2\theta}]/2n^2$, $m = 0, 1, 2, \ldots$

Chapter 22

1. (a) 6.0×10^{19} J; (b) 1.7×10^{12} dollars = 1700 billion dollars.
3. (a) (14 m, 0, 0, -28 ns), (32 m, 0, 0, -73 ns); (b) A, B
5. 200 MeV
7. $t' = 55$ ns, $\mathbf{r}' = (-3.5\hat{\mathbf{i}} + 6.0\hat{\mathbf{j}})$ m, $\mathbf{u}' = (-0.43\hat{\mathbf{i}} + 2.3\hat{\mathbf{j}}) \times 10^8$ m/s
9. 370 million years
11. $(2 - \sqrt{3})c \approx 0.27c$
13. 1.6×10^{-8} m
15. S: (0, 240 m, 0, 0.80 μs); S': (-320 m, 240 m, 0, 1.3 μs)
17. (a) $v_x = c/2$; (b) 1.9 ns; (c) 5.0 m, 4.8 m; (d) No.
19. (a) 20 m, 10/3 m; (b) 25 ns, 95 ns
21. (a) 4.8×10^{-27} kg
23. $(\sqrt{33}/12) M$, $(4/7)c$
25. (a) $M = (4\sqrt{3}/3)m$, $u_3 = c/2$;

(b) $u_1' = (40/41)c$, $u_2' = 4c/5$;
(c) $M = (4\sqrt{3}/3)m$, $u_3' = (13/14)c$;
(d) $\Delta K = \Delta K' = -[(4\sqrt{3}/3) - 2]mc^2$
27. (a) 3.0 m/s^2; (b) 3.0 m/s^2; (c) 3.0 m/s^2; (d) 1.2 m/s^2
29. (a) Yes; (b) No
33. $V = 207c/305$, $v = 527c/625$
35. (a) $71°$; (b) 0.92 m

Chapter 23

1. 0.38 μeV
3. (a) 3.3×10^{15} Hz; (b) 14 eV
5. $\lambda_e = 73$ nm, $\lambda_p = 0.040$ nm
7. $n = 2$, $n = 5$
9. $\frac{81}{16}$
11. (a) 1.5×10^{15} Hz; (b) 4.1 nm; (c) 1.2×10^{-7} eV; (d) 3.3×10^{-19} J; (e) 4.1–1.8 eV
13. (a) 1.0×10^{15} Hz; (b) 8.4×10^5 m/s
15. (a) 2.1×10^{-10} m; (b) 5.2×10^{-9} N; (c) 1.1×10^6 m/s; (d) -3.4 eV
17. (a) 1.1 eV; (b) 2.7×10^{14} Hz
19. 120 nm, 100 nm
21. (a) 3.6×10^{-3} m; (b) 6.6×10^{-30} m
23. (a) Yes; (b) 0.40 eV
25. (c) $\frac{1}{4}$
27. (a) 0.025 nm; (b) 0.021 nm, 0.019 nm, 0.14 nm
29. (a) 12 cm; (b) 0.35 cm^{-1}; (c) 0.082

Chapter 24

1. $A = 239$, $Z = 94$, $N = 145$
3. $\frac{1}{64}$
5. 14
7. 32
9. 1200 nm, 4.5 mW
11. 1800 MeV
13. 35 MeV

Chapter 25

1. 0.69
3. $E = -7.6$ MeV; $\psi = A e^{-\alpha|x|}$, $\alpha = 1.2 \times 10^{15}$ m^{-1}
5. (a) $E = \hbar^2 k^2/2m$; (b) $\Psi(x, t) = A e^{i(kx - \hbar k^2 t/2m)}$
7. (a) 3.1 nm; (b) 2.1×10^{-25} kg·m/s; (c) 0.15 eV; (d) 0.36
9. (a) 0.14 nm; (b) 0.021 eV; (c) 0.50
11. (a) $U_o a/2$; (b) $(-4/3)U_o a$; (c) $2U_o a$
13. -1.9 MeV
17. 0.32 nm
19. $E_1 = -0.14$ MeV, $E_2 = -0.087$ MeV

Index